NATURAL PRODUCTS:
Research Reviews

— Volume 1 —

NATURAL PRODUCTS:
Research Reviews

— *Volume 1* —

Editor
V.K. Gupta
Deputy Director & Head,
Central Animal Facility
Indian Institute of Integrative Medicine (CSIR),
Canal Road, Jammu – 180 001,
India

2012
DAYA PUBLISHING HOUSE®
New Delhi - 110 002

© 2012 EDITOR
ISBN 9789351241249

Published by	:	**Daya Publishing House®** **A Division of** **Astral International Pvt. Ltd.** **– ISO 9001:2008 Certified Company –** 4760-61/23, Ansari Road, Darya Ganj, New Delhi - 110 002 Phone: 23245578, 23244987 Fax: (011) 23260116 e-mail : dayabooks@vsnl.com website : www.dayabooks.com
Laser Typesetting	:	**Classic Computer Services** Delhi - 110 035
Printed at	:	**Chawla Offset Printers** Delhi - 110 052

PRINTED IN INDIA

Editorial Board

-Editor-

Dr. V.K. Gupta

Deputy Director & Head Central Animal Facility,
Indian Institute of Integrative Medicine (CSIR),
Canal Road, Jammu-180 001, INDIA
Tel: +91-0191-2520039 ; 09419228396 (M) Fax: +91-0191-2569017, 2569023
E-mail: vgupta_rrl@yahoo.com; vkguptaiiim@gmail.com

-Members-

Prof. Yu ZHAO

Cheung-Kung Chief Professor of Chinese Ministry of Education,
Assistant to President of Dali University,
Honorary Dean of College of Pharmacy,
Dali University, Wanhua Road, Yunnan Province, CHINA
Tel: +86-0872-2214251; Fax: +86-0872-2257401
E-mail: dryuzhao@126.com ; dryuzhao@hotmail.com

Prof. Yukihiro Shoyama

Faculty of Pharmaceutical Science,
Nagasaki International University,
His Ten Bosh, Sasebo, Nagasaki, JAPAN
Tel: +81-956-20-5653
E-mail: shoyama@niu.ac.jp

Prof. Ian Fraser Pryme

Department of Biomedicine,
University of Bergen,
Jonas Lies vei 91, N-5009 Bergen, NORWAY
Tel: +47-55-586438; Fax: +47-55-586360
E-mail: ian.pryme@biomed.uib.no

Dr. George Qian Li

Herbal Medicines Research and Education Centre,
Faculty of Pharmacy, The University of Sydney,
NSW 2006, AUSTRALIA
Tel: 612 9351; Fax: 9351 8638
E-mail: george.li@sydney.edu.au

Dr. Julia Serkedjieva

Institute of Microbiology,
Bulgarian Academy of Sciences,
26, Academician Georgy Bonchev St.,
1113 Sofia, BULGARIA
Tel: +359 2 979 31 85; Fax: +359 2 870 01 09
E-mail: jserkedjieva@microbio.bas.bg

UNIVERSITETET I BERGEN
UNIVERSITY OF BERGEN

INSTITUTT FOR BIOMEDISIN   DEPARTMENT OF BIOMEDICINE

Jonas Lies vei 91
5009 Bergen
Telefon (55) 586438
Mobil 91345296
Telefax (55) 586360

Jonas Lies vei 91
N0-5009 Bergen
Norway
Telephone +47 55 586438
Mobile +47 91345296
Telefax +47 55 586360

E-mail: ian.pryme@biomed.uib.no

Foreword

The importance of natural products in health and medicine is now well accepted and there is a world-wide desire to increase the pace of replacing pharmaceuticals and drugs with naturally existing materials. In comparison to highly purified drugs natural products contain a multifactorial complement of molecules and it is becoming increasingly likely that many of these work in concert and it is this complex interplay that provides health benefits. In this first volume of " *Natural Products: Research Reviews*" edited by Dr. V.K.Gupta, we meet a wide spectrum of interesting materials from natural sources that in the future can have a major impact on our health and well-being. This is exemplified by the first article: Diabetes and Medicinal Plants in Portugal written by Fernanda M. Ferreira and colleagues. The incidence of diabetes is on the increase in all countries and alternatives to treatment with insulin would be certainly very welcome. Cancer is also a global problem and in two articles: Immunotherapeutic Targeting of Multiple Dysregulated Immune Functions in Cancer by Neem Leaf Glycoprotein (Anamika Bose et al) and Misteltoe Lectins in Cancer Therapy: Administration by the Oral Route or by Subcutaneous Injection? - A Review (Ian F. Pryme), cover different aspects of how plant derived proteins can be useful in cancer treatment. Alzheimers disease is becoming increasingly prevalent and remedies for both prevention and therapeutics are desperately required. This topic is addressed by N. Suganthy and coworkers in the article entitled Anti-Amyloidogenic Effect of Natural Products: Implications for the Prevention and Therapeutics of Alzheimer's Disease. Another mental problem that is also on the increase is that of anxiety. How medicinal plants can be used in its control is the subject of the chapter Medicinal Plants for Management of Anxiety written by Richa Shri. Onions and garlic are well known in folk medicine as being beneficial for our health and this is the subject addressed by Dheeraj Singh and colleagues in: Review on Medicinal Properties of Onions and Garlic. There is currently great interest concerning the immunomodulatory effect of plant-derived products. This area is discussed in the article written by Amit Gupta and colleagues: Effect of Natural Plant Based and Non-Steroidal Anti-Inflammatory Drugs on the Immune System. Two chapters are devoted to phytochemistry. Firstly, the article written by Bhuwan B. Mishra *et al* has the title Phytochemical Profile of *Aegle marmelos* (Family-Rutaceae). Bengal quince (*Aegle marmelos*) is a mid-sized, slender, aromatic, gum-bearing tree growing up to 18 meters tall, and various parts of the plant are used for medicinal purposes. Secondly, that by Vivianne Marcelino de Medeiros and colleagues where the Phytochemical profile of the Genus Croton is discussed. *Acmella oleracea*, a flowering herb, native to the tropics of Brazil, is known to soothe toothache, and this plant is the theme of the article by Ana Claudia F. Amaral *et al*: An Overview of *Acmella oleracea* a Multipurpose Species. In two chapters of analytical nature S. Koul and coworkers discuss Supercritical Fluid Extraction as a Tool for Value Addition of Natural Products, while C. C. Silva and coworkers describe the Inorganic Composition of Medicinal Plants.

I complement the editor as well as the contributors for their valuable contributions. The chapters contained within this volume of "*Natural Products: Research Reviews*" provide the reader with an excellent in-depth overview of several current areas of major interest in the field.

Ian F. Pryme

Preface

The term 'natural products' is applied to materials derived from plants, microorganisms, invertebrates and vertebrates, which are fine biochemical factories for biosynthesis of both, primary and secondary metabolites. Natural products offer a virtually unlimited source of potential new pharmaceuticals and agrochemicals, largely because of the remarkable diversity of both chemical structures and biological activities of naturally occurring secondary metabolites. Bioactive compounds from natural products have attracted the attention of biologists and chemists throughout the world. The importance of natural products in modern medicine has been well recognized. More than 20 new drugs, launched world over between 2000 and 2005, originate from natural products. Scrutiny of medical indications by source of compounds has demonstrated that natural products and related drugs are used to treat 87% of all categorized human diseases (infectious and non-infectious).

Over 50% of prescription drugs on the US market were derived from natural products. Additionally, the utility of novel bioactive natural products as biochemical probes, the development of novel and sensitive techniques to detect biologically active natural products, improved techniques to isolate, purify, and structurally characterize these active constituents, and advances in solving the demand for supply of complex natural products collectively provide a compelling justification to search for bioactive natural products with potential pharmaceutical and agricultural applications.

A potential explanation beyond the success of natural product as drugs is the classification of natural compounds as so-called privileged structures. This is because chemical agents produced by living organisms (particularly the secondary metabolites) have evolved over millennia under the evolutionary pressure, and are therefore more likely to have a specific biological activity than "randomly" assembled, man-made synthetic chemicals. Despite the enormous potential, only a minor fraction of globe's living species has ever been tested for any bioactivity. For instance, approximately only 10% of the existing 350,000 plant species have been investigated from a phytochemical and pharmacological point of view, and in the case of microbes the value is even lower.

In USA, the botanical market, including herbs and medicinal plants is estimated around US$1.6 billion per annum. The dominating countries are China with exports of over 120,000 tonnes annually and India with some 32,000 tonnes annually. It is estimated that Europe imports medicinal plant from Asia and Africa is about 400,000 tonnes annually which cost approximately US$ 1 billion. With the growing awareness about this new commodity towards the foreign-exchange reserves, a number of national economies are beginning to emerge. Surveys are being conducted to unearth new plant sources of herbal remedies and medicines to satisfy this growing demands.

The present volume of the book series entitled, "*Natural Products: Research Reviews*" illustrates the types of critical discoveries that emerge from the interface of chemistry and biology, contains 12 review articles written by leading authorities in their respective fields of research. The aim is the enthuse the reader with this active and exciting area of research and to lay a solid foundation on which further study of its various facts may be based and the editor will consider himself to be amply rewarded if this humble piece of work proves to be useful for those it is meant. The book series shall hopefully provide an outlet for research work being carried out in these fields. Finally, I am thankful to the publisher for their unstinted support for maintaining the high publishing standards of the book.

Dr. V.K. Gupta

Contents

Editorial Board   *v*

Foreword   *vii*

Preface   *ix*

1. **Diabetes and Medicinal Plants in Portugal**   1

   *Fernanda M. Ferreira, Francisco P. Peixoto, Raquel Seiça and
   Maria S. Santos (**Portugal**)*

2. **Immunotherapeutic Targeting of Multiple Dysregulated Immune
   Functions in Cancer by Neem Leaf Glycoprotein**   41

   *Anamika Bose, Shyamal Goswami, Soumyabrata Roy, Koustav Sarkar,
   Krishnendu Chakraborty, Tathagata Chakraborty, Enamul Haque,
   Subrata Laskar and Rathindranath Baral (**India, USA**)*

3. **Mistletoe Lectins in Cancer Therapy: Administration by the Oral
   Route or by Subcutaneous Injection?–A Review**   79

   *Ian F. Pryme (**Norway**)*

4. **A Review on Medicinal Properties of Onions and Garlic**   99

   *Dheeraj Singh, M.K. Chaudhary, H. Dayal, M.L. Meena and A. Dudi (**India**)*

5. **Phytochemical Profile of *Aegle marmelos* (Family-Rutaceae)**   127

   *Bhuwan B. Mishra, Navneet Kishore, Vinod K. Tiwari and
   Vyasji Tripathi (**India**)*

6. **Supercritical Fluid Extraction as a Tool for Value Addition of
   Natural Products**   161

   *S. Koul, D.K. Gupta, V.K. Gupta and S.C. Taneja (**India**)*

7. **Phytochemistry of the Genus *Croton***　　　　221

   Vivianne Marcelino de Medeiros, Josean Fechine Tavares,
   Jackson Roberto Guedes da Silva Almeida, Vicente Toscano de Araújo-Junior,
   Petrônio Filgueiras de Athayde-Filho, Emídio Vasconcelos Leitão da Cunha,
   José Maria Barbosa-Filho and Marcelo Sobral da Silva (Brazil)

8. **Anti-Amyloidogenic Effect of Natural Products: Implications**
   for the Prevention and Therapeutics of Alzheimer's Disease　　　371

   N. Suganthy, S. Karutha Pandian and K. Pandima Devi (India)

9. **An Overview of *Acmella oleracea*: A Multipurpose Species**　　　409

   Ana Claudia F. Amaral, José Luiz P. Ferreira, Aline de S. Ramos,
   Deborah Q. Falcão, Cristina B. Viana, Jane S. Inada, Sílvia L. Basso,
   and Jefferson Rocha de A. Silva (Brazil)

10. **Effect of Natural Plant Based and Non-Steroidal Anti-Inflammatory**
    Drugs on the Immune System　　　439

    Amit Gupta, Anamika Khajuria, Jaswant Singh, Surjeet Singh and
    V.K. Gupta (Taiwan, India)

11. **Inorganic Composition of Medicinal Plants**　　　461

    C.C. Silva, J.E.M. Gai, L.A.A. Freitas, L.S. Freire and V.F. Veiga-Junior (Brazil)

12. **Medicinal Plants for Management of Anxiety**　　　471

    Richa Shri (India)

    Index　　　501

Natural Products: Research Reviews Vol. 1 (2012)
Editor: **V.K. Gupta**
Published by: **DAYA PUBLISHING HOUSE, NEW DELHI**

Pages **1–39**

1

Diabetes and Medicinal Plants in Portugal

Fernanda M. Ferreira[1,2]*, Francisco P. Peixoto[2,3],
Raquel Seiça[4] and Maria S. Santos[5]

ABSTRACT

Due to the global adaptation of western lifestyles and consequent increase in childhood and adult obesity, type 2 diabetes has become an epidemic on global scale. According to the World Health Organization, type 2 diabetes mellitus is the most common endocrine disorder, currently affecting more than 173 million people around the world and about 9 per cent of global mortality is closely related to diabetes mellitus.

Currently, the use of anti-diabetic oral drugs (ADOs) to control hyperglycaemia does not promote a satisfactory goal for most diabetic patients.

In recent years, despite the advances in the diagnosis and treatment of diabetes in the western medicine, an increasing interest in traditional anti-diabetic plants has been observed. Indeed, medicinal plants seem to be a useful alternative to synthetic drugs used in diabetes therapy and several active compounds of some of these synthetic drugs (such as metformin or

1. Environmental Sciences Department (CERNAS)– Agricultural College of Coimbra, Coimbra, Portugal
2. Center for Animal and Veterinary Sciences (CECAV), University of Trás-os-Montes and Alto Douro, Vila Real, Portugal
3. Chemistry Department, University of Trás-os-Montes and Alto Douro, Vila Real, Portugal
4. Institute of Physiology and Institute of Biomedical Research in Light and Image, Faculty of Medicine, University of Coimbra, Portugal
5. Department of Life Sciences, Center for Neurosciences and Cell Biology of Coimbra, University of Coimbra, Portugal
* *Corresponding author*: E-mail: fmlferreira@gmail.com

guanidine) are extracted from plants or have similar effects. In Portugal where the prevalence of diabetes is also increasing, there has been noticed a remarkable search for anti-diabetic medicinal plants to be used alone or in combination with prescribed medication. In this narrative review, we describe and characterize aromatic and medicinal plants as, for instance, Artium minus Bernh., Cistus ladanifer L., Cytisus multiflorus Sweet, Geranium robertianum L., Hypericum androsaemum L., Lupinus albus L., Pterospartum tridentatum (L.) Willk, Salvia officinallis L. and Vaccinium myrtillus L., used in Portuguese folk medicine for type 2 diabetes treatment.

Keywords: Medicinal and aromatic plants, Type 2 diabetes mellitus, Phytotherapy.

Abbreviations

ADO: Antidiabetic oral drug

NO: Nitric oxide

PPAR-γ: Peroxisome proliferator-activated receptor γ

ROS: reactive oxygen species.

Introduction

Plants have been used since immemorial times for treatment of human complaints. However, in the last century, due to the pharmaceutical industry progress, herbal medicine in the developed countries have been relegated to a lesser position and, often, patients who still employed medicinal plants were correlated with ignorance and witchcraft.

Nevertheless, this situation has been reversed in the last few decades and an increasing demand for herbal medicine as a complement for traditional therapy has occurred (Izzo and Ernst, 2009). Simultaneously, there has been an exponential growth in aromatic and medicinal plant research. Actually, medicinal plants are a good source for pharmaceutical industry and many popular medicines in use, as the common aspirin, or the antidiabetic oral drug metformin are derived from plants.

Medicinal plants seem to be an important and useful alternative (or complementation) to the synthetic drugs used in type 2 diabetes' therapy. In fact, several of these synthetic drugs, such as metformin or guanidine, are based on active compounds previously extracted from medicinal plants (Petlevski *et al.*, 2001; Mueller and Jungbauer, 2009). Despite the increasing use of medicinal plants in the treatment of diabetes mellitus (Halestrap *et al.*, 2000; Ryan *et al.*, 2001), there is still little knowledge about the mechanism of action and the therapeutic effects of several of these plants with attributed anti-diabetic action by folk medicine (Naga Raju *et al.*, 2006).

In Portugal, an increasing interest of phytotherapy treatments as a complement for traditional diabetes therapy (oral anti-diabetic medication or insulin) in order to reach normal glycemic levels and prevent later complications has been observed (Barata, 2008).

The current review focuses on herbal drug preparations and plants used in Portugal since ancient times for the treatment of diabetes mellitus.

Diabetes and Significance

Diabetes mellitus is a complex and a multifarious group of disorders characterized by all metabolism of carbohydrates, fat and protein disturbs, with one common manifestation – hyperglycemia (WHO, 1980). While type 1 diabetes is caused by an abolishment of insulin production, type 2 diabetes, the most common form, and accounting for 85-90 per cent of the occurrences, is caused by hepatic and peripheral tissue insulin resistance and pancreatic beta-cell dysfunction (William and Pickup, 2003). Due to western lifestyles of developing countries and to the consequent increasing rates of childhood and adult obesity, type 2 diabetes mellitus has become an epidemiologic problem (WHO, 2006).

Hyperglycaemia, as a result of uncontrolled glucose regulation, promotes oxidative stress and causes severe diabetic complications, such as nephropathy, retinopathy, neuropathy and cardiovascular diseases, which severely impairs diabetic patients' life quality (Engelgau and Geiss, 2000; Nuorooz-Zadeh, 2000; Yorek, 2003). Hence, the major goal of diabetes therapy is to maintain normal glycaemia levels (Agius, 2007; Yu *et al.*, 2010). Nontheless, this target usually remains unreachable using regular therapies and chemical anti-hyperglycaemic agents.

Medicinal plants are being rediscovered for the treatment of chronic diseases, including diabetes. As a matter of fact, many conventional drugs have been derived from prototypic molecules in medicinal plants. Metformin illustrates an efficacious oral glucose-lowering agent. Its development was based on the use of *Galega officinalis* L., which is rich in guanidine, a hypoglycemic component. Nevertheless, as guanidine is too toxic for clinical use, the alkyl biguanides synthalin A and synthalin B were introduced as oral anti-diabetic agents in Europe in the 1920s (Dey *et al.*, 2002). Despite its properties, their use was discontinued after insulin became more readily available. However, the experience with guanidine and biguanides prompted the development of metformin (Dey *et al.*, 2002). To date, over 400 traditional plant treatments for diabetes have been reported, although only a small number of these have received scientific and medical evaluation to assess their efficacy (Modak *et al.*, 2007). Even so, the hypoglycemic effect of some herbal extracts has been corroborated in humans or in type 2 diabetes animal models (Shukia *et al.*, 2000).

The World Health Organization Expert Committee on Diabetes has recommended traditional medicinal herbs to be further investigated (Day, 1989; Dey *et al.*, 2002). In the last two decades innumerous scientific works were published reporting the effectiveness and promising effects of several plants and their chemical constituents in diabetes mellitus therapy.

List of Medicinal and Aromatic Plants Focused in this Chapter

1. Family Asteraceae

   (a) *Arctium minus* (Hill) Bernh

2. Family Cistaceae

   (a) *Cistus ladanifer* L.

3. Family Clusiaceae

   (a) *Hypericum androsaemum* L.

4. Family Ericaceae

   (a) *Vaccinium myrtillus* L.

5. Family Fabacea

   (a) *Cytisus multiflorus* (L' Hér.) Sweet

   (b) *Lupinus albus* L.

   (c) *Pterospartum tridentatum* (L.) Willk. subsp. *tridentatum*

6. Family Gentianaceae

   (a) *Centaurium erythraea* RAFN

7. Family Geraneaceae

   (a) *Geranium robertianum* L.

8. Family Lamiaceae

   (a) *Rosmarinus officinalis* L.

   (b) *Salvia officinalis* L.

1. Family Asteraceae

(a) Arctium minus (Hill) Bernh

Arctium minus (Hill) Bernh (lesser burdock or common burdock), is a species of the genus *Arctium*, tribe *Cynareae*, family *Asteracea*. This genus comprises some other species such as *A. lappa*, *A. pubens* and *A. tomentosum*, being the species *A. lappa* being the mostly used in phytotherapy.

A. minus, a Portuguese autochthonous plant, is locally known as "*pegamasso*", "*pegamaço*" or "*bardana*". Lesser burdock is a biennial thistle spontaneous plant, found mainly in country paths edges and uncultivated grounds. *A. minus* is native to Europe (alternatively to *A. lappa*, usual in eastern countries); however, nowadays it is widespread throughout most of the United States, as a common weed. It can grow up to 1.5 meters tall and forms multiple branches. Flowers are prickly and pink to lavender in colour. Flower heads are about 2 cm wide. The plant flowering season is from July through October. Leaves are long and ovate. It grows an extremely deep taproot, and roots attain up to 30 cm.

A. minus roots decoctions are commonly used in Portugal to treat type 2 diabetes mellitus (Cunha *et al.*, 2006). In fact, *Arctium* roots contain inulin, a fructan compound, indigestible in the upper gastro-intestinal tract, with inhibitory action over the absorption of substances in small intestine and leading to intestinal microflora improvement. Therefore, *A. minus'* roots are used by diabetic patients to slow carbohydrate digestion, to reduce the absorption and control glucose intolerance (Kardosová *et al.*, 2003, Li *et al.*, 2008; Lou *et al.*, 2009).

Some controversial results were found about the antihyperglycemiant activity of *A. minus* roots in animal models. Cavalli and collaborators found that *A. minus* leaves and roots reduce alloxan-treated rats' plasmatic glucose level, in a similar way to the ADO glibenclamide (Cavalli *et al.*, 2007). These authors suggest that these extracts increase both insulin production and insulin cells sensitivity. However, using Goto-Kakizaki (GK) rats, we did not observe an improvement in hyperglycaemia control (Ferreira *et al.*, 2010a). Indeed the plant extract prepared from *A. minus* available in Portuguese herb shops, contained nickel (Ni) and cadmium (Cd) that could inhibit insulin release (Dormer *et al.*, 1974; Gupta *et al.*, 2000; Lei *et al.*, 2007) and possess toxic effects (Wang *et al.*, 2004; Ferreira *et al.*, 2010a). Nevertheless, as most aromatic and medicinal plants have the ability to bioaccumulate several heavy metals (Broadley *et al.*, 2001), the variations observed in both studies may be attributed both to the different animal models of diabetes and to the different chemotypes studied.

Arctium roots also contain arctiin, a lignane glucoside compound (Yakhontova and Kibal'chich, 1971), with anticarcinogenic activity (Wang *et al.*, 2005). Furthermore, ethanolic and aqueous extracts from *A. minus* leaves were found to possess antioxidant and anti-inflammatory properties, being relevant to diabetic patients in order to decrease oxidative stress and common low-grade of inflammation associated to the disease (Erdemoglu *et al.*, 2009).

2. Cistaceae Family

(a) Cistus ladanifer L.

Roots and aerial parts of *Cistaceae* plants have been used since ancient times in the Mediterranean cultures for its medicinal properties. The species *Cistus ladanifer* L. (rock-rose or gum rockrose), is an indigenous plant of the western Mediterranean region, being widely distributed over western Iberia and northwest Africa (Sosa *et al.*, 2005; Belmokhtar *et al.*, 2009). It is a shrub growing 1-2.5 m tall. The leaves are evergreen, dark green above and paler underneath, lanceolate, with dimensions varying from 3-10 cm long and 1-2 cm broad. The flowers are 5-8 cm diameter, with 5 papery white petals, usually with a red to maroon spot at the base, surrounding the yellow stamens and pistils. The whole plant is covered with the sticky exudates of a fragrant resin known as labdanum gum, being particularly appreciated for their therapeutic applications as well as for their balsamic odor and fixative properties (Teixeira *et al.*, 2007; Barrajón-Catalán *et al.*, 2010).

In Portugal, *C. ladanifer* L., locally known as "esteva das cinco chagas" or simply "esteva", is widely distributed, being one of the most abundant species in the southern part of the country, occurring in large areas as pure dense stands. This shrub colonizes degraded areas and inhibits the growth of other plants (Dias and Moreira, 2002), either by restricting aerial growth of plants or by inhibiting germination of other species, due to its phytotoxicity over other plants and soil, a phenomenon known as allelopathy (Chaves *et al.*, 2001; Alías *et al.*, 2006). The genus *Cistus* easily adapts to wildfires that destroy large forest areas, as their seeds resist fire and rapidly repopulate in the following season (Ferrandis *et al.*, 1999). This could explain why *C. ladanifer* is very abundant along the Portuguese forest and landscape, becoming their overgrowth an environmental problem.

In Portugal, rock-rose is used for treating diabetes and for respiratory and rheumatic problems, among other applications (Castro, 1998; Camejo-Rodrigues, 2001; Carvalho, 2005). In fact, *C. ladanifer* is used as a panacea or "a remedy for all diseases" (Carvalho, 2005). Fruits and aerial parts (harvested before flowering season) infusions are usually used with therapeutical applications in Portugal (Castro, 1998; Camejo-Rodrigues, 2001; Carvalho, 2005; Andrade *et al.*, 2009). Nevertheless, in Morocco, the use of leaf infusions or decoctions (Bnouham*et al.*, 2002), seed infusions are also used in diabetes therapy (Merzouki *et al.*, 2003).

Despite its widespread use, at the present, no scientific reports concerning the effectiveness of *C. ladanifer* water extracts on diabetes mellitus were found. In recent years, several publications noticing high antioxidant properties either in water and ethanol extracts and essential oils from *C. ladanifer* were achieved (Andrade *et al.*, 2009; Barrajón-Catalán *et al.*, 2010; Guimarães *et al.*, 2010). Actually, exudates from this plant present characteristic aglycone flavonoid compounds, such as apigenin (and derivatives), several kaempferol derivatives, quercetins and ellagitannins (Chaves *et al.*, 1998; Barrajón-Catalán *et al.*, 2010). Phenolic content of *C. ladanifer* water extract was higher than that of other aqueous extracts previously reported (Dudonné *et al.*, 2009; Barrajón-Catalán *et al.*, 2010) and almost similar to ethanolic *C. ladanifer* extracts contents (Andrade *et al.*, 2009; Barrajón-Catalán *et al.*, 2010). As the inhibition of reactive oxygen species (ROS) is associated with a positive impact on human health, throughout pathogenesis modulation of many diseases associated to oxidative stress, such as atherosclerosis, hypertension, cardiovascular disease, ischemia/reperfusion injury, diabetes mellitus, cancer and neurodegenerative diseases (Ceriello, 2003; Valko *et al.*, 2006; Seifried, 2007; Valko *et al.*, 2007; Barrajón-Catalán *et al.*, 2010), the high antioxidant activity of *C. ladanifer* is probably associated with the widespread therapeutic utilization of this plant.

C. ladanifer extract also demonstrated to possess anticancer activity against several tumour cell lines (Barrajón-Catalán *et al.*, 2010), namely breast and pancreas tumour cells. This anticarcinogenic potential is probably related to the combination of elligitannins present in the extract. Moreover, this plant showed antihypertensive effects (Belmokhtar *et al.*, 2009), improving vascular reactivity and induced an endothelium- and NO-dependent relaxation of vascular smooth muscle.

In conclusion, *C. ladanifer* extracts possess a significant amount and variety of polyphenolic compounds with an important antioxidant activity, being worthy to prevent or reduce the development of diabetic complications.

3. Family Clusiaceae

(a) Hypericum androsaemum L.

Hypericum androsaemum L., commonly known as tutsan, is a *Clusiaceae* (or *Guttiferea*) plant, native of open woods and hillsides in Europe. Usually requires shadowy places to grow (Cunha *et al.*, 2009). In Portugal, it is usually known as "Hipericão do Gerês", "androsemo" or "erva-da-pedra" and, is a common plant in northern and central areas of the country, mainly at the highest altitudes of Minho, Deiras and Estremadura (Valentão, 2002; Cunha *et al.*, 2009). *H. androsaemum* is a perennial shrub, between 30-120 centimeters in height. Branches are separated in two longitudinal lines, with oval and green leaves. It flowers from June to September, and flowers are small (2 cm) and yellow (Coutinho, 1939). The berries turn from white or green, to red and to black during the maturation process.

Infusions (or decoctions) of aerial parts of *H. androsaemum* are traditionally used in folk medicine, being one of the most consumed medicinal plants in Portugal (Valentão, 2002; Cunha *et al.*, 2009). The common name tutsan appears to be a corruption of "*toute saine*", literally meaning all healthy and probably in reference to its healing properties. Usually, dry aerial parts, harvested just before or during flowering season are used for treating kidney and liver ailments and as diuretic (Valentão, 2002; Novais *et al.*, 2004; Cunha *et al.*, 2009). Further, leaves are applied topically to wounds or burns, due to its healing properties (Lavagna *et al.*, 2001; Valentão, 2002; Cunha *et al.*, 2009). Information collected in several ethnobotanical studies, confirms also the use of *H. androsaemum* as sedative, antidepressant, antihypertensive and for digestive ailments treatment (Novais *et al.*, 2004; Carvalho, 2005) as well as for diabetes treatment (Castro, 1998).

Chemically, *H. androsaemum* contain many polyphenolic compounds, namely flavonoids and phenolic acids (Valentão *et al.*, 2002; Valentão, 2002), that show seasonal variations (Guedes *et al.*, 2004) and chemical polymorphism (Valentão *et al.*, 2003). The compounds mainly described are caffeic acid, chlorogenic acid, luteolin, kaempferol, quercetins and several xanthones (Valentão, 2002) and references therein). However, rutin (quercetin-3-rutinoside) and hypericin, an anthraquinone-derivative, present in other *Hypericum* species (*H. perforatum, H. undulatum*), seem to be present in the *H. androsaemum* preparations previously studied.

H. androsaemum extracts present high antioxidant potential, against hypochlorous acid and oxygen and nitrogen free radicals (Valentão *et al.*, 2002; Valentão *et al.*, 2004; Almeida *et al.*, 2009), probably due to the presence of several quercetin glycosides in the tutsan extract (Valentão *et al.*, 2002). The important antioxidant capacity, *in vitro*, seems to be responsible, in part, for the hepatoprotective properties attributed to this plant as observed in hepatocytes submitted to oxidative stress (Valentão *et al.*, 2004). Nevertheless, *in vivo* studies demonstrate that *H. androsaemum* water extract increase hepatotoxicity induced by tert-butyl hydroperoxide (t-BHP) (Valentão *et al.*, 2004). t-BHP is metabolized into free radical intermediates by cytochrome P450 in hepatocytes, which initiate lipid peroxidation, glutathione depletion and cell damage. Histopathological evaluation of the mice livers revealed that *H. androsaemum* infusion raised the incidence of liver lesions induced by t-BHP. Hence, this study, involving cytochrome P450 activity, does not corroborate the effectiveness of *H. androsaemum* infusion as hepatoprotector but rather its effect as hepatotoxicity potentiator (Valentão *et al.*, 2004).

H. androsaemum is commonly referred as a species in which xanthonoids biosynthesis play an important role (Schmidt and Beerhues, 1997; Schmidt *et al.*, 2000), mainly 1,3,5,6 and 1,3,6,7 oxygenated xanthones (Dias *et al.*, 2000).

Concerning to *H. androsaemum* employment as an antihyperglycemiant plant (Castro, 1998), mangiferin (1,3,6,7-tetrahydroxy-2-[3,4,5-trihydroxy-6-(hydroxymethyl)oxan-2-yl]xanthen-9-one), one of the modified xanthones present in *H. androsaemum*, of major importance. Indeed, this glucosylxanthone shows antihyperglycemic activity (Miura *et al.*, 2001; Muruganandan *et al.*, 2005) and also presents important antiatherogenic properties (Muruganandan *et al.*, 2005), preventing diabetic nephropathy progression in streptozotocin-induced diabetic rats and improving renal function in diabetic rats (Li *et al.*, 2010) and presents a high antioxidant activity (Prabhu *et al.*, 2006), counteracting oxidative stress associated to diabetes. Furthermore, luteolin, a flavone present in *H. androsaemum*, also displays antidiabetic activity (Zarzuelo *et al.*, 1996), increasing insulin sensitivity in adipocytes (Ding *et al.*, 2009), presents an important antioxidant power (Torel *et al.*, 1986; Miean and Mohamed, 2001) and a significant anti-inflammatory action (Chen *et al.*, 2007), reducing the impairment of endothelium-dependent relaxation in rat aorta, commonly observed in diabetic patients, by reducing oxidative stress (Qian *et al.*, 2010). Together with the potential antidiabetic activity, *H. androsaemum* possesses an elevated antioxidant activity and undoubtedly plays an important role in avoiding (or decreasing) diabetic complications.

4. Family Ericaceae

(a) Vaccinium myrtillus L.

Vaccinium myrtillus L. (bilberry or European blueberry), is an *Ericaceae*. This genus is widespread over the world and comprises over 200 species of evergreen woody plants varying from dwarf shrubs to trees (Jaakola, 2003). *V. myrtillus* is a shrubby perennial plant that can be found in mountains and forests both in Europe and in the northern United States. In fact, bilberry is found in very acidic and nutrient-poor soils, throughout the temperate and subarctic regions of the world. In Portugal, bilberry (*V. myrtillus*) is commonly known as "mirtilo","mirtilho","arando" or "uva-do-monte".

Bilberries produce single or paired dark berries on the bush, instead of clusters. Berries are dark, near black, with a slight shade of blue. The bilberry fruit is smaller than the blueberry, but with a fuller taste. While the blueberry's fruit pulp is light green, the bilberry's is red or purple, heavily staining the fingers and lips of consumers eating the raw fruit.

V. myrtillus is extremely difficult to grow being seldom cultivated in Portugal. However, both due to its high economic value and growing demand in the last years, the agriculture of several counties close to Vouga River, namely Sever do Vouga, deeply depend on bilberry production. Indeed, this province possesses the climatic requisites to produce high quality berries and this activity is partially supported by Portuguese Agricultural Ministry. Grândola, in the southern Portuguese seacoast, is another region where the bilberry production became a significant economic activity (Sousa *et al.*, 2007). In addition to the cultivated bilberry, we can also find wild bilberries in the northern Portugal, mainly in Trás-os-Montes (Neves *et al.*, 2009) and Alto Minho.

Actually, coupled to its utilization in food industry, mainly in jams, pies or yogurt and ice-creams preparation, this plant is widely used since ancient times in folk medicine, due to its therapeutic properties (Canter and Ernst, 2004; Valentová *et al.*, 2007; Bao *et al.*, 2008). Bilberry's history of medicinal use dates back to the medieval times, but it did not become widely known to herbalists until the 16th century, when it was used for treating biliary disorders, bladder stones, scurvy, coughs and lung tuberculosis. Lately, bilberry fruit extracts have also been used for the treatment of diarrhea, dysentery and mouth and throat inflammations (Anonymous, 2001, Valentová *et al.*, 2007).

V. myrtillus berries contain a large amount of anthocyanines and quercetins (and also pectins and fibers; conversely, present a low amount of glucides) (Häkkinen *et al.*, 1999; Erlund *et al.*, 2003; Sousa *et al.*, 2007). Nevertheless, the amount of phenolic compounds in berries largely depends upon the tissue analysed, the cultivar type and the edaphoclimatic conditions (Häkkinen and Törrönen, 2000; Witzell *et al.*, 2003). Usually, wild bilberry content of anthocyanins is substantially higher than the cultivated one (Kraus *et al.*, 2010).

Owing its high content in antioxidant compounds, namely flavonoids and phenolic acids, bilberry fruit is commonly used in folk medicine for micro-and-macrovascular system protection (Valentová *et al.*, 2007; Bao *et al.*, 2008). In fact, bilberry fruit extract is freely available in the market as a pharmaceutical preparation for the treatment of vascular diseases and diabetic retinopathy (Kalt and Dufour, 1997; Fraunfelder, 2004; Bao *et al.*, 2008). Furthermore, there is an increasing use of pure flavonoids to treat many important common diseases, due to their proven ability to inhibit specific enzymes, to simulate some hormones and neurotransmitters and to scavenge free radicals (Prior *et al.*, 1998; Martín-Aragón *et al.*, 1999; Havsteen, 2002).

Bilberry leaf decoctions and infusions are used by Portuguese people both in the treatment of diabetes mellitus (Castro, 1998) and hypercholesterolemia (Neves *et al.*, 2009). Indeed, all over the world, *V. myrtillus* leaf tea is used due to their antihyperglycemic properties (Cignarella *et al.*, 1996; Jaric *et al.*, 2007), sometimes in herbal mixtures (Petlevski *et al.*, 2001). Neomyrtillin, a glucoside compound found in *V. myrtillus* leaves, also known as "plant insulin" (Edgars, 1936; Bever, 1980), is associated to the attributed antidiabetic properties of bilberry leaf tea. Nevertheless, the decrease in blood glucose levels promoted by this neomyrtillin has no consensual opinions (Helmstädter and Schuster, 2010). Our results show that diabetic Goto-Kakizaki rats treated during 4 weeks with *V. myrtillus* leaf decoction lead to a slight decrease of occasional glycaemia and an improved intraperitoneal glucose tolerance test, mainly during the initial 60 minutes (Ferreira *et al.*, 2010b). Moreover, metal ions analysis showed that *V. myrtillus* extracts have an appreciable content of chromium (Cr) (Castro, 1998; Ferreira *et al.*, 2010c), an ion known to be altered in the diabetic state (Kim *et al.*, 2004; Zhao *et al.*, 2009). In effect, Cr has been described as a potential therapy of insulin resistance, a feature of type 2 diabetes (Kim *et al.*, 2004).

Furthermore, leaves also possess a large amount of flavonoid compounds, mainly procyanidins and the flavonols quercetin and kaempferol (Jaakola, 2003). This high antioxidant capacity, due to presence of flavonoids, is responsible for *V. myrtillus* health-promoting effects. In fact, it is well known that oxidative stress in diabetic patients leads to multiple cellular dysfunctions and chronic complications associated to the disease (Baynes and Thorpe, 1999; 2000; Ceriello, 2003). In consequence, an enhancement in dietary antioxidants possesses advantageous effects to diabetic patients (Prior *et al.*, 1998; Catoni *et al.*, 2008). Previous studies report cytoprotective effects of bilberry anthocyanins against oxidative damage in hepathocytes (Valentová *et al.*, 2007; Bao *et al.*, 2008). Further, protective effects of *V. myrtillus* anthocyanin rich extracts against mitochondrial dysfunction were described (Yao and Vieira, 2007; Ferreira *et al.*, 2010b), probably correlated to the great content of quercetins that can improve mitochondrial biogenesis (Davis *et al.*, 2009). Also, a decrease of potential neurotoxic activity with an improvement of neuronal and cognitive brain functions was also reported (Yao and Vieira, 2007; Zafra-Stone *et al.*, 2007). Indeed, berries as a potential source of natural anthocyanins, have demonstrated a broad spectrum of biomedical functions, in cardiovascular disorders, advancing age-induced oxidative stress, inflammatory responses and diverse degenerative diseases (Zafra-Stone *et al.*, 2007). Berry anthocyanins also protect genomic DNA integrity and, in fact, the intake of these flavonoid compounds seems to be correlated with the growth suppressing-effect observed in several types of cancer cells (Madhavi *et al.*, 1998; Cooke *et al.*, 2005; Zafra-Stone *et al.*, 2007; Nandakumar *et al.*, 2008; Milbury, 2009; Kraus *et al.*, 2010). The chemopreventive effects of flavonoid compounds in tumoral cells are undoubtedly important for diabetic patients, since it is widely proved the relationship between cancer and diabetes (Psarakis, 2006; Giovannucci *et al.*, 2010).

Dyslipidaemia is usually correlated to diabetic condition. Previous studies showed that bilberry leaf and fruit flavonoid compounds exhibited either a dose-dependent lipid-lowering activity in genetically hyperlipidemic Yoshida rats (Cignarella *et al.*, 1996) and an antihypercholesterolemic activity in hamsters, thus preventing atherosclerosis (Zafra-Stone *et al.*, 2007; Rouanet *et al.*, 2010). Additional, there are some reports showing that anthocyanines prevent angiogenesis and are also responsible for collagen stabilization (Roy *et al.*, 2002; Matsunaga *et al.*, 2009). Moreover, some studies reveal that bilberry fruit extracts present antibacterial properties (Rauha *et al.*, 2000; Puupponen-Pimiä *et al.*, 2008).

For these reasons, *V. myrtillus* is usually considered as a nutraceutical (Espín *et al.*, 2007; Garzón *et al.*, 2010) or a functional food (Katsube *et al.*, 2002). Additionally to lowering blood glucose levels ability of bilberry leaf teas, the valuable effects of anthocyanins against cellular damage induced by hyperglycaemic condition associated with high antimicrobial activity, makes this plant a worthy food supplement to consider for diabetic patients.

5. Family Fabacea

(a) Cytisus multiflorus (L' Hér.) Sweet

Cytisus multiflorus (L' Hér.) Sweet (white Spanish broom) is a species of the genus *Cytisus*, locally known as "giesta branca" (or "gesta branca") and widely spread in Portugal, where it invades a wide range of fertile soils. This plant can fix nitrogen and form a dense scrub layer that outcompetes with native species and becoming an environmental problem. White Spanish broom is native to Portugal, Spain and France and it has been introduced as an ornamental plant in India,

Australia, Italy, United States, New Zealand and Argentina (Weed Management Guide, 2003).

This is a shrub growing up to 3 or 4 meters in sprawling height with leaves appearing mainly on lower branches, each of them made up of three leaflets. Some leaves grow on the upper branches, generally made up of a single leaflet. Each leaflet is under a centimetre-long and may have a shape varying from linear to oblong. The white, pea-like flower is up to a centimetre long. The fruit is a hairy legume pod up to 3 centimetres long. The pods turn black when mature and release explosively their four to six seeds away from the parent plant (Weed Management Guide, 2003).

In Portugal, *C. multiflorus* flowers are usually used in popular medicine (sometimes in herbal mixtures) for the treatment of type 2 diabetes, headaches or for controlling hypertension and hypercholesterolemia (Castro, 1998; Camejo-Rodrigues, 2001; Camejo-Rodrigues *et al.*, 2003; Carvalho, 2005).

Some recent studies suggest the validity of ethnical use of *C. multiflorus* in hyperglycaemia control. The effect of aqueous extracts of *C. multiflorus* was studied in the third inbreeding generation of (Wistar) rodents, showing abnormal glucose tolerance and following oral glucose tolerance test, female glucose intolerant rats were selected. A significant dose-dependent decrease of the postprandial blood glucose levels was observed, in response to treatment with the plant extract, possibly due to an increase of insulin release, while fasting glycaemia was not significantly altered in treated rats (Areias *et al.*, 2008; Vieira *et al.*, 2010). Furthermore, the maximum effect of the plant extract was similar to the glycazide treated group of the experimental assay (Areias *et al.*, 2008; Antunes *et al.*, 2009).

The enhanced insulin secretion can be due to the presence of spartein, an alkaloid present in *Cytisus* genus, that blocks K_{ATP} channels and decreases β-cell K^+ permeability (García López *et al.*, 2004). Nevertheless, some intoxication cases are related to the ingestion of *Cytisus* genus plants prepared infusions (Nunes, 1999), due to the presence of spartein.

Despite the chemical characterization available for *C. multiflorus* is far from being complete, water extract from this plant contain a great amount of alkaloids, hydrolysed tannins and triterpenes (or steroids) and flavonoids (Antunes *et al.*, 2009).

This plant also possesses an elevated antioxidant activity, due to its high content in flavonoids and, therefore, plays an important role in preventing or reducing the development of diabetic complications (Gião *et al.*, 2007; Pereira *et al.*, 2009).

(b) Lupinus albus L.

White lupin (*Lupinus albus* L.) is a species of the genus *Lupinus*, tribe *Luppineae*, family *Fabacea*. Four species of this genus (*L. albus*, *L. angustifolius*, *L. luteus* and *L. mutabilis*) are cultivated worldwide with three main uses: human consumption, green manure and as forage (Huyghe, 1997). *L. albus* is an annual, endemic, traditional and widespread legume in Portugal where is commonly used as a snack. The white lupin is an annual, more or less pubescent plant, with 30–120 cm high, and exists in many distinct forms, as a result of adaptation to different edaphoclimatic (soil and climate) conditions, throughout the country (Carmali *et al.*, 2010; Vaz *et al.*, 2004). The plant has a single tap root system with threadlike portions reaching down to 70 cm and owns its name to the flower colour (white). Each plant can produce primary, secondary, and tertiary pods with each pod containing 3 to 7 seeds. Seeds with high protein content (around 30-35 per cent) and rich in dietary fibers, present a bitter taste due to the presence of quinolizdone alkaloids, and therefore, for consumption these alkaloids must be removed (Carmali *et al.*, 2010). It has been previously described that the architecture and behaviour of the plants changes from the North (where tall and late flowering types are common) to the South of Portugal (with short and early flowering types predominant), reflecting diversity in plants phylogenetic origin (Martins, 1994; Vaz *et al.*, 2004). Moreover, the alkaloid composition also varies due to genetic and environmental differences (Carmali *et al.*, 2010).

In Portugal (as well as in other Mediterranean countries) *L. albus*, locally known as "tremoço", is used in folk medicine due to its hypoglycemiant action (Pereira *et al.*, 2001; Eddouks *et al.*, 2002; Sheweita *et al.*, 2002). Traditionally, *L. albus* wastewaters, containing high amounts of alkaloids, mainly lupanine, are used intending to control blood glycaemia (Camejo-Rodrigues, 2001). In fact, scientific studies showed that lupanine (as 13-α-lupanine and 17-oxo-lupanine) stimulate insulin secretion in a glucose-dependent manner (and only at high glucose concentrations, ≥ 7 mM) (Pereira *et al.*, 2001; García López *et al.*, 2004). Spartein, another lupine alkaloid, showed similar effect, blocking K_{ATP} channels, decreasing β-cell K^+ permeability (García López *et al.*, 2004). Moreover, alkaloids due to their phenolic nature, also have antioxidant activity, worthy in diabetes mellitus therapy (Tsaliki *et al.*, 1999).

Likewise, in certain Portuguese regions, *L. albus* debittered seeds are also used to counteract type 2 diabetes (Pereira *et al.*, 2001). White lupine seeds contain a large amount of conglutin-γ (around 2 per cent of dry weight), a glycoprotein composed of two disulfide bridges subunits (Mr $\sim$ 47 kDa) (Magni *et al.*, 2004). Conglutin-γ displays unique features, as an unusual primary structure, with high resistance to *in vitro* proteolysis, and the ability to bind divalent ions, such as Zn^{2+} and Ni^{2+} (Magni *et al.*, 2004 and articles therein). This globular glycoprotein also mimetizes insulin action in myoblasts, playing an important role in vesicular transport of glucose carrier GLUT 4, influencing cell differentiation and controlling muscle growth (Terruzzi *et al.*, 2010).

L. albus seeds also have antioxidant activity, being important in counteracting oxidative stress induced by hyperglycaemia (Tsaliki *et al.*, 1999). *L. albus* is also used in traditional medicine in some diseases associated with diabetes mellitus, such as dyslipidemia, hypercholesterolemia and hypertension (Camejo-Rodrigues, 2001; Sirtori *et al.*, 2009). It has been proven that white lupin alkaloids, 13-α-lupanine and 17-oxo-lupanine, reduce blood pressure (Yovo *et al.*, 1984). Moreover, white lupin seeds significantly decrease blood pressure in diabetic and hypertensive rats (Pilvi *et al.*, 2006). This is a possible consequence of their high content in arginine, leading to an increase in NO production (Duke, 1992; Sirtori *et al.*, 2009). Several scientific reports indicate a substantial reduction of hypercholesterolemia induced by lupin seed proteins and moderate changes in triglycerides (Sirtori *et al.*, 2004), that appears to depend on a down regulation of liver SREBP 1c (sterol regulatory element-binding protein), a transcription factor that regulates the expression on lipogenic enzymes (Spielmann *et al.*, 2007).

Concerning to toxicological studies, main alkaloid compounds (lupanine and ehydroxilupanine (Duke, 1992)) do not pose a health problem for man, since LD_{50} values for oral administration are elevated (around 1500 mg/Kg), being rapidly cleared from the body (Petterson *et al.*, 1987). Spartein, another alkaloid found in *Lupinus* genus, with low LD_{50}, is classified as antiarrhythmic agent and sodium channel blocker (Yovo *et al.*, 1984; Pothier *et al.*, 1988). Some anticholinergic effects of lupine alkaloids have also been observed in rodents (Pothier *et al.*, 1988) and there are two case reports in humans associated with *L. albus* ingestion, due to sparteine intoxication (Tsiodras *et al.*, 1999; Litkey and Dailey, 2007). Moreover, during a 3 months lasting study with rodents, lupin alkaloids ingestion (500mg/kg/day) produced nonlethal hematological effects (Butler *et al.*, 1996).

(c) Pterospartum tridentatum (L.) Willk. subsp. tridentatum

Pterospartum tridentatum (L.) Willk., known as "prickled broom" (and previously known as *Chamaespartum tridentatum*) is an autochthonous plant, found commonly in Portugal. In fact, *P. tridentatum* is usual in the Norwest part of Iberian Peninsula and in Morocco. *P. tridentatum*, locally known as "carqueja" or "carqueija", grows in acidic soils, in brushwood's, thickets and is a shrub, with characteristic yellow flowers with a typical odour, that are traditionally harvested in Spring (from March to June). The yellow flowers are used in popular medicine (sometimes in herbal mixtures) for the treatment of throat irritation conditions, diabetes (Castro, 1998; Vitor *et al.*, 2004; Grosso *et al.*, 2007) or for controlling hypertension and hypercholesterolemia (Camejo-Rodrigues, 2001; Carvalho, 2005). As a matter of fact, in Portugal, the *P. tridentatum* flowers tea is used as a panacea, being a potential cure for all illnesses of the body (Camejo-Rodrigues, 2001; Carvalho, 2005). The leaves (steams) are normally used in culinary applications, to flavour rice, roast meat or hunting animals (Carvalho, 2005).

Complete chemical characterization of the plant is not yet available, since only in the last decade this plant became a subject of scientific research. However, a recent study in *P. tridentatum* essential oils showed that chemical composition of the analyzed oils are less a consequence of climatic factors in different years than due to differences in genetic heritage and/or other environmental factors (Grosso *et al.*, 2007).

Despite the information collected in several ethnobotanical studies, confirming the use of *P. tridentatum* extracts in diabetes therapy (Camejo-Rodrigues, 2001; Carvalho, 2005), at the present time, there is only one scientific report concerning the effects of *P. tridentatum* water extracts on the blood glucose levels (Paulo *et al.*, 2008). In this work, normal Wistar rats' glycaemia was investigated in a situation of oral glucose challenge. Water extract (300 mg/kg) showed an antihyperglycaemic effect in the initial 30 min after glucose challenge, but then the blood glucose levels rose above those of the control group, indicating the presence of compounds with different effects on glucose tolerance. Probably, these opposite effects were due to the presence of two different compounds: the isoflavone sissotrin and the flavonol derivative isoquercitrin. Isoquercitrin (100 mg/kg) showed time-dependent anti-hyperglycaemic activity by delaying the post-oral glucose load glycaemia peak at 30 min, similarly to phloridzin (100 mg/kg), a sodium-dependent glucose transporters inhibitor (Paulo *et al.*, 2008). In contrast, sissotrin (100 mg/kg) showed an opposite effect, impairing glucose tolerance (Paulo *et al.*, 2008).

Water extracts prepared from *P. tridentatum* aerial parts possess a strong antioxidant activity, with a high content of phenolic compounds and flavonoids (Luís *et al.*, 2009). Ethanolic extracts prepared from the same plant samples own lower antioxidant activity, despite having higher flavonoid content; even so, its antioxidant activity is similar to the standard antioxidant BHT (2,6-bis(1,1-dimethylethyl)-4-methylphenol or butylated hydroxytoluene) (Luís *et al.*, 2009). Indeed, Vitor and collaborators (Vitor *et al.*, 2004) suggested that flavonoids present in *P. tridentatum* water extracts exhibit endothelial protection against oxidative injury and, thus, may prevent or reduce the development of diabetic vascular complications.

6. Gentianaceae Family

(a) Centaurium erythraea Rafin

Centaurium erythraea Rafin, previously known as *Erythraea centaurium*, *Centaurium minus* and *Centaurium umbellatum*, belongs to the *Gentianaceae* family and is usually known as "common centaury" or "European centaury". In Portugal, it is known as "centáurea-menor" (or commonly "fel-da-terra", due to its bitter taste) (Cunha *et al.*, 2009). This is an erect annual or biennial herb, reaching half a meter in height.

C. erythraea inflorescences contain many small pinkish-lavender flowers, of about a centimeter across, flat-faced with yellow anthers and usually flowers from June to September. The fruit is a cylindrical capsule. *C. erythraea* is a widespread plant of Europe, Western Asia and Northern Africa. It has also been introduced in parts of North America and throughout Eastern Australia.

Infusions (and decoctions) of aerial parts of common centaury are traditionally used in folk medicine either in mild dyspeptic and/or gastrointestinal disorders and in temporary loss of appetite (Bnouham *et al.*, 2002; El-Hilaly *et al.*, 2003; Jaric *et al.*, 2007; CMHP 2009; Cunha *et al.*, 2009). This usage was approved by the "Committee on herbal medicinal products" of European Medicines Agency (Ref.: EMEA/HMPC/105535/2008) (CMHP 2009; Cunha *et al.*, 2009). In Portugal, *C. erythraea* is also used, as antipyretic, in hypercholesterolemia, hepatobiliary problems and diabetes mellitus therapy and as vermifuge (Camejo-Rodrigues *et al.*, 2003; Carvalho, 2005; Cunha *et al.*, 2009; Neves *et al.*, 2009). Besides, in some provinces of Morocco, *C. erythraea* is used for kidney disorders treatment (El-Hilaly *et al.*, 2003) and as diuretic (Haloui *et al.*, 2000). Reduction of blood pressure and a decrease in smooth muscle spasms of the gastrointestinal tract and sedative action over the central nervous system have also been reported (Loizzo *et al.*, 2008; Cunha *et al.*, 2009). *C. erythraea* utilization is discouraged both in young people and in the presence of peptic ulcers (CMHP 2009; Cunha *et al.*, 2009). Due to its bitter constituents, common centaury should also be avoided by lactating women (Cunha *et al.*, 2009).

C. erythraea is widely used in folk medicine due to its antihyperglycemic properties (Bnouham *et al.*, 2002; Cunha *et al.*, 2009; Hamza *et al.*, 2010), sometimes in herbal mixtures (Petlevski *et al.*, 2001).

Common centaury presents a high xanthone 6-hydroxylase activity, leading to 1,3,5-trihydroxyxanthone intramolecular cyclization (Schmidt *et al.*, 2000). The aerial parts of this plant possess a large content and variety of methoxylated xanthone derivatives (Schimmer and Mauthner, 1996; Valentão *et al.*, 2000; Valentão *et al.*, 2002; Valentão *et al.*, 2003). These xanthones, which contain a distinctive polyphenolic structure, show many pharmacological effects (Singh, 2008; Shekarchi *et al.*, 2010), such as antioxidant (Valentão *et al.*, 2001), antitumor (Schimmer and Mauthner, 1996), anti-diabetes (Petlevski *et al.*, 2001; Hamza *et al.*, 2010), bactericidal (Kumarasamy *et al.*, 2003) and hepatoprotective properties (Jaishree and Badami, 2010). Monotherpenes, β-sitosterol and some flavonoids are also present (Loizzo *et al.*, 2008; Cunha *et al.*, 2009). The gentianaceae plants also present secoiridoid glycosides, responsible for the characteristic bitter taste (Singh, 2008; Cunha *et al.*, 2009).

Recently interest among medicinal potential gentianaceae plants has been revived and phytochemicals, like swerchirin and swertiamarin have been rediscovered (Singh, 2008). Swerchirin (1,8-dihydroxy-3,5-dimethoxyxanthone) decreased high blood glucose, by stimulating insulin release in streptozotocin-induced type 2 diabetes rats (35 mg/kg i.v.) (Saxena *et al.*, 1991; 1993). Furthermore, *C. erythraea* hydroethanolic extract exhibited an antihyperglycemiant effect, decreased insulin resistance and triglycerides. These studies were conducted with C57BL mice with standardized high fat diet induced type 2 diabetes and no weight or caloric differences were noticed, when compared to controls (Hamza *et al.*, 2010). Swertiamarin is another important constituent, to which several medicinal properties are also attributed. Swertiamarin is a bitter secoiridoid glycoside, with high antimicrobial activity (Kumarasamy *et al.*, 2003) and both with high antioxidant and hepatoprotective potential (Jaishree and Badami, 2010). Swertiamarin metabolites also present anti-inflammatory properties (Jun *et al.*, 2008).

In summary, *C. erythraea* extracts stimulate pancreatic β-cells insulin release and decrease insulin resistance. Moreover, common centaury extracts possess a large content and variety of xanthones with uncommon polyphenolic structures, presenting a significant antioxidant and anti-inflammatory activities and therefore may prevent (or diminish) the development of diabetic chronic complications.

7. Geraniaceae Family

(a) Geranium robertianum L.

The genus *Geranium* encompasses more than 400 different species of flowering plants.

G. robertianum, a Geraniaceae commonly known as Herb Robert or Red Robin, is a common species found in Europe, Asia, North Africa, and introduced in North America (Cunha *et al.*, 2009). In Portugal, it is commonly known as "erva de S. Roberto" or "erva roberta". It can grow in shadowy and wet lands, as an annual or biennial plant, and are common at altitudes of up to 1,500 meters. *G. robertianum* produces small, pink, five-petalled flowers, from April until the autumn. The leaves are fern-like, sometimes resembling parsley, turn red at the end of the flowering season, the stems often reddish and possess little roots structure.

In Portugal, there is some confusion concerning to "erva de S. Roberto" plant. Indeed, several etnobotanical studies show that different species from the genus *Geranium*, as *G. dissectum* L. (Carvalho, 2005), *G. lucidum* L. (Carvalho, 2005), *G. purpureum* Vill. (Camejo-Rodrigues *et al.*, 2003; Novais *et al.*, 2004) and *G. molle* L. (Neves *et al.*, 2009) also share the same popular name.

Infusions (and decoctions) of aerial parts of Herb Robert are traditionally used in herbal medicine with several applications: stop bleeding (as nose bleeding) and, thus, are used topically in wounds, accelerating the healing process (Cunha *et al.*, 2009). Further, it is employed in folk medicine either due to its anti-inflammatory properties and anti-cancer potential activity (Amaral *et al.*, 2009). In Lebanon, *G. robertianum* is also used due to their anti-rheumatic properties (Marc *et al.*, 2008).

An infusion made from the aerial parts is usually used for its diuretic and tonic effects and as a remedy for dysentery and digestive problems (Carvalho, 2005; Cunha *et al.*, 2009). *G. robertianum* is also used in Portugal due to its anti-diabetic (Castro, 1998; Braga and Pontes, 2005; Cunha *et al.*, 2009; Ferreira *et al.*, 2010e) and antihypertensive properties (Braga and Pontes, 2005).

Despite the common use of *G. robertianum* as antihyperglycemiant, as well as some other *Geranium* species (Rodriguez *et al.*, 1994), specialized literature regarding the specific effects of this plant is not easy to achieve. In a previous study, we observed a significant decrease in occasional glycaemia of diabetic Goto-Kakizaki rats, treated during 4 weeks with a *G. robertianum* decoction (Nunes *et al.*, 2006; Ferreira *et al.*, 2010d). Furthermore, a decrease in the blood glucose values was observed in the intraperitoneal glucose tolerance test (Nunes *et al.*, 2006). However, this was a preliminary investigation and several studies are still required in order to clarify the mechanisms of action of *G. robertianum* water extract.

Regarding its chemical composition, *G. robertianum* possesses a high content of polyphenols and flavones (Amaral *et al.*, 2009; Neagu *et al.*, 2010). Among these bioactive components were 3,4-dimethoxyflavone, homoeriodictyol and kaempferol. Kaempferol is known to be an excellent antioxidant, since it exhibits the 3-OH and 5-OH groups with the 4-oxo group in the C-ring and the C_2–C_3 double bond, despite having just one OH group in the B-ring (Amaral *et al.*, 2009). In vitro studies suggest that this compound, in association with quercetin, can improve glucose uptake in 3T3-L1 cells. Thus, kaempferol potentially acts at multiple targets to ameliorate hyperglycemia, including by acting as partial agonist of PPAR-γ (Fang *et al.*, 2008).

Moreover, kaempferol also show an important activity as anti-inflammatory agent (Mahat *et al.*, 2010) and, together with quercetin, possesses an important inhibitory effect in the osteoclast bone reabsorption (Wattel *et al.*, 2003). Furthermore, several studies point the kaempferol proapoptotic effect in tumoural cells (Leung *et al.*, 2007; Yoshida *et al.*, 2008; Kang *et al.*, 2009). Hence, the high content of this polyphenolic compound seems to be correlated to several medicinal properties attributed to this plant.

Syringic acid, acetovanillion, ferulic methyl ester and ferulic ethyl ester were also identified. The ferulic acid derivatives also possess a high antioxidant activity and can explain the anti-cancer potential and the anti-inflammatory properties of the plant (Amaral *et al.*, 2009) and also decrease diabetic complications. Accordingly to the exposed, *G. robertianum* seem to be a very promising medicinal plant; nevertheless, further studies are required, in order to determine both the bioactive compounds responsible for its health effects and the underlying biological mechanisms of action.

8. Lamiaceae Family

(a) Rosmarinus officinalis L.

Rosemary or *Rosmarinus officinalis* L. is a woody, perennial herb with fragrant evergreen needle-like leaves (green above and white below) with dense short woolly hair, native to the Mediterranean region. Flowers are terminal and usually blue-coloured, and bloom in summer in the north; nevertheless, *R. officinalis* can be everblooming in warm-winter climates, as in Portugal, where it is commonly known as "alecrim". This lamiaceae possesses a very strong and pleasant odor, being widely used in culinary and perfumery applications. Frequently, in Portugal, bee hives are placed close to rosemary lands and honey show an exquisite and appreciated flavor (Carvalho, 2005).

R. officinalis is used since ancient times in herbal medicine in Portugal and other Mediterranean countries. Indeed, it is often one of the medicinal plants with wide use, sometimes also used in herbal mixtures (Camejo-Rodrigues, 2001). Different therapeutic applications include treatment of respiratory problems, headaches, hypercholesterolemia, hypertension (Camejo-Rodrigues, 2001; Carvalho, 2005; Neves *et al.*, 2009), rheumatism (Carvalho, 2005; Neves *et al.*, 2009), digestive problems, anxiety, as antipyretic (Camejo-Rodrigues, 2001; Neves *et al.*, 2009) and heal wounds and burns (Camejo-Rodrigues, 2001; Neves *et al.*, 2009; Abu-Al-Basal, 2010), and also showing a high antimicrobial activity (Rasooli *et al.*, 2008). Furthermore, is used due to its diuretic (Haloui *et al.*, 2000; Camejo-Rodrigues, 2001; Neves *et al.*, 2009) and anti-diabetic properties (Castro, 2001; Tahraoui *et al.*, 2007; BakIrel *et al.*, 2008) and also due to hepatoprotective properties (Amin and Hamza, 2005). Usually, aerial parts are therapeutically used dried or green, as a decoction (or infusion) or externally as an ointment (Camejo-Rodrigues, 2001; Carvalho, 2005) or ethanolic extract. Essential oils usage is also reported (BakIrel *et al.*, 2008), as well as other applications (Camejo-Rodrigues, 2001; Neves *et al.*, 2009).

Chemically, rosemary plant extracts contain several phenolic compounds, as caffeic acid and its derivative, rosemarinic acid. Carnosic acid and carnosol are also important chemical constituents (Duke 1992; Moreno *et al.*, 2006; Pérez-Fons *et al.*, 2009). Essential oils present α-pinene, 1,8-cineole, camphor, verbenone and borneol, that constitute around 80 per cent of the total oil (Atti-Santos *et al.*, 2005; Santoyo *et al.*, 2005), despite the variations due to different environmental conditions and harvesting time.

Concerning type 2 diabetes therapy, *R. officinalis* water extracts (at doses of 100 or 200 mg/kg) can decrease hyperglicaemia in alloxan-treated rabbits (BakIrel *et al.*, 2008), in a dose-dependent way and similarly to glibenclamide. The observed lower glycaemia can be either produced by a diuretic higher activity (Haloui *et al.*, 2000) and a decrease in pancreatic amylase promoted by rosmarinic acid (McCue and Shetty, 2004), among other features. In fact, this anti-diabetic activity of rosemary water extracts also seem to be related to PPAR-γ activation, induced by carnosol and carnosic acid (Rau *et al.*, 2006).

A different study referred that essential oil administration decreased insulin production and increased hyperglycaemia in alloxan-treated rabbits during the intraperitoneal glucose tolerance test (Al-Hader *et al.*, 1994). Nevertheless, a recent study showed that alloxan-treated rats' occasional glycaemia presented a decrease after an ingestion of rosemary essential oil during 7 days (Benkhayal *et al.*, 2009). Moreover, after a period of 21 days, the ingestion of *R. officinalis* essential oil (0.1 ml/ kg body weight) led to normal glycaemias, compared to control non-diabetic group (Benkhayal *et al.*, 2009). These results, apparently contradictory, probably reflect a *R. officinalis* specific effect in carbohydrate absorption or metabolization.

Several scientific studies point out the high antioxidant capacity of *R. officinalis* water and ethanolic extracts, and are usually referred as one of the most antioxidant aromatic plants (Inatani *et al.*, 1983; Celiktas *et al.*, 2007; Gachkar *et al.*, 2007; Erkan *et al.*, 2008). This high antioxidant capacity is of major importance to counteract oxidative impairments that attain all cells in diabetic patents.

(b) Salvia officinalis L.

Salvia officinalis L. (common sage or sage) is native to the Mediterranean region, though nowadays it is widely widespread throughout the world. *S. officinalis* is a small perennial evergreen small shrub that plant flowers in late spring or summer. The leaves are oblong, ranging in size up to 6 long by 2.5 cm wide. Leaves are grey-green, rough on the upper side, and nearly white underneath due to the many short soft hairs or trichoma.

Sage has a long history of medicinal and culinary use, and in modern times also as an ornamental garden plant. In Portugal it is typically known as "sálvia" or "salva", and commonly leaves (fresh or dried) are used in folk medicine (sometimes also in herbal mixtures) for the treatment of throat irritation (amygdalitis, laryngitis) and respiratory problems (asthma and bronchitis), as an oral antiseptic (against aphthas and ulcers), for high blood pressure increase, for digestive problems treatments (as gases, diarrhea, indigestion or stomach aches), against animal bites, as sweat inhibitor, for inhibiting lactation and for diabetes mellitus therapy (Castro, 1998; Carvalho, 2005; Lima *et al.*, 2006; Neves *et al.*, 2009).

This plant is chemically well characterized, even considering some chemotypes or seasonal variations (Duke, 1992; Chalchat *et al.*, 1998; Kintzios, 2000; Lu and Yeap Foo, 2000; Lu and Yeap Foo, 2002; Dob *et al.*, 2007; Glisic *et al.*, 2010). The major leaf active constituents are tannic acid, oleic acid, ursonic acid, ursolic acid, caffeic acid, thujones, niacin, nicotinamide, flavones and flavonoid glycosides. Probably, due to the toxicity of thujones (Scientific Committee on Food, 2003), this plant is stated as "abortive", and accordingly to folk medicine, its use should be avoided by pregnant woman (Carvalho, 2005).

S. officinalis leaf water extracts (and essential oils) have a great antioxidant activity, due to the presence of large amounts of flavonoids and phenolic compounds (Lu and Yeap Foo, 2001; Miura *et al.*, 2002; Glisic *et al.*, 2010). It was observed that ingestion of sage infusions improves glutathione-S-transferase and/or glutathione reductase status in (Wistar) rats and in (Balb/c) mice (Lima *et al.*, 2005) and thus protect from liver damage (Amin and Hamza, 2005; Lima *et al.*, 2005).

Regarding diabetes therapy, there are several reports confirming *S. officinalis* beverages hypoglycemiant effects on alloxan- or streptozotocin-induced diabetic animal models (Alarcon-Aguilar *et al.*, 2002; Eidi *et al.*, 2005; Lima *et al.*, 2006). Alarcon-Aguilar and collaborators (Alarcon-Aguilar *et al.*, 2002) found that ethanolic extracts of *S. officinalis* significantly reduced blood glucose levels in fasting normal mice 120 and 240 min (15.7 per cent and 30.2 per cent, respectively) following intraperitoneal administration. It also, significantly diminished hyperglycaemia in mildly alloxan-induced diabetic mice 240 min after glucose load (32.6 per cent and 22.7 per cent, respectively). These results seemed to indicate an enhanced insulin release in the presence of sage extract and have good correlation with some other recent studies.

Eidi and collaborators studied the effects of methanolic sage extracts and essential oils on streptozotocin-induced diabetic rats. They found that, in the presence of methanolic extracts, blood glucose concentration only decreased in streptozotocin-induced diabetic fasted rats, but not in healthy fasted rats (Eidi *et al.*, 2005). However, the extract did not affect insulin release from the pancreas of both animal groups. In this work, intraperitoneal administration of sage essential oil did not change serum glucose (Eidi *et al.*, 2005). Conversely, Lima and co-workers, using the same animal model - streptozotocin-induced diabetic rats, demonstrated that sage water extracts and essential oil affected liver glucose uptake and gluconeogenesis (Lima *et al.*, 2006). In this work, primary cultures of hepatocytes from healthy, sage-tea-drinking rats showed, after stimulation, a high glucose liver uptake capacity and decreased gluconeogenesis in response to glucagon. Moreover, sage essential oil both increased hepatocyte sensitivity to insulin and inhibited gluconeogenesis. The authors suggest that these sage effects are similar to metformin, a known inhibitor of gluconeogenesis used in the treatment and prevention of type 2 diabetes mellitus. Nevertheless, in primary cultures of rat hepatocytes isolated from streptozotocin (STZ)-induced diabetic rats, none of these activities was observed. This was probably because STZ-induced diabetic rats used in both research works were not a type 2 diabetes animal model but a type 1 diabetes model (Sitasawad *et al.*, 2001). As a matter of fact, the stimulation of insulin release requires some functional pancreatic beta-cells that are commonly completely destroyed with a dose of 50 mg STZ/kg (or higher) used in both studies. We performed some preliminary experiments in young Goto-Kakizaki (GK) rats, a good animal model for the initial stages of type 2 diabetes mellitus (Goto and Kakizaki, 1981; Portha *et al.*, 2009). We observed that GK rats drinking *S. officinalis* water extracts during 4 weeks showed an improved response in glucose tolerance tests (Nunes *et al.*, 2006). Conversely, comparing sage effects between GK and STZ rats, we did not observe any hypoglycemiant effect in STZ rats (Nunes *et al.*, 2004).

However, there is no major information about the molecular mechanisms of action of sage extract over insulin release. Concerning this point, our results showed that sage decoction used in this study contained a large amount of some insulinotropic agents, namely amino acids (Nunes *et al.*, 2006). Indeed, our sage extract contained mainly alanine and arginine (Milner, 1969; Robert *et al.*, 1982; Sener and Malaisse, 2002; Nunes *et al.*, 2006). Smaller amounts of other insulinotropic amino acids - lysine and leucine – were also found (Milner, 1969; Robert *et al.*, 1982; Welsh *et al.*, 1982; Nunes *et al.*, 2006).

Furthermore, the anti-diabetic activity of sage water extracts also seem to be related to PPAR-γ activation, induced by carnosol and carnosic acid (Rau *et al.*, 2006).

Consequently, fasting glucose levels decrease in normal animals and its metformin-like effects on rat hepatocytes suggest that *S. officinalis* may be useful as a food supplement in the prevention of type 2 diabetes mellitus by lowering the plasma glucose of individuals at risk (Lima *et al.*, 2006). This research team has also performed a pilot trial (non-randomized crossover trial) with six healthy female volunteers (aged 40-50) demonstrating the beneficial properties of sage tea consumption on lipid profile and transaminase activity in humans (Sá *et al.*, 2009). Although not demonstrating positive effects on glucose regulation in human healthy individuals, this study corroborate both the beneficial use of *S. officinalis* extracts in diabetic patients (since lipid profile is usually altered in type 2 diabetic patients). Nonetheless, further studies seem to be necessary to elucidate the sage extracts molecular mechanisms of action in diabetes.

Some other useful properties of *S. officinalis* may be interesting for diabetic patients, as many scientific works since ancient times describe the antimicrobial activities of common sage (Dobrynin *et al.*, 1976; Horiuchi *et al.*, 2007; Longaray Delamare *et al.*, 2007; Pinto *et al.*, 2007; Bouaziz *et al.*, 2009). Furthermore, anti-inflammatory activity is attributed to ursodecolic acid, present in sage extracts (Baricevic *et al.*, 2001).

In conclusion, *S. officinalis* extracts seem to inhibit gluconeogenesis and stimulate insulin release. Moreover, sage extracts possess a large content and variety of polyphenolic compounds with significant antioxidant activity against ROS and therefore may prevent or reduce the development of chronic complications associated with the disease.

Conclusion

This review indicates that for almost all aromatic and medicinal plants used in Portugal for (type 2) diabetes mellitus therapy there are some scientific evidences pointing to anti-hyperglycemiant attributes and antioxidant power, which prevent or delays the onset of associated diseases to diabetes. Nonetheless, for most aromatic and medicinal plants cited, the mechanisms of action or the active chemical compounds remain unclear and further studies are still required.

Furthermore, we cannot forget that usually, and unlike synthetic drugs, different chemical constituents that despite their lower amounts act in a synergistic manner potentiating the effectiveness of medicinal plants.

Hence, the secondary effects are usually reduced, as compared to synthetic drugs. Even so, long lasting treatments to a single medicinal plant should be avoided, since conversely to the popular knowledge that "medicinal plants" are safe, long-lasting treatments with aromatic plants are prone to induce also several pathological conditions.

Different abiotic conditions lead to different chemotypes in the same species and thus a chemical characterization concerning both the active constituents and noxious components of commercialized aromatic and medicinal plants are of major importance but usually is absent. Besides, it should be noticed that phenolic compounds undergo chemical modifications *in vivo*, which may change some of their biological effects, including the antioxidant properties. Thus, further studies concerning these modifications are also of major importance.

In conclusion, medicinal plants commonly used in Portuguese folk medicine for diabetes treatment seem to have scientific support and, thus, seem feasible to be used together with synthetic oral anti-diabetic drugs, not only due to their anti-hyperglycemiant properties but also due to their ability to prevent several pathologies associated to diabetes and as a result, to improve diabetic patients daily life.

Acknowledgements

This work was supported by FCT (Portuguese Research Council).

References

Abu-Al-Basal, M. A. (2010). Healing potential of *Rosmarinus officinalis* L. on full-thickness excision cutaneous wounds in alloxan-induced-diabetic BALB/c mice. *Journal of Ethnopharmacology,* 131: 443-450.

Agius, L. (2007). New hepatic targets for glycaemic control in diabetes. *Clinical Endocrinology and Metabolism,* 21: 587-605.

Alarcon-Aguilar, F. J., Roman-Ramos, R., Flores-Saenz, J. L. and Aguirre-Garcia, F. (2002). Investigation on the hypoglycaemic effects of extracts of four Mexican medicinal plants in normal and alloxan-diabetic mice. *Phytotherapy Research,* 16: 383-386.

Al-Hader, A. A., Hasan, Z. A. and Aqel, M. B. (1994). Hyperglycemic and insulin release inhibitory effects of *Rosmarinus officinalis*. *Journal of Ethnopharmacology,* 43: 217-221.

Alías, J., Sosa, T., Escudero, J. and Chaves, N. (2006). Autotoxicity against germination and seedling emergence in *Cistus ladanifer* L. *Plant and Soil,* 282: 327-332.

Almeida, I. F., Fernandes, E., Lima, J. L. F. C., Costa, P. C. and Bahia, M. F. (2009). *In vitro* protective effect of *Hypericum androsaemum* extract against oxygen and nitrogen reactive species. *Basic and Clinical Pharmacology and Toxicology,* 105: 222-227.

Amaral, S., Mira, L., Nogueira, J. M. F., Silva, A. P. D. and Helena Florêncio, M. (2009). Plant extracts with anti-inflammatory properties- A new approach for characterization of their bioactive compounds and establishment of structure-antioxidant activity relationships. *Bioorganic and Medicinal Chemistry,* 17: 1876-1883.

Amin, A. and Hamza, A. A. (2005). Hepatoprotective effects of *Hibiscus, Rosmarinus* and *Salvia* on azathioprine-induced toxicity in rats. *Life Sciences,* **77**: 266-278.

Andrade, D., Gil, C., Breitenfeld, L., Domingues, F. and Duarte, A. P. (2009). Bioactive extracts from *Cistus ladanifer* and *Arbutus unedo* L. *Industrial Crops and Products,* **30**: 165-167.

Anonymous (2001). *Vaccinium myrtillus* (bilberry). *Alternative Medicine Review,* **6**: 500-504.

Antunes, C. M., Areias, L. R., Vieira, I. P., Costa, A. C., Tinoco, M. T. and Cruz-Morais, J. (2009). Efeito Hipoglicemiante de um Extracto Aquoso de *Cytisus multiflorus. Revista de Fitoterapia,* **9** (Supl. 1): 91.

Areias, L. R., Vieira, I. P., Tinoco, M. T., Antunes, C. M. and Cruz-Morais, J. (2008). Effect of *Cytisus multiflorus* in the control of type-2 diabetes. *XVI Congresso Nacional de Bioquímica. Ponta Delgada – S. Miguel - Açores (Portugal)*

Atti-Santos, A. C., Rossato, M., Pauletti, G. F., Rota, L. D., Rech, J. C., Pansera, M. R., Agostini, F., Serafini, L. A. and Moyna, P. (2005). Physico-chemical evaluation of *Rosmarinus officinalis* L. essential oils. *Brazilian Archives of Biology and Technology,* **48**: 1035-1039.

BakIrel, T., BakIrel, U., Keles, O. Ü., Ülgen, S. G. and Yardibi, H. (2008). In vivo assessment of antidiabetic and antioxidant activities of rosemary (*Rosmarinus officinalis*) in alloxan-diabetic rabbits. *Journal of Ethnopharmacology,* **116**: 64-73.

Bao, L., Yao, X.-S., Yau, C.-C., Tsi, D., Chia, C.-S., Nagai, H. and Kurihara, H. (2008). Protective effects of bilberry (*Vaccinium myrtillus* L.) extract on restraint stress-induced liver damage in mice. *Journal of Agricultural and Food Chemistry,* **56**: 7803-7807.

Barata, J. (ed.) (2008). Terapêuticas alternativas de origem botânica - Efeitos adversos e interacções medicamentosas. Lidel Edições Técnicas, Lda. Lisboa, Portugal (*in* Portuguese).

Baricevic, D., Sosa, S., Della Loggia, R., Tubaro, A., Simonovska, B., Krasna, A. and Zupancic, A. (2001). Topical anti-inflammatory activity of *Salvia officinalis* L. leaves: the relevance of ursolic acid. *Journal of Ethnopharmacology,* **75**: 125-132.

Barrajón-Catalán, E., Fernández-Arroyo, S., Saura, D., Guillén, E., Fernández-Gutiérrez, A., Segura-Carretero, A. and Micol, V. (2010). Cistaceae aqueous extracts containing ellagitannins show antioxidant and antimicrobial capacity, and cytotoxic activity against human cancer cells. *Food and Chemical Toxicology,* **48**: 2273-2282.

Baynes, J. W. and Thorpe, S. R. (1999). Role of oxidative stress in diabetic complications: a new perspective on an old paradigm. *Diabetes,* **48**: 1-9.

Baynes, J. W. and Thorpe, S. R. (2000). Oxidative stress in diabetes. *in* Antioxidants in diabetes management, L. Packer, P. Rosen, H. J. Tritschler, G. L. King and A. Azzi (eds.), Basel: Marcel Dekker, Inc., New York, U.S.A., pp 77-91

Belmokhtar, M., Bouanani, N. E., Ziyyat, A., Mekhfi, H., Bnouham, M., Aziz, M., Matéo, P., Fischmeister, R. and Legssyer, A. (2009). Antihypertensive and endothelium-dependent vasodilator effects of aqueous extract of *Cistus ladaniferus. Biochemical and Biophysical Research Communications,* **389**: 145-149.

Benkhayal, F. A., EL-Ageeli, W. h., Ramesh, S. and Farg hamd, M. (2009). Anti-hyperglycemic effects of volatile oils extracted from *Rosemarinus officinalis* and *Artemisia cinae* in diabetic rats. *Tamilnadu J. Veterinary and Animal Sciences,* **5**: 216-218.

Bever, B. O. (1980). Oral hypoglycaemic plants in West Africa. *Journal of Ethnopharmacology,* **2**: 119-127.

Bnouham, M., Mekhfi, H., Legssyer, A. and Ziyyat, A. (2002). Medicinal plants used in the treatment of diabetes in Morocco. *International Journal of Diabetes and Metabolism,* **10**: 33-50.

Bouaziz, M., Yangui, T., Sayadi, S. and Dhouib, A. (2009). Disinfectant properties of essential oils from *Salvia officinalis* L. cultivated in Tunisia. *Food and Chemical Toxicology,* **47**: 2755-2760.

Braga, T. and Pontes, G. (eds.) (2005). Plantas usadas na medicina popular. EGA - Empresa Gráfica Açoreana, Lda (*in* Portuguese).

Broadley, M. R., Willey, N. J., Wilkins, J. C., Baker, A. J. M., Mead, A. and White, P. J. (2001). Phylogenetic variation in heavy metal accumulation in angiosperms. *New Phytologist,* **152**: 9-27.

Butler, W. H., Ford, G. P. and Creasy, D. M. (1996). A 90-day feeding study of lupin (*Lupinus angustifolius*) flour spiked with lupin alkaloids in the rat. *Food and Chemical Toxicology,* **34**: 531-536.

Camejo-Rodrigues, J. (2001). Contributo para o estudo etnobotânico das plantas medicinais e aromáticas no Parque Natural da Serra de S. Mamede, Faculty of Sciences of University of Lisboa, Graduation Dissertation (*in* Portuguese).

Camejo-Rodrigues, J., Ascensão, L., Bonet, M. À. and Vallès, J. (2003). An ethnobotanical study of medicinal and aromatic plants in the Natural Park of Serra de São Mamede (Portugal). *Journal of Ethnopharmacology,* **89**: 199-209.

Canter, P. H. and Ernst, E. (2004). Anthocyanosides of *Vaccinium myrtillus* (Bilberry) for night vision - a systematic review of placebo-controlled trials. *Survey of Ophthalmology,* **49**: 38-50.

Carmali, S., Alves, V. D., Coelhoso, I. M., Ferreira, L. M. and Lourenço, A. M. (2010). Recovery of lupanine from *Lupinus albus* L. leaching waters. *Separation and Purification Technology,* **74**: 38-43.

Carvalho, A. M. P. (2005). Etnobotánica del Parque Natural de Montesinho. Departamento de Biologia, Universidad Autónoma de Madrid, PhD Dissertation (in Spanish).

Castro, V. R. O. (1998). Chromium in a series of portuguese plants used in the herbal treatment of diabetes. *Biological Trace Element Research,* **62**: 101-106.

Castro, V. R. O. (2001). Chromium and zinc in a series of plants used in Portugal in the herbal treatment of non-insulinized diabetes. *Acta Alimentaria*, **30**: 333-342.

Catoni, C., Peters, A. and Martin Schaefer, H. (2008). Life history trade-offs are influenced by the diversity, availability and interactions of dietary antioxidants. *Animal Behaviour*, **76**: 1107-1119.

Cavalli, V. L. D. L. O., SordiI, C., ToniniI, K., GrandoI, A., MuneronI, T., GuigiI, A. and Júnior, W. A. R. (2007). Avaliação *in vivo* do efeito hipoglicemiante de extratos obtidos da raiz e folha de bardana*Arctium minus* (Hill.) Bernh. *Revista Brasileira de Farmacognosia*, **17**: 64-70.

Celiktas, O. Y., Bedir, E. and Sukan, F. V. (2007). *In vitro* antioxidant activities of Rosmarinus officinalis extracts treated with supercritical carbon dioxide. *Food Chemistry*, **101**: 1457-1464.

Ceriello, A. (2003). New insights on oxidative stress and diabetic complications may lead to a causal antioxidant therapy. *Diabetes Care*, **26**: 1589-1596.

Chalchat, J. C., Michet, A. and Pasquier, B. (1998). Study of clones of *Salvia officinalis* L. Yields and chemical composition of essential oil. *Flavour and Fragrance Journal*, **13**: 68-70.

Chaves, N., Ríos, J. J., Gutierrez, C., Escudero, J. C. and Olías, J. M. (1998). Analysis of secreted flavonoids of *Cistus ladanifer* L. by high-performance liquid chromatography-particle beam mass spectrometry. *Journal of Chromatography A*, **799**: 111-115.

Chaves, N., Sosa, T. and Escudero, J. C. (2001). Plant growth inhibiting flavonoids in exudate of *Cistus ladanifer* and in associated soils. *Journal of Chemical Ecology*, **27**: 623-631.

Chen, C.-Y., Peng, W.-H., Tsai, K.-D. and Hsu, S.-L. (2007). Luteolin suppresses inflammation-associated gene expression by blocking NF-kB and AP-1 activation pathway in mouse alveolar macrophages. *Life Sciences*, **81**: 1602-1614.

Cignarella, A., Nastasi, M., Cavalli, E. and Puglisi, L. (1996). Novel lipid-lowering properties of *Vaccinium myrtillus* L. leaves, a traditional antidiabetic treatment, in several models of rat dyslipidaemia: a comparison with ciprofibrate. *Thrombosis Research*, **84**: 311-322.

CMHP (Committee on Herbal Medicinal Products) (2009). Assessment report on *Centaurium erythraea* Rafn S. L. including *C. majus* (H. et L.) Zeltner and *C. suffruticosum* (Griseb.) Ronn., herba, for the development of a community herbal monograph. *in* Evaluation of Medicines for Human Use. European Medicines Agency. London, U.K.

Cooke, D., Steward, W. P., Gescher, A. J. and Marczylo, T. (2005). Anthocyans from fruits and vegetables - Does bright colour signal cancer chemopreventive activity? *European Journal of Cancer*, **41**: 1931-1940.

Coutinho, A. X. P. (ed.) (1939). Flora de Portugal (2nd Edition), Lisboa, Portugal (*in* Portuguese).

Cunha, A. P., Silva, A. P. and Roque, A. R. (2006). Plantas e Produtos Vegetais em Fitoterapia. Fundação Calouste Gulbenkian, Lisboa, Portugal (*in* Portuguese).

Cunha, A. P., Silva, A. P. and Roque, A. R. (2009). Plantas e Produtos Vegetais em Fitoterapia (3rd Edition). Fundação Calouste Gulbenkian, Lisboa, Portugal (*in* Portuguese).

Davis, J. M., Murphy, E. A., Carmichael, M. D. and Davis, B. (2009). Quercetin increases brain and muscle mitochondrial biogenesis and exercise tolerance. *American Journal of Physiology*, **296**: R1071-1077.

Day, C. (1989). Hypoglycaemic compounds from plants. *in* New Antidiabetic Drugs C. J. Bailey and P. R. Flatt (eds.), pp: 267-278, Smith-Gordon, London, U.K.

Dey, L., Attele, A. S. and Yuan, C.-S. (2002). Alternative therapies for type 2 diabetes. *Alternative Medicine Review*, 7: 45-58.

Dias, A. C. P., Seabra, R. M., Andrade, P. B., Ferreres, F. and Fernandes-Ferreira, M. (2000). Xanthone biosynthesis and accumulation in calli and suspended cells of *Hypericum androsaemum*. *Plant Science*, **150**: 93-101.

Dias, L. S. and Moreira, I. (2002). Interaction between water soluble and volatile compounds of *Cistus ladanifer* L. *Chemoecology*, **12**: 77-82.

Ding, L., Jin, D. and Chen, X. (2010). Luteolin enhances insulin sensitivity via activation of PPAR-γ transcriptional activity in adipocytes. *The Journal of Nutritional biochemistry*, **21**: 941-947.

Dob, T., Berramdane, T., Dahmane, D., Benabdelkader, T. and Chelghoum, C. (2007). Chemical composition of the essential oil of *Salvia officinalis* from Algeria. *Chemistry of Natural Compounds*, **43**: 491-494.

Dobrynin, V., Kolosov, M., Chernov, B. and Derbentseva, N. (1976). Antimicrobial substances from *Salvia officinalis*. *Chemistry of Natural Compounds*, **12**: 623-624.

Dormer, R. L., Kerbey, A. L., McPherson, M., Manley, S., Ashcroft, S. J. H., Schofield, G. J. and Randle, P. J. (1974). The effect of nickel on secretory systems. Studies on the release of amylase, insulin and growth hormone. *Biochemical Journal*, **140**: 135-142.

Dudonné, S. P., Vitrac, X., Coutiére, P., Woillez, M. and Mérillon, J.-M. (2009). Comparative study of antioxidant properties and total phenolic content of 30 plant extracts of industrial interest using DPPH, ABTS, FRAP, SOD, and ORAC assays. *Journal of Agricultural and Food Chemistry*, **57**: 1768-1774.

Duke, J. A. (1992). Handbook of phytochemical constitutens of GRAS herbs and other economic plants. CRC Press, Boca Raton, U.S.A.

Eddouks, M., Maghrani, M., Lemhadri, A., Ouahidi, M. L. and Jouad, H. (2002). Ethnopharmacological survey of medicinal plants used for the treatment of diabetes mellitus, hypertension and cardiac diseases in the south-east region of Morocco (Tafilalet). *Journal of Ethnopharmacology*, **82**: 97-103.

Edgars, N. K. (1936). A new glucoside from blueberry leaf. *Journal of the American Pharmaceutical Association,* **25**: 288-291.

Eidi, M., Eidi, A. and Zamanizadeh, H. (2005). Effect of *Salvia officinalis* L. leaves on serum glucose and insulin in healthy and streptozotocin-induced diabetic rats. *Journal of Ethnopharmacology,* **100**: 310-313.

El-Hilaly, J., Hmammouchi, M. and Lyoussi, B. (2003). Ethnobotanical studies and economic evaluation of medicinal plants in Taounate province (Northern Morocco). *Journal of Ethnopharmacology,* **86**: 149-158.

Engelgau, M. M. and Geiss, L. S. (2000). The burden of diabetes mellitus. In Medical management of diabetes mellitus. J. L. Leahy, N. G. Clark and W. T. Cefalu (eds.), pp. 1-17. Mark Dekker, Inc., New York, U.S.A.

Erdemoglu, N., Turan, N. N., Akkol, E. K., Sener, B. and Abacloglu, N. (2009). Estimation of anti-inflammatory, antinociceptive and antioxidant activities on *Arctium minus* (Hill) Bernh. ssp. minus. *Journal of Ethnopharmacology,* **121**: 318-323.

Erkan, N., Ayranci, G. and Ayranci, E. (2008). Antioxidant activities of rosemary (*Rosmarinus officinalis* L.) extract, blackseed (*Nigella sativa* L.) essential oil, carnosic acid, rosmarinic acid and sesamol. *Food Chemistry,* **110**: 76-82.

Erlund, I., Marniemi, J., Hakala, P., Alfthan, G., Meririnne, E. and Aro, A. (2003). Consumption of black currants, lingonberries and bilberries increases serum quercetin concentrations. *European Journal of Clinical Nutrition,* **57**: 37-42.

Espín, J. C., García-Conesa, M. T. and Tomás-Barberán, F. A. (2007). Nutraceuticals: Facts and fiction. *Phytochemistry,* **68**: 2986-3008.

Fang, X.-K., Gao, J. and Zhu, D.-N. (2008). Kaempferol and quercetin isolated from *Euonymus alatus* improve glucose uptake of 3T3-L1 cells without adipogenesis activity. *Life Sciences,* **82**: 615-622.

Ferrandis, P., Herranz, J. and Martínez-Sánchez, J. (1999). Effect of fire on hard-coated *Cistaceae* seed banks and its influence on techniques for quantifying seed banks. *Plant Ecology,* **144**: 103-114.

Ferreira, F. M., Peixoto, F. P., Nunes, E., Sena, C., Seiça, R. and Santos, M. S. (2010a). Inhibitory effect of *Arctium minus* on mitochondrial bioenergetics in diabetic Goto-Kakizaki rats. *Scientific Research and Essays* (In Press).

Ferreira, F. M., Peixoto, F. P., Nunes, E., Sena, C., Seiça, R. and Santos, M. S. (2010b). *Vaccinium myrtillus* improves liver mitochondrial oxidative phosphorylation of diabetic Goto-Kakizaki rats. *Journal of Medicinal Plants Research,* **4**: 692–696.

Ferreira, F. M., Peixoto, F. P., Nunes, E., Sena, C., Seiça, R. and Santos, M. S. (2010c). Diabetic Goto-Kakizaki rats improved liver mitochondrial oxidative phosphorylation by *Vaccinium myrtillus*. In 6th Conference on Aromatic and Medicinal Plants of Southeast European Countries, Endorgan-Orhan, Ý. (ed.), Antalya, Turkey, pp. 452-462.

Ferreira, F. M., Peixoto, F., Nunes, E., Sena, C., Seiça, R. and Santos, M. S. (2010d). Mito'l'eas: *Vaccinium myrtillus* and *Geranium robertianum* decoctions improve diabetic Goto-Kakizaki rats hepatic mitochondrial oxidative phosphorylation. *Biochimica et Biophysica Acta*, **1797**(Suppl. 1): 79-80.

Ferreira, F. M., Peixoto, F., Nunes, E., Sena, C., Seiça, R. and Santos, M. S. (2010e). "MitoTea": *Geranium robertianum* L. decoctions decrease blood glucose levels and improve liver mitochondrial oxidative phospharylation in diabetic Goto-Kakizaki rats. *Acta Biochimica Polonica*, **57**: 399-402.

Fraunfelder, F. W. (2004). Ocular side effects from herbal medicines and nutritional supplements. *American Journal of Ophthalmology*, **138**: 639-647.

Gachkar, L., Yadegari, D., Rezaei, M. B., Taghizadeh, M., Astaneh, S. A. and Rasooli, I. (2007). Chemical and biological characteristics of *Cuminum cyminum* and *Rosmarinus officinalis* essential oils. *Food Chemistry*, **102**: 898-904.

García López, P. M., de la Mora, P. G., Wysocka, W., Maiztegui, B., Alzugaray, M. E., Del Zotto, H. and Borelli, M. I. (2004). Quinolizidine alkaloids isolated from *Lupinus* species enhance insulin secretion. *European Journal of Pharmacology*, **504**: 139-142.

Garzón, G. A., Narváez, C. E., Riedl, K. M. and Schwartz, S. J. (2010). Chemical composition, anthocyanins, non-anthocyanin phenolics and antioxidant activity of wild bilberry (*Vaccinium meridionale* Swartz) from Colombia. *Food Chemistry*, **122**: 980-986.

Gião, M. S., González-Sanjosé, M. L., Rivero-Pérez, M. D., Pereira, C. I., Pintado, M. E. and Malcata, F. X. (2007). Infusions of Portuguese medicinal plants: Dependence of final antioxidant capacity and phenol content on extraction features. *Journal of the Science of Food and Agriculture*, **87**: 2638-2647.

Giovannucci, E., Harlan, D. M., Archer, M. C., Bergenstal, R. M., Gapstur, S. M., Habel, L. A., Pollak, M., Regensteiner, J. G. and Yee, D. (2010). Diabetes and cancer: A consensus report. *Diabetes Care*, **33**: 1674-85.

Glisic, S., Ivanovic, J., Ristic, M. and Skala, D. (2010). Extraction of sage (*Salvia officinalis* L.) by supercritical CO_2: Kinetic data, chemical composition and selectivity of diterpenes. *The Journal of Supercritical Fluids*, **52**: 62-70.

Goto, Y. and Kakizaki, M. (1981). The spontaneous-diabetes rat: A model of noninsulin dependent diabetes mellitus. *Proceedings of Japanese Academy*, **57**: 381-384.

Grosso, A. C., Costa, M. M., Ganço, L., Pereira, A. L., Teixeira, G., Lavado, J. M. G., Figueiredo, A. C., Barroso, J. G. and Pedro, L. G. (2007). Essential oil composition of *Pterospartum tridentatum* grown in Portugal. *Food Chemistry*, **102**: 1083-1088.

Guedes, A. P., Amorim, L. R., Vicente, A. and Fernandes-Ferreira, M. (2004). Variation of the essential oil content and composition in leaves from cultivated plants of *Hypericum androsaemum* L. *Phytochemical Analysis*, **15**: 146-151.

Guimarães, R., Sousa, M. J. and Ferreira, I. C. F. R. (2010). Contribution of essential oils and phenolics to the antioxidant properties of aromatic plants. *Industrial Crops and Products*, **32**: 152-156.

Gupta, S., Ahmad, N., Husain, M. M. and Srivastava, R. C. (2000). Involvement of nitric oxide in nickel-induced hyperglycemia in rats. *Biochemical Journal,* 4: 129-138.

Häkkinen, S. H. and Törrönen, A. R. (2000). Content of flavonols and selected phenolic acids in strawberries and *Vaccinium* species: influence of cultivar, cultivation site and technique. *Food Research International,* 33: 517-524.

Häkkinen, S., Heinonen, M., Kärenlampi, S., Mykkänen, H., Ruuskanen, J. and Törrönen, R. (1999). Screening of selected flavonoids and phenolic acids in 19 berries. *Food Research International,* 32: 345-353.

Halestrap, A. P., Doran, E., Gillespie, J. P. and O'Toole, A. (2000). Mitochondria and cell death. *Biochemical Society Transactions,* 28: 170-177.

Haloui, M., Louedec, L., Michel, J.-B. and Lyoussi, B. (2000). Experimental diuretic effects of *Rosmarinus officinalis* and *Centaurium erythraea. Journal of Ethnopharmacology,* 71: 465-472.

Hamza, N., Berke, B., Cheze, C., Agli, A.-N., Robinson, P., Gin, H. and Moore, N. (2010). Prevention of type 2 diabetes induced by high fat diet in the C57BL/6J mouse by two medicinal plants used in traditional treatment of diabetes in the east of Algeria. *Journal of Ethnopharmacology,* 128: 513-518.

Havsteen, B. H. (2002). The biochemistry and medical significance of the flavonoids. *Pharmacology and Therapeutics,* 96: 67-202.

Helmstädter, A. and Schuster, N. (2010). *Vaccinium myrtillus* as an antidiabetic medicinal plant - research through the ages. *Pharmazie,* 65: 315-321.

Horiuchi, K., Shiota, S., Hatano, T., Yoshida, T., Kuroda, T. and Tsuchiya, T. (2007). Antimicrobial activity of oleanolic acid from *Salvia officinalis* and related compounds on vancomycin-resistant Enterococci (VRE). *Biological and Pharmaceutical Bulletin,* 30: 1147-1149.

Huyghe, C. (1997). White lupin (*Lupinus albus* L.). *Field Crops Research,* 53: 147-160.

Inatani, R., Nakatani, N. and Fuwa, H. (1983). Antioxidative effect of the constituents of rosemary (*Rosmarinus officinalis* L.) and their derivatives. *Agricultural Biology Chemistry,* 47: 521-528.

Izzo, A. A. and Ernst, E. (2009). Interactions between herbal medicines and prescribed drugs: An updated systematic review. *Drugs,* 69: 1777-1798.

Jaakola, L. (2003). Flavonoid biosynthesis in bilberry (*Vaccinium myrtillus* L.). Academic Dissertation. Faculty of Science, University of Oulu, Finland.

Jaishree, V. and Badami, S. (2010). Antioxidant and hepatoprotective effect of swertiamarin from *Enicostemma axillare* against d-galactosamine induced acute liver damage in rats. *Journal of Ethnopharmacology,* 130: 103-106.

Jaric, S., Popovic, Z., Macukanovic-Jocic, M., Djurdjevic, L., Mijatovic, M., Karadzic, B., Mitrovic, M. and Pavlovic, P. (2007). An ethnobotanical study on the usage of wild medicinal herbs from Kopaonik Mountain (Central Serbia). *Journal of Ethnopharmacology,* 111: 160-175.

Jun, C., Xue-Ming, Z., Chang-Xiao, L. and Tie-Jun, Z. (2008). Structure elucidation of metabolites of swertiamarin produced by *Aspergillus niger*. *Journal of Molecular Structure*, **878**: 22-25.

Kalt, W. and Dufour, D. (1997). Health functionality of blueberries. *HortTechnology*, **7**: 216-221.

Kang, G.-Y., Lee, E.-R., Kim, J.-H., Jung, J. W., Lim, J., Kim, S. K., Cho, S.-G. and Kim, K. P. (2009). Downregulation of PLK-1 expression in kaempferol-induced apoptosis of MCF-7 cells. *European Journal of Pharmacology*, **611**: 17-21.

Kardosová, A., Ebringerová, A., Alföldi, J., Nosál'ová, G., Franová, S. and Hríbalová, V. (2003). A biologically active fructan from the roots of *Arctium lappa* L., var. Herkules. *International Journal of Biological Macromolecules*, **33**: 135-140.

Katsube, N., Iwashita, K., Tsushida, T., Yamaki, K. and Kobori, M. (2002). Induction of apoptosis in cancer cells by bilberry (*Vaccinium myrtillus*) and the anthocyanins. *Journal of Agricultural and Food Chemistry*, **51**: 68-75.

Kim, D.-S., Kim, T.-W. and Kang, J.-S. (2004). Chromium picolinate supplementation improves insulin sensitivity in Goto-Kakizaki diabetic rats. *Journal of Trace Elements in Medicine and Biology*, **17**: 243-247

Kintzios, S. E. (2000). Sage: The Genus Salvia. Kintzios, S. E. (Ed) CRC Press, Harwood Academic Publishers, Amsterdam, The Netherlands.

Kraus, M., Kahle, K., Ridder, F., Schantz, M., Scheppach, W., Schreier, P. and Richling, E. (2010). Colonic availability of bilberry anthocyanins in humans. *in* Flavor and health benefits of small fruits. American Chemical Society. Washington, DC, U.S.A. pp. 159-176.

Kumarasamy, Y., Nahar, L., Cox, P. J., Jaspars, M. and Sarker, S. D. (2003). Bioactivity of secoiridoid glycosides from *Centaurium erythraea*. *Phytomedicine: International Journal of Phytotherapy and Phytopharmacology*, **10**: 344-347.

Lavagna, S. M., Secci, D., Chimenti, P., Bonsignore, L., Ottaviani, A. and Bizzarri, B. (2001). Efficacy of *Hypericum* and *Calendula* oils in the epithelial reconstruction of surgical wounds in childbirth with caesarean section. II. *Farmaco*, **56**: 451 - 453.

Lei, L.-J., Chen, L., Jin, T.-Y., Nordberg, M. and Chang, X.-L. (2007). Estimation of benchmark dose for pancreatic damage in cadmium-exposed smelters. *Toxicology Science*, **97**: 189-195.

Leung, H. W. C., Lin, C. J., Hour, M. J., Yang, W. H., Wang, M. Y. and Lee, H. Z. (2007). Kaempferol induces apoptosis in human lung non-small carcinoma cells accompanied by an induction of antioxidant enzymes. *Food and Chemical Toxicology*, **45**: 2005-2013.

Li, D., Kim, J. M., Jin, Z. and Zhou, J. (2008). Prebiotic effectiveness of inulin extracted from edible burdock. *Anaerobe*, **14**: 29-34.

Li, X., Cui, X., Sun, X., Li, X., Zhu, Q. and Li, W. (2010). Mangiferin prevents diabetic nephropathy progression in streptozotocin-induced diabetic rats. *Phytotherapy Research*, **24**: 893-899.

Lima, C. F., Andrade, P. B., Seabra, R. M., Fernandes-Ferreira, M. and Pereira-Wilson, C. (2005). The drinking of a *Salvia officinalis* infusion improves liver antioxidant status in mice and rats. *Journal of Ethnopharmacology,* **97**: 383-389.

Lima, C. F., Carvalho, F., Fernandes, E., Bastos, M. L., Santos-Gomes, P. C., Fernandes-Ferreira, M. and Pereira-Wilson, C. (2006). Metformin-like effect of *Salvia officinalis* (common sage): is it useful in diabetes prevention? *British Journal of Nutrition,* **96**: 326-333.

Litkey, J. and Dailey, M. W. (2007). Anticholinergic toxicity associated with the ingestion of lupini beans. *The American Journal of Emergency Medicine,* **25**: 215-217.

Loizzo, M. R., Saab, A. M., Tundis, R., Menichini, F., Bonesi, M., Piccolo, V., Statti, G. A., de Cindio, B., Houghton, P. J. and Menichini, F. (2008). *In vitro* inhibitory activities of plants used in Lebanon traditional medicine against angiotensin converting enzyme (ACE) and digestive enzymes related to diabetes. *Journal of Ethnopharmacology,* **119**: 109-116.

Longaray Delamare, A. P., Moschen-Pistorello, I. T., Artico, L., Atti-Serafini, L. and Echeverrigaray, S. (2007). Antibacterial activity of the essential oils of *Salvia officinalis* L. and *Salvia triloba* L. cultivated in South Brazil. *Food Chemistry,* **100**: 603-608.

Lou, Z., Wang, H., Wang, D. and Zhang, Y. (2009). Preparation of inulin and phenols-rich dietary fibre powder from burdock root. *Carbohydrate Polymers,* **78**: 666-671.

Lu, Y. and Yeap Foo, L. (2000). Flavonoid and phenolic glycosides from *Salvia officinalis. Phytochemistry,* **55**: 263-267.

Lu, Y. and Yeap Foo, L. (2001). Antioxidant activities of polyphenols from sage (*Salvia officinalis*). *Food Chemistry,* **75**: 197-202.

Lu, Y. and Yeap Foo, L. (2002). Polyphenolics of *Salvia* - a review. *Phytochemistry,* **59**: 117-140.

Luís, Â., Domingues, F., Gil, C. and Duarte, A. P. (2009). Antioxidant activity of extracts of Portuguese shrubs: *Pterospartum tridentatum, Cytisus scoparius* and *Erica* spp. *Journal of Medicinal Plants Research,* **3**: 886-893.

Madhavi, D. L., Bomser, J., Smith, M. A. L. and Singletary, K. (1998). Isolation of bioactive constituents from *Vaccinium myrtillus* (bilberry) fruits and cell cultures. *Plant Science,* **131(1)**: 95-103.

Magni, C., Sessa, F., Accardo, E., Vanoni, M., Morazzoni, P., Scarafoni, A. and Duranti, M. (2004). Conglutin γ, a lupin seed protein, binds insulin *in vitro* and reduces plasma glucose levels of hyperglycemic rats. *The Journal of Nutritional Biochemistry,* **15**: 646-650.

Mahat, M. Y. A., Kulkarni, N. M., Vishwakarma, S. L., Khan, F. R., Thippeswamy, B. S., Hebballi, V., Adhyapak, A. A., Benade, V. S., Ashfaque, S. M., Tubachi, S. and Patil, B. M. (2010). Modulation of the cyclooxygenase pathway via inhibition of nitric oxide production contributes to the anti-inflammatory activity of kaempferol. *European Journal of Pharmacology*, **642**: 169-76.

Marc, E. B., Nelly, A., Annick, D.-D. and Frederic, D. (2008). Plants used as remedies antirheumatic and antineuralgic in the traditional medicine of Lebanon. *Journal of Ethnopharmacology*, **120**: 315-334.

Martín-Aragón, S., Basabe, B., Benedí, J. M. and Villar, A. M. (1999). *In vitro* and *in vivo* antioxidant properties of *Vaccinium myrtillus*. *Pharmaceutical Biology*, **37**: 109-113.

Martins, J. M. N. (1994). Numerical taxonomy on the study of *Lupinus albus* accessions. *in* J. M. N. Martins and M. L. Beirão da Costa (eds.), Advances in Lupin Research - VIIth International Lupin Conference, ISA Press. Évora, Portugal, pp. 84-89.

Matsunaga, N., Tsuruma, K., Shimazawa, M., Yokota, S. and Hara, H. (2009). Inhibitory actions of bilberry anthocyanidins on angiogenesis. *Phytotherapy Research*, **24**(S1): S42 - S47.

McCue, P. P. and Shetty, K. (2004). Inhibitory effects of rosmarinic acid extracts on porcine pancreatic amylase in vitro. *Asia Pacific Journal of Clinical Nutrition*, **13**: 101-106.

Merzouki, A., F., E.-D. and Molero-Mesa, J. (2003). Contribution to the knowledge of Rifian traditional medicine III: Phytotherapy of diabetes in Chefchaouen province (North of Morocco). *Ars Pharmaceutica*, **44**: 59-67.

Miean, K. H. and Mohamed, S. (2001). Flavonoid (myricetin, quercetin, kaempferol, luteolin, and apigenin) content of edible tropical plants. *Journal of Agricultural and Food Chemistry*, **49**: 3106-3112.

Milbury, P. E. (2009). Berries and cancer. *in* Complementary and alternative therapies and the aging population. Ross W. R. (ed.), Academic Press, San Diego, U.S.A. pp. 347-370.

Milner, R. D. G. (1969). Stimulation of insulin secretion in vitro by essential aminoacids. *The Lancet*, **293**: 1075-1076.

Miura, K., Kikuzaki, H. and Nakatani, N. (2002). Antioxidant activity of chemical components from sage (*Salvia officinalis* L.) and thyme (*Thymus vulgaris* L.) measured by the oil stability index method. *Journal of Agricultural and Food Chemistry*, **50**: 1845-1851.

Miura, T., Ichiki, H., Hashimoto, I., Iwamoto, N., Kao, M., Kubo, M., Ishihara, E., Komatsu, Y., Okada, M., Ishida, T. and Tanigawa, K. (2001). Antidiabetic activity of a xanthone compound, mangiferin. *Phytomedicine*, **8**: 85-87.

Modak, M., Dixit, P., Londhe, J., Ghaskadbi, S. and Devasagayam, T. P. A. (2007). Indian herbs and herbal drugs used for the treatment of diabetes. *Journal of Clinical Biochemistry and Nutrition*, **40**: 167-173.

Moreno, S., Scheyer, T., Romano, C. S. and Vojnov, A. A. (2006). Antioxidant and antimicrobial activities of rosemary extracts linked to their polyphenol composition. *Free Radical Research*, **40**: 223-231.

Mueller, M. and Jungbauer, A. (2009). Culinary plants, herbs and spices - A rich source of PPAR-γ ligands. *Food Chemistry*, **117**: 660-667

Muruganandan, S., Srinivasan, K., Gupta, S., Gupta, P. K. and Lal, J. (2005). Effect of mangiferin on hyperglycemia and atherogenicity in streptozotocin diabetic rats. *Journal of Ethnopharmacology*, **97**: 497-501.

Naga Raju, G. J., Sarita, P., Ramana Murty, G. A. V., Ravi Kumar, M., Seetharami Reddy, B., John Charles, M., Lakshminarayana, S., Seshi Reddy, T., Reddy, S. B. and Vijayan, V. (2006). Estimation of trace elements in some anti-diabetic medicinal plants using PIXE technique. *Applied Radiation and Isotopes*, **64**: 893-900.

Nandakumar, V., Singh, T. and Katiyar, S. K. (2008). Multi-targeted prevention and therapy of cancer by proanthocyanidins. *Cancer Letters*, **269**: 378-387.

Neagu, E., Roman, G. P., Radu, G. L. and Nechigor, G. (2010). Concentration of the bioactive principles in *Geranium robertianum* extracts through membranare procedures (ultrafiltration). *Romanian Biotechnological Letters*, **15**: 5042-5048.

Neves, J. M., Matos, C., Moutinho, C., Queiroz, G. and Gomes, L. R. (2009). Ethnopharmacological notes about ancient uses of medicinal plants in Trás-os-Montes (northern of Portugal). *Journal of Ethnopharmacology*, **124**: 270-283.

Novais, M. H., Santos, I., Mendes, S. and Pinto-Gomes, C. (2004). Studies on pharmaceutical ethnobotany in Arrabida Natural Park (Portugal). *Journal of Ethnopharmacology*, **93**: 183-195.

Nunes, E., Ferreira, F. M., Peixoto, F. P., Louro, T., Seiça, R. and Santos, M. S. (2006). The anti-diabetic effects of plants extracts on a type 2 diabetic rat model. *XV National Congress of Biochemistry*. Aveiro, Portugal.

Nunes, E., Louro, T., Sena, C., Peixoto, F. P., Santos, M. S. and Seiça, R. (2004). Plantas anti-diabéticas: efeitos em modelos animais de diabetes tipos 1 e 2., IV Congresso de Investigação em Medicina, Coimbra, Portugal (*in* Portuguese).

Nunes, J. R. (1999). Medicina Popular. Tratamento pelas plantas medicinais. Lisboa-Porto: Litexa Editora (*in* Portuguese).

Nuorooz-Zadeh, J. (2000). Plasma lipid hydroperoxide and vitamin E profiles in patients with diabetes mellitus. *in* Antioxidants in diabetes management. L. Packer, P. Rosen, H. J. Tritschler, G. L. King and A. Azzi (eds.), Basel: Marcel Dekker, Inc., New York, U.S.A. pp. 53-64.

Paulo, A., Martins, S., Branco, P., Dias, T., Borges, C., Rodrigues, A. I., Costa, M. D. C., Teixeira, A. and Mota-Filipe, H. (2008). The opposing effects of the flavonoids isoquercitrin and sissotrin, isolated from *Pterospartum tridentatum*, on oral glucose tolerance in rats. *Phytotherapy Research*, **22**: 539-543

Pereira, F.C., Ouedraogo, R., Lebrun, P., Barbosa, R. M., Cunha, A. P., Santos, R. M. and Rosário, L. M. (2001). Insulinotropic action of white lupine seeds (*Lupinus albus* L.): Effects on ion fluxes and insulin secretion from isolated pancreatic islets. *Biomedical Research*, **11**. 103-109.

Pereira, O., Domingues, M. and Cardoso, S. (2009). Characterization of the phenolic constituents and the antioxidant activity of *Cytisus multiflorus*. In: *4th International Conference on Polyphenols and Health*. Harrogate, England.

Pérez-Fons, L., Garzón, M. A. T. and Micol, V. (2009). Relationship between the antioxidant capacity and effect of rosemary (*Rosmarinus officinalis* L.) polyphenols on membrane phospholipid order. *Journal of Agricultural and Food Chemistry*, **58**: 161-171.

Petlevski, R., Hadzija, M., Slijepcevic, M. and Juretic, D. (2001). Effect of 'antidiabetis' herbal preparation on serum glucose and fructosamine in NOD mice. *Journal of Ethnopharmacology*, **75**: 181-184.

Petterson, D. S., Ellis, Z. L., Harris, D. J. and Spadek, Z. E. (1987). Acute toxicity of the major alkaloids of cultivated *Lupinus angustifolius* seed to rats. *Journal of Applied Toxicology*, **7**: 51-53.

Pilvi, T. K., Jauhiainen, T., Cheng, Z. J., Mervaala, E. M., Vapaatalo, H. and Korpela, R. (2006). Lupin protein attenuates the development of hypertension and normalises the vascular function of NaCl-loaded Goto-Kakizaki rats. *Journal of Physiology and Pharmacology*, **57**: 167-176.

Pinto, E., Salgueiro, L. R., Cavaleiro, C., Palmeira, A. and Gonçalves, M. J. (2007). In vitro susceptibility of some species of yeasts and filamentous fungi to essential oils of *Salvia officinalis*. *Industrial Crops and Products*, **26**: 135-141.

Portha, B., Lacraz, G., Kergoat, M., Homo-Delarche, F., Giroix, M. H., Bailbé, D., Gangnerau, M. N., Dolz, M., Tourrel-Cuzin, C. and Movassat, J. (2009). The GK rat beta-cell: A prototype for the diseased human beta-cell in type 2 diabetes? *Molecular and Cellular Endocrinology*, **297**: 73-85.

Pothier, J., Cheav, S. L., Galand, N., Dormeau, C. and Viel, C. (1988). A comparative study of the effects of sparteine, lupanine and lupin extract on the central nervous system of the mouse. *Journal of Pharmacy and Pharmacology*, **50**: 945-954.

Prabhu, S., Jainu, M., Sabitha, K. E. and Devi, C. S. S. (2006). Role of mangiferin on biochemical alterations and antioxidant status in isoproterenol-induced myocardial infarction in rats. *Journal of Ethnopharmacology*, **107**: 126-133.

Prior, R. L., Cao, G., Martin, A., Sofic, E., McEwen, J., O'Brien, C., Lischner, N., Ehlenfeldt, M., Kalt, W., Krewer, G. and Mainland, C. M. (1998). Antioxidant capacity as influenced by total phenolic and anthocyanin content, maturity, and variety of *Vaccinium* species. *Journal of Agricultural and Food Chemistry*, **46**: 2686-2693.

Psarakis, H. M. (2006). Clinical challenges in caring for patients with diabetes and cancer. *Diabetes Spectrum*, **19**: 157-162.

Puupponen-Pimiä, R., Nohynek, L., Ammann, S., Oksman-Caldentey, K.-M. and Buchert, J. (2008). Enzyme-assisted processing increases antimicrobial and antioxidant activity of bilberry. *Journal of Agricultural and Food Chemistry*, **56**: 681-688.

Qian, L.-B., Wang, H.-P., Chen, Y., Chen, F.-X., Ma, Y.-Y., Bruce, I. C. and Xia, Q. (2010). Luteolin reduces high glucose-mediated impairment of endothelium-dependent relaxation in rat aorta by reducing oxidative stress. *Pharmacological Research*, **61**: 281-287.

Rasooli, I., Fakoor, M. H., Yadegarinia, D., Gachkar, L., Allameh, A. and Rezaei, M. B. (2008). Antimycotoxigenic characteristics of *Rosmarinus officinalis* and *Trachyspermum copticum* L. essential oils. *International Journal of Food Microbiology*, **122**: 135-139.

Rau, O., Wurglics, M., Paulke, A., Zitzkowski, J., Meindl, N., Bock, A., Dingermann, T., Abdel-Tawab, M. and Schubert-Zsilavecz, M. (2006). Carnosic acid and carnosol, phenolic diterpene compounds of the labiate herbs rosemary and sage, are activators of the human peroxisome proliferator-activated receptor gamma. *Planta Medica*, **72**: 881-887.

Rauha, J.-P., Remes, S., Heinonen, M., Hopia, A., Kähkönen, M., Kujala, T., Pihlaja, K., Vuorela, H. and Vuorela, P. (2000). Antimicrobial effects of Finnish plant extracts containing flavonoids and other phenolic compounds. *International Journal of Food Microbiology*, **56**: 3-12.

Robert, J.-J., Bier, D. M., Zhao, X. H., Matthews, D. E. and Young, V. R. (1982). Glucose and insulin effects on de novo amino acid synthesis in young men: Studies with stable isotope labeled alanine, glycine, leucine, and lysine. *Metabolism*, **31**: 1210-1218.

Rodriguez, J., Loyola, J. I., Maulén, G. and Schmeda-Hirschmann, G. (1994). Hypoglycaemic activity of *Geranium core-core, Oxalis rosea* and *Plantago major* extract in rats. *Phytotherapy Research*, **8**: 372-374.

Rouanet, J.-M., Décordé, K., Rio, D. D., Auger, C., Borges, G., Cristol, J.-P., Lean, M. E. J. and Crozier, A. (2010). Berry juices, teas, antioxidants and the prevention of atherosclerosis in hamsters. *Food Chemistry*, **118**: 266-271

Roy, S., Khanna, S., Alessio, H. M., Vider, J., Bagchi, D., Bagchi, M. and Sen, C. K. (2002). Anti-angiogenic property of edible berries. *Free Radical Research*, **36**: 1023-1032.

Ryan, E. A., Pick, M. E. and Marceaux, C. (2001). Use of alternative medicines in diabetes mellitus. *Diabetic Medicine*, **18**: 242-245.

Sá, C. M., Ramos, A. A., Azevedo, M. F., Lima, C. F., Fernandes-Ferreira, M. and Pereira-Wilson, C. (2009). Sage tea drinking improves lipid profile and antioxidant defences in humans. *International Journal of Molecular Sciences*, **10**: 3937-3950.

Santoyo, S., Cavero, S., Jaime, L., Ibañez, E., Señoráns, F.J., and Reglero, G. (2005). Chemical composition and antimicrobial activity of *Rosmarinus officinalis* L. essential oil obtained via supercritical fluid extraction. *Journal of Food Protection*, **68**: 790-795.

Saxena, A. M., Bajpai, M. B. and Mukherjee, S. K. (1991). Oxcitidinin induced blood sugar lowering of streptozotocin treated hyperglycemic rats. *Indian Journal of Experimental Biology*, **29**: 674-675.

Saxena, A. M., Bajpai, M. B., Murthy, P. and Mukherjee, K. (1993). Mechanism of blood sugar lowering by a swerchirin-containing hexane fraction (SWI) of *Swertia chirayita*. *Indian Journal of Experimental Biology*, **31**: 178-181.

Schimmer, O. and Mauthner, H. (1996). Polymethoxylated xanthones from the herb of centaurium erythraea with strong antimutagenic properties in *Salmonella typhimurium*. *Planta Medica*, **62**: 561-564.

Schmidt, W. and Beerhues, L. (1997). Alternative pathways of xanthone biosynthesis in cell cultures of *Hypericum androsaemum* L. *FEBS Letters*, **420**: 143-146.

Schmidt, W., Peters, S. and Beerhues, L. (2000). Xanthone 6-hydroxylase from cell cultures of *Centaurium erythraea* RAFN and *Hypericum androsaemum* L. *Phytochemistry*, **53**: 427-431.

Scientific Committee on Food (2003). Opinion of the Scientific Committee on Food on Thujone. *in* Health and Consumer Protection Directorate-General, European Commission (SCF/CS/FLAV/FLAVOUR/23 ADD2 Final), Brussel, Belgium.

Seifried, H. E. (2007). Oxidative stress and antioxidants: a link to disease and prevention? *The Journal of Nutritional Biochemistry*, **18**: 168-171.

Sener, A. and Malaisse, W. J. (2002). The stimulus-secretion coupling of amino acid-induced insulin release. Insulinotropic action of -alanine. *Biochimica et Biophysica Acta*, **1573**: 100-104.

Shekarchi, M., Hajimehdipoor, H., Khanavi, M., Adib, N., Bozorgi, M. and Akbari-Adergani, B. (2010). A validated method for analysis of Swerchirin in *Swertia longifolia* Boiss. by high performance liquid chromatography. *Pharmacognosy Magazine*, **6**: 13-18.

Sheweita, S. A., Newairy, A. A., Mansour, H. A. and Yousef, M. I. (2002). Effect of some hypoglycemic herbs on the activity of phase I and II drug-metabolizing enzymes in alloxan-induced diabetic rats. *Toxicology*, **174**: 131-139.

Shukia, R., Sharma, S., Puri, D., Prabhu, K. and Murthy, P. (2000). Medicinal plants for treatment of diabetes mellitus. *Indian Journal of Clinical Biochemistry*, **15**: 169-177.

Singh, A. (2008). Phytochemicals of Gentianaceae: A Review of Pharmacological Properties. *International Journal of Pharmaceutical Sciences and Nanotechnology*, **1**: 33-36.

Sirtori, C. R., Galli, C., Anderson, J. W. and Arnoldi, A. (2009). Nutritional and nutraceutical approaches to dyslipidemia and atherosclerosis prevention: Focus on dietary proteins. *Atherosclerosis*, **203**: 8-17.

Sirtori, C. R., Lovati, M. R., Manzoni, C., Castiglioni, S., Duranti, M., Magni, C., Morandi, S., D'Agostina, A. and Arnoldi, A. (2004). Proteins of white lupin seed, a naturally isoflavone-poor legume, reduce cholesterolemia in rats and increase LDL receptor activity in HepG2 cells. *Journal of Nutrition*, **134**: 18-23.

Sitasawad, S., Deshpande, M., Katdare, M., Tirth, S. and Parab, P. (2001). Beneficial effect of supplementation with copper sulfate on STZ-diabetic mice (IDDM). *Diabetes Research and Clinical Practice*, **52**: 77-84.

Sosa, T., Alías, J. C., Escudero, J. C. and Chaves, N. (2005). Interpopulational variation in the flavonoid composition of *Cistus ladanifer* L. exudate. *Biochemical Systematics and Ecology*, 33: 353-364.

Sousa, M. B., Curado, T., Vasconcellos, F. N. and Trigo, M. J. (2007). Mirtilo - Qualidade pós-colheita. *in* Diversificação da produção frutícola com novas espécies e tecnologias que assegurem a qualidade agro-alimentar, Edição no âmbito do Projecto PO AGRO DE&D Nº 556: (INRB/DPA), Portugal (*in* Portuguese)

Spielmann, J., Shukla, A., Brandsch, C., Hirche, F., Stangl, G. I. and Eder, K. (2007). Dietary lupin protein lowers triglyceride concentrations in liver and plasma in rats by reducing hepatic gene expression of sterol regulatory element-binding protein-1c. *Annals of Nutrition and Metabolism*, 51: 387-392.

Tahraoui, A., El-Hilaly, J., Israili, Z. H. and Lyoussi, B. (2007). Ethnopharmacological survey of plants used in the traditional treatment of hypertension and diabetes in south-eastern Morocco (Errachidia province). *Journal of Ethnopharmacology*, 110: 105-117.

Teixeira, S., Mendes, A., Alves, A. and Santos, L. (2007). Simultaneous distillation-extraction of high-value volatile compounds from *Cistus ladanifer* L. *Analytica Chimica Acta*, 584: 439-446.

Terruzzi, I., Senesi, P., Magni, C., Montesano, A., Scarafoni, A., Luzi, L. and Duranti, M. (2010) Insulin-mimetic action of conglutin-γ, a lupin seed protein, in mouse myoblasts. *Nutrition, Metabolism and Cardiovascular Diseases* (in Press)

Torel, J., Cillard, J. and Cillard, P. (1986). Antioxidant activity of flavonoids and reactivity with peroxy radical. *Phytochemistry*, 25: 383-385.

Tsaliki, E., Lagouri, V. and Doxastakis, G. (1999). Evaluation of the antioxidant activity of lupin seed flour and derivatives (*Lupinus albus* ssp. Graecus). *Food Chemistry*, 65(1): 71-75.

Tsiodras, S., Shin, R. K., Christian, M., Shaw, L. M. and Sass, D. A. (1999). Anticholinergic toxicity associated with lupine seeds as a home remedy for diabetes mellitus. *Annals of Emergency Medicine*, 33: 715-717.

Valentão, P. C. R. (2002). Limonete, Hipericão-do-Gerês, Cardo-do-Coalho, Fel-da-Terra: Metodologias de controlo de qualidade com base na fracção fenólica: Estudos de acção antioxidante e hepatoprotectora. PhD Dissertation, Faculdade de Farmácia, Universidade do Porto (*in* Portuguese).

Valentão, P., Andrade, P. B., Silva, A. M. S., Moreira, M. M. and Seabra, R. M. (2003). Isolation and structural elucidation of 5-formyl-2,3-dihydroiscoumarin from *Centaurium erythraea* aerial parts. *Natural Product Research*, 17: 361-364.

Valentão, P., Andrade, P. B., Silva, E., Vicente, A., Santos, H., Bastos, M. L. and Seabra, R. M. (2002). Methoxylated xanthones in the quality control of small centaury (*Centaurium erythraea*) flowering tops. *Journal of Agricultural and Food Chemistry*, 50: 460-463.

Valentão, P., Areias, F., Amaral, J., Andrade, P. and Seabra, R. (2000). Tetraoxygenated xanthones from *Centaurium erythraea*. *Natural Product Letters*, 14: 319-323.

Valentão, P., Carvalho, M., Carvalho, F., Fernandes, E., Neves, R. P. d., Pereira, M. L., Andrade, P. B., Seabra, R. M. and Bastos, M. L. (2004b). *Hypericum androsaemum* infusion increases tert-butyl hydroperoxide-induced mice hepatotoxicity *in vivo*. *Journal of Ethnopharmacology*, **94**: 345-351.

Valentão, P., Carvalho, M., Fernandes, E., Carvalho, F., Andrade, P. B., Seabra, R. M. and Bastos, M. d. L. (2004a). Protective activity of *Hypericum androsaemum* infusion against tert-butyl hydroperoxide-induced oxidative damage in isolated rat hepatocytes. *Journal of Ethnopharmacology*, **92**: 79-84.

Valentão, P., Dias, A., Ferreira, M., Silva, B., Andrade, P. B., Bastos, M. L. and Seabra, R. M. (2003). Variability in phenolic composition of *Hypericum androsaemum*. *Natural Product Research*, **17**: 135-140.

Valentão, P., Fernandes, E., Carvalho, F., Andrade, P. B., Seabra, R. M. and Bastos, M. L. (2001). Antioxidant activity of *Centaurium erythraea* infusion evidenced by its superoxide radical scavenging and xanthine oxidase inhibitory activity. *Journal of Agricultural and Food Chemistry*, **49**: 3476-3479.

Valentão, P., Fernandes, E., Carvalho, F., Andrade, P. B., Seabra, R. M. and Bastos, M. L. (2002). Antioxidant activity of *Hypericum androsaemum* infusion: Scavenging activity against superoxide radical, hydroxyl radical and hypochlorous acid. *Biological and Pharmaceutical Bulletin*, **25**: 1320-1323.

Valentová, K., Ulrichová, J., Cvak, L. and Simánek, V. (2007). Cytoprotective effect of a bilberry extract against oxidative damage of rat hepatocytes. *Food Chemistry*, **101**: 912-917.

Valko, M., Leibfritz, D., Moncol, J., Cronin, M. T. D., Mazur, M. and Telser, J. (2007). Free radicals and antioxidants in normal physiological functions and human disease. *The International Journal of Biochemistry and Cell Biology*, **39**: 44-84.

Valko, M., Rhodes, C. J., Moncol, J., Izakovic, M. and Mazur, M. (2006). Free radicals, metals and antioxidants in oxidative stress-induced cancer. *Chemico-Biological Interactions*, **160**: 1-40.

Vaz, A. C., Pinheiro, C., Martins, J. M. N. and Ricardo, C. P. P. (2004). Cultivar discrimination of Portuguese *Lupinus albus* by seed protein electrophoresis: the importance of considering glutelins and glycoproteins. *Field Crops Research*, **87**: 23-34.

Vieira, I. P., Costa, A. C., Teixeira, D. M., Antunes, C. M. and Cruz-Morais, J. (2010). Efeito Hipoglicemiante de um Extracto Aquoso de *Cytisus multiflorus*. In Jornadas 2010 do Departamento de Química., J. V. N. Júlio Cruz Morais, António Candeias, António Teixeira, and C. G. e. J. Teixeira. (eds.), FLM, Fundação Luís de Molina. Universidade de Évora, Portugal, p. 59 (in Portuguese).

Vitor, R. F., Mota-Filipe, H., Teixeira, G., Borges, C., Rodrigues, A. I., Teixeira, A. and Paulo, A. (2004). Flavonoids of an extract of *Pterospartum tridentatum* showing endothelial protection against oxidative injury. *Journal of Ethnopharmacology*, **93**: 363-370.

Wang, X., Li, F., Sun, Q., Yuan, J., Jiang, T. and Zheng, C. (2005). Application of preparative high-speed counter-current chromatography for separation and purification of arctiin from *Fructus arctii*. *Journal of Chromatography A*, **1063**: 247-251.

Wang, Y., Fang, J., Leonard, S. S. and Rao, K. M. K. (2004). Cadmium inhibits the electron transfer chain and induces Reactive Oxygen Species. *Free Radical Biology and Medicine*, **36**: 1434-1443.

Wattel, A., Kamel, S., Mentaverri, R., Lorget, F., Prouillet, C., Petit, J.-P., Fardelonne, P. and Brazier, M. (2003). Potent inhibitory effect of naturally occurring flavonoids quercetin and kaempferol on *in vitro* osteoclastic bone resorption. *Biochemical Pharmacology*, **65**: 35-42.

Weed Management Guide (2003). White Spanish broom –*Cytisus multiflorus*. in Alert List for Environmental Weeds: Weed Management Guide (ed.), CRC Weed Management, Natural Heritage Trust, Australia.

Welsh, M., Hellerström, C. and Andersson, A. (1982). Respiration and insulin release in mouse pancreatic islets: Effects of l-leucine and 2-ketoisocaproate in combination with d-glucose and l-glutamine. *Biochimica et Biophysica Acta*, **721**: 178-184.

WHO (World Health Organization) (1980). Expert committee on diabetes mellitus: second report. World Health Organization Technical Report Series. **646**: 1-80.

WHO (World Health Organization) (2006). Definition and diagnosis of diabetes mellitus and intermediate hyperglycemia : report of a WHO/IDF consultation.

William, G. and Pickup, J. C. (2003). Handbook of Diabetes (3rd Edition), Blackwell Publishing, Oxford, U.K.

Witzell, J., Gref, R. and Näsholm, T. (2003). Plant-part specific and temporal variation in phenolic compounds of boreal bilberry (*Vaccinium myrtillus*) plants. *Biochemical Systematics and Ecology*, **31**: 115-127.

Yakhontova, L. and Kibal'chich, P. (1971). The question of the content of arctiin in the seeds of *Arctium leiospermum*. *Chemistry of Natural Compounds*, **7**: 287-288.

Yao, Y. and Vieira, A. (2007). Protective activities of *Vaccinium* antioxidants with potential relevance to mitochondrial dysfunction and neurotoxicity. *NeuroToxicology* **28**: 93-100.

Yorek, M. A. (2003). The role of oxidative stress in diabetic vascular and neural disease. *Free Radical Research*, **37**: 471-480.

Yoshida, T., Konishi, M., Horinaka, M., Yasuda, T., Goda, A. E., Taniguchi, H., Yano, K., Wakada, M. and Sakai, T. (2008). Kaempferol sensitizes colon cancer cells to TRAIL-induced apoptosis. *Biochemical and Biophysical Research Communications* **375**: 129-133.

Yovo, K., Huguet, F., Pothier, J., Durand, M., Breteau, M. and Narcisse, G. (1984). Comparative pharmacological study of sparteine and its ketonic derivative lupanine from seeds of *Lupinus albus*. *Planta Medica*, **50**: 420-424.

Yu, P. C., Bosnyak, Z. and Ceriello, A. (2010). The importance of glycated haemoglobin (HbA1c) and postprandial glucose (PPG) control on cardiovascular outcomes in patients with type 2 diabetes. *Diabetes Research and Clinical Practice*, **89**: 1-9.

Zafra-Stone, S., Yasmin, T., Bagchi, M., Chatterjee, A., Vinson, J. A. and Bagchi, D. (2007). Berry anthocyanins as novel antioxidants in human health and disease prevention. *Molecular Nutrition and Food Research*, **51**: 675-683.

Zarzuelo, A., Jiménez, I., Gámez, M. J., Utrilla, P., Fernadez, I., Torres, M. I. and Osuna, I. (1996). Effects of luteolin 5-O-β-rutinoside in streptozotocin-induced diabetic rats. *Life Sciences*, **58**: 2311-2316.

Zhao, P., Wang, J., Ma, H., Xiao, Y., He, L., Tong, C., Wang, Z., Zheng, Q., Dolence, E. K., Nair, S., Ren, J. and Li, J. (2009). A newly synthetic chromium complex - Chromium (D-phenylalanine)$_3$ activates AMP-activated protein kinase and stimulates glucose transport. *Biochemical Pharmacology*, **77**: 1002-1010.

Natural Products: Research Reviews Vol. 1 (2012)
Editor: V.K. Gupta
Published by: DAYA PUBLISHING HOUSE, NEW DELHI

Pages 41–78

2

Immunotherapeutic Targeting of Multiple Dysregulated Immune Functions in Cancer by Neem Leaf Glycoprotein

Anamika Bose[1], Shyamal Goswami[2], Soumyabrata Roy[2],
Koustav Sarkar[3], Krishnendu Chakraborty[4],
Tathagata Chakraborty[2], Enamul Haque[2],
Subrata Laskar[5] and Rathindranath Baral[2*]

ABSTRACT

Non-specificity and collateral toxicity of conventional therapeutics attracts researchers to the immunotherapy for cancerous diseases. Several forms of non-toxic, biological therapeutics are housed under an umbrella of immunotherapy. In cancer, generally immunotherapy intervenes therapeutic modalities, instead of prophylactic setting, which initially aimed to activation and functional optimization of immune cells, like macrophages, NK cells, T cells and dendritic cells. With advent of knowledge in last decade it is known that tumor instructed suppressor cells are dynamically active in cancer to paralyze host immune surveillance. And unfortunately, many promising prophylactic vaccines show only modest efficacy when translated into therapeutic settings because of suppressor cells are either promoted or unaffected

1   University of Pittsburgh Medical Center, W1004 BST, Pittsburgh, PA, USA

2   Department of Immunoregulation and Immunodiagnostics; Chittaranjan National Cancer Institute, 37, S. P. Mukherjee Road, Kolkata 700026; India

3   Childrens Hospital, Pittsburgh, PA, USA

4   Louisiana State University, Baton Rough, LA, USA

5   Department of Chemistry, University of Burdwan, Burdwan, West Bengal, India

*   Corresponding author. E-Mail: baralrathin@hotmail.com; baral.rathindranath@gmail.com; Phone: 91-033-2476-5101; Fax: 91-033-2475-7606

under influence of these vaccines. Accordingly, there is a definite need for an alternative candidate that might be able to stimulate immune effectors in concert with suppressing regulatory cells.

In our continuous experimental effort with neem leaf glycoprotein (NLGP), a natural immunomodulator, we have experienced that NLGP can effectively break the T cell anergy in tumor condition and prevent the IDO mediated apoptosis of T cells. At the same time, T cells acquire greater cytotoxic efficacy to kill tumors utilizing IFNγ dependent perforin-granzyme B pathway. NLGP also mobilizes activated T cells to tumor site by maintaining CXC receptor-ligand homeostasis. In such normalization of T cell functions, maturation of dendritic cells by NLGP and subsequent antigen delivery to T cells plays a critical role. NLGP matured DCs effectively prime antigen specific T cells by optimizing co-stimulatory signaling and histocompatibility in vivo to obtain robust anti-tumor response translated into tumor growth restriction in mice. Elevated level of suppressor regulatory T cells in cancer condition may interfere with NLGP rescued T cell functions. Fortunately, at the sametime, NLGP prevents the tumor promoted conversion of T cells to regulatory T cells, thereby, partially withdraws the suppression of regulatory T cells on T cells and other immune components. Additionally, NLGP efficiently facilitate the trafficking of APCs (DC and macrophages), while limiting the migration of regulatory cells towards tumor microenvironment by differentially regulating CCR7/CCR5 and CCR4 (counter-regulatory) expression, thereby, optimizing tumor cell lysis by T cells. Mechanism of targeting multiple dysregulated immune functions in cancer by a single non-toxic natural agent NLGP is discussed here.

Keywords: Chemokines, Dendritic cells, Macrophages, NK cells, Neem Leaf Glycoprotein, Regulatory T cells, T cells.

Introduction

Non-specificity and collateral toxicity of conventional therapeutics attracts researchers to the immunotherapy for cancerous diseases. Several forms of non-toxic, biological therapeutics are housed under an umbrella of immunotherapy (Figure 2.1). To maintain an individual in good health, host's defense is always alert, which is composed of several innate and adaptive immune components (McCullough and Summerfield, 2005). Optimally effective state of these cellular and extracellular components denotes the health of a physiological system. In disease, immune functions are altered. Relief from a disease depends on the extent of immune recovery after elimination of the responsible etiological agent. With the onset of carcinogenesis, immune system is vigorously reactive to eliminate potential malignant cells and success in such elimination denotes maintenance of 'cancer free' state. Sometime immune system cannot eliminate the malignant cells, but, maintains dormancy on its propagation. Such state of equilibrium helps to be asymptotic after initiation of carcinogenesis. Unfortunately, in few cases malignant cells are escaped from the immune surveillance and malignant clones are expanded to be metastasized. Immune system appears as well as reacts differently in these three 'E' conditions, namely, Elimination, Equilibrium and Escape (Dunn *et al.*, 2002).

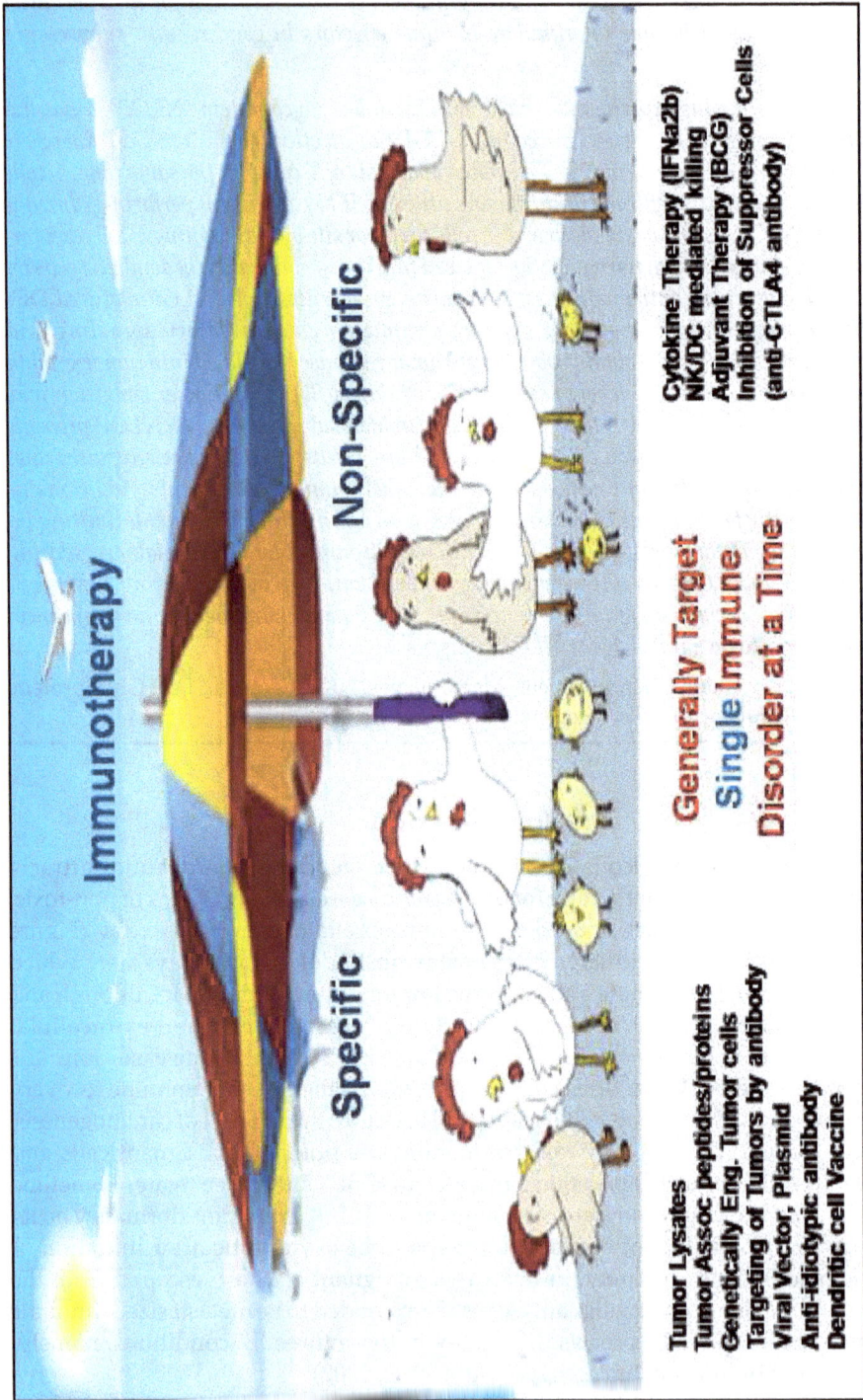

Immunotherapy

Specific

Tumor Lysates
Tumor Assoc. peptides/proteins
Genetically Eng. Tumor cells
Targeting of Tumors by antibody
Viral Vector, Plasmid
Anti-idiotypic antibody
Dendritic cell Vaccine

Non-Specific

Generally Target
Single Immune
Disorder at a Time

Cytokine Therapy (IFNa2b)
NK/DC mediated killing
Adjuvant Therapy (BCG)
Inhibition of Suppressor Cells
(anti-CTLA4 antibody)

Figure 2.1: Immunotherapy consists of several branches. It can be broadly divided into specific and non-specific immunotherapy. Each branch also consists of several forms of therapy under an umbrella of immunotherapy.

In cancer, generally immunotherapy intervenes therapeutic modalities, instead of prophylactic setting, which initially aimed to activation and functional optimization of various immune cells (Sonia *et al.*, 2008). With advent of knowledge in last decade it is known that tumor instructed suppressor cells, are dynamically active in cancer to paralyze host immune surveillance (Rabinovich *et al.*, 2007). And unfortunately, many promising prophylactic vaccine shows only modest efficacy when translated into therapeutic settings because of suppressor cells are either promoted or unaffected under influence of these vaccines (Ha, 2009). Accordingly, there is a definite need for an alternative candidate that might be able to stimulate immune effectors in concert with suppressing regulatory cells.

Immune Dysfunctions and Cancer

In cancer, particularly when a malignant tumor escaped from the immune surveillance, a series of immune alterations (Table 2.1) may be exhibited in tumor bearers to create an immunosuppressive condition. We and several others reported immune alterations in cancer and associated downregulation in functions as well as activation of cytotoxic T cells, NK cells, NK-T cells and macrophages (Block and Markovic, 2009; Hamaï *et al.*, 2010). Moreover, migratory ability of T cells and monocytes obtained from cancer patients was significantly less than healthy cells because of the dysregulated chemokine signaling (Ben-Baruch *et al.*, 2006; Mantovani *et al.*, 2010). This situation causes poor tumor cell lysis by cytotoxic T lymphocytes. Again, poor tumor killing is directly related to the reduced antigen presentation by macrophages and dendritic cells (DCs), those are dysregulated due to lack of optimum co-stimulation and expression of MHC molecules in cancer (Martin-Orozco and Dong, 2007). In addition to these dysregulated cellular and extracellular functions in cancer, immune suppression is contributed by a variety of other mechanisms and creation of immunoavoidance by tumor is a dominant reason for 'immune escape' in cancer. Regulatory T cells (Tregs) and myeloid derived suppressor cells (MDSC) play significant role in conversion and maintenance of type 2/type 3 tumor microenvironment favors tumor progression and appear as a major barrier to effective cancer immunotherapy (Rivoltini *et al.*, 2005) (Figure 2.2).

Implications of Therapeutic Cancer Vaccines to Target Immune Dysfunctions

The majority of cancer vaccines are targeted within the therapeutic setting as opposed to the prophylactic setting proposed for pathogen vaccines (Pashov *et al.*, 2007). The exceptions are cancer vaccines for specific viral infections, recognized as predisposing for the development of malignancies. Three such preventive vaccines have till date been FDA approved - Hepatitis B virus (HBV) vaccine (Lim *et al.*, 2009), for hepatocellular carcinoma (HCC), GARDASIL (Agius *et al.*, 2010) and CERVARIX (Szarewski, 2010) for cervical carcinoma.

Producing effective therapeutic vaccines has proved much more difficult and challenging than developing cancer preventive vaccines. Limited success with therapeutic cancer vaccines has nurtured many speculations. While various components of the innate and adaptive immune responses are able to mediate tumor

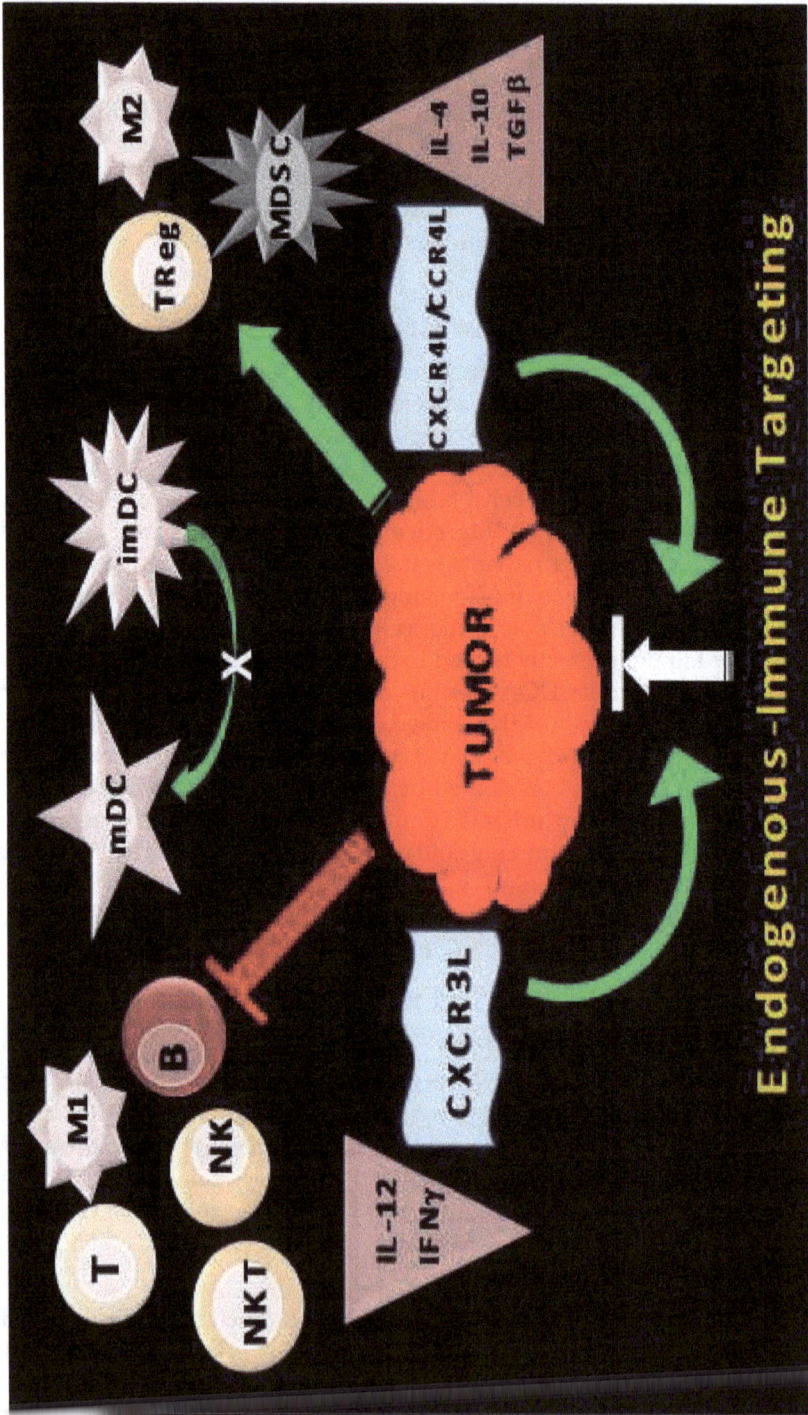

Figure 2.2: Immune targets for immunotherapy. Within tumor microenvironment several immune alterations take place. These alterations influence immune functions within tumors as well as in periphery. Components involved in altered immune functions in cancer may be targeted by immunotherapy by various ways.

Table 2.1: Dysregulated immune components as a target of cancer immunotherapy.

Sl.No.	Dysregulated Immune Components	Adverse Functions	Ref.
1.	Regulatory T (Treg) Cells	Expansion in tumor and peripheral blood of cancer patients suppress the host protective immune responses	Liyanage *et al.*, 2002
2.	Tumor Associated Macrophages (TAMs)	Release a vast diversity of growth factors, proteolytic enzymes, cytokines, and inflammatory mediators, act as key agents in cancer metastasis	Chen *et al.*, 2005
3.	Type 2 Helper (TH2) T-cells	Secrete cytokines (IL-10, TGFβ etc) to favor the establishment of tumor	Keilholz *et al.*, 2002
4.	Myeloid Derived Suppressor Cells (MDSCs)	Impairs effector T cell functions, induces T cell apoptosis and promotes expansion of Treg cells	Gabrilovich and Nagaraj, 2009
5.	Suboptimally Matured Dendritic Cells (iDCs)	Hinders effective antigen presentation and co-stimulation to T cells, release different anti-tumor cytokines	Vieweg and Jackson, 2005
6.	Cancer cells themselves	Promoting the proliferation of suppressor cells by secreting several soluble factors, *e.g.*, TGFβ, IL-10, IL-13, IDO and VEGF	Rabinovich *et al.*, 2007
7.	Tumor associated stromal cells	Secretes growth factor, hampers immune effector cell infiltration	Shurin *et al.*, 2006

cell destruction, specific types of immune cells can also induce a protumor environment that favors tumor growth and the development of metastasis (Palena and Schlom, 2010). These cells create an immunosuppressive tumor environment by accumulating at the site of the tumor, thereby negatively impacting the establishment of antitumor T-cell responses. Different factors may interfere in clinical outcome of various therapeutic vaccines (Table 2.2). A therapeutic cancer vaccine must override these cancer associated immunesuppressions to mount an effective antitumor response.

Several preclinical and clinical trials are initiated with a variety of immunotherapeutic regimens. These trials have shown partial success or could not meet their expectation, due to above discussed reasons. Most of these vaccines occasionally target single altered immune component, instead of several dysregulated factors. For example, anti-CTLA-4 antibody can suppress regulatory T cells without promoting normalization of dendritic cell functions or angiogenesis. Table 2.2 describes the outcome of some clinical trials using therapeutic cancer vaccines. Results obtained from these studies clearly pointing out that these vaccine approaches are functional in *in vivo* system, but durable response is limited due to predominance of nontargeted immune components, still exhibiting cancer associated dysregulations (Ribas *et al.*, 2003). Thus, targeting of multiple dysregulated immune components might appear more consequential to achieve the desired therapeutic outcome. A molecule having effects on multiple targets would have greater possibility of success in clinical setting.

Table 2.2: Various tested immunotherapeutic strategy for cancer.

Sl.No.	Immunotherapeutic Strategy (Target)	Disease	Outcome	Response Status		Ref.
				Immune Response	Clinical Response	
I.	**Peptide and Protein**					
	gp100, tyrosinase+ IL-12	High-risk resected stage III or IV melanoma	Significant proportion of patients with resected melanoma mount an antigen-specific immune response against a peptide vaccine and indicate that IL-12 may increase the immune response and supporting further development of IL-12 as a vaccine adjuvant	33/38 + ELISA	24/48 relapses at 20 months	Lee et al., 2001
	gp100+tetanous tox+Montamide ISA-51 or QS-21	High-risk resected melanoma (stage IIB-IV)	Although this Phase I study was not intended to evaluate clinical benefit, the excellent survival of patients on this protocol suggests the possibility of a benefit that should be assessed in future studies	14 per cent with gp100+ CTL	75 per cent survival at 4.5 years	Slingluff et al., 2001
	gp100, IL-2, IL-12	Melanoma	This study shows that a peptide-based vaccine can effectively generate a quantifiable T cell-specific immune response in the PBMC of cancer patients, though such a response does not associate with a clinically evident regression of metastatic melanoma	7/7 alone 4/5 with IL-12 1/11 with IL-2	Greater tumor regression with IL-2	Lee et al., 1999
	MAGE-3, PADRE, IFA	Resected high-risk melanoma	The data suggest that immune responses can be detected against PADRE and MAGE-3 in vaccinated melanoma patients, albeit with a low frequency of effector cells	5/14 MAGE+ Immunity	NR	Weber et al., 1999

Contd...

Table 2.2–*Contd...*

Sl.No.	Immunotherapeutic Strategy (Target)	Disease	Outcome	Response Status		Ref.
				Immune Response	*Clinical Response*	
	MART-1 + IFA	High-risk resected stages IIB, III and IV melanoma	The data suggest a significant proportion of patients with resected melanoma mount an antigen-specific immune response against a peptide vaccine and support further development of peptide vaccines for melanoma.	13/25 DTH⁺ 10/22 Elispot⁺	9/25 relapsed at 16 mo	Wang et al., 1999
	HER2/neu helper peptides	Breast Cancer (Stage III and IV) Ovarian cancer (Stage III)	Results demonstrate that HER-2/neu MHC class II epitopes containing encompassed MHC class I epitopes are able to induce long-lasting HER-2 specific IFNγ producing CD8+ T cells.	14/18 with antigen specific proliferation	NR	Knutson et al., 2001
	Idiotype + GM-CSF	Lymphoma	The demonstration of molecular remissions, analysis of CTLs against autologous tumor targets, and addition of GM-CSF to the vaccine formulation provide principles relevant to the design of future clinical trials of other cancer vaccines administered in a minimal residual disease setting	19/20 tumor+ T cells	8/11 became PCR- in PB	Bendandi et al., 1999
II.	**Tumor Cells**					
	Melanoma transduced with Ad- GM-CSF	Melanoma	The data suggest that repeated vaccinations with irradiated autologous GM-CSF-producing tumor cells were well tolerated by patients and led to the activation of an antitumor immune response in some patients.	5/9 + CTL activity	1/9 minor response	Kusumoto et al., 2001

Contd...

Table 2.2–Contd...

Sl. No.	Immunotherapeutic Strategy (Target)	Disease	Outcome	Response Status		Ref.
				Immune Response	Clinical Response	
	Allogenic pan-creatic tumor secreting GM-CSF	Pancreatic Cancer	This vaccine approach seems to induce dose-dependent systemic anti-tumor immunity as measured by increased postvaccination DTH responses against autologous tumors. Further clinical evaluation of this approach in patients with pancreatic cancer is warranted	3/14 DTH+	3 patients with DFS >25 months	Jaffee et al., 2001
	CancerVax (melanoma)	Melanoma	PMCV therapy greatly enhances serum CDC against melanoma cells. This enhancement is directly correlated with DFS following initiation of vaccine therapy	82 per cent + complement dependent cyto-toxicity (CDC)	Median Survival 54 months	Hsueh et al., 1998
	Autologous colon CA + BCG	Stage II and stage III colon cancer	Adjuvant active specific immuno-therapy provided significant clinical benefits in patients with stage II colon cancer and appears to bean important new adjuvant treatment for these patients	Increased DTH in all patients	No overall survival benefit	Hanna et al., 2001
	Autologous GBM+ Newcastle virus	Glioblastoma Multiforme	Active specific immunization with NDV-modified glioblastoma cells produced a noticeable peripheral immune response. In this preliminary series survival of patients was not significantly longer after active specific immunization than after combined treatment of surgery, radio-therapy and chemotherapy. As there were no side effects, however, active specific immunization may be considered an alternative in the management of glioblastoma	DTH increased from 1.67 to 4.05 cm^2	Median survival was 46 weeks	Schneider et al., 2001

Contd...

Table 2.2–*Contd...*

Sl.No.	Immunotherapeutic Strategy (Target)	Disease	Outcome	Response Status		Ref.
				Immune Response	Clinical Response	
III. Viral Vector and Plasmid						
	ALVAC-CEA B7.1	Advanced or metastatic CEA-expressing Adenocarcinoma	This pilot study, in which patients were vaccinated with advanced CEA-expressing adenocarcinomas, has demonstrated that ALVAC-CEA B7.1 is safe when used alone or in combination with GM-CSF	5/9 CEA+ T cell precursor frequency	0 PR; 4/23 with decreased CEA	von Mehren et al., 2001
	Vaccinia CEA/ avipox-CEA + GMCSF	Advanced tumors expressing CEA	rV-CEA was more effective in its role as a primer of the immune system; avipox-CEA could be given up to eight times with continued increases in CEA T-cell precursors. Future trials should use rV-CEA first followed by avipox-CEA. Vaccines specific to CEA are able to generate CEA-specific T-cell responses in patients without significant toxicity	Greater immune response than reverse order	Minimal	Marshall et al., 2001
	Vaccinia CEA	Metastatic adenocarcinoma	No objective clinical responses to the rV-CEA vaccine were observed among this population of patients with widely metastatic adenocarcinoma	No CEA+ responses	4/20 with stable disese	Conry et al., 1999
	PSMA/CD86 plasmid	Prostate Cancer	No immediate or long-term side effects following immunizations have been recorded	67-100% DTH+	NR	Mincheff et al., 1999

Contd...

Table 2.2—Contd...

Sl.No.	Immunotherapeutic Strategy (Target)	Disease	Outcome	Response Status		Ref.
				Immune Response	Clinical Response	
IV. Dendritic Cells (DCs)						
	CD34+DC+gp100 MART-1, MAGE-3, tyrosinase, gp100	Metastatic melanoma	Results indicate that vaccination of stage IV melanoma patients with antigen-pulsed CD34-DCs is well tolerated and results in enhanced immunity to a viral antigen as well as to several MelAgs	16/18 with Ag+ T cell reactivity	Regression of >1 melanoma met in 7/18	Banchereau et al., 2001
	DC+MAGE-3 peptide	Gastrointestinal carcinomas	Results suggested that DC vaccination with MAGE-3 peptide is a safe and promising approach in the treatment of gastrointestinal carcinomas	4/8 MAGE-specific CTL	3/12 with minor responses	Sadanaga et al., 2001
	Flt3L mobilized DC+ CAP1-6D (CEA+)	Colon or nonsmall cell lung cancer	After vaccination, two of 12 patients experienced dramatic tumor regression, one patient had a mixed response, and two had stable disease. Clinical response correlated with the expansion of CD8+ tetramer1 T cells, confirming the role of CD8 T cells in this treatment strategy	5/12 tetramer+ T cells	2/12 PR, 2/12 MR	Fong et al., 2001
	DC+ MAGE1,3	MAGE-A1 and/or -A3 tumors	It can be concluded that anti-tumor vaccination using DC pulsed with MAGE peptides induces a potent but transient anti-MAGE, IFN-gamma secretion that is not influenced by the additional delivery of a nonspecific, T-cell help	12/24 IFNγ secreting PBL	5/17 with stable disease	Toungouz et al., 2001

Contd...

Table 2.2–*Contd...*

Sl.No.	Immunotherapeutic Strategy (Target)	Disease	Outcome	Response Status		Ref.
				Immune Response	Clinical Response	
	DC + tumor RNA, KLH	Colon Cancer	The therapy was well tolerated. Dendritic cells were verified by phenotype and in vitro function. The positive keyhole limpet hemocyanin skin test confirms in vivo function by effective vaccination to keyhole limpet hemocyanin. Demonstration of any anticancer efficacy will require further follow-up	11/13 KLH+ responses	7/13 decresed CEA; 0 PRs	Rains *et al.,* 2001
	DC + mouse PAP	Metastatic prostate cancer	Results suggest that while activated DC can prime T cell immunity regardless of route, the quality of this response and induction of Ag-specific Abs may be affected by the route of administration	100 per cent IFNγ+ T cells	NR	Fong *et al.,* 2001

Ad: Adenovirus; CDC: Complement-Dependent Cytotoxicity; DFS: Disease-Free Survival; DTH: Delayed-Type Hypersensitivity; NR: No response; PMCV: Polyvalent Melanoma Cell Vaccine; PR: Primary response; rV: recombinant Vaccinia.

Neem Leaf Glycoprotein to Target Multiple Immune Dysregulations

In our venture in an unexplored field of cancer immunology and immunotherapy with a novel glycoprotein (identified in neem (*Azadirachta indica* leaf), efficacy of it as a prophylactic tumor vaccine in murine model is established (Baral and Chattopadhyay, 2004). Therapeutic potential of NLGP with or without chemotherapy is presently under investigation (*data not shown*). Biochemical characteristics of this glycoprotein, termed neem leaf glycoprotein (NLGP), are shown in Figure 2.3. Carbohydrate portion of NLGP is important for its biological activity, as neuraminidase treatment abolishes tumor growth restriction, observed after prophylactic treatment of mice with NLGP. NLGP has excellent role in promoting tumor antigen specific immunity and the total work is reviewed in this year (Baral *et al.*, 2010). In course of our study on role of NLGP in restoration of various dysregulated innate and adaptive immune functions in cancer, several encouraging data were obtained and published.

NK, NK-T and Macrophages

To encounter carcinogenic threat, innate immune system reacts initially to eliminate tumor with help of macrophages and NK cells. However, altered/ dysregulated functions of these cells may not tackle the situation. In initial phase of our venture, we have used neem leaf preparation (NLP) and tested its role in prevention of the murine tumor growth [Ehrlich's carcinoma (EC) and B16 melanoma] (Baral and Chattopadhyay, 2004). Using adoptive cell transfer technology, it was established that NLP mediated activation of immune cells may be involved in tumor growth restriction (Haque and Baral, 2006). Mononuclear cells (MNC) from blood and spleen of NLP activated Swiss and C57BL/6 mice caused enhanced cytotoxicity to murine EC cells in vitro. Fractionation of spleen cells exhibited greater percentage of tumor cell lysis in macrophage and B cell depleted NK and T cell rich fraction. Flow cytometric analysis revealed in both blood and spleen, NK cells (DX5$^+$ or NK1.1$^+$) and NK-T cells (CD3$^+$/DX5$^+$ or CD3$^+$/NK1.1$^+$) were increased in number in Swiss, C57BL/6 and athymic nude mice after pretreatment with NLP. NLP stimulated spleen cells showed greater secretion of IFNg and TNFa (Haque and Baral, 2006). Thus, NLP activated NK and NK-T cells in mice may regulate tumor cell cytotoxicity by enhancing the secretion of different cytotoxic cytokines (Table 2.3; adopted from Haque and Baral, 2006). In addition to murine system, NLP activates human NK cells (CD56$^+$CD3$^-$) to enhance their cytotoxic ability to tumor cells (NK sensitive K562 cells) and stimulates the release of IL-12 from macrophages from healthy individuals and HNSCC patients. NLP upregulates cytotoxic (CD16$^+$ and CD56^{dim}) NK cells and cytotoxicity of NK sensitive K562 cells by NLP stimulated PBMC was decreased significantly following IL-12 neutralization. Although NLP stimulates the release of IFNγ from NK cells, NK cellular cytotoxicity is not IFNγ dependent. This NK mediated cytotoxicity is manifested by upregulating IL-12 dependent intracellular expression of perforin-granzyme B system. Moreover, NK cytotoxic function was abolished after use of concanamycin A, a perforin inhibitor, but not by brefeldin A, a Fas inhibitor, confirming the participation of perforin-granzyme B system. Moreover, NLP

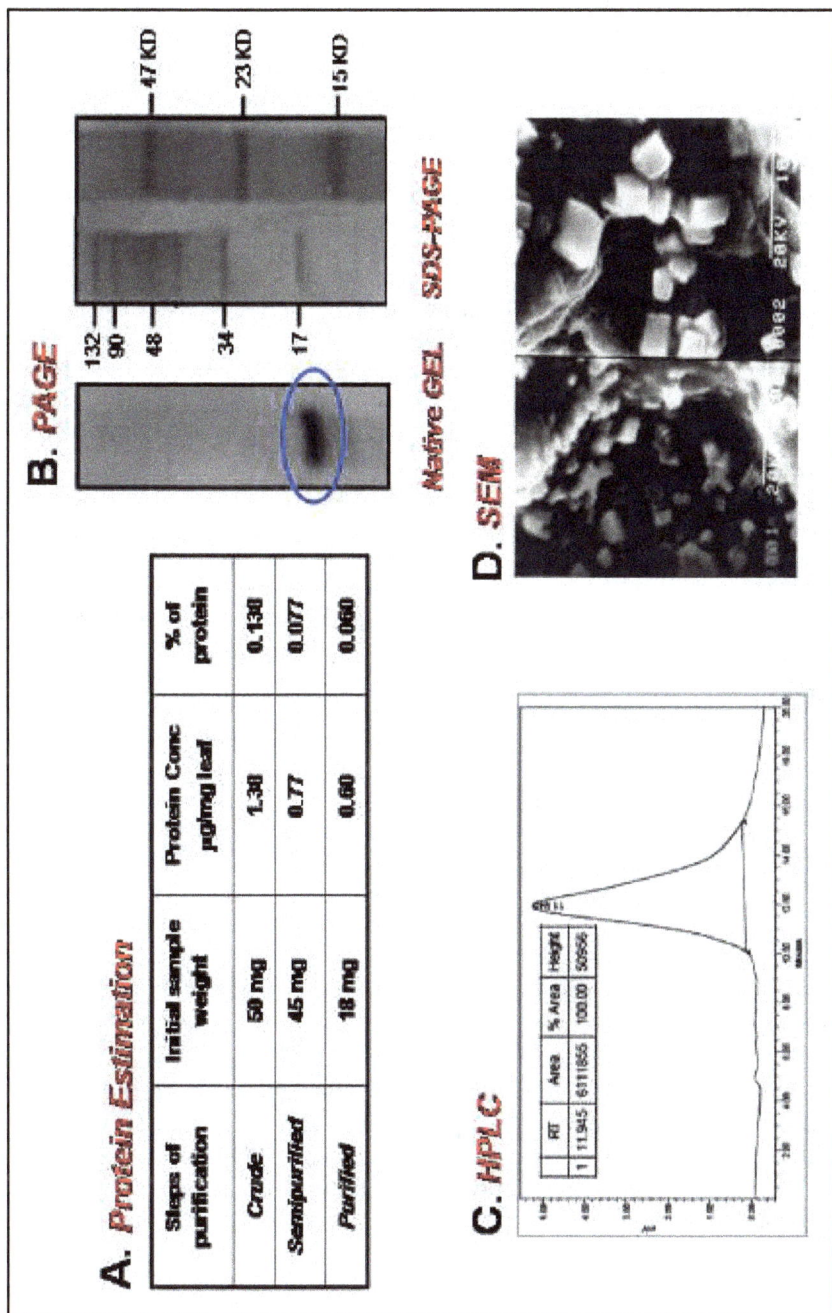

Figure 2.3: Biochemical features of Neem Leaf Glycoprotein (NLGP). A. Yield of NLGP at different stages of purification. B. Non-denatured and denatured electrophoretic profile of NLGP. C. High performance liquid chromatographic analysis for NLGP. D. Scanning electron microscopic picture of NLGP (*Courtesy*, Baral *et al.*, 2010).

upregulates the expression of CD40 on CD14$^+$ monocytes and CD40L on CD56$^+$ lymphocytes (NK cells). Neutralization of CD40, CD40L in NLP stimulated PBMC culture causes significant downregulation of IL-12 release and cytotoxicity of NK cells, proving the role of CD40-CD40L interaction in the observed functions. Signals involved in the NLP induced release of IL-12, thereby, induction of the NK cell cytotoxicity are mediated by activating p38MAPK pathway, but not through ERK1/ 2 signaling pathway (Bose and Baral, 2007). Overall results suggest that NLP operates NK cellular cytotoxicity by CD40-CD40L mediated endogenous production of IL-12, which critically control the perforin dependent tumor cell cytotoxicity (Figure 2.4; adopted from Bose *et al.*, 2007). After obtaining purified NLGP from NLP, similar line of results were obtained.

Table 2.3: Increase in NK cells and NK-T cells in the Blood and Spleen of C57BL/6 and Swiss Mice.

	NK		NKT	
	NK1.1 $^+$	DX5$^+$	CD3$^+$NK1.1$^+$	CD3+DX5$^+$
C57BL/6				
Blood	2.26	1.70	1.36	2.0
	N=10	N=3	N=5	N=3
Spleen	1.67	2.60	5.00	8.50
	N=7	N=3	N=3	N=3
Swiss				
Blood		2.9		2.63
		N=3		N=3
Spleen		1.3		4.2
		N=3		N=3

Data are presented as fold increase in per cent positive cells from NLP-treated mice over PBS-treated control.

T Cells

Along with the demonstration of NLP/NLGP influenced CD3$^-$CD56$^+$ NK cell mediated tumor cell cytotoxicity, CD8$^+$CD56$^-$ T cell mediated tumor cell cytotoxicity by NLGP was also observed (Bose *et al.*, 2009). This T cells were isolated from the peripheral blood of HNSCC patients with a state of immunosuppression. NLGP induces TCRαβ associated cytotoxic T lymphocyte (CTL) reaction to kill oral cancer (KB) cells. This CTL reaction is assisted by NLGP mediated upregulation of CD28 on T cells and HLA-ABC, CD80/86 on monocytes and dendritic cells. CTL mediated killing of KB cells is associated with activation of these cells by NLGP. This activation is evidenced by increased expression of early activation marker CD69 with altered expression of CD45RO/CD45RA. NLGP is a strong inducer of IFNγ from T cells. Unlike to the NK cells, IFNγ regulates the T cell mediated cytotoxicity. Reason of this differential regulation may lies within upregulated expression of IFNγ receptor on T cell surface, not on NK cells. This NLGP induced cytotoxicity is dependent on

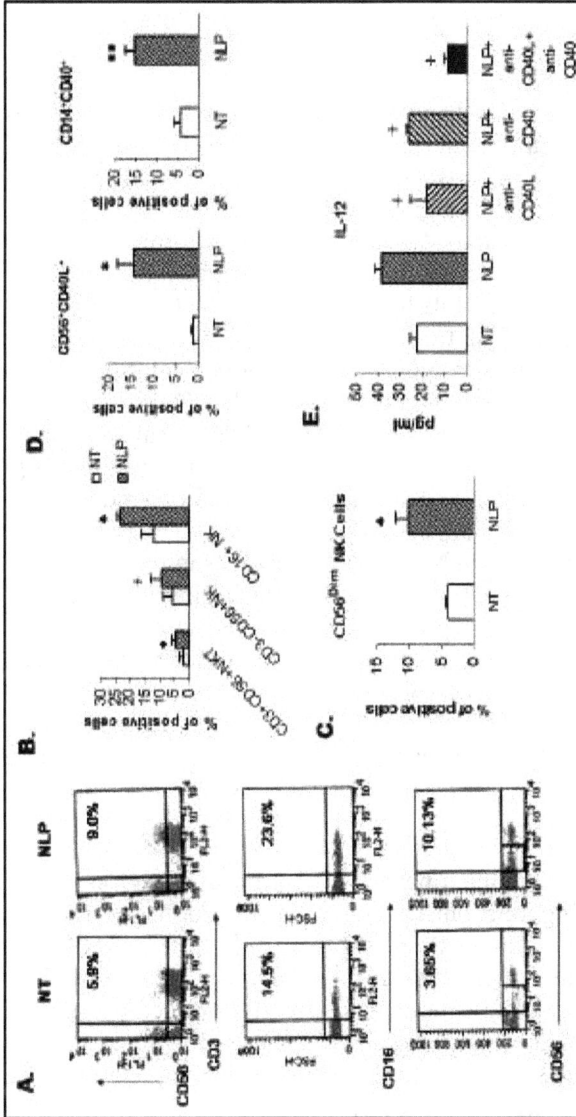

Figure 2.4: A. Enhancement of cytotoxic NK cell population by NLP. PBMC isolated from HNSCC patients (n=8) were cultured in vitro for 48 hrs with/without NLP. Cultured cells were then washed and either double stained with anti-CD3-PE and anti-CD56-FITC antibodies or single stained with either anti-CD16-PE or anti-CD56-PE antibodies and incubated for 30 mins. Percentage of positive cells was monitored by flow cytometric analysis. Representative figures from each case before and after NLP treatment are presented (A). Values of CD56+CD3+, CD56+CD3- and CD16+ cells are monitored by quadrant statistics. Bar diagram shows the average expression as percent positive cells ± SD from 8 individual experiments (B). CD56+ dim cells are presented by region statistics (C). NLP enhances CD40-CD40L interaction for IL-12 production. PBMC were stimulated with NLP for 48 hrs and doubled stained with either CD14/CD40 (CD14-FITC and CD40-PE) or CD56/ CD154 (CD40L) (CD56-PE and CD154-FITC). Staining status was assessed by flow cytometric analysis. Bar diagram shows the percent positive cells ± SD from 6 individual experiments (D). PBMC from HNSCC patients (n=4 in each case) was stimulated in vitro with NLP in presence of either anti-CD40L, anti-CD40 or both (anti-CD40 + anti-CD40L) and culture supernatants were collected after 48 hours to measure the levels of IL-12 by ELISA (E). NT= No Treatment; NLP= Neem Leaf Preparation. (*Courtesy,* Bose and Baral, *Human Immunol,* 2007).

upregulated perforin/granzyme β expression in killer cells, which is again IFN-γ dependent in T cells. Although, FasL expression is increased by NLGP, it may not truly linked with the cytotoxic functions, as brefeldinA could not block such NLGP mediated cytotoxicity, like, concanamycin α, a perforin inhibitor. Based on these results, we conclude that NLGP might be effective to recover the suppressed cytotoxic functions of T cells from HNSCC patients (Figure 2.5, adopted from Bose *et al.*, 2009a).

Regulatory T Cells

NLGP activated macrophage, NK cell and T cell mediated antitumor functions may be negatively influenced by suppressor cells, like, regulatory T (Treg) cells and myeloid derived suppressor cells (MDSC). Involvement of Treg cells in NLGP mediated tumor growth restriction and associated immune mechanisms were investigated. NLGP downregulates CD4+CD25+Foxp3+ Treg cells within tumor and this downregulation may directly translate into tumor growth restriction (Chakraborty *et al.*, 2009). Downregulation of CCR4 and its ligand CCL22 by NLGP restrict Treg migration at tumor site. As a result, NLGP treatment retarded Treg promoted tumor growth and increased survivability of mice. NLGP does not induce Treg apoptosis *in vitro*, but, significantly downregulates Foxp3 expression, along with decreased expression of: CD25, CTLA4, GITR and increased CD8+ T cells. Functional impairment of T-effector cells by Tregs, in terms of IFNγ secretion, proliferation, and tumor cell cytotoxicity, is also reversed by NLGP. NLGP immunization results remodeling of tumor microenvironment by increasing IFNγ, IL-12, but decreasing IL-10, TGFβ, VEGF and IDO, thus, favoring tumor inhibition. NLGP also intervenes in the interaction between Foxp3, *p*-NFATc3 and *p*-Smad2/3 to overcome the Foxp3 induced anergy of T cells (Chakraborty *et al.*, 2010). NLGP might be an excellent tool as a T cell anergy breaker by suppressing the suppressor functions of Treg cells in cancer (Figure 2.6). In line with this, another suppressor cells, MDSC, are also downregulated with NLGP as shown by downregulation of CD11b+Gr1+ cells in NLGP treated B16 melanoma bearing mice (*unpublished observation*), however, this preliminary observation warrant further investigation.

Antigen Presenting Cells (APCs)

APCs are essential link between the innate and adaptive immune responses. Besides their potent Ag-presenting function, APC, including macrophages and DCs, were also act as tumor killer cell. Tumoricidal functions of both these APCs are seriously dysregulated in a variety of cancers. Additionally, maturation of DCs as well as macrophages is subnormal in cancer. Accordingly, antigen presenting functions to B cells and T cells are immensely disturbed. We have attempted to know the role of NLGP in different levels of functions of antigen presenting cells.

Macrophage as a Tumor Killer and APC

Killing by macrophages to B16 melanoma and Ehrlich's carcinoma cells (in murine system) and U937 cells (in human) were studied in vitro and significant upregulation in killing efficacy was noted after NLGP treatment. Macrophages with adjuvant help from NLGP can effectively present carcinoembryonic antigen (CEA) to B cells to generate CEA specific antibody (IgG2a) response, having role in induction

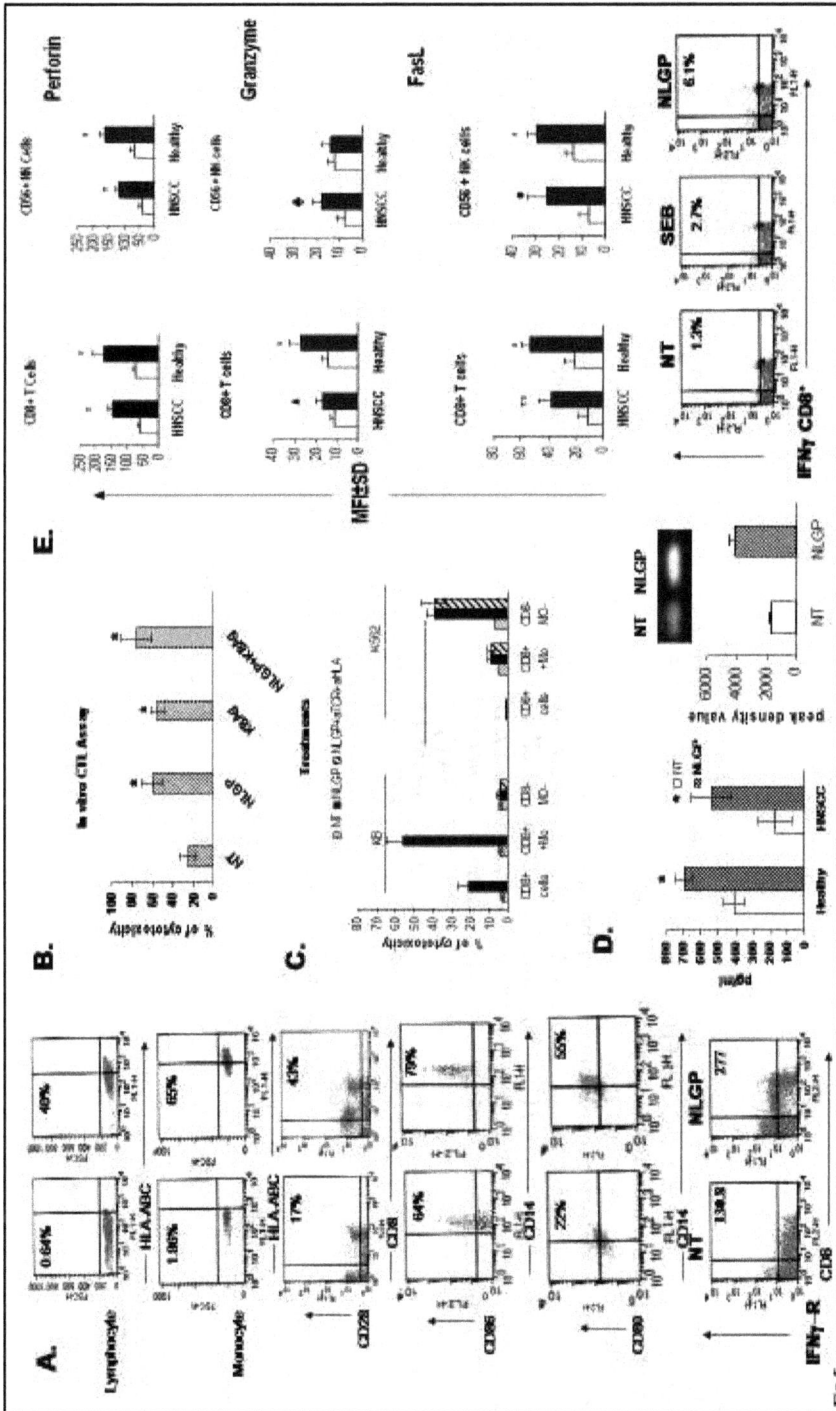

Figure 2.5: NLGP efficiently participate in CTL generation A. PBMC isolated from healthy individuals (n=10 in each case) were cultured in vitro for 72 hrs with/without NLGP. Cultured cells were then washed and stained either single labeled with anti-HLA-ABC-FITC or doubled stained with anti-CD8-PE/anti-CD28-FITC, anti-CD14-FITC/anti-CD80-PE, anti-CD14-FITC/anti-CD86-PE and anti-IFNγR-FITC/anti-CD8 and incubated for 30 mins. Percentage of positive cells was monitored by flow cytometric analysis. A representative figure from each case is presented. B. CD8+ cell population isolated from healthy individuals (n= 6) were cultured *in vitro* for 7 days with/without NLGP, KBAg, KBAg+NLGP in presence of macrophages and then incubated with KBAg positive KB cells. Cytotoxicity of CTLs towards cancer cells was determined by LDH release assay. * *p*<0.001. C. CD8+ T cells were in vitro cultured for 7 days with/without NLGP in presence or absence of anti¯CRαβ+HLA-ABC antibody (1µg/ml). After 7 days of culture, effector cells were co-cultured with different cancer cells (KB and K562) for another 4 hrs. Cytotoxicity was determined by LDH release assay. D. NLGP increases the secretion and synthesis of IFNγ. PBMC from HNSCC patients and healthy individuals (n=10 in each case) were stimulated in vitro with/without NLGP and culture supernatants were collected after 72 hours to measure the levels of IFNγ by ELISA (*p*<0.0001 in comparison to no treatment values). PBMC from HNSCC patients (n= 6 in each case) were stimulated in vitro with NLGP for 4 hrs and RNA was extracted and RT-PCR was carried out using IFNγ specific primer. HNSCC-PBMC were incubated with media, SEB and NLGP for 6 hrs along with golgistop (monensin). Cells were surface stained with FITC conjugated anti-CD8 antibody along with anti-human IFNγ-PE. Percentage of single or double positive cells was assessed by flow cytometric analysis. Values show the percent of positive cells of the indicated figure. E. NLGP increases the expression of cytotoxic molecules Perforin, Granzyme B and FasL. PBMC isolated from HNSCC patients and healthy individuals (n=8 in each case) were cultured in vitro for 24 hrs with/without NLGP. Cells were washed and stained for intracellular Perforin or GranzymeB (by anti-human Perforin-FITC or anti-human GranzymeB-FITC) along with PE conjugated anti-CD8 or anti-CD56 antibodies. Similar cells were also stained with anti-human FasL along with PE conjugated anti-CD8 or anti-CD56 antibodies. Staining status was assessed by flow cytometric analysis. Bar diagrams show the average expression of Perforin, GranzymeB and FasL on CD8+ T cells and CD56+ NK cells, as mean fluorescence intensity (MFI ±SD) from 8 individual experiments in each case. (*p*<0.0001, ᐞ*p*=0.35; °*p* =0.0047; **p*=0.0002; •*p*=0.0005 in comparison to pretreatment values) (*Courtesy, Bose et al., J Immunother,* 2009).

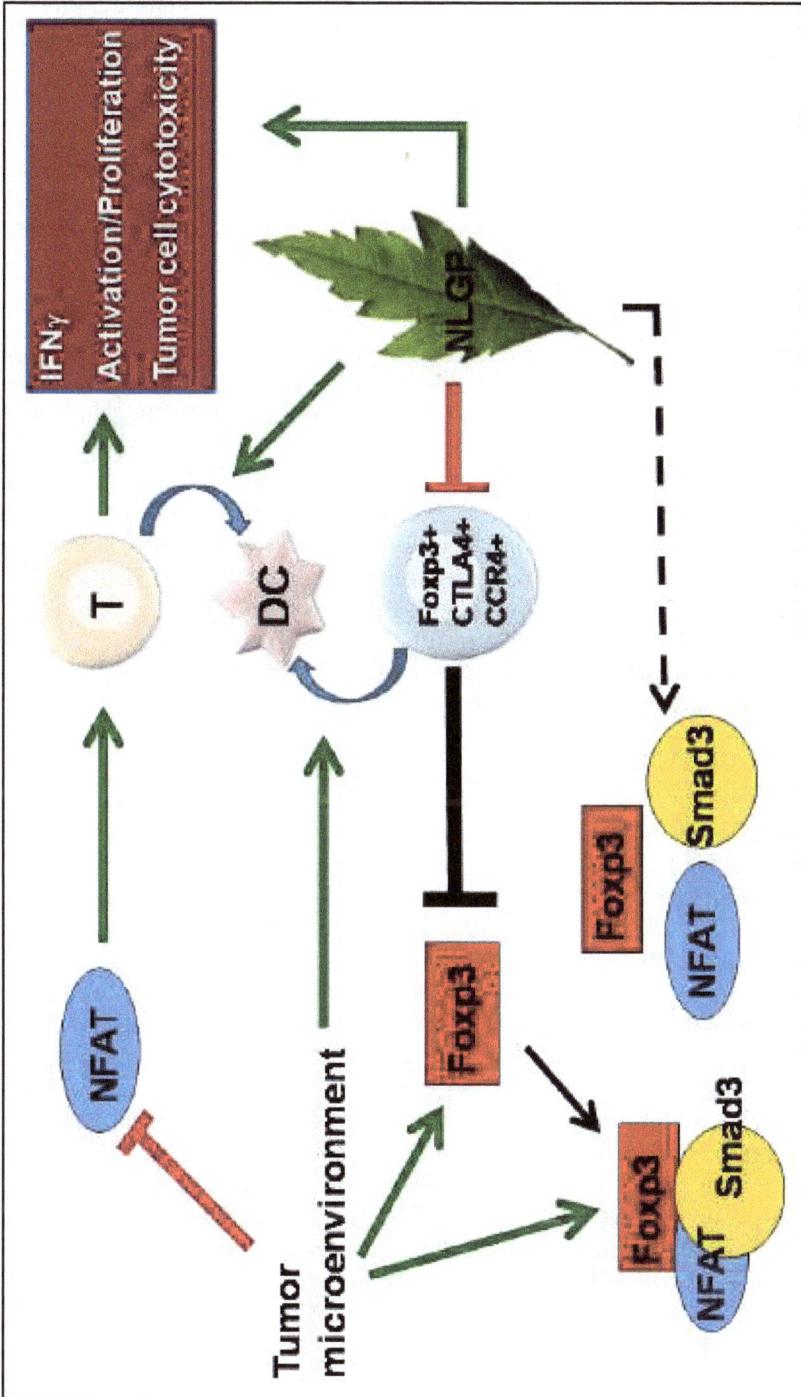

Figure 2.6: Possible mechanism of NLGP mediated eradication of T cell anergy induced by T reg cells. Tumor microenvironment stimulates Foxp3 and promotes its interaction with NFAT and Smad3. This situation causes T cell anergy. NLGP inhibits Foxp3 and other Treg associated molecules, thereby, releases NFAT from Foxp3-NFAT-Smad3 complex. NLGP also matures dendritic cells to type 1 polarity to facilitate antigen presentation to T cells not to Treg cells. These altered immune events ultimately help to withdraw the T cell anergy.

of anti-tumor immune response to prevent the growth of CEA$^+$ tumors *in vitro* and *in vivo* (Sarkar *et al.*, 2008). NLGP also controls the function of B cells by altering expressions of various regulatory molecules, like, CD19, CD11b etc. (Figure 2.7A, adopted from Sarkar *et al.*, 2008).

DC as a Tumor Killer

In addition to macrophages, NLGP matured DCs also shown to be exhibited greater cytotoxicity of lymphoma (U937) cells in vitro and this cytotoxic function is associated with NLGP mediated upregulation of perforin and granzyme B content within DCs (*unpublished observation*).

Human Myeloid DC as APC

Most significant function of DCs in induction of anti-tumor immunity is antigen presentation. Altered DC maturation results partial antigen processing and poor presentation-costimulation to T cells, thereby, causing T cell anergy. Thus, *in vivo* or *ex vivo* DC maturation is a primary task in designing DC based vaccine. We have tested NLGP as a nontoxic DC maturating agent for human use. NLGP matured DCs (NLGP-DCs) generated from myeloid cells (CD14$^+$CD3$^-$) show upregulated expression of CD83, CD80, CD86, CD40 and MHCs, in a comparable extent of control (LPS matured DCs). NLGP-DCs secrete high amount of IL-12p70 with low IL-10. NLGP also upregulated CD28 and CD40L on T cells, thus, promoting the scope of DC-T interactions (Goswami *et al.*, 2010). As a result, T cells secrete high amount of IFNγ with low IL-4 and generates anti-tumor type 1 immune microenvironment (Figure 2.7B, adopted from Goswami *et al.*, 2010).

Mouse Bone Marrow DCs as APC

NLGP also promote maturation of mouse bone marrow derived DCs (BmDCs). Vaccination with NLGP matured CEA pulsed DCs enhances antigen specific humoral and cellular immunity against CEA and restricts the growth of CEA$^+$ murine tumors. Indeed, NLGP helps better CEA uptake, processing and presentation to T/B cells. This vaccination (DCNLGPCEA) elicits mitogen induced and CEA specific T cell proliferation, IFNγ secretion and induces specific cytotoxic reactions to CEA$^+$ colon tumor cells. In addition to T cell response, DCNLGPCEA vaccine generates anti-CEA antibody response, which is principally IgG2a in nature. This observation is similar to the macrophage mediated CEA delivery system. This antibody participates in cytotoxicity of CEA$^+$ cells in antibody dependent manner. This strong anti-CEA cellular and humoral immunity protects mice from tumor development and these mice remained tumor free following second tumor challenge, indicating generation of effector memory response. Evaluation of underlying mechanism suggests vaccination generates strong CEA specific CTL and antibody response that can completely prevent the tumor growth following adoptive transfer. In support, significant upregulation of CD44 on the surface of lymphocytes from DCNLGPCEA immunized mice was noticed with a substantial reduction in L-selectin (CD62L) (Sarkar *et al.*, 2010) (Figure 2.7C, adopted from Sarkar *et al.*, 2010).

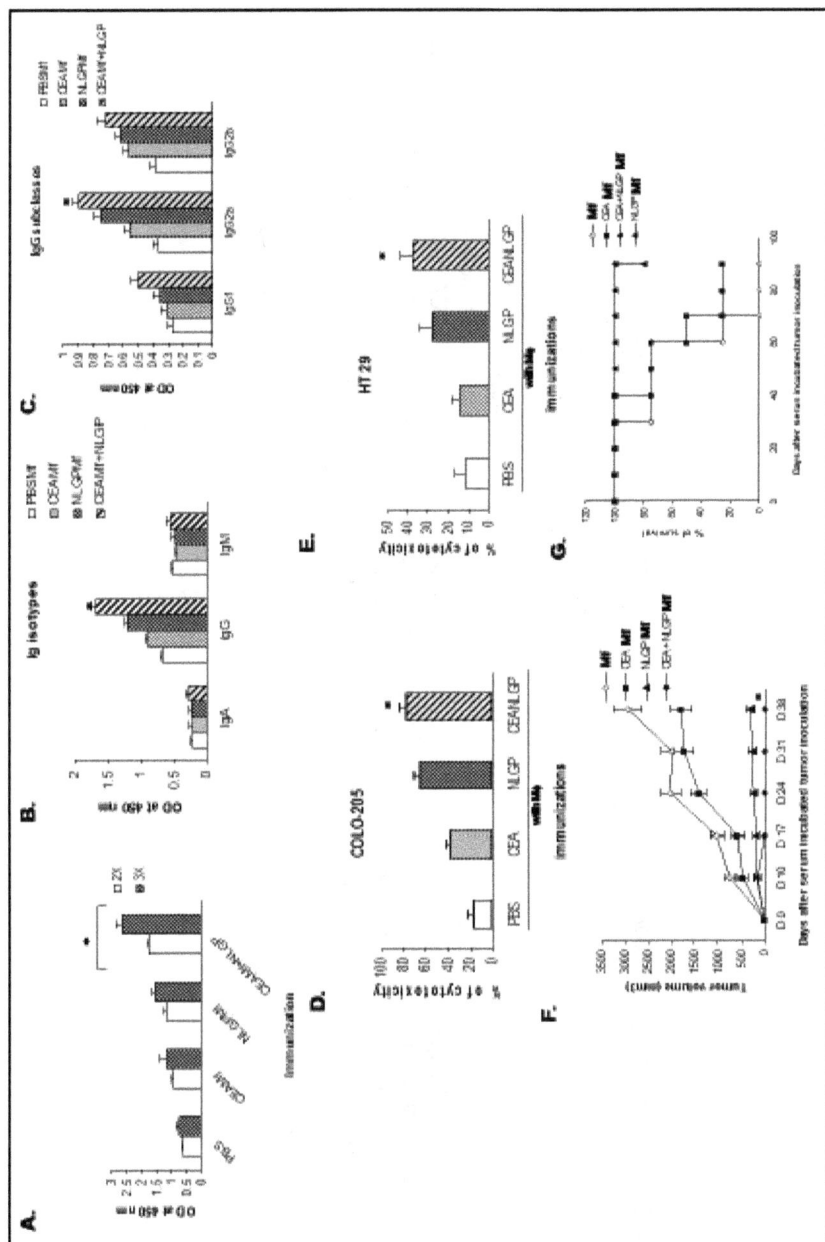

Figure 2.7A: NLGP enhances the generation of CEA reactive antibodies in immunized mice. Four groups of mice (n=6 in each group) were injected with Mφ, CEAMφ, NLGPMφ and CEAMφNLGP weekly for 3 weeks. The presence of anti-CEA antibodies in the sera of the immunized animals after second and third immunizations was assessed by ELISA (A). *$p<0.001$, in comparison to CEAMφ treated group. Same sera were analyzed for the immunoglobulin isotypes (B) and IgG subclasses (C) as indicated. Each data represents the mean value ± SD. *$p<0.001$ in comparison to either IgA/IgM or IgG1/IgG2b.

Sera from CEAMφNLGP immunized mice exhibits enhancement in ADCC to CEA$^+$ cells. Four groups of mice (n=6 in each group) were injected with Mφ, CEAMφ, NLGPMφ and CEAMφNLGP weekly for 3 weeks. Seven days after the last injection, sera from individual mice were collected and tested in ADCC. Splenocytes isolated from a syngenic mice were used as effector and CEA$^+$ colon tumor cells - COLO 205 (D) and HT 29 (E) were used as target. Data obtained by using an E:T ratio of 10:1 and immune sera at a dilution of 1:100 is presented. Each data represents the mean value ± SD. *$p<0.0001$, in comparison to CEAMφ group.

Immune sera from CEAMφNLGP vaccinated mice restrict the growth of CEA$^+$ tumors in mice. CEA$^+$ colorectal tumor cells were incubated with immune sera, generated by immunizing mice with Mφ, CEAMφ, NLGPMφ, and CEAMφNLGP weekly for 3 weeks. Treated tumor cells were then injected subcutaneously to four groups of mice (n = 6 in each group). *$p<0.0001$, in comparison to CEAMφ group (F). Survivability of mice was recorded by regular observation till day 90 after tumor inoculation (G). Data presented is a representative of two identical experiments (*Courtesy, Sarkar et al., Vaccine,* 2008).

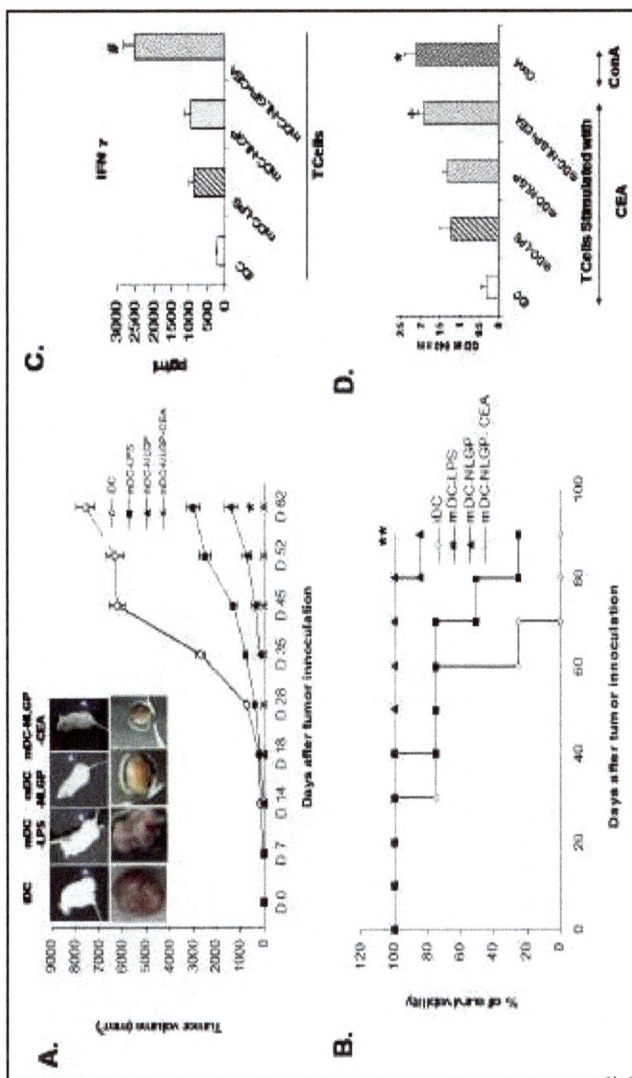

Figure 2.7B: Tumor growth restriction, survivability and T cell functions of dendritic cell (DC) immunized mice. A. Four groups of Swiss mice (n=6 in each group) were immunized with iDCs, LPS matured BmDCs, NLGP matured BmDCs and NLGP matured CEA pulsed BmDCs, in alternate weeks, for three times in total. Three days following completion of the immunization, mice were inoculated with mouse colon carcinoma cells (1×10^7) subcutaneously on right hind leg quarter. A. Tumor growth curve, as monitored by caliper measurement using the formula: (width^{2}xlength)/2. *inset*. Representative mice and isolated tumors from each group. B. Survivability curve of mice. $*p<0.0001$, in comparison to iDC and $* p<0.01$ in comparison to mDC-LPS and mDC-NLGP. C. T cells were cultured in various conditions as mentioned in figure and stimulated with CEA to measure the release of IFNγ by ELISA. In comparison to, T cell + iDC, $\#p<0.0001$ and T cell + mDC-LPS + CEA, $\#p<0.01$. D. CD8$^+$ T cells were cultured in various conditions as mentioned in figure and stimulated with CEA/ConA for 72 hours and proliferation was assessed by MTT colorimetric assay. In comparison to T cell + iDC, $^+p<0.0001$ and T cell + mDC-LPS/mDC-NLGP, $^+p<0.01$ (*Courtesy, Goswami et al., Vaccine, 2010*).

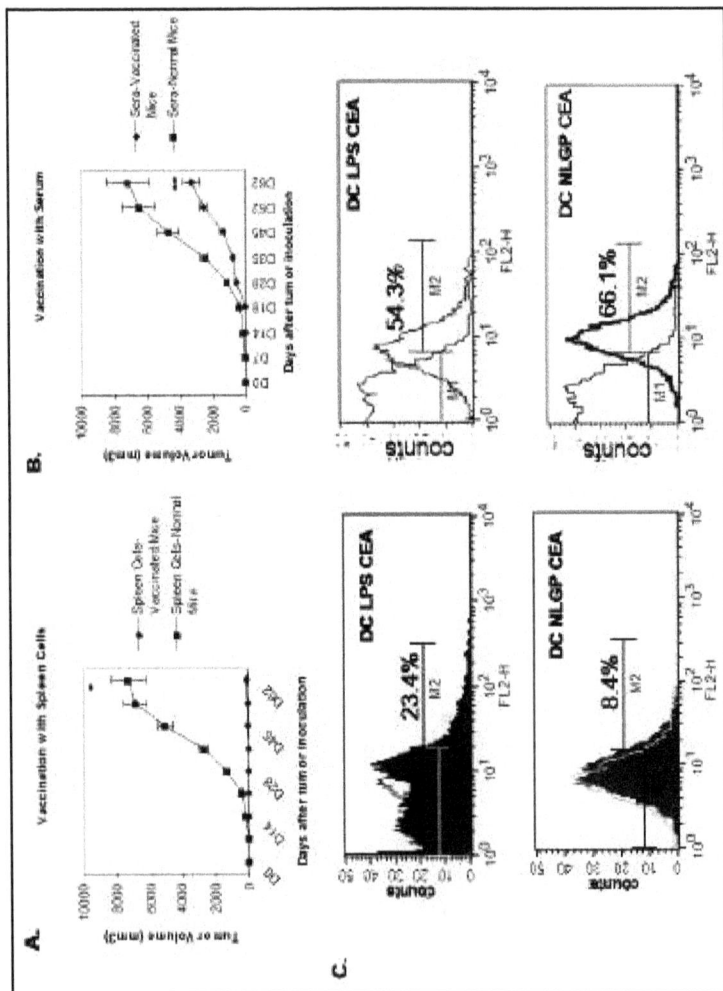

Figure 2.7C: Adoptive transfer of CTLs/serum from DCNLGPCEA immunized tumor free mice offers complete tumor growth restriction. A. Spleen cells were collected from tumor restricted mice following fifth week of second tumor inoculation and stimulated in vitro with CEA for seven days. These cells were adoptively transferred to normal mice and tumor inoculation (mouse CEA⁺) was given after a day, along with a control group received normal spleen cells + tumor. B. Another group of mice was adoptively transferred with immune sera from tumor restricted mice and tumor inoculation (mouse CEA⁺) was given after a day, along with a control group received normal sera + tumor. C. Expression of CD62L and CD44 on spleen cells from DCLPSCEA and DCNLGPCEA immunized mice as studied by flow cytometry (*Courtesy*, Sarkar *et al.*, *Int Immunopharmacol*, 2010).

Altered Antigen Presenting Function of DCs in Cancer

Endogenous DCs in cancer-bearing patients are a target of tumor-associated suppressive factors, resulting in their aberrant functions, impaired maturation and sub-optimal expression of costimulatory molecules. Consistent with these reports, we observed myeloid DCs generated from monocytes of Cervical Cancer (CaCx), Stage IIIB patients, show a dysfunctional maturation, in comparison to age matched healthy females as controls. As a consequent, suboptimal functional repertoire of CaCx-DC leading to impairment of many anti-tumor T cell functions was observed. In this context, NLGP was examined to optimize dysregulated T cell functions and creation of anti-tumor type 1 immune milieu by maturating myeloid derived DCs from CaCx patients. *In vitro* NLGP treatment of immature DCs (iDCs) from CaCx patients results upregulated expression of various cell surface markers (CD40, CD83, CD80, CD86, HLA-ABC), which indicates DC maturation. Consequently, NLGP matured DCs displayed a balanced cytokine secretion with type 1 bias and noteworthy functional property. These DCs displayed substantial T cell allo-stimulatory capacity and promoted generation of CTLs. Although, NLGP matured DCs derived from CaCx monocytes is generally subdued from those of normal monocyte origin, considerable revival of the suppressed DC based immune functions is noted *in vitro* at a fairly advanced stage of CaCx and, thus, further exploration for *ex vivo* and *in vivo* DC based vaccine is proposed. Moreover, DC maturating efficacy of NLGP might be much effective in earlier stages of CaCx, where extent of immune dysregulation is less, thus, scope of further investigation may be explored.

Chemokine Signaling in Lymphocytes

For optimum tumor cell killing a balanced crosstalk between effector T cells and APCs is required, along with reduced interaction between regulatory T cells and APCs at the site of tumor. For immune cell mediated eradication of tumor desired movement (trafficking and homing) of effector cells to tumor site is essential. As movement of immune cells is primarily regulated by chemokines, chemokine signaling in lymphocytes and monocytes under influence of NLGP was studied.

We have observed interaction between IFNγ dependent chemokine, CXCL10 and CXCR3 is dysregulated in HNSCC that impaired chemotaxis of cytotoxic cells at tumor site. Effect of NLGP in rectification of the dysregulated CXCL10 and its receptor CXCR3 splice variants, A and B, was investigated. Upregulated expression of CXCR3B in lymphocytes from HNSCC patients were downregulated following in vitro NLGP treatment. Unchanged expression of CXCR3A+B by NLGP with downregulation of the CXCR3B indirectly suggests the upregulation of the CXCR3A, responsible for cellular migration. However, stimulation of healthy-PBMC with NLGP maintains physiological homeostasis of CXCL10 and increases IFNγ secretion. The suppressed chemotaxis of HNSCC-lymphocytes could be restored either by *in vitro* treatment with NLGP or during use of NLGP stimulated PBMC supernatant as a chemoattractant. Neutralization studies confirmed that the chemoattraction process is guided by both receptor (CXCR3A) and its ligand (CXCL10). Neutralization of the IFNγ in PBMC culture in presence of NLGP unexpectedly increases the intracellular release of CXCL10, suggesting the NLGP mediated IFNγ independent release of

CXCL10. Interestingly, downregulation of the CXCL10 release was detected after IFNγ neutralization in absence of NLGP and IFNγ receptor neutralization in presence of NLGP (Chakraborty, 2008) (Figure 2.8A, adopted from Chakraborty *et al.,* 2008).

Chemokine Signaling in Monocytes

After demonstration of NLGP induced optimum migration of lymphocytes, migration of monocytes (considering its important role in phagocytosis and antigen presentation for priming of tumor specific T cells) was also studied. Downregulated expression of a CC chemokine receptor, CCR5 on monocytes/macrophages (MO/Mφ), surfaces from HNSCC patients (Stage IIIB) was demonstrated. Ligands (RANTES, MIP1α, MIP1β) of this chemokine receptor were also secreted in lesser quantity from MO/Mφ of HNSCC patients, in comparison to healthy individuals. In an objective to rectify this dysregulated receptor-ligand status, we have used NLGP. NLGP upregulated CCR5 expression, as evidenced from studies on MO/Mφ of peripheral blood from HNSCC patients as well as healthy individuals. Intracellular secretory status of RANTES, MIP1α, MIP1β was also rectified upon NLGP treatment of these cells in vitro. These rectifications in receptor-ligand level were reflected in improved CCR5 dependent, *p*38MAP kinase mediated migration of MO/Mφ after NLGP treatment against a standard chemoattractant. NLGP also potentiates MO/Mφ for better antigen presentation and simultaneous co-stimulation to effector T cells by upregulating HLA-ABC, CD80 and CD86 on CCR5+ macrophages. NLGP treated MO/Mφ primed T cells can effectively lyses tumor cells in vitro (Chakraborty *et al.,* 2010) (Figure 2.8B, adopted from Chakraborty *et al.,* 2010).

Induction of Type 1 Immune Response

Anti-tumor immune functions are associated with type 1 immune polarity. Induction of immune activation instead of tolerization and type 1 immune polarization by NLGP was studied *in vitro* and *in vivo* system. NLGP induced activation is reflected in upregulation of early activation marker CD69 on lymphocytes, monocytes and dendritic cells. Activation is also denoted by CD45RO enhancement, with a decrease in CD45RA phenotype and CD62L (L-selectin) on T cell population. NLGP activated T cells secrete greater amount of signature type 1 cytokines IFNγ and a lower amount of type 2 cytokine, IL-4. Similar type 1 directiveness is also observed in antigen presenting monocytes and dendritic cells by upregulation of IL-12, TNFα and downregulation of IL-10. Similar result was obtained in *in vivo* studies, where mice were immunized with NLGP either tumor bearing or tumor free condition. Creation of type 1 microenvironment is also assisted by NLGP induced downregulation of Foxp3+ Treg cells. A type 1 specific transcription factor, T-bet, is upregulated in circulating immune cells after their stimulation with NLGP. In the creation of type 1 immune network, increased NLGP induced phosphorylation of STAT1/STAT4 with decreased phosphorylation of STAT3 might have significance (Bose *et al.,* 2009b) (Figure 2.9, adopted from Bose *et al.,* 2009b).

In tumor related creation of type 2/type 3 immunity, tryptophan breaking enzyme, IDO played a pivotal role, which is released from macrophages and DCs after exposure into tumor microenvironment consisting of Treg cells. Interestingly,

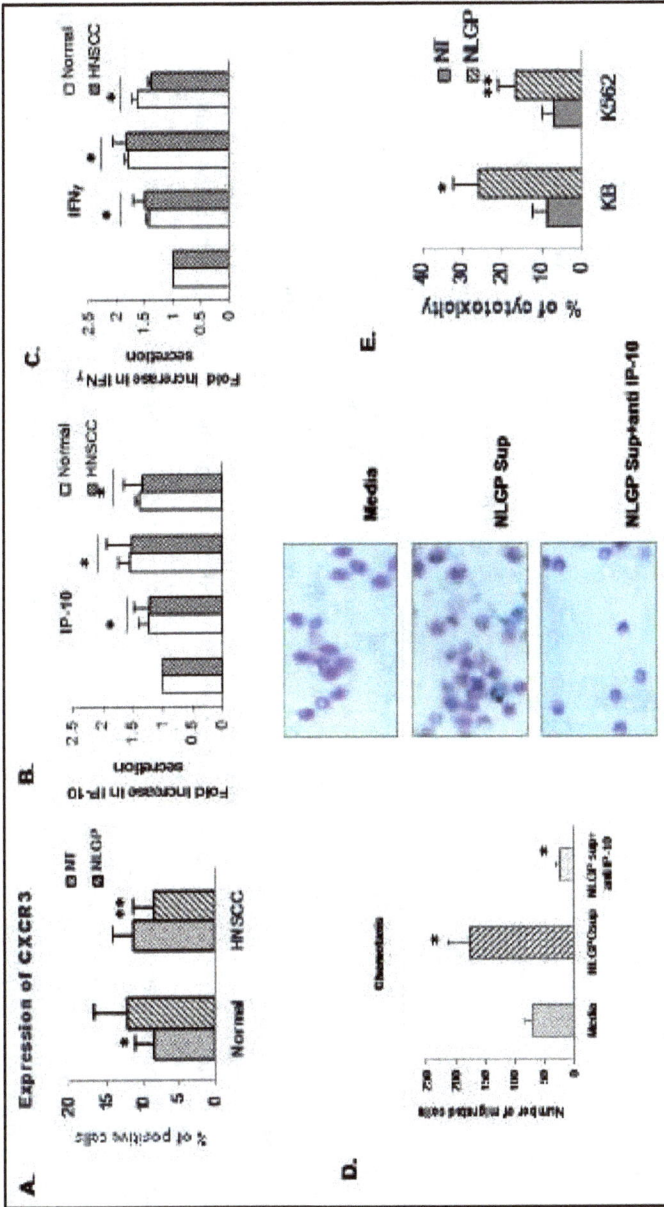

Figure 2.8A: A. NLGP reduces CXCR3 expression on HNSCC-PBMC, but not in healthy-PBMC. PBMC were isolated from venous blood of HNSCC patients (n=6) and healthy individuals (n=6) and treated with NLGP for 48 hrs. A. Total PBMC were then surface labeled with anti-CXCR3 antibody for flow cytometric analysis. Mean±SD is presented in bar diagram. * p=0.01; ** p=0.017. B, C. NLGP enhances intracellular and extracellular release of IP10 and IFNγ from healthy and HNSCC-PBMC after NLGP treatment, in comparison to values obtained from PBMC having no treatment. * p<0.0001. D. NLGP-Supernatant mediated migration of PBMC is dependent on IP10. PBMC were allowed to migrate against NLGP-Supernatant in presence or absence of antibody against IP10. Mean number of migrated cells is presented in bar diagram and a representative figure of PBMC migration with or without IP10 neutralization is presented. * p<0.0001. E. NLGP enhances cytotoxic ability of PBMC towards tumor cells (*Courtesy*, Chakraborty *et al.*, *Int Immunopharmacol*, 2008).

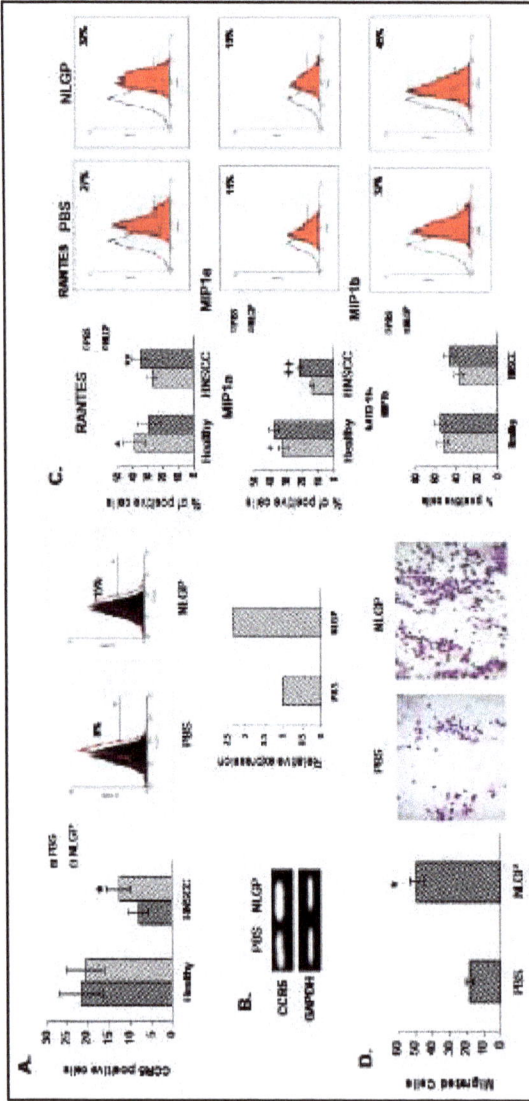

Figure 2.8B: NLGP upregulates CCR5 expression in HNSCC-monocytes. PBMC were isolated from venous blood of HNSCC patients (n=12) and healthy individuals (n=12) and adherent fractions were taken and incubated with NLGP (1.5mg/ml) for 48 hrs.

A. Cells were surface labeled with anti-CCR5 antibody for flow cytometric analysis. Mean ± SD is presented in bar diagram (*p=0.001, in comparison to PBS-HNSCC) and a representative figure on CCR5 expression from HNSCC-monocytes is also presented.; B. RT-PCR analysis shows NLGP upregulates mRNA expression of CCR5 in HNSCC-monocytes in respect to GAPDH. A representative figure (*left panel*) and relative expression of CCR5 mRNA in relation to GAPDH (*right panel*) are presented. C. NLGP upregulates expression of CCR5 ligands in HNSCC monocytes. Monocytes were isolated from PBMC of HNSCC patients (n=6) and healthy individuals (n=6) and incubated with NLGP (1.5mg/ml) for 48 hrs. Intracellular expression of RANTES, MIP1α, and MIP1β was assessed by flow cytometry. RANTES, *p=0.009, **p=0.02, when compared with PBS-HNSCC. MIP1α,*p<0.001, ++p=0.0023, when compared with PBS-HNSCC. MIP1β, #p=0.019, ##p=0.05, when compared with PBS-HNSCC (*left panel*). A representative figure in each case generated from HNSCC-monocytes is presented (*right panel*). D. Suppressed CCR5 dependent chemotaxis of HNSCC-monocytes can be rectified by NLGP. HNSCC monocytes (n=6) were incubated with NLGP (1.5 µg/ml) for 48 hrs and migration was checked against a standard chemoattractant. Mean ± SD of migrated cells is presented in bar diagram, *p<0.001, in comparison to other groups and representative figures of migrated cells are shown (*Courtesy*, Chakraborty *et al.*, *Cellular Mol Immunol*, 2010).

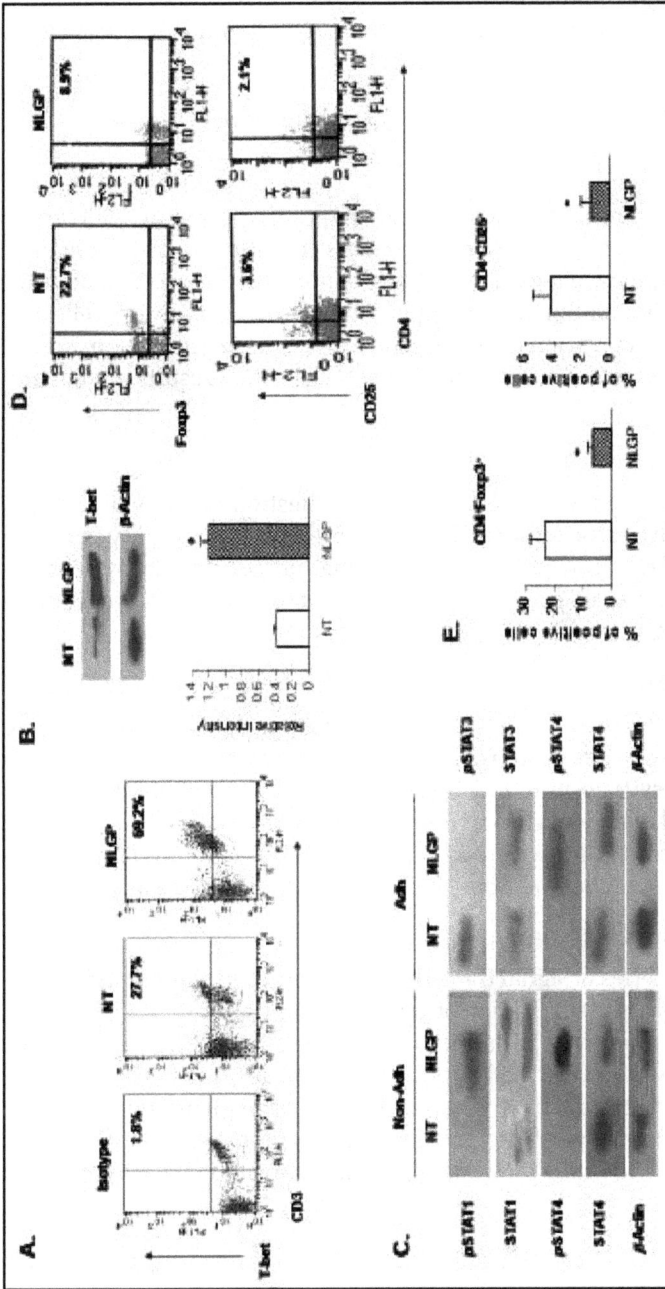

Figure 2.9: I. Induction of Th1 commitment by NLGP. T-bet expression is augmented by NLGP on human PBMC (n=6) as determined by flow cytometry (A) and immunoblotting (B) *p<0.001. NLGP upregulates the phosphorylation of STAT1 on non-adherent cells, STAT4 on both adherent and non-adherent cells and downregulates STAT3 on adherent cells (C). Result of a representative experiment is presented from six individual experiments in each case. II. Downregulation of Regulatory T cell population by NLGP. PBMC isolated from tongue cancer patients (n=8) were cultured in vitro for 48 hrs with NLGP. Cultured cells were then washed, and doubled stained with either anti-CD4-FITC/ anti-CD25-PE or anti-CD4-FITC/anti-Foxp3-PE. Percentage of positive cells was monitored by flow cytometric analysis after gating of lymphocyte rich zones respectively from FSC/SSC plots. A representative figure from each case is presented (D). Values show the percentage of positive cells of the indicated figure. Bar diagram shows the average expression as percent positive cells ± SD from 8 individual samples (*p<0.0001, in comparison to no treatment values) (E) (*Courtesy*, Bose *et al., Human Immunol*, 2009).

supplementation of NLGP (exactly mimic the results found with IDO inhibiting agent, 1-MT) results diminution of IDO releases thereby prevention of T cell anergy and apoptosis. In this way, NLGP can save the life of effector T cells, which is incredibly important in induction of anti-tumor immunity. Thus, type 1 tumor microenvironment is created, where level of TGFβ, IL-10, iNOS, Arginase is decreased significantly by NLGP (*unpublished observation*).

Further Questions to be Answered

Angiogenesis

These NLGP mediated immune normalization requires appropriate delivery of immune cells as well as drugs. This delivery demands conditioned vascularization within tumor. Anti-angiogenic drugs restrict the blood supply to tumor to limit the supply of nutrition. Such situation also hampers the supply of immune cells and drugs. Thus, vascular normalization is required instead of vaso-constriction. Role of NLGP in such direction is under evaluation and some positive results are already obtained.

Autoimmunity

Type 1 immune polarization by NLGP, asks a question whether it induces autoimmunity (either Th1 associated or Th 17 associated, but possibility is less as NLGP downregulates STAT3 phosphorylation). During course of our study no tissue damage in any organ was noticed by histological studies. Further studies are required to validate this hypothesis.

Therapeutic Outcome

Our elaborate study with NLGP in prophylactic setting, asked its feasibility in therapeutic setting. With demonstration of *in vitro* and *in vivo* immunomodulation in cancer, it is required to test its applicability in cancer therapy. Preliminary study with Sarcoma, Carcinoma and Melanoma model revealed the therapeutic efficacy of NLGP at least in 50 per cent cases. Extensive study in this direction is required along with supplementation of low dose chemotherapy for rapid translation into clinical setting.

Signal Transduction Gateway

Demonstration of a series of signal transduction events for immunomodulation emphasized the necessicity of identification of gateway(s) for signaling. In this context, involvement of toll like receptors (TLRs) needs to be revealed. Series of evidences proved the dissimilarity of NLGP with lipopolysaccharides, thus, possibility of signaling through TLR4 is excluded. Apart from that we have identified binding of NLGP on the surface of monocytes and lymphocytes, suggestive of the presence of specific receptor for NLGP. Identification and molecular characterization are essential at this point.

Concluding Remarks

In vitro and *in vivo* studies as summarized above generated bulk of evidences in favor of implication of NLGP as a therapeutic cancer vaccine. NLGP may appear as a single missile to target multiple immune dysregulations (Figure 2.10). Unlike to the

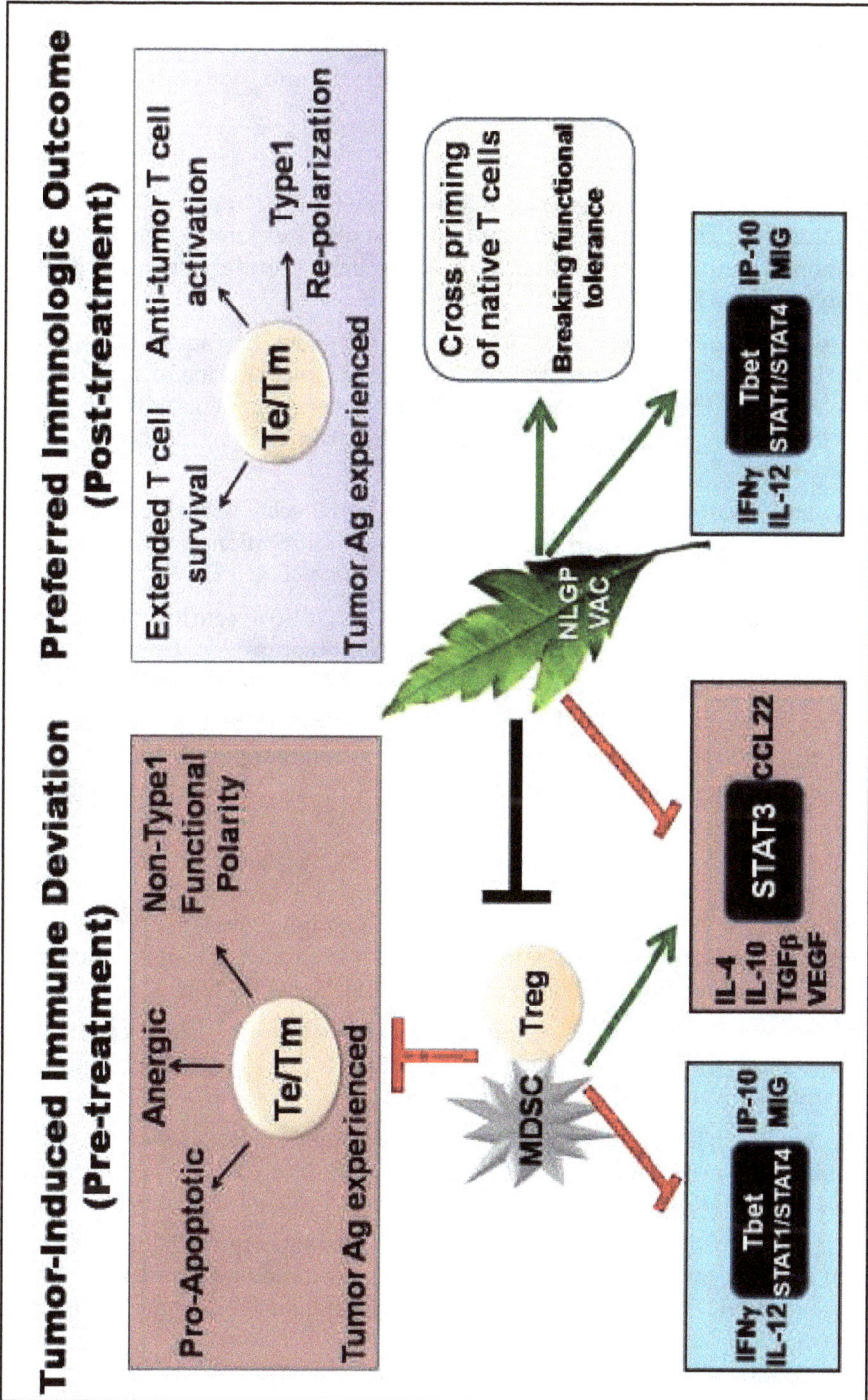

Figure 2.10: Diagrammatic presentation on correction of tumor induced immune deviation by NLGP.

vaccine designed to target single immune dysfunction, NLGP can normalize several immune abnormalities that contribute significantly in remodeling tumor microenvironment in favor of the host. Emergence of this nontoxic bioregimen as a therapeutic vaccine is very important and warrants the initiation of preclinical and clinical trial.

References

Agius, P.A., Pitts, M.K., Smith, A.M., and Mitchell, A. (2010). Human papillomavirus and cervical cancer: Gardasil vaccination status and knowledge amongst a nationally representative sample of Australian secondary school students. *Vaccine*, **28:** 4416-4422.

Banchereau, J., Palucka, A.K., Dhodapkar, M., Burkeholder, S., Taquet, N., Rolland, A., Taquet, S., Coquery, S., Wittkowski, K.M., Bhardwaj, N., Pineiro, L., Steinman, R., and Fay, J. (2001). Immune and clinical responses in patients with metastatic melanoma to CD34(+) progenitor-derived dendritic cell vaccine. *Cancer Res.,* **6:** 6451-6458.

Baral, R., and Chattopadhyay, U. (2004). Neem (*Azadirachta indica*) leaf mediated immune activation causes prophylactic growth inhibition of murine Ehrlich carcinoma and B16 melanoma. *Int Immunopharmacol.,* **4:** 355-366.

Baral, R.N., Sarkar, K., Mandal-Ghosh, I., and Bose, A. (2010). Neem leaf glycoprotein as a new vaccine adjuvant for cancer immunotherapy. *In:* Comprehensive Bioactive Natural Products: Immune-modulation and Vaccine Adjuvants, Vol. 5, Ed. By Gupta, V.K., Studium Press LLC, USA, pp. 21-45.

Ben-Baruch, A. (2006). Inflammation-associated immune suppression in cancer: the roles played by cytokines, chemokines and additional mediators. *Semin Cancer Biol.,* **16:** 38-52.

Bendandi, M., Gocke, C.D., Kobrin, C.B., Benko, F.A., Sternas, L.A., Pennington, R., Watson, T.M., Reynolds, C.W., Gause BL, Duffey, P.L., Jaffe, E.S., Creekmore, S.P., Longo, D.L., and Kwak, L.W. (1999). Complete molecular remissions induced by patient-specific vaccination plus granulocyte-monocyte colony-stimulating factor against lymphoma. *Nat Med.,* **5:** 1171-1177.

Block, M.S., and Markovic, S.N. (2009). The Tumor/Immune Interface: Clinical Evidence of Cancer Immunosurveillance, Immunoediting and Immunosubversion. *Am J Immunol.,* **5:** 29-49.

Bose, A., and Baral, R. (2007). Natural killer cell mediated cytotoxicity of tumor cells initiated by neem leaf preparation is associated with CD40-CD40L-mediated endogenous production of interleukin-12. *Hum Immunol.,* **68:** 823-831.

Bose, A., Chakraborty, K., Sarkar, K., Goswami, S., Chakraborty, T., Pal, S., and Baral, R. (2009). Neem leaf glycoprotein induces perforin-mediated tumor cell killing by T and NK cells through differential regulation of IFNgamma signaling. *J Immunother.,* **32:** 42-53.

Bose, A., Chakraborty, K., Sarkar, K., Goswami, S., Haque, E., Chakraborty, T., Ghosh, D., Roy, S., Laskar, S., and Baral, R. (2009). Neem leaf glycoprotein directs T-bet-associated type 1 immune commitment. *Hum Immunol.*, **70**: 6-15.

Chakraborty, K., Bose, A., Chakraborty, T., Sarkar, K., Goswami, S., Pal, S., and Baral, R. (2010). Restoration of dysregulated CC chemokine signaling for monocyte/macrophage chemotaxis in head and neck squamous cell carcinoma patients by neem leaf glycoprotein maximizes tumor cell cytotoxicity. *Cell Mol Immunol.*, **7**: 396-408.

Chakraborty, K., Bose, A., Pal, S., Sarkar, K., Goswami, S., Ghosh, D., Laskar, S., Chattopadhyay, U., and Baral, R. (2008). Neem leaf glycoprotein restores the impaired chemotactic activity of peripheral blood mononuclear cells from head and neck squamous cell carcinoma patients by maintaining CXCR3/CXCL10 balance. *Int Immunopharmacol.*, **8**: 330-340.

Chakraborty, T., Bose, A., Chakraborty, K., Sarkar, K., Goswami, S., Haque, E., Ghosh, D., Mandal-Ghosh, I., Pal, S., Laskar, S., and Baral, R.N. (2009). Suppression of Suppressors: Neem leaf glycoprotein Guided Crosstalk Between Regulatory T cells and T cells/NK Cells/Macrophages in Cancer. *In:* Treatments of Advanced Stage Cancers: current status and emerging frontiers, Ed. By Talwar, G.P., and Sood, O.P. Narosa Publishing House Pvt. Ltd, India, pp 153-157.

Chakraborty, T., Bose, A., Ghosh, D., Goswami, S., Chakraborty, K., Sarkar, K., Banerjee, S., Roy, S., and Baral, R. (2010). Neem leaf glycoprotein restricts murine tumor growth by suppressing migration, tumor induced conversion and functions of CD4$^+$CD25$^+$Foxp3$^+$ regulatory T cells. *J Immunother (in press)*.

Chen, J.J., Lin, Y.C., Yao, P.L., Yuan, A., Chen, H.Y., Shun, C.T., Tsai, M.F., Chen, C.H., and Yang, P.C. (2005). Tumor associated macrophages: the double-edged sword in cancer progression. *J Clin Oncol.*, **23**: 953-964.

Conry, R.M., Khazaeli, M.B., Saleh, M.N., Allen, K.O., Barlow, D.L., Moore, S.E., Craig, D., Arani, R.B., Schlom, J., and LoBuglio, A.F. (1999). Phase I trial of a recombinant vaccinia virus encoding carcinoembryonic antigen in metastatic adenocarcinoma: comparison of intradermal versus subcutaneous administration. *Clin Cancer Res.*, **5**: 2330-2337.

Dunn, G.P., Bruce, A.T., Ikeda, H., Old, L.J., and Schreiber, R.D. (2002). Cancer immunoediting: from immunosurveillance to tumor escape. *Nat Immunol*, **3**: 991-998.

Fong, L., Brockstedt, D., Benike, C., Wu, L., and Engleman, E.G. (2001). Dendritic cells injected via different routes induce immunity in cancer patients. *J Immunol.*, **166**: 4254-4259.

Fong, L., Hou, Y., Rivas, A., Benike, C., Yuen, A., Fisher, G.A., Davis, M.M., and Engleman, E.G. (2001). Altered peptide ligand vaccination with Flt3 ligand expanded dendritic cells for tumor immunotherapy. *Proc Natl Acad Sci U S A.*, **98**: 8809-8814.

Gabrilovich, D.I., and Nagaraj, S. (2009). Myeloid-derived suppressor cells as regulators of the immune system. *Nat Rev Immunol.*, **9:** 162-174.

Goswami, S., Bose, A., Sarkar, K., Roy, S., Chakraborty, T., Sanyal, U., and Baral, R. (2010). Neem leaf glycoprotein matures myeloid derived dendritic cells and optimizes anti-tumor T cell functions. *Vaccine*, **28:** 1241-1252.

Ha, T.Y. (2009) The role of regulatory T cells in cancer. *Immune Netw.*, **9:** 209-235.

Hamaï, A., Benlalam, H., Meslin, F., Hasmim, M., Carré, T., Akalay, I., Janji, B., Berchem, G., Noman, M.Z., and Chouaib, S. (2010). Immune surveillance of human cancer: if the cytotoxic T-lymphocytes play the music, does the tumoral system call the tune? *Tissue Antigens*, **75:** 1-8.

Hanna, M.G. Jr., Hoover, H.C. Jr., Vermorken, J.B., Harris, J.E., and Pinedo, H.M. (2001). Adjuvant active specific immunotherapy of stage II and stage III colon cancer with an autologous tumor cell vaccine: first randomized phase III trials show promise. *Vaccine*, **19:** 2576-2582.

Haque, E., and Baral, R. (2006). Neem (*Azadirachta indica*) leaf preparation induces prophylactic growth inhibition of murine Ehrlich carcinoma in Swiss and C57BL/6 mice by activation of NK cells and NK-T cells. *Immunobiology*, **211:** 721-731.

Hsueh, E.C., Famatiga, E., Gupta, R.K., Qi, K., and Morton, D.L. (1998). Enhancement of complement-dependent cytotoxicity by polyvalent melanoma cell vaccine (CancerVax): correlation with survival. *Ann Surg Oncol.*, **5:** 595-602.

Jaffee, E.M., Hruban, R.H., Biedrzycki, B., Laheru, D., Schepers, K., Sauter, P.R., Goemann, M., Coleman, J., Grochow, L., Donehower, R.C., Lillemoe, K.D., O'Reilly, S., Abrams, R.A., Pardoll, D.M., Cameron, J.L., and Yeo, C.J. (2001). Novel allogeneic granulocyte-macrophage colony-stimulating factor-secreting tumor vaccine for pancreatic cancer: a phase I trial of safety and immune activation. *J Clin Oncol.*, **19:** 145-156.

Keilholz, U., Weber, J., Finke, J.H., Gabrilovich, D.I., Kast, W.M., Disis, M.L., Kirkwood, J.M., Scheibenbogen, C., Schlom, J., Maino, V.C., Lyerly, H.K., Lee, P.P., Storkus, W., Marincola, F., Worobec, A., and Atkins, M.B. (2002). Immunologic monitoring of cancer vaccine therapy: results of a workshop sponsored by the Society for Biological Therapy. *J Immunother.*, **25:** 97-138.

Knutson, K.L., Schiffman, K., and Disis, M.L. (2001). Immunization with a HER-2/neu helper peptide vaccine generates HER-2/neu CD8 T-cell immunity in cancer patients. *J Clin Invest.*, **107:** 477-484.

Kusumoto, M., Umeda, S., Ikubo, A., Aoki, Y., Tawfik, O., Oben, R., Williamson, S., Jewell, W., and Suzuki, T. (2001). Phase 1 clinical trial of irradiated autologous melanoma cells adenovirally transduced with human GM-CSF gene. *Cancer Immunol Immunother.*, **50:** 373-381.

Lee P., Wang, F., Kuniyoshi, J., Rubio, V., Stuges, T., Groshen, S., Gee, C., Lau, R., Jeffery, G., Margolin, K., Marty, V., and Weber, J. (2001). Effects of interleukin-12 on the immune response to a multipeptide vaccine for resected metastatic melanoma. *J Clin Oncol.*, **19:** 3836-3847.

Lee, K.H., Wang, E., Nielsen, M.B., Wunderlich, J., Migueles, S., Connors, M., Steinberg, S.M., Rosenberg, S.A., and Marincola, F.M. (1999). Increased vaccine-specific T cell frequency after peptide-based vaccination correlates with increased susceptibility to *in vitro* stimulation but does not lead to tumor regression. *J Immunol.*, **163:** 6292-6300.

Lim, S.G., Mohammed, R., Yuen, M.F., and Kao, J.H. (2009). Prevention of hepatocellular carcinoma in hepatitis B virus infection. *J Gastroenterol Hepatol.*, **24:** 1352-1357.

Liyanage, U.K., Moore, T.T., Joo, H.G., Tanaka, Y., Herrmann, V., Doherty, G., Drebin, J.A., Strasberg, S.M., Eberlein, T.J., Goedegebuure, P.S., Linehan, D.C. (2002). Prevalence of regulatory T cells is increased in peripheral blood and tumor microenvironment of patients with pancreas or breast adenocarcinoma. *J Immunol.*, **169:** 2756-2761.

Mantovani, A., Savino, B., Locati, M., Zammataro, L., Allavena, P., and Bonecchi, R. (2010). The chemokine system in cancer biology and therapy. *Cytokine Growth Factor Rev.*, **21:** 27-39.

Marshall, J.L., Hoyer, R.J., Toomey, M.A., Faraguna, K., Chang, P., Richmond, E., Pedicano, J.E., Gehan, E., Peck, R.A., Arlen, P., Tsang, K.Y., and Schlom, J. (2000). Phase I study in advanced cancer patients of a diversified prime-and-boost vaccination protocol using recombinant vaccinia virus and recombinant nonreplicating avipox virus to elicit anti-carcinoembryonic antigen immune responses. *J Clin Oncol.*, **18:** 3964-3973.

Martin-Orozco, N., and Dong, C. (2007). Inhibitory costimulation and anti-tumor immunity. *Semin Cancer Biol.*, **17:** 288-298.

McCullough, K.C. and Summerfield, A. (2005) Basic concepts of immune response and defense development. *ILAR J.*, **46:** 230-240.

Mincheff, M., Tchakarov, S., Zoubak, S., Loukinov, D., Botev, C., Altankova, I., Georgiev, G., Petrov, S., and Meryman, H.T. (2000). Naked DNA and adenoviral immunizations for immunotherapy of prostate cancer: a phase I/II clinical trial. *Eur Urol.*, **38:** 208-17.

Palena, C., and Schlom, J. (2010). Vaccines against human carcinomas: strategies to improve antitumor immune responses. *J Biomed Biotechnol.*, **2010:** 380697.

Pashov, A., Monzavi-Karbassi, B., Chow, M., Cannon, M., and Kieber-Emmons, T. (2007). Immune surveillance as a rationale for immunotherapy? *Hum Vaccin.*, **3:** 224-228.

Rabinovich, G.A., Gabrilovich, D., and Sotomayor, E.M. (2007). Immunosuppressive strategies that are mediated by tumor cells. *Annu Rev Immunol.*, **25:** 267-96.

Rains, N., Cannan, R.F., Chen, W., and Stubbs, R.S. (2001). Development of a dendritic cell (DC)-based vaccine for patients with advanced colorectal cancer. *Hepatogastroenterology*, **48:** 347-351.

Ribas, A., Butterfield, L.H., Glaspy, J.A., Economou, J.S. (2003). Current developments in cancer vaccines and cellular immunotherapy. *J Clin Oncol.*, **21:** 2415-2432.

Rivoltini, L., Canese, P., Huber, V., Iero, M., Pilla, L., Valenti, R., Fais, S., Lozupone, F., Casati, C., Castelli, C., and Parmiani, G. (2005). Escape strategies and reasons for failure in the interaction between tumour cells and the immune system: how can we tilt the balance towards immune-mediated cancer control? *Expert Opin Biol Ther.*, **5:** 463-476.

Sadanaga, N., Nagashima, H., Mashino, K., Tahara, K., Yamaguchi, H., Ohta, M., Fujie, T., Tanaka, F., Inoue, H., Takesako, K., Akiyoshi, T., and Mori, M. (2001). Dendritic cell vaccination with MAGE peptide is a novel therapeutic approach for gastrointestinal carcinomas. *Clin Cancer Res.*, **7:** 2277-2284.

Sarkar, K., Bose, A., Chakraborty, K., Haque, E., Ghosh, D., Goswami, S., Chakraborty, T., Laskar, S., and Baral, R. (2008). Neem leaf glycoprotein helps to generate carcinoembryonic antigen specific anti-tumor immune responses utilizing macrophage-mediated antigen presentation. *Vaccine*, **26:** 4352-4362.

Sarkar, K., Goswami, S., Roy, S., Mallick, A., Chakraborty, K., Bose, A., and Baral, R. (2010). Neem leaf glycoprotein enhances carcinoembryonic antigen presentation of dendritic cells to T and B cells for induction of anti-tumor immunity by allowing generation of immune effector/memory response. *Int Immunopharmacol.*, **10:** 865-874.

Schneider, T., Gerhards, R., Kirches, E., and Firsching, R. (2001). Preliminary results of active specific immunization with modified tumor cell vaccine in glioblastoma multiforme. *J Neurooncol.*, **53:** 39-46.

Shurin, M.R., Shurin, G.V., Lokshin, A., Yurkovetsky, Z.R., Gutkin, D.W., Chatta, G., Zhong, H., Han, B., and Ferris, R.L. (2006). Intratumoral cytokines/chemokines/ growth factors and tumor infiltrating dendritic cells: friends or enemies? *Cancer Metastasis Rev.*, **25:** 333-356.

Slingluff, C.L. Jr., Yamshchikov, G., Neese, P., Galavotti, H., Eastham, S., Engelhard, V.H., Kittlesen, D., Deacon, D., Hibbitts, S., Grosh, W.W., Petroni, G., Cohen, R., Wiernasz, C., Patterson, J.W., Conway, B.P., and Ross, W.G. (2001). Phase I trial of a melanoma vaccine with gp100(280-288) peptide and tetanus helper peptide in adjuvant: immunologic and clinical outcomes. *Clin Cancer Res.*, **7:** 3012-24.

Sonia, A., Perez, P., and Michael, P. (2008). Cancer Immunotherapy: Perspectives and Prospects. *Adv Exp Med Bio.*, **622:** 235-253.

Szarewski, A. (2010). HPV vaccine: Cervarix. Review. *Expert Opin Biol Ther.*, **10:** 477-487.

Toungouz, M., Libin, M., Bulté, F., Faid, L., Lehmann, F., Duriau, D., Laporte, M., Gangji, D., Bruyns, C., Lambermont, M., Goldman, M., and Velu, T. (2001). Transient expansion of peptide-specific lymphocytes producing IFN-gamma after vaccination with dendritic cells pulsed with MAGE peptides in patients with mage-A1/A3-positive tumors. *J Leukoc Biol*, 69: 937-940.

Vieweg, J., and Jackson, A. (2005). Modulation of antitumor responses by dendritic cells. *Springer Semin Immunopathol.*, **26:** 329-341.

von Mehren, M., Arlen, P., Gulley, J., Rogatko, A., Cooper, H.S., Meropol, N.J., Alpaugh, R.K., Davey, M., McLaughlin, S., Beard, M.T., Tsang, K.Y., Schlom, J., and Weiner, L.M. (2001). The influence of granulocyte macrophage colony-stimulating factor and prior chemotherapy on the immunological response to a vaccine (ALVAC-CEA B7.1) in patients with metastatic carcinoma. *Clin Cancer Res.*, **7:** 1181-1191.

Wang, F., Bade, E., Kuniyoshi, C., Spears, L., Jeffery, G., Marty, V., Groshen S, and Weber, J. (1999). Phase I trial of a MART-1 peptide vaccine with incomplete Freund's adjuvant for resected high-risk melanoma. *Clin Cancer Res.*, **5:** 2756-2765.

Weber, J.S., Hua, F.L., Spears, L., Marty, V., Kuniyoshi, C., and Celis, E. (1999). A phase I trial of an HLA-A1 restricted MAGE-3 epitope peptide with incomplete Freund's adjuvant in patients with resected high-risk melanoma. *J Immunother.*, **22:** 431-440.

Natural Products: Research Reviews Vol. 1 (2012)
Editor: **V.K. Gupta**
Published by: **DAYA PUBLISHING HOUSE, NEW DELHI**

Pages 79–98

3

Mistletoe Lectins in Cancer Therapy: Administration by the Oral Route or by Subcutaneous Injection?–A Review

Ian F. Pryme[1]

ABSTRACT

Various types of mistletoe preparations have been in use in cancer treatment for many years, particularly in the German-speaking countries. Despite the fact that much is now known about the individual components in such preparations, especially the lectins, mistletoe extracts used for subcutaneous injection are still essentially in their original form. Thus in spite of major scientific advances during the last 20-30 years little has changed with regard to the manner of use and composition of mistletoe-based preparations. Three major lectins have been isolated and purified from mistletoe. Following binding to appropriate receptors in the intestinal or nasal mucosa the lectins initiate an immune response, and production of anti-angiogenic factors and endorphins has been reported. In extensive in vitro and in vivo studies anti-cancer properties have been described – the so-called RIP-effect, leading to apoptosis and cell death. This effect is dependent on lectin binding to specific receptors in the cell membrane through the B-chain, this then enabling the internalization of the A-chain, followed by activation of its ribosome-inactivating properties. The mistletoe lectins are highly glycosylated molecules resulting in resistance to the low pH in the stomach and also to protection from proteolysis by the digestive enzymes present in the gastrointestinal tract. These properties mean that the mistletoe lectins can be presented by the oral route and that a biological response can be instigated. Tumours in the oral cavity or gastrointestinal tract bearing mistletoe lectin binding receptors would be susceptible to the RIP effect which could ultimately result in apoptosis and tumour cell death. In a situation where tumours are at other sites then following receptor binding in the small intestine a biological response would occur, resulting in stimulation of

---

1   Department of Biomedicine, University of Bergen, Jonas Liesvei 91, N-5009 Bergen, Norway.

*   E-mail: ian.pryme@biomed.uib.no

the body's own defence mechanisms against tumour growth. In this paper the advantages of this approach are compared to the traditional subcutaneous injection manner of administration of a mistletoe lectin-containing preparation. It is argued that this approach can prove to be more effective than the currently established procedure.

Keywords: Apoptosis, Anti-angiogenesis, Cytokines, Cytotoxicity, Immunostimulation, NK-cells, Peyer's patch, Tumour.

Introduction

Mistletoe extracts have been used in cancer therapy for more than 80 years (see Ostermann *et al.*, 2009 for references), particularly in clinics in Austria, Switzerland and Germany. Use of these extracts has been heavily criticized by practitioners of traditional "school medicine" due to the lack of knowledge concerning the actual nature and characterisation of possible active anti-cancer components from mistletoe. A further complicating factor is that the results of more than 40 randomised clinical trials where preparations such as Iscador, Helixor, abnoba VISCUM have been used in cancer treatment are largely inconclusive such that the efficacy of mistletoe treatment is still in doubt. The manner of administration of these preparations is essentially through subcutaneous injection.

There has been much debate on the use of such preparations because individual mistletoe extracts prepared by different manufacturers contain a host of largely undefined components including lectins, viscotoxins, peptides, amino acids, amines, lipids, cyclitoles, thiols, lipids, phytosterols, triterpenes, flavonoids, phenylpropanes, polysaccharides and various minerals (Franz *et al.*, 1981; Pfüller, 2000). The cytotoxic and immunomodulatory effects of mistletoe extracts that have been described appear to depend upon the host tree, the manufacturing process and also the actual composition of the different components present in the extracts. Furthermore, in a study designed to compare the biological effects of 12 different clinically applied mistletoe preparations both apoptosis and cytokine production were shown to be induced differentially in leukocyte cultures by the mistletoe preparations tested (Elsasser-Beile *et al.*, 1998). Zarkovich *et al.* (1998) suggested that the beneficial therapeutic effects of Isorel in cancer therapy may well be a result of the combined biological activity of high and low molecular weight components and thus not due to a single class of molecule alone. This suggests, therefore, that clinical studies performed with different preparations are not comparable either. Studies using purified mistletoe components are thus warranted in order to determine whether or not active anti-cancer molecules can indeed be identified.

Eifler *et al.* (1993) have isolated three lectins from mistletoe extracts (ML-I, ML-II and ML-III). These have now been subject to rigorous study (see Pryme *et al.*, 2007). It has been clearly established that these purified carbohydrate-binding lectins possess cytotoxic properties. The best studied mistletoe lectin is ML-I, a type-2 RIP (ribosome-inactivating protein) which is composed of two chains - the A-chain which has N-glycosidase activity (accounting for the property of inactivating ribosomes), and a B-chain possessing galactose-specific binding properties, this subunit being responsible for binding and cellular uptake of the molecule. The individual subunits do not appear to possess cytotoxic properties (Vervecken *et al.*, 2000).

Not only do the mistletoe lectins impart cytotoxic activity through their type-2 RIP property but they also have the ability to initiate a profound immunomodulating response. Stein *et al.* (1998) demonstrated a strong initial proliferation of peripheral blood mononuclear cells following the exposure of healthy individuals to an aqueous mistletoe extract, accompanied by increased TNFα and IL-6 production. A less pronounced release of IFN-γ and IL-4 was also observed. In breast carcinoma patients treated with galactose-specific lectin standardized mistletoe extract, Heiny *et al.* (1998) reported increased β-endorphin plasma levels, stimulation of T-lymphocytes exhibiting expression of CD25/interleukin-2 receptors and HLA/DR-antigens and enhanced activity of peripheral blood natural killer cells (NK). Baxevanis *et al.* (1998) also observed enhanced NK activity following the incubation of cultures of PBL for 3 days with ML-I. They reported an expansion and activation of T-cells which demonstrated both NK and LAK-like cytotoxicity. ML-I preferentially stimulated and expanded CD8+ T cells which mediated the cytotoxic effect. An activation of PBL with both ML-I and IL-2 resulted in simultaneous induction of T and CD56+ cell-mediated NK and LAK cytotoxicity. Büssing *et al.* (1999) have analysed mitochondrial alterations in human lymphocytes incubated with ML-I and demonstrated generation of reactive oxygen intermediates and the induction of expression of newly described mitochondrial membrane proteins Apo2.7. Part of the cytotoxic response thus appears to be a distinct 'death signal' resulting in an induction of Apo2.7 molecules within 24 hr. Stein *et al.* (2000a,b) have shown that intracellular expression of IL-4 and inhibition of IFN-γ production are processes related to the induction of apoptosis in U-266 plasmacytoma cells. It has now been established that serum glycoproteins, particularly haptoglobin, but also α(1)-acid glycoprotein and transferrin are able to inhibit the apoptosis-inducing properties of ML lectins (Frantz *et al.*, 2000). Deglycosylated haptoglobin, however, did not exhibit a protective effect. These results explain why MLs do not exert harmful effects when administered to patients.

Subcutaneous Injection of Mistletoe Extracts in Oncology

A number of commercial mistletoe preparations have been in use in cancer treatment in Europe for more than 80 years. Since these differ with respect to both processing and molecular composition, their biological effects also vary and this had made it very difficult to interpret the results of clinical trials conducted over a period of several decades. One of the most widely used preparations is Iscador®, a product of Weleda AG (Arlesheim, Switzerland). It is a fermented aqueous extract of the fresh, leafy shoots and fruits of the mistletoe plant. This mistletoe extract is the one that has been most commonly reported in clinical trials. The typical manner of administration is subcutaneous (sc) injection 2-3 times per week. In a recent article Ostermann *et al.* (2009) performed a systematic literature review, including a statistical analysis of the data available, with respect to determining the effectiveness of the preparation in patients suffering with cancer.

The following databases were searched: MEDLINE, Excerpta Medica (EMBASE), the Cochrane Library, Deutsches Institut für Medizinische Dokumentation und

Information (DIMDI), and CAMbase for published, controlled clinical trials where Iscador was used in the treatment of cancer patients. In addition they hand-searched reference lists of published articles and, furthermore, contacted experts and also the manufacturers of the mistletoe extract for additional published clinical information. Excluded from their analysis were case reports, uncontrolled studies, unpublished manuscripts, and trials using a combination of Iscador with other types of mistletoe products. Only articles written in English or German were included in the study. Two members of the team independently assessed the methodological quality utilized in the various studies and extracted the data meeting the required criteria.

Their programme of study resulted in the identification of 40 publications that met the criteria set out for analysis. Data were analyzed separately for studies with a placebo control, an active control, and where standard medical care was used as a control. Standard medical care served as the control in 22 studies, providing sufficient data to extract hazard ratios. Of these studies (published in the period 1963 to 2008) 12 were prospective, and 10 had a matched-pair design. The studies encompassed 7,253 patients that received standard medical care and 3,388 patients who were treated with Iscador in addition to standard medical care.

Ostermann *et al.* (2009) judged the quality of these trials as ranging from poor to moderate. Interestingly, a meta-analysis gave an overall hazard ratio of 0.59 ($P <$ 0.0001), meaning that patients who received Iscador were at a lower risk of dying (*i.e.* had better survival rates). Both funnel plots and statistical analysis indicated a possible publication bias, where studies with statistically significant results were more likely to be published than studies without statistically significant results. Of the 22 studies only 5 were randomised, and these latter studies showed a lower effect of Iscador on survival than results of non-randomised studies. Studies with a matched-pair design showed greater effects of Iscador on survival than the prospective design studies.

Three studies constituted randomised trials where active controls were used. In these 3 trials, 462 patients were treated with Iscador, while 450 patients were treated with an active control (such as interferon or radiation therapy). Combined analysis of these 3 trials showed no effect of Iscador on survival. Only 1 placebo-controlled trial was identified. In this trial Iscador proved to have no effect on survival in 224 patients suffering from advanced lung cancer.

According to Ostermann *et al.* (2009) adjuvant treatment with Iscador is associated with a significant improvement in survival in people with cancer. They point out that the findings varied according to study design, with randomised trials showing a lower effect of Iscador on survival than that observed in matched-pair trials. In their article the authors discuss results from previous systematic analyses of a variety of mistletoe preparations. A Cochrane review (Horneber *et al.*, 2008) suggested that mistletoe extracts may improve the quality of life in women being subjected to chemotherapy for breast cancer. In general, however, the evidence for improved survival in cancer patients is rather poor. The authors concluded that the strongest evidence were benefits related to improved quality of life and a reduction of the adverse side effects commonly experienced by cancer patients undergoing chemotherapy. The analysis performed by Ostermann *et al.* (2009) demonstrated a wide variability in data from the individual studies and the possibility of publication bias. Despite these limitations the number of studies reporting positive benefits of Iscador clearly indicates that further investigation using mistletoe preparations is warranted. It is envisioned that in future studies where improved study design is utilized then a better insight into the clinical value of mistletoe preparations in improving the quality of life, outcome, and survival rate of cancer patients will be gained.

In their recent studies Huber *et al.* (2010) have shown strong inter-individual differences in the pharmacokinetics of mistletoe lectins after sc injection. This was the case despite the fact that all the 15 volunteers in the trial (healthy males aged 18-42 years) were injected by the same investigator in the same abdominal quadrant. The authors attributed this to the likelihood of different patterns of binding to carbohydrates in skin tissue thus affecting the kinetics of release from the subcutaneous tissue. This would then be partially responsible for major individual differences in the availability of mistletoe lectin with respect to initiating a biological response. Another complicating factor would be the fact that as mentioned earlier serum proteins (haptoglobin, $\alpha(1)$-acid glycoprotein and transferrin) can inhibit ML activity (Frantz *et al.*, 2000). Again differences in the levels of these serum proteins could impart large differences in the amounts of ML made readily available for eliciting a biological response. The levels of these serum proteins are almost certainly extremely variable when comparing individual cancer patients, more so than in normal individuals, and this together with differences in release of lectin from the tissue at the site of sc injection will contribute to a high degree of variability in a population study. It is thus not surprising that many of the trials performed with mistletoe lectin injected sc have given such widely varying results.

When the use of mistletoe extracts was first introduced as a form of cancer treatment knowledge of the macromolecular content of such preparations was extremely limited. In recent years our understanding of one of the biologically active class of molecules in mistletoe, namely the lectins, has become extensive. It is surprising that after more than 80 years of use little attention has been drawn toward a "modernisation" of the traditional mistletoe extract. In the following sections our current knowledge of certain aspects of mistletoe lectins will be discussed and a case for the oral use of mistletoe lectin-containing preparations in cancer therapy will be outlined. This represents a new alternative to the traditional administration of mistletoe extracts by sc injection.

Effects of Purified Mistletoe Lectins on Tumour Cells in vitro

A series of workers have shown that the addition of mistletoe lectins to various tumour cell lines in culture results in growth impairment and often entry into apoptosis with ensuing cell death (Table 3.1). Some of these findings are discussed below.

Mockel *et al.* (1997) examined the effects of ML-I on the human T-cell leukemia line MOLT-4, the monocytic line THP-1 and on human peripheral blood mononuclear cells (PBMC) with regard to general cell viability and induction of apoptosis. Using a sensitive serum-free cytotoxicity assay, the time- and concentration-dependent direct

Table 3.1: Mistletoe lectins : apoptosis/reduced growth induced in various tumour cell lines *in vitro*.

Cell Line	Reference
Molt-4 (T-cell leukemia)	Mockel *et al.*, 1997
Molt-4	Ribereau-Gayon *et al.*, 1997
Leukemic B- and T-cell lines	Bantel *et al.*, 1999
HL-60 human leukemia	Pae *et al.*, 2000a
Human myeloleukemic U937 cells	Kim *et al.*, 2000, 2003
Human myeloleukemic cells	Park *et al.*, 2000
HL-60	Park *et al.*, 2000
HL-60 human leukemia	Lyu *et al.*, 2001
U937 human monoblastic leukemia	Pae *et al.*, 2001b
Calu-1 (human lung carcinoma)	Köteles *et al.*, 1998
Human lung carcinoma (Calu-1)	Kubasova *et al.*, 1996
Human lung carcinoma (A549)	Siegle *et al.*, 2001
Hepatocarcinoma cells	Yoon *et al.*, 1999
Human hepatocarcinoma SK-Hep1 cells	Pae *et al.*, 2001a
Human hepatocarcinoma (SK-Hep-1, Hep 3B)	Lyu *et al.*, 2002
B16-BL6 melanoma	Park *et al.*, 2001
Human melanoma cells	Thies *et al.*, 2005
Human colon cancer HT29	Valentiner *et al.*, 2002
Human colon cancer (COLO)	Khil *et al.*, 2007
F-98 anaplastic glioma	Lenartz *et al.*, 1998
HeLa (human cervix carcinoma)	Pae *et al.*, 2000b
MCF-7 breast carcinoma	Pae *et al.*, 2000b
Jurkat T-cells; RAW 264.7; DLD-1 cells	Park *et al.*, 2000
U-266 plasmacytoma	Stein *et al.*, 2000
Human A253 cells	Choi *et al.*, 2004
Various human and murine tumour cells	Yoon *et al.*, 1999
Human tumor models	Kelter *et al.*, 2007

toxicity towards MOLT-4 cells was determined with IC_{50}-values ranging from 20-40 pg/ml (300-600 fmol/l). Investigations on the time course of the toxic effect using selected concentrations of ML-I revealed distinct response curves for concentrations of high, low and intermediate toxicity, respectively. The ratio of apoptotic to viable MOLT-4 cells was determined after treatment with ML-I for 24hr. Apoptosis and cytotoxicity were correlated at low and intermediate concentrations. The data showed that in the concentration range of low cytotoxicity ML-I - induced cell death is quantitatively due to apoptotic processes. The immunomodulatory activity of ML-I was investigated *in vitro* by measuring cytokine release. At concentrations of low cytotoxicity ML-I showed immunostimulatory activity on PBMC and THP-1. RT-

PCR with THP-1 cells confirmed that cytokine induction by ML-I is regulated at the transcriptional level. These findings suggest that in the blood cells investigated both apoptosis and cellular signaling are induced by the same concentration range of ML-I. Ribereau-Gayon *et al.* (1997) also demonstrated an inhibition of Molt 4 cell growth. This occurred with lectin concentrations in the pg/ml range. The first events which were observed were membrane perforation and protrusions typical of apoptosis. They showed that ML-111 was about 10 times more cytotoxic than ML-I. Treatment of leukemic T- and B-cell lines with ML-I also triggered cell death through the induction of apoptosis (Bantel *et al.*, 1999). A peptide cascade inhibitor was almost completely able to prevent the effect of ML-I. These authors showed that ML-I potentiated the effect of chemotherapeutic drugs.

The cytotoxic activity of ML-I towards the anaplastic glioma cell line (F98) using a three dimentional spheroid model was studied by Lenartz *et al.* (1998). F98 glioma cell spheroid growth was significantly inhibited after incubation with defined ML-I concentrations of 10 and 100 ng/ml, demonstrating dose dependent cytotoxicity to the lectin.

Yoon *et al.* (1999) isolated cytotoxic lectins (KML-C) from an extract of Korean mistletoe [*Viscum album* C. (coloratum)] by affinity chromatography on a hydrolysed Sepharose 4B column, and the chemical and biological properties of KML-C were examined, partly by comparing them with a lectin (EML-I) from European mistletoe [*Viscum album* L. (loranthaceae)]. The isolated lectins showed strong cytotoxicity against various human and murine tumour cells, and the cytotoxic activity of KML-C was higher than that of EML-I. Tumour cells treated with KML-C exhibited typical patterns of apoptotic cell death, such as typical morphological changes and DNA fragmentation, and its apoptosis-inducing activity was blocked by addition of Zn^{2+}, an inhibitor of Ca^{2+}/Mg^{2+}-dependent endonucleases, in a dose-dependent manner. These results suggest that KML-C is a novel lectin and that its cytotoxic activity against tumour cells is due to apoptosis mediated by Ca^{2+}/Mg^{2+}-dependent endonucleases. Pae *et al.* (2000a) have shown that incubation of human leukemia HL-60 cells with various doses of Korean mistletoe lectin (ML-II) results in apoptosis. Activation of PKA or PKC, however, appeared to convey protection against apoptosis induced by the lectin.

There is thus a wealth of information clearly demonstrating that mistletoe lectins when added to cultures of *e.g.* leukemia, melanoma, lung carcinoma, hepatocarcinoma and colon cancer cell lines can exert cytotoxic effects. There are, however, serious differences between the *in vivo* and *in vitro* situations where the obvious question can be raised as to whether it is likely that sufficient amounts of lectin injected sc will arrive at *e.g.* the stomach, breast, lung etc. and there trigger a cytotoxic response in a tumour.

Effects of Purified Mistletoe Lectins on Tumours in vivo

A number of investigations have demonstrated that mistletoe lectins, when applied directly in the *in vivo* situation, can have profound effects on the growth of various types of tumours (Table 2). Two such cases are discussed below.

Table 3.2: Mistletoe lectins : reduced tumour growth observed *in vivo*.

Tumour	Reference
Non-Hodgkin lymphoma (mice)	Ewen *et al.*, 1998, Pryme *et al.*, 1998, Ewen *et al.*, 1999, Pryme *et al.*, 2002, 2004, 2006, 2007
Plasmacytoma (mice)	Pryme *et al.*, 1996b
F-98 glioma (rat)	Lenartz *et al.*, 1998
Malignant glioma (human study*)	Lenartz *et al.*, 2000
B16-BL6 melanoma (mice)	Yoon *et al.*, 1998
B16-BL6 melanoma (mice)	Park *et al.*, 2001
B16-BL6 melanoma (mice)	Duong van Huyen *et al.*, 2006
MV3 melanoma (human)	Thies *et al.*, 2008
Colon 26-M3.1 carcinoma (mice)	Yoon *et al.*, 1998
Lewis lung tumour (mice)	Kubasova *et al.*, 1998
MB49 urinary bladder carcinoma (mice)	Mengs *et al.*, 2000
Acute lymphoblastic leukemia (human)	Seifert *et al.*, 2008

*: Prolongation of overall survival rate.

To investigate the *in vivo* cytotoxic efficacy of ML-I, Fischer 344 rats were intracerebrally implanted with F98 glioma cells and subjected to both local and systemic ML-I treatment (Lenartz*et al.*, 1998). Histological and immunohistochemical evaluation showed a reduction in tumour volume for both treatment modalities, most pronounced and statistically significant after systemic (immunomodulating) administration of the optimal ML-I dosage (1 ng/kg body weight, subcutaneously) and after low dose (10 ng ML-I per application) local treatment. High dose ML-I administration (10 ng/kg body weight; systemically; 100 ng/application, locally) was less effective than low (optimal) dose treatment and apparently the systemic/ immunomodulating approach resulted in a greater benefit for glioma bearing rats.

Yoon *et al.* (1998) have demonstrated the prophylactic effect of lectins (KM-110) from *Viscum album coloratum*, a Korean mistletoe, on tumour metastasis produced by highly metastatic tumour cells, colon 26-M3.1 carcinoma, B16-BL6 melanoma and L5178Y-ML25 lymphoma cells, using experimental models in mice. Intravenous (i.v.) administration of KM-110 (100 µg/mouse) 2 days before tumour inoculation significantly inhibited lung metastasis of B16-BL6 and colon 26-M3.1 cells, and liver and spleen metastasis of L5178Y-ML25 cells. Furthermore, mice given KM-110 (100 µg) 2 days before tumour inoculation showed significantly prolonged survival rates compared with the untreated mice. In a time course analysis of NK activity, i.v. administration of KM-110 (100 µg) significantly augmented NK cytotoxicity to Yac-a tumour cells from 1 to 3 days after KM-110 treatment. Furthermore, depletion of NK by injection of rabbit anti-asialo GM1 serum completely abolished the inhibitory effect of KM-110 on lung metastasis of colon 26-M3.1 cells. These results suggest that KM-110 possesses immunopotentiating activity which enhances the host defence system against tumours, and that its prophylactic effect on tumour metastasis is mediated by NK activation.

Oral Intake of Mistletoe Plant does not Induce Toxicity

There are reports that following ingestion of mistletoe plant material, or its white berries, there can be an induction of symptoms of nausea, and occasionally vomiting may be caused. Two studies where the outcome of over 2000 exposures to accidental intake of the plant has been examined did not show any evidence of toxicity in humans. Hall *et al.* (1986) reviewed a total of 318 cases of mistletoe ingestion that were reported to the FDA poison control case reporting system between 1978 and 1983 (n=177) and 1984 (n=141). The majority remained asymptomatic and no deaths were recorded. In their study Krenzelok *et al.* (1997) followed the outcome of a total of 1754 exposures to mistletoe plant ingestion. Their conclusion was quite clear in that the accidental intake of mistletoe material was not associated with any signs of profound toxicity. It is thus evident that the unfortunate historical reputation that mistletoe is toxic following oral intake has no support from clinical findings.

The addition of large doses of purified mistletoe lectin to diets (67 or 200 mg ML-I/kg body weight) and fed to rats for 10 days did not result in the induction of any observable toxic reactions (Pusztai *et al.,* 1998). It is thus evident that large amounts of mistletoe lectins are well tolerated when presented orally. Rather than initiating toxic effects dietary mistletoe lectins have in fact been shown to exert growth-stimulatory effects on crypt cells of the small intestine, both in rats (Pusztai *et al.,* 1998) and mice (Pryme *et al.,* 2002).

Oral Mistletoe Lectins in Rats and Mice and Their Effect on Growth of a Non-Hodgkin Lymphoma

Pusztai *et al.* (1998) observed a significant increase in plasma TNFα levels 30 hr after feeding an ML-I-containing diet to rats. Plasma interleukin-1β was also elevated. Thus the release of cytokines that is considered to be an essential step in immunomodulation leading to tumour depression (Männel *et al.,* 1991), was promoted when the mistletoe lectin was given orally. An avid binding of ML-I to M cells of Peyer's patch was demonstrated in the rat small intestine. It is likely that this is implicated in the observed increase in plasma cytokines. The obvious advantage of providing mistletoe lectins by the oral route is that large amounts of the lectins, through their binding to the gut mucosa, are in due course presented to lymphocytes of Peyers patches and thereby able to induce a major cytokine response. We have observed a 14 per cent reduction in the weight of the spleen within 24 hr of feeding ML-I to mice (Christiansen and Pryme, unpublished observations). This can be attributed to a major release of lymphocytes into the blood circulation following a burst of cytokine release from stimulated cells of Peyers patches. It is highly unlikely that such a response would be evoked by the small amounts of lectins which are likely to reach the lymphatic tissue when mistletoe extracts or lectins are injected subcutaneously. Interestingly, Lavelle *et al.* (2000) have demonstrated that oral delivery of ML-I stimulated the production of specific serum IgG and IgA antibody after three oral doses.

Based on their observations Pusztai *et al.* (1998) were able to conclude that ML-I provided in the diet was sufficiently without detrimental effects that it should be tested for anti-tumour properties. The lectin was incorporated into lactalbumin-based semi-synthetic diets (Pryme *et al.*, 1996c). The effect of feeding mice the LA control diet, or one containing three different concentrations of ML-I lectin (0.42, 0.83 and 1.67mg/g diet) were tested on tumour growth after 10 days following the subcutaneous injection of 2.10^6 non-Hodgkin lymphoma tumour cells. The results (see Figure 1 in Pryme *et al.*, 2006) showed that even the lowest lectin concentration had a marked effect on tumour growth, tumour mass being reduced by about 25 per cent. At the highest concentration of ML-I tested tumour mass was reduced by about 40 per cent (Pryme *et al.*, 1998a; 2004). The observations showed that a reduction in tumour growth occurred in an apparent dose-dependent manner. Interestingly, the observations were similar to those seen earlier when the effects of another lectin (phytohaemagglutinin – PHA) on growth of a NHL tumour in NMRI mice fed PHA in the range 0.45 - 7.0 µg/g diet were studied (Pryme *et al.*, 1996a), and in Balb/c mice where the same PHA concentrations were tested on the growth of the MPC-11 plasmacytoma tumour (Pryme *et al.*, 1996b).

In earlier experiments PHA was shown to cause hyperplasia of the small intestine (Pryme *et al.*, 1998b) and the results indicated a relationship between the stimulation of "normal" growth and the depressed growth of the tumour, suggesting a competition between the two types of cell proliferation for nutrients and growth factors from a common body pool (Bardocz *et al.*, 1999; Pryme *et al.*, 1999). Based on tissue mass measurements ML-I was shown to cause hyperplasia of the gut in a similar manner to that previously observed with PHA (Pryme *et al.*, 1998a). These results were confirmed by histological analysis of the small intestine (Pryme *et al.*, 2002). It was evident that the level of hyperplasia initiated by the two lowest concentrations of the lectin used in the first experiments (Pryme *et al.*, 1998a) was only of a minor degree. The effect on tumour growth seen at the lower doses, therefore, may well have been imparted essentially through an immunomodulatory response. In contrast, at the highest dose, where a major proliferation of the small intestine was observed, a nutritional restraint on tumour growth may have been incurred, working in concert with the immunomodulatory effect. The data from this first study suggested that the mistletoe lectin exhibited similar effects both on tumour growth and on the gut as earlier described for PHA.

Further detailed studies have been performed to examine the characteristics of the NHL tumour at the microscopic level following the feeding of ML-I-containing diets to mice (Ewen *et al.*, 1998, 1999). An intense lymphoid host response within the NHL tumour was evoked by ML-I, compared to a patchy and sparse lymphoid reaction in the control-fed animals (see Figure 2a,b in Pryme *et al.*, 2006). Accelerated cellular turnover within the transplanted NHL tumour as a response to oral intake of ML-I was seen as increased numbers of apoptotic cells with an increased area of serpiginous irregular dead cells, and the non-viable cells occupied a two fold increased area in the mice fed the lectin (Ewen *et al.*, 1999). Apoptoses were more numerous in the tumours of mice fed ML-I compared to control diet (see Figure 3a,b in Pryme *et al.*, 2006). These were identified around areas of non-viable tumour cells, at the advancing edge of the tumour and within intense lymphoid aggregates. Pryme *et al.* (2004) demonstrated a decrease in the nuclear area of the tumour cells following feeding ML-I (mean decrease 21 per cent). The number of tumour cell mitoses (measured as mitoses/high power field) was reduced from 7.4 (control) to 1.7 in tumours from mice with the highest daily intake of ML-I (Pryme *et al.*, 2004). There was a simultaneous increase in crypt length of the jejunum with increasing intake of ML-I (Ewen *et al.*, 1998: Pryme *et al.*, 2004).

The NHL tumour in LA or PHA fed mice was characterised as a tumour with an extremely well-developed capillary system and thus was bloody in appearance upon dissection. Interestingly, the NHL tumours removed from ML-I fed mice were not only much smaller but were white in nature with evidence of a less profusely developed vascularisation. Morphological studies of tumour sections showed a greatly reduced incidence of tumour vascularisation (see Figure 4a, b in Pryme *et al.*, 2006). Quantification showed that tumour vascularisation was reduced by about 90 per cent, indicating a strong anti-angiogenic response. These observations strongly suggest that one of the responses to oral ML-I is an induction of the production of anti-angiogenic factors. These results corroborated well with the initial "naked eye" observations of the NHL tumours mentioned above.

Yoon *et al.* (1995) suggested that the anti-metastatic effect of an extract of Korean mistletoe on a series of tumour cell lines in mice was in part due to an inhibition of tumour-induced angiogenesis. Importantly, as a result of feeding ML-I the NHL tumour was histologically completely ablated in 4 of 14 mice after a period of 11 days (Pryme *et al.*, 2004).

An experiment was performed with purified ML-III to establish whether this lectin also has anti-cancer properties. Two groups of mice were injected s.c. with NHL tumour cells and fed the LA–based control diet for 3 days before one group was switched to a diet containing ML-III (2.5 mg lectin/g diet). Eight days later mice were sacrificed and tumours excised. The results showed that addition of the lectin to the diet of mice bearing established NHL tumours was extremely effective in reducing further tumour growth (Pryme *et al.*, 2002). The tumour mass in mice fed ML-III was about 50 per cent of that in control mice demonstrating that the addition of lectin to the diet must have had a rapid deleterious effect on tumour growth.

Inhibition of Mistletoe Lectins in the Blood Circulation

On reaching the blood it seems likely that mistletoe lectins would be very rapidly inactivated through two mechanisms. Firstly, it is well known that mistletoe lectins exhibit an immuno-stimulatory response in the body resulting in the production of anti-lectin antibodies (Hajto *et al.*, 1989). Stettin *et al.* (1990) demonstrated that anti-mistletoe lectin antibodies produced in patients during therapy with an aqueous mistletoe extract caused neutralization of lectin-induced cytotoxicity *in vitro*. Oral intake of a mistletoe lectin-containing preparation was seen to induce antibodies against the lectins (Wing *et al.*, 2010). Additionally, Stein *et al.* (1997) showed that anti-mistletoe antibodies were able to neutralise the cytotoxic effect of mistletoe lectins on peripheral blood mononuclear cells. Secondly, Frantz *et al.* (2000) established that serum glycoproteins, particularly haptoglobin, but also α(1)-acid glycoprotein and transferrin, had the ability to inhibit the apoptosis-inducing properties of ML lectins.

It is thus evident that there must be a limitation as to the period of how long mistletoe lectins presented by the sc route can remain active before being bound by either circulating antibodies or serum glycoproteins. This is in contrast to mistletoe lectins provided orally where a biological response is initiated following binding to cell surface receptors in the small intestine. Sharma*et al.* (1996), using histochemistry on human biopsies of small intestine sections, identified binding of mistletoe lectins to both enterocytes and M-cells of the follicle-associated epithelium, which would then result in a biological response (Bardocz *et al.,* 1995). Since cells of the intestinal epithelium are replaced every 48-72 hr it is thus evident that mistletoe lectin supplementation in the diet will be required in order to sustain a biological response. In contrast to the subcutaneous injection of a mistletoe lectin-containing preparation it is extremely unlikely that lectin antibodies in the blood, or serum glycoproteins, will interfere with the biological effects of mistletoe lectins that are presented by the oral route.

Oral Preparation Enriched in Mistletoe Lectins

As mentioned above mistletoe extracts contain nausea-inducing components. Winge *et al.* (2010) have made a preparation for oral use that is enriched in mistletoe lectins but free of the undesired nausea-inducing components. The potential immunostimulatory effect of this preparation was evaluated in a pilot study. Eight healthy individuals took the preparation on a daily basis for three months. Blood samples were taken before intake of the preparation and after 28 days and 84 days during the course of the study. All individuals had anti-lectin antibodies in the blood at day 84, showing that an immune response had indeed occurred. No antibodies against mistletoe lectins were identified in the initial blood sample. During the observation period four cell types showed a marked response: activated IL-2 helper cells, memory cells, activated T-cells and activated natural killer (NK) cells. In general on day 28 the levels of activated IL-2 helper cells and memory cells were elevated with a return to normal levels on day 84. The levels of activated T-cells and the activated NK cells, however, were still elevated on day 84. Surprisingly at the onset of the pilot study four of the individuals showed NK levels that were well below normal values. Three of these, however, at 84 days had values that were now acceptably within the normal range. The observations taken together show that the mistletoe lectin preparation not only functions as an unspecific immunogenic trigger but can also activate the specific immune system. In a short-term study it was shown that there were increases in serum levels of endostatin, TNFα and β-endorphin values within 1-2 days of ingesting the preparation. The oral mistletoe preparation enriched in lectins thus behaved as an extremely effective dietary supplement for immunostimulation.

Because of the nature of the effects of current cancer treatment (such as radiotherapy and chemotherapy) many patients suffer from immunoinsufficiency. In order to test whether or not terminal cancer patients, where further treatment for their condition was no longer applicable, could benefit from immunostimulation

such patients on their own initiative tested the preparation. The results were encouraging and there were positive indications that immunostimulation, and presumably other effects such as induced endostatin and β-endorphin production, were contributory factors to reported improved quality of life (Pryme *et al.*, 2007).

Suggested Stages in the Effect of Orally Fed Mistletoe Lectins

Based on observations from our animal model system, published data and our unpublished work the following sequence of events following the oral intake of mistletoe lectins is suggested:

1. ML´s bind strongly to the gut mucosa.
2. ML´s are endocytosed through the mucosa of the small intestine.
3. Binding of endocytosed ML´s to lymphocytes of Peyers patches.
4. Stimulation of cytokine release and increased plasma levels.
5. Activation and release of splenic lymphocytes, activation of NK and macrophages.
6. Production and release of anti-angiogenic factors.
7. The anti-angiogenic effect reduces the availability of nutrients for tumour growth and oxygen supplies stagnate.
8. Cytotoxic (RIP) effects exerted on tumour cells.
9. Induction of apoptosis.
10. Tumour cell death.

Comparison of Oral Intake Compared with Subcutaneous Injection

Table 3.3 presents a comparison between the expected effects of oral intake of a mistletoe lectin-containing preparation with the traditional form of mistletoe treatment *i.e.* sc injection. The advantages of choosing the oral route of administration are several, while the limitations of sc injection are evident.

It is envisaged that a major future use of oral ML in cancer therapy would be in the treatment of throat, oesophageal, stomach, colon and rectal cancer *i.e.* through a direct cytotoxic (RIP) effect of the lectin. This would require that these tumour types bear receptors on their surface that are enable to recognise and bind the B-chain which is necessary for internalisation of the A-chain into cells. By utilizing the oral route large amounts of lectin can be delivered directly to the tumour. This is not possible when choosing sc injection as the method of application.

For those tumours where oral presentation would not result in direct contact with lectin molecules then other forms of biological response such as immuno-stimulation and initiation of an anti-angiogenic response would be important lines of anti-tumour activity.

Table 3.3: A comparison of the administration of mistletoe lectins orally with subcutaneous injection of mistletoe preparations.

Oral Intake of Mistletoe Lectins	Mistletoe Preparation : Subcutaneous Injection
Large amounts can be taken	Only very small amounts can be injected
Side effects : none evident	Side effects : nausea, fever, flu-like symptoms, soreness at injection site
Binding to receptors in the small intestine and generation of a biological response : release of cytokines, anti-angiogenic factors and endorphins	*Not applicable*
Direct contact with GALT (80 per cent of the body's immuno-system)	No contact with GALT. Limited contact with immuno-system (only local stimulation)
Activation of receptors in Peyer's patches	*Not applicable*
Not affected by generated antibodies	Subject to inactivation by circulating antibodies
Not affected by inhibitory serum glycoproteins	Subject to inhibition by serum glycoproteins
Can come into direct contact with cancer cells in the oral cavity/GIT, bind to appropriate receptors, *exert* RIP effect after internalization, induce apoptosis and cause tumour cell death	*Not applicable*
Following binding to receptors in the small intestine release of cytokines, anti-angiogenic factors and endorphins induced	

Concluding Remarks

Taken together the extremely promising results obtained using mistletoe lectin-containing preparations with respect to reduced growth of a series of transplantable tumours may provide a new window of opportunity for establishing a novel form of cancer treatment. Taken from the point of view of the cancer patient the oral ingestion of an anti-cancer preparation with a minimum of discomforting side-effects would be highly preferable to conventional present-day anti-cancer treatment which is very often associated with major trauma.

References

Bantel, H., Engels, I.H., Voelter, W., Schulze-Osthoff, K. and Wesselborg, S. (1999). Mistletoe lectin activates caspase-8/FGLICE independently of death receptor signaling and enhances anticancer drug-induced apoptosis. *Cancer Research*, **59** : 2083-90.

Bardocz, S., Grant, G., Ewen, S.W.B., Duguid, T.J., Brown, D.S., Englyst, K. and Pusztai, A. (1995). Reversible effect of phytohaemagglutinin on the growth and metabolism of rat gastrointestinal tract. *Gut.*, **37**: 353-60.

Bardocz, S., Grant, G., Brown, D.S. and Pusztai, A (1999). Uptake, inter-organ distribution and metabolism of dietary polyamines in the rat. *In*: Polyamines in Health and Nutrition, *Ed.* by Bardocz, S. and White, A., Kluwer Academic Publishers Boston/Dordrecht/London, pp. 241-58.

Baxevanis, C.N., Voutsas, I.F., Soler, M.H.J., Gritzapis, A.D., Tsitsilonis, O.E., Stoeva, S., Voelter, W., Arsenis, P. and Papamichail, M. (1998). Mistletoe lectin I- induced effects on human cytotoxic lymphocytes. I. Synergism with IL-2 in the induction of enhanced LAK cytotoxicity. *Immunopharmacology and Immunotoxicology*, **20** : 355-72.

Büssing, A., Wagner, M., Wagner, B., Stein, G.M., Schietzel, M., Schaller, G. and Pfüller U. (1999). Induction of mitochondrial Apo2.7 molecules and generation of reactive oxygen-intermediates in cultured lymphocytes by the toxic proteins from *Viscum album* L. *Cancer Letters*, **139**: 79-88.

Choi, S.H., Lyu, S.Y. and Park,W.B. (2004). Mistletoe lectin induces apoptosis and telomerase inhibition in human A253 cancer cells through dephosphorylation of Akt. *Archives of Pharmacal Research*, **27** : 68-76.

Duong van Huyen, J.P., Delignat, S., Bayry, J., Kazatchkine, M.D., Bruneval, P., Nicoletti, A. and Kaveri, S.V. (2006). Interleukin 12 is associated with the *in vivo* anti-tumor effect of mistletoe extracts in B16 mouse melanoma. *Cancer Letters*, **243** : 32-37

Eifler, R., Pfüller, K., Göckeritz, W. and Pfüller, U. (1993). Improved procedures for isolation of mistletoe lectins and their subunits : lectin pattern of the European mistletoe. *In* : Lectins : Biology, Biochemistry, Clinical Biochemistry, Vol. 9,*Ed.* by Basu, J., Kundu, M. and Chakrabarty, P., New Delhi : Wiley Eastern Limited, pp. 144-51.

Elsasser-Beile, U., Lusebrink, S., Grussenmeyer, T., Wetterauer, U. and Schultze-Seemann, W. (1998). Comparison of the effects of various clinically applied mistletoe preparations on peripheral blood leukocytes. *Drug Research*, **48**: 1185-9.

Ewen, S.W.B., Bardocz, S., Grant, G., Pryme, I.F. and Pusztai, A. (1998). The effects of PHA and mistletoe lectin binding to epithelium of rat and mouse gut. *In* : COST 98, Effects of antinutrients on the nutritional value of legume diets, Vol. 5. *Ed.* by Bardocz, S., Pfüller, U. and Pusztai, A., Luxembourg pp. 221-5.

Ewen, S.W.B., Pryme, I.F., Bardocz, S. and Pusztai, A. (1999). Does oral ingestion of ML-I lectin ablate murine transplanted non-Hodgkin lymphoma by apoptosis or necrosis? *In* : COST 98, Effects of antinutrients on the nutritional value of legume diets, Vol. 6, *Ed.* by Bardocz, S., Hajos, G. and Pusztai, A., Luxembourg. pp. 126 –33.

Franz, H., Ziska, P. and Kindt, A. (1981) Isolation and properties of three lectins from mistletoe. *Biochemical Journal*, **195** : 481-84.

Frantz, M., Jung, M.L., Ribereau-Gayon, G. and Anton, R. (2000). Modulation of mistletoe (*Viscum album* L.) lectins cytotoxicity by carbohydrates and serum glycoproteins. *Drug Research*, **50**: 471-79

Hajtó, T., Hostanska, K. and Gabius, H.J. (1989). Modulatory potency of the betagalactoside-specific lectin from mistletoe extract (Iscador) on the host defense system *in vivo* in rabbits and patients. *Cancer Research*, **49**: 4803–8.

Hall, A.H., Spoerke, D.G. and Rumack, B.H. (1986). Assessing mistletoe toxicity. *Annals of Emergency Medicine*, **151**: 320–3.

Heiny, B.M., Albrecht, V. and Beuth, J. (1998). Correlation of immune cell activities and beta-endorphin release in breast carcinoma patients treated with galactose-specific lectin standardized mistletoe extract. *Anticancer Research*, **18** : 583-6.

Horneber, M., Bueschel, G., Huber, R., Linde, K. and Rostock, M. (2008). Mistletoe therapy in oncology. *Cochrane Database of Systematic Reviews*, Issue 2. Art. No.: CD003297. DOI: 10.1002/14651858.CD003297.pub2.

Huber, R., Eisenbraun, J., Miletzki, B., Adler, M., Scheer, R., Klein, R. and Gleiter, C.H. (2010). Pharmacokinetics of natural mistletoe lectins after subcutaneous injection. *European Journal of Clinical Pharmacology*, **66** : 889-97.

Kelter, G., Schierholz, J.M., Fischer, I.U. and Fiebig, H.H. (2007). Cytotoxic activity and absence of tumor growth stimulation of standardized mistletoe extracts in human tumor models *in vitro*. *Anticancer Research*, **27** : 223-33.

Khil LY, Kim W, Lyu S, Park WB, Yoon JW, Jun HS. (2007). Mechanisms involved in Korean mistletoe lectin-induced apoptosis of cancer cells. *World Journal of Gastroenterology*, **13** : 2811-8.

Kim, M.S., So, H.S., Lee, K.M., Park, J.S., Lee, J.H., Moon, S.K., Ryu, D.G., Chung, S.Y., Jung, B.H., Kim, Y.K., Moon, G. and Park, R. (2000). Activation of caspase cascades in Korean mistletoe (*Viscum album* var. coloratum) lectin-II-induced apoptosis of human myeloleukemic U937 cells. *General Pharmacology*, **34** : 349-55.

Kim, M.S., Lee, J., Lee, K.-M., Yang, S.-H., Choi, S., Chung, S.-Y., Kim, T.-Y., Moon, Jeong, W.-H. and Park, R. (2003). Involvement of hydrogen peroxide in mistletoe-11 induced apoptosis of myeloleukemic U937 cells. *Life Sciences*, **73** : 1231-43.

Köteles. G.J., Kubasova, T., Hurná, E., Horváth, Gy. and Pfüller, U. (1998). Cellular and cytogenetic approaches in testing toxic and safe concentrations of mistletoe lectins. *In*: COST 98, Effects of antinutrients on the nutritional value of legume diets, Vol.5. *Ed.* by Bardocz, S., Pfüller, U. and Pusztai, A. Luxembourg, pp 81-6.

Krenzelok, E.P., Jacobsen, T.D. and Aronis, J. (1997). American mistletoe exposures. *American Journal of Emergency Medicine*, **15**: 516–20.

Kubasova, T., Pfüller, U., Köteles. G.J., Csollák, M. and Eifler, R. (1996). Study of some cellular and immune parameters after the effect of mistletoe lectins *in vitro*. In: COST 98, Effects of antinutrients on the nutritional value of legume diets, Vol. 1, *Ed.* by Bardocz S., Gelencsér. É. and Pusztai A. Luxembourg, pp 55-9.

Kubasova, T., Pfüller, U., Bojtor, I. and Köteles, G.J. (1998). Modulation of immune response by mistletoe lectin 1 as detected on tumour model *in vivo*. *In*: COST 98,

Effects of antinutrients on the nutritional value of legume diets, Vol. 5, *Ed.* by Bardocz S., Pfüller U. and Pusztai A., Luxembourg, pp 202-7.

Lavelle, E.C., Grant, G., Pusztai, A., Pfüller, U. and O'Hagan, D.T. (2000). Mucosal immunogenicity of plant lectins in mice. *Immunology*, 99: 30-7.

Lenartz, D., Andermahr, J., Plum, G., Menzel, J. and Beuth, J. (1998). Efficiency of treatment with galactoside-specific lectin from mistletoe against rat glioma. *Anticancer Research*, 18: 1011-14.

Lenartz, D., Dott, U., Menzel, J., Schierholz and Beuth, J. (2000). Survival of glioma patients after complementary treatment with galactoside-specific lectin from mistletoe. *Anticancer Research*, 20 : 2073-76.

Lyu, S.Y., Park, W.B., Choi, K.H. and Kim, W.H. (2001). Involvement of caspase-3 in apoptosis induced by *Viscum album* var.coloratum agglutinin in HL-60 cells. *Bioscience Biotechnology Biochemistry*, 65 : 534-41.

Lyu, S.Y., Choi, S.H. and Park, W.B. (2002). Korean mistletoe lectin induced apoptosis in hepatocarcinoma cells is associated with inhibition of telomerase via mitochondrial controlled pathway independent of p53. *Archives Pharmacology Research*, 25 : 93-101.

Männel, D.N., Becker, H., Gundt, A., Kist, A. and Franz, H. (1991). Induction of tumour necrosis factor expression by a lectin from *Viscum album. Cancer Immunology Immunotherapy*, 33 : 177-82.

Mengs, U., Schwarz, T., Bulitta, M. and Weber, K. (2000). Antitumoral effects of an intravesically applied aqueous mistletoe extract on urinary bladder carcinoma MB49 in mice. *Anticancer Research*, 20 : 3565-8.

Mockel, B., Schwarz, T., Zinke, H., Eck, J., Langer, M. and Lentzen, H. (1997). Effects of mistletoe lectin I on human blood cell lines and peripheral blood cells. Cytotoxicity, apoptosis and induction of cytokines. *Drug Research*, 47: 1145-51.

Ostermann, T., Raak, C. and Büssing, A. (2009). Survival of cancer patients treated with mistletoe extract (Iscador) : a systematic literature review. *BMC Cancer*, 9: 451- 60.

Pae, H.O., Seo, W.G., Shin, M.K., Lee, H.S., Kim, S.B. and Chung, H.T. (2000a). Protein kinase A or C modulates the apoptosis induced by lectin II isolated from Korean mistletoe, *Viscum album* var. Coloratum, in the human leukemic HL-60 cells. *Immunopharmacology and Immunotoxicology*, 22 : 279-95

Pae, H.O., Seo, W.G., Oh, G.S., Shin, M.K., Lee, H.S., Lee, H.S., Kim, S.B. and Chung H.T. (2000b). Potentiation of tumour necrosis factor alpha-induced apoptosis by mistletoe lectin. *Immunopharmacology and Immunotoxicology*, 22 : 697-709.

Pae, H.O., Oh, G.S., Seo WG, Shin, M.K., Hong, S.G., Lee, H.S. and Chung. H.T. (2001a). Mistletoe lectin synergizes with paclitaxel in human SK-hep1 hepatocarcinoma cells. *Immunopharmacology and Immunotoxicology*, 23: 531- 40.

Pae, H.O., Oh, G.S., Kim, N.Y., Shin, M.K, Lee, H.S., Yun, Y.G., Oh, H., Kim, Y.M. and Chung, H.T. (2001b). Roles of extracellular signalregulated kinase and p38 mitogen-activated protein kinase in apoptosis of human monoblastic leukemia U937 cells by lectin-II isolated from Korean mistletoe. *In Vitro Molecular Toxicology,* **14** : 99-106.

Park, R., Kim, M.S., So, H.S., Jung, B.H., Moon, S., Chung, S.Y., Ko, C.B., Kim, B. and Chung, H.T. (2000). Activation of c-Jun N-terminal kinase 1 (JNK1) in mistletoe lectin II-induced apoptosis of human myeloleukemic U937 cells. *Biochemical Pharmacology,* **60** : 1685-91.

Park W.B., Lyu S.Y., Kim J.H., Choi S.H., Chung H.K., Ahn S.H., Hong S.Y., Yoon T.J. and Choi M.J. (2001). Inhibition of tumour growth and metastasis by Korea mistle- toe lectin is associated with apoptosis and antiangiogenesis. *Cancer Biotherapy Radiopharmacology,* **16** : 439-47.

Pfüller, U. (2000). Chemical constituents of European mistletoe (*Viscum album* L.). Isolation and characterization of the main relevant ingredients: lectins, viscotoxins, oligo-/polysaccharides, flavonoids. *In*: Mistletoe. The Genus Viscum, *Ed.* by Büssing, A., Harwood Academic Publishers, Amsterdam, pp. 101-22.

Pryme, I.F., Bardocz, S. and Pusztai, A. (1994). A diet containing the lectin phyto-haemagglutinin (PHA) slows down the proliferation of Krebs II cell tumours in mice. *Cancer Letters,* **76** : 133-7.

Pryme, I.F., Pusztai, A., Grant, G. and Bardocz, S. (1996a). Phytohaemagglutinin induced gut hyperplasia and the growth of a mouse lymphosarcoma tumour. *Journal of Experimental and Therapeutic Oncology,* **1** : 171-6.

Pryme, I.F., Pusztai, A., Grant, G. and Bardocz, S. (1996b). Dietary phytohaemagglutinin slows down the proliferation of a mouse plasmacytoma (MPC-11) tumour in Balb/c mice. *Cancer Letters,* **103**: 151-55.

Pryme, I.F., Bardocz, S., Grant, G., Ewen, S.W.B., Pusztai, A. and Pfüller, U. (1998a) The plant lectins PHA and ML-I suppress the growth of a lymphosarcoma tumour in mice. *In*: COST 98, Effects of antinutrients on the nutritional value of legume diets, Vol. 5, *Ed.* by Bardocz, S., Pfüller, U. and Pusztai, A., Luxembourg, pp. 215-20.

Pryme, I.F., Pusztai, A., Grant, G. and Bardocz, S. (1998b). The induction of gut hyperplasia by phytohaemagglutinin in the diet and limitation of tumour growth. *Histology and Histopathology,* **13**: 575-83.

Pryme, I.F., Grant, G., Pusztai, A. and Bardocz, S. (1999). Limiting the availability of polyamines for a developing tumour : an alternative approach to reducing tumour growth. *In*: Polyamines in Health and Nutrition. *Ed.* by Bardocz, S. and White, A., Kluwer Academic Publishers Boston/Dordrecht/London, pp. 283-91.

Pryme, I.F., Bardocz, S., Pusztai, A. and Ewen, S.W.B. (2002). Dietary mistletoe lectin supplementation and reduced growth of a murine non-Hodgkin lymphoma. *Histology and Histopathology,* **17**: 261–71.

Pryme, I.F., Bardocz, S., Pusztai, A., Ewen, S.W.B. and Pfüller, U. (2004). A mistletoe lectin (ML-I)-containing diet reduces the viability of a murine non-Hodgkin lymphoma tumour. *Cancer Detection and Prevention*, **28** : 52–6.

Pryme, I.F., Bardocz, S., Pusztai, A. and Ewen, S.W.B. (2006). Suppression of growth of tumour cell lines *in vitro* and tumours *in vivo* by mistletoe lectins. *Histology and Histopathology*, **21**: 285–99.

Pryme, I.F., Dale, T.M. and Tilrem, J.P.P. (2007). Oral mistletoe lectins : A case for their use in cancer therapy. *Cancer Therapy*, **5**: 287-300.

Pusztai, A., Grant, G., Gelencsér, É., Ewen, S.W.B., Pfüller, U., Eifler, R. and Bardocz, S. (1998). Effects of an orally administered mistletoe (type-2 RIP) lectin on growth, body composition, small intestinal structure, and insulin levels in young rats. *Nutritional. Biochemistry*, **9** : 31-6.

Ribereau-Gayon, G., Jung, M.L., Frantz, M. and Anton, R. (1997). Modulation of cytotoxicity and enhancement of cytokine release induced by *Viscum album* L. extracts or mistletoe lectins. *Anticancer Drugs*, **8 Suppl 1**: S3-8.

Seifert, G., Jesse, P., Laengler, A., Reindl, T., Lüth, Lobitz, S., Henze, G., Prokop, A. and Lode, H.N. (2008). Molecular mechanisms of mistletoe plant extract-induced apoptosis in acute lymphoblastic leukemia *in vivo* and *in vitro*. *Cancers Letters*, **264** : 218-28.

Siegle, I., Fritz, P., McClellan, M., Gutzeit, S. and Mürdter, T.E. (2001). Combined cytotoxic action of *Viscum album* agglutinin-1 and anticancer agents against human A549 lung cancer cells. *Anticancer Research*, **21**(4A) : 2687-91.

Sharma, R., van Damme, E.J.M., Peumanns, W.J., Sarsfield, P. and Schumacher, U. (1996). Lectin binding reveals divergent carbohydrate expression in human and mouse Peyer's patches. *Histochemistry and Cell Biology*, **105** : 459–65.

Stein, G.M., Stettin, A., Schultze, J. and Berg, P.A. (1997). Induction of anti-mistletoe lectin antibodies in relation to different mistletoe-extracts. *Anticancer Drugs*, **8 (Suppl.1)**: 57–89.

Stein, G., Henn, W., von Laue, H. and Berg, P. (1998). Modulation of the cellular and humoral immune responses of tumor patients by mistletoe therapy. *European Journal of Medical Research*, **3** : 194-202.

Stein, G.M., Pfüller, U., Schietzel, M. and Büssing, A. (2000a). Toxic proteins from European mistletoe (*Viscum album* L.): Increase of intracellular IL-4 but decrease of IFN-gamma in apoptotic cells. *Anticancer Research*, **20** : 1673-78.

Stein, G.M., Pfüller, U., Schietzel., M. and Büssing, A. (2000b). Intracellular expression of IL-4 and inhibition of IFN-gamma by extracts from European mistletoe is related to induction of apoptosis. *Anticancer Research*, **20** : 2987-94.

Stettin, A., Schultze, J.L., Stechemesser, E. and Berg, P.A. (1990). Anti-mistletoe lectin antibodies are produced in patients during therapy with an aqueous mistletoe extract derived from *Viscum album* L. and neutralize lectin-induced cytotoxicity in nitro. *Klinische Wochenschrift*, 60. 096–900.

Thies, A., Nugel, D., Pfüller, U., Moll, I. and Schumacher, U. (2005). Influence of mistletoe lectins and cytokines induced by them on cell proliferation of human melanoma cells *in vitro*. *Toxicology,* **207**: 105-16.

Thies, A., Dautel, P., Meyer, A., Pfüller, U. and Schumacher, U. (2008). Low-dose mistletoe lectin-1 reduces melanoma growth and spread in a scid mouse xenograft model. *British Journal of Cancer*, **98** : 106-12.

Valentiner, U., Pfüller, U., Baum, C. and Schumacher, U. (2002). The cytotoxic effect of mistletoe lectins I, II and III on sensitive and multidrug resistant human colon cancer cell lines *in vitro*. *Toxicology*, **171**: 187-99

Vervecken, W., Kleff, S., Pfüller, U. and Büssing, A. (2000). Induction of apoptosis by mistletoe lectin 1 and its subunits. No evidence for cytotoxic effects caused by isolated A- and B-chains. *International Journal of Biochemistry and Cell Biology*, **32**: 317-26.

Winge, I., Dale. T.M., Tilrem, P. and Pryme, I.F. (2010). A mistletoe lectin-containing preparation for oral use provokes an immune response and induces an increase in the population of activated natural killer cells. *In* : Comprehensive Bioactive Natural Products, Vol. **5**, Immune-modulation and Vaccine Adjuvants, *Ed.* by Gupta, V.K. Studium Press LLC, USA, pp. 283-99.

Yoon, T.J., Yoo, Y.C., Choi, O.B., Do, M.S., Kang, T.B., Lee, K.H., Azuma, I. and Kim, J.B. (1995). Inhibitory effect of Korean mistletoe (*Viscum album* coloratum) extract on tumor angiogenesis and of hematogenous and non-hematogenous tumor cells in mice. *Cancer Letters*, **97**: 83-91.

Yoon, T.J., Yoo, Y.C., Kang, T.B., Baek, Y.J., Huh, C.S., Song, S.K., Lee, K.H., Azuma, I. and Kim, J.B. (1998). Prophylactic effect of Korean mistletoe (*Viscum album* coloratum) extract on tumour metastasis is mediated by enhancement of NK cell activity. *International Journal of Immunopharmacology*, **20** : 163-72.

Yoon, T.J., Yoo, Y.C., Kang, T.B., Shimazaki, K., Song, S.K., Lee, K.H., Kim, S.H., Park, C.H., Azuma, I. and Kim, J.B. (1999). Lectins isolated from Korean mistletoe (*Viscum album* coloratum) induce apoptosis in tumour cells. *Cancer Letters*, **136**: 33-40.

Zarkovic, N., Kalisnik, T., Loncaric, I., Borovic, S., Mang, S., Kissel, D., Konitzer, M., Jurin, M. and Grainza, S. (1998). Comparison of the effects of *Viscum album* lectin ML-1 and fresh plant extract (Isorel) on the cell growth *in vitro* and tumorigenicity of melanoma B16F10. *Cancer Biotherapy and Radio-pharmaceuticals*, **13** : 121-31.

Natural Products: Research Reviews Vol. 1 (2012)
Editor: V.K. Gupta
Published by: DAYA PUBLISHING HOUSE, NEW DELHI

Pages **99–125**

4

A Review on Medicinal Properties of Onions and Garlic

Dheeraj Singh[1]*, M.K. Chaudhary[1], H. Dayal[1],
M.L. Meena[1] and A. Dudi[1]

ABSTRACT

Garlic and onion are both found in the Allium family and contain organosulphur compounds with antioxidant, anti-inflammatory, and antimicrobial properties. Garlic (Allium sativum) and onion (Allium cepa) are two very important food ingredients widely used in our gastronomy. Moreover, garlic and onion extracts have been recently reported to be effective in cardiovascular disease, because of their hypocholesterolemic, hypolipidemic, anti-hypertensive, anti-diabetic, antithrombotic and anti-hyperhomocysteinemia effects, and to possess many other biological activities including antimicrobial, antioxidant, anticarcinogenic, antimutagenic, antiasthmatic, immunomodulatory and prebiotic activities. Garlic, has many purported benefits and a long medicinal history dating back to Aristotle, Hippocrates, and Aristophane. Garlic typically contains three times greater levels of organosulphur compounds than onion. After chopping or crushing, the enzyme allinase converts alliin (a cysteine-sulphoxide) in garlic to allicin (a thiosulphate). The latter compound is thought to confer many of garlic's medicinal effects, but garlic has also been shown to be metabolized to a number of additional organosulphur compounds. The potential medical benefits of these include possible roles in lowering the risk for atherosclerosis, cardiovascular disease, cancer and diabetes. While onion have been used for their presumed health benefits for thousands of years that also include purported roles in preventing earaches, hair loss, and treating warts, new medical research is now uncovering the underlying physiological and molecular mechanisms of their action as well as providing some scientific evidence of their effectiveness.

1   Krishi Vigyan Kendra, Central Arid Zone Research Institute, Pali, Rajasthan, India

*   Corresponding author. E-mail. dheerajthakurala@yahoo.com

Onion is a useful herb for the prevention of cardiovascular disease, especially since they diminish the risk of blood clots. Onion also protects against stomach and other cancers, as well as protecting against certain infections. Onion can improve lung function, especially in asthmatics.While both alliums have specific beneficial effects, e.g. in aiding digestion by increasing the production of saliva and gastric juices, this paper will focus on more general and fundamental features, i.e. the antioxidant, anti-inflammatory, and antimicrobial properties of the two spices.

Keywords: Garlic, Onion, Organosulphur compounds, Medicinal, Antioxidant, Anti-Inflammatory, Antimicrobial properties.

Introduction

Onions and allied alliums are among the most produced vegetables worldwide, just after watermelon and tomato (FAO Stat, 2006). Produced worldwide, they are consumed by most of cultures as a staple food. Garlic (*Allium sativum*) and onion (*Allium cepa*) are two food ingredients widely used for medicinal purposes as well as for added flavor and aroma in many food dishes worldwide. Traditional wisdom and scientific literature (over 3000 publications) have confirmed the health benefits of onion and garlic (Amagase *et al.*, 2001). These benefits include reduction of risk factors for cardiovascular diseases (Ali *et al.*, 2000; Milner, 2001; Bazzano *et al.*, 2002), reduction in cancer incidence (Fleischauer and Arab, 2001), reduction of inflammatory response (Srivastava, 1986; Kim *et al.*, 2001), enhanced xenobiotic detoxification (Munday *et al.*, 2003), antidiabetic (Srinivasan, 2005), antioxidant (Prasad *et al.*, 1995), antibiotic (Sivam, 2001) and antifungal properties (Lancaster and Kelly, 1983; Rose *et al.*, 2005), etc. Garlic, *Allium sativum*, is a widely studied plant with immence benefits and a long medicinal history dating back to Aristotle, Hippocrates, and Aristophane (Ali *et al.*, 2000). Garlic, besides to be used like food, has been used as medicinal plant for over 4000 years for a variety of ailments including headache, bites, intestinal worms and tumors (Block, 1985). In India, garlic has been used for centuries as an antiseptic lotion for washing wounds and ulcers; for fever, headache, cholera and dysentery; garlic is still being employed in folk medicine all over the world for the treatment of a variety of diseases (Ali *et al.*, 2000). Onions are one of the most ancient cultivated plants and were well known in pharaonic Egypt, 7000 years ago (Brewster, 1994).

Sulphur Containing Compounds

Garlic and onion are both found in the Allium family and contain organosulphur compounds with antioxidant, anti-inflammatory, and antimicrobial properties. Evidence from several investigations suggests that the biological and medical functions of garlic and onions are mainly due to their high organo-sulphur compounds content (Augusti and Mathew, 1974). Sulphur containing compounds are integral part of *Allium* metabolism and they provide the characteristic flavour and odour of onion. Garlic typically contains three times greater levels of organosulphur compounds than onion (Benkeblia, 2004). After chopping or crushing, the enzyme allinase converts alliin (a cysteine-sulphoxide) in garlic to allicin (a thiosulphate) (Benkeblia, 2004; Banerjee *et al.*, 2003). The latter compound is thought to confer many of garlic's medicinal effects, but garlic has also been shown to be metabolized to a number of additional organosulphur compounds (Khanum *et al.*, 2004). Cysteine sulphoxides in some *Alliums* represent close to 1 per cent of their fresh weight (Kubec *et al.*, 2000). According to Lancaster and Kelly (1983), nonprotein cysteine and glutathione and their derivatives account for almost 5 per cent of the plant's dry weight. As for glucosinolates in *Brassicas*, sulphur compounds in *Alliums* are believed to participate in defense protection against pathogens and herbivores (Brewster, 1994). They play a crucial role in many central metabolisms like sulphur assimilation by plants, redox homeostasis and xenobiotic detoxification (Noctor *et al.*, 1998). The biosynthesis of sulphur compounds in *Alliums* is complex due to the large variety of chemicals involved. Thorough reviews on the biosynthesis of these compounds have recently been published (Jones *et al.*, 2004; Rose *et al.*, 2005. Four major nonvolatile cysteine sulphoxides are the precursors of the volatile compounds found in *Allium*. The first and most ubiquitous is S-allyl-cysteine sulphoxide (ACSO) (Alliin) found in garlic; S-methyl cysteine sulphoxide (MCSO) (Methiin) found in *Alliums* and some *Brassicaceae*; S-trans-prop-1-enyl cysteine sulphoxide (PeCSO) (Isoalliin) found in onions; and S-propyl cysteine sulphoxide (PCSO) (Propiin) also found in onions. The primary sulphur- containing constituents in both whole vegetables are the S-alk(en)yl-L-cysteine sulphoxides (ACSOs), such as alliin, and γ-glutamylcysteines, which, besides to serve as important storage peptides, are biosynthetic intermediates for corresponding ACSOs from which, and by different metabolic pathways in each vegetable, volatile, such as allicin, and lipid-soluble sulphur compounds, such as diallyl sulphide (DAS), diallyl disulphide (DADS) and others, are originated (Lancaster and Shaw, 1989). These compounds provide to garlic and onion their characteristic odour and flavour, as well as most of their biological properties (Lanzotti, 2006). Flavonoids, abundant in onion but practically absent in garlic, and a small amount of non-volatile water-soluble sulphur compounds found in garlic, as S-allyl cysteine (SAC), (coming from enzymatic transformation of g-glutamylcysteines when garlic is extracted with an aqueous solution) (Amagase *et al.*, 2001), are also responsible for a great part of the health benefits of both vegetables. The onion, *Allium cepa*, is another food with medicinal properties as well as uses for flavor and aroma. A major active ingredient of onion is S-propenylcysteine sulphoxide (Ali *et al.*, 2000). Onion furthermore contains cepaenes that are best known for their inhibition of pro-inflammatory messengers (Ali *et al.*, 2000). Onions possess antioxidant and antibacterial properties, but their antioxidant activity is less than that of garlic (Shobana and Naidu, 2000). The antioxidant activity of onion is reduced after cooking, and onion is thus most effective in its raw form (Ali *et al.*, 2000). Interestingly, different types of onions were found to vary in their properties, with highest total antioxidant activities as well as greatest *in vitro* tumor cell inhibition seen in shallots and the onion variety Western Yellow (Yang *et al.*, 2004).

The biological effects of additional constituents of intact garlic and onion, such as lectins (the most abundant proteins in garlic and onion), prostaglandins, fructan, pectin, adenosine, vitamins B1, B2, B6, C and E, biotin, nicotinic acid, fatty acids, glycolipids, phospholipids and essential amino acids, have been studied for over several decades (Fenwick and Hanley, 1985) and the importance of biological and pharmacological activities, such as antifungal, antibacterial, antitumor, anti-inflammatory, antithrombotic and hypocholesterolemic properties of certain steroid saponins and sapogenins, such as β-chlorogenin, has been recently demonstrated (Lanzotti, 2006). Other characteristic chemical constituents of garlic include allixin and organo-selenium compounds. These chemical compounds are reported to exhibit several biological effects, including cholesterol reduction, cancer prevention and others, and probably work synergistically with organo-sulphur compounds (Amagase, 2006).

Onion is usually consumed as fresh, in powder or as essential oil and its commercial products is less abundant than those of garlic. Like garlic, organo-sulphur compounds present in onion preparations depend on the variety (Yang *et al.*, 2004) and the extraction and/or processing conditions. Given the importance of these vegetables as much in feeding as in therapeutic, in the present work, the main biological activities of garlic and onions have been reviewed, indicating the responsible compounds for each one of them.

Regulation of Metabolism by Oxidation/Antioxidation

Spices contain the classic antioxidant vitamins ascorbic acid (vitamin C) and tocopherols (vitamin E group) but also other, very potent antioxidants, such as phenols, thiols (as sulphur compounds), and carotenoids (Sharma, 2005; Yang *et al.*, 2004). As antioxidants, all of these are compounds able to slow down, stop, or reverse oxidation processes by scavenging oxidizing agents, such as reactive oxygen species (ROS), and recycling oxidized lipids, proteins, and nucleic acids. Oxidation of lipids can cause specific, direct effects, such as destabilization of (lipid) membranes resulting, *e.g.* in decreased survival of red blood cells (Yang *et al.*, 2004; Kempaiah and Srinivasan, 2004). Allicin has been shown to act as an antioxidant by scavenging ROS and preventing lipid oxidation and production of proinflammatory messengers (Banerjee *et al.*, 2003), and similar results were obtained for garlic and onion extracts (Shobana and Naidu, 2000).

A key mechanism for the multiple effects of ROS is the activation of redox-regulated gene regulatory proteins (Lavrovsky *et al.*, 2000) that turn on genes for proinflammatory enzymes such as cyclooxygenase (COX) and lipoxygenase (LOX). Redoxregulated genes are controlled by reduction (via antioxidants) and oxidation (via ROS) of components of the signal transduction pathways that control their expression. Expression of COX is upregulated by a surplus of ROS and downregulated by antioxidants (such as those present in garlic and onion). How much of these pro-inflammatory enzymes (COX and LOX) are synthesized is regulated by gene regulatory factors (transcription factors). One of these is nuclear factor kappa B or NFkB, a master control gene of the immune/inflammatory response (Janssen-Heininger *et al.*, 2000). Under normal conditions, NFkB remains inactivated by another factor, its inhibitor IkB. When NFkB is stimulated, more COX/LOX is synthesized and inflammation is triggered. This transcription factor is, in turn, strongly regulated by dietary factors; it is activated under insufficient levels of antioxidants, particularly sulphur-containing ones (Janssen-Heininger *et al.*, 2000). In a study by Kempaiah and Srinivasan (2004), rats were given a high-fat diet with or without garlic, and blood levels of triglycerides (lipids with three fatty acids known to increase atherosclerosis risk) and thiols such as glutathione (amino acids or peptides with sulphur groups that recycle, or re-reduce, oxidized proteins, scavenge ROS, and have a potent effect on redox-regulated signaling pathways, such as that involving NFkB) were assessed. Food intake per se was not affected by garlic in this study. The high-fat diet increased the levels of blood triglycerides, decreased the levels of thiols such as glutathione, and increased lipid oxidation. Kempaiah and Srinivasan (2004) found that all of these adverse effects of the high-fat diet were effectively reduced by regular addition of garlic to the diet, thus presumably reducing the risk of atherosclerosis. When garlic was added to the high-fat diet, total endogenous thiols increased by 16 per cent, glutathione increased by 28 per cent, and the level of an endogenous antioxidant enzyme, catalase, which is depleted under oxidative stress, also increased (Kempaiah and Srinivasan, 2004). The sulphur compounds in garlic are thus able to protect the endogenous thiol pool (by re-reducing thiols that become oxidized). Other studies (Ali *et al.*, 2000; Ashraf *et al.*, 2005; Srinivasan, 2005b; Rahman and Lowe, 2006) support the effect of garlic in improving cardiovascular health, *e.g.*, via decreases in platelet aggregation, a lowering blood pressure and cholesterol levels, and inhibition of several steps in the inflammation process as described in the present review.

Antiparasitic Activity

Due to the great antimicrobial activity that garlic and onion possess, both vegetables could be used as natural preservatives, to control the microbial growth (Pszczola, 2002). In folk medicine, garlic and onion have been used for centuries in several societies against parasitic infections. Recent chemical characterization of their sulphur compounds has allowed stating that they are the main active antimicrobial agents (Rose *et al.*, 2005). However, some proteins, saponins and phenolic compounds can also contribute to this activity (Griffiths *et al.*, 2002). Regarding the activity that garlic and onion and their constituents exert on parasitic protozoa, some of them have demonstrated that garlic extracts are effective against *Opalina ranarum, Opalina dimidicita, Balantidium entozoon, Entamoeba histolytica, Tripanosoma brucei, Leishmania, Leptomonas* and *Crithidia* (Reuter *et al.*, 1996). Due to the occurrence of unpleasant side effects and increasing resistance to the synthetic pharmaceuticals recommended for the treatment of giardiasis, there has been an increasing interest to explore natural alternatives. Results of a clinical study (Lun *et al.*, 1994) demonstrated that garlic is effective against *Giardia lamblia* and *Giardia intestinalis*. In China, DATS, an allicin breakdown product, easily synthesised and more stable than the extremely volatile allicin, is commercially available as a preparation, called Dasuansu, prescribed for the treatment of giardiasis and infections by *E. histolytica* and *Trichomonas vaginalis* (Lun *et al.*, 1994). Allicin, ajoene and other organo-sulphur compounds from garlic are also effective antiprotozoals. Antiparasitic properties of onion extracts towards different strains of *Leishmania* and *T. vaginalis* have been reported as well (Saleheen *et al.*, 2004).

Anti-inflammatory Effect

Chronic over-production of either COX or LOX (and also NFkB itself) causes excess inflammation and contributes to chronic pro-inflammatory diseases such as cardiovascular disease, diabetes, and others (Goodsell, 2005). The messengers produced by LOX can also either stimulate or prevent programmed cell death. Excessive cell death is involved in *e.g.* neurodegenerative disease, while insufficient cell death can lead to cancer (Hannun, 1997; Tatton and Olanow, 1999). In addition to limiting how much of these inflammatory enzymes is manufactured (see above), spices can also dampen the actual activity of existing the pool of inflammatory enzymes such as COX and LOX. Both COX and LOX convert oxidized lipids, such as arachidonic acid (AA), to pro-inflammatory, hormone-like messengers. COX produces prostaglandins that signal pain and trigger inflammation and LOX produces a related group of messengers, leukotrienes (Goodsell, 2005). Spices inhibit the activity of both COX and LOX (Goodsell, 2005). Onion, apparently via its thiosulphinate and cepaene content, inhibits the production of AA as well as its conversion to pro-inflammatory prostaglandins and leukotrienes (Ali *et al.*, 2000). More specifically, onion cepaenes were shown to inhibit COX and LOX activity as well as blood platelet aggregation (Ali *et al.*, 2000). The same study also showed that onion extract can decrease the onset and development of tumors as well as have antiasthmatic effects (the latter again via COX inhibition). Allicin inhibited the production of pro-inflammatory cytokine messengers in a study of inflammatory bowel disease, apparently by inactivating the pro-inflammatory factor NFkB via its IkB inhibitor (Lang *et al.*, 2004). By virtue of sulphur-based antioxidants found in garlic, NFkB was maintained in its inactive state, thus preventing synthesis of excess COX/LOX.

Antifungal Activity

Antifungal activity of garlic and related species was first established in 1936 by Schmidt and Marquardt (1936) studying epidermophyte cultures. Onions and garlic extracts have been shown to inhibit growth of more than 80 species of plant pathogenic fungi (Fenwick and Hanley, 1985). *In vitro* and *in vivo* studies have shown a great effectiveness of garlic and its derivatives against a broad spectrum of fungi and yeasts, including *Candida, Trichophyton, Torulopsis, Rhodotorula, Cryptococcus, Aspergillus* and *Trichosporon* (Davis and Perrie, 2003), as well as a synergistic activity with amphotericin B *in vitro*, one of the main antifungal drugs (Shen *et al.*, 1996). Onion extracts are also effective against many yeast species and their essential oil inhibits the dermatophytic fungi (Zohri *et al.*, 1995). The active compounds of garlic and onion destroy fungal cells decreasing the oxygen uptake, reducing cellular growth, inhibiting the synthesis of lipids, proteins and nucleic acids, changing the lipid profile of the cell membrane and inhibiting the synthesis of the fungal cell wall (Gupta and Porter, 2001; Tansey and Appleton, 1975). Like for the antibacterial activity, the main active antifungal agents from onion and garlic extracts are the breakdown products of allicin, including diallyl trisulphide (DATS), DADS, DAS and ajoene, which have a greater antifungal effect than allicin (Tansey and Appleton, 1975). An antifungal compound, fistulosin (octadecyl 3-hydroxyindole), has been isolated from welsh onion (*A. fistulosum*), which shows a high activity towards *Fusarium oxysporium* inhibiting primarily the protein synthesis (Phay *et al.*, 1999). In addition to sulphur compounds, a great variety of antifungal proteins and peptides have been isolated from several *Allium* species (Lam *et al.*, 2000; Wang and Ng, 2001) such as allicepin, a novel isolated antifungal peptide from onion bulbs (Wang and Ng, 2004). Finally, it is necessary to consider certain steroid saponins, such as eruboside-B, isolated from the garlic bulb that also exhibit antifungal activity for *Candida albicans* (Matsuura *et al.*, 1988).

Antibacterial Activity

In addition to being antioxidants and anti-inflammatory agents, alliums also have antibacterial/antimicrobial properties (Lai and Roy, 2004). The antibacterial properties of garlic can be eliminated by inhibition of the allinase enzyme and prevention of allicin formation (Jonkers *et al.*, 1999). The antibacterial effect garlic apparently results from interaction of sulphur compounds, like allicin, with sulphur (thiol) groups of microbial enzymes (such as trypsin and other proteases), leading to an inhibition of microbial growth (Jonkers *et al.*, 1999; Bakri and Douglas, 2005). Many bacterial strains, both gram-positive and gram-negative, can be inhibited with garlic, and some strains were inhibited much more strongly by allicin or garlic extract compared to antibiotics (Bakri and Douglas, 2005; Lai and Roy, 2004). The bacterial strain *Staphylococcus aureus* causes pus-producing infections, such as boils, as well as pneumonia and urinary tract infections (Todar, 2005). Cultures of this strain (as well as *Salmonella enteritidis*, the bacterium responsible for salmonella food poisoning, and several fungi) are effectively inhibited by garlic and onion oil or extracts (Benkeblia, 2004). Aqueous extracts of onions were shown to be active against several gram-negative bacteria (Zohri *et al.*, 1995). Moreover, onion extracts inhibit oral bacteria and may thus reduce the incidence of cavities (Kim, 1997). According to Kyung and Lee (2001), PeCSO is the compound involved in the inhibition of microbial metabolism. Garlic has been proven to inhibit the growth of gram-positive, gram-negative and acid-fast bacteria, as well as toxin production. Bacteria against which garlic is effective include strains of *Pseudomonas, Proteus, Escherichia coli, Staphylococcus aureus, Klebsiella, Salmonella, Micrococcus, Bacillus subtilis, Mycobacterium*, and *Clostridium* (Delaha and Garagusi, 1985), some of which are resistant to penicillin, streptomycin, doxycilline and cephalexin, among other antibiotics. Other microbes inhibited by garlic include *Bacillus subtilis*, a gram-positive bacterium found in soil, *Escherichia coli*, a common toxin-producing, food-borne bacterium, and *Saccharomyces cerevisiae*, a yeast species (Lai and Roy, 2004). Garlic also inhibits beneficial intestinal microflora, but it is more effective against potentially harmful enterobacteria, probably due to a greater sensitivity of enterobacteria to allicin (Miron *et al.*, 2000). Remarkably, mouthwash containing garlic significantly reduced total salivary bacteria, including *Porphyromonas gingivalis*, the bacterium causing gingivitis (Bakri and Douglas, 2005). Recently, it has been reported that onion and garlic extracts exert bactericidal effects towards *Streptococcus mutans, Streptococcus sobrinus, Porphyromonas gingivalis* and *Prevotella intermedia* (gram-positive bacteria), considered as the main bacteria responsible of dental caries and adult periodontitis, respectively (Bakri and Douglas, 2005). Onions possess antibacterial properties as well. Although less research is available on the antibacterial activity of onion, it is suggested that S-propenylcysteine sulphoxide is the compound that inhibits antibacterial metabolism by the same mechanism as garlic (Kyung and Lee, 2001). Onion extract, the activity of which remained stable for 48 h, inhibited *Streptococcus mutans*, a bacterium that causes strep throat, tonsillitis, bacterial pneumonia, as well as other diseases (Ali *et al.*, 2000).

In both vegetables, the major active antibacterial components *in vivo* are the allicin-derived organo-sulphur compounds, such as DAS, DADS and ajoene, as well as other thiosulfinates isolated from oil-macerated garlic (Tsao and Yin, 2001). Epidemiological studies have demonstrated that DAS and DADS from garlic can protect against the *Helicobacter pylori* infection and, therefore, to reduce the risk of gastric neoplasia, since *H. pylori* is deeply involved in stomach cancer development (You *et al.*, 1998). In addition to organo-sulphur compounds, it has been recently reported that certain quercetin oxidation products found in onion also present antibacterial activity against *H. pylori* and MRSA (multidrug-resistant *S. aureus*) (Ramos *et al.*, 2006).

Antiviral Activity

The antiviral activities of various commercial garlic products, including garlic powder tablets and capsules, oil-macerated garlic, steam-distilled garlic oils, garlic aged in aqueous alcohol and fermented garlic oil, against herpes simplex virus Types 1 and 2, influenza A and B viruses (Fenwick and Hanley, 1985), human cytomegalovirus (Meng *et al.*, 1993), vesicular stomatitis virus, rhinovirus, human immunodeficiency virus (HIV), viral pneumonia and rotavirus, have been studied. Antiviral activities of these commercial products seem to be dependent on their preparation process and those products with the highest levels of allicin and other thiosulfinates, mainly DADS, DATS and ajoene, have the best antiviral activities (Weber *et al.*, 1992). In addition to sulphur compounds, it has been reported that quercetin, the major onion flavonoid, also possesses antiviral activity and enhances the bioavailability of some antiviral drugs (Wu *et al.*, 2005). Lectins are a very heterogeneous group of glycoproteins with the ability to recognize and bind specifically to carbohydrate ligands. Onion lectins, unlike the garlic lectins, have a pronounced anti-HIV activity (Van Damme *et al.*, 1993).

Antioxidant Activity

Oxidation of DNA, proteins and lipids by reactive oxygen species (ROS) plays an important role in aging and in a wide range of common diseases, including cancer and cardiovascular, inflammatory and neurodegenerative diseases, such as Alzheimer's disease and other age-related degenerative conditions (Borek, 1997; Richardson, 1993). plant-based diets, in particular those rich in vegetables and fruits, provide a great amount of antioxidant phytochemicals, such as vitamins C and E, glutathione, phenolic compounds (flavonoids) and vegetable pigments, which offer protection against cellular damage (Dimitrios, 2006). Onion is one of the major sources of dietary flavonoids in many countries, which are present either as sugar conjugates or as aglycones. The major flavonoid found in onion is quercetin, present in conjugated form, as quercetin 40-O-b-glycopyranoside, quercetin 3,40-O-b-diglycopyranoside, and quercetin 3,7,40-O-b-triglycopyranoside (Sellappan and Akoh, 2002). The dry outer layers of onion, which are wasted before food processing such as cooking, contain large amounts of quercetin, quercetin glycoside and their oxidative products (Gu'lsen *et al.*, 2007), which are effective antioxidants against non enzymatic lipid peroxidation and oxidation of low density lipoproteins (LDL). Quercetin and its dimerized compound show the highest antioxidative activity, which is comparable to that of a-tocopherol. Therefore, the outer layer extract of onion is expected to be a resource for food ingredients (Ly *et al.*, 2005).

Among garlic-derived products, AGE is the preparation with higher antioxidant activity, even more than fresh garlic and other commercial garlic supplements. This is due to its own extraction procedure, which increases stable and highly bioavailable water-soluble organo-sulphur compounds content, such as SAC and S-allylmercaptocysteine (SAMC), with potent antioxidant activity (Imai *et al.*, 1994). SAC and SAMC are the major organo-sulphur compounds found in AGE, nevertheless, this garlic preparation has other compounds with antioxidant effect. Another recently identified antioxidant compounds of AGE are N-fructosyl glutamate, N-fructosyl arginine (Ryu *et al.*, 2001) (whose antioxidant activity is comparable to that of ascorbic acid) and N-fructosyl lysine (Moreno *et al.*, 2006). Phytochemicals in AGE may act in synergistic or additive way and exert their antioxidant action by scavenging ROS (Borek, 2001), by enhancing the cellular antioxidant enzymes superoxide dismutase (SOD), catalase and glutathione peroxidase, and by increasing glutathione in the cells (Liu *et al.*, 1992), important defence mechanism in living cells (Borek, 1997). Particularly, due to its antioxidant action, AGE decreases the risk of cardiovascular and cerebrovascular disease inhibiting the lipid peroxidation and oxidation of LDL (Lau, 2006). Moreover, AGE has radioprotective effects (Lau, 1989), protecting against ionising radiation and UV light-induced damage. In addition, it protects the erythrocytes membrane against oxidative stress inhibiting the formation of abnormally dense erythrocytes, which are believed to play an important role in the clinical manifestations of sickle cell anaemia patients (Ballas and Smith, 1992) and protects against cardiotoxicity and liver toxicity induced by several oxidant environmental, chemical and medicinal substances (Wang *et al.*, 1998). Finally, AGE has also anti-aging effects, since recent studies have demonstrated that it promotes neuronal cells survival, increasing cognitive functions, memory and longevity and slowing down age-related impairment of learning behaviour and memory. Due to this neurotrophic activity attributed to AGE, the garlic potential as natural alternative for the treatment of neurodegenerative diseases, such as Alzheimer's disease or dementia, is being studied (Chauhan, 2006).

Anticarcinogenic and Antimutagenic Activities

Many prospective and epidemiological studies have shown that the regular consumption of *Alliums* could have protective effects against cancer (Lampe, 1999). For instance, there appears to be a strong link between the consumption of onions and the reduced incidence of stomach and intestine cancers (You *et al.*, 2005). Epidemiological studies also show correlation between the consumption of onions and a reduced incidence of cancers (Griffiths *et al.*, 2002). A synthesis of case-control studies carried in Italy and Switzerland reveals that consumption of one to seven portions of onions per week reduces the risks of colon, ovary, larynx and mouth cancers (Galeone *et al.*, 2006). Similar correlations are also observed for brain and stomach cancers in a case-control study in China (Hu *et al.*, 1999). Dutch researchers have shown an inverse relationship between onion consumption and the incidence

of stomach cancers (Dorant *et al.*, 1996). Mortality due to prostate cancer also appears to be reduced by a diet making a large place for onions (Grant, 2004). The risk of breast cancer was shown to decrease as consumption of *Alliums* was increased in a French case-control study (Challier *et al.*, 1998). The Epic Prospective Study, conducted on more than half a million subjects, shows clear correlation between onion consumption and reduction in intestinal and stomach cancers (Gonzalez and Riboli, 2006). Studies have reported that garlic and onion intake diminishes the risk of sarcoma and carcinoma in various tissues and organs, such as stomach, colon, oesophagus, prostate, bladder, liver, lungs, mammas, skin and brain (Hu *et al.*, 1999; Lau *et al.*, Tosk, 1990; Le Marchand *et al.*, 2000; You *et al.*, 1998). It is thought that more important direct anticarcinogen action of garlic is the potentiation of the immune system (Lamm and Riggs, 2001). Several investigations have shown that both water- and lipid-soluble sulphur compounds from garlic and onion provide their anticarcinogen benefits. DAS, diallyl sulphoxide (DASO), diallyl sulfone (DASO2), DADS, DATS, and SAC (Sigounas *et al.*, 1997) (from garlic), and dipropyl sulphide (DPS) and dipropyl disulphide (DPDS) (from onions) (Guyonnet *et al*, 1999), can inhibit both early and late stages of carcinogenesis. Other sulphur compounds, as SAMC (Sigounas *et al.*, 1997), ajoene (Dirsch *et al.*, 1998) and methiin (more abundant in onion than in garlic), along with DADS and DATS (Sakamoto *et al.*, 1997), can inhibit the cellular proliferation by inducing apoptosis in human cell cultures, for example, in human leukaemic cells. In addition to organo-sulphur compounds, eruboside-B, a steroid saponin isolated from garlic bulb, and organo-Se compounds are largely responsible for the anticarcinogenic activity of garlic and onion. Se-enriched garlic and onion have higher anticarcinogenic activity than the common plants (El-Bayoumy *et al.*, 2006; Matsuura, 1997). Onions probably act at different stages of the aetiology of cancers (Sengupta *et al.*, 2004). Some studies say that onion extracts can inhibit the mutation process (Shon *et al.*, 2004) and reduce the proliferation of cancer cells (Yang *et al.*, 2004). This effect is being attributed to quercetin in particular. Quercetin and kaempferol, from onion, also possess anticarcinogenic properties. Particularly, they have antineoplastic effects by inhibiting bioactivating enzymes (Lautraite *et al.*, 2002), by inducing detoxifying enzymes, by inducing apoptosis (Brisdelli *et al.*, 2007), and due to their antioxidant and anti-inflammatory activities (Raso *et al.*, 2001). Several epidemiological studies have found inverse associations between lung cancer risk and onion intake, probably due to its high content of flavonoids (Le Marchand *et al.*, 2000). Moreover, recent studies have reported that quercetin enhances bioavailability of some anticancer drugs, as Tamoxifen, a non-steroidal antiestrogen for treating and preventing breast cancer, by promoting their intestinal absorption and reducing their metabolism (Shin *et al.*, 2006; Wu *et al.*, 2005).

Effects on Lipid Metabolism

Diseases related to arteriosclerosis, such as ischaemic heart disease and stroke, are associated with elevated serum lipids. Many studies using garlic essential oil and raw garlic have reported that garlic consumption decreases significantly the content of total serum cholesterol (Chang and Johnson, 1980), LDL and very low density lipoproteins (VLDL) and also increases significantly the level of high density lipoproteins (HDL) in rats and rabbits. In another study with cholesterol-fed rabbits, it was shown that AGE reduces vessel wall cholesterol accumulation and arteriosclerotic plaques development in arterial wall (Effendy et al., 1997). Studies on onion are less advanced. Some investigations have demonstrated that onion also has compounds with capacity to reduce blood triglycerides levels and to inhibit rat hepatic cholesterol biosynthesis in vitro (Effendy et al., 1997).

Several clinical reports and meta-analyses have revealed the cholesterol-lowering effects of raw garlic and some garlic supplements, such as garlic essential oil and AGE (Lau et al., 1987; Neil et al., 1996). Allicin and its derivative compounds are the main active substances responsible for the hypolipidemic and hypocholesterolemic effects of onion and garlic, as much in humans as in experimentation animals (Liu and Yeh, 2002; Yeh et al., 1997). Some allicin-derived compounds in garlic that have demonstrated to possess a beneficial effect on cardiovascular variables are ajoene, methyl ajoene, DAS, DATS, 2-vinyl-4H-1,3-dithiin and SAC. Methiin and flavonoid quercetin (Glasser et al., 2002), both more abundant components in onion than in garlic, have also shown to have the ability to reduce serum cholesterol levels and arteriosclerosis severity. Moreover, other non-sulphur components of garlic, such as steroid saponins, have also demonstrated to be able to reduce serum cholesterol concentrations (Koch, 1993).

All these compounds may exert their hypocholesterolemic effect by inhibiting hepatic cholesterol biosynthesis (Gupta and Porter, 2001; Singh and Porter, 2006), enhancing cholesterol turnover to bile acids and its excretion through gastrointestinal tract (Srinivasan and Sambaiah, 1991), or, in the case of plant saponins, by inhibiting cholesterol absorption from intestinal lumen without changing HDL cholesterol levels in hypercholesterolemic animal models (Slowing et al., 2001).

Anti-hypertensive Effect

Anti-hypertensive effect of garlic has been determined in multiple studies with hypertensive rats using AGE, aqueous garlic extracts and garlic powder (Al-Qattan et al., 1999; Harauma and Moriguchi, 2006). In contrast, other investigations carried out with ethanolic extracts of onion and garlic in hypertensive rats reported that oral administration of extracts during a normal salt diet or during a high salt diet do not influence blood pressure (Kiviranta et al., 1989). Studies in humans are few and contradictory; in a work involving the administration of the garlic essential oil to 70 hypertensive patients, it was observed that 47 per cent of the patients considerably improved, 20 per cent hardly improved and 33 per cent did not improve. These results suggest that garlic essential oil may be effective in reducing blood pressure in hypertensive patients (Mansell and Reckless, 1991). Several investigations have allowed the determination of the mechanism by which garlic exerts its anti-hypertensive action. Some studies of garlic effect on muscular contraction in vitro have concluded that its hypotensive action may be, at least partly, due to a direct relaxant effect on smooth muscles (Aqel et al., 1991). On the other hand, other studies have suggested that garlic may also exert an indirect vasodilator effect, inducing the nitric oxide and hydrogen sulphide synthesis, both potent vasodilators. The latter is synthesised from sulfhydryl- containing amino acids, which are present in large amounts in garlic extracts, such as cysteine and the S-alk(en)yl derivatives as SAC, SEC (S-ethylcysteine) and SPC (S-propylcysteine) (Liu and Yeh, 2002). Likewise, a recent study with several rat models of hypertension has indicated that quercetin and its methylated metabolite isorhamnetin, found in onion, can reduce blood pressure and prevent angiotensin II-induced endothelial dysfunction by inhibiting the overexpression of p47 (phox), a regulatory subunit of the membrane NADPH oxidase, and the subsequent increased superoxide production, resulting in a highest nitric oxide bioavailability (Sanchez *et al.*, 2007). A novel drug assayed in hypertensive rats has been recently synthesised through the reaction of the pharmaceutical drug Captopril with allicin. The reaction product, called allylmercaptocaptopril (CPSSA), provides better protection against hypertension, since it has the Captopril ability to inhibit the angiotensin-converting enzyme (ACE) and the allicin ability to reduce serum cholesterol and triglycerides levels (Miron *et al.*, 2004).

Anti-hyperglycaemic or Anti-diabetic Potential

The onion and garlic effectiveness as hypoglycaemic agents has been scarcely investigated. Babu and Srinivasan (1997) observed that dietary onion intake for 8 weeks produced significant hypolipidemic effect besides hypoglycaemic influence in diabetic rats. Recently, it has been reported that long-term absorption of natural flavonoids as quercetin could be useful to prevent advanced glycation of collagens, which contributes to development of cardiovascular complications in diabetic patients (Urios *et al.*, 2007). Owing to the presence of prebiotic polysaccharides (inulin), which are poorly degraded by the gut enzymes, and the presence of flavonoids, onions have been shown to possess antidiabetic potential (Srinivasan, 2005). In a recent study on the use of natural remedies for Type II diabetes mellitus treatment in a diabetic women group from United States, garlic appeared among the most used vegetables (Johnson *et al.*, 2006). The bioactive constituents from onion and garlic, such as methiin and S-allyl cysteine sulphoxide (SACS), exert their anti-diabetic action by stimulating the insulin production and secretion by pancreas, interfering with dietary glucose absorption, and favouring the insulin saving (Srinivasan, 2005). Sharma *et al.* (1977) showed that onions had antihyperglycemic effects. Such effects were confirmed by Tjokroprawiro *et al.* (1983) who conducted a crossover comparative study with twenty diabetic patients to assess the effect of a diet comprising onions and green beans on serum glucose levels. They showed that the consumption of 20 g fresh onion three times daily significantly reduced blood sugar levels.

Antiasthmatic Effect

Compounds found in onions (thiosulfinates) appear to have antiasthmatic properties. This anti-inflammation activity is mediated through a suppression of cyclooxygenase reaction cascades, initiating once again the eicosanoid metabolism, leading to bronchial restriction (Wagner *et al.*, 1990). Onions contain cepaenes that are best known for their inhibition of proinflammatory messengers like arachidonic acid.

Antiplatelet or Antithrombotic Effect

The major function of blood platelets is to maintain the haemostatic integrity of blood vessels and to stop bleeding after injury (Ali *et al.*, 2000). Both onion and garlic inhibit platelet aggregation *in vitro* (Ali *et al.*, 1999; Lawson *et al.*, 1992) and several platelet inhibitors have been isolated and characterized from these vegetables. Studies on the antithrombotic action of onion have reported that its aqueous extracts inhibit thromboxane formation, potent inducers of platelet aggregation (Moon *et al.*, 2000). Aqueous and organic garlic extracts are also able to inhibit platelet aggregation induced by a number of physiologically important aggregating agents, as collagen and adrenaline, and the thromboxanes synthesis *in vivo* (Mohammad and Woodward, 1986) by several mechanisms, such as inhibition of several steps of the arachidonic acid pathway in platelets (Ali *et al.*, 2000), which is the thromboxanes precursor. Due to the variations in methods of preparation, the different garlic products available at the market may show different inhibitory effect on platelet aggregation (Lawson *et al.*, 1992). It has been reported that, in onions and garlic, the antiplatelet activity is determined, in part, by the native concentration of organo-sulphur compounds and genotypically determined sulphur content of the bulb (Goldman *et al.*, 1996), being garlic 13 folds more potent than onion (Effendy *et al.*, 1997). Several epidemiologic studies have reported that antiplatelet activity of onion and garlic is considered to be a property of organo-sulphur compounds. In particular, a class of a-sulphinyl-disulphides (cepaenes) found in onion extracts has demonstrated antithrombotic activity (Block *et al.*, 1997). These compounds have structural similarity to ajoene, considered the major antiplatelet compound in garlic extracts. In addition, other non-sulphur compounds, such as b-chlorogenin and quercetin, have also been shown to inhibit platelet aggregation (Rahman *et al.*, 2006). A recent study on pigs reveals that feeding raw onions for six weeks did not have an impact on platelet aggregation but caused a significant reduction in blood triglycerides, another biomarker of cardiovascular diseases (Gabler and Osrowska, 2003). The quantity fed to the animals was equivalent to the daily consumption of one onion by a human. A preliminary study conducted this time on humans showed that the consumption of the equivalent of three onions in a soup was sufficient to significantly reduce the blood platelet aggregation (Hubbard *et al.*, 2006b). This activity appears to be less important after cooking (Janssen *et al.*, 1998) and more important in pungent onions (Osmont *et al.*, 2003). It is attributed to quercetin and alkyl-propenyl cysteine sulphoxyde molecules, but the exact mode of action remains elusive. It is suggested that these compounds stimulate the release of arachidonic acid from membrane phospholipids, which initiates eicosanoid metabolism in mammals leading to the inhibition of thromboxane A synthesis and a significant reduction in platelet aggregation and vasoconstriction (Moon *et al.*, 2000).

Effect on Hyperhomocysteinemia

Homocysteine is a sulphur-containing amino acid formed during metabolism of methionine, an essential amino acid derived from the diet. The determination of total plasma homocysteine (Hyc) has become a very useful tool because moderately elevated values of circulating homocysteine constitute an important risk factor for the development and progress of occlusive vascular affections and of ischaemic heart disease in diabetic patients (Fischer *et al.,* 2000). The commonest cause of acquired hyperhomocysteinemia is the folate, vitamin B6 and/or B12 deficiency (Durand *et al.,* 1996; Ubbink *et al.,* 1996) and the drugs consumption that interfere with these vitamins metabolism. Because garlic contains as much vitamins B6 and B12 as a large amount of aminothiol compounds, such as SAMC, DAS, diethyl disulphide (DEDS) and dipropyl disulphide (DPDS) (Liu and Yeh, 2000), it was thought that garlic intake may be an effective way to reduce plasma homocysteine levels.

Effects on the Respiratory System

Certain onion-derived compounds, in particular thiosulfinates and cepaenes, show a remarkable *in vitro* inhibitory effect of cyclooxygenase and lipoxygenase mediated reactions which initiate eicosanoid metabolism and lead to bronchial restriction. Therefore, these compounds have antiasthmatic activity (Wagner *et al.,* 1990). It has been described that, in general, saturated thiosulphinates are less active than unsaturated ones and that cepaenes are more active than thiosulphinates. Likely, these effects *in vitro* are responsible, at least in part, for onion extracts anti-inflammatory and antiasthmatic properties observed *in vivo* (Breu and Dorsch, 1994).

Immunomodulatory Effect

Recent investigations are beginning to clarify important roles of immune functions modulation in some diseases progression. Currently, available data suggest that garlic may be a promising candidate as a biological immune response modifier, being able to maintain the homeostasis of immune function, stimulating necessary functions and suppressing unnecessary functions (Kyo *et al.,* 2001). A great variety of immunomodulatory effects have been studied in different garlic-derived products, mainly in AGE, whereas in onion no immune-stimulating properties have been reported. It has been demonstrated that AGE exerts an anti-allergic (Kyo *et al.,* 2001) and antitumor effect (Lamm and Riggs, 2001) through direct and/or indirect modification of immune function, stimulating the lymphocytes and antibodies proliferation and the antibodies production (Zhang *et al.,* 1997), among other mechanisms. The maintenance of immune stimulation offers protection against cancer and impairment of immune defences, as occurs with acquired immunodeficiency syndrome (AIDS) (Lamm and Riggs, 2001), and improves aging-related cognitive deterioration (Zhang *et al.,* 1997). Modification of immune function by garlic may contribute to the treatment and prevention of certain diseases caused by immune dysfunction, for example, the invasive fungal disease. Several studies (Capel *et al.,* 1979) have demonstrated that garlic may be used synergistically in conjunction with other antifungal agents due to its capacity to enhance the host cellular immunity. Nakata and Fujiwara (1975) identified a carbohydrate in the garlic extract that appeared to be responsible for the antitumor immunity.

Other Beneficial Effects

In addition to the above mentioned biological activities, it has been observed that AGE may protect the small intestine against antitumor drugs-induced damage, for example, nauseas, vomits, diarrhoea, stomatitis and gastrointestinal ulceration, and, consequently, intestinal dysfunction (Capel *et al.,* 1979; Horie *et al.,* 2001). Onion and garlic prebiotic activity is also being investigated (Benkeblia and Shiomi, 2006; Sharma *et al.*, 2006) due to their high soluble fibre content, specially inulin and fructooligosaccharides which stimulate in the colon the growth of specific microorganisms, as *bifidobacteria* and *lactobacilli*, with a general positive health effect (Ernst and Feldheim, 2000; Gibson, 1998). As previously discussed, onions are a rich source of dietary fibers and especially of inulin, a polyfructosan. The health benefits of inulintype fructans to human health have now been studied for more than a decade (Ritsema and Smeekens, 2003). It has prebiotic properties as it is preferably fermented by beneficial bowel bacteria like *Lactobacilli* and *Bifidobacteria*, thereby altering the bacterial mycoflora of the intestine in such a way that pathogenic, or harmful bacteria become less abundant (Kruse *et al.*, 1999). It has been reported that onion stimulates the digestive process, accelerating digestion and reducing food transit time in the gastrointestinal tract (Platel and Srinivasan, 2001). Both, inulin and fructooligosaccharides of onion and garlic may be used as functional ingredients to enrich many processed foods without any negative impact on their taste (Causey *et al.*, 2000). Neokestose, another fructan found in onion, has recently been shown to be an excellent promoter of the growth of beneficial bacteria (Kilian *et al.*, 2002). Fructans also promote the absorption of calcium and could thus be useful in the prevention of osteoporosis (Scholz-Ahrens *et al.*, 2001). High fructan diets have also been shown to lower concentration of cholesterol, tryacylglycerol, phospholipids, glucose and insulin in the blood of middle-aged men and women (Jackson *et al.*, 1999).

Conclusions

Garlic and onion each possess antioxidant, anti-inflammatory, and antibacterial properties. The effectiveness of these spices in decreasing pro-inflammatory diseases is rooted in their nature as modulators of metabolism, for example as COX and LOX inhibitors. While the available evidence is encouraging, controlled human trails are needed to establish the effectiveness of these spices in disease prevention. Many of the available studies utilized relatively high doses of the effective compounds in garlic and onion, and it remains to be seen whether a moderate level of consumption, that avoids the toxic effects of excessive doses, is effective. Until such trials are available, it seems safe to conclude that garlic and onion should be included in the human diet as whole foods and spices, while high-dose extracts should be used with caution. In conclusion, both vegetables onion and garlic should be consumed, within an equilibrated diet, not only for their organoleptic characteristics but also for the potential and/or verified biological activities. The research summarized here supports the ancient wisdom of Aristotle and Hippocrates who recommended garlic for medicinal purposes and generally promoted the use of food as medicine. This would allow enriching the processed foods with onion and garlic extracts or their active compounds to fulfill a therapeutic or beneficial function on human health. The efforts of research and industry should be directed towards the improvement of processing and extraction methods in order to obtain garlic and onion and their derivatives with high quality preserving and/or improving the particular biological properties. Thus, it is necessary to find the best way to elaborate high added-value foods derived from garlic and onion.

References

Ali, M., Bordia, T., and Mustafa, T. (1999). Effect of raw versus boiled aqueous extract of garlic and onion on platelet aggregation. *Prostaglandins Leukotrienes and Essential Fatty Acids*, **60**: 43-47.

Ali, M., Thomson, M. and Afzal, M. (2000). Garlic and onions: their effect on eicosanoid metabolism and its clinical relevance. *Prostaglandins, Leukotrienes and Essential Fatty Acids*, **6(2):** 55-73.

Al-Qattan, K. K., Alnaqeeb, M. A., and Ali, M. J. (1999). The antihypertensive effect of garlic (*Allium sativum*) in the rat two-kidney-one-clip Goldblatt model. *Ethnopharmacology*, **66**: 217-222.

Amagase, H. (2006). Clarifying the real bioactive constituents of garlic. *Journal of Nutrition*, **136**: 716-725.

Amagase, H., Petesch, B. L., Matsuura, H., Kasuga, S., and Itakura, Y. (2001). Intake of garlic and its bioactive components. *The Journal of Nutrition*, **131**: 955-962.

Aqel, M. B., Gharaibah, M. N., and Salva, A. S. (1991). Direct relaxant effects of garlic juice on smooth and cardiac muscles. *Journal Ethnopharmacology*, **33**: 13-19.

Arnault, I., and Auger, J. (2006). Seleno-compounds in garlic and onion. *Journal of Chromatography A*, **1112**: 23-30.

Ashraf, M.Z., Hussain, M.E. and Fahim, M. (2005). Antiatherosclerotic effects of dietary supplementations of garlic and turmeric: restoration of endothelial function in rats. *Life Sciences*, **77(8):** 837-57.

Augusti, K. T., and Mathew, P. T. (1974). Lipid lowering effect of allicin (diallyl disulfide oxide) on long-term feeding in normal rats. *Experientia*, **30**: 468-470.

Babu, P. S., and Srinivasan, K. (1997). Influence of dietary capsaicin and onion on the metabolic abnormalities associated with streptozotocin induced diabetes mellitus. *Molecular and Cellular Biochemistry*, **175**: 49-57.

Bakri, I. M., and Douglas, C. W. I. (2005). Inhibitory effect of garlic extract on oral bacteria. *Archives of Oral Biology*, **50(7):** 645-651.

Ballas, S. K., and Smith, E. D. (1992). Red cell changes during the evolution of the sickle cell painful crisis. *Blood*, **79**: 2154-2163.

Banerjee, S.K., Mukherjee, P.K. and Maulik, S.K. (2003). Garlic as an antioxidant: the good, the bad and the ugly. *Phytotherapy Research*, **17**: 97-106.

Bazzano, L.A., He, J., Ogden, L.G., Loria, C.M., Vupputuri, S., Myers, L. and Whelton, P.K. (2002). Fruit and vegetable intake and risk of cardiovascular disease in US adults: the first National Health and Nutrition Examination Survey Epidemiologic Follow-up Study. *Am. J. Clin. Nutr.*, **76**: 93-99.

Benkeblia, N. (2004). Antimicrobial activity of essential oil extracts of various onions (*Allium cepa*) and garlic (*Allium sativum*). *Lebensmittel-Wissenschaft und Technologie-Food Science and Technology*, **37(2):** 263-268.

Benkeblia, N., and Shiomi, N. (2006). Hydrolysis kinetic parameters of DP 6, 7, 8 and 9-12 fructooligosaccharides (FOS) of onion bulb tissues. Effect of temperature and storage time. *Journal of Agricultural and Food Chemistry*, 54(7): 2587-2592.

Block, E. (1985). The chemistry of garlic and onions. *Scientific American*, 252: 114-119.

Block, E., Gulati, H., Putman, D., Sha, D., Niannian, Y., and Zhao, S.-H. (1997). *Allium* chemistry: synthesis of 1-[alk(en)ylsulfinyl]-propyl alken(en)yl disulfides (cepaenes), antithrombotic flavorants from homogenates of onion (*Allium cepa*). *Journal of Agricultural and Food Chemistry*, 45: 4414 - 4422.

Borek, C. (1997). Antioxidants and cancer. *Science and Medicine*, 4: 51-62.

Borek, C. (2001). Antioxidant health effects of age garlic extract. *The Journal of Nutrition*, 131: 1010-1015.

Breu, W., and Dorsch, W. (1994). *Allium cepa* L. (onion), chemistry, analysis and pharmacology. In H.Wagner, and N. R. Farnsworth (Eds.), Economic and Medicinal Plant Research, Vol. 6 (pp. 115-147). London: Academic Press.

Brewster, J.L. (1994). Onions and other vegetable alliums. Ed 1st. Vol 3. CAB International, Wallingford.

Brisdelli, F., Coccia, C., Cinque, B., Cifone, M. G., and Bozzi, A. (2007). Induction of apoptosis by quercetin: different response of human chronic myeloid (K562) and acute lymphoblastic (HSB-2) leukemia cells. *Molecular and Cellular Biochemistry*, 296: 137-149.

Capel, I. D., Pinnock, M. H., and Williams, D. C. (1979). An *in vitro* assessment of the effect of cytotoxic drugs upon the intestinal absorption of nutrients in rats. *European Journal of Cancer*, 15: 127-131.

Causey, J. L., Feirtag, J. M., Gallaher, D. D., Tungland, B. C., and Slavin, J. L. (2000). Effects of dietary inulin on serum lipids, blood glucose and the gastrointestinal environment in hypercholesterolemic men. *Nutrition Research*, 20: 191-201.

Challier, B., Perarnau, J.-M. and Viel, J.-F. (1998). Garlic, onion and cereal fibre as protective factors for breast cancer: A French case-control study. *Eur. J. Epidemiol.*, 14: 737-747.

Chang, M. L. W., and Johnson, M. A. (1980). Effect of garlic on lipid metabolism and lipid synthesis in rats. *The Journal of Nutrition*, 110: 931-936.

Chauhan, N. B. (2006). Effect of aged garlic extract on APP processing and tau phosphorylation in Alzheimer's transgenic model Tg2576. *Journal of Ethnopharmacology*, 108(3): 385-394.

Davis, S. R. and Perrie, R. (2003). The *in-vitro* susceptibility of *Cryptococcus neoformans* to allitridium. In Program and Abstracts of the 15th Congress of ISHAM (abstract 113). San Antonio, TX, USA, May 25-29.

Davis, S. R. (2005). An overview on the antifungal properties of allicin and its breakdown products e the possibility of a safe and effective antifungal properties. *Mycoses*, 48(2): 95-100.

Delaha, E. C., and Garagusi, V. F. (1985). Inhibition of mycobacterial by garlic extract (*Allium sativum*). *Antimicrobial Agents and Chemotherapy*, **27**: 485-486.

Dimitrios, B. (2006). Sources of natural phenolic antioxidants. *Trends in Food Science and Technology*, **17(9):** 505-512.

Dirsch, V. M., Gerbes, A. L., and Vollmar, A. M. (1998). Ajoene, a compound of garlic, induces apoptosis in human promyieloleukemic cells, accompanied by generation of reactive oxygen species and activation of nuclear factor kB. *Molecular Pharmacology*, **53**: 402-407.

Dorant, E., van Den Brandt, P.A., Goldbohm, R.A. and Sturnmans, F. (1996). Consumption of onions and a reduced risk of stomach carcinoma. *Gastroenterology*, **110**: 12-20.

Durand, P., Fortin, L. J., Lussier-Cacan, S., Davignon, J., and Blanche, D. (1996). Hyperhomocysteinemia induced by folic acid deficiency and methionine load-applications of a modified HPLC method. *Clinical Chemistry Acta*, **252**: 83-93.

Effendy, J. L., Simmons, D. L., Campbell, G. R., and Campbell, J. H. (1997). The effect of aged garlic extract "Kyolic", on the development of experimental atherosclerosis. *Atherosclerosis*, **132**: 37-42.

El-Bayoumy, K., Sinka, R., Pinto, J. T., and Rivlin, R. S. (2006). Cancer chemoprevention by garlic and garlic-containing sulfur and selenium compounds. *Journal of Nutrition*, **136(3):** 864-869.

Ernst, M., and Feldheim, W. J. (2000). Fructans in higher plants and in human nutrition. *Angewandte Botanik*, **74**: 5-9.

FAO Stat. (2006). Vegetable production statistics.

Fenwick, G. R., and Hanley, A. B. (1985). The genus *Allium. CRC Critical Reviews in Food Science and Nutrition*, **22**: 199-377.

Fenwick, G.R. and Hanley, A.B. (1985). The genus *Allium* Part 3. *Crit. Rev. Food Sci. Nutr.*, **23**: 1-73.

Fischer, P. A., Falcon, C., and Masnatta, L. D. (2000). Hiperhomocisteinemia moderada: fisiopatologia de la lesion endotelial e implicancia clinica. *Revista de la Federacio´n Argentina de Cardiologý´a*, **29**: 57-66.

Fleischauer, A.T. and Arab, L. (2001). Garlic and cancer: A critical review of the epidemiologic literature. *J.Nutr.*, **131**: 1032-1040.

Gabler, N.K. and Osrowska, E. (2003). Dietary onion intake as part of a typical high fat diet improves indices of cardiovascular health using the mixed pig model. *Plant Foods Hum. Nut.*, **61**: 179-185.

Galeone, C., Pelucchi, C., Levi, F., Negri, E.,Franceschi, S., Talamini, R., Giacosa, A. and La Vecchia, C. (2006). Onion and garlic use and human cancer. *Am. J. Clin. Nutr.*, **84**: 1027-1032.

Gibson, G. R. (1998). Dietary modulation of the human gut microflora using prebiotics. *The British Journal of Nutrition*, **80**: 209-212.

Glasser, G., Graefe, E. U., Struck, F., Veit, M., and Gebhardt, R. (2002). Comparison of antioxidative capacities and inhibitory effects on cholesterol biosynthesis of quercetin and potential metabolites. *Phytomedicine*, **9**: 33-40.

Goldman, I. L., Kopelberg, M., Debaene, J. E. P., and Schwartz, B. S. (1996). Antiplatelet activity of onion (*Allium cepa*) is sulphur dependent. *Thrombosis and Aemostasis*, **76**: 450-453.

Gonzalez, C.A. and Riboli, E. (2006). Diet and cancer prevention: Where we are, where we are going. *Nutr. Cancer*, **56**: 225-231.

Goodsell, D. (2005). Cyclooxygenase. Protein data bank, online, available at: www.rcsb.org/pdb/molecules/pdb17_1.html (accessed 19 October 2005).

Grant, W.B. (2004). A multicountry ecologic study of risk and risk reduction factors for prostate cancer mortality. *Eur. Urol.*, **45**: 271-279.

Griffiths, G., Trueman, L., Crowther, T.E., Thomas, B. and Smith, B. (2002). Onions - A global benefit to health. *Phytother. Res.*, **16**: 603-615.

Gu¨lsen, A., Makris, D. P., and Kefalas, P. (2007). Biomimetic oxidation of quercetin: isolation of a naturally occurring quercetin heterodimer and evaluation of its in vitro antioxidant properties. *Food Research International*, **40**: 7-14.

Gupta, N., and Porter, T. D. (2001). Garlic and garlic-derived compounds inhibit human squalene monooxygenase. *The Journal of Nutrition*, **131**: 1662-1667.

Guyonnet, D., Siess, M. H., Le Bon, A. M., and Suschetet, M. (1999). Modulation of phase II enzymes by organosulfur compounds from allium vegetables in rat tissues. *Toxicology and Applied Pharmacology*, **154(1)**: 50-58.

Hannun, Y.A. (1997). Apoptosis and the dilemma of cancer chemotherapy. *Blood*, **89(6)**: 1845-1853.

Harauma, A., and Moriguchi, T. (2006). Aged Garlic Extract improves blood pressure in spontaneously hypertensive rats more safely than raw garlic. *Journal of Nutrition*, **136**: 769-773.

Horie, T., Awazu, S., Itakura, Y., and Fuwa, T. (2001). Alleviation by garlic of antitumor drug-induced damage to the intestine. *The Journal of Nutrition*, **131**: 1071-1074.

Hu, J., La Vecchia, C., Negri, E., Chatenoud, L.,Bosetti, C., Jia, X., Liu, R.H., Huang, G., Bi, D. and Wang, C. (1999). Diet and brain cancer in adults: a case-control study in northeast China. *Int. J.Cancer*, **81**: 20-23.

Hubbard, G.P., Wolffram, S., de Vos, R., Bovy, A., Gibbins, J.M. and Lovegrove, J.A. (2006a). Ingestion of onion soup high in quercetin inhibits platelet aggregation and essential components of the collagen-stimulated platelet activation pathway in man: a pilot study. *British J. Nutr.*, **96**: 482-488.

Imai, J., Ide, N., Nagae, S., Moriguchi, T., Matsuura, H., and Itakura, Y. (1994). Antioxidants and free radical scavenging effects of aged garlic extract and its constituents. *Planta Medica*, **60**: 417-420.

Jackson, K.G., Taylor, G.R., Clohessy, A.M. and Willieams, C.M. (1999). The effect of the daily intake of inulin on fasting lipid, insulin and glucose concentrations in middle-aged men and women. *Br. J. Nutr.*, **82**: 23-30.

Janssen-Heininger, Y.M.W., Poynter, M.E. and Baeuerle, P.A. (2000). Recent advances towards understanding redox mechanisms in the activation of nuclear factor kB. *Free Radical Biology and Medicine*, **28(9)**: 1317-1327.

Jones, M.G., Hughes, J., Tregova, A., Milne, J.,Tomsett, A.B. and Collin, H.A. (2004). Biosynthesis of the flavour precursors of onion and garlic. *J.Exp. Bot.*, **55**: 1903-1918.

Jonkers, D., van den Broek, E., van Dooren, I., Thijs, C., Dorant, E., Hageman, G. and Stobberingh, E. (1999). Antibacterial effect of garlic and omeprazole on Helicobacter pylori. Journal of Antimicrobial Chemotherapy, **43**: 837-9.

Kempaiah, R.K. and Srinivasan, K. (2004). Influence of dietary curcumin, capsaicin and garlic on the antioxidant status of red blood cells and the liver in high-fat-fed rats. *Annals of Nutrition and Metabolism*, **48**: 314-320.

Khanum, F., Anilakumar, K.R. and Viswanathan, K.R. (2004). Anticarcinogenic properties of garlic: a review. *Critical Reviews in Food Science and Nutrition*, **44(6)**: 479-88.

Kilian, S., Kritzinger, S., Rycroft, C., Gibson, G.R. and du Preez, J. (2002). The effects of the novel bifidogenic trisaccharide, neokestose, on the human colonic microbiota. *World Journal of Microbiology and Biotechnology*, **18**: 637-644.

Kim, J.H. 1997. Anti-bacterial action of onion (*Allium cepa* L.) extracts against oral pathogenic bacteria. *J. Nihon Unive. School Dentistry*, **39**: 136-141.

Kim, K.-M., Chun, S.-B., Koo, M.-S., Choi, W.-J.,Kim, T.-W., Kwon, Y.-G., Chung, H.-T., Billiar, T.R.and Kim, Y.-M. (2001). Differential regulation of NO availability from macrophages and endothelial cells by the garlic component S-allyl cysteine. *Free Rad. Biol. Med.*, **30**: 747-756.

Kiviranta, J., Huovinen, K., Seppanen-Laakso, T., Hiltunen, R., Karppanen, H., and Kilpelainen, M. (1989). Effects of onion and garlic extracts on spontaneously hypertensive rats. *Phytotherapy Research*, **3**: 132-135.

Koch, H. P. (1993). Saponine in knoblauch und kü˝chenzwiebel. *Deutsche Apotheker Zeitung*, **133**: 3733-3743.

Kruse, H.-P., Kleessen, B. and Blaut, M. (1999). Effects of inulin on faecal bifido bacteria in human subjects. *Br. J. Nutr.*, **82**: 375-382.

Kubec, R., Svobodova, M. and Velisek, J. Norat, T. and Riboli, E. (2001). Meat consumption and colorectal cancer: a review of epidemiologic evidence. *Nutr. Rev.*, **59**: 37-47.

Kyo, E., Uda, N., Kasuga, S., and Itakura, Y. (2001). Immunomodulatory effects of aged garlic extract. *The Journal of Nutrition*, **131**: 1075-1079.

Kyung, K.H. and Lee, Y.C. (2001). Antimicrobial activities of sulfur compounds derived from S-alk(en)yl-Lcysteine sulfoxides in *Allium* and *Brassica*. *Food Reviews International*, **17**: 183-198.

Lai, P.K. and Roy, J. (2004). Antimicrobial and chemopreventive properties of herbs and spices. *Current Medicinal Chemistry*, **11(11)**: 1451-1460.

Lam, Y. W., Wang, H. X., and Ng, T. B. (2000). A robust cysteine deficient chitinase-like antifungal protein from inner shoots of the edible chive *Allium tuberosum*. *Biochemical and Biophysical Research Communications*, **279**: 74-80.

Lamm, D. L., and Riggs, D. R. (2001). Enhanced immunocompetence by garlic: role in bladder cancer and other malignancies. *The Journal of Nutrition*, **131**: 1067-1070.

Lampe, J.W. (1999). Health effects of vegetables and fruit: assessing mechanisms of action in human experimental studies. *Am. J. Clin. Nutr.*, **70**: 475-490.

Lancaster, J. E., and Shaw, M. L. (1989). G-Glutayl peptides in the biosynthesis of S-alk(en)yl-L-cysteine sulfoxides (flavor precursors) in *Allium*. *Phytochemistry*, **28**: 455-460.

Lancaster, J.E. and Kelly, K.E. (1983). Quantitative analysis of the S-alk(en)yl-L-cysteine sulphoxides in onion (*Allium cepa* L.). *J. Sci. Food Agric.*, **34**: 1229-1235.

Lang, A., Lahav, M., Sakhnini, E., Barshack, I., Fidder, H.H., Avidan, B., Bardan, E., Hershkoviz, R., Bar-Meir, S. and Chowers, Y. (2004). "Allicin inhibits spontaneous and TNF- induced secretion of proinflammatory cytokines and chemokines from intestinal epithelial cells", Clinical Nutrition, Vol. 5, pp. 1199-208.

Lanzotti, V. (2006). The analysis of onion and garlic. *Journal of Chromatography A*, **1112**: 3-22.

Lau, B. H. S. (1989). Detoxifying, radioprotective and phagocyteenhancing effects of garlic. *International Clinical Nutrition Review*, **9**: 27-31.

Lau, B. H. S. (2006). Suppression of LDL oxidation by garlic compounds is a possible mechanism of cardiovascular health benefit. *Journal of Nutrition*, **136**: 765-768.

Lau, B. H. S., Lam, F., and Wang-Cheng, R. (1987). Effects of an odormodified garlic preparation on blood lipids. *Nutrition Research*, **7**: 139-149.

Lau, B. H. S., Tadi, P. P., and Tosk, J. M. (1990). *Allium sativum* (garlic) and cancer prevention. *Nutrition Research*, **10**: 937-948.

Lautraite, S., Musonda, A. C., Doehmer, J., Edwards, G. O., and Chipman, J. K. (2002). Flavonoids inhibit genetic toxicity produced by carcinogens in cells expressing CYP1A2 and CYP1A1. *Mutagenesis*, **17**: 45-53.

Lawson, L. D., Ransom, D. K., and Hughes, B. G. (1992). Inhibition whole blood platelet aggregation by compounds in garlic clove extract and commercial garlic products. *Thrombosis Research*, **65**: 141-156.

Le Marchand, L., Murphy, S. P., Hankin, J. H., Wilkens, L. R., and Kolonel, L. N. (2000). Intake of flavonoids and lung cancer. *Journal of the National Cancer Institute*, **92**: 154-160.

Liu, J. Z., Lin, X. Y., and Milner, J. A. (1992). Dietary garlic powder increases glutathione content and glutathione S-transferase activity in rat liver and mammary tissues. *The FASEB Journal*, **6:** A3230, (abstract).

Liu, L., and Yeh, Y. Y. (2002). S-alk(en)yl cysteines of garlic inhibit cholesterol synthesis by deactivating HMG-CoA reductase in cultured rat hepatocytes. *The Journal of Nutrition*, **132(6):** 1129-1134.

Lun, Z. R., Burri, C., Menzinger, M., and Kaminsky, R. (1994). Antiparasitic activity of diallyl trisulfide (Dasuansu) on human and animal pathogenic protozoa (*Trypanosoma* sp., *Entamoeba histolytica* and *Giardia lamblia*) *in vitro*. *Annales de la Societe Belge de Medecine Tropicale*, **74:** 51-59.

Ly, T. N., Hazama, C., Shimoyamada, M., Ando, H., Kato, K., and Yamauchi, R. (2005). Antioxidative compounds from the outer scales of onion. *Journal of Agricultural and Food Chemistry*, **53(21):** 8183-8189.

Mansell, P., and Reckless, J. P. D. (1991). Garlic: effects on serum lipids, blood pressure, coagulation, platelet aggregation, and vasodilatation. *British Medical Journal*, **303:** 379.

Matsuura, H. (1997). Phytochemistry of garlic horticultural and processing procedures. In P. A. Lachance (Ed.), Neutraceuticals: Designer foods III. Garlic, soy and licorice (pp. 55-59). Trumbull, CT: Food and Nutrition Press.

Matsuura, H., Ushiroguchi, T., Itakura, Y., Hayashi, H., and Fuwa, T. I. (1988). A furostanol glycoside from garlic bulbs of *Allium sativum*. *Chemical and Pharmaceutical Bulletin*, **36:** 3659-3663.

Meng, Y., Lu, D., Guo, N., Zhang, L., and Zhou, G. (1993). Anti-HCMV effect of garlic components. *Virologica Sinica*, **8:** 147-150.

Milner, J.A. (2001). Garlic: The mystical food in health promotion. p.193-207. In: R.E.C. Wildman (ed.), *Handbook of Nutraceuticals and Functional Foods*. CRC Press, Boca Raton.

Miron, T., Rabinkov, A., Mirelman, D., Wilchek, H., and Weiner, L. (2000). The mode of action of allicin: its ready permeability through phospholipid membranes may contribute to its biological activity. *Biochimica and Biophysica Acta*, **1463:** 20-30.

Miron, T., Rabinkov, E., Peleg, T., Rosenthal, D., Mirelman, M., and Wilchek, M. (2004). Allylmercaptocaptopril: a new antihypertensive drug. *American Journal of Hypertension*, **17(1):** 71-73.

Mohammad, S. F., and Woodward, S. C. (1986). Characterisation of a potent inhibitor of platelet aggregation and release reaction isolated from *Allium sativum* (garlic). *Thrombosis Research*, **44:** 793-806.

Moon, C. H., Jung, Y. S., Kim, M. H., Lee, S. H., Baik, E. J., and Park, S. W. (2000). Mechanism for antiplatelet effect of onion: araechidonic acid release inhibition, thromboxane A(2) synthase inhibition and TXA (2)/PGH (2) receptor blockade. *Prostaglandins Leukotrienes and Essential Fatty Acids*, **62:** 277-283.

Moreno, F. J., Corzo-Martý´nez, M., del Castillo, M. D., and Villamiel, M. (2006). Changes in antioxidant activity of dehydrated onion and garlic during storage. *Food Research International*, 39: 891-897.

Munday, R., Munday, J.S. and Munday, C.M. (2003). Comparative effects of mono-, di-, tri-, and tetrasulfides derived from plants of the *Allium* family: redox cycling in vitro and hemolytic activity and Phase 2 enzyme induction *in vivo*. *Free Rad. Biol. Med.*, 34: 1200-1211.

Nakata, T., and Fujiwara, M. (1975). Adjuvant action of garlic sugar solution in animals immunized with Ehrlisch ascites tumors cells attenuated with allicin. *Gann*, 66: 417-419.

Neil, H. A., Silagy, C. A., Lancaster, T., Hodgeman, J., Vos, K., Moore, J. W., *et al.* (1996). Garlic powder in the treatment of moderate hyperlipidemia: a controlled trial and a meta-analysis. *Journal of the Royal College of Physicians of London*, 30: 329-334.

Noctor, G., Arisi, A.-C.M., Jouanin, L., Kuert, K.J.,Renneberg, H. and Foyer, C.H. (1998). Glutathione: biosynthesis, metabolism and relationship to stress tolerance explored in transformed plants. *J. Exp.insulin receptor kinase. Biochemistry*, 44: 8167-8175.

Osmont, K.S., Arnt, C.R. and Goldman, I.L. (2003). Temporal aspects of onion-induced antiplatelet activity. *Plant Foods Hum. Nut.*, 58: 27-40.

Phay, N., Higashiyama, T., Tsuji, M., Matsuura, H., Fukushi, Y., Yokota, A., *et al.* (1999). An antifungal compound from roots of Welsh onion. *Phytochemistry*, 52: 271-274.

Platel, K., and Srinivasan, K. (2001). Studies on the influence of dietary spices on food transit time in experimental rats. *Nutrition Research*, 21: 1309-1314.

Prasad, K., Laxdal, V.A., Yu, M. and Raney, B.L. 1995. Antioxidant activity of allicin, an active principle in garlic. *Mol. Cell. Biochem.*, 48: 183-189.

Pszczola, D. E. (2002). Antimicrobials: setting up additional hurdles to ensure food safety. *Food and Technology*, 56: 99-107.

Rahman, K. and Lowe, G.M. (2006), Garlic and cardiovascular disease: a critical review. *Journal of Nutrition*, 136 (3): 736-740.

Rahman, K., Allison, G. L., and Lowe, G. M. (2006). Mechanisms of inhibition of platelet aggregation by aged garlic extract and its constituents. *Journal of Nutrition*, 136: 782-788.

Ramos, F. A., Takaishi, Y., Shirotori, M., Kawaguchi, Y., Tsuchiya, K., Shibata, H., et al. (2006). Antibacterial and antioxidant activities of quercetin oxidation products from yellow onion (*Allium cepa*) skin. *Journal of Agricultural and Food Chemistry*, 54: 3551-3557.

Raso, G. M., Meli, R., Di Carlo, G., Pacilio, M., and Di Carlo, R. (2001). Inhibition of inducible nitric oxide synthase and cyclooxygenase-2 expression by flavonoids in macrophage J774A.1. *Life Sciences*, 68: 921-931.

Reuter, H. D., Koch, H. P., and Lawson, L. D. (1996). Therapeutic effects and applications of garlic and its preparations. In H. P. Koch, and L. D. Lawson (Eds.), Garlic, the science and therapeutic application of *Allium sativum* L. and related species (pp. 135-213). Baltimore: Williams and Wilkins.

Richardson, S. J. (1993). Free radicals in the genesis of Alzheimer's disease. *Annals of the New York Academy of Sciences,* **695**: 73-76.

Ritsema, T. and Smeekens, S. (2003). Fructans: beneficial for plants and humans. *Current Opinion in Plant Biology,* **6**: 223-230.

Rose, P., Whiteman, M., Moore, P.K. and Zhu, Y.Z. (2005). Bioactive S-alk(en)yl cysteine sulfoxide metabolites in the genus *Allium*: the chemistry of potential therapeutic agents. *Nat. Prod. Rep.,* **22**: 351-368.

Ryu, K., Ide, N., Matsuura, H., and Itakura, Y. (2001). N Alpha-(1-deoxy- D-fructos-1-yl)-L-arginine, an antioxidant compound identified in aged garlic extract. *The Journal of Nutrition,* **131**: 972-976.

Sakamoto, K., Lawson, L. D., and Milner, J. A. (1997). Allyl sulfides from garlic suppress the *in vitro* proliferation of human A549 lung tumor cells.*Nutrition and Cancer,* **29(2)**: 152-156.

Saleheen, D., Ali, S. A., and Yasinzai, M. M. (2004). Antileishmanial activity of aqueous onion extract *in vitro. Fitoterapia,* **75(1)**: 9-13.

Sanchez, M., Lodi, F., Vera, R., Villar, I. C., Cogolludo, A., Jimenez, R., et al. (2007). Quercetin and isorhamnetin prevent endothelial dysfunction, superoxide production, and overexpression of p47(phox) induced by angiotensin II in rat aorta. *The Journal of Nutrition,* **137**: 910-915.

Schmidt, P. W., and Marquardt, U. (1936). U¨ ber den antimykotischen efffekt a¨therischer o¨ le von lauchgewa¨chsen und kreuzblu¨ tlern auf pathogene hautpilze. *Zentralblatt fur Bakteriologie, Parasitenkunde, Infektionskrankheiten und Hygiene. Abteilung,* **138**: 104-128.

Scholz-Ahrens, K.E., Schaafsma, G., van den Heuvel, E.G.H.M. and Schrezenmeir, J. (2001). Effects of prebiotics on mineral metabolism. *Am.J. Clin. Nutr.,* **73**: 459-464.

Sellappan, S., and Akoh, C. C. (2002). Flavonoids and antioxidant capacity of Georgia-grown Vidalia onions. *Journal of Agricultural and Food Chemistry,* **50**: 5338-5342.

Sengupta, A., Gosh, S. and Bhattacharjee, S. (2004). *Allium* vegetables in cancer prevention: an overview. *Asian Pac. J. Cancer Prev.,* **5**: 237-245.

Sharma, A. D., Kainth, S., and Gill, P. K. (2006). Inulinase production using garlic (*Allium sativum*) powder as a potential substrate in *Streptomyces* sp. *Journal of Food Engineering,* **77(3)**: 486-491.

Sharma, K.K., Gupta, R.K., Gupta, S.C. and Samuel, K.C. (1977). Antihyperglycemic effect of onion: effect on fasting blood sugar and induced hyperglycemia in man. *Ind. J. Med. Res.,* **65**: 422-429.

Sharma, R.A., Gescher, A.J. and Steward, W.P. (2005). Curcumin: the story so far. *European Journal of Cancer*, **41(13)**: 1955-1968.

Shen, J. K., Davis, L. E.,Wallace, J. M., Cai, Y., and Lawson, L. D. (1996). Enhanced diallyl trisulfide has *in vitro* synergy with amphotericin B against *Cryptococcus neoformans*. *Planta Medica*, **62**: 415-418.

Shin, S.-C., Choi, J.-S., and Li, X. (2006). Enhanced bioavailability of tamoxifen after oral administration of tamoxifen with quercetin in rats. *International Journal of Pharmaceutics*, **313**: 144-149.

Shobana, S. and Naidu, K.A. (2000). Antioxidant activity of selected Indian spices. *Prostoglandins, Leukotrienes and Essential Fatty Acids*, **62(2)**: 107-110.

Shon, M.-Y., Choi, S.-D., Kahng, G.-G., Nam, S.-H.and Sung, N.-J. (2004). Antimutagenic, antioxidant and free radical scavenging activity of ethyl acetate extracts from white, yellow and red onions. *Food Chem. Toxicol.*, **42**: 659-666.

Sigounas, G., Hooker, J. L., Li, W., Anagnostou, A., and Steiner, M. (1997). S-Allylmercaptocysteine, a stable thioallyl compound, induces apoptosis in erythroleukemia cell lines. *Nutrition and Cancer*, **28(2)**: 153-159.

Singh, D. K., and Porter, T. D. (2006). Inhibition of sterol 4 alpha-methyl oxidase is the principal mechanism by which garlic decreases cholesterol synthesis. *Journal of Nutrition*, **136(3)**: 759-764.

Sivam, G.P. 2001. Protection against *Helicobacter pylori* and other bacterial infections by garlic. *J.Nutr.*, **131**: 1106-1108.

Slowing, K., Ganado, P., Sanz, M., Ruiz, E., Beecher, C., and Tejerina, T. (2001). Effect of garlic in cholesterol-fed rats. *Journal of Nutrition*, **131**: 994-999.

Srinivasan, K. (2005). Plants food in the management of diabetes mellitus: spices as beneficial antidiabetic food adjuncts. *International Journal of Food Science and Nutrition*, **56:** 399-414.

Srinivasan, K. (2005b). Spices as influencers of body metabolism: an overview of three decades of research. *Food Research International*, **38(1):** 77-86.

Srinivasan, K., and Sambaiah, K. (1991). The effect of spices on cholesterol 7-alpha-hydroxilase activity and on serun and hepatic cholesterol levels in rats. International *Journal of Vitamin and Nutrition Research*, **61**: 364-369.

Srivastava, K.C. (1986). Onion exerts antiaggregatory effects by altering arachidonic acid metabolism in platelets. *Prostag. Leukotr. Med.*, **24**: 43-50.

Tansey, M. R., and Appleton, J. A. (1975). Inhibition of fungal growth by garlic extract. *Mycologia*, **67**: 409-413.

Tatton, W.G. and Olanow, C.W. (1999), "Apoptosis in neurodegenerative diseases: the role of mitochondria", Biochimica et Biophysica Acta-Bioenergetics, Vol. 1410 No. 2, pp. 195-213.

Tjokroprawiro, A., Pikir, B.S., Budhiarta, A.A., Pranawa, S.H., Dennosepoetro, M., Budhianto,F.X., Widowu, J.A., Tanuwidjaja, S.J. and Pangemanan, M. (1983).

Metabolic effects of onion and green beans on diabetic patients. *Tohoku J. Exp. Med.*, **141**: 671-676.

Todar, K. (2005). "Staphylococcus", Todar's online textbook of bacteriology, online, available at: http: //textbookofbacteriology.net/staph.html (accessed 20 October 2005).

Tsao, S. M., and Yin, M. C. (2001). *In-vitro* antimicrobial activity of four diallyl sulphides occurring naturally in garlic and Chinese leek oils. *Journal of Medical Microbiology*, **50(7)**: 646-649.

Ubbink, J. B., van der Merwe, A., Delport, R., Allen, R. H., Stabler, S. P., Riezler, R., et al. (1996). The effect of a subnormal vitamin B6 status on homocysteine metabolism. *The Journal of Clinical Investigation*, **98**: 177-184.

Urios, P., Grigorova-Borsos, A.-M., and Sternberg, M. (2007). Flavonoids inhibit the formation of the cross-linking AGE pentosidine in collagen incubated with glucose, according to their structure. *European Journal of Nutrition*, **46**: 139-146.

Van Damme, E. J. M., Smeets, K., Engelborghs, I., Aelbers, H., Balzarini, J., Pusztai, A., et al. (1993). Cloning and characterization of the lectin cDNA clones from onion, shallot and leek. *Plant Molecular Biology*, **23**: 365-376.

Wagner, H., Dorsch, W., Bayer, T., Breu, W. and Willer, F. (1990). Antiasthmatic effects of onions: Inhibition of 5-lipoxygenase and cyclooxygenase in vitro by thiosulfinates and "Cepaenes". Prosta. Leuko. Ess. *Fatty Acids*, **39**: 59-62.

Wagner, H., Dorsch, W., Bayer, T., Breu, W., and Willer, F. (1990). Antiasthmatic effects of onions: inhibition of 5-lipoxygenase and cyclooxygenase in vitro by thiosulfinates and "cepaenes". *Prostaglandins Leukotrienes and Essential Fatty Acids*, **39**: 59-62.

Wang, B. H., Zuel, K. A., Rahaman, K., and Billington, D. (1998). Protective effects of aged garlic extract against bromobenzene toxicity to precision cut rat liver slices. *Toxicology*, **126**: 213-222.

Wang, H. X., and Ng, T. B. (2001). Purification of allivin, a novel antifungal protein from bulbs of the round-cloved garlic. *Life Sciences*, **70**: 357-365.

Wang, H. X., and Ng, T. B. (2004). Isolation of allicepin, a novel antifungal peptide from onion (*Allium cepa*) bulbs. *Journal of Peptide Science*, **10(3)**: 173-177.

Weber, N. D., Anderson, D. O., North, J. A., Murray, B. K., Lawson, L. D., and Hughes, B. G. (1992). *In vitro* virucidal activity of *Allium sativum* (garlic) extract and compounds. *Planta Medica*, **58**: 417- 423.

Wu, C. P., Calcagno, A. M., Hladky, S. B., Ambudkar, S. V., and Barrand, M. A. (2005). Modulatory effects of plant phenols on human multidrug-resistance proteins 1, 4 and 5 (ABCC1, 4 and 5). *FEBS Journal*, **272(18)**: 4725 - 4740.

Yang, J., Meyers, K. J., Van der Heide, J., and Liu, R. H. (2004). Varietal differences in phenolic content and antioxidant and anti proliferative activities of onions. *Journal of Agricultural and Food Chemistry*, **52**: 6787- 6793.

Yang, J., Meyers, K.J., VanderHeide, J. and Liu, R.H. 2004. Varietal differences in phenolic content and antioxidant and antiproliferative activities of onions. *J. Agric. Food Chem.*, **52**: 6787-6793.

Yeh, Y. Y., Lin, R. I., Yeh, S. M., and Evens, S. (1997). Garlic reduced plasma cholesterol in hypercholesterolemic men maintaining habitual diets. In H. Ohigashi, et al. (Eds.), Food factors for cancer prevention. Tokyo: Springer-Verlag. Yeh, Y. Y., and Yeh, S. M. (2006). Homocysteine-lowering action is another potential cardiovascular protective factor of aged garlic extract. *Journal of Nutrition*, **136**: 745-749.

You, W. C., Zhang, L., Gail, M. H., Ma, J. L., Chang, Y. S., Blot, W. J., *et al.* (1998). Helicobacter pylori infection, garlic intake and precancerous lesions in a Chinese population at low risk of gastric cancer. *International Journal of Epidemiology*, **27(6)**: 941-944.

You, W.C., Li, J.Y., Zhang, L., Jin, M.L., Chang, Y.S., Ma, J.L. and Pan, K.F. 2005. Etiology and prevention of gastric cancer: a population study in high risk area of China. *Chi. J. Dig.Dis.*, **6**: 149-154.

Zhang, Y., Moriguchi, T., Saito, H., and Nishiyama, N. (1997). Improvement of age-related deterioration of learning behaviours and immune responses by aged garlic extract. In P. P. Lachance, and P. Paul (Eds.), Nutraceuticals: Designer foods III. Garlic, soy and licorice (pp. 117-129). Trumbell, CT: Food and Nutrition Press.

Zohri, A. N., Abdel-Gawad, K., and Saber, S. (1995). Antibacterial, antidermatophytic and antioxigenic activities of onion (*Allium cepa* L.) oil. *Microbiological Research*, **150**: 167-172.

Natural Products: Research Reviews Vol. 1 (2012)
Editor: V.K. Gupta
Published by: DAYA PUBLISHING HOUSE, NEW DELHI

Pages **127–160**

5

Phytochemical Profile of *Aegle marmelos* (Family-Rutaceae)

Bhuwan B. Mishra[1], Navneet Kishore[1],
Vinod K. Tiwari[1] and Vyasji Tripathi[1]*

ABSTRACT

Aegle marmelos Corr. (Bael), a widely distributed plant containing numerous primary (fatty acids and amino acids) and secondary metabolites (including coumarins, alkaloid-amides, anthraquinones, triterpenoids and sterols etc.) has been considered significant in Homeopathic and Ayurvedic system of medicine for several clinical conditions. Besides tremendous use of root, bark, leaves and fruits, our group have search out new therapeutic potentials from seeds of A. marmelos. Present review highlights a current state of knowledge on phytochemical constituents of A. marmelos.

Keywords: *Aegle marmelos*, Medicinal properties, Phytochemicals.

Introduction

The tree *Aegle marmelos* Correa has been used for millennia in the Indian subcontinent and Indochina as a traditional medicine (Ghate, 1999). Historic mention of *A. marmelos* (also known as *Bael* fruit) has been traced to Vedic period (2000-800 BC) with first mention of *bael* fruit in *Yajurveda* (Roy and Singh, 1979; Mukherjee and Wahile, 2006). In fact, as per *Charaka* (1500 BC) no drug has been longer or better known or appreciated by Indian inhabitants than the *Bael* (Singanan *et al.,*2007). The

---

1   Department of Chemistry, Faculty of Science, Banaras Hindu University, Varanasi – 221 005, India

*   *Corresponding author:* E-mail: vyas_45@rediffmail.com, Phone: 05426704263; Fax: 91-542-2368174.

Bael tree is held sacred by Hindus, and is commonly grown in temple gardens in India (Dubey *et al.*, 2000).

The *Aegle marmelos* (*Bael* tree) is an Indian native plant, also naturalized in Burma, Pakistan, Bangladesh, Sri Lanka, Thailand and various other parts of South-eastern Asia (Kumar and Prabhakar, 1987). The tree grows wild in mixed deciduous and dry dipterocarp forests of former French Indochina up to 1000 m in altitude (Wiart, 2006) and is cultivated throughout the Pacific Rim. It is also found cultivated in Ceylon and northern Malaya, the drier areas of Java and to a limited extent on northern Luzon in the Philippine Islands.

Classification

Kingdom	:	Plantae
Division	:	Magnoliphyta
Class	:	Magnoliopsida
Subclass	:	Rosidae
Order	:	Sapindales
Family	:	Rutaceae
Genus	:	Aegle
Species	:	A. marmelos

Aegle marmelos (L.) Correa

The name *Aegle marmelos* (L) Correa is preferred in GRIN Taxonomy. The botanical name was ascribed to the author Linnaeus in the Transactions of the Linnaean Society (London 5:223) in the year 1800, and subsequently verified in the name of the author Correa by the Systematic Botany Laboratory in 1988. It is usually referred to in other literature as *Aegle marmelos* Correa that belongs to a small genus *Aegle* of family Rutaceae comprising only three species distributed in tropical Asia and Africa. Only one species *i.e. A. marmelos* commonly known as ' *Bel* or *Bael*' is cultivated in India due to its diverse range of biological significance (Dash and Padhy, 2006).

Common Names

Common English names used to refer *A. marmelos* are *viz.*, wood apple, Bengal quince, golden apple, Indian quince, holy fruit, and stone apple (Jauhari *et al.*, 1969). Other vernacular names are *Maredu* (Andra Pradesh), *Bel* (Bengal), *Bil* (Gujarat), *Kumbala* (Karnatak), *Vilwam* (Kerala), *Bilwa* (Sanskrit) and *Kuvalum* (Tamil Nadu) etc.

Botanical Description

Aegle marmelos is a medium sized, armed and deciduous tree growing up to 8.0 m in height, with a short thick trunk and narrow oval head; in the wild state smaller and more irregular, with short, strong, sharp, spiny branches bearing aromatic leaves, sweet scented and greenish-white flowers (Johri and Ahuja, 1957). Complete botanical description of the plant *A. marmelos* has been depicted under followings.

Leaves

Leaves alternate, compound, with one or two (rarely) pairs of shortly stalked opposite leaflets, and a larger long-petioled terminal one, leaflets 1-2 inches long, ovate or oval-ovate, abrupt or tapered at the base, somewhat attenuated towards the blunt apex, very shallowly serratocrenate, smooth, thin, midrib prominent beneath (Nene, 1977).

Flowers

Flowers 3/4 inch wide, sweet-scented, stalked, solitary, or in few-flowered, lax, erect, axillary or terminal cymes. Calyx shallow, with 5 short, broad teeth, pubescent outside. Petals 5 (rarely 4), oblong-oval, blunt, thick, pale greenish-white, dotted with glands, imbricate, spreading. Numerous stamens that is sometimes coherent in bindles, hypogynous with short filaments half as long as the linear anthers (Tilak and Nene, 1978; Srivastava and Singh, 2000). Disk none or very small. Ovary oblong-ovoid, slightly tapering into the thick short style which is again somewhat thickened upward, stigma capitate, axis of ovary wide, cells numerous, 8-20, small, arranged in a circle, with numerous ovules in each cell. The ovary wall in *A. marmelos* has thirteen carpellary dorsals and about 15-20 other wall bundles (Parmar and Kaushal, 1962).

Fruits

Fruit usually globose, 2 1/2 to 3 1/4 inches in diameter, pericarp nearly smooth, greyish-yellow, about 1/8 inch thick, hard, filled with softer tissue becoming very hard and orange-red when dry, cells as in ovary. Rind is about 1/8 inch thick and adherent to a light red pulp, in which there are 10-15 cells, each of which contains several woolly seeds. It has a faint aromatic odor and mucilaginous taste (Parmar and Kaushal, 1962).

Seeds

Seeds very numerous, somewhat compressed, ranged in closely packed tiers in the cells and surrounded by a very tenacious, slimy transparent mucus which becomes hard when dry, testa white, covered with woolly hairs immersed in the mucus, embryo with large cotyledons, and a short superior radicle, no endosperm (Kakiuchi *et al.*, 1991).

Medicinal Properties of Various Plant Parts

The whole plant as well as the separate plant parts *i.e.* roots, bark, stem, leaves and fruits are used in Homeopathic and Ayurvedic system of medicines.In Ayurveda, the plant has been considered significant in clinical conditions like swollen joints, healing of wound (Udupa *et al.*, 1994; Jaswanth *et al.*, 2001) high blood pressure, troubles during pregnancy, snake bites, eye problems, fever, burn feeling on the skin, diarrhea (Shoba and Thomas, 2001), urinary troubles, nocturnal seminal emission, colds and catarrh, illnesses like consumption, jaundice, scurvy and skin problems (Sofowora, 1982). A virucidal agent, marmelide showing efficient activity against human coxsackie viruses B1-B6 has been isolated from *A. marmelos* (Badam *et al.*, 2002).

Leaf

The plant is also known as herbal medicine for the treatment of diabetes mellitus (Narendhirakannan *et al.*, 2006). The leaf extracts have shown hypoglycemic and anti-hyperglycemic effects on rabbits (Nammi *et al.*, 2001). The alkaloid extract prepared from leaves and the crude aqueous leaf extract (1 g/kg for 30 days) exhibit hypoglycaemic effect in alloxanized diabetic rats. Aqueous leaf extract reverses the increase in *Km* values of liver malate dehydrogenase enzyme and improves histopathological alterations in the pancreatic and kidney tissues of STZ induced diabetic rats. The leaf extract has been found effective in the regeneration of damaged pancreas (β-cells) in diabetic rats. Fresh aqueous and alcoholic leaf extracts possess cardio tonic effects like digitalis and decrease the requirement of circulatory stimulants and at 5 μg/mL concentration, protect HPBLs against radiation-induced DNA damage and genomicinstability (Jagetia *et al.*, 2003). The methanol extract of leaves significantly reduce the writhing induced by acetic acid. The leaves of the plant have shown promising anti-spermatogenic, anti-inflammatory, antipyretic and analgesic properties (Arul *et al.*, 2005). The leaf extract at dose levels of 200 and 300 mg/kg exhibits significant analgesic activity in the tail flick test (Shankarananth *et al.*, 2007). The leaves of *A. marmelos* display a high margin of drug safety. The intraperitoneal administration of leaf extracts (total alcoholic, total aqueous, whole aqueous and methanol extracts) at 50, 70, 90 and 100 mg/kg body weight for 14 consecutive days to male and female Wistar rats produce no short-term toxicological effects. Also, doses of up to 1000 mg/kg body weight do not produce mortality, indicating that the leaf extract of *A. marmelos* is devoid of short-term harmful effects (Veerappan *et al.*, 2007).

Root and Bark

The methanol extracts of root bark exhibits preventive effects on myocardial diseases. The root and bark of tree are used in the treatment of fever by making a decoction good against malarial, jaundice and skin diseases such as ulcers, urticaria, and eczema. Oral administration of aqueous decoction prepared from root bark produces hypoglycaemic and anti-diarrhoeal effects (Mazumder *et al.*, 2006).

Fruit

Fruits (pulp and rind) are especially important in Homeopathy and Ayurvedic system of medicine. Significant anti-hyperlipidaemic effect of fruit extracts has been observed in STZ-induced diabetes rats (Kamalakkannan and Prince, 2005). The aqueous extracts of fruits exhibit hypoglycaemic activity. The ripe fruit, tamarind and sugar in mixture are used as laxative to overcome constipation and body heating problems. Sweet drink prepared from the pulp is considered significant for patients who have just recovered from bacillary dysentery. The ripe fruit is a good cure for dyspepsia (Kalaivani *et al.*, 2009). The unripe and half-ripe fruits improve appetite and digestion.

Seed

The *A. marmelos* seeds exhibit the antimicrobial and antihelminthic properties. Seed oil exhibits antibacterial activity against different strains of *Vibrios* and inhibits

Table 5.1 : Organic compounds isolated from *A. marmelos.*

Organic Constituents	Plant Part	References
(i) Fatty acids:		
Stearic acid	Seed	(Singh and Malik, 2000)
	Fruit	(Human Metabolome Database, 2002)
Palmitic acid	Seed	(Singh and Malik, 2000)
	Fruit	(Human Metabolome Database, 2002)
Linoleic acid	Seed	(Singh and Malik, 2000)
	Fruit	(Human Metabolome Database, 2002)
Linolenic acid	Seed	(Singh and Malik, 2000)
	Fruit	(Human Metabolome Database, 2002)
Valencic acid	Leaf	(Ali and Pervez, 2004)
(ii) Amino acids		
Alanine	Fruit	(Barthakur and Arnold, 1989)

Contd...

Table 5.1 – *Contd...*

Organic Constituents	Plant Part	References
L-Arginine	Fruit	(Barthakur and Arnold, 1989)
Aspartic acid	Fruit	(Barthakur and Arnold, 1989)
L-Cystine	Fruit	(Barthakur and Arnold, 1989)
Glutamic acid	Bark	(Barthakur and Arnold, 1989)
Glycine	Fruit	(Barthakur and Arnold, 1989)

Contd...

Table 5.1–*Contd...*

Organic Constituents	Plant Part	References
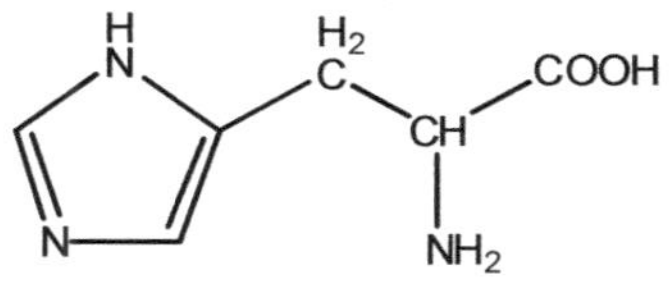 Histidine	Fruit	(Barthakur and Arnold, 1989)
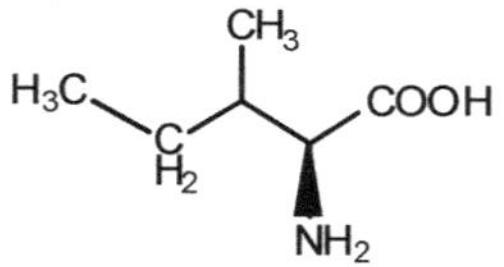 Isoleucine	Fruit	(Barthakur and Arnold, 1989)
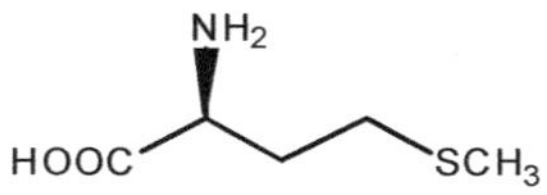 Methionine	Fruit	(Barthakur and Arnold, 1989)
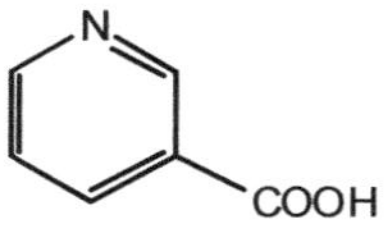 Niacin	Fruit	(Barthakur and Arnold, 1989)
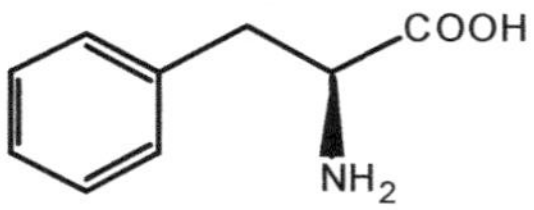 Phenylalanine	Fruit	(Barthakur and Arnold, 1989)

Contd...

Table 5.1–*Contd...*

Organic Constituents	Plant Part	References
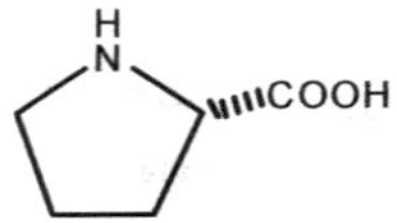 Proline	Fruit	(Barthakur and Arnold, 1989)
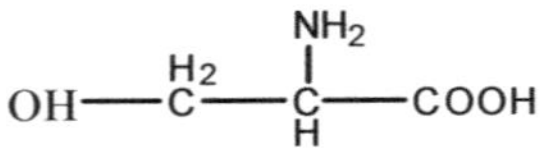 Serine	Fruit	(Barthakur and Arnold, 1989)
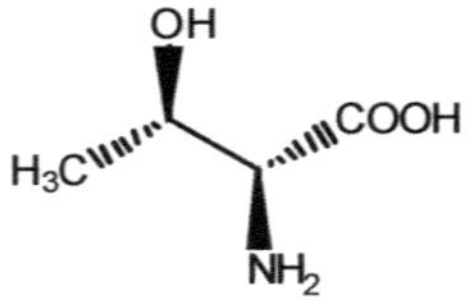 Threonine	Fruit	(Barthakur and Arnold, 1989)
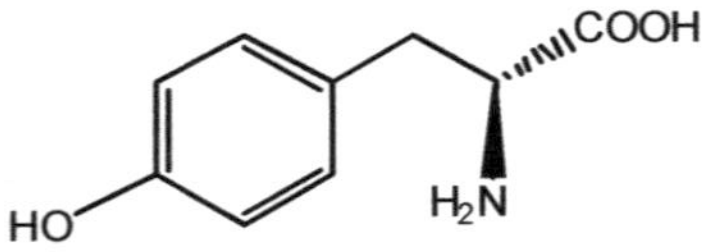 Tyrosine	Fruit	(Barthakur and Arnold, 1989)
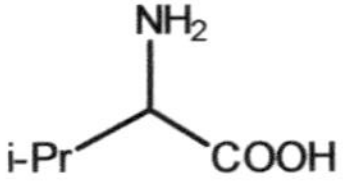 Valine	Fruit	(Barthakur and Arnold, 1989)

Contd...

Table 5.1–*Contd...*

Organic Constituents	Plant Part	References

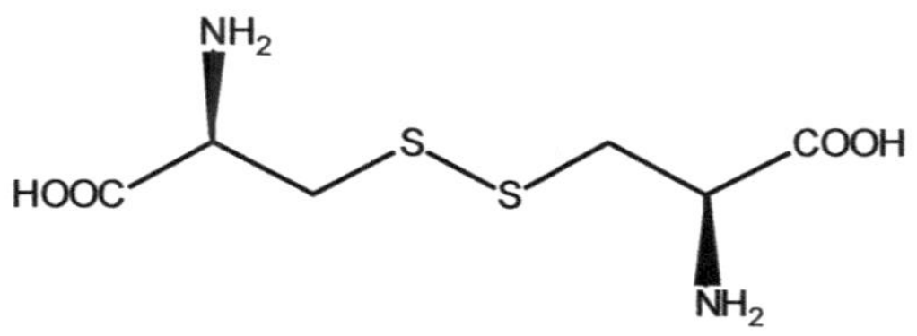

L-Cystine

Fruit

(Barthakur and Arnold, 1989)

(iii) Glycosides:

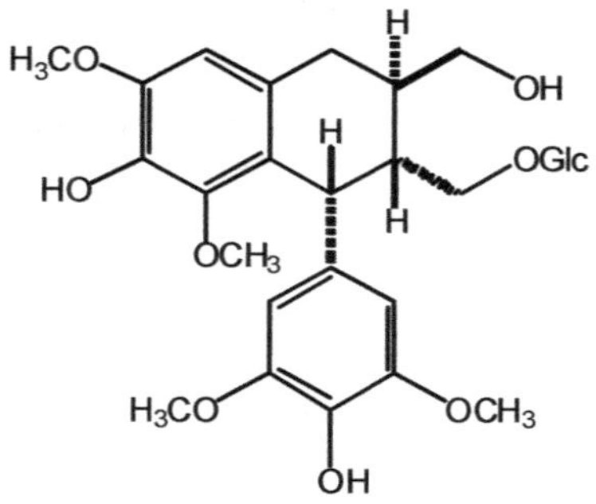

(-)-4-Epi-lyoniresinol-3α-O-β-D-glucopyranoside

Bark

(Ohashi and Watanabe, 1994)

Contd...

Table 5.1–*Contd...*

Organic Constituents	Plant Part	References
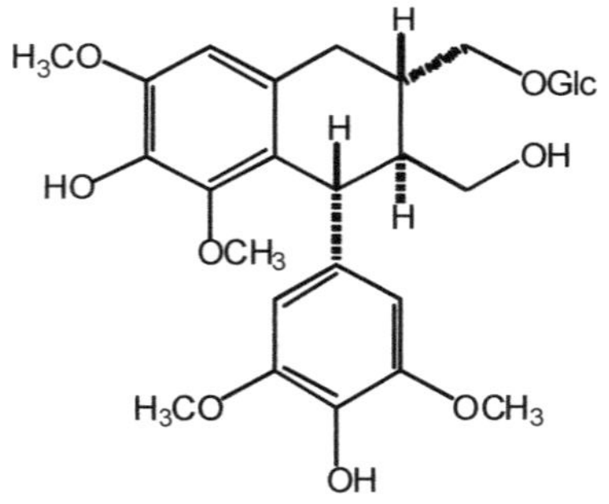 (-)-Lyoniresinol -2α-O-β-D-glucopyranoside	Bark	(Ohashi and Watanabe, 1994)
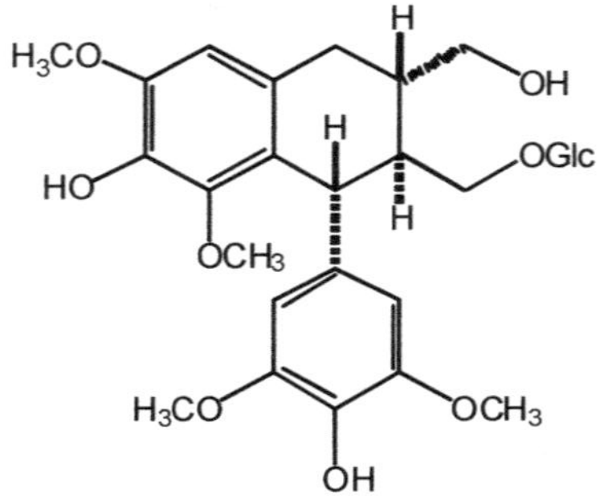 (-)-Lyoniresinol -3α-O-β-D-glucopyranoside	Bark	(Ohashi and Watanabe, 1994)

Contd...

Table 5.1 –*Contd...*

Organic Constituents	Plant Part	References
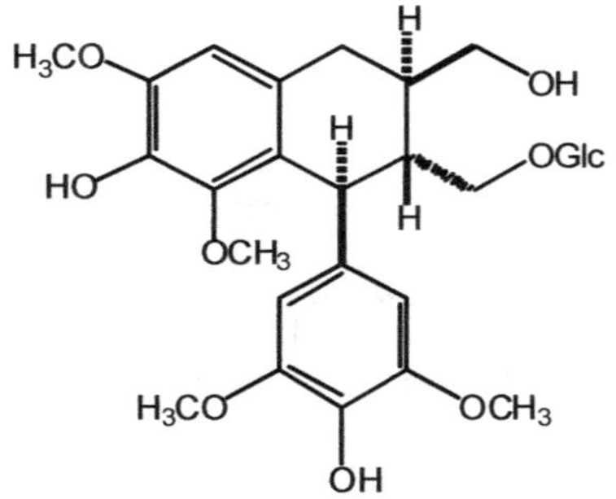 (+)-Lyoniresinol-3α-O-β-D-glucopyranoside	Bark	(Ohashi and Watanabe, 1994)
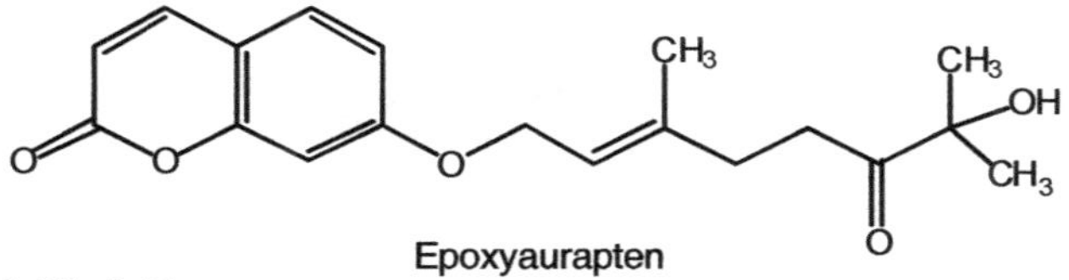 Epoxyaurapten	Root	(Yang et al., 1996a)
(iv) Alkaloids: 4-Methoxy-1-methyl-2-quinolone	Root	(Yang et al., 1996a)

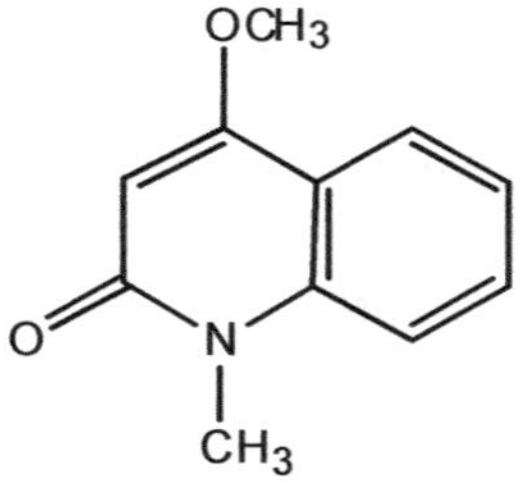

Contd...

Table 5.1–*Contd...*

Organic Constituents	Plant Part	References
Aegeline	Leaf Fruit Leaf	(Kapoor, 1990) (Sharma and Sharma, 1981) (Chatterjee and Bose, 1952)
Dictamine	Root Fruit Wood	(Yang et al., 1996a) (Human Metabolome Database, 2002) (Kapoor, 1990)
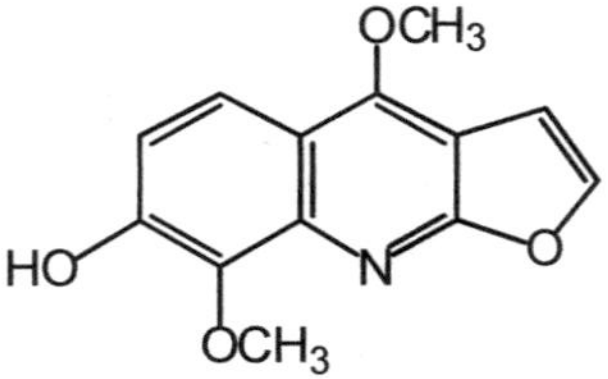 Haplopine	Aerial part	(Banerji et al., 1988)

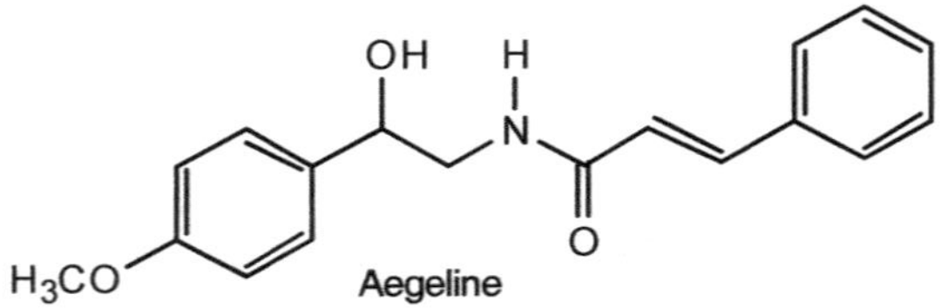

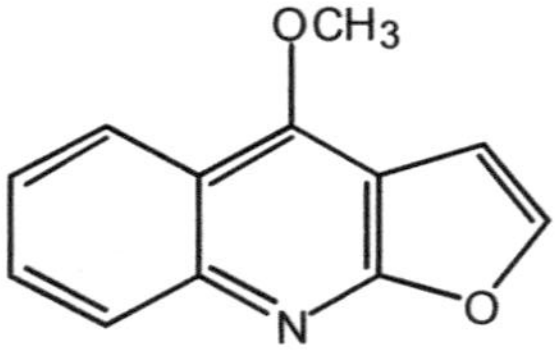

Contd...

Table 5.1–*Contd...*

Organic Constituents	Plant Part	References
OCH₃ ... (structure with furoquinoline nucleus, OCH₃, H₃CO, OCH₃, N, O)	Aerial part Root Leaf	(Banerji *et al.*, 1988) (Chatterjee and Chaudhury, 1960) (Chrubasik *et al.*, 2006)
Skimmianine		
(structure: OCH₃, HO, N–CH₃, O)	Root	(Yang *et al.*, 1996b)
Integriquinolone		
OCH₃ ... (furoquinoline structure, OCH₃, N, O)	Root Fruit Bark	(Yang *et al.*, 1996b) (Human Metabolome Database, 2002) (Karawya *et al.*, 1980)
γ-fagarine		

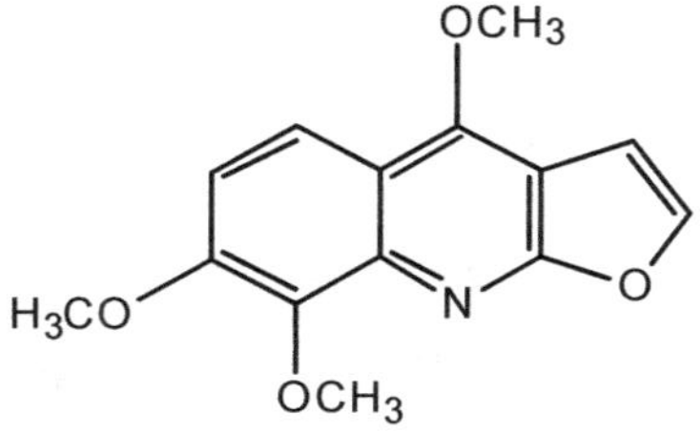

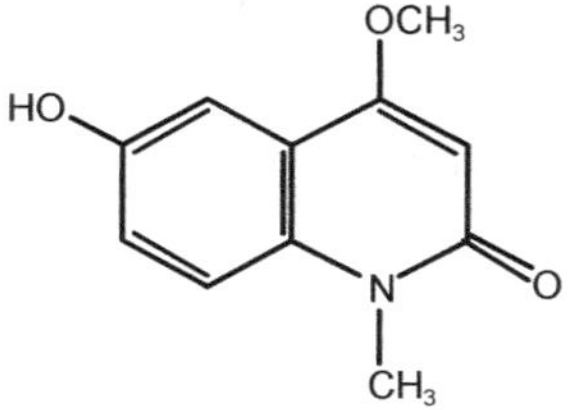

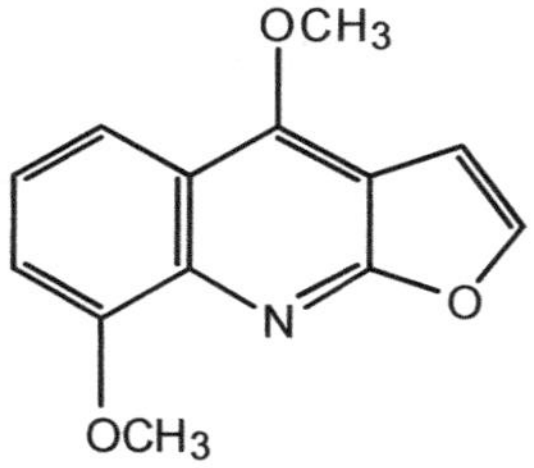

Contd...

Tabl≣ 5.1–*Contd...*

Organic Constituents	Plant Part	References
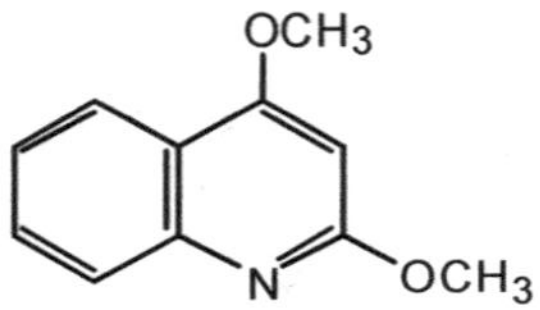 Montanine	Leaf	(Ali and Pervez, 2004)
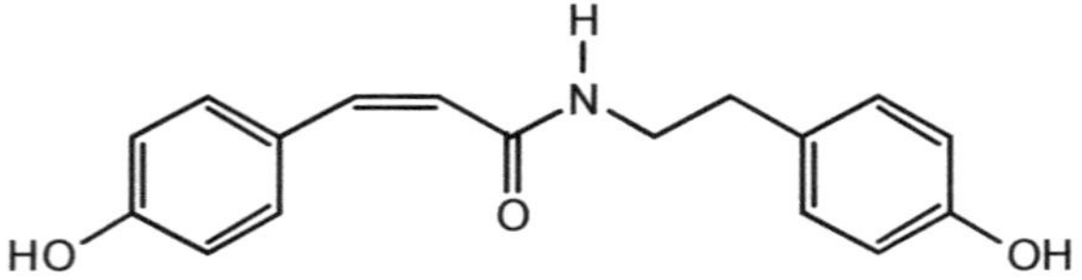 N-p-cis-Coumaryltyramine	Leaf	(Ali and Pervez, 2004)
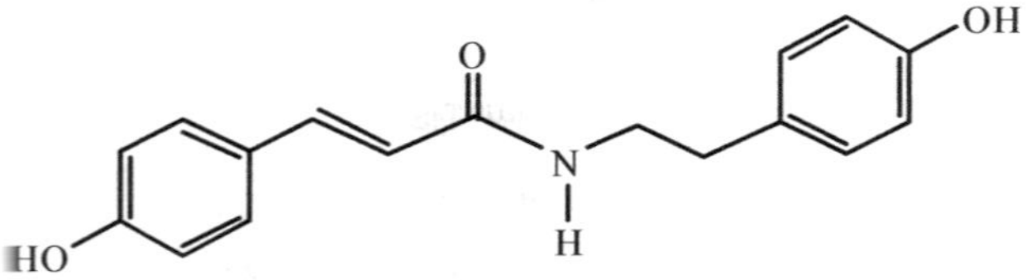 N-p-trans-Coumaryltyramine	Leaf	(Ali and Pervez, 2004)
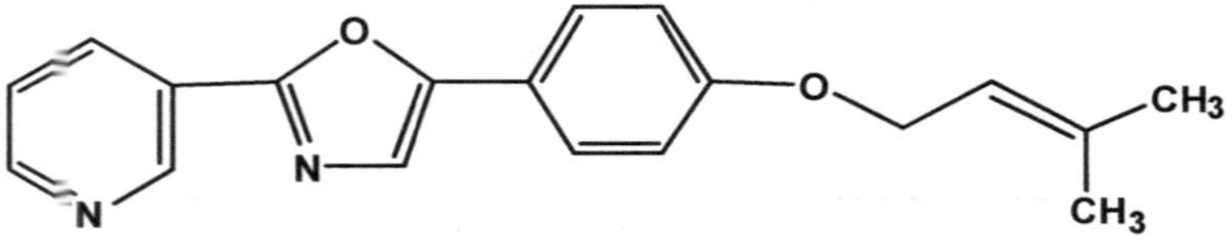 O-isopentinylhalfordinol	Fruit	(Sharma *et al.*, 1981)

Contd...

Table 5.1—*Contd...*

Organic Constituents	Plant Part	References
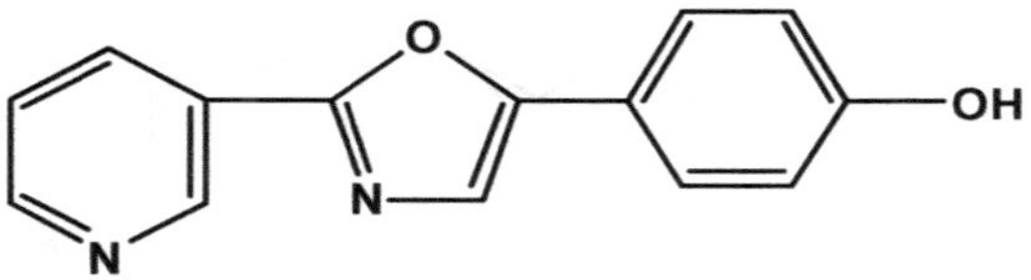 O-methylhalfordinol	Fruit	(Manandhar *et al.*, 1978)
(v) Coumarins:		
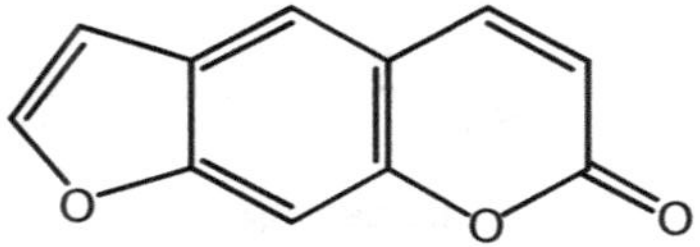 Psoralen	Fruit	(Chrubasik *et al.*, 2006)
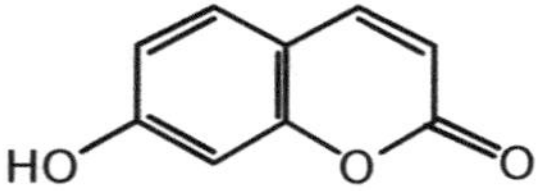 Umbelliferone	Root	(Chatterjee and Chaudhury, 1960)
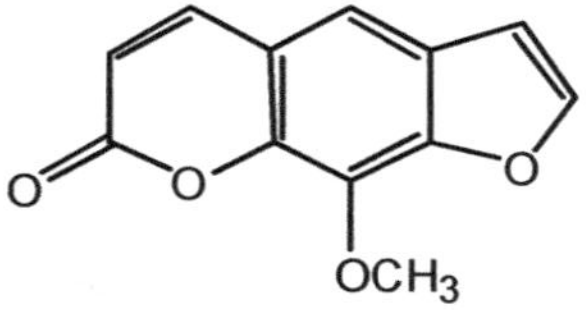 Xanthotoxin	Root	(Yang *et al.*, 1996b)

Contd...

Table 5.1–*Contd...*

Organic Constituents	Plant Part	References
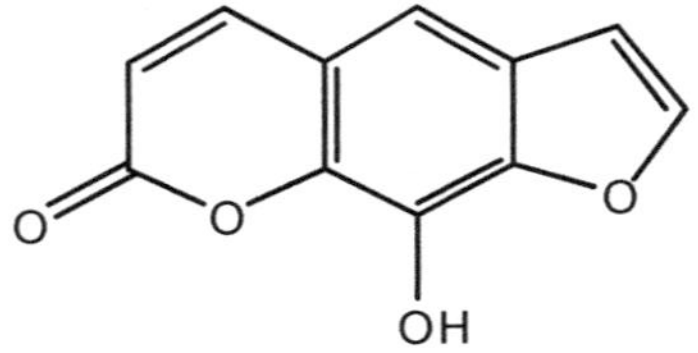 Xanthotoxol	Wood	(Srivastava *et al.*, 1996)
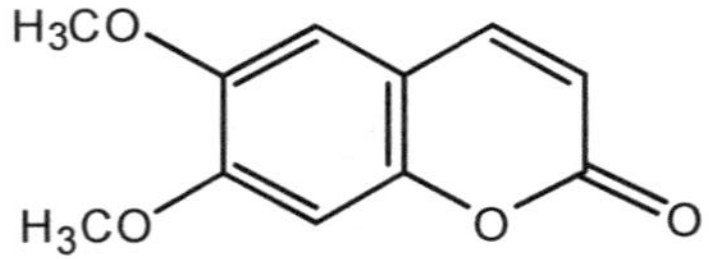 Scoparone	Fruit	(Sharma *et al.*, 1980)
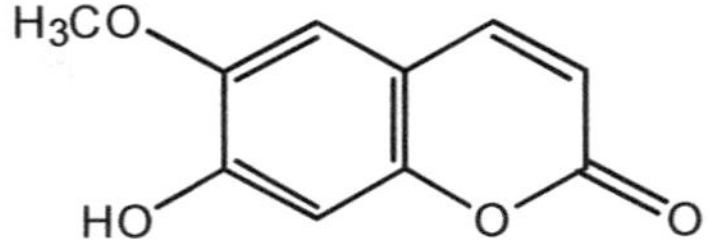 Scopoletin	Fruit Leaf	(Sharma *et al.*, 1980) (Panda and Kar, 2006)
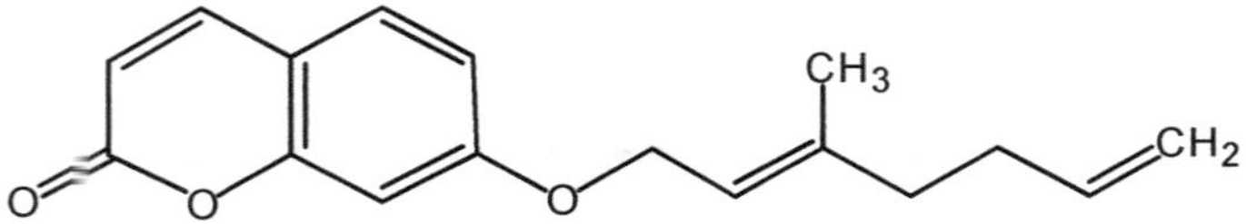 Aurapten	Root	(Chatterjee and Chaudhury, 1960)

Contd...

Table 5.1 *–Contd...*

Organic Constituents	Plant Part	References
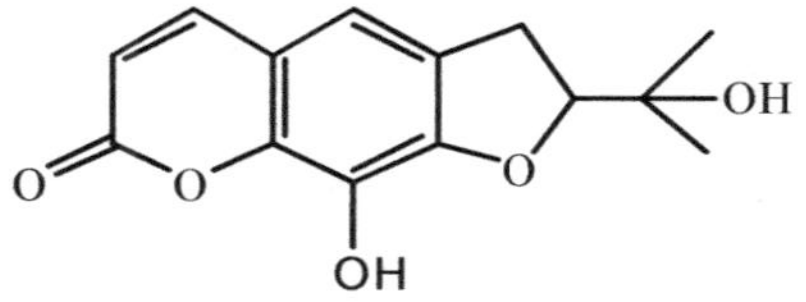 Rutaretin	Leaf	(Ali and Pervez, 2004)
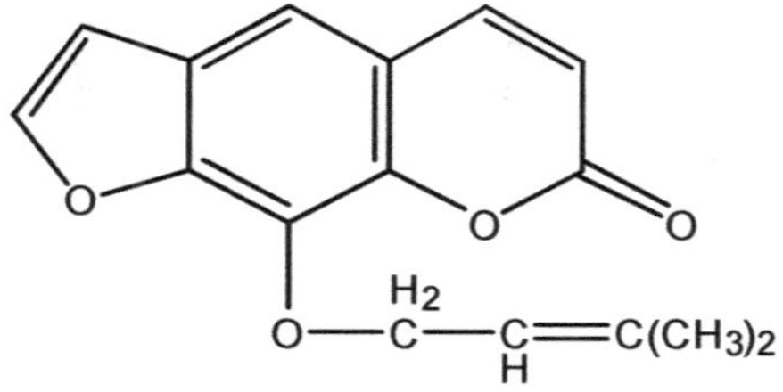 Imperatorin	Root Fruit	(Yang *et al.*, 1996b) (Sharma *et al.*, 1981)
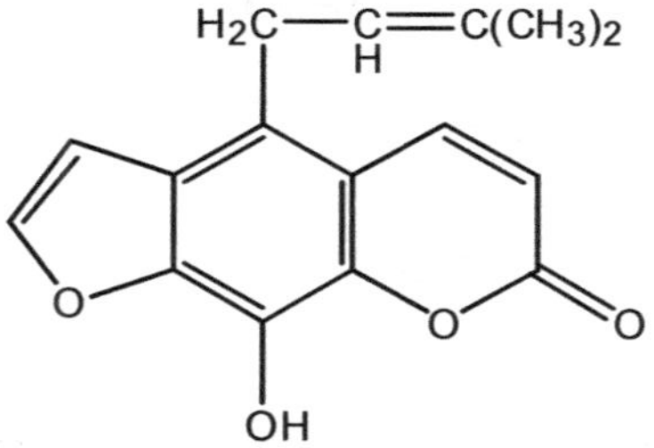 Allo-imperatorin	Fruit	(Saha and Chatterjee, 1957)

Contd...

Table 5.1–Contd...

Organic Constituents	Plant Part	References
 Alloimperatorin methyl ether	Fruit	(Sharma et al., 1981)
 Marmenol	Leaf	(Ali and Pervez, 2004)
 Marmin	Wood Bark Root	(Srivastava et al., 1996) (Chrubasik et al., 2006) (Chatterjee and Chaudhury, 1960)

Contd...

Table 5.1–*Contd...*

Organic Constituents	Plant Part	References
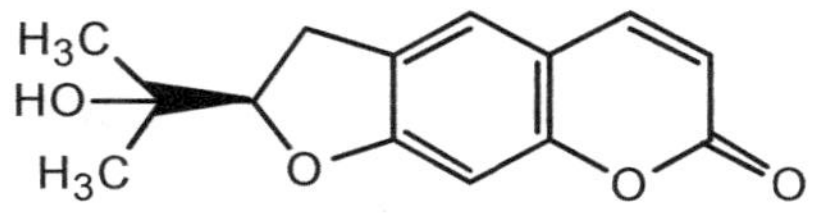 Marmesin	Root Fruit Bark	(Yang *et al.*, 1996b) (Sharma *et al.*, 1981) (Goswami *et al.*, 2005)
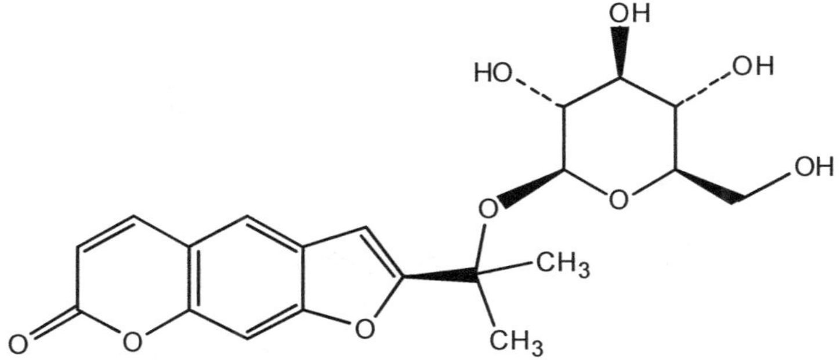 Marmesin	Leaf	(Sharma *et al.*, 1980)
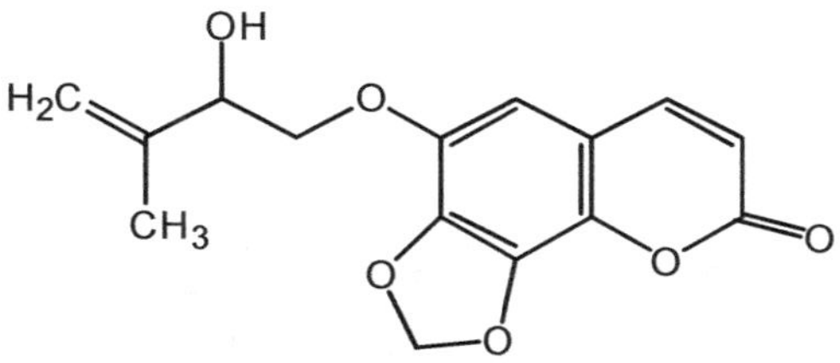 (+)-4-(20-hydroxy-30-methylbut-30-enyloxy)-8H- [1,3]dioxolo[4,5-h]chromen-8-one	Seed	(Mishra *et al.*, 2010a)

Contd...

Table 5.1 – *Contd*...

Organic Constituents	Plant Part	References
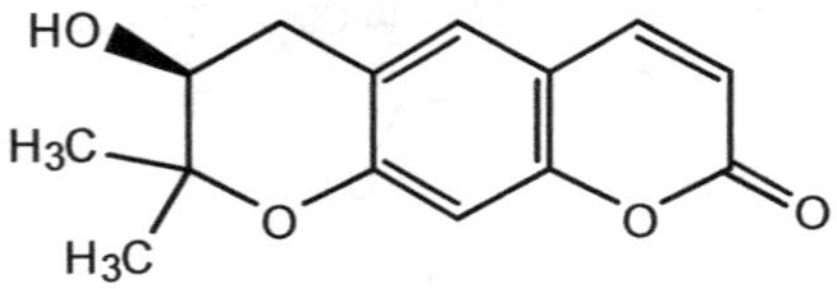 Decursinol	Root	(Basu and Sen, 1974)
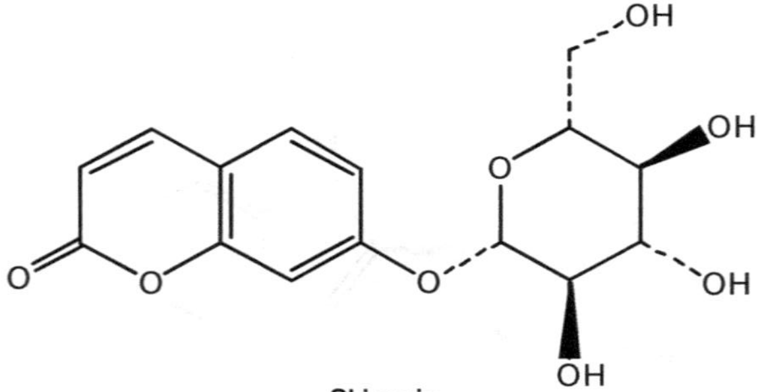 Skimmin	Fruit	(Sharma *et al.*, 1980)
(vi) Anthraquinones: 6-Hydroxy-1-methoxy-3-methylanthraquinone	Wood	(Srivastava *et al.*, 1996)

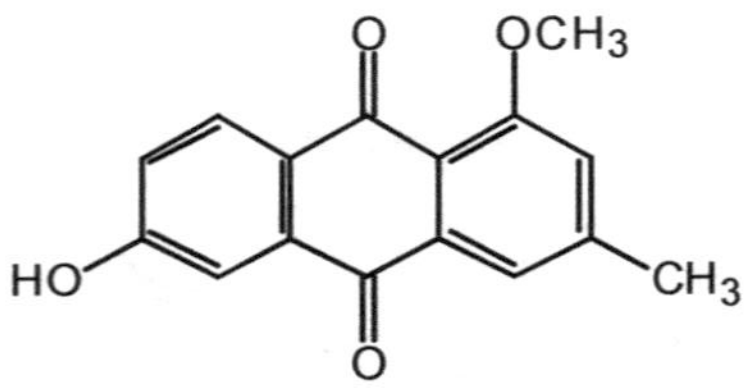

Contd...

Table 5.1—*Contd...*

Organic Constituents	Plant Part	References
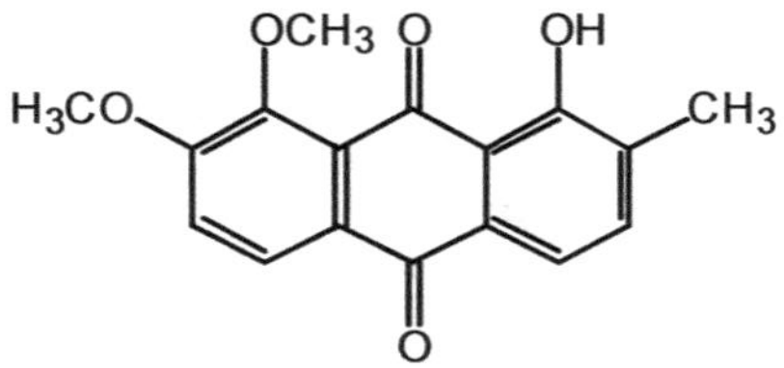 7,8-Dimethoxy-1-hydroxy-2-methylanthraquinone	Wood	(Srivastava et al., 1996)
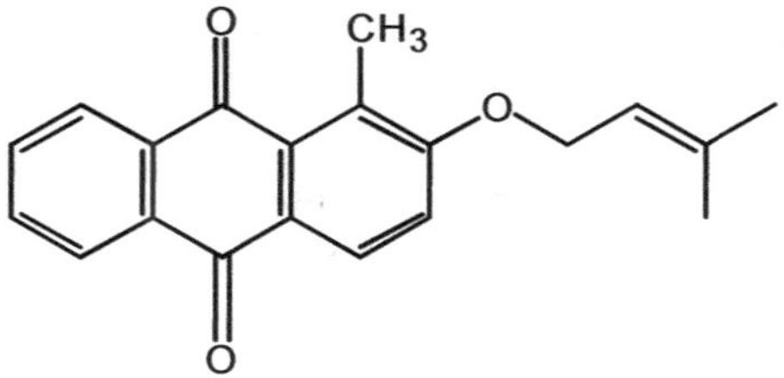 1-Methyl-2-(3'-methyl-but-2'-enyloxy)-anthraquinone	Seed	(Mishra et al., 2010b)
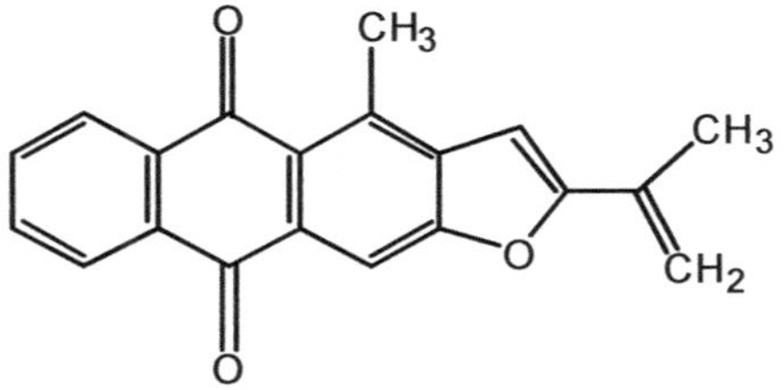 2-Isopropenyl-4-methyl-1-oxa-cyclopenta[b]anthracene -5,10-dione	.Seed	(Mishra et al., 2010b)

Contd...

Table 5.1–*Contd...*

Organic Constituents	Plant Part	References
(vi) Terpenoids:		
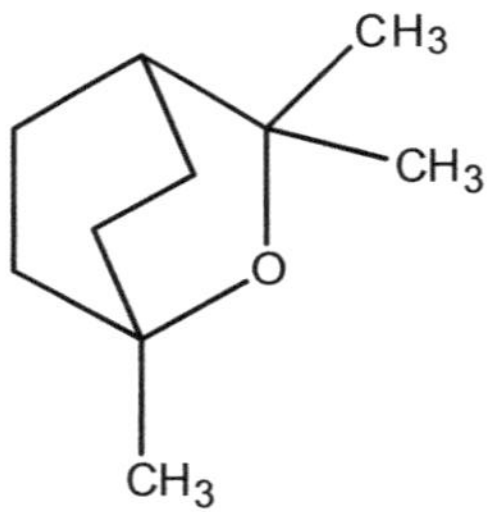 Cineole	Leaf	(Chrubasik *et al.*, 2006)
$C_{10}H_{18}O_2$ *Cis*-linalool oxide	Fruit	(Mac and Pieris, 1981)
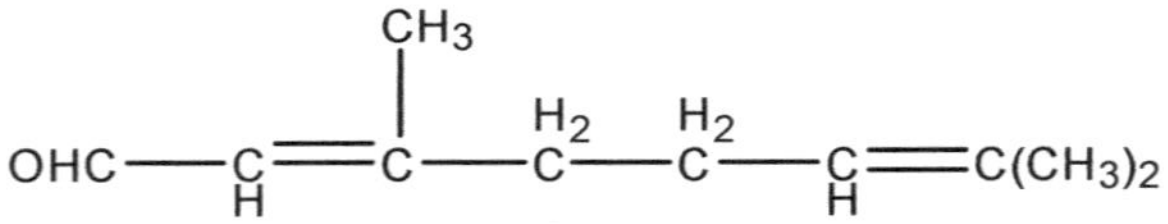 Citral	Leaf	(Chrubasik *et al.*, 2006)
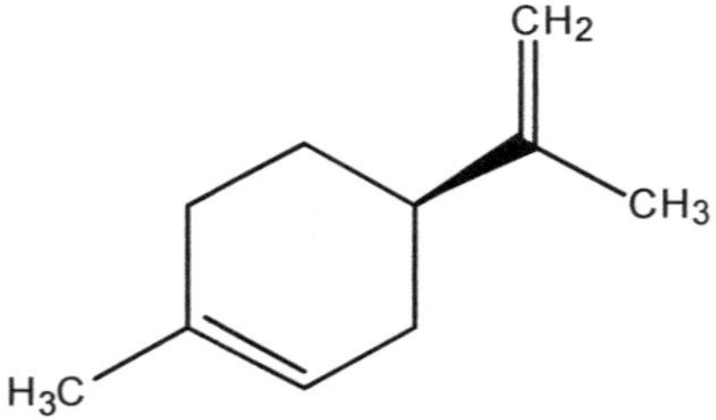 *d*-Limonene	Leaf	(Chrubasik *et al.*, 2006)

Contd...

Table 5.1–*Contd...*

Organic Constituents	Plant Part	References
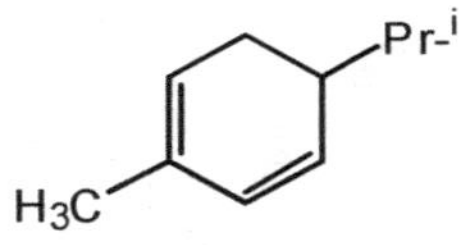 α-Phellandrene	Fruit	(Kapoor, 1990)
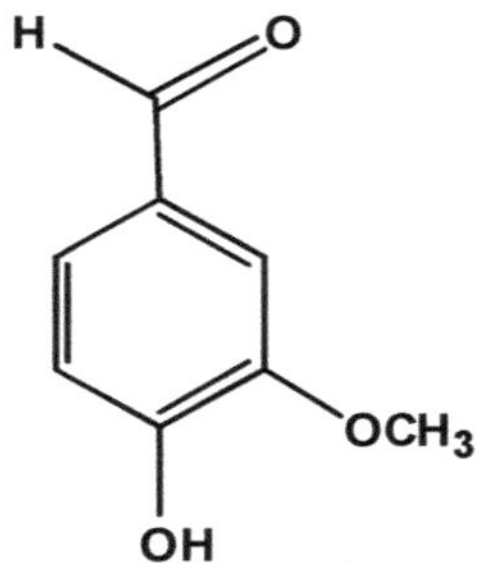 Vanillin	Fruit	(Tokitomo *et al.,*1982)
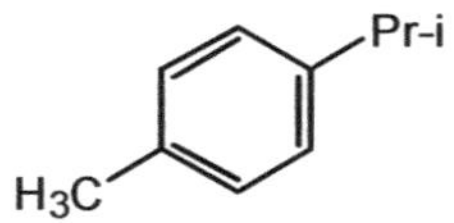 p-Cymene	Leaf	(Chrubasik *et al.,* 2006)

Contd...

Table 5.1 –Contd....

Organic Constituents	Plant Part	References
α-Amyrin	Fruit	(Human Metabolome Database, 2002)
Lupeol	Leaf	(Arul et al., 1999)

Contd...

Table 5.1–*Contd...*

Organic Constituents	Plant Part	References
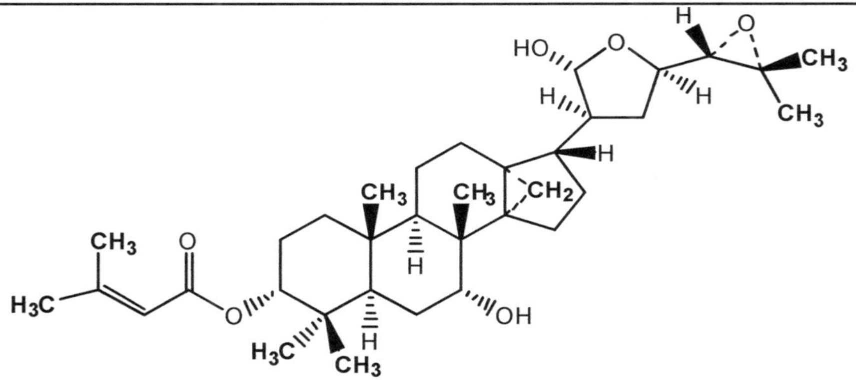 Skimmiarepin-C	Bark	(Samarasekera *et al.*, 2004)
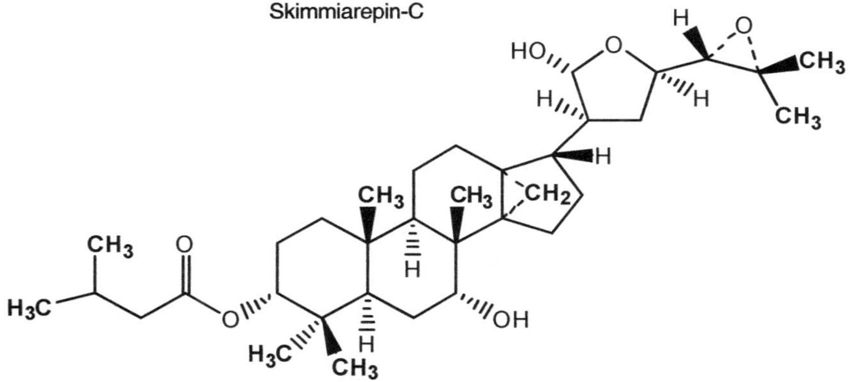 Skimmiarepin-A	Bark	(Samarasekera *et al.*, 2004)

Contd...

Table 5.1–*Contd...*

Organic Constituents	Plant Part	References
(vii) Others:		
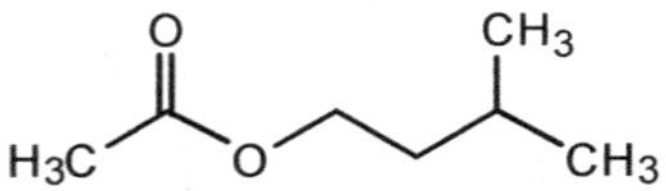 Isoamyl acetate	Fruit	(Mac and Pieris, 1981)
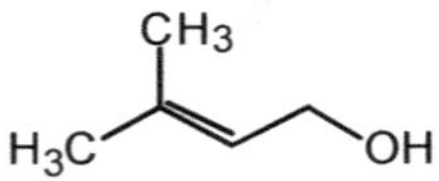 3-Methylbut-2-en-1-ol	Fruit	(Mac and Pieris, 1981)
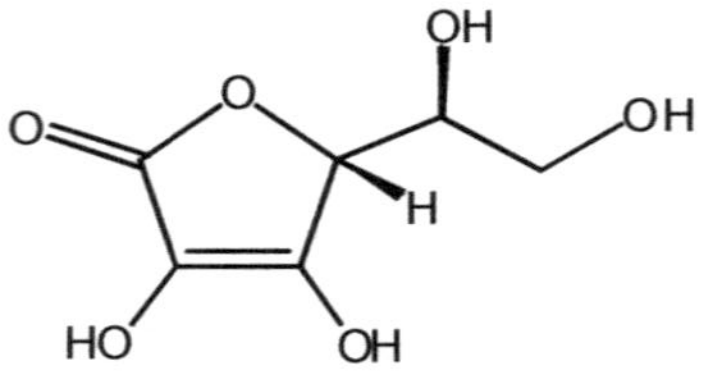 Ascorbic acid	Fruit	(Singh and Malik, 2000)
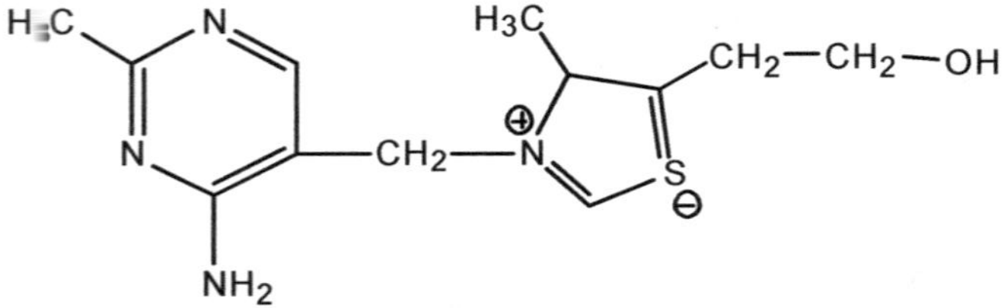 Thiamin	Fruit	(Singh and Malik, 2000)

Contd...

Table 5.1–*Contd...*

Organic Constituents	Plant Part	References
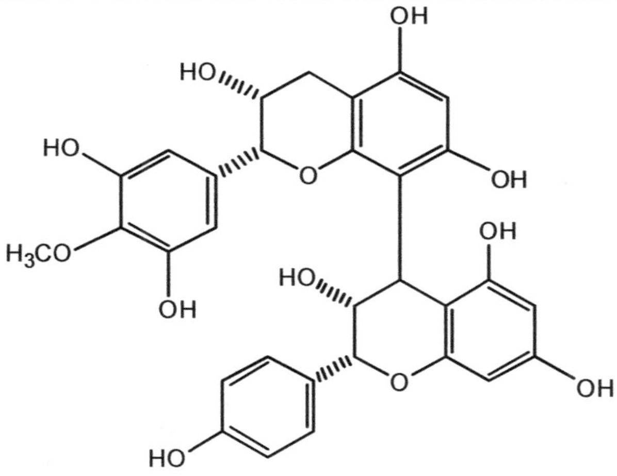 Proanthocynidin	Fruit	(Abeysekera *et al.,* 1996)
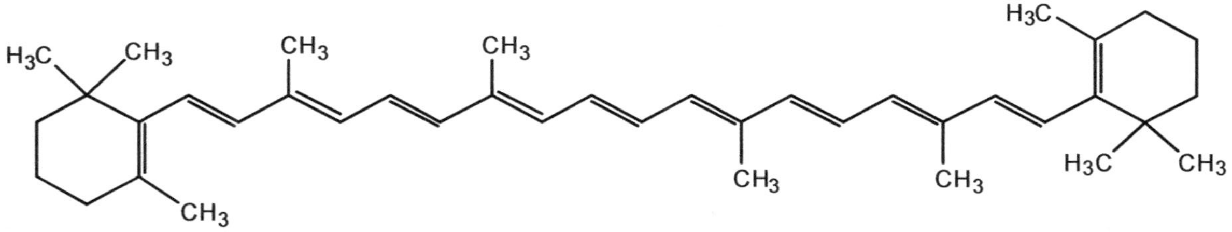 β-carotene	Fruit	(Chrubasik *et al.,* 2006)

Contd...

Table 5.1–*Contd...*

Organic Constituents	Plant Part	References
 β-sitosterol	Leaf Wood Root Fruit	(Chakravarti and Dasgupta, 1958) (Srivastava et al., 1996) (Yang et al., 1996b) (Saha and Chatterjee, 1957)
 β-sitosterol-β-D-glucoside	Leaf Fruit	(Sharma et al., 1980) (Sharma et al., 1980)

Contd...

Table 5.1 *–Contd...*

Organic Constituents	Plant Part	References
 Rutin	Leaf	(Sharma *et al.*, 1980)

the growth of *Vibrio cholerae*, *Staphylococcus aureus*, and *Escherichia coli*. The compound luvangetin isolated from the seeds of *A. marmelos* have shown anti-ulcer activity. The aqueous seed extract of plant displays anti-diabetic and hypolipidemic effects in diabetic rats (Kesari *et al.*, 2006). Essential oil exhibits antifungal activities against fungi *Physalospora tucumanensis*, *Ceratocystis paradoxa*, *Cephalosporium sacchari* and *Sclerotium rolfsii*.

Earlier Work Done on Aegle marmelos

Quantitative reports show that an *A. marmelos* fruit consists of moisture 61.5 per cent, protein 1.8 per cent, fat 0.3 per cent, minerals 1.7 per cent, fibre 2.9 per cent and carbohydrates 31.8 per cent per 100 grams of edible portion. Its mineral and vitamin contents include calcium, phosphorus, iron, carotene, thiamin, riboflavin, niacin and vitamin C. Its calorific value is 137. Earlier, several coumarins (Goswami *et al.*, 2005), anthraquinones (Srivastava *et al.*, 1996), sterols (Karawya *et al.*, 1980), lignan-glucosides, triterpenoids (Samarasekera *et al.*, 2004), alkaloid-amides and carbohydrates have been isolated from root, stem, leaves and fruits of *A. marmelos*. Organic and inorganic chemical constituents so far have been reported from *A. marmelos* are appended below in Tables 5.1 and 5.2, respectively.

Table 5.2: Inorganic compounds isolated from *A. marmelos*.

Inorganic Constituents	Plant Part	References
Boron (compounds)	Fruit	(Barthakur and Arnold, 1989)
Calcium (compounds)	Fruit	(Barthakur and Arnold, 1989)
Chlorine (compounds)	Fruit	(Barthakur and Arnold, 1989)
Copper (compounds)	Fruit	(Barthakur and Arnold, 1989)
Iron (compounds)	Fruit	(Barthakur and Arnold, 1989)
Magnesium (compounds)	Wood	(Barthakur and Arnold, 1989)
Magnesium (compounds)	Fruit	(Barthakur and Arnold, 1989)
Phosphorus (compounds)	Fruit	(Barthakur and Arnold, 1989)
Potassium (compounds)	Fruit	(Barthakur and Arnold, 1989)
Sodium (compounds)	Fruit	(Barthakur and Arnold, 1989)

Acknowledgement

Banaras Hindu University is sincerely acknowledged for providing the infrastructural and research facilities for the present work.

References

Abeysekera, A.M., De-Silva, K.T.D., Samarasinghe, S., Seneviratne, P.A.K., Van-Den-Berg, A.J.J. and Labadie, R.P. (1996). An immunomodulatory *C*-glucosylated propelargonidin from the unripe fruit of *Aegle marmelos*. *Fitoterapia*, **67**: 367-370.

Ali, G.M. and Pur vcz, M.K. (2004). Marmenol: A 7-geranyloxycoumarin from the leaves of *Aegle marmelos* Corr. *Natural Product Research*, **18**: 141-146.

Arul, V., Kumaraguru, S. and Dhananjayan, R. (1999). Effects of aegeline and lupeol, the two cardioactive principles isolated from the leaves of *Aegle marmelos* Corr. *Journal of Pharmacy and Pharmacology*, **51**(S): 252.

Arul, V., Miyazaki, S. and Dhananjayan, R. (2005). Studies on the anti-inflammatory, antipyretic and analgesic properties of the leaves of *Aegle marmelos* Corr. *Journal of Ethno-pharmacology*, **96**: 159-163.

Badam, L., Bedekar, S.S., Sonawane, K.B. and Joshi, S.P. (2002). *In vitro* antiviral activity of bael (*Aegle marmelos* Corr.) upon human coxsackieviruses B1-B6. *Journal of Communicable Diseases*, **34**: 88-99.

Banerji, J., Das, A.K., Ghoshal, N. and Das, B. (1988). Studies on Rutaceae Part VIII, Chemical Investigation on the constituents of *Atalantia wightii* Tanaka, *Aegle marmelos* Correa Ex Koen, *Ruta graveolens* Linn. and *Micromelum pubescens* Blume. *Indian Journal Chemistry*, Section B, **27**: 594-6.

Barthakur, N.N. and Arnold, N.P. (1989). Certain organic and inorganic constituents in Bael (*Aegle marmelos* Correa) fruit. *Tropical Agriculture* (Trinidad), **66**: 65-8.

Basu, D. and Sen, R. (1974). Alkaloids and coumarins from root bark of *Aegle marmelos*. *Phytochemistry*, **13**: 2329-30.

Chakravarti, R.N. and Dasgupta, B. (1958). Sitosterol from the leaves of *Aegle marmelos* Correa. *Journal of Indian Chemical Society*, **35**: 194-6.

Chatterjee, A. and Bose, S. (1952). Studies on the active principles isolated from the leaves of *Aegle marmelos* Correa. *Journal of Indian Chemical Society*, **29**: 425-429.

Chatterjee, A. and Chaudhury, B. (1960). Occurrence of auraptene, umbelliferone, marmin, lupeol and skimmianine in the root of *Aegle marmelos* Correa. *Journal of Indian Chemical Society*, **37**: 334-6.

Chrubasik, C., Duke, R.K. and Chrubasik, S. (2006). The evidence for clinical efficacy of rose hip and seed: A systematic review. *Phytotherapy Research*, **20**: 1-3.

Dash, S.K., and Padhy, S. (2006). Review on ethnomedicines for diarrhoea diseases from Orissa: Prevalence versus culture. *Journal of Human Ecology*, **20**: 59-64.

Dubey, G., Shahu, P. and Sahu, R. (2000). Role of plant in different religious ceremonies common to Bundelkhand region of Madhya Pradesh. *Journal of Medicinal and Aromatic Plant Sciences*, **22-23**: 542-5.

Ghate, V.S. (1999). Bruhat-panchamula in ethno-medico-botany and Ayurved. *Journal of Medicinal and Aromatic Plant Sciences*, **21**: 1099-110.

Goswami, S., Gupta, V.K., Sharma, A.S. and Gupta, B.D. (2005). Supramolecular structure of S-(+)-marmesin-a linear dihydrofuranocoumarin. *Bulletin of Materials Science*, **28**: 725-729.

Human Metabolome Database. (2002). (http://www.hmdb.ca/metabolites/ HMDB03546) IHP, Indian Herbal Pharmacopoeia, Revised New Edition, *Aegle marmelos*, Indian Drug Manufacturers Association, Mumbai, India. 40-8.

Jagetia, G.C., Venkatesh, P. and Baliga, M.S. (2003). Evaluation of the radioprotective effect of _Aegle marmelos_ (L.) Correa in cultured human peripheral blood lymphocytes exposed to different doses of γ-radiation: a micronucleus study. _Mutagenesis_, **18**: 387-393.

Jaswanth, A., Akilandeswari, Loganathan, V., Manimaran, S. and Ruckmani. (2001). Wound healing activity of _Aegle marmelos_. _Indian Journal of Pharmaceutical Sciences_, **63**: 41-4.

Jauhari, O.S., Singh, R.D. and Awasthi, R.K. (1969). Survey of some important varieties of Bael (_Aegle marmelos_ Correa). _Punjab Horticulture Journal_, **9**: 48-53.

Johri, B.M. and Ahuja, M.R. (1957). A contribution to the floral morphology and embryology of _Aegle marmelos_ Correa. _Phytomorphology_, **7**: 10-24.

Kakiuchi, N., Senaratne, L.R., Huang, S.L., Yang, X.W., Hattori, M., Pilapitiya, U. and Namba, T. (1991). Effect of constituents of Belli [_Aegle marmelos_ (L.) Corr.] on spontaneous beating and calcium-paradox of myocardial cells. _Planta Medica_, **57**: 43-6.

Kalaivani, T., Premkumar, N., Ramya, S., Siva, R., Vijayakumar, V., Meignanam, E., Rajasekaran, C. and Jayakumararaj, R. (2009). Investigations on hepatoprotective activity of leaf extracts of _Aegle marmelos_ (L.) Corr. (Rutaceae). _Ethnobotanical Leaflets_, **13**: 47-50.

Kamalakkannan, N. and Prince, P.S.M. (2005). Antihyperlipidaemic effect of _Aegle marmelos_ fruit extract in streptozotocin-induced diabetes in rats. _Journal of the Science of Food and Agriculture_, **85**: 569-73.

Kapoor, L.D. (1990). Handbook of ayurvedic medicinal plants. CRC Press LLC, 21.

Karawya, M.S., Mirhom, Y.W. and Shehata, I.A. (1980). Sterols, triterpenes, coumarins and alkaloids of _Aegle marmelos_ Correa, cultivated in Egypt. _Egyptian Journal of Pharmaceutical Sciences_, **21**: 239-248.

Kesari, A.N., Gupta, R.K., Singh, S.K., Diwakar, S. and Watal, G. (2006). Hypoglycemic and antihyperglycemic activity of _Aegle marmelos_ seed extract in normal and diabetic rats. _Journal of Ethno-pharmacology_, **107**: 374-379.

Kumar, D.S. and Prabhakar, Y.S. (1987). On the ethnomedical significance of the arjun tree, _Terminalia arjuna_ (Roxb.) Wight and Arnot. _Journal of Ethno-pharmacology_, **20**: 173-90.

Mac, L.A.J. and Pieris, N.M. (1981). Volatile flavor components of beli fruit (_Aegle marmelos_) and a processed product. _Journal of Agriculture and Food Chemistry_, **29**: 1262-1264.

Manandhar, M.D., Shoeb, A., Kapil, R.S. and Popli, S.P. (1978). New alkaloids from _Aegle marmelos_. _Phytochemistry_, **17**: 1814-15.

Mazumder, R., Bhattacharya, S., Mazumder, A., Pattnaik, A.K., Tiwary, P.M. and Chaudhary, S. (2006). Antidiarrhoeal evaluation of _Aegle marmelos_ (Correa) Linn. root extract. _Phytotherapy Research_, **20**: 82-4.

Mishra B.B., Singh D.D., Kishore N., Tiwari V.K., Tripathi V. (2010a). Antifungal constituents isolated from the seeds of *Aegle marmelos*. *Phytochemistry*, **71**: 230-234.

Mishra B.B., Kishore N., Tiwari V.K., Singh D.D., Tripathi V. (2010b). A novel antifungal anthraquinone from seeds of *Aegle marmelos* Correa (family Rutaceae). *Fitoterapia*, **81**: 104-107.

Mukherjee, P.K. and Wahile, A. (2006). Integrated approaches towards drug development from Ayurveda and other Indian system of medicines. *Journal of Ethno-pharmacology*, **103**: 25-35.

Nammi, S., Ganti, S.S. and Lodagala, D.S. (2001). Hypoglycemic effect of *Aegle marmelos* leaf extract on normal and diabetic model rabbits. *Asia Pacific Journal of Pharmacology*, **15**: 9-11.

Narendhirakannan, R.T., Subramanian, S. and Kandaswamy, M. (2006). Biochemical evaluation of antidiabetogenic properties of some commonly used Indian plants on streptozotocin-induced diabetes in experimental rats. *Clinical and Experimental Pharmacology and Physiology*, **33**: 1150-7.

Nene, P.M. (1977). Placentation in Rutaceae. *Proceedings of the Indian Academy of Sciences*, **85**B: 378-383.

Ohashi, K. and Watanabe, H. (1994). Indonesian medicinal plants: XII, four isomeric lignan-glucosides from the bark of *Aegle marmelos* (Rutaceae). *Chemical and Pharmaceutical Bulletin*, **42**: 1924-1926.

Panda, S. and Kar, A. (2006). Evaluation of the antithyroid, antioxidative and antihyperglycemic activity of scopoletin from *Aegle marmelos* leaves in hyperthyroid rats. *Phytotherapy Research*, **20**: 1103-5.

Parmar, C. and Kaushal, M.K. (1962). *Aegle marmelos*, In: Wild Fruits, Kalyani Publishers, New Delhi, India, 1-5. (http://newcrop.hort.purdue.edu/newcrop/parmar/01.html)

Roy, S.K. and Singh, R.N. (1979). Bael fruit (*Aegle marmelos*) - A potential fruit for processing. *Economic Botany*, **33**: 203-212.

Saha, K.S. and Chatterjee, A. (1957). Isolation of allo-imperatorin and (beta-sitosterol from the fruits of *Aegle marmelos* Correa. *Journal of Indian Chemical Society*, **34**: 228-30.

Samarasekera, J.K.R.R., Khambay, B.P.S. and Hemalal, K.P. (2004). A new insecticidal protolimonoid from *Aegle marmelos*. *Natural Product Research*, **18**: 117-122.

Shankarananth, V., Balakrishnan, N., Suresh, D., Sureshpandian, G., Edwin, E. and Sheeja, E. (2007). Analgesic activity of methanol extract of *Aegle marmelos* leaves. *Fitoterapia*, **78**: 258-259.

Sharma, B.R. and Sharma, P. (1981). Constituents of *Aegle marmelos* II. Alkaloids and coumarin from fruits. *Planta Medica*, **43**: 102-103.

Sharma, B.R., Rattan, R.K. and Sharma, P. (1980). Constituents of leaves and fruits of *Aegle marmelos*. *Indian Journal of Chemistry* B, **19**: 162.

　　　　　　　　　　　　　　　　　Natural Products: Research Reviews Vol. 1

Sharma, B.R., Rattan, R.K. and Sharma, P. (1981). Marmeline, an alkaloid and other components of unripe fruits of *Aegle marmelos*. *Phytochemistry*, **20**: 2606-2607.

Shoba, F.G. and Thomas, M. (2001). Study of antidiarrhoeal activity of four medicinal plants in castor-oil induced diarrhea. *Journal of Ethno-pharmacology*, **76**: 73-6.

Singanan, V., Singanan, M. and Begum, H. (2007). The hepatoprotective effect of bael leaves (*Aegle marmelos*) in alcohol induced liver injury in albino rats. *International Journal of Science and Technology*, **2**: 83-92.

Singh, Z. and Malik, A.U. (2000). The bael. WANATCA Yearbook, **24**: 12-7.

Sofowora, A. (1982). Medicinal plant and traditional medicine in Africa. Second Ed. Wiley, Ibadan. 8-14.

Srivastava, K.K. and Singh, H.K. (2000). Floral biology of bael (*Aegle marmelos*) cultivars, *Indian Journal of Agricultural Sciences*, **70**: 797-8.

Srivastava, S.D., Srivastava S. and Srivastava, S.K. (1996). New anthraquinones from the heartwood of *Aegle marmelos*. *Fitoterapia*, **66**: 83-84.

Tilak, V.D. and Nene, P.M. (1978). Floral anatomy of the Rutaceae. *Indian Journal Botany*, **1**: 83-90.

Tokitomo, Y., Shimono, Y., Kobayashi, A. and Yamanishi, T. (1982). Aroma Components of Bael fruit (*Aegle marmelos* Correa), *Agricultural and Biological Chemistry*, **46**: 1873-1877.

Udupa, S.L., Udupa, A.L. and Kulkarni, D.R. (1994) Studies on the anti-inflammatory and wound healing properties of *Moringa oleifera* and *Aegle marmelos*. *Fitoterapia*, **65**: 119-23.

Veerappan, A., Miyazaki, S., Kadarkaraisamy, M. and Ranganathan, D. (2007). Acute and subacute toxicity studies of *Aegle marmelos* Corr., an Indian medicinal plant. *Phytomedicine*, **14**: 209-215.

Wiart, C. (2006). Medicinal plants of India and Pacific, CRC Press, 212-213.

Yang, X.W., Hattori, M. and Namba, T. (1996a). Two new coumarins from the roots of *Aegle marmelos*. *Journal of Chinese Pharmaceutical Sciences*, **5**: 68-73.

Yang, X., Masao, H. and Tsuneo, N. (1996b). Studies on the anti-lipid peroxidation actions of the methanol extract of the root of *Aegle marmelos* and its constituents *in vivo* and *in vitro*. *Journal of Chinese Pharmaceutical Sciences*, **5**: 132-140.

Natural Products: Research Reviews Vol. 1 (2012)   Pages 161–220
Editor: V.K. Gupta
Published by: DAYA PUBLISHING HOUSE, NEW DELHI

6

Supercritical Fluid Extraction as a Tool for Value Addition of Natural Products

S. Koul[1]*, D.K. Gupta[1]*, V.K. Gupta[1] and S.C. Taneja[1]

ABSTRACT

Since antiquity, the existence of mankind has been dependent on natural products, in particular plants for various purposes such as food, shelter, clothing, medicines, religious ceremonies etc. To obtain these materials, several methodologies have been developed for the extraction, isolation, purification for their end use i.e., food materials, nutraceuticals, cosmetics, flavors and fragrances etc. Supercritical fluid extraction (SCFE) methodology, introduced in recent times, is becoming the method of choice for the extraction of natural products. The use of SCFE for value addition in the area of natural products is discussed, highlighting the key advantages of SCFE in terms of better quality, shelf life, free from solvent residues, better fractionation with minimal transformation of the chemical components as well as harvesting of enriched bioactive extracts and fractions, compared to conventional approaches and lesser issues of safety.

Keywords: Natural products, Value addition, Extraction, Supercritical fluid, Hydrodistillation, Sonication, Solvent extraction, Isolation, LCMS, Quantification, Bioactivity.

Introduction

Man's existence is generally depended on natural products particularly on plants, not only as a source of drugs and nutrition but also as a source of various other requirements such as shelter, religious purposes, colorants, fragrances *etc*. Natural product chemistry has tended to advance regularly on a broad front during this

---

1   Indian Institute of integrative Medicine (CSIR), Canal Road, Jammu – 180 001, India.

*   *Corresponding authors*: E-mail: skoul@iim.ac.in; dkgupta@iiim.ac.in

century. The useful chemicals, drugs and pharmaceuticals have come mainly from plants (terrestrial, aquatic, semi terrestrial), marine and invertebrates. The chemical constituents present in the plant range from non polar (lipids, steroids etc.) to most polar (cellulose, starch etc.) depicting a broad range of their chemical character. With the advancement of knowledge, more and more plants have been brought under use and the process goes unabated. World Health Organization (WHO) report shows dependence of 80 per cent of the world's population on the modern medicine with >25 per cent contribution from plants. The international market of medicinal plants and their downstream products runs to the tune of several billion dollars per year. In recent years, a new term called "mediculture" has come up for sustained production of herbo-chemicals through scientific cultivation and adaptation of post harvest technologies.

In present day scenario, there exists a great scope for value added products – may it be a plant, a plant extract, a fraction, an oil, biomass or the chemical components embedded in these materials. The value addition may be achieved by using various approaches involving unconventional strategies, newer techniques and methodologies. Use of extraction/fractionation, isolation or enrichment of chemical compounds of importance or medicinal use, bypassing the conventional route to achieve the said objective will therefore amount to value addition. For example, use of a plant waste (obtained after the removal of the required/wanted product/s) to get products that add to its value (Ubalua *et al.*, 2007) or removal of moisture, color sediments, solvent impurity, removal of a particular component from the essential oils can prove a very useful value addition. Preparation of a toxin free extract, by removal of toxic part or components, from a bioactive extract also leads to its value addition. Use of a cheap catalyst for conversion of low cost highly abundant compound into a compound of high cost may also be described as a case of value addition such as the conversion of carvone to carvacrol.

Barring few examples, where the extracts are used as such for consumption such as tea, coffee *etc.*, the efforts to isolate pure chemical components for their use aimed at various activities has resulted in establishing the basis of medicinal chemistry, applied chemistry and molecular modeling. To get chemical components from the plants, their extrusion from the plant matrix forms the very first important step (Cannel, 1998), which is not as simple as one would think of. In fact, it differs from plant to plant, and due to variations and complexity of composition, one needs to adopt the right kind of extraction procedures. For the compounds of acidic or basic nature, besides solvents, pH management is also a basic requirement. In a particular class of compounds, a solvent of appropriate polarity becomes a requirement *e.g.*, for the alkaloids one needs polar solvents compared to terpenoids/steroids where extraction with low polarity solvents may be useful. It is important to keep in mind that an extract obtained from a plant, fungi or microorganism may be a simple mixture or a highly complex mixture with many of the secondary metabolites present in micro to nanogram quantities, but having high value as exemplified by Vitamin D3 diol (Sigma Aldich, 2008) in *Cestrum* and *Solanum* species. The selection of the extraction process becomes the most important step to achieve goal of isolating the pure molecules not withstanding the constraints of low abundance. A process that fails to capture all the essential ingredients of an extract from a plant or other natural source shall prove to be inefficient and the reason could be the wrong choice of the extraction method.

Depending upon the nature of chemical components and their abundance present in the plant material/matrix the followings extractions procedures can be adapted:

1. The conventional system of extraction involves either sequential extraction process that starts with low polar solvents such as petroleum ether/hexane and ends with polar solvent *e.g.,* water or hydro alcoholic extraction under ambient (Silva *et al.,* 1998) to hot conditions (soxhlet) (Zymount *et al.,* 2003).

2. Extraction with alcohol, aqueous alcohol mixture and water are recommended as a standard extraction procedure as followed in herbal preparations (Silva *et al.,* 1998).

3. Ultrasonication extraction: herein, diffusivity and solubility of chemical compounds at interface is used to extract compounds, ranging from polysaccharides to esters, fatty acids, volatile oils, sterols, pyrethrins etc. (Luque *et al.,* 2003).

4. Microwave extraction process can be used successfully in the extraction of materials of olive seeds, paprika, pigments, glycrrhizic acid from liquorice root etc. (Hudiab *et al.,* 2003). The method has less solvent requirement, short operational time, higher recovery, reproducibility and minimal sample manipulation.

5. Accelerated solvent extraction or pressurized liquid extraction: works on the principle of static extraction with super heated liquids, that result in high diffusion rates and disruption of strong matrix solute interaction caused by hydrogen bonding, Vander Wall forces and dipole-dipole interactions (Benthin *et al.,* 1999).

6. Medium pressure solid liquid extraction method involves extraction of powdered material in the medium pressure liquid chromatography column. The selected suitable extraction solvent is pumped through the matrix which leads to rapid and exhaustive extraction.

7. Micelle mediated extraction method: Herein along with the organic solvent – a polar solvent such as water in combination with surfactants is used. This methodology has been sucessfully used in the extraction of ginsenosides, tanshinones (Fang *et al.,* 2000; Shi *et al.,* 2004) terpenoids and other class of compounds (Pisacane, 2001).

8. Rotation planar extraction method also called as forced flow solid liquid extraction method involves extraction of single/complex materials by the use of centrifugal force and is affected by the action of force. The extraction procedure is well suited for small size samples and allows to screen a large number of samples due to short extraction time (Vovk *et al.,* 2003; Mesaros *et al.,* 1987; Nyiredy and Botz, 2001).

9. Lastly, one can also use a more recent but highly applicable extraction method *i.e.,* Super Critical Fluid Extraction (SCFE). It exploits the physical

property of certain gases which enables them to behave as liquids, having the properties of a solvent, and thus applicable for extraction of a wide variety of natural products. A supercritical fluid is a substance that exists at a temperature and pressure above its thermodynamic critical point. This property enables it to diffuse through matrix as gases, and dissolve materials behaving as a liquid. Above its critical point, small variations in temperature, pressure result in large changes in density, which can be used in various applications. The supercritical fluids are suitable substitutes for organic solvents. Carbon dioxide, water, nitric oxide, ethane etc. are used as supercritical fluids with best application obtained with CO_2 The use of many co-solvents along with the carbon dioxide (binary/ternary mixtures) exerts a significant effect on the yields and the extraction of the components as well. These solvents called as modifiers are mostly co-used with CO_2 after the first extraction is carried out with carbon dioxide alone and can be introduced with eutrainer pump as mixed fluids kept in a mixing chamber or by the introduction of solvent into the sample before extraction. The process has become a method of choice and finds wide industrial application for the extraction, fractionation and isolation of molecules, especially from the plant sources.

The origin of super critical extraction dates back to 1870's with the classic studies of Hannay and Hogath (1879), describing the phenomenon of solubility of solutes in gases. The phenomenon after remaining dormant for almost 7-8 decades got reactivitated in 1950's. Supercritical fluid extraction (SCFE) was exploited by two groups, one led by Paul and Wise (1971) and the other by Schneider and his coworkers (1980) for the extraction and fractionation of components from natural products. However, the real impetus for adoption of this methodology *i.e.*, supercritical extraction process came with the work of Zosel (1976) described in a US patent, in which as many as 60 examples have been listed along with the application of SCFE in food and other materials laying thereby a solid foundation for the use of SCFs. Reviews and semantic papers by several workers are available in the literature (Wang and weller, 2006; Meireless and Angela, 2003; Lang and Wai 2001; Reverchon 1997; Daintree *et al.*, 2008) that enlist the application of SCFE and effect of various parameters such as matrix size, its nature, effect of pressure, temperature and modifiers, development of computer aided and derived mathematical models on the overall performance of the process. The SCFE is becoming a method of choice and as a promising alternative technique because of the following attributes:

(*i*) Higher yields, (*ii*) no residual solvent/s, (*iii*) high potency of active compounds, (*iv*) selective extraction of chemical components through fractionation, (*v*) biological contaminant free extracts/fractions, (*vi*) shelf life longevity, (*vii*) recycling of CO_2 leading to minimum waste generation, (*viii*) freedom of choice to carryout modification in experimental conditions, (*ix*) safe solvent use [Green chemistry], (*x*) fractionation of the extract in a single step, (*xi*) non-explosive, (*xii*) nontoxic, (*xiii*) relatively inexpensive, (*xiv*) lesser time constraints.

SCFE products are free from solvent residue which is the main advantage for food products. SCFE can be applied from gram scale to pilot plant and industrial scale and is used for the extraction of lipids, essential oils, flavours and herbal medicine (Wang and Weller, 2006; Meireles and Angela, 2003).

Description of Super Critical CO_2 Fluid Extraction

SCFE is a two step process which uses a dense gas such as CO_2 as a solvent above its critical temp (31°C) and critical pressure (74 bar) for extraction by allowing CO_2 to enter into the extractor housing the matrix of appropriate mesh size. The extract is then separated either by depressurizing fully to get the extract in separator-1 and recycle the extract free carbon dioxide back to the CO_2 tank or by depressurizing partially in separator-1 (S-1) and fully in extractor-2 which can bring in fractionation of the extract in S-1 and separator-2 (S-2). This is a continuous process and the extraction is usually completed in 2.5-3.0 hours. In case of use of co-solvents (modifiers), the solvent mixture is pumped into the liquid CO_2 stream via entrainer pump and the process carried out as described above. In Figure 6.1, all the components of the supercritical fluid extraction unit are highlighted.

Some of the factors responsible for the effective extraction of materials using SCFE include:

Analyte Solubility

It is one among several important factors that leads to the efficiency of the SCFE process. Various workers have proposed techniques for modeling analyte solubility (Branahas *et al.*, 1994).

Particle Size

Finer the material better is the extraction. However, it has also been observed that excessive grinding may hinder the extraction due to re-adsorption of the analytes into the matrix surfaces and pressure drop in the extraction chamber (Loudi *et al.*, 2004).

Effect of Pressure and Temperature

As the solvating power of SCFE is related to density and dielectric constant of the fluid used in extraction, which in turn depends upon pressure and temperature, a small increase of pressure can enhances the significantly solubility of solutes (McNally and Wheeler, 1988). Generally speaking, the best suited pressure for SCFE (using CO_2) ranges from 75 to 300 bar, though examples are now pouring in the literature where pressure as high as 800 bar has been used (Reverchon and Marco, 2006). Studies that relate the effect of pressure and temperature in supercritical fluid extraction are available in the literature (Anderson *et al.*, 1990; Gmez-Prieto *et al.*, 2003; Roop *et al.*, 1989; Caude and Theibut, 1999; Baysal and Starmens, 1999; Careri *et al.*, 2001; Reverchon, 1997; Hamburge *et al.*, 2004; Kaiser *et al.*, 2001; Taylor, 1996). Several workers have studied the effects of pressure and temperature for the overall efficiency of SCFE. The better performance at high pressure is said to be due to an increase in the solubility which is also true for conditions that involve fixed pressure and high temperature. Temperature control is more important for volatile materials

Figure 5.1 : Photograph of SCFE plant installed at IIIM, Jammu (India).

where higher temperature can lead to complex mixture or decomposition of the chemical components (Campos *et al.*, 2005; Anderson *et al.*, 1990; Caude and Theibut, 1999; Gmez-Prieto *et al.*, 2003). Studies involving several of these parameters (Jaubert *et al.*, 2000; Berna *et al.*, 2000; Giddings *et al.*, 1969; King, 1989; Roop *et al.*, 1989, 1999; Loudi *et al.*, 2004). Berna *et al.* (2000) has shown that at 80 bar pressure, linalool and limonene contents are maximum even with increasing temperature (upto 55°C). However, in some cases small change in temperature can cause dramatic variation in the solubility *e.g.*, an increase of 5°C (288.15→293.15K) at 66.7 bar pressure lead to 14 per cent increase in solubility but further increment of 5°C *i.e.*, 293.15 to 298.15 K lead to decrease of solute solubility by 42 per cent (Kurnik and Reid, 1981).

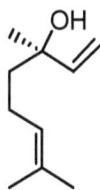

Linalool

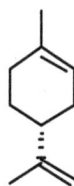

Limonene

By optimizing extraction temperature and pressure conditions, a desirable component in the fraction of the extract can be increased. Mastinellin *et al.* (1994) have highlighted the work regarding the solubility of essential oil components in SCFE based on their chemical structure and polarity of the constituents. Apart from factors like solubility and selectivity for better extraction, mass transfer rate has been studied for effective extraction (Ferriera *et al.*, 1999; Lim *et al.*, 1989; Martinez *et al.*, 2007). Deterpenation of the essential oil of citrus has been achieved by using SCFE. It has been shown that at low temperature and pressure, SCFE carried out oil adsorbed on silica gel get separated into hydrocarbon portion with silica gel retaining the oxygenated compounds, and with the increase in temperature and pressure more polar compounds are extracted along with hydrocarbons (Dugo *et al.*, 1995).

Modifiers

A limitation of SCFE is that, it often fails in quantitative extraction of polar analytes from a matrix because of poor solvating power of CO_2 fluid and insufficient interaction between solvent and the matrix (Pavoliszyn *et al.*, 1993). Modifiers help in overcoming this problem by increasing solubility of the analytes, by interacting with sample matrix and or by inducing matrix modification.

SCFE leads to value addition by minimizing the degradation due to isomerisation and preventing the of oxidation of components present in the plant/plant extract as exemplified by the extraction of *E*-lycopene (a health promoting ingredient). SCFE extraction of tomato affords only trans- lycopene at 40 °C whereas normal extraction affords the mixture of both the isomers (Gmez-Prieto *et al.*, 2003). A similar approach has been used for the extraction of carotene from carrot (Barth *et al.*, 1995; Hart *et al.*,

1995). SCFE processes have wide applications in the field of spices, oils and oleoresin, herbal medicines, flavors and fragrances, food colors and preservatives.

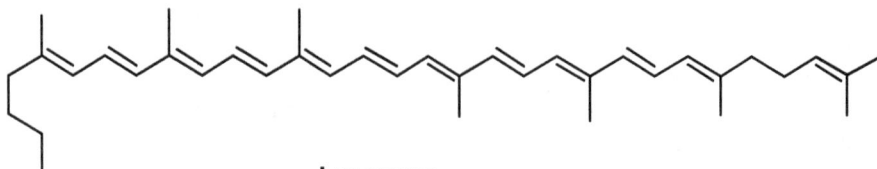

Lycopene

Results and Discussion

In the last few years SCFE has proved to be one of the most interesting techniques for various applications. These include its use in the improvement of diastereoselectivity of a reaction to afford higher yields of one diastreoisomer over the other *e.g.*, in sulfoxidation of cysteine derivatives, SCFE assisted reaction result in high diastreoisomeric ratio (as high as 95 per cent) which in the absence of SCFE shows no selectivity (Oakes *et al.*, 1999). SCFE can replace perchloroethylene or other solvents used for dry cleaning purpose (Science News, online 20th Nov. 2007). Nano and micro particle formation method has become important for better delivery and SCFE can promote nucleation, spinodel decomposition over crystal generation affording very minute and regularly sized particles (Sang and Kirana, 2005) in polymer synthesis (Cooper, 2000). The most widely application of SCFE is the extraction of materials related to spices, herbs, essential oils, pharmaceuticals, food industry, nutraceuticals, decaffeination of green coffee beans and extraction of hops for beer production. SCFE is highly advantageous and desirable alternative to solvent extraction of several classes of natural substances for food, providing high speed and efficiency of extraction, eliminating concentration steps and simplifying the analytical method (Anklam *et al.*, 1998).The interest in SCFE is promoted by legal limitation of conventional solvents for food and pharmaceuticals and is, therefore, most commonly used as extraction method because of its low toxicity, chemical inertness, low cost and easy availability. In addition CO_2 has a low critical temperature value (Tc= 31°C), making it ideal for extraction of thermally labile compounds. For these reasons, SCFE (using CO_2) has been successfully used for isolation of carotenoids from various vegetables in the preparation of herbal fractions and in the extraction of essential oil.

Advantages of SCFE Over Traditional Methods

For investigating the composition of genuine essential oils and related aromatic compounds, hydrodistillation is not very useful because of discrimination and transformation processes, due to high temperature and acidic conditions as seen in the case of lavender oil, which is overcome by supercritical fluid extraction (Richter*et al.*, 2007). SCFE has advantage over steam distillation as the laters can lead to thermal degradation and hydrolysis which in SCFE is avoided *e.g.*, linalyl acetate content obtained by hydrodistillation/steam distillation of lavender is much lower as compared to SCFE (Unpublished results; Figure 6.2). The SCFE extraction of the lavender from Kashmir Valley (J&K State, India) afforded linalyl acetate and linalool in the ratio of ~1:9 (Table 6.1), whereas by hydrodistillation the ratio is reduced to 1:2. Hydrodistillation of samples from Himachal Pradesh showed the presence of two compounds in near equal proportion as identified by gas chromatography.

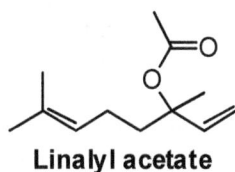

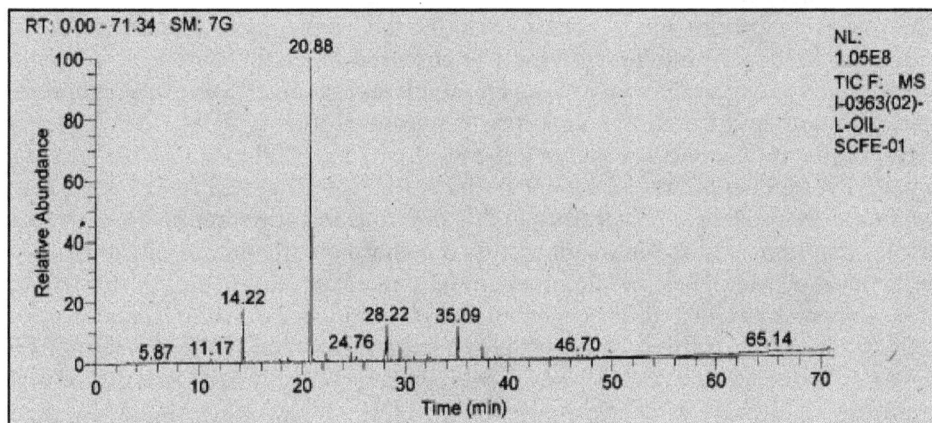

Linalyl acetate

Figure 6.2: GCMS chromatogram of SCFE product of Lavander.

Table 6.1: Comparative study on the Extraction of essential oil from Lavender using SCFE and hydrodistillation extraction method.

Sl.No.	Name of Constituent	SCFE Product Area per cent	Hydrodistillation Product Area per cent
1	CAMPHENE	–	0.39
2	3-OCTANONE	–	0.53
3	P-CYMENE	–	0.35
4	LIMONENE	0.43	0.58
5	CINEOLE	–	1.31
6	LINALOOL	7.51	22.49
7	BORNEOL	0.50	1.20
8	p-MENTH-1-EN-4-OL	0.24	0.72
9	TERPINYL ACETATE	0.23	2.31
10	LINALYL ACETATE	62.81	52.43
11	TERPENDIOL	1.20	0.72
12	LIMONENE-DIEPOXIDE	–	0.60
13	CARYOPHYLLENE	5.39	0.40
14	CARYOPHYLLENE OXIDE	5.70	4.11

For the determination of amount and composition of essential oils and related compounds from aroma plants, different extraction methods have been reported. Hydro distillation is a time consuming method and therefore not useful especially for the screening of very large quantities of plant samples for their aroma compound composition. Moreover transformations of genuine aroma active compounds can also occur during processing due to the influence of heat, steam and pH (Fischer *et al.*, 1987, 1988; Jimenez-Carmina *et al.*, 1999; Cornwell *et al.*, 1999; Oszagyan *et al.*, 1996). High volatile as well as water soluble components can get lost during hydrodistillation/steamdistillation (Langer *et al.*, 1996; Ronyai *et al.*, 1999; Maldao-Martius *et al.*, 2002). SCFE has gained more attention in recent years because of legal limitations of solvents. This makes the process more economical in the food, beverage and pharmaceutical industries. Enhanced inhibitory activity is observed for SCFE extract compared to solvent derived extract. Example to cite for the advantages of SCFE include marjorana, sage, caraway and thyme. Extraction by different processes show that *Origanum majorana* (having the most sensitive terpenes) by hydrodistillation (HD), effects the sensitive components like E-sabinene hydrate and sabinene hydrate acetate. These are found in much lesser quantities in HD due to transformation processes in the acidic conditions and at high temperature. Furthermore, the increased contents of terpinene-4-ol, γ-terpinene and limonene in HD are said to be due to the transformation of E-sabinene hydrate and its acetate into these products. In terms of its biological activity, one finds a contrast between the extracts and essential oils obtained by SCFE to those obtained by the rest of the extraction processes. Analysis of essential oil and antimicrobial activity of *O. majorana* extracts, SCFE extracts have shown stronger antimicrobial properties in comparison to the poor inhibitory effects of the extract obtained by other extraction methods. Fresh and dried leaves and flowering tops of *O. majorana* are widely used to flavor many foods. Its essential oil and extracts are used in pharmaceuticals, perfumes and cosmetics (Vagi *et al.*, 2005). The essential oil of *O.majorana* has been shown to exhibit strong antifungal properties (Deans and Svoboda, 1990). SCFE extracts has been shown to inhibit growth of the filamentous fungi at 4 mg/liter while ethanolic extract even at 25mg/ml showed no inhibition. For antibacterial activity, the SCFE extract at 2 to 4mg/ml concentration showed higher inhibitory activity against bacteria than extract obtained from soxhlet extraction. These results suggest and support the notion that extracts obtained by SCFE might have a role as flavouring, nature colorants as well as preservatives in food and cosmetics. Likewise, the adverse effect on chemical composition of other plant materials like sage, caraway and thyme by conventional extraction or even by accelerated solvent extraction method can be avoided by resorting to SCFE extraction method. Antioxidant activity of *O. majorana* essential oils and its purified substance has been reported. Its volatile oil possesses antimicrobial properties against food born bacteria and mycotoxigenic fungi. The SCFE extracts exhibit strong inhibitory activity as compared to alcoholic extracts (prepared by conventional extraction method). The SCFE extracts in the concentration of 0.4 w/v, showed 96.4, 98.8 and 98.7 per cent inhibition against *E.coli, P. fluorscenes* and *Bacillus cereus* respectively. In contrast to this, alcoholic extract showed inhibition of 21.5, 56.6 and 76.8 per cent respectively against these three microbial strains. In 0.2 w/v extract concentration, inhibitory property of SCFE extract remained virtually unaffected but alcoholic extract showed poor inhibitory activity. Essential oils by this method gave 20 per cent terpine-4-ol, (+) cis-sabinene hydrate (3-18 per cent) which attribute flavor fragrance to the oil besides α and r-terpinene,terpinolene, thymol and cravacrol (Unpublished results).

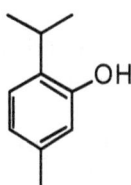 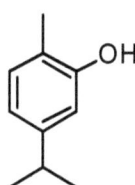

Thymol   **Cravacrol**

SCFE protocols for the extraction of medicinal plants of repute have been developed *e.g.*, preparation of anti-migraine extract from *Tanacetum parthenium* (Beuscher and Willigmann, 1993) and Spasmolytic and anti-asthamatic phytomedicine from *Petasites hybridus* (Steiner *et al.*, 1988, 1999).

From the flowers of *Chamomilla recutita*, α- bisabolol and chamazulene are important bioactive phytomolecules (Kim *et al.*, 2007; Mikhova *et al.*, 2004) and have been obtained in high yields by SCFE as compared to conventional extraction method (Povh, 2001; Kotnik and Knez, 2007). The laboratory scale method of extraction has been transferred to pilot scale (Kotnik and Knez, 2007; Scalia *et al.*, 1999).

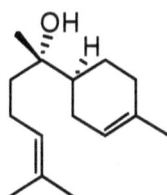 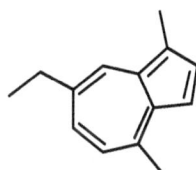

Chamazulene

α-bisabolol

It has been shown that SCFE method is advantageous in terms of oil content, flavor and recovery of the components in nondegradative form over other methods of extraction of the plants namely *Origanum majorana, Carum carvi, Salvia officinalis* and *Thymus vulgaris* (Richter and Schellenberg, 2007).

Fractional extraction of compounds from Grape seeds by SCFE and analysis for antimicrobial and agrochemical activation (Palma *et al.*, 1999) allows for the isolation of compounds without interference from air and light thereby guaranting conservation of their antioxidant properties. This has not been achieved by other conventional approaches as evident from the reports in the literature (Tipsrisnkond *et al.*, 1998).

In the extraction of medicinal plants, the application of SCFE has taken over the conventional solvent extraction method (Hamburger *et al.*, 2004) because of some of the inherent properties that are associated with SCFE (List and Schmidtt, 1984). In this direction, extraction of herbs and various natural products has been documented by various workers (Castioni *et al.*, 1995; Modey *et al.*, 1996; Kaiser *et al.*, 2001) encompossing various class of compounds like alkaloids (Heaton *et al.*, 1993), free and glycosidic phenolics (Tsuda *et al.*, 1995; Moraes *et al.*, 1997; Tena *et al.*, 1998; Palma *et al.*, 1999; Murga *et al.*, 2000) and cardiac glycosides (Moore and Taylor, 1996).

In the extraction of grape seed oil, three different kinds of compounds are isolated in the oil fraction *i.e.*, aliphatic aldehydes, fatty acids and their derivatives and sterols. These results agree with the data from the list employing both SCFE (Molero-Gomez *et al.*, 1996) and a coventional liquid extraction (Rao, 1994). The two extracted fractions obtained by SCFE were shown to retain the main characteristics previously described for both oils and phenolic compounds that are obtained by using conventional extraction procedures for grape seed. In SCFE, in the consecutive extracts two different set of compounds were obtained bearing activities that varied quite a bit. The steroidal fraction A showed much better activity than fraction B (containing mainly catechins) when tested against bacterial strains like *Bacillus cereus*, *Staphylococcus coagulans*, *A. niger*, *E. cloacae* and *E. coli* etc. Fraction A was also found to be more active than fraction B on human pathogens.

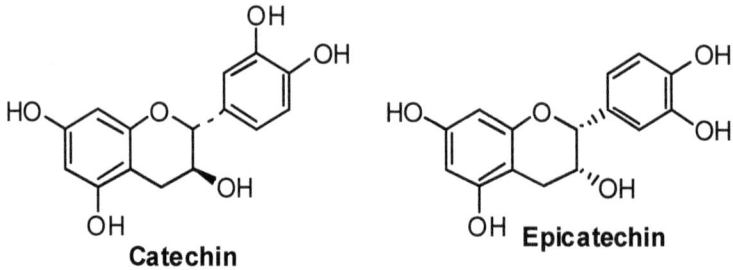

Catechin   Epicatechin

Therefore, it may be concluded that it is possible to fractionate compounds from grape seeds wherein activity enrichment of its fractions/extracts can be achieved.

The advantageous extraction of essential oil from dry celery fruit by SCFE has been shown responsible for the high antimicrobial activity (MIC 40µg/ml) when tested against *Staphylococcus aureus* and *Listeria monocytogenes*. The active components sedanolide, sedanenolide and 3-n-butylphthalide were higher in oil as compared to hydrodistillation *e.g.*, SCFE gave 2.74 per cent oil with thymol 67.6 per cent and hydrodistillation afforded 1.63 per cent oil with thymol content of 44.6 per cent. (Misic *et al.*, 2008; Scalia *et al.*, 1999).

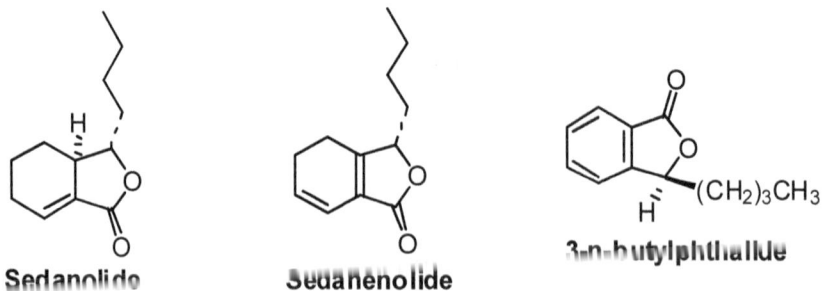

Sedanolide   Sedanenolide   3-n-butylphthalide

Essential oil from *Myristica fragrans* Houtt by SCFE and steam distillation showed presence of 38 compounds from SCFE and 48 compounds from steam distillation. The major components of oil obtained from steam distillation were alpha and beta pinene, terpine-4-ol, gamma-terpinene and beta-phellandrene and major components from SCFE were myristic acid, myristicin, terpinen-4-ol, alpha pinene and safrole, showing. Thereby a lot of variation in the chemical composition of the two extracts/oils. To conclude, one should not rule out the role of high temperature and the pH conditions during steam distillation in bringing a transformation in the components of the oil during steam distillation (Qiu *et al.*, 2004).

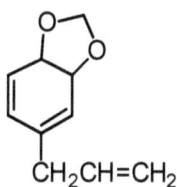

Safrole

A comparative study on the extraction process of black cumin using SCFE and hydro distillation revealed a major contrast in the chemical composition. SCFE extract showed major components namely r-terpinene ~38 per cent, cinnamaldehyde 11.5 per cent alpha-methylbenzene methanol 25.5 per cent whereas hydro distillation extract showed the presence of cuminyl alcohol and p-cymene as the major components (Pourmortazavi *et al.*, 2005).

SCFE and hydrodistillation extraction studies on Chinese plant *Schizonepeta tenuifolia* Briq show many of the chemical constituents present in SCFE absent in hydrodistillation and vise versa (Qiu *et al.*, 2005). SCFE of *Foeniculum vulgare* Mill seeds (Table 6.2) has been found to be advantageous over hydrodistallation in terms of yields and intense aroma (Damjanovic *et al.*, 2005).

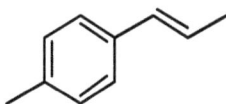

trans-Anethole   **Fenchone**

Table 6.2: Major percent composition of the extracts obtained by SCFE and Hydrodistillation from *Foeniculum vulgare* Mill seeds.

Chemical Component	SCFE (per cent)	Hydrodistillation (per cent)
t-anethole	~ 75 per cent	~62
methyl chavicol	~9.0 per cent	<5 per cent
fenchone	~14.7	20 per cent

Humulus lupulus is a plant of repute used over the centuries in brewery, as preservative and as antibiotic (Gerhauser *et al.*, 2005; Teuber *et al.*, 1973). The extract is reported to contain humulon and lupulon, their derivatives and flavonoids as major components. The single molecule isolated from the plant have been shown to exhibit anti bacterial activity with minimum inhibitory concentration that range from 6.25 µg/ml and above.

Lupulon

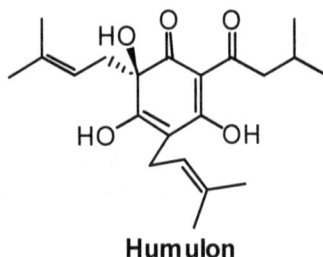

Humulon

In our SCFE and solvent extraction experiments directed towards the extraction and bioevaluation of *Humulus lupulus* extracts (collected from Ladakh region at 17,000 ft height, J and K, India) against *Staphylococcus aureus* and methicillin resistant *Staphylococcus aureus* (MRSA), the observed MIC of SCF extract was as low as 8µg/ml compared to solvent extracts where MIC was 125 and 250 µg/ml (Table 6.3). The variation in the activity was also conspicuous by their respective HPLC chromatograms (Figure 6.3) the HPLC analysis of the SCFE-1 extracts, by using column: RP-18 (Merck, 5µM, 4 × 250mm) and solvent system: MeOH (B), 0.25 per cent phosphoric acid in water (A) at flow rate: 0.8ml/min., revealed the presence of only four components that contribute towards higher activity while the rest of the extracts/fractions were found more complex in the chemical composition.

Table 6.3: Antimicrobial activity profile of extracts of *Humulus lupulus*.

Sl.No.	Tested Samples	MIC (mg/ml)	
		S. aureus ATCC-29213	MRSA-15187
1	Alcholic extract(A001)	125	125
2	Hydroalcholic Extract(A002)	125	125
3	Water extract (A003)	250	500
4	SCFE 1	8.0	8.0
5	SCFE 2	32	32
6	SCFE 3	64	64
	CC Fraction 1	>128	>128
	CC Fraction 2	16	32
	CC Fraction 3	128.0	128.0
	Ciprofloxacin	0.25	8.0

Figure 6.3: HPLC profile of extracts from *Humulus lupulus*.

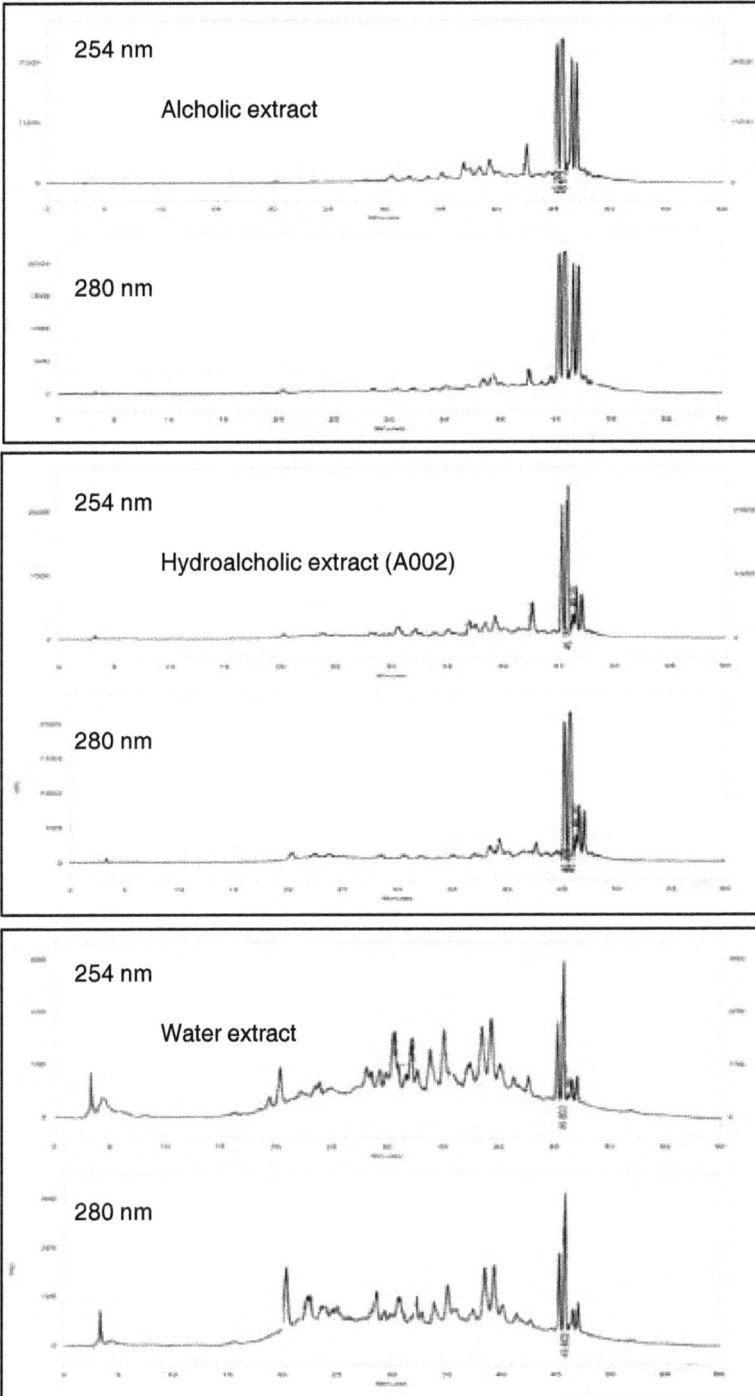

Contd...

Figure 6.3–*Contd...*

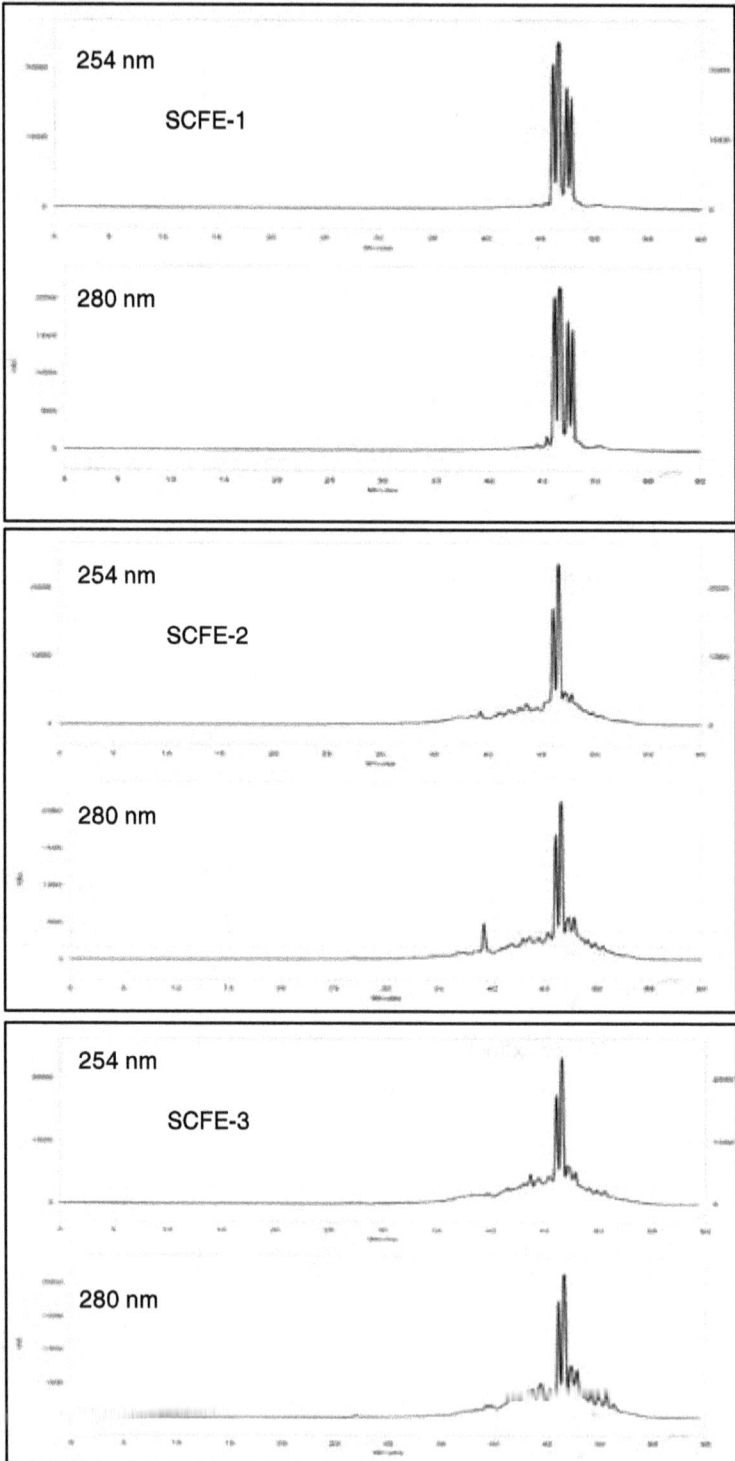

In SCFE, higher concentration of sesquiterpenes in (stem part) *Spilanthes americana* has been observed in SCFE than in simultaneous distillation extraction (SDE). The SD extract of the leaf, flowers and stem, however, exhibited more enrichment of oxygenated and monoterpene content than SCFE. The SCFE extract showed antimicrobial activity against several bacterial/fungal strains such as *Staphylococcus aureus, Bacillus cereus, Pseudomonas aerugnosa* and *Candida albicans* (Piggot *et al.*, 1997).

Eikani *et al.* (1999) and Pourmartazavi *et al.* (2004) observed distinct differences in the physical properties and chemical composition of SCFE and hydrodistillation extracts of *Cuminum cyminum*. It was shown that the desired chemical compound p-mentha-1, 4-dien-7-al (because of close odor resemblance with the extract) was higher in SCFE (41 per cent) than in hydrodistillation (27.4 per cent). Preparation and analysis of extracts from *Chamomile* flowers and fruits by SCFE and by conventional methods showed that SCFE have 4.4 times higher extractive value than that of steam distillation method and 71.4 per cent more apigenin content than soxhlet method and 124.6 per cent more than maceration method, besides the disadvantage of the non SCFE method/s involving long duration time extending upto 3 days. However, one disadvantage with the SCFE process has been the low isolation of glucoside of apigenenin.

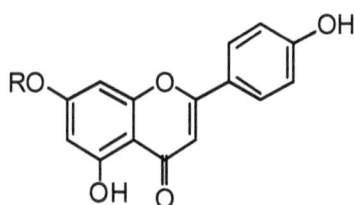

Apigenin R = H
Apigenin-7-D glucoside R = Glucose

Tandem process of SCFE and vacuum distillation for the separation of oxygenated compounds of high quality with excellent recovery rates has been developed. With the management of pressure, optimum conditions have been achieved by Sonsuzer *et al.* (2004) for the minimization of monoterpenes in the oil extraction of *Thymbra spicata*. Ranalli *et al.* (2004) have shown carrot root SCFE extract richer in carotene, phenolics, phytosterol, wax and sesquiterpene content as compared to commercial (solvent extracted) carrot oil. Among several extraction processes for the essential oil of Western Australia sandalwood, yields of extractable material and total volatile components in SCFE outnumbered the rest of the extraction processes (Piggott *et al.*, 1997). Better selectivity and efficiency in the extraction and isolation of sesquiterpenes, high mol. weight hydrocarbons and nitrogenated compounds from *Spilanthes americana* has been found better in SCFE as compared to simultaneous distillation extraction process (Stashenko *et al.*, 1996). Preferrential isolation of oxygenated compounds (sequiterpenes) from *Baccharis* leaves in high yields has been achieved by SCFE which however could not be achieved with hydrodistillation method, wherein high content of monoterpenes was osbserved (Cassel *et al.*, 2000). Extraction of important biomolecules like thymol and cravacrol (antioxidant active) has been achieved in higher contents by SCFE as compared to steam distillation from the *Zataria multiflora* (Ebrahimzadeh *et al.*, 2003).

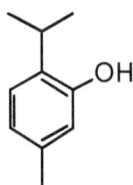

Thymol

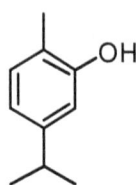

Cravacrol

SCFE of the fruits of *Archangelica officinalis,* at an optimum pressure and temperature has been successful for extracting the wax and triglycerides free, essential oil, which however could not be achieved by the conventional extraction processes (Gawdzile *et al.*, 1996; Modey *et al.*, 1996).

SCFE has been successfully applied in the extraction of antioxidant components from the leaves of rosemary. The two step operation at 40°C/100 bar afforded potent antioxidant components like carsonic acid, carnosol, rosmanol, epirosemanol and isorosemanol while at 60°C/400 bar, essential oil obtained showed lower antioxidant activity (Ibanez *et al.*, 1997).

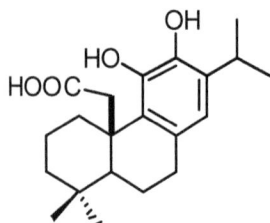

Carsonic acid

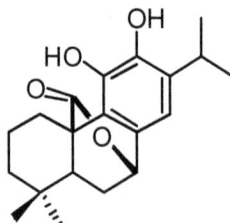

Carnosol

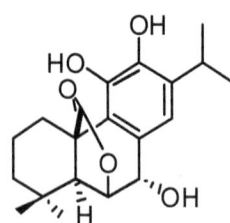

Rosmanol

SCFE has been applied to *Dancus carrota* for the extraction of carotol (major component) with low composition of high mol. weight molecules which has not possible with conventional extraction methods (Glisic *et al.*, 2007; Boceuka and Sovova, 2007; Yamini *et al.*, 2008). Extraction of aniseed (*Pimpinella anisum* L.) by SCFE has an edge over conventional hydrodistillation (HD) extraction process. The extraction by SCFE not only gave double the extractive value of the oil (7.5 per cent) as compared to HD (3.1 per cent) (Rodrigues *et al.*, 2003; Yamini *et al.*, 2008) but alos extracted anethole in >90 per cent yield where as in hydrodistillation it was only 60 per cent (Quii *et al.*, 2004).

Del Valle and Aguilera (1999) reviewed the role of high pressure CO_2 for the selective extraction of essential oils, pungent principles, carotenoid pigments, antioxidants, antimicrobials and related substances from spices, herbs and other plant materials. Conventional methods for the extraction of carotenoids from natural matrices are time consuming and require large amounts of organic solvents which are often expensive and potentially harmful (FDA 1992, Official methods AOAC 13th

Ed. AOAC, Washington DC 1992). Owing to problems associated with traditional processes, interest is growing in the development of simpler, faster and more efficient methods for carotenoid extraction from food and other natural products (Favti *et al.*, 1988; Cygnarowicz *et al.*, 1990; Spanos *et al.*, 1993). SCFE and solvent extraction methods have been applied to *Pyrethrum* plant material to study issues like color, yield and product quality (Otterback and Wenclawiak, 1999; Bunzenberger *et al.*, 1984; Stahl *et al.*, 1997; Wynn *et al.*, 1995; Kiriamiti *et al.*, 2003). The SCFE method has been found to be most effective and affords 0.55g of light brown product (pyrethrins) per gram of *Pyrethrum* whereas greenish black product in 0.16g/g yield is obtained by solvent extraction.

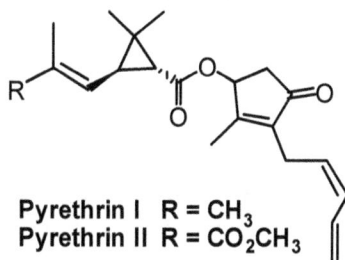

Pyrethrin I R = CH$_3$
Pyrethrin II R = CO$_2$CH$_3$

Extraction of lipids from *Sorhgum* distillers grains with soluble (DDGS) have been found to give not only higher yields of lipids (150g/kg) by SCFE against 85g/kg by solvent extraction method, but also contained higher contents of policosanols, phytosterols, free fatty acids and toropherols (Wang *et al.*, 2007). These compounds find various medicinal uses *e.g.*, policosanols as beta blockers (Castano *et al.*, 2004). Beneficial unsaturated fatty acids from cranberry seeds have been obtained by SCFE, contrary to solvent extraction product where higher content of saturated fatty acids is obtained (Mansi, 2004). The value addition of flax waste (obtained after the removal of fiber) has been exploited for valuable products. The extractive value by SCFE found 7.4 per cent compared to solvent extraction where it is reported to be 4 per cent. The SCFE extract was enriched with polycosanols which are known for their LDL lowering and anti-aggregating effects. These compounds also find use as nutraceuticals (Taylor, 1996). Comparison of essential oils from clove buds extracted by SCFE and other traditional extraction methods (Guan *et al.*, 2007) showed that best results are obtained with SCFE. The SCFE extract is less colored and have higher eugenol (57.77 per cent) and eugenol acetate (~22 per cent) contents, while in other methods, lower yields of chemical constituents are obtained (Morrison *et al.*, 2006).

SCFE/GC is seen as better choice for rapid quantitative method for capsiacinoids obtained from *Capsicum annuum* L. The findings showed that SCFE method needs no pretreatment or have minimal use of organic solvents, which, however was a must for usual extraction-HPLC method (Sato *et al.*, 1999).

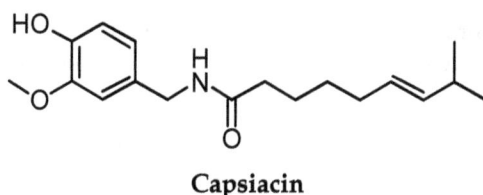

Capsiacin

Catchpole and his coworkers (2003) have shown that volatile contents of ginger, chilli powder and black pepper are best obtained with SCFE rather than by the solvent extraction method. Dramatic increase in the extractive values of scopolamine hydrochloride and hyoscyamine from *Datura arobrea* has been achieved by SCFE method (Choi *et al.*, 1999).

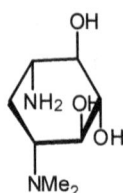 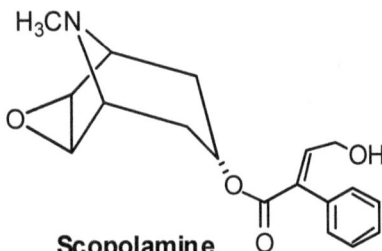

Hyoscyamine   **Scopolamine**

In the extraction of stevia glycosides from *Stevia*, the influence of co-solvents in SCFE has been studied and the combination of CO_2 with water: ethanol (1:1) found to be the effective extraction solvent mixture for achieving the best extractive results (Pasquel *et al.*, 2000). Selective extraction of alkaloid vindoline - a precursor of vinblastine and vincristine has been obtained by SCFE from the leaves of *C. roseus* at 300 bar and 35°C temperature which has not been possible by using normal extraction process (Song *et al.*, 1992). Application of SCFE in removing pesticides from natural products has been documented in the literature (Lehotay, 1997).

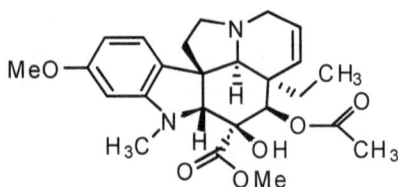

Vindoline

The method is well suited for the extraction of most of the pesticides and therefore provides an effective means to clean up any pesticide residues from herbs or other plant materials without introducing any toxic solvent residues. Ling *et al.* (1999) have successfully removed the pesticides from Chinese herbal plant *Glycyrrhizia radix* by water modified Super Critical CO_2. Reports for the removal of herbicides like diuron from various herbs by SCFE method has been acheived, whereas the other extraction methods failed to do the job (Gurdial *et al.*, 1991). This extraction method can be, therefore, used to make the materials safe for human use.

Lycopene, a natural product with strong antioxidant properties and protective effect against cancer and coronary heart has been obtained in solvent free state from tomato (Gmez-Prieto *et al.*, 2003; Sabio *et al.*, 2003).

Astaxanthin has been extracted from red yeast *Phaffia rhodozyma* by SCFE method in two step process at higher pressure of 500 bars whereas extraction at lower pressure of 300 bars afforded the lipid compounds with very low presence of astaxanthin (Lim *et al.*, 2002).

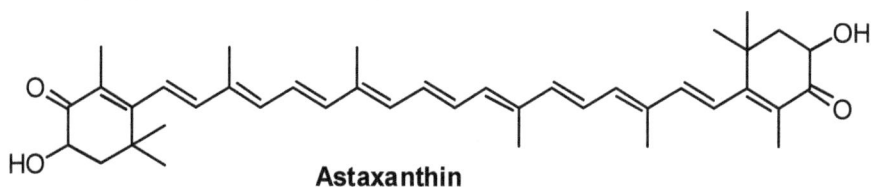

Astaxanthin

Over 30 million tons of soybeans are extracted each year for domestic use by using hexane as preferred solvent but complete removal of hexane from the extract is a problem. This issue has been addressed by using SCFE at 100 bars pressure and hexane from the product is removed completely as it is soluble in CO_2 and then triglycerides are extracted using pressure of > 200 bar (Friedrich *et al.*, 1982).The SCFE methodology has also been employed successfully in the extraction of lipids from wheat flour (Hubbard *et al.*, 2004). An efficient fractionation method to extract fatty acids, aliphatic aldehydes and sterols fatty acids followed by phenolic compounds (mainly catechin, epicetechin and gallic acid) has been developed by using SCFE method (Palma *et al.*, 1999).

Hypericin and hyperforin, the important components of *Hypericum perforatum* are manifested with many biological activity *e.g.*, hypericin has been reported as antiviral/antitumor (Miskovsky, 2002; Wills *et al.*, 2001) and hyperforin as antidepressant and antimalarial/antibacterial (Reichling *et al.*, 2001; Vollmer and Rosenson, 2004). We have carried out comparative studies on the extraction processes of *H. perforatum* leaves using soxhlet, accelerated solvent extraction, sonification and SCFE extraction methods. Among all the extraction methods, SCFE method proved to be the best choice for the extraction of hyperforin. Figures 6.4 and 6.5 depicts the chromatogram of the extracts obtained by following the above mentioned four extraction methods and Table 6.4 summarizes the results that enlist the extractive value and the content of hypericin and hyperforin present in the extracts.

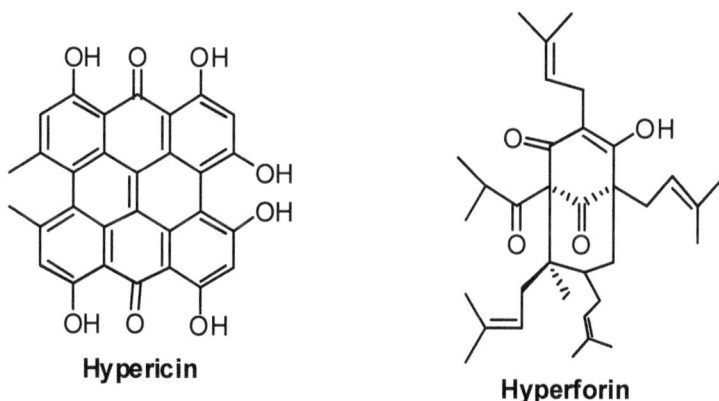

Hypericin

Hyperforin

Figure 6.4: HPLC profiles of Hypericin in the extracts of *Hypericum perforatum* (aerial parts) using different extraction methods.

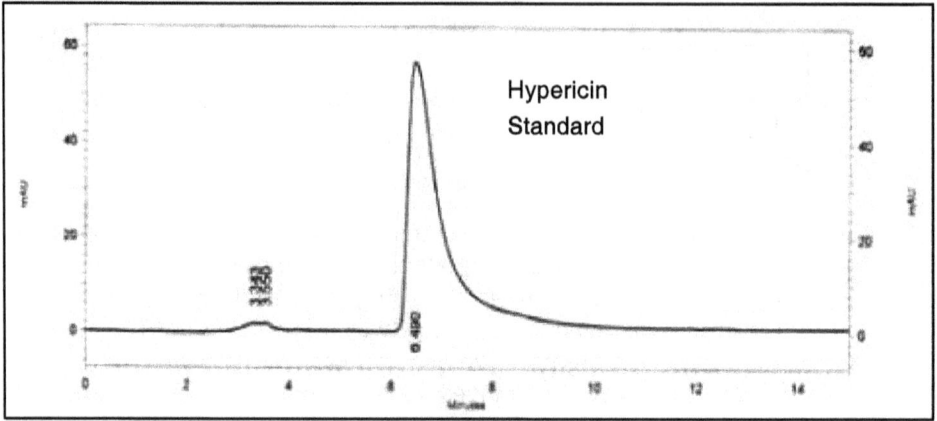

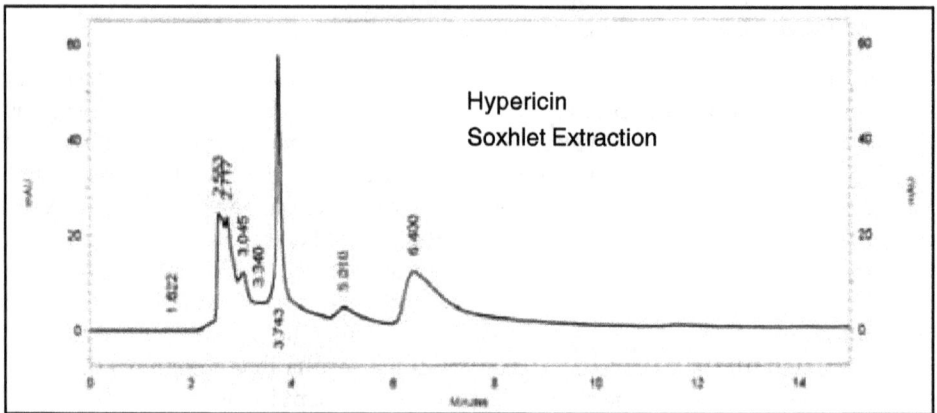

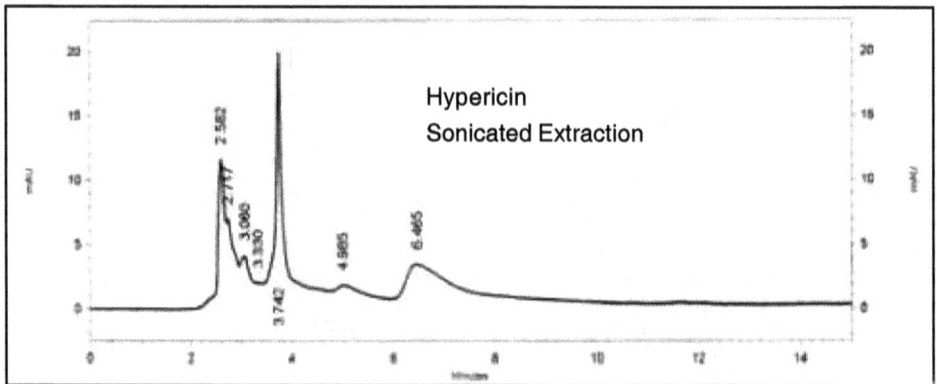

Contd...

Figure 6.4–*Contd...*

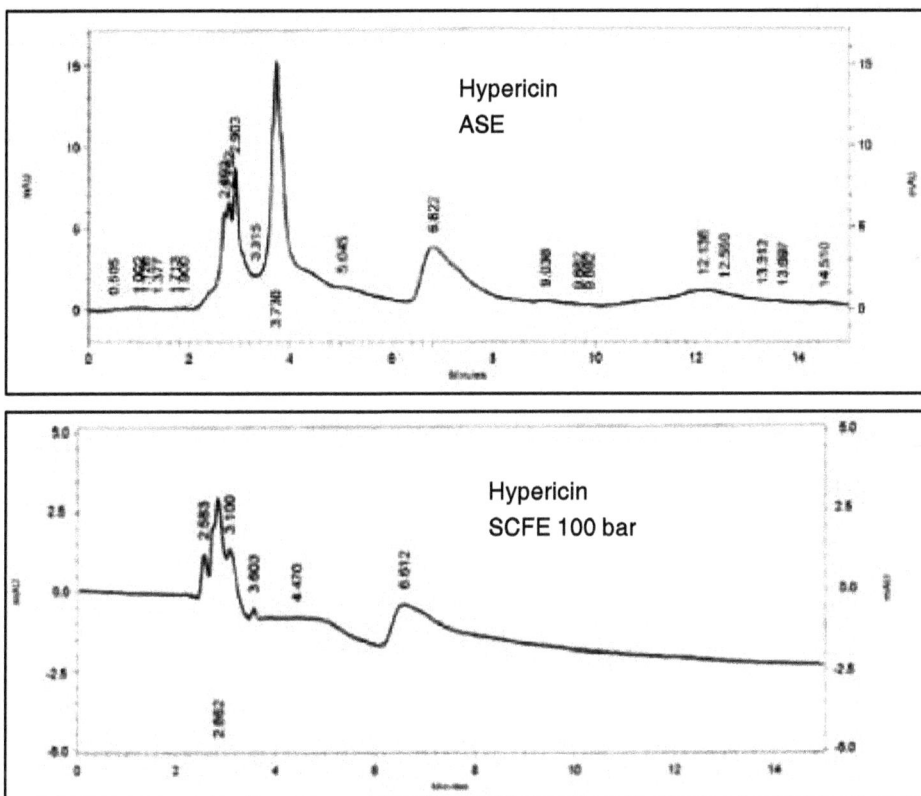

HPLC conditions for Hypericin:

Column: C-18 (Merck, 5µm, 4 x 100mm) Flow rate: 0.3 ml/min

Solvent System: MeOH: EtOAc:Buffer (67:16:17)

Table 6.4: Hypericin and Hyperforin content in the extracts obtained by four different methods.

	Hypericin		Hyperforin	
Method of Extraction	Extractive Value	Hypericin in the Extract	Extractive Value	Hyperforin in the Extract
SCFE	3.65	0.17 per cent	1.85	56.72 per cent
ASE	24.8	0.39 per cent	3.57	13.58 per cent
Sonication	10.53	6.20 per cent	2.25	0.25 per cent
Soxhlet	24.8	0.39 per cent	5.20	11.6 per cent

Figure 6.5: HPLC profiles of Hyperforin in the extracts of *Hypericum perforatum* (aerial parts) using different extraction methods.

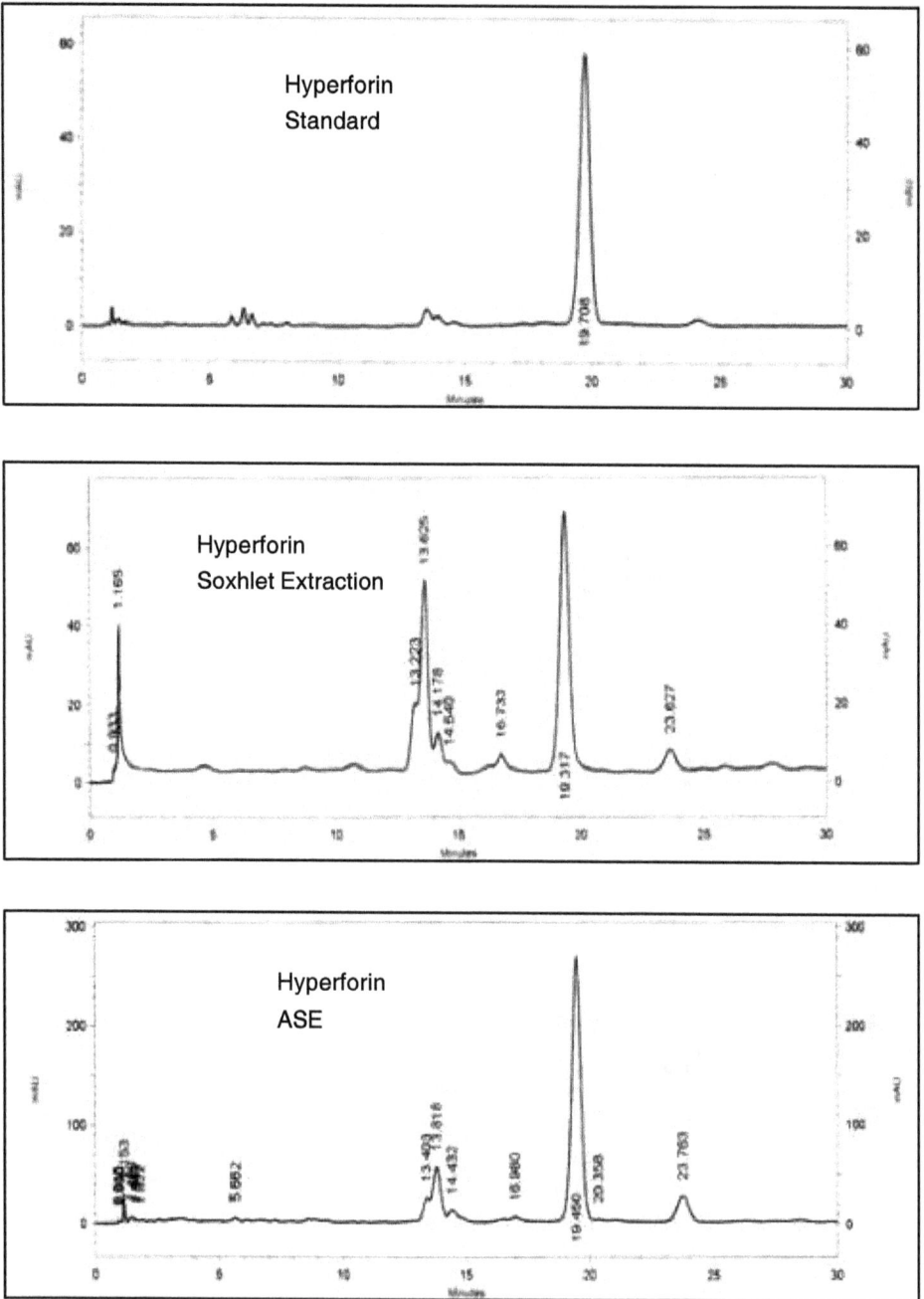

Contd...

Figure 6.5–*Contd...*

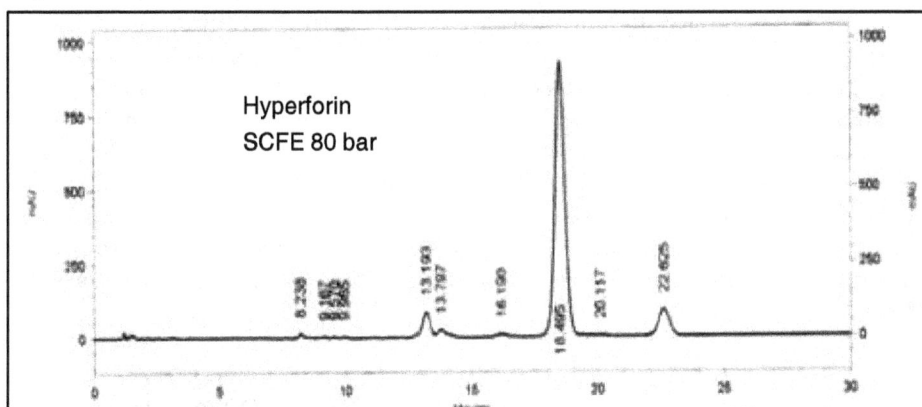

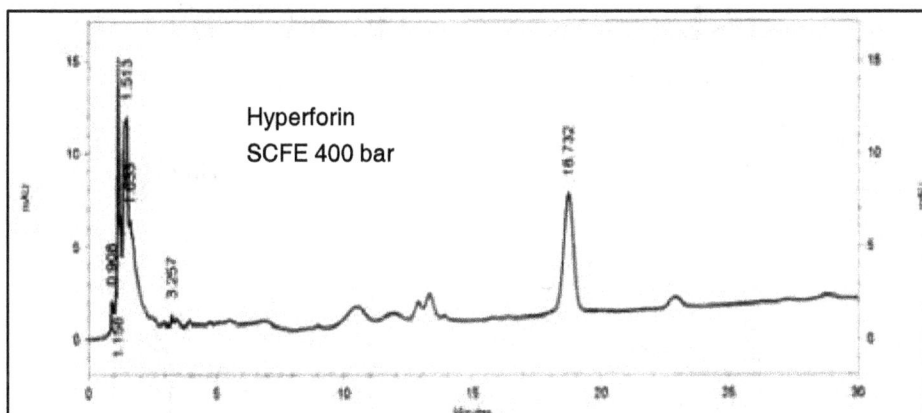

HPLC conditions for Hyperforin:

Column: RP-18 (Merck, 5µm, 4 x 250mm)

Flow rate: 1.5 ml/min

Solvent System: Acetonitrile: Buffer (85:15)

SCFE method has also been used for the isolation of phloroglucinols from *Hypericum perforatum*. The extraction has been reported at pressure of 367 bar and temperature of 50°C (Cui and Ang, 2002).The SCFE extract obtained from black sesame seed has been shown to exhibit more significant antioxidant activity compared to the extracts obtained by solvent extraction. The comparative studies were done by using known antioxidants like α-tocopherol and trolox (Hu *et al.*, 2004). Azadirachtins (Salanin, Cjemudin, Nimbin) manifested with valuable medicinal properties have been extracted by SCFE after removal of its seed oil (that contains very small quantities of Azadirachtins) at pressure of 300 bar and temperature upto 50 °C to afford the

Azadiractins upto 10,000 ppm. Its application on industrial scale is being worked up by an Italian company Essences Sri. Salerno (Tonthubthimthong *et al.*, 2001; Ambrosino *et al.*, 2004; Johnson and Morgan, 1997). Azadirachtin A has been selectively extracted by SCFE method from neem seed kernels (Ambrosino *et al.*, 1999; Walsh *et al.*, 1987). Pacilitaxel and baccatin III have been extracted from the Taxus plant pacific yew tree by SCFE (at 300 bar and 40°C temperature) using modifiers like dichloromethane and diethylether (Chun *et al.*, 1994). Santonin- a biologically active phytomolecule has been obtained by SCFE (using water as modifier) with recovery of > 92 per cent (Smith and Burford, 1992). The extraction of 1-hydroxy pinoresinol from *Fraxinus japonica* and other *Fraxinus* species has been effectively achieved by SCFE (Miyachi *et al.*, 1987).

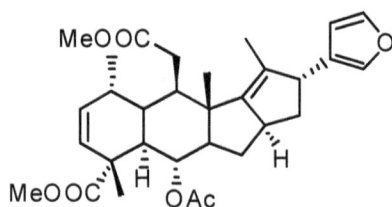

Nimbin

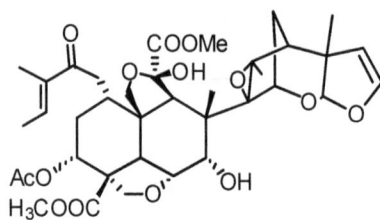

Azadirachtin A

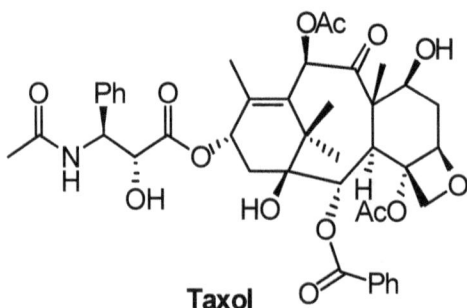

Taxol

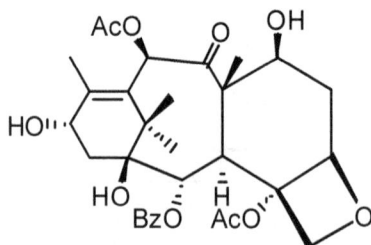

Baccatin III

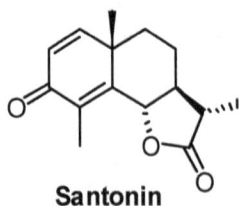

Santonin

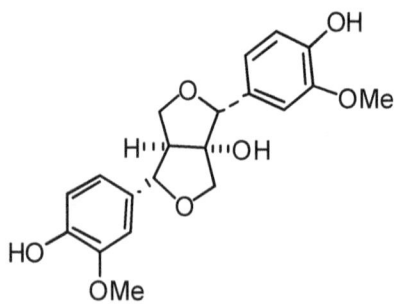

1-Hydroxypiniresinol

The extraction of flavonoids from the *Scutellariae radix* by SCFE has resulted in higher yields of bioactive compounds such as baicalin, baicalein and woginin giving 137.6, 8.6 and 2.2 mg/g compared to 113.5, 5.7 and 2.3 mg/g respectively obtained by sonication method (Ling *et al.*, 1999). These flavonoids are reported to have inhibitory effect on osteolytic bone metastasis of breast cancer (Won Ji-Hee *et al.*, 2008; Chen *et al.*, 2001).

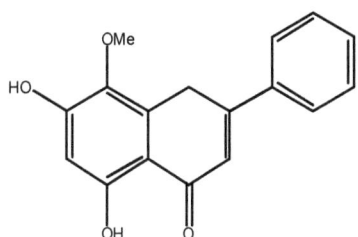

Wogonin

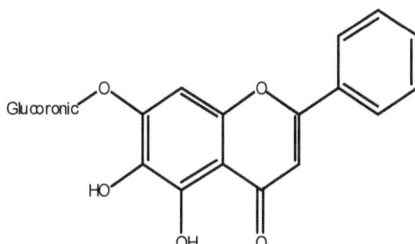

Baicalin

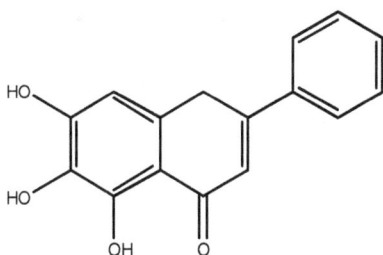

Baicalein

Docosahexaenoic acid (DHA) and ecosapentaenoic acid obtained from fish oil, are important molecules associated with several biological activities that include anti atherosclerosis and bioenhancer activity (Hirafugi *et al.*, 2003). However, fermentation process is being exploited as a viable technology both in terms of yield and quality product. The solvent aided downstream processing of fermented material affords brownish extract with 6.46 per cent of DHA which could be enriched by three fold by extraction of the fermented biomass by SCFE method and the process furnishes a pale yellow product overcoming thereby the issue of color problem. Figures 6.6A and 6.6B shows the chromatographs and GCMS of DHA enriched fractions (as methyl ester prepared by diazomethane esterification of the extract) obtained by solvent and SCF extraction.

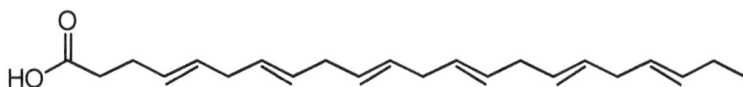

Docosahexaenoic acid

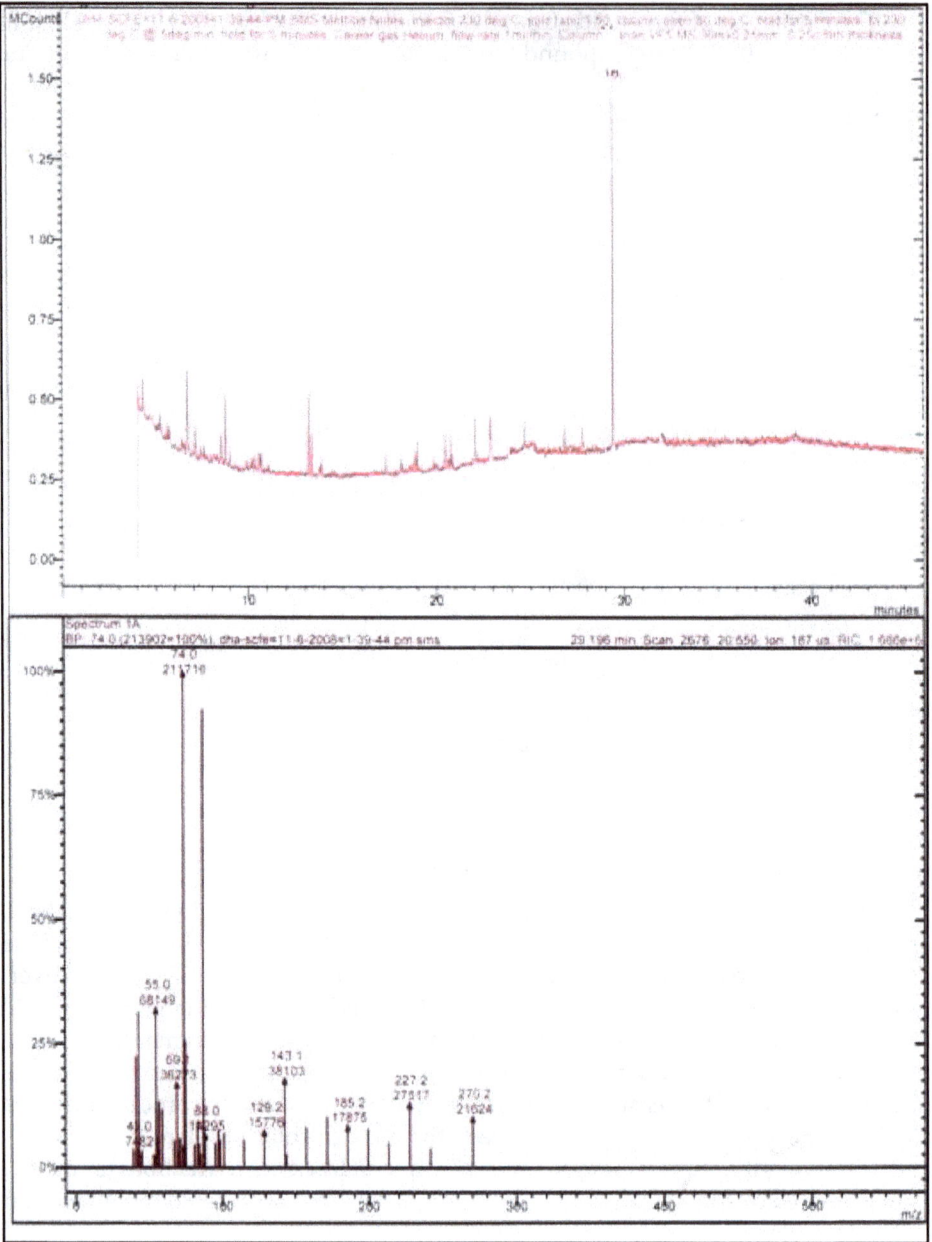

Figure 6.6A: SCFE extracted DHA fraction (GC: DHA Methyl ester).

Apart from the examples cited in the foregone text, which demonstrate the advantages associated with SCFE over both conventional and non conventional extraction methods in the area of natural products, the data in Table 6.5 shows the distinct advantage of SCFE over other methods in vogue-may it be essential oils, flavouring agents, perfumary chemicals, insecticides, antioxidants, anti cancer agents

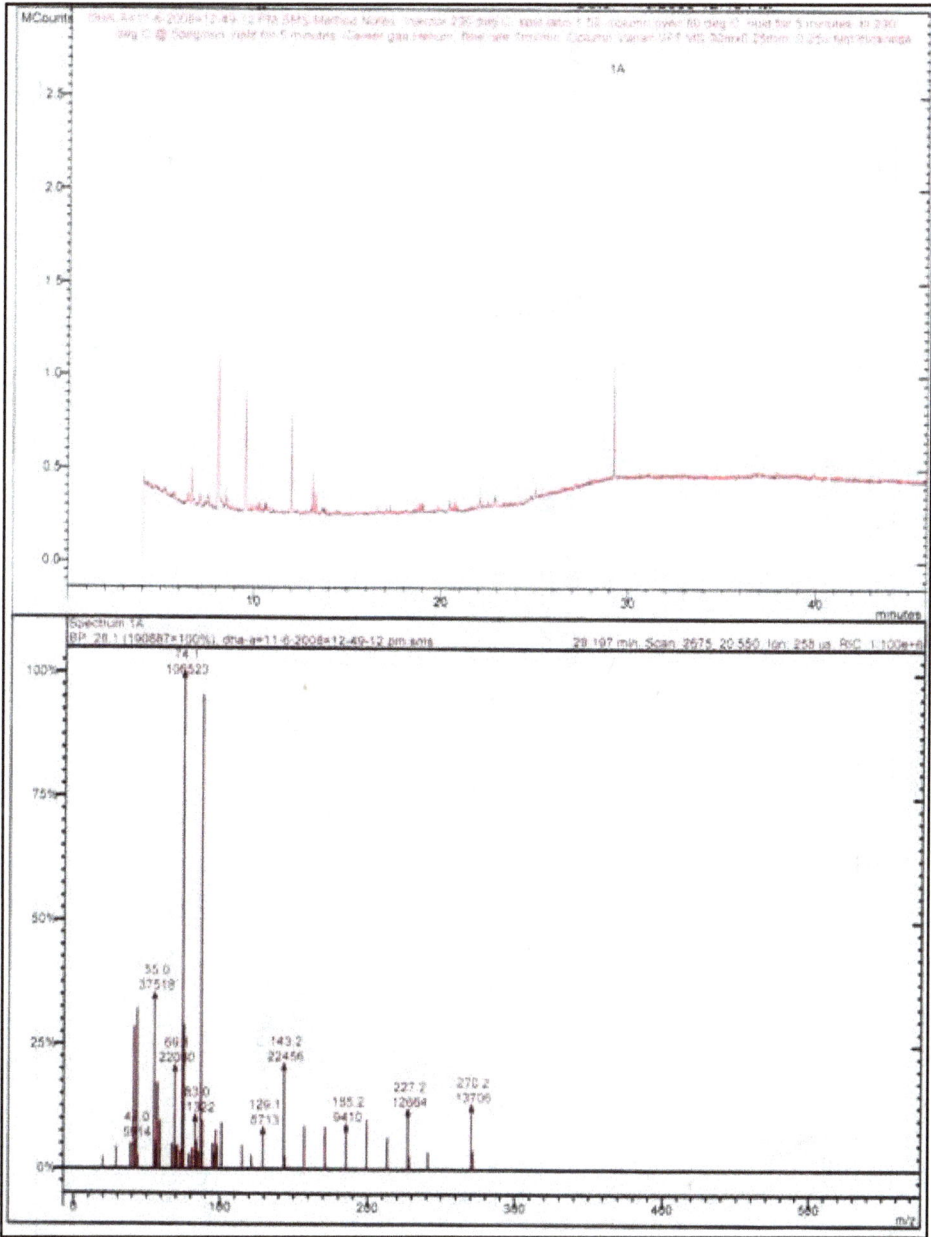

Figure 6.6B: Solvent extracted DHA fraction (GC: Methyl Ester).

etc. However, there are several examples where the method of choice does not matter much - may it be SCFE or solvent extraction or steam/hydrodistillation extraction. Having said that, one can not ignore the classical methods of extraction and both classical as well as non classical methods of extractions, being complimentary to each other should find use in industry and pharma companies (Hamburger *et al.*, 2004).

Table 6.5: Comparative studies on the chemical composition of essential oils derived by SCFE and Conventional extraction methods.

Sl. No.	Botanical Name	Major Constituents of SCFE Product	Major Constituents of Conventional Extraction Product	References
1.	*Mentha pulegium* L.	Pulegone 52.0%, Menthone 30.3%.	Pulegone 37.8%, menthone 20.3%, piperitenone 6.8%.	Aghel *et al.*, 2004
2.	*Zataria multiflora* Boiss.	Thymol 67.6%, λ-terpinene 19.5%, ρ-cymene 12.0%.	Thymol 44.6%, λ-terpinene 21.5%, ρ-cymene 13.7%.	Ebrahimzadeh *et al.*, 2003
3.	*Cuminum cyminum* L.	p-Mentha-1,4-dien-7-al 41.0%, p-Mentha-1,3-dien-7-al 0.2%, Cuminaldehyde 13.0%, γ-Terpinene 23.2%, Perillaldehyde 0.5%, o-Cymene 3.7%, Myrcene 0.4%, β-Pinene 16.4%, Sabinene 0.9%, α-Pinene 0.8%.	p-Mentha-1,4-dien-7-al 27.4%, p-Mentha-1,3-dien-7-al 8.2%, Cuminaldehyde 15.7%, γ-Terpinene 23.9%, Perillaldehyde 0.8%, o-Cymene 5.1%, Myrcene 1.1%, β-Pinene 16.3%, Sabinene 0.8%, α-Pinene 0.7%.	Eikani *et al.*, 1999
4.	*Pelargonium* sp.	Guaia-6,9-diene 9.58%, Geranyl formate 8.56%, Isomenthone3.67%, Geraniol 9.11%, Linalool 0.08%, Citronellyl formate 10.20%, Germacrene-D 4.99%, Citronellol 24.78%, Geranyl tiglate 4.61%, Phenylethyl tiglate 2.24%, Rose oxide 0.39%.	Guaia-6,9-diene 5.9%, Geranyl formate 5.5%, Isomenthone 5.6%, Linalool 2.7, Citronellyl formate 13.2%, Germacrene-D 2.4%, Citronellol 26.9%, Geraniol 8.1%, Geranyl tiglate 3.3%, Phenylethyl tiglate 1.8%, Rose oxide 0.5%.	Gomes *et al.*, 2007
5.	*Eucalyptus citriodora*	Neral 9.01%, Caryophyllene oxide 10.10%, Caryophyllene 2.77%, Citronellal 78.50%.	Neral 5.23%, Citronellal 87.97%.	Rozzi *et al.*, 2002
6.	*Melissa officinalis*	Geranial 63.23%, Caryophyllene 8.47%, Neral 23.30%, Neral acetate 1.57%, Caryophyllene oxide 2.63%.	Geranial 47.06%, Citro-nellal 5.00%, Neral 33.63%, Caryophyllene 1.23%, Caryophyllene oxide 3.56%.	Rozzi *et al.*, 2002
7.	*Cymbopogon citratus*	Caryophyllene 27.87%.Neral 21.27%, Geranial 42.03%.	Caryophyllene 0.67%, Neral 35.30%, Geranial 52.73%.	Rozzi *et al.*, 2002
8.	*Menarda citriodora*	Thymol 90.43%, Thymol methyl ester 1.93%.	Thymol 65.50%, α-Terpine 12.23%.	Rozzi *et al.*, 2002
9.	*Salvia officinalis* L.	1,8-Cineole 9.54%, α-Thujone 26.52%, β-Thujone 4.23%, Camphor 27.26%, α-Thujene traces, α-Pinene 4.28%, Camphene 7.16%, β-Pinene 0.96%, β-Myrcene1.06%, α-Phellandrene traces, α-Terpinene 0.83%, p-Cymene 3.14%, Borneol	8-Cineole 7.96%, α-Thujone 24.29%, β-Thujone 4.03%, Camphor 23.72%, Borneol 2.21%, Terpinen-4-ol 0.39%, Bornyl acetate 2.73%, β-Caryophyllene 1.8%, 2.25%, Aroma-dendrene 0.33%, α-Humulene 2.83%, Manool	Aleksovski *et al.*, 2007

Contd...

Table 6.5–*Contd...*

Sl.No.	Botanical Name	Major Constituents of SCFE Product	Major Constituents of Conventional Extraction Product	References
		traces, Terpinen-4-ol tr., Bornyl acetate2.25%, β-Caryophyllene 3.83%, Aromadendrene tr., α-Humulene3.82%, Manool 0.65%, Sclareol 1.52%, Heneicosane 2.04%.	4.07%,Sclareol 0.19%, α-Thujene 0.27%, α-Pinene 4.35%,Camphene 7.61%, β-Pinene 0.94%, β-Myrcene0.88%, α-Phellandrene 0.15%, α-Terpinene0.78%, p-Cymene 2.77%.	
10.	Polygala cyparissias	Methyl salicylate 29.3%, hexanoic acid 24.78%, cyclohexanone, 2-ethyl-oxime 33.14%, octadecane 5.42%, docosane 1.73%, squalene 5.6%, p-Benzoquinone 28.311%, Acetyl eugenol 6.082%, 11,14-Octadecanoic methyl ester acid 2.425%, Cadinene 2.482%, β-Caryophyllene oxide 2.573%, N,N-bis (2-hydroxyethyl) dodecanamide 0.925%, α-Spinasterol 0.866%, Hexadecanoic acid 38.632%, Dodecane 3.203%, n-Tridecane 8.358%, Tetradecane 15.834%, Pentadecane 40.752%, Hexadecane 3.873%, Heptadecane 15.726%, Octadecane 6.575%, Nonadecane 0.530%.	β-Caryophyllene oxide 51.97% (With DCM).	Weinhold et al., 2008
11.	Schizandra chinensis	Schizandrin A 23.8%, Gomisin A 28.8%.	Schizandrin A 5.1%, Gomisin A 16.8%.	Lojková et al., 1998
12.	Chamomile	Matricin 7.3%, β-Farnesene 9.9%, Spathulenol 1.0%, α-Bisabolol oxide B 4.6%, α-Bisabolone oxide 2.5%, α-Bisabolol 2.3%, α-Bisabolol oxide A 28.5%, trans-en-in-dicycloether 3.9%, waxes 4.8%.	Cis-en-in-dicycloether 2.7%,β-Matricin 0.5% Farnesene 12.8%,Spathulenol 2.6%, α-Bisabolol oxide B 7.8%, α-Bisabolone oxide 9.2%, α-Bisabolol 3.6%, α-Bisabolol oxide A 36.6%.	Scalia et al., 1999
13.	Piper nigrun L.	δ-3-carene 14.34%, β-Caryophyllene 21.76%, limonene 19.81%, α-Thujene1.00%, α-Pinene 4.09%, sabinene 11.64%, myrecene 7.70%, β-Pinene 2.61%, α-phelandrene 1.85%, p-cymene and α-terpinene 1.77%, β-phelandrene 0.36%,	Hydrocarbons 89% (The hydrocarbon fraction is composed of monoterpenes 70–80% and sesquiterpenes 20–30%), oxygenated terpenes and aromatic compounds 10%.	Debrawere et al., 1976; Sandra et al., 1999

Contd...

Table 6.5–Contd...

Sl.No.	Botanical Name	Major Constituents of SCFE Product	Major Constituents of Conventional Extraction Product	References
14.	Lavandula angustifolia	α-terpinolene 0.73%, linalool 0.42%, δ-elemene 0.78%, α-copaene 1.31%, β-elemene 1.34%, α-humulene 1.46%, γ-murolene 0.43%, β-selinene 2.50%, α-selinene 3.08%, δ-cadinene 0.46%, calamenene 0.46%. Linalyl acetate 34.7%.	Linalyl acetate 12.1%.	Reverchon et al., 1995
15.	Scutellaria baicalensis	Baicalin 33.4%, baicalein 2.1%, wogonin 0.5%.	Baicalin 14.3%, baicalein 4.1%, wogonin 1.7%.	Lin et al., 1999
16.	Juniperus communis L.	Terpene-4-ol 9.8%, bornyl acetate 1.3%, terpinenyl acetate 3.3%, sabinene 34.9%, α-Thujene 25.1%, α-Pinene 22.4%, β-Pinene 1.4%, myrecene 3.8%, α-phelandrene 0.6%, 3- carene 36.8%, α-terpinene 2.8%, p-cymene 1.5%, limonene 24.2%, γ-terpinene 4.1%, trans- sabinene hydrate 1.1%, terpinolene 2.7%, cis- sabinene hydrate 2.4%, α-copaene 1.2%, β-elemene 1.2%, γ-murolene 3.2%, δ-cadinene 4.4%, γ-cadinene 2.8%.	Camphene 0.9%, β-phelandrene 25.1%, α-Pinene 24.5%, sabinene 0.4%, β-Pinene 1.7%, myrecene 3.4%, delta-3- carene 39.4%, α-terpinene 0.6%, p-cymene 0.5%, γ-terpinene 0.4%, terpinolene 3.1%.	Pourmortazavi et al., 2004
17.	Foeniculum vulgare	(E)-anethol 90.14%, α-Pinene 1.32%, sabinene 0.89%, limonene 9.34%, (z)-β-ocimene 1.39%, γ-terpinene 1.94%, fenchone 9.25%, estragole 3.09%, (E)- germecene 2.9%.	(E)-anethol 69.41%, α-Pinene 0.89% camphene 0.09%, sabinene 0.15%, 1.4%, myrecene 10.58%, p-cymene 0.24%, limonene 10.0%, (z)-β-ocimene 0.96%, γ-terpinene 0.72%, fenchone11.0%, estragole 4.45%, p-anisaldehyde 0.57%, (z)- anethol 0.27%, germecene D 0.25%.	Yamini et al., 2002
18.	Valeriana officinalis L.	Acetoxy valeranone 5.6-9.6%, Isovaleric acid 18.7-41.8%, valerenic acid 8.2-11.8%, (Z)-valernyl acetate 4.5-6.5%, bornyl acetate 2.3-7.7%, valerenol 3.7-5.2%.	Acetoxy valeranone 7.6%, Bornyl acetate 11.6%, valerenic acid 8.0%, (Z)-valernyl acetate 7.9%.	Safaralie et al., 2008

Contd...

Table 6.5—Contd...

Sl.No.	Botanical Name	Major Constituents of SCFE Product	Major Constituents of Conventional Extraction Product	References
19.	Artemisia sieberi	Camphor 77.4%, cis-Chrysanthenol 3.17%, trans-Pineocorveol 1.55%, Chrysanthenone 4.46%, γ-Terpinene 4.35%, cis-Arbusculone 4.47%, 1,8-Cineol 7.03%, Camphene 3.36%.	Camphor 54.7%, camphene 11.8%, 1,8-cineol 9.9%, β-thujone 5.7%, α-pinene 2.5%.	Ghasemi et al., 2007
20.	Origanum vulgare	Thymol 21.7%, Carvacrol 6.4%, myrecen 1.2%, Linalool 43.6%, Linalyl acetate 3.1%, γ-Terpinene 2.5%, trans-Thujan-4-ol 4.7%,4-Terpineol 3.0%, α-Terpineol 3.2%, Methyl thymol ether 1.9%, trans-Caryophyllene 1.7%.	Thymol 15.3%, Carvacrol 4.6%, Sabinene 4.9%, Linalyl acetate 1.5%, myrecene 1.7%, α-Terpinene 3.8%, p-cymene 2.6%, γ-Terpinene 7.7%, trans-Thujan-4-ol 3.9%, Linalool 31.7%, 4-Terpineol 10.7%, α-Terpineol 3.3%, Methyl thymol ether 1.8%, trans-Caryophyllene 1.1%.	Díaz-Maroto et al., 2002
21.	Romanian Mentha	Z- Sabinene hydrate acetate 43.5%, Sabinene 1.88%, 4-Terpineol 0.83%, γ-Terpinene 0.28%, Terpineolene 0.07%, α-Terpinene 0.12%.	Z- Sabinene hydrate37.3%, Z- Sabinene hydrate acetate 0.42%, Sabinene 2.93%, 4-Terpineol 19.16%, γ-Terpinene 6.30%, Terpineolene 1.60%, α-Terpinene 4.24%.	Eugenia et al., 2001
22.	Nigella sativa	α-Methyl-benzene methanol 25.6%, Dillapiole 3.5%,, cuminaldehyde 11.5%, α-Pinene 0.8%, β-Pinene 1.5%, Myrecene 0.6%, o-cymene 7.8%, Limonene 6.8%, γ-terpinene 38.0%, α-Terpinolene 0.4%, β-Selinene 0.6%, Germacene-B 0.9%.	Methyl-benzene methanol 3.5%, α-Pinene 2.8%, Sabinene 1.2%, β-Pinene 3.7%, Myrecene 1.0%, p-cymene 5.6%, o-cymene 0.1%, Limonene 10.6%, γ-terpinene 45.7%, α-Terpinolene 0.4%, α-Terpinolene 1.7%, cuminaldehyde 12.7%, Cuminyl alcohol 6.4%, α- and β-Selinene 0.1%.	Pourmortazavi et al., 2005
23.	Marchantia convoluta	Cedrol 4.60%, Benzothiazole 11.82%, 2-ethyl-hexanoic acid 9.82%, ethylphenoxy benzene 8.99%, acetic acid octadecyl ester 8.82%, 4-cyanothiophenol 5.49%, 9,12-octadecadienoic acid ethyl ester 3.25%, 2(3H)-benzothiazolone 2.79%, octadecanoic acid ethyl ester 2.39%, n-hexadecanoic acid 2.08%, 1,1'-(3-methyl-1-propene-1,3-diyl) bis-benzene 2.07%. The total content of organic acids and esters was 32.19%.	Ester 57.21%, phytol 6.32%.	Chen et al., 2008

Contd...

Table 6.5—Contd...

Sl. No.	Botanical Name	Major Constituents of SCFE Product	Major Constituents of Conventional Extraction Product	References
24	Eucalyptus globulus	α-Pinene 10.5%, 1,8-Cineole 62.6%, p-Mentha-1,3,5-triene 1.6%, Guaiol 4.2%, α-Eudesmol 0.8%, n-Nonacosane 28.5%, n-Dotriacontane 24.8%, n-Pentatriacontane 11.8%, n-Hexatriacontane 21.9%, Camphor 1.4%, Aromadendrene 8.0%, allo-Aromadendrene 1.8%, Valencene 0.7%, Spathulenol 1.3%.	α-Pinene 6.9%, 1,8-Cineole 48.2%, p-Mentha-1,3,5-triene 3.8%, Camphor 1.7%, Aromadendrene 13.7% allo-Aromadendrene 2.9%, Valencene 1.0%, Spathulenol 2.0%, Guaiol 7.6%, α-Eudesmol 1.2%.	Porta et al., 1999
25.	Coriandrum sativum L.	Geranyl acetate 2.4%, α-Thujene 0.1%, α-Pinene 2.8%, Camphene 1.5%, β-Pinene 0.9%, Sabinene 0.9%, β-Myrcene 1.0%, p-Cymene 4.0%, Limonene 2.7%, δ-Terpinene 3.5%, Linalool 61.9%, Camphor 5.6%, p-Cymen-8-ol 0.3%, Carvone 1.0%, Geraniol 2.2%, Eugenol 1.4%, β-Caryophylene 0.8%.	Geranyl acetate 1.8%, α-Thujene traces, α-Pinene 2.3%, Camphene 0.4%, β-Pinene 0.3%, Sabinene 0.3%, β-Myrcene 0.8%, p-Cymene 4.0%, Limonene 2.3%, δ-Terpinene 3.5%, Linalool 62.8%, Camphor 5.6%, p-Cymen-8-ol 0.1%, Carvone 1.0%, Geraniol 2.8%, Eugenol 2.6%, β-Caryophylene 2.1%.	Anitescu et al., 1997
26.	Eugenia caryophyllata	Eugenol 56.81%, Eugenol acetate 20.91%, α-Copaene 0.95%, Caryophyllene 17.77%, α-Humulene 2.32%, Cadinene 0.71%, Diethyl Phthalate 1.43%.	Eugenol 48.82%, Eugenol acetate 3.89%, α-Copaene 1.93%, Caryophyllene 20.59%, α-Humulene 4.41%, Cadinene1 46%.	Gyan et al., 2007
27.	Ridolfia segetum	α-Phellandrene 19.4%, terpinolene 20.5%, piperitenone oxide 11.6%, β-phellandrene 8.2%, (Z)-β-ocimene 7.8%, myristicin 7.5%, p-cymene 4.4%.	α-Phellandrene 12.9%, terpinolene 11.6%, myristicin 11.0%, p-cymene 9.9%, β-phellandrene 8.2%, (Z)- β-ocimene 6.0% while the main components of the fruits were found to be myristicin 70.8%, piperitenone oxide 19.9%, dill apiole 4.2%.	Bruno et al., 2007
28.	Thymbra spicata	Oxygenated monoterpenes 85.25%, Monoterpenes 2.75% and Sesquiterpenes 0.04%.	Oxygenated monoterpenes 53.10%, Monoterpenes 39.80% and Sesquiterpenes 4.40%.	Sonsuzer et al., 2004
29.	Laurus nobilis	1,8-Cineole 22.8%, linalool 12.5%, α-terpinyl acetate 11.4%, and methyleugenol 8.1%. Hydrocarbon monoterpenes 12.67%, oxygenated monoterpenes 79.43, hydrocarbon	Hydrocarbon monoterpenes 15.51%, oxygenated monoterpenes 70.28%, hydrocarbon sesquiterpenes 7.19%, oxygenated sesquiterpenes 7.03%.	Pourmortazavi et al., 2007

Contd...

Table 6.5—Contd...

Sl.No.	Botanical Name	Major Constituents of SCFE Product	Major Constituents of Conventional Extraction Product	References
		sesquiterpenes 7.45%, oxygenated sesqui-terpenes 0.5%.		
30.	*Daucus carota* L.	Total carotenes 0.1850%, total polyphenols 0.0623%, tyrosol 0.00 23%, hydroxytyrosol 0.00018%, chlorophylls 0.00 23%, tocopherols 0.0160%.	Total polyphenols 0.0045%, tyrosol 0.0007%, hydroxytyrosol 0.0007%, chlorophylls 0.0017%, total carotenes 0.0170%, tocopherols 0.0182%.	Ranalli *et al.*, 2004
31.	*Thymus vulgaris* L.	Thymol 60.91%, carvacrol 8.90%. α-Pinene 0.20%, camphene 0.13%, 1-octen-3-ol 0.11%, p-cymene 8.49%, 1,8-cineole + limonene 2.26%, δ-terpinene 1.60%, linalool 3.29%, camphor 1.38%, endo-borneol 3.95%, 4-terpineol 1.08%, Verbenone 1.43%, linalyl acetate 1.03%.	Thymol 31.06%, carvacrol 5.18%. α-Pinene 2.13%, camphene 1.49%, verbenene 0.07%, 1-octen-3-ol 0.23%, 1-octen-3-one 0.11%, p-cymene 33.00%, 1,8-cineole + limonene 2.75%, δ-terpinene 4.75%, linalool 3.57%, camphor 0.70%, endo-borneol 3.55%, 4-terpineol 1.00%, Verbenone 1.67%, linalyl acetate 0.20%.	Diaaz-Maroto *et al.*, 2005
32.	*Foeniculum vulgare* Mill.	Trans-anethole 63.80%, 1,8-Cineole + limonene 0.87%, fenchone 12.71%, estragole 20.33%, p-anisaldehyde 0.99%.	Trans-anethole 49.71%. 1,8-Cineole + limonene 1.01%, fenchone 19.33%, estragole 25.84%, p-anisaldehyde 1.90%.	Diaaz-Maroto *et al.*, 2005
33.	*Lippia alba* Mill.	Monoterpene hydrocarbons 37.9%, Monoterpene 47.4%, Sesquiterpene Carvone 45.1%, limonene 36.90%, Piperitone 0.8%, Piperitenone 1.1%, β-Bourbonene 1.0%, Bicyclosesquiphellandrene 8.9%, hydrocarbons 13.9%.	Carvone 51.0%, limonene 32.60%, Piperitone 0.93%, Piperitenone 1.47%, β-Bourbonene 0.7%, Bicyclosesquiphellandrene 7.3%, Monoterpene hydrocarbons 33.3%, Monoterpene 54.0%, Sesquiterpene hydrocarbons 11.3%.	Stashenko *et al.*, 2004
34.	*Santalum spicatum*	Epi-α-bisabolol 3.1%, (Z)-α-2(E),6(E)-farnesol 5.3%, (Z)-β-Santalol 3.8%, (Z)-nuciferol 3.6%, Santalol 10.0%.	Epi-α-bisabolol 6.6%, (Z)-α-2(E),6(E)-farnesol 11.0%, (Z)-β-Santalol 8.1%, (Z)-nuciferol 6.9%, Santalol 21.6%.	Piggott *et al.*, 1997
35.	*Spilanthes americana*	Sesquiterpenes (α-and β-bisabolenes, Caryophyllene and Cadinens) 40.0%, 5-Phenyl-2,4-Pentadienyl acetate 2.6%.	Sesquiterpenes 32.0%, Oxygenated compounds 28.0%, Monoterpenes 27.0%, 5-Phenyl-2,4-Pentadienyl acetate 3.2%.	Stashenko *et al.*, 1996

Contd...

Table **6.5**–Contd...

Sl.No.	Botanical Name	Major Constituents of SCFE Product	Major Constituents of Conventional Extraction Product	References
36.	*Spilanthes americana*	Nitrogenated compounds 43.0%, oxygenated compounds 36.0%, Spilanthol 17.03%, 5-phenyl-2,4-pentadienyl acetate 7.3%.	Nitrogenated compounds not detected, sesqui-terpenes 28.0%, oxygenated compounds 52.0%, monoterpenes 10.0%, 5-phenyl-2,4-pentadienyl acetate 13.9%.	Stashenko et al., 1996
37.	*Spilanthes americana*	Nitrogenated compounds 27.0%, Oxygenated compounds 23.0%, Spilanthol 27.0%, 5-Phenyl-2,4-Pentadienyl acetate 4.7%.	Sesquiterpenes 20.0%, Oxygenated compounds 32.0%, Monoterpenes 42.0%, 5-Phenyl-2,4-Pentadienyl acetate 4.7%.	Stashenko et al., 1996
38.	*Baccharis dracunculifolia*	Oxygenated compounds 52.2%, spathulenol 16.4%, (E)-nerolidol 47.5%.	Oxygenated compounds 35.9%, (E)-spathulenol 9.8%, nerolidol 13.5%, β-Pinene, 28.2%, limonene 10.6%.	Cassel et al., 2000
39.	*Juniperus virginiana* L.	Cedrene 0.187%, cedrol 2.250%.	Cedrene 0.6%, cedrol 14.6%.	Hu et al., 2004
40.	*Bunium persicum* Boiss.	α-Methyl-benzenemethanol 25.55%, γ-Terpinene 37.98%, dillapiole 3.5%, cuminaldehyde 11.48%, α-Pinene 0.8%, β-Pinene 1.5%, Myrcene 0.6%, o-Cymene 7.8%, Limonene 6.8%, α-Terpinolene 0.4%.	α-Methyl-benzenemethanol 3.5% γ-Terpinene 45.7%, cuminaldehyde 12.7%, α-Pinene 2.8%, Sabinene 1.2%, β-Pinene 3.7%, Myrcene 1.0%, p-Cymene 5.6%, o-Cymene 0.1%, Limonene 10.6%, α-Terpinolene 1.5%, α-Terpineol 1.7%, Cuminyl alcohol 6.4%.	Pourmortazavi et al., 2005
41.	*Achillea millefolium* L.	Borneol 4.68%, bornyl acetate 16.73%, tricyclene 0.43%, α-thujene 0.24%, α-pinene 0.53%, camphene 0.92%, β-pinene 0.15%, myrcene 3.14%, p-cymene 2.76%, 1,8-cineole 9.59%, γ-terpinene 8.97%, terpinolene 7.58%, linalool 2.32%, β-terpineol 0.795, camphor 26.36%, terpinen-4-ol 0.32%, α-terpineol 0.66%, sabinyl acetate 0.58%, α-terpinyl acetate 0.58%, α-copaene 0.46%, β-caryophyllene 0.31%, γ-cadinene 0.69%.	Borneol 3.53%, bornyl acetate 4.26%, α-thujene 0.44%, α-pinene 1.24%, camphene 2.02%, myrcene 4.93%, p-cymene 5.40%, 1,8-cineole 16.19%, γ-terpinene 9.42%, terpinolene 3.89%, linalool 3.37%, β-terpineol 0.795, camphor 38.44%, terpinen-4-ol 0.96%, α-terpineol 1.92%.	Bocevska et al., 2007
42.	*Rosa canin* L.	Carotene 36.3μg g^{-1}, unsaturated fatty acids 92.7%, poly unsaturated fatty acids 76.25%, Pheophytin 45.8μg/g.	Unsaturated fatty acids 90%, poly unsaturated fatty acids 60%, carotene 145.3μg/g.	Szentminhalyi et al., 2002

Contd...

Table 6.5–*Contd...*

Sl.No.	Botanical Name	Major Constituents of SCFE Product	Major Constituents of Conventional Extraction Product	References
43.	Elaeis guineensis	Myristic acid (C14) 15.60%, C16 7.80%, C18 2.00%, C18:1 15.10%, C18:2 2.70% Caprylic acid (C8) 4.4%, capric acid (C10) 3.70%, lauric acid (C12)48.30%.	Myristic acid (C14) 16.70%, C16 6.80%, C18 1.86%, C18:1 9.63%, C18:2 1.03% Caprylic acid (C8) 5.44%, capric acid (C10) 4.90%, lauric acid (C12) 53.65%.	Norulaini et al., 2004
44.	Capsicum annuum	Capsorubin 5.32%, capsanthin 39.10%, zeaxanthin 14.96%, β-cryptoxanthin 8.78%, β-carotene 3.02%.	Capsorubin 3.29%, capsanthin 23.3%, zeaxanthin 9.01%, β-cryptoxanthin 9.02%, β-carotene 4.92%.	Jaren-Galan et al., 1999
45.	Secale cereale	Alkylresorcinolos C17:0 26.0%, C19:0 34.0%, C21:0 24.0%, C23:0 9.0%, C25:0 7.0%.	Alkylresorcinolos C17:0 23.0%, C19:0 32.0%, C21:0 25.0%, C23:0 11.0%, C25:0 9.1%.	Landberg et al., 2007
46.	Pyrethrum	Pyrethrin 55%.	Pyrethrin 16%.	Kiriamiti et al., 2003
47.	Maclura pomifera	Osajaxanthone 0.203%, euchrestaflavanone B 0.215%, euchrestaflavanone C 0.241%, alvaxanthone 0.553%, macluraxanthone 1.054%, 8-prenyltoxyloxanthone 0.175%.	Osajaxanthone 0.192%, euchrestaflavanone B 0.217%, euchrestaflavanone C 0.235%, alvaxanthone 0.586%, macluraxanthone 1.054%, 8-prenyltoxyloxanthone 0.182%.	da Costa et al., 1999
48.	Calendulae	Faradiol-3-O-myristate 0.648%, isorhamnetin-3-O-(2''-rhamnosyl) glucoside 0.0006%, Faradiol-3-O-palmitate 0.0557%. Isorhamnetin-3-O-(2',6'-dirhamnosyl)glucoside 0.0017%.	Faradiol-3-O-myristate 0.410%,Isorhamnetin-3-O-(2'',6''-dirhamnosyl)glucoside 0.3%, isorhamnetin-3-O-(2''-rhamnosyl) glucoside 0.502%, Faradiol-3-O-palmitate 0.281%.	Hamburger et al., 2004
49.	Crataegi folium	Epigallocatechin gallate 0.268%, vitexin-2''-O-rhamnoside 0.0064%, gallocatechin gallate 0.0832%.	Epigallocatechin gallate 0.141%, vitexin-2''-O-rhamnoside 0.524%, gallocatechin gallate 1.381%.	Hamburger et al., 2004
50.	Matricariae	Hyperoside 0.0017%, luteolin 0.0032%, Matricaria spiroketal 0.436%.	Hyperoside 0.061%, luteolin 0.078%.	Hamburger et al., 2004
51.	Spirulina pacifica	Zeaxanthin 0.48%, β-Cryptoxanthin 0.075%, β-Carotene 1.18%.	Zeaxanthin 0.50%, β-Cryptoxanthin 0.08%, β-Carotene 1.20%.	Careria et al., 2001

Contd...

Table 6.5– *Contd...*

Sl.No.	Botanical Name	Major Constituents of SCFE Product	Major Constituents of Conventional Extraction Product	References
52.	Sorghum	Policosanols 3.41%, phytosterols 2.03%, palmitic acid 2.16%, oleic acid 2.77%, linoleic acid 4.49%, α-tocopherol 7.43%, γ-tocopherol 6.14%.	Policosanols 2.6%, phytosterols 1.42%, palmitic acid 2.2%, oleic acid 2.7%, linoleic acid 4.48%, α-tocopherol 8.18%, γ-tocopherol 9.1% δ-tocopherol 9.1%	Wang et al., 2008
53.	Anacardium occidentale	Cardanol 70–90%.	Cardanol 5.0%, anacardic acid 70.0%, cardol 18.0%.	Patel et al., 2006
54.	Olea europaea L.	Tocopherol 97.0%.	Tocopherol 0.00%.	de Lucas et al., 2002
55.	Pelargonium sp.	Geranyl formate 7.9%, Geraniol 8.5%, Geranyl tiglate 3.3%, 2-Phenylethyl tiglate1.8%, rose oxide 0.4%, Isomenthone 3.5%, Linalool 0.1%, Guaia-6,9-diene 8.8%, Citronellyl formate 10.2%, Germacrene-D 4.6%, Citronellol 24.8%.	Geranyl formate 3.7%, rose oxide 0.5%, Geraniol 8.4%, Geranyl tiglate 3.1%, 2-Phenylethyl tiglate1.9%,Isomenthone 4.8%, Linalool 4.4%, Guaia-6,9-diene 7.2%, Citronellyl formate 11.1%, Germacrene-D 2.5%, Citronellol 26.5%.	Gomes et al., 2007
56.	Nepta tuberose	Benzaldehyde 0.60%, β-pinene 1.5%, p-cymene 0.76%, 1,8-cineole 8.22%, eugenyl acetate 36.70%, benzyl benzoate 4.06%, p-cymenene 0.25%, methyl benzoate 16.47%, 2-methylbenzonitrile 1.13%, benzyl acetate 0.35%, α-terpineol 1.25%, methyl salicylate 6.68%, ethyl benzoate 0.49%, methylanthranilate 1.36%, benzyl butyrate 0.24%, eugenol 1.13%, cis-isoeugenol 7.52%, cis-methyl-isoeugenol 0.77%, cis-β-farnesene 1.05%, valencene 4.93%, nepetalactone 14.18%, dehydro-nepetalactone 3.23%, trans-methylisoeugenol 44.79%, α-farnesene 3.08%, α-farnesol 6.54%, $C_{14}H_{12}O_2$ 5.92%, n-nonadecane 3.75%, n-heneicosane 0.39%, n-tricosane 4.88%, n-tetracosane 0.94%, n-tricosan-1-ol 2.4%, n-pentacosane 30.69%, n-hexacosane 3.02%,	Benzaldehyde 0.03%, β-pinene 0.26%, p-cymene 0.13%, 1,8-cineole 3.41%, eugenyl acetate 6.09%, benzyl benzoate 0.72%, p-cymenene, methyl benzoate 3.57%, 2-methylbenzonitrile 0.24%, benzyl acetate, α-terpineol, methyl salicylate 3.57%, ethyl benzoate 0.17%, methylanthranilate 0.66%, benzyl butyrate 0.07%, eugenol 0.48%, cis-isoeugenol, cis-methylisoeugenol 2.73%, cis-β-farnesene 0.30%, valencene 2.87%, nepeta-lactone 4.08%, dehydronepetalactone 0.90%, trans-methylisoeugenol 15.21%, α-farnesene 1.44%, α-farnesol 0.90%, $C_{14}H_{12}O_2$ 0.81%, n-nonadecane 0.02%, n-heneicosane 0.08%, n-tricosane 0.90%, n-tetracosane 0.39%, n-tricosan-1-ol 0.36%, n-pentacosane 10.79%,	Reverchon et al., 1997

Contd...

Table 6.5—Contd...

Sl.No.	Botanical Name	Major Constituents of SCFE Product	Major Constituents of Conventional Extraction Product	References
		n-tetracosan-1-ol 13.32%, n-heptacosane 43.37%, n-octacosane 1.20%, n-pentacosan-1-ol 8.38%, n-nonacosane 14.97%, n-hentriacontane 3.20%.	n-hexacosane 0.98%, n-tetracosan-1-ol 1.52%, n-heptacosane 20.25%, n-octacosane 0.61%, n-pentacosan-1-ol 1.62%, n-nonacosane 10.91%, n-hentriacontane 1.46%.	
57.	Scutellariae radix	Baicalin 23.76%, Baicalein0.86%, Wogonin 0.22%.	Baicalin 2.29%, Baicalein 0.66%, Wogonin 0.28%.	Lin et al., 1999
58.	Artemisia annua L.	Artemisinin 4.0%, artemisinic acid 10.20%.	Artemisinin 7.5%, artemisinic acid 4.0%.	Kohler et al., 1997
59.	Lippia alba	Carvone 80%, limonene 17%.	Carvone 24%, limonene 13.4%.	Braga et al., 2005
60.	Peumus boldus M.	Boldine 0.74%.	Boldine 5.97%.	del Valle et al., 2005
61.	Smilax china	Sapogenins 0.454%.	Sapogenins 0.385%.	Shu et al., 2004
62.	Mauritia flexuosa	Carotene 80.0%.	Carotene 1.0%.	De Franca et al., 1999.
63.	Sesamum indicum L.	Vitamin E 4.667%.	Vitamin E 2.467%.	Hu et al., 2004.
64.	Azadirachta indica	Azadirachtin A 8810 mg/kg of oil.	Azadirachtin A 550 mg/kg of oil.	Ambrosino et al., 1999
65.	Rosa damascena L.	Phenyl ethyl alcohol 50.01%, Paraffins 15.12%, Paraffinic alcohols 3.69%.	Phenyl ethyl alcohol 10.37%, Paraffins 22.69%, Paraffinic alcohols 3.69%.	Reverchon et al., 1996
66.	Rosa damascena Mill	2-phenylethanol 50.0% citronellol 26.1%, 2-phenylethyl acetate 7.5%, n-Nonadecane 7.8%, n-Heneicosane 3.4%, n-Tricosane 0.9%.	2-phenylethanol 10.4%, citronellol 11.4%, 2-phenylethyl acetate 14.8%, n-Nonadecane 10.0%, n-Heneicosane 1.7%, n-Tricosane 1.1%.	Reverchon et al., 1997
67.	Thymus vulgaris L.	α-pinene 0.20%, camphene 0.13%, p-cymene 8.49% 1,8-cineolea thymol 60.91%, carvacrol 8.90%, + limonene 2.26% γ-terpinene 1.60%, linalool 3.29%, camphor 1.38%, endo-borneol 3.95%, verbenone 1.43%, linalyl acetate 1.03%.	α-pinene 2.13%, camphene 1.49%, p-cymene 33.0%, 1,8-cineolea + limonene 2.75%, γ-terpinene 4.75%,%, thymol 31.06%, carvacrol 5.18% linalool 3.57%, camphor 0.70%, endo-borneol 3.55%, verbenone1.67%, linalyl acetate 0.20.	Diaaz-Maroto et al., 2005.

Contd...

Table 6.5—Contd...

Sl.No.	Botanical Name	Major Constituents of SCFE Product	Major Constituents of Conventional Extraction Product	References
68.	*Botryococcus braunii, Chlorella vulgaris, Dunaliella salina, Arthrospira maxima*, microalgae	γ-linolenic acid 0.44%, lipids 3.1%.	γ-linolenic acid 0.01%, lipids 2.6%.	Mendes *et al.,* 2003
69.	*Nannochloropsis gaditana*	Carotenoid 0.0343%, chlorophyll 0.0223%.	Carotenoid 0.08%, chlorophyll 1.85%.	Macýas-Sanchez *et al.,* 2005
70.	*Silybum marianum*	α-Tocopherol 0.085%.	α-Tocopherol 0.043%.	Hadolin *et al.,* 2001
71.	*Olea europaea* L.	Total phenols 0.076%.	Total phenols 0.018%.	Le Floch *et al.,* 1998
72.	*Olea europaea* L.	Tocopherol 97.10%.	Tocopherol 0.00%.	de Lucas *et al.,* 2002
73.	*Pandanus amaryllifolius*	2-acetyl-1-pyrroline 1.734%.	2-acetyl-1-pyrroline 0.2312%.	Bhattacharjee *et al.,* 2005
74.	*Hypericum perforatum* L.	Hyperforin 35.0%.	Hypericin 0.1–0.3%, hyperforin 0.6–11%.	Römpp *et al.,* 2004
75.	Shark liver	Squalene 95.0%	Squalene 50.0%, pristine 0.1%.	Catchpole *et al.,* 1997
76.	*Glycine max* L.	Isoflavones 0.086%, (Genistin 0.00536%, Genistein 0.00017%, Daidzein 0.00309%).	Isoflavones 0.0213%, (Genistin 0.0205%, Genistein 0.00042%, Daidzein 0.00035%).	Rostagno *et al.,* 2002
77.	*Phaffia rhodozyma*	Astaxanthin 10 times and carotenoids 13 times more in the extract.	Astaxanthin and carotenoids.	Lim *et al.,* 2002
78.	*Scutellaria lateriflora* L.	Dihydrobaicalin 0.167%, oroxylin A 0.356%, baicalein 0.293%, wogonin 0.085%.	Baicalin 1.314%, dihydrobaicalin 0.744%, laterifIorin 0.135%, ikonnikoside I 0.073%, scutellarin 0.089%, oroxylin A 7-O- glucuronide 0.023%, oroxylin 0.004%, baicalein 0.066%, wogonin 0.001%.	Bergeron *et al.,* 2005

Contd...

Table 6.5—Contd...

Sl.No.	Botanical Name	Major Constituents of SCFE Product	Major Constituents of Conventional Extraction Product	References
79.	Curcuma longa L.	α-pinene tr, 1,8-cineole 0.18%, *trans*-caryophyllene 0.5%, Ar-curcumene 2.3%, α-zingiberene 2.4%/%, β-bisabolene 0.5%, β-sesquiphellandrene 2.6%, Ar-turmerol1.2%, Ar-turmerol isomer 1.3%, Ar-turmeron 28.1%, (Z)- γ-atlantone 39.5%, (E)- γ-atlantone 20.3%, dihydro-Ar-turmerone 0.4%, 1-epi-cubeno 0.7%, 6S,7R-bisabolone 1.18%, (Z)-α-atlantone0.5%, (E)- α-atlantone 0.7%.	α-pinene 2.7%, 1,8-cineole 1.4%, *trans*-caryophyllene traces, Ar-curcumene 1.0%, α-zingiberene 2.4%, β-bisabolene traces, β-sesquiphellandrene 1.9%, Ar-turmerol 1.1%, Ar-turmerol isomer 0.7%, Ar-turmeron 18.0%, (Z)-γ-atlantone44.0%, (E)-γ-tlantone18.3%, dihydro-Ar-turmerone tr, 1-epi-cubeno 0.6%, 6S,7R-bisabolone 0.6%, (Z)-α-atlantone 0.6%, (E)- α-atlantone 0.9%.	Mara et al., 2003
80.	Pimpinella anisum L.	Methyl chavicol 91.7%, limonene 3.6%.	Anethole 90%, γ-himachalene 2-4%, p-anisaldehyde <1%, methylchavicol 1.5%, cis-pseudoisoeugenyl 2-methylbutyrate 3%, trans-pseudoisoeugenyl 2-methylbutyrate 1.3%.	Rodrigues et al., 2003; Mallavarapu et al., 2004
81.	Xylopia aromatica	β-Phellandrene 60%, β-myrcene 9.1%, α-pinene 8.1%.	β-Phellandrene 40%, β-myrcene 5.1%, α-pinene 5.9%.	Stashenko et al., 2004
82.	Xylopia aromatica	β-Phellandrene 61%, β-myrcene 9.1%, α-pinene 8.1%.	α-pinene 5.9%, β-Phellandrene 40%, β-myrcene 5.1%.	Stashenko et al., 2004
83.	Origanum majorana L.	Terpinen-4-ol 30.3%, γ-terpinene 14.0%, α-terpineol 4.4%, α-terpinolene 1.8%, α-terpinene 3.2%, β-caryophyllene1.8%, α-pinene 5.9%, p-Cymol 9.8%, linalool 12.1%.	Spathuleno 9.9%, Terpinen-4-ol 30.6%, α-terpineol 4.2%, α-terpinolene 1.1%, α-terpinene 2.7%, β-caryophyllene 2.2%, α-pinene 5.9%, cis-Sabinene hydrate 1.1%, Terpenyl-ester 0.9%, Neophytadiene 1.9%, γ-terpinene 5.3%, linalool 1.1%, p-Cymol 1.8%.	Vági et al., 2005
84.	Schizandra chinensis	Schizandrin A 96.2%,Gomisin A 96.8%, Gomisin N 97.7%, Deoxy Schizandrin 97.6%, Wuweizisu C 97.6%.	Schizandrin A 93.4%, Gomisin A 97.4%, Gomisin N 100%, Deoxy Schizandrin 95.2%, Wuweizisu C 92.0%.	Lojková et al., 1998
85.	Salvia triloba L.	1,8-Cineol38.9%, camphor 8.4%, α-terpineol 4.9%, α-pinene 5.9%, α-pinene 4.4%.	1,8-cineol 33.1%, camphor 8.1%, α-terpineol 3.7%, α-pinene 4.3%, α-pinene 3.2%.	Rónyai et al., 2002

Contd...

Table 6.5–Contd...

Sl.No.	Botanical Name	Major Constituents of SCFE Product	Major Constituents of Conventional Extraction Product	References
86.	*Mentha spicata*	Limonene 0.00117%, Cineole 0.0118%, Dihydrocarvone 0.00282%, Carvone 0.1481%.	α-Pinene0.000 912%, Limonene 0.0103%, Cineole 0.0167%, Dihydrocarvone 0.0103%, Carvone 0.1431%, Menthyl acetate 0.00 143%.	Ali *et al.,* 2007
87.	*Ocimum basilicum*	Eugenol 6.12%, trans- methyl cinnamate 5.96%, Methyl eugenol 1.07%, 1,8-Cineol 10.94%, inalool 35.99%, α-Terpineol 1.07%, Estragol 22.59%, Cis-methyl cinnamate 0.85%, α-Bergamotene 2.35%, β-Selinene 0.64%, δ-Guaiene 0.94%, trans-Cadinol 3.09%.	Eugenol 8.22%, trans- methyl cinnamate 8.71%, Methyl eugenol 1.18%, 1,8-Cineol 5.85%, Linalool 30.73%, α-Terpineol 0.82%, Estragol 21.80%, Cis- methyl cinnamate 1.17%, α-Berga-motene 5.67%, β-Selinene 1.57%, δ-Guaiene 1.99%, trans-Cadinol 3.59%.	Díaz-Maroto *et al.,* 2002
88.	*Mentha piperita*	1,8-Cineole 4.0%, *trans*-Menthone 37.6%, *cis*-Menthone 5.0%, Neomenthol 1.3%, *trans*-Menthol 24.3%, Piperitone 0.9% *trans*-Menthyl acetate 11.1%, β-Caryophyllene 4.4%, β-Elemene 1.1%, γ-Cadinene 3.2%.	1,8-cineole 8.1%, trans-Menthone 39.3%, cis-Menthone 4.6%, Neomenthol 1.0%, trans-Menthol 23.3%, Piperitone 1.6%, trans-Menthyl acetate 11.6%, β-Caryophyllene4.2%, β-Elemene 1.0%, γ-Cadinene 4.1%.	Vilcu *et al.,* 2002
89.	*Borago officinalis* L.	Palmitic acid 13.32%,palmitoleic acid 0.19%, stearic acid 4.58%, oleic acid 19.78%, linoleic acid 39.57%, γ-linolenic acid 22.29%, α-linolenic acid 0.27%.	Palmitic acid 13.28%, palmitoleic acid 0.18%, stearic acid 5.06%, oleic acid 20.77%, linoleic acid 39.02%, γ-linolenic acid 21.04%, α-linolenic acid 0.65%.	Gómez *et al.,* 2002
90.	*Platonia insignis*	B-Bisabolene 1.07%, 3,7-Dimethyloct-1-en-3,7-diol 4.41%, Linalool 9.12%, α-Terpineol 1.39%, Stearic acid 2.68%, Linolenic acid 3.81%, Oleic acid 15.27%, Palmitic acid 26.68%, eugenol 1.69%.	Methyl benzene 4.38%, 2-Methylheptane 4.49%, Nonacosane 4.77%, 3,7-Dimethyloct-1-en-3,7-diol 1.29%, Linalool 51.03%, α-Terpineol 13.74%, Caprylic acid 1.41%, cis-linalool oxide 9.16%, trans-linalool oxide 4.11%.	Monteiro *et al.,* 1997
91.	*Vitis vinifera*	Linoleic acid 55.9%,Oleic acid 12.9%, Palmitic acid 2.8%, Stearic acid 1.4%.	Linoleic acid 57.1%,Oleic acid 13.8%, Palmitic acid 3.1%, Stearic acid 1.4%.	Cao *et al.,* 2003
92.	*Laurus nobilis* L.	α-pinene 8.0%, camphene 2.6%, sabinene 1.8%, β-pinene 4.2%, 1,8-cineole 8.8%, (Z)-β-ocimene 2.0%, (E)-β-ocimene 20.9%, linalool 2.2%, para-mentha-1,5-dien-8-ol 1.5%, linalyl acetate 4.5%,	α-pinene 10.3%, camphene 3.8%, sabinene 2.6%, β-pinene 5.8%, 1,8-cineole 8.1%, (Z)-β-ocimene 3.0%, (E)-β-ocimene 23.7%, linalool 4.2%, para-mentha-1,5-dien-8-ol 1.5%, linalyl	Marzouki *et al.,* 2008

Contd...

Table 6.5–Contd...

Sl.No.	Botanical Name	Major Constituents of SCFE Product	Major Constituents of Conventional Extraction Product	References
		bornyl acetate 2.9%, α-terpinyl acetate 3.8%, β-cubebene 2.2%, β-longipinene 7.1%, methyl eugenol 1.4%, (E)-caryophyllene 2.5%, germacrene D 2.7%, viridiflorene 1.5%, α-bulnesene 3.5%, trans-cadinene 2.7%, δ-cadinene 4.7%, spathulenol 2.3%, α-cadinol 2.0%, 5-isocedranol 2.1%.	acetate 1.3%, bornyl acetate 2.1%, α-terpinyl acetate 3.0%, β-cubebene 1.9%, β-longipinene 6.8%, methyl eugenol 1.0%, (E)-caryophyllene 1.9%, germacrene D 1.8%, viridiflorene 1.0%, α-bulnesene 2.7%, trans-cadinene 2.1%, δ-cadinene 3.9%, spathulenol 1.4%, α-cadinol 1.1%, 5-isocedranol 1.1%.	
93.	Schizandria chinesis	Schisandrol A 0.580%, Schisandrol B 0.164%, Schisandrin A 0.134%, Schisandrin B 0.597%, Schisandrol C 0.113%.	Schisandrol A 0.500%, Schisandrol B 0.146%, Schisandrin A 0.145%, Schisandrin B 0.584%, Schisandrol C 0.112%.	Choi et al., 1998
94.	Peumus boldus M.	Boldine 2.9%.	Boldine 3.05%.	del Valle et al., 2004
95.	Theobroma cacao	Saturated fatty acids 60.96%, Unsaturated fatty acids 39.03%.	Saturated fatty acids 59.36%, Unsaturated fatty acids 40.63%.	Marleny et al., 2002
96.	Foeniculum vulgare Mill.	Fenchone 12.71%, estragole 20.33%, p-anisaldehyde 0.99%, trans-anethole 49.71%.	Fenchone 19.33%, estragole 25.84%, p-anisaldehyde 1.90%, trans-anethole 49.71%.	Diaaz-Maroto et al., 2005
97.	Capsicum annuum L.	β-Carotene 0.0791%, cis β-carotene 0.0252%, Diesters 0.0559%.	β-Carotene 0.1981%, cis β-carotene 0.0589%, Diesters 110% (Subcritical methods).	Daood et al., 2002
98.	Capsicum annuum L.	Capsorubin 1.10%, Capsanthin 0.91%, Capsanthin 5,6-epoxide 1.04%, Zeaxanthin 2.41%, Cryptocapsin 0.62%, β-Cryptoxanthin 0.89%, β-Carotene 0.69%.	Capsorubin 1.14%, Capsanthin 1.07%, Capsanthin 5,6-epoxide 1.24%, Zeaxanthin 2.61%, Cryptocapsin 0.83&, β-Cryptoxanthin 1.11%, β-Carotene 1.76%.	Uquiche et al., 2004
99.	Hierochloe odorata	5,8-dihydroxycoumarin, 7.8%, 5-hydroxy-8-O-β-D-glucopyranosyl-benzopyranone 1.0%	5,8-dihydroxycoumarin 7.6%, 5-hydroxy-8-O-β-D-glucopyranosyl-benzopyranone 1.3%	Grigonis et al., 2005
100.	Urtica dioica L.	Chlorophyll a 0. 073%, Chlorophyll b 0.10%, Pheophytin a 0.278%, Lutein 0.039%, β-Carotene 0.024% (yield from dry matter).	Chlorophyll a 0.173%, Chlorophyll b 0.032, Pheophytin a 0.0%, Lutein 0.046%, β-Carotene 0.020% (yield from dry matter).	Sovova et al., 2004

Conclusions

In the present chapter, we have tried to put in place the advantages of SCFE as a tool for value addition and have cited numerous examples from the literature where in the SCFE method has been proved to be a very efficient extraction method, may it be for herbals, medicinal plants, essential oils or preparation of products with lesser color problems or removal of unwanted/health hazard materials such as pesiticides etc. We have also listed examples wherein parameters like pressure, temperature and modifiers have been optimised for better extractive values or better extrusion of bioactive components.The examples that are referenced in this chapter touch various type of materials and class of compounds which include alkaloids, sterols, terpenoids, polysols, flavonoids, insecticides etc. The use of SCFE as a tool for extraction of useful materials, enrichment of fractions/compounds should be thought of as a need in present day scenario because of its environmental friendly approach, thereby reducing the level of toxins which are responsible for various diseases and ailments. There is a great scope for the SCFE to be used as a tool for various activities in the pharma, food and herbal industries. All the issues of budgeting etc. can be overridden when the health and environment becomes a priority. FDA approval for SCFE and processes alike makes the task of the pharma, food and other related industries much easier for wide acceptance of their products.

References

Aghel, N., Yamini, Y.,Hadgiakhoondi, A. and Pourmortazavi, S.M. (2004). Supercritical extraction of *Mentha pulegium* L. essential oil. *Talanta*, 62: 407-411.

Aleksovski, S. A. and Sovova, H. (2007). Supercritical CO_2 extraction of *Salvia officinalis* L. *J. Supercrit. Fluids*, 40: 239-245.

Ali, H. A., Rao, M. V. and Jobe, B. (2007).Comparative Evaluation of SFE and Steam Distillation on the Yield and Composition of essential oil extracted from Spearmint (*Mentha spicata*). *Journal of Liquid Chromatography and Related Technologies*, 30: 463–475.

Ambrosino, P., Fresa, R., Fogliano, V., Monti, S.M. and Ritieni, A. (1999). Extraction of azadirachtin A from neem seed kernels by supercritical fluid and its evaluation by HPLC and LC/MS. *J.Agric. chem.*, 47: 5252-5256.

Andreson, M.R., King, J.W. and Hawthorne, S.B. (1990). In: Lee, M.L., Markides, K.E. (Eds.) Chromatography conferences, Provo Ut, p313.

Andreson, M. R., Swamson, J.T., Porter, N.L. and Richter, B.E. (1989). Supercritical fluid extraction as a sample introduction method for chromatography. *J.Chromatogr. Sci.* 27: 371.

Anitescu, G., Doneanu C. and Radulescu, V. (1997). Isolation of *Coriander* oil, comparison between steam distillation and supercritical CO_2 extraction. *Flavour and Fragrance*, 12: 173-176.

Anklam, E., Berg, H., Mathiasson, L., Shjarma, M. and Ulberth, F.(1998). Supercritical fluid extraction for liquid chromatographic determination of carotenoids in Spirulina Pacifica algae: a chemometric approach. *Food Add. Contam.*, 15: 729.

Barth, M.M., Zhou, C., Kute, K. M. and Rosenthal, G.A. (1995). Determination of optimum conditions for supercritical fluid extraction of carotenoids from carrot (*Dausus carota* L.). *J. Agric. Food Chem.*, **43**: 28-76.

Bartle, K. D., Clifford, A. A. and Shilstone, G.F. (1992). Estimation of solubilities in supercritical carbon dioxide: A correlation for the Peng-Robinson –interaction parameters. *J.Supercrit fluids*, **5**: 220.

Baysal, T. and Starmens, D.A.J. (1999).Supercritical carbon dioxide extraction of carvone and limonene of caraway seeds. *J. Supercritical Fluids*, **14**: 225.

Benthin, B., Danz, H. and Hamburger, M.J. (1999). Pressurized liquid extraction of medicinal plants. *J.Chromatogr. A.*, **837**: 211-219.

Bergeron, C., Gainer, S., Clausen, E. and Carrier, D. J. (2005).Comparison of the chemical composition of extracts from *Scutellaria lateriflora* using accelerated solvent extraction and supercritical fluid extraction.*J.Agric.Food Chem.*,**53**: 3076-3080.

Berna, A., Tarrega, A., Blasco, M. and Subirats, S. (2000). Supercritical CO_2 extraction of essential oil from orange peel: effect of the height of the bed. *J.Supercrit.Fluids*, **18**: 222-237.

Bernardo- Gil, M.G., Grentra, J. Santos, J. and Cardoso, P. (2002). Supercritical fluid extraction and characterization of oil from hazelnut. *Eur. J.Lipid Sci. Technol.*, **104**: 402-409.

Beuscher, N. and Willigmann, I. (1993). Pharmaceutically active composition extracted from *Tenacetum parthenium*, process for extraction and medicinal composition. *EP* 0553658.

Beveridge, T. H., Li, T. S. and Drover, J. C. (2002). Phytosterol content in American ginseng seed oil. *J. Agric. Food Chem.*, **50**: 744-750.

Bhattacharajee, P., Kshirsagar, A. and Singal, R. S. (2005). Supercritical carbon dioxide extraction of 2-acetyl-1-pyrroline from*Pandanus amaryliofolius* Roxb.*Food Chem.*, **91**: 255-259.

Boceuka, M. and Sovova, H. (2007). Supercritical CO_2 extraction of essential oil from *Achillea millefoluim. J. Supercrit. Fluids*, **40(3)**: 360-367.

Braga, M.E.M., Ehlertb, P. A.D., Ming, L.C. and Meireles, M. A.A. (2005). Supercritical fluid extraction from*Lippia alba*: global yields, kinetic data, and extract chemical composition. *J. Supercrit. Fluids*, **34(20)**: 149-156.

Braga, M.E.M.; Leal, P.F., Carvalho, J. E. and Meireles, M.A.A.(2003).Comparison of yield,composition and antioxidant activity of turmeric (*Curcuma longa* L) extracts using various techniques. *J.Agric. Food Chem.*, **51**: 6604-6611.

Branabas, I. J., Dean, J.R. and Owen, S. P. (1994). Critical review: Supercritical fluid extraction of analytes from environment samples. *Analyst*, **119**: 2381-2394.

Bruno, M., Alessandra, P., Silvia, P., Enrica, T. and Andra, M. (2007). Comparative analysis of the oil and supercritical CO_2 extract of *Ridolfia segetum* (L) Moris. *Nat. Prod. Res.***21**: 412-417.

Bunzenberger, G., Lack, E. and Marr, R. (1984). Co2 –Extraction: Comparison of super – and subcritical extraction conditions. Ger. *Chem. Engg.*, 7: 25-31.

Butterbar, S. R., Tratz, W., Brattstrom, A. and Debrunner, B. (1998). High pressure extraction and fractionation of pharmacologically active components from petasites, hybrids. *In proceedings of the 5th meeting on supercritical fluids materials and natural products processing. Nice* perrut M, Subra P (Eds.) Institute natural polytechnique de Lorraine; Vaudocuvre, 12: 509-514.

Campos, L.M.A.S., Miechielin, E.M.Z., Danielski, L. and Ferreira, S.R.S. (2005). Experimental data and modeling the supercritical fluid extraction of marigold (*Calendula officinalis*) oleoresin. *J. Supercrit. Fluids*, 34: 163.

Cannel, R.J.P. (1998). Methods in Biotechnology 4. How to approach the isolation a natural product p 1-51, Natural products isolation. Humana Press, Totowa, New Jersey, USA.

Cao, X., Tian, Y., Zhang, T.Y. and Ito,Y. (2003). Supercritical fluid extraction of catechins from *Cratoxylum prumifolium* Dyer and subsequent purification by high speed cunter current chromatography. *J.Chromatogr. A*, 898: 75-81.

Cao, X. and Ito, Y. (2003). Supercritical fluid extraction of grape seed oil and subsequent separation of free fatty acids by high speed counter current chromatography. *J.Chromatogr. A*, 1021: 117-124.

Careri, M., Furlattini, L., Mangia, A., Musci, E, M., Anklam, E., Thepbald, A. and Holst, C.V. (2001). Experimental design for the optimization of supercritical fluid extraction of carotenoids from Spirulina Pacifica algae. J.Chromatogr. A, 912: 61.

Cassel, E., Frizzo, C.D., Vanderlinde, R., Atti-Serafini, L., Lorenzo, D. and Dellacassa, E. (2000). Extraction of *Baccharis* oil by supercritical CO_2. *Ind. Eng. Chem. Res.*, 39: 4803-4805.

Castano, G., Mas, R., Gamez, R., Fernandez, J., Illnait, J., Fernandez, L., Mendoza, S., Mesa, M., Gutierraz, J.A. and Lopez, E. (2004). Concomitant use of policosanols and beta blockers in older patients. *Int. J. Clin.Pharmacol.Res.*, 24(2-3): 65-77.

Castioni, P., Christen, P. and Veuthey, J.L. (1995). L'extraction enphase supoercritique des substances d'origine vegetale. *Analusis*, 23: 95-106.

Catchpole, O. J., Grey, J. B., Perry, N.B., Burgess, E.J., Redmond, W.A. and Porter, N.G. (2003). Extraction of chili, black pepper and ginger with near-critical CO_2, propane, and dimethyl ether: analysis of the extracts by quantitative nuclear magnetic resonance. *J. Agric. Food Chem.*, 51(17): 4853–4860.

Catchpole, O. J., Tallon, S.J., Grey, J. B., Fletcher, K. and Fletcher, A.J. (2008). Extraction of lipids from a specialist dairy stream. *J.Supercrit. Fluids*, 45: 314-321.

Catchpole, O. J., von Kamp, J. C. and Grey, J. B. (1997). Extraction of squalene from shark liver oil in a packed column using supercritical carbon dioxide. *Ind. Eng. Chem. Res.*, 36: 4318-4324.

Caude, M. and Theibut, D. (1999). *Practical supercritical fluid chromatography and* extraction. Harwood, Amsterdam.

Chafer, A., Berna, A., Monton, J.B. and Mulet, A. (2001). High pressure solubility data of the system limonene + linalool+ CO_2 *J.Chem. Eng Data*, **46**: 1145.

Chen, Yen-chou., Shen, Shing-chung., Chou, Lih-Geen., Lee-Tony, J.F. and Yang, Ling-Ling. (2001). Wogonin, baicalin, and baicalein inhibition of inducible nitric oxide synthase and cyclooxygenase-2 gene expressions induced by nitric oxide synthase inhibitors and lipopolysaccharide. *Biochemical pharmacology*, **61(11)**: 1417-1427.

Chen, Y., Jiang, X. Y., Tong, X. and Chen, X. Q. (2008). Comparison of volatile components from *Marchantia convoluta* obtained by microwave. *J. Chil. Chem. Soc.*, **53**: 15-18.

Choi, Y.H., Kim, J., Noh, M.J., Choi, E.S. and Yoo, K.P. (1997). Comparison of supercritical carbon dioxide extraction with solvent of nonacosan-10-ol, α-amyrin acetate, squalene and stigmasterol from medicinal plants. *Phytochem. Anal.*, **8**: 233-237.

Choi, Y.H., Kim, J., Jeon, S.H., Yoo, K. P. and Lee, H. K. (1998). Optimum SFE condition for lignans of *Schisendra chinesis* fruits.*Chromatographia*, **48**: 695-699.

Choi, Y. H., Chin, Y.W., Kim, J., Jeon, S.H. and Yoo, K.P. (1999). Strategies for supercritical fluid extraction of hyoscyamine and scolopamine salts using basified modifiers. *J. Chromatogr. A*, **863**: 47-55.

Chun, M.K., Shin, H.W. and Lee, H. (1994). Supercritical fluid extraction of taxol and baccatin III from needles of *Taxus cuspidate. Biotechnology Techniques.*, **8**: 543.

Cocero, M.J. and Garcia, J. (2001) Mathematical model of Supercritical extraction applied to oil seed extraction by CO_2 + saturated alcohol-1. Desorption Model. *J.Supercrit.Fluids*, **22(3)**: 229-243.

Coelho J.A.P., Periera, A.P., Mendes, R.L. and Palavra, A.M.F. (2003). Supercritical carbon dioxide extraction of *Foeniculum vulgare* volatile oil. *Flavour Fragr. J.*, **18**: 316-319.

Cooper, A. I. (2000). Synthesis and processing of polymers using supercritical carbon dioxide. *J. Mater. Chem.*, **10**: 207.

Cornwell, C.P., Leach, D.N. and Wyllie, S.G. (1999). The origin of terpinen-4-ol in the steam distillates of *Melaleuca argentea, M. dissitiflora* and *M. linariifolia. J. Essent. Oil Res.*, **11**: 49-53.

Cristina, T., da Costa, Sam A., Margolis, Bruce A., Benner Jr. and Horton, D. (1999). Comparison of methods for extraction of flavanones and xanthones from the root bark of the Osage orange tree using liquid chromatography. *J. Chromatogr.A*, **831**: 167-178.

Cui, Y. and Ang, C.Y.W. (2002). Supercritical fluid extraction and high-performance liquid chromatographic determination of phloroglucinols in St. John's Wort (*Hypericum perforatum* L.). *J.Agric Food. Chem.*, **50**: 27-55.

Cygnarowicz, M.L., Maxwell, R.J. and Seider, W.D. (1990). Equilibrium solubilities of β-carotene in supercritical carbon dioxide. *Fluid Phase Equilibria*, **59**: 57–71.

Damjanovic, Z., Lepojevic, V. and Zivkovic, Tolic A. (2005). Extraction of fennel (*Foeniculum vulgare* Mill.) seeds with supercritical CO$_2$: Comparison with hydrodistillation. *Food Chemistry*, **92(1)**: 143-149.

Daood, H.G., Illés, V., Gnayfeed, M.H., Mészáros, B., Horváth, G. and Biacs, P.A. (2002). Extraction of pungent spice paprika by supercritical carbon dioxide and subcritical propane. *J. Supercritc. Fluids*, **23**: 143–152.

De Franca, L.F., Reber, G., Meirles, M. A. M., Maehado, N. T. and Brunner, G. (1999). Supercritical exraction of carotenoids and lipids from buriti (*Mauritia flexuosa*), a fruit from the Amazon region. *J.Supercrit. Fluids*, **14**: 247-256.

Deans, S.G. and Svoboda, K. P. (1990). The antimicrobial properties of marjoram (*Origanim majorana* L.) volatile oil. *Flavour and Fragrance Journal*, **5**: 187–190.

Debrawere, J., and Verzele, M. (1976). Constituents of pepper, the hydrocarbons of pepper essential oil. *J. Chromatogr. Sci.*, **14**: 296.

Del valle, J. M., Godoy, C., Asencio, M. and Aguilera, J. M. (2004). Recovery of antioxidants from boldo (*Peumus boldus* M.) by conventional and supercritical CO$_2$ extraction water. *Food Research Intern.*, **37**: 695-702.

Del Valle, J. M. and Aguilera, J. M. (1999). Review: High pressure CO$_2$ extraction. Fundamemtals and applications in the food industry. *Food Science Intern.*, **5(1)**: 1-24.

Del Valle, J.M., de la Fuente., Juan, C. and Cardarelli, D. A. (2005). Contributions to supercritical extraction of vegetable substrates in Latin America. *J.Food Eng.*, **67**: 35-57.

Del Valle, J. M., Rogalinski, T., Zentl, C. and Brunner, G. (2005). Extraction of boldo (*Peumus boldus* M.) leaves with supercritical CO$_2$ and hot pressurized water. *Food Research Intern.*, **38**: 203–213.

Della Porta, G., Porcedda, S., Marongiu, B. and Reverchon, E. (1999). Isolation of eucalyptus oil by supercritical fluid extraction. *Flavour Fragr. J.*, **14**: 214-218.

Díaz-Maroto, M. C., Pérez-Coello, M. S. and Cabezudo, M. D. (2002). Supercritical carbon dioxide extraction of volatiles from spices: Comparison with simultaneous distillation extraction. *J. Chromatogr.A.*, **947**: 23-29.

Djarmati, Z., Jankov, R. M. E., Schwirtlich, julinac, B.D. and Djordjevic, A. (1991). High antioxidant activity of oleoresins obtained from sage by supercritical CO$_2$ extraction. *J.Am. Oil Chem. Soc.*, **68**: 731.

Doriaswamy, L.K., Ferriera, S.R.S., Meireless, M., Angela, A., Nikolov, Z.L. and Petenate, A. (1999). Supercritical fluid extraction of black pepper (*Piper nigurm* L.). *J.Supecrit. Fluids*, **14**: 235.

Dugo, P., Monedello, L., Bartle, K.D., Clifford, A.A., Breen, D.G.P.A. and Dugo, G. (1995). Deterpenation of sweet orange and lemon essential oils with supercritical carbon dioxide using silica gel as an adsorbent. *Flavour Fragrance J.*, **10**: 51–58.

Ebrahimzadeh, E., Yamini, Y. Sefidkon, F., Chaloosi, M. and Pourmortazavi, S.M. (2003). Chemical composition of the essential oil and supercritical CO$_2$ extracts of *Zataria multiflora* Boiss. *Food Chem.*, **83**: 357.

Eikani, L.M.H., Goodarznia, I. and Mirza, M. (1999). Supercritical carbon dioxide extraction of *Cumin Cymimum*. *Flavour Fragr. J.*, **14**: 29-31.

Eugenia, G.H.P. and Danielle, B. (2001). Supercritical fluid extraction of Z-sabinene hydrate-rich essential oils from Romanian*Mentha hybrids*. *Pure Appl. Chem.*,**73**: 287-1291.

Fang, Q., Yeung, H.W., Leung, H.W. and Huie, C.W. (2000). Micelle-mediated extraction and preconcentration of ginsenosides from Chinese herbal medicine. *J.Chromatogr.A*, **904**: 43-55.

Favti. F., King, F.W., Freidrich, J.P. and Eskoins, K. (1988). Supercritical CO_2 Extraction of carotene and lutein from leaf protein concentrates. *J. Food Sci.*, **53**: 1532.

Ferriera, S.R.S., Nikolov, Z. L., Doriswamy, L.K., Meireles, M.A.A. and Patenate, A.J. (1999). Supercritical fluid extraction of black pepper (*P.nigrum* L) essential oil. *J.Supercrit. Fluids*, **14**: 235-245.

Fischer, N., Nitz, S. and Drawert, F. (1987). Original flavour compounds and the essential oil composition of marjoram (*Marojana hortensis* Moench). *Flavour Fragr. J.*, **2**: 55-61.

Fischer, N., Nitz, S. and Drawert, F. (1988). Original composition of marjoram flavor and its changes during processing. *J. Agric. Food Chem.*, **36**: 996-1003.

Friedrich, J.P., List, G.R. and Heakin, A. J. (1982). Petroleum free extraction of oil from Soybeans with supercritical CO_2. *J.Am.Oil.Chem. Soc.*, **59**: 288-292.

Gãinar, I., Vilcu, R. and Mocan, M. (2002). Supercritical fluid extraction and fractional separation of essential oils. *www.chimie.unibuc.ro/biblioteca/anale/a/63-67.pdf*

Gawdzile, J., Mardarowicz, M. and Wolski, T. (1996). Supercritical fluid extraction of essential oils from the fruits of *Archangelica officinalis*. Hoffm and their characterization by GC/MS. *Journal of high resolution chromatography*, **19**: 237.

Gerhauser, C. (2005). Broad spectrum antiinfective potential of xanthohumol from hop (*Humuluslupulus* L.) in comparison with activities of other hop constituents and xanthohumol metabolites. *Mol. Nutr. Food. Res.*, **49**: 827-831.

Ghasemi, E., Yamini, Y., Bahramifar, N. and Sefidkon, F. (2007). Comparative analysis of the oil and supercritical CO_2 extract of *Artemisia sieberi*. *J. Food Engineering*, **79**: 306-311.

Giddings, J.C., Myres, M.N. and King, J.W. (1969). Dense gas chromatography in micro columns to 2000 atmospheres. *J.Chromatogr. Sci.*, **I**: 276.

Glisic, S.B., Stamenic, M.D., Zozovic, I.T., Asanin, R.M. and Skala, D.U. (2007). Supercritical carbon dioxide extraction of carrot fruit essential oil composition and antimicrobial activity. *Food chemistry*, **105**: 346-352.

Gmez-Prieto, M.S., Caja, M.M., Herraiz, M. and Mariaa, G.S. (2003). Supercritical fluid extraction of all –trans-lycopene from Tomato. *J. Agric. Food Chem.*, **51**: 3.

Gomes, P.B. and Mata, V.G. (2007). Production of rose geranium oil using supercritical fluid extraction. *J. Supercrit. Fluids,* **41**: 50-60.

Gomez, A.M. and Martinez, D.L.O. (2002). Quality of borage seed oil extracted by liquid and supercritical carbon dioxide. *Chem.Eng. J.,* **88**: 103-109.

Grigonis, D., Venskutonis, P.R., Sivik, Sandahl, B. M. and Eskilsson, C.S. (2005). Comparison of different extraction techniques for isolation of antioxidants from sweet grass (*Hierochloë odorata*). *J. Supercrit. Fluids,* **33**: 223–233.

Guan, W., Li, R., Yan, R., Tang, S. and Quan, C. (2007). Comparison of essential oils of clove buds extracted with supercritical carbon dioxide and otherthree traditional extraction methods. *Food chemistry*, **101**(4): 1558-1564.

Gurdial, G. S., Foster, N. R. and Yun, J. S. L. (1991). Chromatography conference, M.L. Lee and K.E. Markides (Eds.). Provo,UT, p 68.

Hadolin, M., kerget, M., Knez, Zeljko. and Bauman, D. (2001). High pressure extraction of vitamin E-rich oil from *Silybum maria*. *Food Chem.,* **74**(3): 355-364.

Hamburger, M., Baumann, D. and Adler, S. (2004). Supercritical carbon dioxide extraction of selected medicinal plants - effects of increasing pressure and added ethanol on the yield of extracted substances. *Phytochem. Anal.,* **15**: 46-54.

Hannay, J.B. and Hogard, J. (1879). On the solubility of solids in gases. *Proc. R. Soc London*, **29**: 324.

Hart, D.J. and Scott, K. J. (1995). Development and evaluation of an HPLC method for the analysis of carotenoids in foods and measurement of the carotenoid content of vegetables and fruits commonly consumed in the UK. *Food Chem.,* **54**: 101.

Heaton, D. M., Bartle, K. D., Rayner, C. M. and Clifford, A. A. (1993). Application of super-critical fluid extraction and supercritical fluid chromatography to the production of taxanes as anti-cancer drugs. *J. High Resolut. Chromatogr.,* **16**: 666-670.

Hirafugi, M., Machida, T., Hamane, N. and Minami, M.J. (2003). Cardiovasecular protective effects of n-3 doosahexaenoic acid with special emphasis on docohexaenoic acid. *Pharmacol.Sci.,* **92**(4): 308-316.

Hu, Q., Xu, J., Chem, S. and Yang, F. (2004). Antioxidant activity of extracts of black sesame seed (*Sesamum indicum* L.) by supercritical dioxide extraction. *J. Agric. Food Chem.,* **52**: 943.

Hubbard, J.D., Dowing, J.M., Ram, M.S. and Chung, O.K. (2004). Lipid extraction of wheat flour using supercritical fluid extraction.*Cereal Chemistry*, **81**(6): 693-698.

Hudaib, M., Gott, R., Pomonio, R. and Cavrini, V. (2003). Recovery evaluation of lipophilic markers from *Echinacea purpurea* roots applying microwave-assisted solvent extraction versus conventional methods. *J.Sep. Sci.,* **26**: 97-104.

Ibanez, E., Lopez-Sebastain, S., Ramos, E., Tabera, J. and Regeiro, G. (1997). Analysis of highly volatile components of foods by off line sFe/GC. *J.Agric Food Chem.,* **45**: 39-40.

Jaren-Galan, M., Nienaber, U. and Schnartz, S.S. (1999). Paprika (*Capsicum annum*) oleoresin extraction with supercritical crbon dioxide. *J. Agric. Food Chem.*, **47**: 3558-3564.

Jaubert, J.N., Conclaves, M.M. and Barth, D.A. (2000). A theoretical model to simulate supercritical fluid extraction: application to the extraction of terpenes by supercritical carbon dioxide. *Ind. Eng. Chem. Res.*, **39**: 49-91.

Jimenez-Carmma, M.M., Ubera, J.L. and L.de Castro M.D. (1999). Comparison of continuous subcritical water extraction and hydrodistillation of marjoram essential oil. *J. Chromatogr. A*, **855**: 625-632

Johnson, S. and Morgan, E.D. (1997). Supercritical fluid extraction of oil and triterpenoids from Neem seeds. *Phytochem. Anal.*, **8**: 228-232.

Kaiser, C. S., Rompp.H. and Schmidt, P.C. (2001). Pharmaceutical applications of supercritical carbon dioxide. *Pharmazie*,**56**: 907-926.

Kim, J.T., Ren, C.J., Fielding, G.A., Pihi, A., Kasum,T. and Wajada, M. (2007).Treatment with lavender aromatherapy in the post-anesthesia care unit reduces opioid requirements of morbidly obese patients undergoing laparosic adjustable gastric banding.*Obesity Surgery*,**17(7)**: 920-925.

King, J.W. (1989). Fundamentals and applications of sueprcritical fluid extraction in chromatographic science. *J. Chromatogr. Sci.*, **27**: 355.

Kiriamiti H., Camy, S., Gourdon, C. and Condoret, J.S. (2003). Supercritical carbon dioxide processing of pyrethrum oleoresin and pale. *J. Agric. Food Chem.*, **51**: 880-884.

Kohler, M., Haerdi, W., Christen. P. and Veuthey, J.L. (1997). Extraction of artemisinin and artemisinic acid from *Artemisia annua* L. using supercritical carbon dioxide. *J. Chromatogr.A*, **785**: 353-360.

Kotnik, M. S. and Knez, Z.(2007). Supercritical fluid extraction of chamomile flower heads. *J. Supcrit. Fluids*, **43**: 192.

Kurnik, R.T. and Reid, R.C. (1981). Solubility extrema in solid-fluid equilibria. *AIChE J.* **27**: 861.

Landberg, R., Dey, E.S., Francisco, José Da C., Åman, P. and Kamal-Eldin, A. (2007). Comparison of supercritical carbon dioxide and ethyl acetate extraction of alkylresorcinols from wheat and rye. *J. Food Comp. and Anal.*, **20**: 534-538.

Lang,Q.Y. and Wai, C.M. (2001). Supercritical fluid extraction in herbal and natural product studies-a practical review. *Talanta*, **53**: 771-782.

Langer, R., Mechtler, C. and Jurenitch, J. (1996). Composition of the essential oils of commercial samples of *Salvia officinalis* L. and *S. fruticosa* Miller: acomparison of oils obtained by extraction and steam distillation.*Phytochemical Anal.*, **7**: 289-293.

Lea Lojková, Jií Slanina, Milena Mike ová and Eva Táborská, JiVejrosta. (1998). Supercritical fluid extraction of lignans from seeds and leaves of *Schizandra chinensis. Phytochemical Anal.*, **8**: 261–265.

Lehotay, S.J. (1997). Supercritical fluid extraction of pesticides in foods. *J.Chromatogr. A*, **785**: 289.

Li, H.W., Li, T. and Wang, J.Z. (2003). Supercritical carbon dioxide extraction of essential oil from *Cinnamomum migao*. *Hong Yao Cai.*, **26**: 178-80.

Lim, G.B., Holder, G.D. and Shah, Y.T. (1989). K.P Johnson and JML Penninger Editors. *Supercritical fluid science and Technology* Acs, Washington, D.C. 379.

Lim, G.B., Lee. S.Y., Lee. E.K., Haam, S.J. and Kim, W.S. (2007). Supercritical fluid extraction and determination of lutein in heterotrophically cultivated *Chlorella pyrenoidosa*. *J.Food process Eng.*, **30**: 174.

Lim, G.B., Lee. S.Y., Lee. E.K., Haam, S.J. and Kim, W.S. (2002). Separation of astaxanthin from red yeat *Phaffia rhodozyma* by supercritical carbon dioxide extraction. *Biochem.Eng.J.*, **11**: 181-187.

Lin, Mei-Chih., Tsai, Ming-Jer. and Wen, Kuo-Ching. (1999). Supercritical fluid extraction of flavonoids from Scutellariae Radix. *J. Chromatogr. A*, **830**: 387-395.

Ling, Y.C., Teng, H.C. and Cartwright, C. (1999). Supercritical fluid extraction and clean-up of organochlorine pesticides in Chinese herbal medicines. *J.Chromatogr.A*, **835**: 145.

List, P.H. and Schmidt, P.C. (1984).Technologie pflanlicher Arzneizubereitungen Wissens- chaftliche Verlagsgesellschaft: *Sttugart*, 99-232.

Lorenzo, T., Schwartz, S.J. and Kila, P.K. (1991). In: MA McHugh (Ed.). Proceedings of the 2nd International symposium on SCFE Department of chemical Engineering. John Hophier Univ. Baltimore MD.

Llefloch F.,Tena, M. T., Rios, A. and Valcarcel, M. (1998). Supercritical fluid extraction of phenol compounds from olive leaves. *Talanta*, **46**: 1123.

Loudi,V., Folas, G., Voutass, E. and Magoulas, K. (2004). Extraction of parsley seed oil by supercritical CO_2. *J. Supercrit. Fluids*, **30**: 163.

Lucas, A. De., Martinez de la Ossa, E., Rincón, J., Blanco, M.A. and Gracia, I.(2002). Supercritical fluid extraction of tocopherol concentrates from olive tree leaves. *J. Supercritical Fluids*, **22**: 221–228.

Luque-Garcia, J. L. and Luque De Castro, M. D. (2003). Ultrasound: a powerful tool for leaching. *Trends Anal-Chem.*, **22**: 41-47.

Macias-Sanchez, M.D., Mantell, C., Rodriguez, M., De La Ossa, E.M., Lubian, L.M. and Montero, O. (2005). Supercritical fluid extraction of carotenoids and chlorophyll a from *Nannochloropsis gaditana*. *J. Food Eng.*, **66**: 245–251.

Mallavarapu, G.R., Kulkarni, R.N. and Ramesh, S. (2004). The essential oil composition of anise hyssop grown in India. *Flavour and Fragrance J.*, **19 (4)**: 351 – 353.

Mansi, B. (2004). M.Sc Thesis,Virginia Polytechnic Institute and State University,USA, p1- 32.

Manuel J-G., Nienaber,U. and Schwartz, S.J. (1999). Paprika (*Capsicum annuum*) oleoresin extraction with supercritical carbon dioxide. *J. Agric. Food Chem.*, **47**: 3558-3564.

Maria de Lourdes, L., Moraes, Janete H. Y. and Vilegas Fernando, M. (1997). Super critical fluid extraction of glycosylated flavonoids from *Passiflora* leaves. *Phytochemical Anal.*, **8**:257.

Marleny, D.A., Saldaña, R.S. and Mohamed, P.M. (2002). Extraction of cocoa butter from Brazilian cocoa beans using supercritical CO_2 and ethane. *Fluid Phase Equilibria*, **194–197**: 885–894.

Marongiu, B., Porcedda, S., Caredda, A., De Gioannis, B., Vargiu, L. and La Colla, P. (2003). Extraction of *Juniperus oxycedrus* sp. *oxycedrus* essential oil by supercritical carbon dioxide: Influence of some process parameters and biological activity. *Flavour Fragr. J.*, **18**: 390-397.

Marsili, D.R. and Callahan, D. (1993). Comparison of a liquid solvent extraction technique and supercritical fluid extraction for the determination of α- and β-carotene in vegetable. *Journal of Chromatography Science*, **31**: 422–428.

Martinelte/mastinelli, E., Shulz, K. and Mansoori, G.A. (1994). In T.J Bruno, J.F Ely (Eds.), SCF Technology. *CRC Press*, Boca Raton, Florida USA, 451-478.

Martinez, J., Rosa, P. T. V. and Meireles, M. A. A. (2007). Extraction of clove and vetiver oils with supercritical carbon dioxide: Modeling and simulation. *The Open Chemical Engineering Journal*, **1**: 1-7.

Marzouki, H., Piras, A. Marongiu, B., Rosa, A., and Dessi, M.A. (2008). Extraction and separation of volatile and fixed oils from berries of *Laurus nobilis* L by supercritical CO_2. *Molecules*, **13**: 1702-1711.

McNally, M.E. and Wheeler, J.R. (1988). Increasing extraction efficiency in supercritical fluid extraction from complex matrices Predicting extraction efficiency of diuron and linuron in supercritical fluid extraction using supercritical fluid chromatographic retention. *J.Chromatogr.A*, **447**: 53.

Meireless, A.A. M. (2003). Supercritical extraction from solid process design data (2001-2003). *Curr. Opin. Solid State Mater. Sci.*, 7: 321-330.

Mendes, R.L., Nobre, B., Cardoso, M., Pereira, A. and Palavra, A. (2003). Supercritical carbon dioxide extraction of compounds with pharmaceutical importance from microalgae, *Inorganica Chimica Acta.*, **356**: 328–334.

Mesaros.S., Verzar-Petri.G., Nyiredy, M.K., Tyihak, E., Nyiredy,S.Z., Meier, B., Sticher, O. and Dallenbach, T.K. (1987). Planar Centrifugal Chromatography Device *US Patent* US4, 678-570.

Mikhova, B., Duddeck, H., Taskova, R., Mitova, M. and Alipicva, K. (2004). Oxygenated bisabolone fucoside from *Carthus lanatus* L. *Z. Naturforsh*, **59**C: 244-48.

Misic, D., Zizovic, I., Stamenic, M., Asanin, R., Ristic, M., Petrivic, S.D. and Skala, D. (2008). Antimicrobial activity of celery fruit isolates and SFE process modeling. *Biochem. Eng. J.*, **42**: 148-152.

Miskovsky, P. (2002). Hypericin - A New Antiviral and Antitumor Photosensitizer: mechanism of Action and Interaction with Biological Macromolecules. *Curr. Drug Targets*, **3(1)**: 55-84.

Miyachi, H., Manabe,A., Tokumori, T., Sumida, Y., Yoshida, T., Nishibo, S., Agate, T. and Okuda,T. (1987). Application of supercritical fluid extraction to components of crude drugs and plants. *Yakagaku Zasshi*, **107:** 435.

Modey, W.K., Mulholland, D.A., Mahomed, H. and Raynor, M.W. (1996). Analysis of extracts from *Cedrela toona* (meliaceae) by on-line and off-line supercritical fluid extraction-capillary gas chromatography. *J.Microcolumn Sep.*, **8:** 67.

Mohacsi-Farkas, C., Tulok, M. and Balogh, B. (1994). SCF/RC with application in Biotech.In: T.J Bruno and J.F Ely. Editors SCFT CRC press raton, *FL,* 451-478.

Moldano-Martins, M., Bernando-Gil, M., Gabriela., D.A. and Costa, A. M. L.(2002). Sensory and chemical evaluation of *Thymus zygis* L. essential oil and compressed CO_2 extracts. *Eur Food Res. Technol.*, **214**: 207-211.

Molero-Gomez, A.,Pereyra-Lopez, C. and Martinez de la, O. (1996). Recovery of grape seed oil by liquid supercritical carbon dioxide extraction: a comparison with convention solvent extraction. *Chem. Eng. J.*, **61:** 227-231.

Monteiro, A. R., Meirles, M.A., Marques, M.O.M. and Patente, A. J. (1997). Extraction of the soluble material from the shells of the bacuri fruit (*Platonia insignis* Mart) with pressurized CO_2 and other solvents. *J.Supercrit. Fluids,* **11:** 91-102.

Moore, W.N. and Taylor, L.T.(1996).Extraction and quantitation of diogixin and acetyldigoxin from the digitalis lanata leaf via near supercritical methanol modified carbon dioxide. *J.Nat. Prod.*, **59:** 690-693.

Moreas, M.D.L.,Vilegas, J.H.Y. and Lancas, F.M. (1997). SCFE of glycosylated flavonoids from *Passiflora* leaves. *Phytochem.Anal.B*, 257-260.

Morrison, W.H., Holser, R., and Akin, D.E. (2006). Cuticular wax from flax processing waste with hexane and super critical carbon dioxide extractions. *Industrial Crops and Products*, **24**(2): 119-122.

Murga, R., Ruíz, R., Beltrán, S. and Cabezas, J.L. (2000). Extraction of natural complex phenols and tannins from grape seeds by using supercritical mixtures of carbon dioxide and alcohol. *Journal of Agricultural and Food Chemistry*, **48:** 3408–3412.

Norulaini, N. A., Zaidul, N. Md., Anuar, I. S. O. and Omar, A. K. M. (2004). Supercritical enhancement for separation of lauric acid and oleic acid in palm kernel oil (PKO). *Separ.Purif.Technol.*, **39(3)**: 133-138.

Nyiredy, S.Z. and Botz, L. (2001). Medium-pressure solid-liquid extraction: A new preparative method based on the principle of counter-current. *J. Planar Chromatogr.*, **14:** 393-395.

Oakes, R.S., Clifford, A.A., Bartle, K.D., Pett, M.T. and Rayner, C.M. (1999). Sulfur oxidation in supercritical carbon dioxide : Dramatic pressure dependant enhancement of diastereo selectivity for sulfoxidation of cysteine derivatives. *Chem. Commun.*, 247-248.

Oszagyan, M., Simandi, B. and Sawinsky, J. (1996). Supercritical fluid extraction of volatile compounds from lavandin and thyme. *Flavour Fragr. J.* **11:** 157-165

Otterback, A. and Wenclawiak, B.W. (1999). Ultrasonic/soxhlet/supercritical fluid extraction kinetics of pyrethrins from flowers and of allethrin from paper strips. *J. Anal. Chem.*, **365:** 472-474.

Palma, M., Taylor, L.T, Zoecklein, B.W. and Douglas, L.S. (2000). Supercritical fluid extraction of grape glycosides. *J.Agric. Food Chem.*, **48**: 775-779.

Palma, M., Taylor, L.T. and Varela, R.M. (1999). Fractional extraction of compounds from grape seeds by Supercritical fluid extraction and analysis for antimicrobial and agrochemical activities. *J.Agric. Food Chem.*, **47**: 5044-5048

Pan Wynn, H.T., Chang, C.C., Lee, F. and Fuh, M.R.S. (1995). Preparative supercritical extraction of pyretherin 1 from Pyrethrum flowers. *Talanta*, **42:** 1745-1749.

Pasquel, A., Meirless, M.A. A., Marques, M.O.M. and Pentenate A.J. (2000). Extraction of stevia glycosides with CO_2+water, CO_2+ethanol, and CO_2+water+ethanol. *Braz.J.Chem.Eng.*, **17**: 271.

Patel, R.N., Bandyopadhyay, S. and Ganesh, A. (2006). Extraction of Cashew (*Anacardium occidentale*) Nut Shell Liquid Using Supercritical Carbon Dioxide. *Bioresource Technology*, **97(6)**: 847-853.

Paul, P. M. F. and Wise, W.S. (1971). The principles of gas extraction, London Mills &Boon Ltd.

Pavoliszyn, J. (1993). Kinetic model of supercritical fluid extraction. *J. Chromatogr. Sci.*, **31**: 31-37.

Piggott, M. J., Ghisalbeti, E. L., R.D. and Trengrove, E. L. (1997). Western Australian Sandalwood oil: extraction by different techniques and variations of the major components in different Sections of a Single Tree. *Flavour Frag. J.*, **12**: 43.

Pisacane, A. (2001). Extraction of materials from Plants,WO 01/07135,2001.

Pourmartazavi, S.M., Baghaee, P. and Mirhosseini, M. (2004). Extraction of volatile compounds from *Juniperus communis* L. leaves with supercritical fluid carbon dioxide: comparison with hydrodistillation.*Flavour Frag. J.*, **5**: 417.

Pourmortazavi, S. M., Ghadiri, S.M. and Hajimirasadeghi, S.S. (2005). Supercritical fluid extraction of volatile components from *Bunium persicum* Boiss. (black cumin) and *Mespilus germanica* L. (medlar) seeds.*J.Food Comp. and Anal.*, **18(5)**: 439-446.

Pourmortazavi, S. M. and Hazimirsadeghi, S. S. (2007). Supercriical fluid extraction in plant essential and volatile oil analysis. *J. Chromatogr. A*, **1163(1-2)**: 2-24.

Povh, N. P., Marques, M. D. M. and Meireles, A. A. (2001). Supercritical CO_2 exraction of essential oils and concretes from Floweres. *J. Supercrit. Fluids*, **21**: 245-256.

Qui, Q., Zhang, G., Sun, X. and Liu, X. (2003). Supercritical carbon dioxide extraction of essential oil from *Cinnamomum migao. Zhong Yao Cai*, **26(3)**: 178-80.

Qiu. Q., Ling, J., Ding, Y., Wang, H. L. and Liu, T. (2005). Comparison of supercritical fluid extraction and steam distillation methods for the extraction of essential oils from *Schizonepeta tenuifolia* Briq. *Se PU.*, **23(6)**: 646-50.

Quii, Q., Zhang. G., Sun, X. and Liu, X. (2004). Study on chemical constituents of the essential oil from *Myristica fragrans* Houtt by supercritical fluid extraction and steam distillation. *Zhang- Yao Cai*, **27(11)**: 823-826.

Ranalli, A., Contento,S., Lucera, L., Pavone, G Giacorno, G., Alosio, L. and Gregorio, D.I. (2004). Characterization of carrot root oil arising from supercritical fluid carbon dioxide extraction. *J. Agric. Food Chem.*, **52(15)**: 4795–4801.

Rao, P. U. (1994). Nutrient composition of some less familiar oil seeds. *Food Chem.*, **50**: 379-382.

Reichling, L., Weseler, A. and Saller, R. (2001). A current review of the antimicrobial activity of *Hypericum perforatum* L. *Pharmacopsychiatry Supplement*, **34**: 116-118.

Reverchon, E. (1997). SCFE and Fractionation of essential oil. *J.Supercrit. Fluids*,**10** : 1-37.

Reverchon, E. and Marco, I. De. (2006). Supercritical fluid extraction and fractionation of natural matter. *J. Supercrit. Fluids*, **38**: 146–166.

Reverchon, E. and Poletto, M. (1996). Mathematical modelling of supercritical CO_2 fractionation of flower concretes. *Chem. Eng. Sci.Chem. Eng. Sci.*, **51**:3741-3753.

Reverchon, E., Porta, G. D. and Gorgoglione, D. (1997). Supercritical CO_2 extraction of volatile oil from rose concrete. *Flavour Fragr. J.*, **12**: 37-39.

Reverchon, E. and Senatora, F. (1995). Supercritical CO_2 extraction and fractionation of Lavender essential oil and waxes. *J Agric. Food Chem.*, **43**: 1654 – 1658.

Richter, J. and Schellenberg, I. (2007). Comparison of different extraction methods for the determination of essential oils and related compounds from aromatic plants and optimization of solid-phase microextraction/gas chromatography. *Anal.Bioanl.Chem.*, **387**: 2207-2217.

Rodrigues, V.M., Rosa, P.T., Marques, M.O., Petervate, A. J. and Meireles, M.A. (2003). Supercritical extraction of essential oil from aniseed (*Pimpinella anisum* L.) using CO_2: solubility, kinetics, and composition data. *J.Agric. Food Chem.*, **51**(6): 1518-23.

Römpp, H., Seger, C., Kaiser, C., Haslinger, S. and Schmidt, P.C. (2004). Enrichment of hyperforin from St. John's Wort (*Hypercum perforatum*) by pilot scale supercritical carbon dioxide extraction. *Eur. J. Pharm. Sci.*, **21**: 443–451.

Rónyai, E., Simandi, B., Veress, T., Lemberkovics, E. and Patika, D. (1999). Essential oil obtained by hydrodistillation from different *Salvia* spp. and GCMS analysis. *J.Ess.Oils Res.*, **11**: 499-502.

Roop, R.K., Akermann, A., Dexter, B.J. and Irvin, T.R. (1989). Extraction of Phenol from Water with Supercritical Carbon Dioxide. *J.Supercrit fluids*, **2**:51-56.

Roop, R.K., German, A.A.K., Dexter, B.J. and Irvin, T.R. (1999). In: Caude, M. and Thiebaut, D., Editors practical supercritical fluid chromatography and extraction, Harwood, Amsterdam.

Rosa, P.T.V. and Meireles M.A. (2005). Supercritical Technology in Brazil: System Investigation. *J. Supercrit. Fluids*, **34**: 109-117.

Rostagno, M. A., Araujo, J. M. A. and Sandi, D. (2000). Supercritical fluid extraction of isoflavones from soybean flour. Food Chem., **78**: 11-117.

Roy, B. B., Goto, M., Kodama, A. and Hirose, T. (1996). SCFE CO_2 extraction of essential oils and cuticular waxes from peppermint leaves. *J.Chem. Tech. Biotechnol.*, **67**: 21-26.

Rozzi, N. L., Phippen, W., Simon, J. E. and Singh, R. K. (2002). Supercritical fluid extraction of essential oil components from lemon-scented botanicals. *Lebensm.-Wiss.U.-Technol.*, **35**: 319-324.

Sabio, E., Lozano, M., Montero de Espimosa, V., Mendes, R.L., Periera, A.P., Palavra, A.P. and Coelho, J. A. (2003). Lycopene and other carotenoids extraction from tomato waste using supercritical CO_2 *Ind. Eng. Chem. Res.*, **42**: 6641-6646.

Safaralie, A., Fatemi, S. and Sefidkon, F. (2008). Essential oil composition of *Valeriana officinalis* L. roots cultivated in Iran. Comparative analysis between supercritical CO_2 extraction and hydrodistillation. *J. Chromatogr. A*, **1180(2)**: 159-164.

Sandra, R.S., Ferreira, Z.L., Nikolov, L.K., Doraiswamy, M.A.A., Meireless, M.A. and Peteniate. A. J. (1999). Supercritical fluid extraction of black pepper (*Piper nigrum* L.) essential oil. *J. Supercrit. Fluids*, **14**: 235-245.

Sang, D. Y. and Kirana, E. (2005). Formation of polymer particles with supercritical fluids: A Review. *J. Supercrit. Fluids*, **34(3)**: 287- 308.

Sato, K., Sasaki, S. S., Goder, Y., Yamarda, T., Nunomura, O., Ishikanoa, K. and Maitani, T. (1999). Direct connection of supercritical fluid extraction and supercritical fluid chromatography as a rapid quantitative method for capsaicinoids in placentas of capsicum. *J. Agric. Food Chem.*, **47**: 4665.

Scalia, S., Guiffreda, L. and Pallado, P. (1999). Analytical and preparative supercritical fluid extraction of Chamomile flowers and its comparison with conventional methods. *J. Pharm. and Biomedical Analysis*, **21**: 549-558.

Schneider, G. M., Stahl, E. and Wilke, G. (1980). Extraction with supercritical gases. Deerfield Beach Fl. Verlagchemie.

Shi, Z., He, J. and Chang, W. (2004). Micelle-mediated extraction of tanshinones from*Salvia miltiorrhiza bunge* with analysis by high-performance liquid chromatography. *Talanta*, **64**: 401-407.

Shu, X.S., Gao, Z.H. andYang, X.L. (2004). Supercritical fluid extraction of sapoenins from fibers of *Smilax china*. *Fitoterapia*, **75**: 656– 661.

Sigma-Aldrich chemicals. 2008-2009,p 661, D-1530.

Silva,G.L., Lee,I-S. and Kinghorn, A.D. (1998). Special problems with the extraction of plants pp 343-363 in Cannel R.J.P(Ed.) : Methods in Biotechnology 4. How to approach the isolation a natural product p 1-51, Natural products isolation, Humana Press, Totowa, New Jersey, USA.

Smith, R.M. and Burford, M.D. (1992). Optimization of supercritical fluid extraction of volatile constituents from a model plant matrix. *J.Chromatography*, **600**: 175.

Song, K.M., Park, S.W., Hong, W.H., Lee, H., Kwak, S.S. and Liu, J.R. (1992). Isolation of vindoline from *Catharanthus roseus* by supercritical fluid extraction. *Biotechnol Prog.*, **8**: 583.

Sonsuzer, S., Sachin, S. and Yilmaz, L.(2004). Optimization of supercritical CO_2 extraction of *Thymbra spicata* oil. *J.Supercrit. Fluids*, **30**: 189.

Sovova, H., Sajfrtova, M., Bartlova, M. and Opletal, L. (2004). Near critical extraction of pigments and oleoresin from stinging nettle leaves. *J. Supercrit. Fluids*, **30**: 213–224.

Spanos, G.A., Chen, H. and Schwartz, S.J. (1993). Supercritical CO_2 extraction of carotene from sweet potatoes. *J.Food Sci.*, **58**: 817-820.

Stahl, E., Quinn, K. W., and Gerard, D. (1987).Verdichtete Gase zur extraction and Raffination, Springer, Berlin: 1-260.

Stashenko, E. E., Jaramillo, B. E. and Martinez, J.R. (2004). Comparison of different extraction methods for the analysis of volatile secondary metabolites of *Lippia alba* (Mill), Nie brown grown in Colombia and evaluation of its in vitro antioxidant activity. *J. Chromatogr. A*, **1025**: 93-104.

Stashenko, E.E., Puertas, M.A. and Combariza, M.Y. (1996). Volatile secondary metabolites from *Spilanthes americana* obtained by simultaneous steam distillation-solvent extraction and supercritical fluid extraction.*J.Chromatogr. A*, **752**: 223.

Stashenko, E. E., Jaramillo, B. E. and Martinez, J. R. (2004). Analysis of volatile secondary metabolites from *Colombian xylopia* aromatic (Lamarck) by different extraction and headspace methods and gas chromatography. *J. Chromatogr.A*, **1025**: 105-113.

Steiner, R., Hauk. A. and Tratz, W. (1999). Method for producing medicinal plant extracts, WO 99/18984.

Steiner, B., Trat, W., Brattstrom, A. and Debrunner, B.(1998). High pressure extraction and fractionation of Pharmacologically active components from petasites, hybrids. In proceedings of the 5th meeting on supercritical fluids materials and natural products processing, Nice: Vol. 2, Perrut M, Subra P (Eds.) Institute natural polytechnique de Lorraine; Vaudocuvre, 509-514.

Steiner, R., Tratz, W., Brattstrom, A. and Debrunner, B. (1998). Proceedings of the 5th meeting on SCF-Materials and Natural products Processing Nice Vol.2 perrut M subra P (Eds.), 509- 514.

Szentihalyi, K., Peter, V., Bela, L., Venel, I. and Maria,T. (2002). Rose hip (*Rosa cania* L) oil obtained from waste hip seeds by different extraction methods. *Bioresource Technology*, **82**: 195-201.

Taylor, L.T. (1996). Supercritical fluid extraction.John Wiley and Sons : New York, p. 1-181.

Teuber, M. and Schmalreck, A.F. (1973). Membrane leakage in *Bacillus subtilis* 168 induced by the hop constituents lupulone, humulone, isohumulone and humulinic acid. *Arch Microbiol.*, **94**: 159-171.

Tena, M.T., Rios, A. and Valcarcel, M. (1998). Supercritical fluid extraction of t-resveratrol and other phenolics from a spiked solid. *J.Anal.Chem.*, **361**: 143-148.

Tietz,U., Thomann, R. and Forstner, S. (1991). High pressure extraction of marjoram leaves. *Die Nahrung*, **35**: 101-102.

Tipsrisnkond, N., Fernando, L. N., and Clarke, A.D. (1998). Antioxidant effects of essential and oleoresin of black pepper from supercritical carbon dioxide extractions in ground pork. *J. Agric. Food Chem.*, **46**: 4329-4333.

Tonthubthimthong, P., Chuaprasert, S., Douglas, P. and Lnewisuthichat, W. (2001). Supercritical CO_2 extraction of nimbin from neem seeds-an experimental study. *J. Food. Eng.*, **47**: 289-293

Tsuda, T., Mizuno, K., Oshima, K., Kawaishi, S. and Osawa, T. (1995). Supercritical carbon dioxide extraction of antioxidative components from tamarind (*Tamarindus indica* L.) seed coat *J. Agric. Food Chem.*, **43**: 2803-2806.

Ubalua, O. A. (2007). Cassava Wastes: treatment options and vaue addition alternatives. *African J.Biotech.*, **6(18)**: 2065-2073.

Uquiche, E., deValle, J. M. and Ortiz, J. (2004). Supercritical carbon dioxide extraction of red pepper (*Capsicum annum* L.) oleoresin. *J. Food Eng.*, **65**: 55–66.

Vági, E., Rapavi, E., Hadolin, M., Vasarhelyin,P., Balazs, A. and Blazovics, Simandi.B. (2005). Phenolic and triterpenoid antioxidants from *Origanum majorana* L. herb and extracts obtained with different solvents. *Food Res.Intern.*, **38**: 51-57.

Vilcu,G.I. and Mocan, M.(2002). Supercritical Fluid extraction and fractional separation of essential oils, Bucharest. Univ. www.chimie.unibuc.ro/biblioteca/anale/2002a/63-67

Vivok, I., Simonovska, B., Andernsek, S., Vuorela, H. and Vuorela, P. (2003). Rotation planar extraction and rotation planar chromatography of Oak (*Quercus robur* L). *J.Chromatogr. A*, **991**: 267-274.

Vollmer, J. J. and Rosenson J. (2004). Chemistry of St. John's Wort: Hypericin and Hyperforin. *J.Chem. Edu.*, **81(10)**: 1450.

Walsh, J. M., Ikonomou, G. D. and Donohue, M. D.(1987). Supercritical Phase behavour: The entrainer effect. *Fluid Phase Equileb.*, **287**: 33.

Wang, J. and Li,T. (2003). Supercritical carbon dioxide extraction of essential oil from *Cinnamomum migao. Zhong Yao Cai.*, **26(3)**: 178-80.

Wang, L.J. and Weller, C.L. (2006). Recent advances in extraction of natural products from plants. *Trends Food Sci. Technol.*, **17**: 300-312.

Wang, L.,Weller, C. L., Schlegel, V. L., Carr, T. P. and Cuppett, S. L. (2007). Comparison of supercritical CO_2 and hexane extraction of lipids from Sorghum distillers grains. *Eur.J. Lipid Sci.Technol.*, **109**: 567-574.

Wang, L., Weller, C. L., Schlegel, V. L., Carr, T.P. and Cuppett, S.L. (2008). Supercritical CO_2 extraction of lipids from grain sorghum dried distillers grains with solubles. *Bioresource Technology*, **99**: 1373-1382.

Weinhold, T. de., Bresciani, L.F.V.,Tridapalli,C.W.,Yunes, H. and Herreira, S.R.S. (2008). *Polygala cyparissias* oleoresin: Comparing CO_2 and classical organic solvent extractions Chemical Engineering and Processing. *Process Intensification*, **47**: 109-117.

Wills, N. J., Pork, J., Wen, J., Kesavan, S., Kraus, G. A., Petrich, J.W. and Carpenter, S. (2001). Tumor cell toxicity of hypericin and related analogs.*Photochem. Photobiol.*, **74**: 216-220.

Won, Ji-Hee., Park, Kwang-Kyon, and Chung, won-Yoon. (2008).*Cancer treatment reviews*, 3421-22.

Wynn Pan, H.T., Chang, C-C, Sn,T.T., Lee, F. and Fuh, M.R.S. (1995). Preparative supercritical fluid extraction of pyrethrin I and II from pyrethrum flower.*Talanta*, **42**: 1745-1749.

Yamini, Y., Bahramifarz, N. and Sefidkon, F. (2008). Extraction of essential oil from *Pimpinella anisum* during supercritical carbon dioxide and comparison with hydrodidstillation. *Nat. Prod. Res.*, **22(3)**: 212-218.

Yamini,Y., Sefidkon, F. and Pourmortazavi, S. M. (2002).Comparisin of essential oil composition of Iranian Fennel (*Foeniculum vulgare*) obtained by supercritical carbon dioxide extraction and hydrodistillation. *Flavour Fragr. J.*, **17(5)**: 345-348.

Yeo, S. and Kiran, E. (2005). Formation of polymer particles with supercritical fluids: a Review. *J.Supercrit.Fluids*, **34**(3): 287- 308.

Zosel, K. (1976). Process for the separation of mixture of substances. *US patent.* 3,969,196.

Zygmunt, B., Jastrzebska, A. and Namiesinik, J. (2001). Solid Phase Micro extraction– A convenient tool for the determination of organic pollutants. In: Environmental matrices. *Critical reviews in Analytical Chemistry*, **31(1)**: 1-18.

Natural Products: Research Reviews Vol. 1 (2012)   *Pages* **221–370**
Editor: **V.K. Gupta**
Published by: **DAYA PUBLISHING HOUSE, NEW DELHI**

7

Phytochemistry of the Genus *Croton*

Vivianne Marcelino de Medeiros[1], Josean Fechine Tavares[1],
Jackson Roberto Guedes da Silva Almeida[2],
Vicente Toscano de Araújo-Junior[3], Petrônio Filgueiras de
Athayde-Filho[1], Emídio Vasconcelos Leitão da Cunha[1,4],
José Maria Barbosa-Filho[1] and Marcelo Sobral da Silva[1*]

ABSTRACT

A literature review on 108 species of the genus Croton has been carried out and presented in this chapter. The species belonging to this genus are rich in alkaloids, phenylpropanoids, monoterpenes, sesquiterpenes, diterpenes and lignans, which demonstrates the importance of the phytochemical study of this genus.

Keywords: *Croton*, Euphorbiaceae, Phytochemistry, Review.

Introduction

The family Euphorbiaceae, is comprised by 300 genera and some 7500 species, widely distributed over the world, specialy in tropical and subtropical regions of the

1   Universidade Federal da Paraíba, Laboratório de Tecnologia Farmacêutica, C.P. 5009, 58051-900, João Pessoa, PB, Brazil.

2   Fundação Universidade Federal do Vale do São Francisco, Avenida José de Sá Maniçoba, s/n, Centro, C.P. 252, 56304-205, Petrolina, PE, Brasil.

3   Universidade Federal do Rio Grande do Norte, Departamento de Tecnologia Farmacêutica e de Alimentos, 59010-180, Natal, RN, Brazil.

4   Universidade Estadual da Paraíba, Departamento de Farmácia e Biologia, CCBS, 58100-000, Campina Grande, PB, Brazil.

*   *Corresponding author*: E-mail: marcelosobral@ltf.ufpb.br

Americas, Africa and Asia. The most important genera are: *Euphorbia* (1500 species), *Croton* (700 species), *Phyllantus* (400 species), *Acalypha* (400 species), *Macaranga* (400 species), *Antidesma* (150 species), *Drypetes* (150 species), *Jatropha* (150 species), *Manihot* (150 species) and *Tragia* (150 species) (Cronquist, 1981; Webster, 1994).

There are two previous reviews of the literature on the phytochemistry of the genus *Croton*. The first was published by Farnsworth and collaborators in 1969 (Farnsworth *et al.*, 1969) and soon after by Stuart in 1970 (Stuart, 1970). The fact that the last review was published some 39 years ago presented an obvious challenge, given the large volume of publications on the phytochemistry of this genus that have appeared in the ensuing years.

The richness of the chemical composition of the plants of the genus *Croton*, led this research group to carry out a complete literature review, covering its phytochemistry, and the use of some species in folk medicine.

Methodology

The review was carried out through searches in data-bases such as NAPRALERT (Natural Products Alert) and ISI/Web of Sciences. Chemical Abstracts was also an important source of informations used in this work, updated to December 2009. The references found in the search were then studied in detail.

The obtained data are organized separately by plants and within each species by class of compounds, compounds, part of the plant from where the compound was isolated and also part of the world where the work was done.

Results and Discussion

The word *Croton* derives from the Greek word crston meaning "tick" (*Ixodes ricinus* L.) an allusion to the resemblance of the seeds of its representants to this insect (Di Stasi *et al.*, 1989). In this genus it is possible to find herbs, shrubs and even big trees with more than 30 m high. *Croton* plants have been used in many countries as a folk remedy, and claimed to be effective for the treatment of a variety of illnesses amongst which are wounds, cancers, inflammations and infections. However, many of them contain a milky latex which can in some cases be poisonous. Detailed informations are shown in Table 7.1. The commercially available oil of Croton derives from the Asian species *Croton tiglium* L. and is a source of phorbol esters, whose cocarcinogenic properties have been object of numerous research efforts.

A survey of the present available chemical data suggests that terpenoids, especially diterpenes and alkaloids are the main classes of substances of interest because of their structural variety and pharmacological activities. Also abundant are the lower molecular weight terpenoids (mono and sesquiterpenes), which together with arylpropanoids form another important class of natural products, the essential oils, responsible for the characteristic odour of many species of *Croton*. Other compounds, also isolated from different *Croton* species have been classified under twenty one categories, *viz.*, alicyclics, aliphatics, alkaloids, benzenoids, carbohydrates, diterpenes, flavonoids, lignans, lipids, monoterpenes, oxygen heterocycles, peptides, phenanthrene, phenylpropanoids, proteids, quinone, sesterpenes, sesquiterpenes,

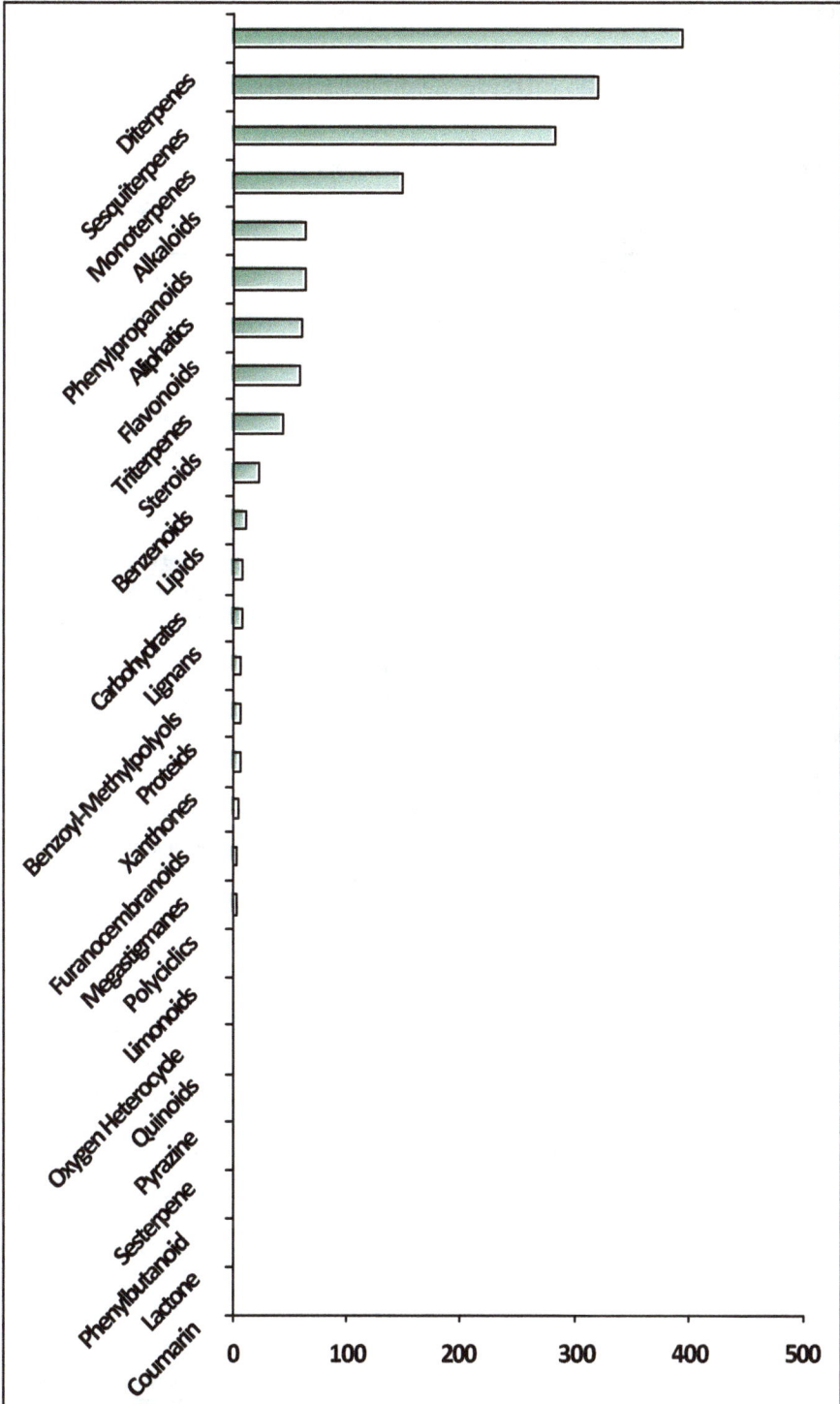

Chart 7.1: Phytochemical profile of the genus *Croton.*

steroids, triterpenes and xanthones. Compounds are listed in Table 7.1 and structures in Figure 7.1.

Chart 7.1 shows a graphical representation of the classes of substances isolated from plants of the genus *Croton*.

Table 7.1: Chemical constituents of the plants of the genus *Croton*.

Compound Type Chemical Name	Place (Part Used)	References
Croton adenocalyx A. DC		

Popularly known as "catinga de porco", the barks are used in the treatment of some skin diseases (Lima, 2000).

Compound Type Chemical Name	Place (Part Used)	References
Monoterpenes		
Camphene	Brazil (EO, L)	Craveiro *et al.*, 1990
Cineole, 1,8	Brazil (EO, L)	Craveiro *et al.*, 1990
Cymene, *para*	Brazil (EO, L)	Craveiro *et al.*, 1990
Limonene	Brazil (EO, L)	Craveiro *et al.*, 1990
Myrcene	Brazil (EO, L)	Craveiro *et al.*, 1990
Pinene, α	Brazil (EO, L)	Craveiro *et al.*, 1990
Pinene, β	Brazil (EO, L)	Craveiro *et al.*, 1990
Terpinene, α	Brazil (EO, L)	Craveiro *et al.*, 1990
Thujene, α	Brazil (EO, L)	Craveiro *et al.*, 1990
Sesquiterpenes		
Bourbonene, β	Brazil (EO, L)	Craveiro *et al.*, 1990
Cadinene, δ	Brazil (EO, L)	Craveiro *et al.*, 1990
Cadinene, γ	Brazil (EO, L)	Craveiro *et al.*, 1990
Caryophyllene, β	Brazil (EO, L)	Craveiro *et al.*, 1990
Cubebene, α	Brazil (EO, L)	Craveiro *et al.*, 1990
Elemene, β	Brazil (EO, L)	Craveiro *et al.*, 1990
Elemene, δ	Brazil (EO, L)	Craveiro *et al.*, 1990
Elemene, γ	Brazil (EO, L)	Craveiro *et al.*, 1990
Germacrene D	Brazil (EO, L)	Craveiro *et al.*, 1990
Gurjunene, α	Brazil (EO, L)	Craveiro *et al.*, 1990
Humulene, α	Brazil (EO, L)	Craveiro *et al.*, 1990
Croton affinis var. mucronifolius		

Infusion of the leaves is used for the treatment of influenza, rheumatism and syphilis (Lemos *et al.*, 1992).

Compound Type Chemical Name	Place (Part Used)	References
Monoterpenes		
Ascaridol	Brazil (EO, L)	Craveiro *et al.*, 1981a
Ascaridol, iso	Brazil (EO, L)	Craveiro *et al.*, 1981a
Borneol	Brazil (EO, L)	Craveiro *et al.*, 1981a

Contd...

Table 7.1 – *Contd...*

Compound Type Chemical Name	Place (Part Used)	References
Camphor	Brazil (EO, L)	Craveiro *et al.,* 1981a
Car-3-ene	Brazil (EO, L)	Craveiro *et al.,* 1981a
Cineole, 1,8	Brazil (EO, L)	Craveiro *et al.,* 1981a
Cymene, *para*	Brazil (EO, L)	Craveiro *et al.,* 1981a
Phellandrene, α	Brazil (EO, L)	Craveiro *et al.,* 1981a
Pinene, α	Brazil (EO, L)	Craveiro *et al.,* 1981a
Terpinene, g	Brazil (EO, L)	Craveiro *et al.,* 1981a

Croton arboreous Millsp.

The aerial parts of *Croton arboreous*, common name "cascarillo", are the source of a popular beverage used in Tabasco and Chiapas, Mexico, as an auxiliary anti-inflammatory agent in the treatment of respiratory ailments (Guadarrama and Rios, 2004).

Diterpene

Junceic acid	Mexico (AP)	Guadarrama and Rios, 2004

Flavonol

Flavone, 5, 5', 7-trihidroxy-3, 4'-dimethoxy: 2'-O-β-D-glucoside	Mexico (AP)	Guadarrama and Rios, 2004
Quercetin-3, 4'-dimethyl ether	Mexico (AP)	Guadarrama and Rios, 2004

Steroids

Daucosterol	Mexico (AP)	Guadarrama and Rios, 2004

Sesquiterpenes

Eudesm-3-en-1β,6α-diol, 5α,10β	Mexico (AP)	Guadarrama and Rios, 2004
Eudesm-4(15)-en-1β,6β-diol, 5α,10β	Mexico (AP)	Guadarrama and Rios, 2004
Eudesm-4(15)-ene-1β,6α-diol, 5α,10β	Mexico (AP)	Guadarrama and Rios, 2004
Germacran-6β-ol, 7α-(H):1α,4β,5α, 10β-diepoxy	Mexico (AP)	Guadarrama and Rios, 2004
Germacran-6β-ol, 7α-(H):1β,4β,5α, 10α-diepoxy	Mexico (AP)	Guadarrama and Rios, 2004
Guaiane, 9α,10β-dihydroxy-2β,4β-peroxy-1α,5β,7αH	Mexico (AP)	Guadarrama and Rios, 2004
Guai-9-ene, 1α,5β-(H): 4β,6β-dihydroxy	Mexico (AP)	Guadarrama and Rios, 2004
Patchoul-4(14)-en-2α-ol, 5α,7α,10β-(H)	Mexico (AP)	Guadarrama and Rios, 2004
Patchoul-3-en-2-one, 5α,7α,10β-(H)	Mexico (AP)	Guadarrama and Rios, 2004
Spathulenol	Mexico (AP)	Guadarrama and Rios, 2004
Teucladiol	Mexico (AP)	Guadarrama and Rios, 2004

Triterpene

Lupenone	Mexico (AP)	Guadarrama and Rios, 2004

Contd...

Table 7.1–*Contd...*

Compound Type Chemical Name	Place (Part Used)	References

Croton argyrophylloides Muell. Arg.

Occurs in Northeastern Brazil, where it is popularly known as "marmeleiro branco" (Webster, 1994; Craveiro *et al.*, 1981a). Argyrophilic acid, a diterpene isolated from the roots, has antibiotic activity (Albuquerque *et al.*, 1974).

Diterpenes

Argyrophylic acid	Brazil (RB, EP)	Albuquerque *et al.*, 1974
Cleroda-13(16),14-diene, 3,12-dioxo-15,16-epoxy-4-hydroxy	Brazil (RB)	Monte *et al.*, 1988
Croton argyrophylloides diterpene 3	Brazil (RB)	Monte *et al.*, 1983
Croton argyrophylloides diterpene A	Brazil (TW)	Monte *et al.*, 1984
Croton argyrophylloides diterpene B	Brazil (TW)	Monte *et al.*, 1984
Kaur-16-en-15-oxo-18-oic acid: *ent*	Brazil (RB)	Monte *et al.*, 1988
Kaur-16-en-19-oic acid: ent	Brazil (RB)	Monte *et al.*, 1988
rel-(1*R*,4a*R*,5*R*,8*R*)-Methyl-7-(1-(methoxycarbonyl)vinyl)-5,8-diacetoxy-1,2,3,4a,5,6,7,8,9,10,10a-dodecahydro-1,4a-dimethyl-2-oxophenanthrene-1-carboxylate.	Brazil (B)	Santos *et al.*, 2009
rel-(1*S*,4a*S*,7*S*,8a*S*)-7-(1-vinyl)-Tetradeca-hydro-1,4a dimethyl-phenanthrene-7,8a-carbolactone-1-carboxylic acid	Brazil (B)	Santos *et al.*, 2009

Monoterpenes

Cineole, 1,8	Brazil (EO, L)	Craveiro *et al.*, 1978a Craveiro *et al.*, 1981a
Pinene, α	Brazil (EO, L)	Craveiro *et al.*, 1978a Craveiro *et al.*, 1981a
Sabinene	Brazil (EO, L)	Craveiro *et al.*, 1978a Craveiro *et al.*, 1981a

Sesquiterpenes

Caryophyllene, β	Brazil (EO, L)	Craveiro *et al.*, 1978a Craveiro *et al.*, 1981a
Elemene, β	Brazil (EO, L)	Craveiro *et al.*, 1978a Craveiro *et al.*, 1981a
Elemene, γ	Brazil (EO, L)	Craveiro *et al.*, 1978a Craveiro *et al.*, 1981a
Humulene, α	Brazil (EO, L)	Craveiro *et al.*, 1978a Craveiro *et al.*, 1981a

Croton aromaticus L.

Widely distributed plant in Sri Lanka, is used in ethnomedical preparations and in traditional agriculture, which showed insecticidal activity against *Apis craccivora* (Albuquerque *et al.*, 1974).

Diterpene

Hardwickiic acid, (-)	Sri Lanka (R)	Bandara *et al.*, 1987 Bandara et al., 1990

Contd...

Table 7.1 –*Contd...*

Compound Type Chemical Name	Place (Part Used)	References
Sesquiterpene		
Cyperenoic acid	Sri Lanka (R)	Bandara *et al.*, 1990

Croton betulaster Muell. Arg.

Croton betulaster Müll. Arg. is a shrub that is found in the northern part of the Cadeia do Espinhaço from Grão Mogol to the Serra do Sincorá in the Chapada, Diamantina, Bahia, Brazil (Cordeiro, 1995).

Benzoyl-methylpolyols		
1-Benzoyloxy-4-para-hydroxy-benzoyloxy- 2-3-5-trihydroxy-5-methylhexane	Brazil (L)	Barbosa *et al.*, 2004
1,5-Dibenzoyloxy-4-*para*-hydroxy-benzoyloxy- 5-methylhex-2-ene	Brazil (L)	Barbosa *et al.*, 2004
3-Benzoyloxy-4-parahydroxy-benzoyloxy-2- ethoxy-1-hydroxy-5-methylhexane	Brazil (L)	Barbosa *et al.*, 2004
Flavonoid		
Flavone, 5-hydroxy-7,4'-dimethoxy	Brazil (AP)	Barbosa *et al.*, 2003
5,3'-Dihydroxy-3,6,7,4'-tetramethoxyflavone	Brazil (L)	Barbosa *et al.*, 2004
Triterpenes		
Cycloart-24-en-26-oic-3-oxo acid	Brazil (AP)	Barbosa *et al.*, 2003
Cycloart-trans-24-en-26-oic acid, 3-oxo	Brazil (AP)	Barbosa *et al.*, 2003
Hopane, 22-hydroxy-3-oxo	Brazil (AP)	Barbosa *et al.*, 2003
Lupane, 20-hydroxy-3-oxo	Brazil (AP)	Barbosa *et al.*, 2003
Lupenone	Brazil (AP)	Barbosa *et al.*, 2003
Lupeol	Brazil (AP)	Barbosa *et al.*, 2003
Olean-12-en-28-oic-3-oxo acid	Brazil (AP)	Barbosa *et al.*, 2003
Olean-18-en-28-oic-3-oxo acid	Brazil (AP)	Barbosa *et al.*, 2003
Taraxastane, 3-oxo-20β-hydroxy	Brazil (AP)	Barbosa *et al.*, 2003

Croton bonplandianus Baill.

In the southern Taiwan *C. bonplandianus* is found as a weed commonly in wastelands, roadsides, and orchards. It is locally abundant in the areas of sandy or sandy clay soils. It blooms and fruits all the year round. This accidentally introduced weed has a potential to become an agricultural pest in Taiwan.

Alkaloids		
Crotosparine	India (L)	Farnsworth *et al.*, 1969
Crotsparine	India (AP, EP)	Casagrande *et al.*, 1975 Bhakuni, 1984
Crotsparine, *N*-methyl	Not stated (BR)	Bhakuni and Jain, 1981
Crotsparinine	India (EP, AP)	Casagrande *et al.*, 1975 Bhakuni, 1984
Crotsparinine, iso	Not stated (AP)	Casagrande *et al.*, 1975

Contd...

Table 7.1 – *Contd...*

Compound Type Chemical Name	Place (Part Used)	References
Crotsparinine, iso: *N*-methyl	Not stated (AP)	Casagrande *et al.*, 1975
Crotsparinine, *N*-methyl	Not stated (AP, BR)	Bhakuni and Jain, 1981
Crotsparinol, *N*-methyl	India (EP)	Bhakuni, 1984
Glaziovine, tetrahydro: (DL)	Not stated (AP)	Casagrande *et al.*, 1975
Morphinandien-7-one, 4, 6-dihydroxy-3-methoxy	India (EP)	Tiwari *et al.*, 1981
Nuciferine 1, nor	India (EP)	Bhakuni, 1984
Nuciferine I, nor	India (BR)	Bhakuni *et al.*, 1979
Nuciferine, pro: (+)	India (L)	Farnsworth, 1969
Sinoacutine, nor	India (EP)	Tiwari *et al.*, 1981
Sparsiflorine	India (L, AP, EP)	Farnsworth, 1969 Bhakuni, 1984 Casagrande *et al.*, 1975
Sparsiflorine, *N*-methyl	Not stated (BR)	Bhakuni and Jain, 1981
Diterpenes		
Phorbol myristate acetate-20-linolenate	India (SE)	Upadhyay and Hecker, 1976
Phorbol, 12-*O*-dodecanoate: 13-acetate	India (S)	Upadhyay and Hecker, 1976
Flavonoids		
Quercetin-3-*O*-α-L-rhamnosyl-glucoside	India (L)	Subramanian *et al.*, 1971
Lipids		
Fatty acids unsaturated	India (SE, O)	Kapoor *et al.*, 1986
Croton sparsiflorus (Fixed oil)	India (SE)	Upadhyay and Hecker, 1976
Steroids		
Sitosterol, β	India (L, S)	Bhakuni *et al.*, 1971
Sesquiterpene		
Vomifoliol	India (L, S)	Satish and Bhakuni, 1972
Triterpenes		
Taraxerol	India (L, S)	Bhakuni *et al.*, 1971
Ursolic acid	India (L, S)	Satish and Bhakuni, 1972

Croton cajucara Benth.

Known in the Amazon area as "sacaca", it is a medicinal plant used in the form of a tea for ailments such as diarrhea, diabetes and inflammation of the liver (Maciel *et al.*, 2000; Itokawa *et al.*, 1989).

Diterpenes

Cajucarin A	Brazil (B)	Itokawa *et al.*, 1990
Cajucarin B	Brazil (B)	Itokawa *et al.*, 1990 Maciel *et al.*, 1998
Cajucarin B, *trans*	Brazil (B)	Maciel *et al.*, 1998

Contd...

Table 7.1–*Contd...*

Compound Type Chemical Name	Place (Part Used)	References
Cajucarinolide	Brazil (B, L)	Ichihara *et al.*, 1992 Maciel *et al.*, 1998 Maciel *et al.*, 2000
Cajucarinolide, iso	Brazil (B)	Ichihara *et al.*, 1992
Crotonin, dehydro	Brazil (B, C)	Simões *et al.*, 1979 Itokawa *et al.*, 1989
Crotonin, *cis*-dehydro	Brazil (B)	Kubo *et al.*, 1991
Crotonin, *trans*	Brazil (B)	Maciel *et al.*, 1998 Grynberg *et al.*, 1999
Crotonin, *trans*-dehydro	Brazil (B, L, SB, R)	Kubo *et al.*, 1991 Farias *et al.*, 1997 Hiruma-Lima *et al.*, 1999 Agner *et al.*, 1999 Agner *et al.*, 2001 Brito *et al.*, 1998 Grynberg *et al.*, 1999 Costa *et al.*, 1999 Carvalho *et al.*, 1996 Maciel *et al.*, 2000
Crotonin, T	Brazil (C)	Itokawa *et al.*, 1989
Crotonin, T-dehydro	Brazil (B)	Maciel *et al.*, 1998
Sacacarin	Brazil (B)	Maciel *et al.*, 1998
Essential Oil		
Essential oil	Brazil (L)	Araújo *et al.*, 1972
Flavonoids		
Kaempferol 3,7-dimethyl-ether	Brazil (L)	Maciel *et al.*, 2000
Kaempferol 3,5,7-trimethyl-ether	Brazil (L)	Maciel *et al.*, 2000
Kaempferol 3,4',7-trimethyl ether	Brazil (L)	Maciel *et al.*, 2000
Kumatakenin	Brazil (L)	Maciel *et al.*, 2000
Monoterpenes		
Linalool	Brazil (EO, L)	Araújo *et al.*, 1972 Lopes *et al.*, 2000 Rosa *et al.*, 2003
Sesquiterpenes		
Caryophyllene, β	Brazil (EO, L)	Lopes *et al.*, 2000
Nerolidol, *trans*	Brazil (EO, L)	Lopes *et al.*, 2000
Steroids		
Daucosterol	Brazil (L)	Maciel *et al.*, 2000
Sitosterol, β	Brazil (L)	Maciel *et al.*, 2000
Stigmasterol	Brazil (L)	Maciel *et al.*, 2000
Triterpene		
Aleuritolic acid, acetyl	Brazil (B)	Maciel *et al.*, 1998

Contd...

Table 7.1–*Contd...*

Compound Type Chemical Name	Place (Part Used)	References
Croton californicus **Müll. Arg**		

American Indians prepare a hot poultice of its powdered leaves as a pain reliever for rheumatism (Wilson *et al.*, 1976). The leaf and stem showed antimalarial activity (Spencer *et al.*, 1947).

Compound Type / Chemical Name	Place (Part Used)	References
Aliphatic		
Triacontan-1-ol	USA (EP)	Luzbetak *et al.*, 1979
Tridecane, n	Mexico (F, L)	Williams III *et al.*, 2001
Diterpenes		
Barbascoate methyl, (-)	USA (L)	Wilson *et al.*, 1976
Hadwickiic acid, (-)	USA (EP)	Luzbetak *et al.*, 1979
Phorbol, 12-deoxy:13,20-*O*-didecanoyl	USA (EP)	Chavez *et al.*, 1982
Phorbol, 12-deoxy-13-*O*-decanoyl-20-*O*-dodecanoyl	USA (EP)	Chavez *et al.*, 1982
Phorbol, 12-deoxy:13-*O*-decanoyl-20-*O*-hexadecanoyl	USA (EP)	Chavez *et al.*, 1982
Phorbol, 12-deoxy:13-*O*-decanoyl-20-tetradecanoyl	USA (EP)	Chavez *et al.*, 1982
Monoterpenes		
Thujene, 3	Mexico (F, L)	Williams III *et al.*, 2001
Pinene, β	Mexico (F, L)	Williams III *et al.*, 2001
Myrcene, β	Mexico (F, L)	Williams III *et al.*, 2001
Terpinene, δ	Mexico (F, L)	Williams III *et al.*, 2001
Sesquiterpenes		
Caryophyllene, β	Mexico (F, L)	Williams III *et al.*, 2001
Caryophyllene, α	Mexico (F, L)	Williams III *et al.*, 2001
Xanthone		
Xanthone-1,2,3,4,6,7-hexamethoxy	USA (EP)	Tammami *et al.*, 1977

Croton campestris **St. Hill.**		

Popularly known as "velame do campo, velame verde, velame verdadeiro and curraleira", it is used in the Brazilian folk medicine as internal and external use preparations as a potent purgative, and to drain and treat syphilis. It has been successfully employed in the treatment of bile duct infections (Ribeiro *et al.*, 1993; El Babili *et al.*, 1998).

Compound Type / Chemical Name	Place (Part Used)	References
Alkaloids		
Crotosinoline-*N*-oxide, 1,2,10-trihydroxy	Brazil (L)	Ribeiro *et al.*, 1993
Taspine	Brazil (L)	Ribeiro *et al.*, 1993
Aliphatic		
Butyric acid	Brazil (EO)	Freise, 1935
Valeric acid	Brazil (EO)	Freise, 1935

Contd...

Table 7.1–*Contd...*

Compound Type Chemical Name	Place (Part Used)	References
Diterpenes		
Cleroda-3,13(16)-14-trien-15,16-epoxy-2-oxo: *ent*	Brazil (RB)	El Babili *et al.*, 1998
Cleroda-3,13(16)-14-trien-15,16-epoxy-20-acetoxy-2-oxo: *ent*	Brazil (RB)	El Babili *et al.*, 1998
Cleroda-3,13(16)-14-triene,15,16-epoxy-20-hydroxy-2-oxo: *ent*	Brazil (RB)	El Babili *et al.*, 1998
Lipid		
Tetracosanoic acid butyl ester	Brazil (L)	Ribeiro *et al.*, 1993
Monoterpenes		
Angelic acid	Brazil (EO)	Farnsworth *et al.*, 1969 Freise, 1935
Croton cascarilloides **Raeusch.**		
Alkaloid		
Crotonosine	Taiwan (S)	Farnsworth *et al.*, 1969
Julocrotine	Vietnam (R)	Cuong *et al.*, 2002
Linearisine, homo	Taiwan (S)	Farnsworth *et al.*, 1969
Benzenoids		
Anisic acid	China (NS)	Chen *et al.*, 2006
Vanillin	China (NS)	Chen *et al.*, 2006
Diterpenes		
ent-8,9-7α-Hydroxy-11β-acetoxykaura-8(14),16-dien-9,15-dione	China (NS)	Chen *et al.*, 2006
ent-8,9-seco-7α-Hydroxykaura-8(14),16-dien-9,15-dione	China (NS)	Chen *et al.*, 2006
ent-8,9-seco-8,14-Epoxy-7α-hydroxy-11β-acetoxy-16-kauren-9,15-dione	China (NS)	Chen *et al.*, 2006
Flavonoids		
Luteolin-7-*O*-α-L-rhamnoside	China (NS)	Chen *et al.*, 2006
Quinoid		
Rubiadin-1-methyl ether	Vietnam (R)	Cuong *et al.*, 2002
Steroids		
Sitosterol, β	China (NS)	Chen *et al.*, 2006
Daucosterol, β	China (NS)	Chen *et al.*, 2006
Triterpene		
Aleuritolic acid, 3-acetyl	Vietnam (R)	Cuong *et al.*, 2002

Contd...

Table 7.1–*Contd...*

Compound Type Chemical Name	Place (Part Used)	References

Croton caudatus **Geiseler**

Occurs in Eastern Himalayan range at altitudes of about 6000 ft. This plant is popular for its use in the treatment of stomach disorders. Its extracts have been found to possess some insecticidal and insect repelling properties (Banerjee *et al.,* 1988).

Diterpenes

Crotocaudin	India (SB)	Chaterjee *et al.,* 1977 Chaterjee *et al.,* 1978
Crotocaudin, iso	India (SB)	Chaterjee *et al.,* 1978
Teucvidin	India (SB)	Fujita *et al.,* 1976 Chaterjee *et al.,* 1977 Chaterjee *et al.,* 1978

Steroids

Sitosterol, β	India (SB)	Chaterjee *et al.,* 1977 Chaterjee *et al.,* 1978 Banerjee *et al.,* 1988
Stigmastane-3,6-dione, 5α	India (SB)	Banerjee *et al.,* 1988

Triterpenes

Taraxerol	India (SB)	Chaterjee *et al.,* 1977 Chaterjee *et al.,* 1978
Taraxerone	India (SB)	Chaterjee *et al.,* 1977 Banerjee *et al.,* 1988
Taraxeryl acetate	India (SB)	Chaterjee *et al.,* 1978
Tataxerol	India (SB)	Banerjee *et al.,* 1988

Croton celtidifolius **Baillon**

Known as "arvore de sangue", occurs in the Southern part of Brazil in the State of Santa Catarina. The latex is known as skin irritant (Mukherjee and Axt, 1984).

Alkaloids

Boldine, iso	Brazil (L, S)	Amaral and Barnes, 1997
Laudanidine, (+)	Brazil (L, S)	Amaral and Barnes, 1997
Reticuline, (-)	Brazil (L, S)	Amaral and Barnes, 1997

Carbohydrates

Bornesitol, D: (-)	Brazil (L, T)	Mukherjee and Axt, 1984
Inositol, neo	Brazil (L, T)	Mukherjee and Axt, 1984
Bornesitol, L: (+)	Brazil (L, T)	Mukherjee and Axt, 1984

Steroid

Sitosterol, β	Brazil (L, T)	Mukherjee and Axt, 1984

Contd...

Table 7.1–*Contd...*

Compound Type Chemical Name	Place (Part Used)	References

Croton chilensis Muell. Arg.

Shrub endemic to the region of Antogasta, Chile, where it is popularly known as "hirriquerilla de paposo", it is considered a rare and endangered species (Bittner *et al.*, 1997).

Aliphatic

Crotonic acid	Chile (EP)	Borquez *et al.*, 1995

Alkaloids

Coridine, iso	Chile (EP)	Bittner *et al.*, 1997
Flavinantine	Chile (EP)	Bittner *et al.*, 1997
Flavinantine, *O*-methyl	Chile (EP)	Bittner *et al.*, 1997
Salutaridine, iso	Chile (EP)	Bittner *et al.*, 1997

Benzenoids

Salidroside	Chile (EP)	Bittner *et al.*, 1997
Tyrosol	Chile (EP)	Bittner *et al.*, 1997

Sesquiterpene

Vomifoliol	Chile (EP)	Bittner *et al.*, 1997

Croton ciliatoglanduliferus Ort.

Croton ciliatoglanduliferus Ort. (Euphorbiaceae) is a wild plant, found in the Tehuacan region (Puebla State) and in Guerrero State, in Mexico. It has been used as a repellent against insects and for herbal medicine to cure some diseases (Morales-Flores *et al.*, 2007).

Diterpenes

Labdane-8a,15-diol	Mexico (S)	Morales-Flores *et al.*, 2007
Labdane-8a,15-diol, acetyl	Mexico (S)	Morales-Flores *et al.*, 2007
12-*O*-[(2*R*)-*N*,*N*-Dimethyl-3-methyl-butanoyl]-4-deoxyphorbol 13-Acetate	Mexico (AP)	Rios and Aguillar-Guadarrama, 2006
12-*O*-[(2*S*)-*N*,*N*-Dimethyl-3-methyl-butanoyl]-4-deoxyphorbol 13-Acetate	Mexico (AP)	Rios and Aguillar-Guadarrama, 2006
12-*O*-[3-Methyl-2-butenoyl]-4-deoxyphorbol 13-acetate	Mexico (AP)	Rios and Aguillar-Guadarrama, 2006
12-*O*-[(2*R*)-*N*,*N*-Dimethyl-3-methyl-butanoyl]phorbol 13-acetate		

Flavonoids

5-Hydroxy-3,7,3',4'-tetramethoxy-flavone	Mexico (L)	González-Vázquez *et al.*, 2006
5,4'-Dihydroxy-3,7,3'-trimethoxy-flavone	Mexico (L)	González-Vázquez *et al.*, 2006

Contd...

Table 7.1–*Contd...*

Compound Type Chemical Name	Place (Part Used)	References

Croton columnaris Airy. Slaw.

According to the Flora of Thailand it grows in dry deciduous oak forest, rocky hardwood forest slopes, open sandy places and over sandstone. Flowering and fruiting in Jan., Mar., Apr., Sept., Nov. It is popularly known as "Plao kham" (Esser, 2003).

Diterpene

Geranyl-geraniol, all-*trans*:8-hydroxy	Japan (EP)	Mishima *et al.,* 1977
Plaunol	Thailand (EP)	Mishima *et al.,* 1977
Plaunol monoacetate	Thailand (EP)	Mishima *et al.,* 1977

Croton cortesianus H.B.K. (Kunth)

It is a shrub 1-3 m high, growing in Texas and Mexico (vernacular names "chilpatli", "palillo", "pozual", "pinolillo"). The leaves, twigs and roots, are used in folk medicine for venereal diseases and uterine hemorrhages (Siems *et al.,* 1992; Dominguez and Alcorn, 1985).

Lipids

Linolenic acid	Mexico (AP)	Siems *et al.,* 1992
Octadeca-10-*trans*-12-*cis*-15-*cis*-trienoic acid, 9-hydroxy	Mexico (AP)	Siems *et al.,* 1992
Octadeca-9-*cis*-11-*trans*-15-*cis*-trienoic acid, 13-hydroxy	Mexico (AP)	Siems *et al.,* 1992
Octadeca-9-*cis*-13-*trans*-15-*cis*-trienoic acid-12-hydroxy	Mexico (AP)	Siems *et al.,* 1992

Carbohydrates

Inositol, L: (-)	Mexico (AP)	Siems *et al.,* 1992
Inositol, muco: 1-*O*-methyl	Mexico (AP)	Siems *et al.,* 1992
Quebrachitol	Mexico (AP)	Siems *et al.,* 1992

Diterpenes

Hoffmanniaaldehyde	Mexico (AP)	Siems *et al.,* 1992
Hoffmanniaaldehyde, 1-10(5)-dehydro: 5-deformyl	Mexico (AP)	Siems *et al.,* 1992
Printziane,10-β-(H): 5,10-dihydro:5α-hydroxy	Mexico (AP)	Siems *et al.,* 1992
Strigyllanoic acid B	Mexico (AP)	Siems *et al.,* 1992

Oxygen heterocycle

Tocopherol, α	Mexico (AP)	Siems *et al.,* 1992

Sesquiterpene

Caryophyllene, oxide	Mexico (AP)	Siems *et al.,* 1992

Triterpenes

Squalene	Mexico (AP)	Siems *et al.,* 1992
Oleanolic acid	Mexico (R)	Dominguez and Alcorn, 1985

Contd...

Table 7.1–*Contd...*

Compound Type Chemical Name	Place (Part Used)	References

Croton corylifolius LAM

In the West Indies FL and leaves are used in folk medicine for amenorrhea (Napralert, 2002).

Diterpenes

Corylifuran	Jamaica (L, T)	Burke *et al.*, 1976
Crotofolin A	Jamaica (EP, L, T)	Chan *et al.*, 1975 Burke *et al.*, 1976
Crotofolin E	Jamaica (EP)	Burke *et al.*, 1979

Croton crassifolius Geiseler

It is a small herb commonly found throughout northeastern Thailand (Boonyaratanakornkit *et al.*, 1988). The MeOH ext of the roots showed HIV-1 reverse transcriptase inhibition (Tan *et al.*, 1991).

Diterpene

Chettaphanin I	Thailand (R)	Boonyaratanakornkit *et al.*, 1988

Sesquiterpene

Cyperenoic acid	Thailand (R)	Boonyaratavej and Roengsumran, 1988 Boonyaratanakornkit *et al.*, 1988

Triterpenes

Aleuritolic acid, 3-*O*-acetyl	Thailand (R)	Boonyaratanakornkit *et al.*, 1988
Amyrin, β	Thailand (R)	Boonyaratanakornkit *et al.*, 1988

Croton cuneatus Klotzsch.

Croton cuneatus is a medicinal plant used by the natives of the Amazonian to treat inflammations, gastrointestinal disturbances and as an analgesic (Suárez, *et al.*, 2004). It is a medium-size tree which grows widely in the Amazons of Venezuela, Peru and Brazil, and it is known by the names *arapurina, caferana* and *reventillo* (Webster, *et al.*, 1999).

Alkaloids

Julocrotine	Venezuela (AP)	Suarez *et al.*, 2004
Julocrotol	Venezuela (AP)	Suarez *et al.*, 2004
Julocrotol, iso	Venezuela (AP)	Suarez *et al.*, 2004
Julocrotone	Venezuela (AP)	Suarez *et al.*, 2004

Monoterpenes

Cymene, *p*	Venezuela (EO, SB, L)	Suarez *et al.*, 2005
Limonene	Venezuela (EO, SB, L)	Suarez *et al.*, 2005
Linalil-oxide, *trans*-	Venezuela (EO, SB, L)	Suarez *et al.*, 2005
Linalool	Venezuela (EO, SB, L)	Suarez *et al.*, 2005
Norboneol	Venezuela (EO, SB, L)	Suarez *et al.*, 2005

Contd...

Table 7.1–*Contd...*

Compound Type Chemical Name	Place (Part Used)	References
Terpinol, α	Venezuela (EO, SB, L)	Suarez *et al.*, 2005
Verbenone	Venezuela (EO, SB, L)	Suarez *et al.*, 2005
Carvacrol	Venezuela (EO, SB, L)	Suarez *et al.*, 2005
Phenylpropanoids		
Eugenol	Venezuela (EO, SB, L)	Suarez *et al.*, 2005
Capric acid	Venezuela (EO, SB, L)	Suarez *et al.*, 2005
Hexyl-hexanoate, n-	Venezuela (EO, SB, L)	Suarez *et al.*, 2005
Methyleugenol	Venezuela (EO, SB, L)	Suarez *et al.*, 2005
Methylisoeugenol	Venezuela (EO, SB, L)	Suarez *et al.*, 2005
Elemicine	Venezuela (EO, SB, L)	Suarez *et al.*, 2005
3,4,5-Trimethoxybenzaldehyde	Venezuela (EO, SB, L)	Suarez *et al.*, 2005
Cedryl-propyl-ether	Venezuela (EO, SB, L)	Suarez *et al.*, 2005
Isoelemicine	Venezuela (EO, SB, L)	Suarez *et al.*, 2005
Sesquiterpene		
Selin-11-en-4α-ol	Venezuela (AP)	Suarez *et al.*, 2004
Gurjunene, α	Venezuela (EO, SB, L)	Suarez *et al.*, 2005
Caryophyllene, β	Venezuela (EO, SB, L)	Suarez *et al.*, 2005
Valencene	Venezuela (EO, SB, L)	Suarez *et al.*, 2005
Veratral	Venezuela (EO, SB, L)	Suarez *et al.*, 2005
Cadinene, γ	Venezuela (EO, SB, L)	Suarez *et al.*, 2005
Bisabolene, β	Venezuela (EO, SB, L)	Suarez *et al.*, 2005
Cadinene, δ	Venezuela (EO, SB, L)	Suarez *et al.*, 2005
Cadinene, α	Venezuela (EO, SB, L)	Suarez *et al.*, 2005
Calacorene	Venezuela (EO, SB, L)	Suarez *et al.*, 2005
Spathulenol	Venezuela (EO, SB, L)	Suarez *et al.*, 2005
Eudesmol, γ	Venezuela (EO, SB, L)	Suarez *et al.*, 2005
Cadinol, γ	Venezuela (EO, SB, L)	Suarez *et al.*, 2005
Asaraldehyde	Venezuela (EO, SB, L)	Suarez *et al.*, 2005
Eudesmol, α	Venezuela (EO, SB, L)	Suarez *et al.*, 2005
Carotol	Venezuela (EO, SB, L)	Suarez *et al.*, 2005
Cadinol, α	Venezuela (EO, SB, L)	Suarez *et al.*, 2005
Cubenol	Venezuela (EO, SB, L)	Suarez *et al.*, 2005
Agarospirol	Venezuela (EO, SB, L)	Suarez *et al.*, 2005
Cadinol, *t*	Venezuela (EO, SB, L)	Suarez *et al.*, 2005
Cadinol, δ	Venezuela (EO, SB, L)	Suarez *et al.*, 2005
11-Eudesmol, α	Venezuela (EO, SB, L)	Suarez *et al.*, 2005

Contd...

Table 7.1 –*Contd...*

Compound Type Chemical Name	Place (Part Used)	References
Ledenol	Venezuela (EO, SB, L)	Suarez *et al.*, 2005
Nootkatone	Venezuela (EO, SB, L)	Suarez *et al.*, 2005
Xanthone		
Lichexanthone	Venezuela (AP)	Suarez *et al.*, 2004

Croton diasii Pires ex Secco and P.E. Berry

Diterpene		
Diasin	Brazil (TW)	Alvarenga *et al.*, 1978
Steroid		
Sitosterol, β	Brazil (TW)	Alvarenga *et al.*, 1978

Croton dichogamus Pax

Elephants in the Serengeti-Mara region of Northern Tanzania and Southern Kenya feed on the variety of trees and shrubs, however, they show distinct preferences by *C. dichogamus* (Jogia *et al.*, 1989).

Diterpenes		
Crotoxide A	Kenya (L)	Jogia *et al.*, 1989
Crotoxide B	Kenya (L)	Jogia *et al.*, 1989

Croton discolor Willd.

Alkaloids		
Nuciferine, pro: (+)	Jamaica (EP)	Farnsworth *et al.*, 1969
Salutaridine, 8, 14-dihydro	Jamaica (EP)	Farnsworth *et al.*, 1969

Croton dracco Schlecht (Schltdl. e Cham.)

Commonly known as "sangre de drago", occurs in Mexico, where it is used in traditional medicine to relieve fever and to harden the flesh by which the teeth are surrounded (Rodriguez-Hahn *et al.*, 1975; Hernandez and Delgado, 1992).

Aliphatics		
Castaprenol 11	Mexico (AP)	Hernandez and Delgado, 1992
Benzenoids		
Benzoica cid, 4-(2-hydroxy-ethyl)	Costa Rica (LX)	Kostova *et al.*, 1999
Phenylethanol, 2,5-dihydroxy	Costa Rica (LX)	Kostova *et al.*, 1999
Phloroglucinol	Costa Rica (LX)	Kostova *et al.*, 1999
Diterpenes		
Draconin	Mexico (B)	Rodriguez-Hahn *et al.*, 1975
Flavonoids		
Catechin (+)	Costa Rica (LX)	Kostova *et al.*, 1999
Catechin, epi (-)	Costa Rica (LX)	Kostova *et al.*, 1999
Gallocatechin (+)	Costa Rica (LX)	Kostova *et al.*, 1999
Gallocatechin, epi (-)	Costa Rica (LX)	Kostova *et al.*, 1999
Myricitrin	Costa Rica (LX)	Kostova *et al.*, 1999
Quercitrin	Costa Rica (LX)	Kostova *et al.*, 1999

Contd...

Table 7.1 *–Contd...*

Compound Type Chemical Name	Place (Part Used)	References
Sesquiterpenes		
Vomifoliol	Mexico (AP)	Hernandez and Delgado, 1992
Steroids		
Daucosterol	Costa Rica (LX)	Hernandez and Delgado, 1992 Kostova *et al.*, 1999
Ergosterol, peroxide	Mexico (AP)	Hernandez and Delgado, 1992
Sitosterol, β	Costa Rica (LX), Mexico (AP)	Hernandez and Delgado, 1992 Kostova *et al.*, 1999
Stigmasterol	Mexico (AP)	Hernandez and Delgado, 1992

Croton draconoides Muell. Arg.

Also know as *C. palanostigmina*, tree of several areas of South American jungle, produces a blood-red latex, named "Sangre de Drago", largely used in the popular medicine for wound healing. The blood-red latex is also produced by various South American *Croton* species, including *C. lechleri*, and, *C. erythrochylus* (Aquino *et al.*, 1991; Pieters *et al.*, 1992).

Alkaloids

Taspine	Peru (LX)	Marini-Bettolo and Scarpati, 1979; Aquino *et al.*, 1991
Glaucine	Peru (L, T)	Maini-Bettolo and Scarpati, 1979
Thalyporphine	Peru (L, T)	Marini-Bettolo and Scarpati, 1979
Flavonoids		
Flavan, epi:3,3',5,5',7-pentahydroxy: (-)	Peru (LX)	Aquino *et al.*, 1991
Gallocatechin, (+)	Peru (LX)	Aquino *et al.*, 1991 Mahmood *et al.*, 1993
Gallocatechin, epi (-)	Peru (LX)	Aquino *et al.*, 1991 Mahmood *et al.*, 1993

Croton echinocarpus Muell. Arg. (Baill.)

The resin from this plant shows strong antihelmintic activity (Farnsworth *et al.*, 1969).

Alkaloids

Laurelliptine	Brazil (L, S)	Pereira *et al.*, 1999
Laurelliptine, *N*-methyl	Brazil (L, S)	Pereira *et al.*, 1999
Jacularine	(EP)	Farnsworth *et al.*, 1969
Laurelliptine, dehydro	Brazil (L, S)	Pereira *et al.*, 1999
Salutaridine, nor: 8, 14-dihydro	(EP)	Farnsworth *et al.*, 1969

Croton eluteria Bennet

This species, popularly known as "cascarilla" has been used for years as a tonic, febrifuge and stimulant; for dyspepsia, diarrhea, and as diaphoretic in "eruptive fevers", the grippe and pneumonia (Farnsworth *et al.*, 1969).

Aliphatics

Decan-2-one	Not stated (EO)	Hagedorn and Brown, 1991
Heptan-2-one	Not stated (EO)	Hagedorn and Brown, 1991

Contd...

Table 7.1–*Contd...*

Compound Type Chemical Name	Place (Part Used)	References
Heptan-2-one, 6-methyl	Not stated (EO)	Hagedorn and Brown, 1991
Hexa-2,4-dien-1-al	Not stated (EO)	Hagedorn and Brown, 1991
Hept-5-en-2-one, 6-methyl	Not stated (EO)	Hagedorn and Brown, 1991
Nonan-2-one	Not stated (EO)	Hagedorn and Brown, 1991
Tridecan-2-one	Not stated (EO)	Hagedorn and Brown, 1991
Undec-*cis*-5-en-2-one	Not stated (EO)	Hagedorn and Brown, 1991
Undecan-2-one	Not stated (EO)	Hagedorn and Brown, 1991
Valeric acid	Not stated (EO)	Halberstein and Saunders, 1978
But-3-en-1-ol, 3-methyl: benzoate	Not stated (EO)	Hagedorn and Brown, 1991
Heptan-2-ol	Not stated (EO)	Hagedorn and Brown, 1991
Propanone, *trans*-2-(2'-hexyl-cyclo-propyl)	Not stated (EO)	Hagedorn and Brown, 1991
Prenyl benzoate	Not stated (EO)	Hagedorn and Brown, 1991
Tridec-*cis*-5-en-2-one	Not stated (EO)	Hagedorn and Brown, 1991
Undeca-5,8-dien-2-one	Not stated (EO)	Hagedorn and Brown, 1991
Benzenoids		
Acetophenone, *para*-methyl	Not stated (EO)	Hagedorn and Brown, 1991
Anisole, *orto*-methyl	Not stated (EO)	Hagedorn and Brown, 1991
Benzoic acid, pentyl ester	Not stated (EO)	Hagedorn and Brown, 1991
Toluene, 2,5-dimethoxy	Not stated (EO)	Hagedorn and Brown, 1991
Diterpenes		
Cascarillin B	Ecuador (SB)	Vigor *et al.*, 2001
Cascarillin C	Ecuador (SB)	Vigor *et al.*, 2001
Cascarillin D	Ecuador (SB)	Vigor *et al.*, 2001
Eluterin A	South America (B)	Fattorusso *et al.*, 2002
Eluterin B	South America (B)	Fattorusso *et al.*, 2002
Eluterin C	South America (B)	Fattorusso *et al.*, 2002
Eluterin D	South America (B)	Fattorusso *et al.*, 2002
Eluterin E	South America (B)	Fattorusso *et al.*, 2002
Eluterin F	South America (B)	Fattorusso *et al.*, 2002
Eluterin G	South America (B)	Fattorusso *et al.*, 2002
Eluterin H	South America (B)	Fattorusso *et al.*, 2002
Eluterin I	South America (B)	Fattorusso *et al.*, 2002
Eluterin J	South America (B)	Fattorusso *et al.*, 2002
Cascarillardione	Not stated (B)	Appendino *et al.*, 2003
Cascarillin	South America (B)	Fattorusso *et al.*, 2002

Contd...

Table 7.1—*Contd...*

Compound Type Chemical Name	Place (Part Used)	References
Cascarillin E	Ecuador (B)	Vigor *et al.*, 2002
Cascarillin F	Ecuador (B)	Vigor *et al.*, 2002
Cascarillin G	Ecuador (B)	Vigor *et al.*, 2002
Cascarillin H	Ecuador (B)	Vigor *et al.*, 2002
Cascarillin I	Ecuador (B)	Vigor *et al.*, 2002
Eluterin B, pseudo	Not stated (B)	Appendino *et al.*, 2003
Eluterin K	Not stated (B)	Appendino *et al.*, 2003
Naphthyl-methyl-3-furanyl-ketone, decahydro-7-hydroxy-1,2,4(A)-5-tetramethyl-1	Not stated (EO)	Hagedorn and Brown, 1991
Lipid		
Phorbic acid	Not stated (B)	Nordal *et al.*, 1965
Nonanoic acid, 3-vinyl: methyl ester	Not stated (EO)	Hagedorn and Brown, 1991
Monoterpenes		
Camphene	Not stated (EO)	Hagedorn and Brown, 1991
Camphor	Not stated (EO)	Hagedorn and Brown, 1991
Car-2-ene	Not stated (EO)	Hagedorn and Brown, 1991
Carvacrol, methyl ether	Not stated (EO)	Hagedorn and Brown, 1991
Carvone	Not stated (EO)	Hagedorn and Brown, 1991
Carvotanacetone	Not stated (EO)	Hagedorn and Brown, 1991
Cineole, 1,8	Not stated (EO)	Hagedorn and Brown, 1991
Cuminaldehyde	Not stated (EO)	Hagedorn and Brown, 1991
Cymene, *meta*: 2,5-dimethoxy	Not stated (EO)	Hagedorn and Brown, 1991
Cymene, *para*	Not stated (EO)	Hagedorn and Brown, 1991
Fennelone	Not stated (EO)	Hagedorn and Brown, 1991
Limonene	Not stated (EO)	Hagedorn and Brown, 1991
Linalool, acetate	Not stated (EO)	Hagedorn and Brown, 1991
Myrcene	Not stated (EO)	Hagedorn and Brown, 1991
Ocymene, β-*cis*	Not stated (EO)	Hagedorn and Brown, 1991
Ocymene, β-*trans*	Not stated (EO)	Hagedorn and Brown, 1991
Phellandral	Not stated (EO)	Hagedorn and Brown, 1991
Phellandrene, α	Not stated (EO)	Hagedorn and Brown, 1991
Pinene, α	Not stated (EO)	Hagedorn and Brown, 1991
Pinene, β	Not stated (EO)	Hagedorn and Brown, 1991
Sabinene	Not stated (EO)	Hagedorn and Brown, 1991
Terpinen, 4-ol-acetate	Not stated (EO)	Hagedorn and Brown, 1991

Contd...

Table 7.1–*Contd...*

Compound Type Chemical Name	Place (Part Used)	References
Terpinene, γ	Not stated (EO)	Hagedorn and Brown, 1991
Terpineol, α-acetate	Not stated (EO)	Hagedorn and Brown, 1991
Terpinolene	Not stated (EO)	Hagedorn and Brown, 1991
Thujene, α	Not stated (EO)	Hagedorn and Brown, 1991
Thujone	Not stated (EO)	Hagedorn and Brown, 1991
Thymol, methyl-ether	Not stated (EO)	Hagedorn and Brown, 1991
Styrene, α: *para*-dimethyl	Not stated (EO)	Hagedorn and Brown, 1991
Phenylpropanoids		
Estragole	Not stated (EO)	Hagedorn and Brown, 1991
Eugenol, methyl ether	Not stated (EO)	Hagedorn and Brown, 1991
Sesquiterpenes		
Aromadendrene, allo	Not stated (EO)	Hagedorn and Brown, 1991
Bisabolol, dihydro	Not stated (EO)	Hagedorn and Brown, 1991
Cadalene	Not stated (EO)	Hagedorn and Brown, 1991
Cadinene, δ	Not stated (EO)	Hagedorn and Brown, 1991
Cadinene, γ	Not stated (EO)	Hagedorn and Brown, 1991
Calacorene, α	Not stated (EO)	Hagedorn and Brown, 1991
Calacorene, β	Not stated (EO)	Hagedorn and Brown, 1991
Calamenene	Not stated (EO)	Hagedorn and Brown, 1991
Calamenene, 1,11-oxyde	Not stated (EO)	Hagedorn and Brown, 1991
Caryophyllene, oxide	Not stated (EO)	Hagedorn and Brown, 1991
Caryophyllene, α	Not stated (EO)	Hagedorn and Brown, 1991
Cascarylladiene I	Not stated (EO)	Hagedorn and Brown, 1991
Cedren-10-ol, β	Not stated (EO)	Hagedorn and Brown, 1991
Cedrene, β	Not stated (EO)	Hagedorn and Brown, 1991
Chamigrene, α	Not stated (EO)	Hagedorn and Brown, 1991
Chamigrene, β	Not stated (EO)	Hagedorn and Brown, 1991
Copaene, α	Not stated (EO)	Hagedorn and Brown, 1991
Cubebene, α	Not stated (EO)	Hagedorn and Brown, 1991
Cubebene, β	Not stated (EO)	Hagedorn and Brown, 1991
Cuparene	Not stated (EO)	Hagedorn and Brown, 1991
Curcumene, α	Not stated (EO)	Hagedorn and Brown, 1991
Cycosativene	Not stated (EO)	Hagedorn and Brown, 1991
Elemene, β	Not stated (EO)	Hagedorn and Brown, 1991
Farnesene, β-*trans*	Not stated (EO)	Hagedorn and Brown, 1991
Geranyl, acetone	Not stated (EO)	Hagedorn and Brown, 1991

Contd...

Table 7.1–*Contd...*

Compound Type Chemical Name	Place (Part Used)	References
Germacrene D	Not stated (EO)	Hagedorn and Brown, 1991
Humulene, epoxy isomer	Not stated (EO)	Hagedorn and Brown, 1991
Maaliene, β	Not stated (EO)	Hagedorn and Brown, 1991
Muurolene, α	Not stated (EO)	Hagedorn and Brown, 1991
Naphth-1-ol-1,2,3,5,6,7,8,8-(A)-octahydro-6-iso-propenyl-4-8-(A)-dimethyl	Not stated (EO)	Hagedorn and Brown, 1991
Naphth-2-ol-1,2,6,7,8,8(A)-hexahydro-5,8 (A)-dimethyl-3-iso-propyl	Not stated (EO)	Hagedorn and Brown, 1991
Naphth-2-ol-5,6,7,8-tetrahydro-4,8-dimethyl-1-iso-propyl	Not stated (EO)	Hagedorn and Brown, 1991
Naphthalene, 1,2,3,4-tetrahydro-1,5-dimethyl-7-isopropenyl	Not stated (EO)	Hagedorn and Brown, 1991
Naphthalene, 1,2,3,4-tetrahydro-1,5-dimethyl-8-isopropenyl	Not stated (EO)	Hagedorn and Brown, 1991
Naphthalene, 1,2,3,4-tetrahydro-1,5-dimethyl-8-isopropyl	Not stated (EO)	Hagedorn and Brown, 1991
Naphthalene,1,2-dihydro-4,8-dimethyl-6-isopropyl	Not stated (EO)	Hagedorn and Brown, 1991
Selenene, β	Not stated (EO)	Hagedorn and Brown, 1991
Selenene, γ	Not stated (EO)	Hagedorn and Brown, 1991
Tetralone, 4-isopropyl-6-methyl	Not stated (EO)	Hagedorn and Brown, 1991
Thujopsene	Not stated (EO)	Hagedorn and Brown, 1991
Valencene	Not stated (EO)	Hagedorn and Brown, 1991
Naphthalene, 1,2,3,4-tetrahydro-5-methyl-1-methylene-7-iso-propyl	Not stated (EO)	Hagedorn and Brown, 1991
Triterpene		
Lupeol	South America (B)	Fattorusso *et al.*, 2002

Croton erythrochilus Muell. Arg.

Another plant popularly known as "Sangre de Drago" or "Sangue de Drago", produces a blood-red latex, largely used for wound healing (Pieters *et al.*, 1990).

Lignan

Cedrusin, 3',4-*O*-dimethyl	Peru (LX)	Pieters *et al.*, 1990

Croton essequiboensis Klotzsch

This species (sin *Croton populifolius* var. *essequiboensis* (Klotzsch) Muell. Arg), occurs in Northeast Brazil in the State of Ceara (Craveiro *et al.*, 1981a).

Monoterpenes

Pinene, α	Brazil (EO, L)	Craveiro *et al.*, 1981a
Pinene, β	Brazil (EO, L)	Craveiro *et al.*, 1981a

Contd...

Table 7.1–*Contd...*

Compound Type Chemical Name	Place (Part Used)	References
Phenylpropanoids		
Estragole	Brazil (EO, L)	Craveiro *et al.*, 1981a
Anethole, *trans*	Brazil (EO, L)	Craveiro *et al.*, 1981a
Sesquiterpenes		
Caryophyllene, β	Brazil (EO, L)	Craveiro *et al.*, 1981a
Copaene, α	Brazil (EO, L)	Craveiro *et al.*, 1981a
Cubebene, α	Brazil (EO, L)	Craveiro *et al.*, 1981a
Elemene, β	Brazil (EO, L)	Craveiro *et al.*, 1981a
Humulene, α	Brazil (EO, L)	Craveiro *et al.*, 1981a

Croton flavens L.

On the island of Curaçao and other areas of Central America, the roots are chewed for their "stimulating" effect, and the fresh young leaves and tips of T are used to prepare "bush tea", a popular beverage and folk remedy. To relieve oral inflammation, the leaves of this plant are held in the mouth; also they serve as an insect repellent and as a detergent (Weber and Hecker, 1978).

Alkaloids		
Amuronine, (-)	Colombia (L)	Charris *et al.*, 2000
Coreximine	Barbados (L)	Eisenreich *et al.*, 2003
Flavinantine	Barbados (L) Jamaica (BR)	Eisenreich *et al.*, 2003 Stuart and Graham, 1973
Salutarine	Barbados (L)	Eisenreich *et al.*, 2003
Salutaridine	Barbados (L) Jamaica (EP)	Eisenreich *et al.*, 2003 Farnsworth *et al.*, 1969
Scoulerine	Barbados (L)	Eisenreich *et al.*, 2003
Sebiferine	Barbados (L)	Eisenreich *et al.*, 2003
Sinoaucutine, nor	Barbados (L) Jamaica (BR)	Eisenreich *et al.*, 2003 Stuart and Graham, 1973
Choline	Curacão (L)	Morton, 1968
Crotosparine	Jamaica (EP)	Farnsworth *et al.*, 1969
Histamine	Curacão (L)	Morton, 1968
Sinoacutine	Jamaica (BR)	Stuart and Graham, 1973
Sparsiflorine	Jamaica (EP)	Farnsworth *et al.*, 1969
Diterpenes		
Phorbol, 16-hydroxy-12-hexadecanoate-13-acetate-20-decanoate	Curaçao (R)	Weber and Hecker, 1977
Phorbol-13-acetate, 16-hydroxy-12-myristate	Curaçao (R, L)	Hecker, 1984
Phorbol-13-acetate, 16-hydroxy-12-*O*-Palmitate	Curaçao (R, L)	Hecker, 1984
Phorbol-13-acetate, 16-hydroxy-12-stearate	Curaçao (R, L)	Hecker, 1984

Contd...

Table 7.1–*Contd...*

Compound Type Chemical Name	Place (Part Used)	References
Phorbol-13-acetate, 16-hydroxy-20-decanoate-12-palmitate	Curaçao (R, L)	Hecker, 1984
Phorbol, 4-deoxy-16-hydroxy-12-*O*-hexadecanoate-13-*O*-acetate	Curaçao (R)	Weber and Hecker, 1977
Oblongifoliol	Jamaica (EP)	Farnsworth *et al.*, 1969
Monoterpenes		
Pinene, α	Curaçao (EO, AP)	Woerdenbag *et al.*, 2000
Polycyclic		
Crotoflavol	Barbados (L) Curaçao (L)	Eisenreich *et al.*, 2003 Eisenreich and Bracher, 2001
Sesquiterpenes		
Caryophyllene, iso	Curaçao (EO, AP)	Woerdenbag *et al.*, 2000
Spathulenol	Curaçao (EO, AP)	Woerdenbag *et al.*, 2000

Croton geayi Leandri.

Is a shrub endemic to Madagascar where it is used in traditional medicine for various purposes, particularly as antimalarial (Pallazino *et al.*, 1997).

Diterpenes		
Geayine	Madagascar (W)	Palazzino *et al.*, 1997
Geayine, 7-deoxo	Madagascar (W)	Palazzino *et al.*, 1997
Geayinine	Madagascar (W)	Palazzino *et al.*, 1997
Geayinine,iso	Madagascar (W)	Palazzino *et al.*, 1997

Croton glabellus L.

Grows in the warm areas of Colombia. Its leaves are used as antihypotensive by the Indians and forest populations (Paredes *et al.*, 1985).

Flavonoids		
Ayanin	Colombia (L)	Garcia *et al.*, 1986
Quercitrin	Colombia (L)	Novoa *et al.*, 1979
3-*O*-Methylkaempferol	Mexico (L)	García *et al.*, 2006
5,7,3',4'-Tetrahydroxy 3-methoxy-flavonoid	Mexico (L)	García *et al.*, 2006
Diterpenes		
Austroinulin	Mexico (L)	García *et al.*, 2006
Cajucarinolide	Mexico (L)	García *et al.*, 2006
6-*O*-Acetylaustroinulin	Mexico (L)	García *et al.*, 2006
Marrubiagenin	Mexico (L)	García *et al.*, 2006
Dehydrocrotonin, *trans*	Mexico (L)	García *et al.*, 2006

Contd...

Table 7.1–*Contd...*

Compound Type Chemical Name	Place (Part Used)	References
Croton gossypifolius Vahl		
Alkaloid		
Crotogossamide	Trinidad (LX)	Quintyne-Walcott *et al.*, 2007
Flavonoids		
Kaempferol-3-*O* rhamnopyranoside	Trinidad (LX)	Quintyne-Walcott *et al.*, 2007
Myricitrin	Trinidad (LX)	Quintyne-Walcott *et al.*, 2007
Quercitrin	Trinidad (LX)	Quintyne-Walcott *et al.*, 2007
Croton gratissimus Burch.		

Despite the extensive traditional use of *Croton gratissimus* Burch. var. *gratissimus* (Euphorbiaceae) for medicinal purposes (Watt and Breyer-Brandwijk, 1962; Van Wyk *et al.*, 1997; Van Wyk and Gericke, 2000), scientific studies validating the therapeutic properties of this indigenous plant are lacking. One of the vernacular (Afrikaans) names for *Croton gratissimus* is "koorsbessie" ("koors" = fever) suggesting that the plant is used as a pyrogenic.

Alkaloid		
Nuciferine, pro: (+)	Transvaal (B)	Farnsworth *et al.*, 1969
Croton gubouga S. Moore		

Croton gubouga, S. Moore, is a cmall tree growing on the losw veldt in the Eastern Transvaal near the Sabi and Selati rivers. The bark of the tree has considerable local reputat.ion among the natives as a remedy for malaria, and both the seeds and the bark have been used by Captain Maberley in conjunction with opium in the treatment of malarial fever (*Lancet*, 1899).

Alkaloid		
Turumiquirensine	East Africa (R)	Farnsworth *et al.*, 1969
Diterpene		
Phorbol	East Africa (EO)	Farnsworth *et al.*, 1969
Monoterpene		
Angelic acid	East Africa (EO)	Farnsworth *et al.*, 1969
Croton haumanianus J Léonard		

It is a tropical shrub of which leaves and barks were used in folk medicine against gonorrhea, gastric diseases and also antihypertensive and antiepileptic drug (Tchissambou *et al.*, 1990).

Diterpenes		
Crotocorylifuran	Congo (TB)	Tchissambou *et al.*, 1990
Crotohaumanoxide	Congo (TB)	Tchissambou *et al.*, 1990
Triterpene		
Lupeol	Congo (TB)	Tchissambou *et al.*, 1990
Croton hemiargyreus Muell. Arg.		

Popularly known as "Marmeleiro" it is a small tree and its wood is used to make toothpicks. The barks are known to have antibiotic acitivity (Braga, 1976).

Alkaloids		
Corydine	Brazil (L, S)	Pereira *et al.*, 1999 Lin *et al.*, 2003

Contd...

Table 7.1–*Contd...*

Compound Type Chemical Name	Place (Part Used)	References
Corydine, iso	Brazil (L, S)	Pereira *et al.*, 1999 Lin *et al.*, 2003
Glaucine	Brazil (L, S)	Pereira *et al.*, 1999 Lin *et al.*, 2003 Amaral and Barnes, 1998b
Glaucine, oxo	Brazil (L, S)	Amaral and Barnes, 1998b
Hemiargyrine	Brazil (L, S)	Amaral and Barnes, 1998b Pereira *et al.*, 1999
Reticuline	Brazil (L, S)	Pereira *et al.*, 1999
Reticuline, 3,4-dehydro	Brazil (L, S)	Pereira *et al.*, 1999
Salutaridine	Brazil (L, S)	Amaral and Barnes, 1998b Lin *et al.*, 2003
Salutaridine, nor	Brazil (L, S)	Amaral and Barnes, 1998b
Corydine, nor	Brazil (L)	Lin *et al.*, 2003
Hemiargine A	Brazil (L)	Lin *et al.*, 2003
Hemiargine B	Brazil (L)	Lin *et al.*, 2003
Hemiargine C	Brazil (L)	Lin *et al.*, 2003
Hemiargine D	Brazil (L)	Lin *et al.*, 2003
Laudanosine,nor	Brazil (L)	Lin *et al.*, 2003
Palmatrubine, tetrahydro	Brazil (L)	Lin *et al.*, 2003
Xylopinine	Brazil (L)	Lin *et al.*, 2003
Diterpenes		
Cromiargyne	Brazil (B)	Amaral and Barnes, 1998a
Cromiargyne, 7-acetoxy	Brazil (B)	Amaral and Barnes, 1998a
***Croton hieronymi* Griseb.**		
Aliphatics		
Eicosan-1-ol	Argentina (AP)	Catalan *et al.*, 2003
Alkaloid		
Aurantiamide acetate	Argentina (AP)	Catalan *et al.*, 2003
Phenylalaninol, *N*-benzoyl	Argentina (AP)	Catalan *et al.*, 2003
Phenylalaninyl, *N*-benzoyl-phenylalanianate, *N*-benzoyl	Argentina (AP)	Catalan *et al.*, 2003
Benzenoid		
Acetophenone, dihydroxy-methoxy-acetoxy	Argentina (AP)	Catalan *et al.*, 2003
Xanthoxylin	Argentina (AP)	Catalan *et al.*, 2003
Diterpene		
Phytol	Argentina (AP)	Catalan *et al.*, 2003

Contd...

Table 7.1–*Contd...*

Compound Type Chemical Name	Place (Part Used)	References
Monoterpene		
Cyclohept-2-en-1-one, 5-hydroxy-3,7,7-trimethyl	Argentina (AP)	Catalan *et al.,* 2003
Sesquiterpenes		
Cadinol, T	Argentina (AP)	Catalan *et al.,* 2003
Cubenol, epi	Argentina (AP)	Catalan *et al.,* 2003
Sesterpene		
Eicos-2-1-ol,*trans*- 2,3,7,11,15,19-pentamethyl	Argentina (AP)	Catalan *et al.,* 2003
Steroids		
Campesterol	Argentina (AP)	Catalan *et al.,* 2003
Cholest-4-en-3-one	Argentina (AP)	Catalan *et al.,* 2003
Cholest-8(14)-en-3β-ol	Argentina (AP)	Catalan *et al.,* 2003
Cholesterol	Argentina (AP)	Catalan *et al.,* 2003
Ergosta-4,22-dien-3-one	Argentina (AP)	Catalan *et al.,* 2003
Fucosterol, isso	Argentina (AP)	Catalan *et al.,* 2003
Sitostenone, β	Argentina (AP)	Catalan *et al.,* 2003
Sitosterol, β	Argentina (AP)	Catalan *et al.,* 2003
Stigmast-4-en-3-one	Argentina (AP)	Catalan *et al.,* 2003
Stigmasterol	Argentina (AP)	Catalan *et al.,* 2003
Triterpene		
All-trans-2,6,15,19,23-pentamethyl-tetra-cosa-2,6,10 (28),14,22,28-hexaene-11-ol	Argentina (AP)	Catalan *et al.,* 2003
All-trans-10-methylene-2,6,10,14,18,22-pentamethyl-tetracosa-1,6,10,14,18,22-hexaen-3-ol	Argentina (AP)	Catalan *et al.,* 2003
Amyrin, α	Argentina (AP)	Catalan *et al.,* 2003
Amyrin, β	Argentina (AP)	Catalan *et al.,* 2003
Gramisterol	Argentina (AP)	Catalan *et al.,* 2003
Lophenol	Argentina (AP)	Catalan *et al.,* 2003
Lupeol	Argentina (AP)	Catalan *et al.,* 2003
Mortenol	Argentina (AP)	Catalan *et al.,* 2003

Croton hovarum Leandri

It is a toxic tree, endemic to Madagascar (Krebs and Ramiarantsoa, 1997).

Diterpenes

Cleroda-13(16),14-dien, 9-al-3α,4β-dihydroxy-15,16, epoxy-12-oxo	Madagascar (SB)	Krebs and Ramiarantsoa, 1996
Cleroda-13(16),14-dien, 3α,4β-dihydroxy-15,16, epoxy-12-oxo	Madagascar (SB)	Krebs and Ramiarantsoa, 1996

Contd...

Table 7.1–*Contd...*

Compound Type Chemical Name	Place (Part Used)	References
Cleroda-13(16),14-dien-9-al,3,12-dioxo-15, 16-epoxy	Madagascar (L)	Krebs and Ramiarantsoa, 1997
Cleroda-5(10),13(16),14-triene,19-nor-3α, 4β-dihydroxy-15,16-epoxy-12-oxo	Madagascar (L)	Krebs and Ramiarantsoa, 1997
Flavonoid		
Vitexin	Madagascar (L)	Krebs and Ramiarantsoa, 1997
Proteid		
Hygric acid, 4-hydroxy	Madagascar (SB)	Krebs and Ramiarantsoa, 1996
Triterpenes		
Amyrin, β	Madagascar (SB)	Krebs and Ramiarantsoa, 1996
Friedelin	Madagascar (SB)	Krebs and Ramiarantsoa, 1996
Friedoolean-14-en-28-oic acid, 3-β-acetoxy	Madagascar (SB)	Krebs and Ramiarantsoa, 1996

Croton humilis L.

In Jamaica it is used as an insecticide to kill bed bugs, in Brazil it is used for the treatment for erysipelas (Kutney *et al.,* 1971a).

Aliphatics		
Butyric acid	Brazil (EO)	Freise, 1935
Valeric acid	Brazil (EO)	Freise, 1935
Alkaloids		
Glutarimide, 2-[N-(2-methyl-propanoyl]-N-phenylethyl	Jamaica (EP, L, T)	Kutney *et al.,* 1971b
Glutarimide, 2-[N-(2-methylbutanoyl]-N-phenylethyl	Jamaica (EP)	Kutney *et al.,* 1971b
Glutarimide, 2-[N-(2R-methyl-butanoyl]-N-phenylethyl	Jamaica (L, T)	Stuart *et al.,* 1973
Glutarimide, 2-[N-(2R-methyl-propanoyl]-N-phenylethyl	Jamaica (L, T)	Stuart *et al.,* 1973
Phenylethylamine, 2:N-[N-(2-methyl-propanoyl]-L-glutaminoyl	Jamaica (L, T)	Stuart *et al.,* 1973
Phenylethylamine, 2:N-[N-(2-methyl-butanoyl]-glutaminoyl	Jamaica (EP)	Kutney *et al.,* 1971a
Phenylethylamine, 2:N-[N-2-methyl-propionoyl]-glutaminoyl	Jamaica (EP)	Kutney *et al.,* 1971a
Phenylethylamine, 2: N-(N-2(R)-methyl-butanoyl)-L-glutaminoyl	Jamaica (L, T)	Stuart *et al.,* 1973

Contd...

Table 7.1–*Contd...*

Compound Type Chemical Name	Place (Part Used)	References
Monoterpene		
Angelic acid	Brazil (EO)	Freise, 1935

Croton hutchinsonianus Hosseus

Croton hutchinsonianus HOSSEUS belongs to the family Euphorbiaceae. It is a shrub or small tree reaching 4-5 meters high and native to Thailand (Shaw, 1972), where it is commonly called "Plao phae".

Diterpene		
Plaunol A	Thailand (S)	Ogiso *et al.*, 1981

Croton insularis Baill.

Croton insularis BAILLON is a small tree widespread in New Caledonia, growing to *ca.* 15 m, and characterized by the silvery-white color of its branchlets and under-leaf surface (McPherson and Tirel, 1987).

Diterpenes		
Crotinsularin	New Caledonia (L)	Graikou *et al.*, 2004
Trachyloban-3-one: *ent*	New Caledonia (L)	Graikou *et al.*, 2004
Trachyloban-19-oic acid: *ent*	New Caledonia (L)	Graikou *et al.*, 2004
Trachylobane, 3α,19-dihydroxy: *ent*	New Caledonia (L)	Graikou *et al.*, 2004
Trachyloban-3β-ol: *ent*	New Caledonia (L)	Graikou *et al.*, 2004
Cleroda-13(16), 14-diene	New Caledonia (L)	Graikou *et al.*, 2004
Manoyl oxide, 13-epi: *ent*	New Caledonia (L)	Graikou *et al.*, 2004
Trachylobane, ent: 19-acetoxy	New Caledonia (L)	Graikou *et al.*, 2004
Flavonoids		
Catechin, (+)	New Caledonia (L)	Graikou *et al.*, 2004
Catechin, epi: (-)	New Caledonia (L)	Graikou *et al.*, 2004
Sesquiterpene		
Vomifoliol-3'-*O*-β-D-glucopyranoside	New Caledonia (L)	Graikou *et al.*, 2004
Triterpene		
Friedoolean-14-en-28-oic acid, 3β-acetoxy	New Caledonia (L)	Graikou *et al.*, 2004

Croton jacobinensis Baill.

Grows in Northeastern Brazil in the "Agreste" and "Sertão" areas (Craveiro *et al.*, 1981a). It is said to possess antibiotic activity against Gram positive and Gram negative bacteria (Lima *et al.*, 1973).

Monoterpene		
Terpinolene, α	Brazil (EO, L)	Craveiro *et al.*, 1981a
Sesquiterpenes		
Aromadendrene	Brazil (EO, L)	Craveiro *et al.*, 1981a
Caryophyllene, β	Brazil (EO, L)	Craveiro *et al.*, 1981a
Elemene, β	Brazil (EO, L)	Craveiro *et al.*, 1981a

Contd...

Table 7.1–*Contd...*

Compound Type Chemical Name	Place (Part Used)	References
Elemene, δ	Brazil (EO, L)	Craveiro *et al.*, 1981a
Elemene, γ	Brazil (EO, L)	Craveiro *et al.*, 1981a
Farnesene, α	Brazil (EO, L)	Craveiro *et al.*, 1981a
Farnesene, β	Brazil (EO, L)	Craveiro *et al.*, 1981a
Humulene, α	Brazil (EO, L)	Craveiro *et al.*, 1981a

Croton jatrophoides Pax

The native people of East Africa use the roots as a remedy for colds and stomach ache (Kubo *et al.*, 1990).

Triterpenes

Dumsin	Kenya (RB)	Kubo *et al.*, 1990 Nihei *et al.*, 2002
Zumisin	Kenya (RB)	Nihei *et al.*, 2005
Zumketol	Kenya (RB)	Nihei *et al.*, 2005
Zumsenin	Kenya (RB)	Nihei *et al.*, 2005
Zumsenol	Kenya (RB)	Nihei *et al.*, 2005

Limonoids

Musidunin	East African (RB)	Nihei *et al.*, 2006
Musiduol	East African (RB)	Nihei *et al.*, 2006

Croton jimenezii Standl et. (&) Valerio

Aliphatics

Octanone, 3	Costa Rica (L)	Ciccio and Segnini, 2002
Undecanone, 2	Costa Rica (L)	Ciccio and Segnini, 2002
Decanal	Costa Rica (L)	Ciccio and Segnini, 2002
Methyl hexadecanoate	Costa Rica (L)	Ciccio and Segnini, 2002
Undecanal	Costa Rica (L)	Ciccio and Segnini, 2002
Dodecanal	Costa Rica (L)	Ciccio and Segnini, 2002
Methyl salicylate	Costa Rica (L)	Ciccio and Segnini, 2002
Undecanol	Costa Rica (L)	Ciccio and Segnini, 2002
Dodecanol	Costa Rica (L)	Ciccio and Segnini, 2002
Nonanol	Costa Rica (L)	Ciccio and Segnini, 2002
Tetradecanal	Costa Rica (L)	Ciccio and Segnini, 2002
Octanol, 3	Costa Rica (L)	Ciccio and Segnini, 2002
Tridecane	Costa Rica (L)	Ciccio and Segnini, 2002
1-Octen-3-yl acetate	Costa Rica (L)	Ciccio and Segnini, 2002
Hexyl butyrate	Costa Rica (L)	Ciccio and Segnini, 2002
Pentadecanal	Costa Rica (L)	Ciccio and Segnini, 2002

Contd...

Table 7.1–*Contd...*

Compound Type Chemical Name	Place (Part Used)	References
1-Octen-3-ol	Costa Rica (L)	Ciccio and Segnini, 2002
Nonanal	Costa Rica (L)	Ciccio and Segnini, 2002
Ocatanol	Costa Rica (L)	Ciccio and Segnini, 2002
Diterpenes		
Hardwickiic acid, (-)	Costa Rica (NS)	Murillo and Jakupovic, 2000
Strictic acid	Costa Rica (NS)	Murillo and Jakupovic, 2000
1,3 Cyclodecadiene-1-carboxylic acid	Costa Rica (NS)	Murillo and Jakupovic, 2000
Phytol	Costa Rica (L)	Ciccio and Segnini, 2002
Phytol acetate	Costa Rica (L)	Ciccio and Segnini, 2002
Monoterpenes		
Camphor	Costa Rica (L)	Ciccio and Segnini, 2002
Bornyl acetate	Costa Rica (L)	Ciccio and Segnini, 2002
Linalool	Costa Rica (L)	Ciccio and Segnini, 2002
Camphene	Costa Rica (L)	Ciccio and Segnini, 2002
Pinene, α	Costa Rica (L)	Ciccio and Segnini, 2002
Terpineol, α	Costa Rica (L)	Ciccio and Segnini, 2002
Phellandrene, α	Costa Rica (L)	Ciccio and Segnini, 2002
Terpinene, γ	Costa Rica (L)	Ciccio and Segnini, 2002
Terpinene, α	Costa Rica (L)	Ciccio and Segnini, 2002
Cymene, *p-*	Costa Rica (L)	Ciccio and Segnini, 2002
Geraniol	Costa Rica (L)	Ciccio and Segnini, 2002
Myrcene	Costa Rica (L)	Ciccio and Segnini, 2002
Pinene, β	Costa Rica (L)	Ciccio and Segnini, 2002
Limonene	Costa Rica (L)	Ciccio and Segnini, 2002
Cineole, 1,8	Costa Rica (L)	Ciccio and Segnini, 2002
Borneol	Costa Rica (L)	Ciccio and Segnini, 2002
Tricyclene	Costa Rica (L)	Ciccio and Segnini, 2002
Phellandrene, β	Costa Rica (L)	Ciccio and Segnini, 2002
Terpinen-4-ol	Costa Rica (L)	Ciccio and Segnini, 2002
Terpinolene	Costa Rica (L)	Ciccio and Segnini, 2002
Thujene, α	Costa Rica (L)	Ciccio and Segnini, 2002
(Z)-β-Ocimene	Costa Rica (L)	Ciccio and Segnini, 2002
Sabinene	Costa Rica (L)	Ciccio and Segnini, 2002
(E)-β-Ocimene	Costa Rica (L)	Ciccio and Segnini, 2002

Contd...

Table 7.1–*Contd...*

Compound Type Chemical Name	Place (Part Used)	References
Phenylpropanoids		
Hexadecanoic acid	Costa Rica (L)	Ciccio and Segnini, 2002
Linoleic acid	Costa Rica (L)	Ciccio and Segnini, 2002
Methyl eugenol	Costa Rica (L)	Ciccio and Segnini, 2002
Eugenol	Costa Rica (L)	Ciccio and Segnini, 2002
Octanoic acid	Costa Rica (L)	Ciccio and Segnini, 2002
Methyl chavicol	Costa Rica (L)	Ciccio and Segnini, 2002
Dodecanoic acid	Costa Rica (L)	Ciccio and Segnini, 2002
Methyl linolenate	Costa Rica (L)	Ciccio and Segnini, 2002
Decanoic acid	Costa Rica (L)	Ciccio and Segnini, 2002
Tetradecanoic acid	Costa Rica (L)	Ciccio and Segnini, 2002
Methylisoeugeno, I(E)	Costa Rica (L)	Ciccio and Segnini, 2002
3-Hexanyl-benzoate, (Z)	Costa Rica (L)	Ciccio and Segnini, 2002
Sesquiterpenes		
Caryophyllene, β	Costa Rica (L)	Ciccio and Segnini, 2002
Cadinol, α	Costa Rica (L)	Ciccio and Segnini, 2002
Cadinene, δ	Costa Rica (L)	Ciccio and Segnini, 2002
Viridiflorol	Costa Rica (L)	Ciccio and Segnini, 2002
Caryophyllene oxide	Costa Rica (L)	Ciccio and Segnini, 2002
Cyperene	Costa Rica (L)	Ciccio and Segnini, 2002
Copaene, α	Costa Rica (L)	Ciccio and Segnini, 2002
Bourbonene, β	Costa Rica (L)	Ciccio and Segnini, 2002
Spathulenol	Costa Rica (L)	Ciccio and Segnini, 2002
Humulene, α	Costa Rica (L)	Ciccio and Segnini, 2002
Muurolene, α	Costa Rica (L)	Ciccio and Segnini, 2002
Cubebene, β	Costa Rica (L)	Ciccio and Segnini, 2002
Germacrene B	Costa Rica (L)	Ciccio and Segnini, 2002
Cubebene, α	Costa Rica (L)	Ciccio and Segnini, 2002
Humulene epoxide II	Costa Rica (L)	Ciccio and Segnini, 2002
Cubebol	Costa Rica (L)	Ciccio and Segnini, 2002
Germacrene D	Costa Rica (L)	Ciccio and Segnini, 2002
Cadinene, α	Costa Rica (L)	Ciccio and Segnini, 2002
Bicyclogermacrene	Costa Rica (L)	Ciccio and Segnini, 2002
Alloaromadendrene	Costa Rica (L)	Ciccio and Segnini, 2002
Germacrene A	Costa Rica (L)	Ciccio and Segnini, 2002
Cadina 1,4 diene	Costa Rica (L)	Ciccio and Segnini, 2002

Contd...

Table 7.1–*Contd...*

Compound Type Chemical Name	Place (Part Used)	References
Elemene, β	Costa Rica (L)	Ciccio and Segnini, 2002
Epi-cubebol	Costa Rica (L)	Ciccio and Segnini, 2002
Cadinene, γ	Costa Rica (L)	Ciccio and Segnini, 2002
Chrysanthenol, *Cis*	Costa Rica (L)	Ciccio and Segnini, 2002
Chrysanthenyl acetate, *Cis*	Costa Rica (L)	Ciccio and Segnini, 2002
Gurjunene, β	Costa Rica (L)	Ciccio and Segnini, 2002

Croton joufra Roxb

In traditional medicine of Thailand it is used for blood purification, and in Brazil it is used as an antipyretic (Mokkhasmit *et al.*, 1971a; Mokkhasmit *et al.*, 1971b).

Diterpenes

Labda-8(17),12(13),14(15)-triene,2α, 3α-dihydroxy	Thailand (L)	Sutthivaiyakit *et al.*, 2001
Pimara-8(9),15-dien-7-one, 19-*O*-acetyl- 3β-hydroxy	Thailand (L)	Sutthivaiyakit *et al.*, 2001
Plaunol A	Thailand (S)	Ogiso *et al.*, 1981
Plaunol C	Thailand (S)	Ogiso *et al.*, 1981
Swassin	Thailand (S)	Roengsumran *et al.*, 1982

Croton kerrii A. Shaw

Vernacular name in Thailand is Plao (Sato *et al.*, 1980).

Diterpenes

Hexadeca-2,6,1,1,14-tetraen-1-ol, [*E,E,Z*]-11-hydroxymethyl-3,7,15-trimethyl	Thailand (L)	Sata *et al.*, 1980
Hexadeca-2,6,10,14-tetraen-1-ol, [*E,E,E*]- 11-formyl-3,7,15-trimethyl	Thailand (L)	Sata *et al.*, 1980

Croton lacciferus Linn.

It is a common plant distributed in Sri Lanka and South India it is used in the treatment of fever, colds, dysentery, skin diseases, and lung diseases including tuberculosis (Bandara and Wimalasiri, 1988; Bandara *et al.*, 1988).

Diterpenes

Kaur-15-en-17-hydroxy-3b-yl acetate: *ent*	Sri Lanka (R)	Bandara and Wimalasiri, 1988
Kaur-15-en-3b,17-diol, *ent*	Sri Lanka (R)	Bandara *et al.*, 1988
Kaur-15-en-17-ol, *ent*	Sri Lanka (R)	Bandara and Wimalasiri, 1988
Kauran-17-oic acid,16a-(H): *ent*	Sri Lanka (R)	Bandara *et al.*, 1988
Kauran-17-ol, 15b-16-epoxy: *ent*	Sri Lanka (R)	Bandara *et al.*, 1988
Kauran-3b,16b,17-triol: *ent*	Sri Lanka (R)	Bandara and Wimalasiri, 1988
Kauran-16β-17-diol	Sri Lanka (R)	Bandara and Wimalasiri, 1988
Kauran-3β-16-β-17-triol, *ent*	Sri Lanka (R)	Bandara and Wimalasiri, 1988

Contd...

Table 7.1–*Contd...*

Compound Type Chemical Name	Place (Part Used)	References
Quinone		
Benzoquinone, 2,6-dimethoxy	Sri Lanka (R)	Bandara and Wimalasiri, 1988
Triterpenes		
Friedoolean-14-en-28oic acid, D: 3-β-acetoxy	Sri Lanka (R)	Bandara *et al.*, 1988
Oleanolic acid	Sri Lanka (R)	Bandara *et al.*, 1988

Croton lanjouwensis Jablonski

Grows in the Brazilian Amazon, where it is popularly known as "dima" and "jima" (Leão *et al.*, 1998).

Monoterpenes		
Borneol	Brazil (SE)	Leão *et al.*, 1998
Bornyl acetate	Brazil (F)	Leão *et al.*, 1998
Camphene	Brazil (B)	Leão *et al.*, 1998
Cineole, 1,8	Brazil (F, FL)	Leão *et al.*, 1998
Cymene, *para*	Brazil (L, F)	Leão *et al.*, 1998
Linalool	Brazil (L, F, FL)	Leão *et al.*, 1998
Myrcene, β-oxide	Brazil (F)	Leão *et al.*, 1998
Ocymene, β	Brazil (F, FL)	Leão *et al.*, 1998
Phellandrene, α	Brazil (L)	Leão *et al.*, 1998
Pinene, α	Brazil (L, F, FL, B)	Leão *et al.*, 1998
Pinene, β	Brazil (L, F, FL)	Leão *et al.*, 1998
Sabinene	Brazil (F, FL)	Leão *et al.*, 1998
Terpinene, α	Brazil (L)	Leão *et al.*, 1998
Terpinene, γ	Brazil (L, F)	Leão *et al.*, 1998
Terpinolene	Brazil (L)	Leão *et al.*, 1998
Thujene, α	Brazil (L, F, FL)	Leão *et al.*, 1998
Phenylpropanoid		
Elemicin	Brazil (L)	Leão *et al.*, 1998
Sesquiterpenes		
Cadinene, γ	Brazil (B)	Leão *et al.*, 1998
Caryophyllene, β	Brazil (L, F, FL)	Leão *et al.*, 1998
Caryophyllene, oxide	Brazil (L, F, FL)	Leão *et al.*, 1998
Cubebene, α	Brazil (F, B)	Leão *et al.*, 1998
Cubebene, β	Brazil (L, FL)	Leão *et al.*, 1998
Elemene, β	Brazil (L, F)	Leão *et al.*, 1998
Elemene, γ	Brazil (F)	Leão *et al.*, 1998
Farnesol (2Z,1E)	Brazil (F)	Leão *et al.*, 1998

Contd...

Table 7.1–*Contd...*

Compound Type Chemical Name	Place (Part Used)	References
Germacrene A	Brazil (L, F, FL)	Leão *et al.*, 1998
Germacrene B	Brazil (L, F, FL)	Leão *et al.*, 1998
Germacrene bicycle	Brazil (B)	Leão *et al.*, 1998
Guaiol	Brazil (FL)	Leão *et al.*, 1998
Gurgujene, γ	Brazil (FL)	Leão *et al.*, 1998
Humulene, α	Brazil (L, F, FL)	Leão *et al.*, 1998
Selinene, β	Brazil (FL)	Leão *et al.*, 1998
Sesquiphellandrene, epi, bicycle	Brazil (L, F)	Leão *et al.*, 1998
Seychelene	Brazil (FL)	Leão *et al.*, 1998
Ylangene, α	Brazil (B)	Leão *et al.*, 1998

Croton lechleri L. (Müll. Arg.)

This species, together with *C. draco*, *C.draconoides*, *C. salutaris* and *C. erythrochylus*, are known as "sangue de dragão" and "sangre de drago", due to blood-red latex that exudates from the plants (Pieters *et al.*, 1993). In folk medicine it is used by Peruvian Indians for a few different diseases including rheumatism (Vaisberg *et al.*, 1989; Cai *et al.*, 1991), healing of wounds (Persinos, 1972), against cancer (Cai *et al.*, 1991; Chen *et al.*, 1994).

Alkaloids

Boldine, iso	Peru (L)	Milanowski *et al.*, 2002
Glaucine	Peru (L)	Milanowski *et al.*, 2002
Magnoflorine	Peru (L)	Milanowski *et al.*, 2002
Sinoacutine	Peru (L)	Carlin *et al.*, 1996
Taspine	Peru (LX, SA, L) Ecuador (B,SA,LX) Italy (LX)	Persinos, 1972 Cai *et al.*, 1993a Pieters *et al.*, 1993 Pieters *et al.*, 1995 Vaisberg *et al.*, 1989 Milanowski *et al.*, 2002 Risco *et al.*, 2003 Persinos *et al.*, 1979 Persinos *et al.*, 1974 Marino *et al*, 2008
Thalyporphine	Ecuador (L)	Milanowski *et al.*, 2002
Boldine, iso: nor	Peru (L)	Milanowski *et al.*, 2002

Benzenoids

Benzene, 1,3,5-trimethoxy	Ecuador (B, SA)	Cai *et al.*, 1993a Chen *et al.*, 1994
Benzyl alcohol, 3,4-dimethoxy	Ecuador (B, SA)	Cai *et al.*, 1993a
Phenethyl alcohol, 4-hydroxy	Ecuador (B, SA)	Cai *et al.*, 1993a Chen *et al.*, 1994
Phenol, 3,4-dimethoxy	Ecuador (B, SA)	Cai *et al.*, 1993a

Contd...

Table 7.1–*Contd...*

Compound Type Chemical Name	Place (Part Used)	References
Phenol, 2,4,6-trimethoxy	Ecuador (SA, B)	Cai *et al.*, 1993a Chen *et al.*, 1994
Phenethyl alcohol, 4-hydroxy: acetate	Ecuador (B, SA)	Cai *et al.*, 1993a
Diterpenes		
Bincatriol	Ecuador (B, SA)	Cai *et al.*, 1993a Chen *et al.*, 1994
Crolechinic acid	Ecuador (B, SA)	Cai *et al.*, 1993a Chen *et al.*, 1994
Crolechinol	Ecuador (B, SA)	Chen *et al.*, 1994 Cai *et al.*, 1993a
Hardwickiic acid	Ecuador (B, SA)	Cai *et al.*, 1993a Chen *et al.*, 1994
Korberin A	Ecuador (B, SA)	Cai *et al.*, 1993b Chen *et al.*, 1994
Korberin B	Ecuador (B, SA)	Cai *et al.*, 1993b Chen *et al.*, 1994
Floribundic acid glucoside	Italy (LX)	Marino *et al.*, 2008
Flavonoids		
Catechin, (+)	Ecuador (LX, SA) Italy (LX)	Cai *et al.*, 1991 Chen *et al.*, 1994 Risco *et al.*, 2003 Marino *et al.*, 2008
Catechin, epi (-)	Ecuador (LX SA) Italy (LX)	Cai *et al.*, 1991 Chen *et al.*, 1994 Risco *et al.*, 2003 Marino *et al.*, 2008
Catechin (4α-8)-gallocatechin (4α-6)-gallocatechin	Ecuador (LX)	Cai *et al.*, 1991
Catechin (4α-8)-gallocatechin (4α-8)-gallocatechin, (+)	Ecuador (LX)	Cai *et al.*, 1991
Gallocatechin (4-α-8)-gallocatechin (4-α-8)-epi-gallocatechin	Ecuador (LX)	Cai *et al.*, 1991
Gallocatechin, (+)	Ecuador (LX, SA) Italy (LX)	Cai *et al.*, 1991 Chen *et al.*, 1994 Marino *et al.*, 2008
Gallocatechin, epi (-)	Ecuador (LX, SA) Italy (LX)	Cai *et al.*, 1991 Chen *et al.*, 1994 Risco *et al.*, 2003 Marino *et al.*, 2008
Gallocatechin-(4-a-6)-epi-gallocatechin	Ecuador (LX)	Cai *et al.*, 1991
Gallocatechin-(4-a-8)-epi-catechin	Ecuador (LX)	Cai *et al.*, 1991
Procyanidin, B-1	Ecuador (LX)	Cai *et al.*, 1991 Risco *et al.*, 2003

Contd...

Table 7.1—*Contd...*

Compound Type Chemical Name	Place (Part Used)	References
Procyanidin, B-4	Ecuador (LX, SA) Chen *et al.*, 1994	Cai *et al.*, 1991
SP-303	South America (LX) Peru (LX)	Ubillas *et al.*, 1994 Jones, 2003 Fischer *et al.*, 2004
Procyanidin B-2	Ecuador (LX)	Risco *et al.*, 2003
SB-300	Peru (LX)	Fischer *et al.*, 2004
Lignans		
Cedrusin, 3',4-*O*-dimethyl	Peru (SA) Italy (LX)	Pieters *et al.*, 1992 Pieters *et al.*, 1993 Pieters *et al.*, 1995 Marino *et al.*, 2008
Cedrusin, 4-*O*-methyl	Peru (SA)	Pieters *et al.*, 1993
Benzofuran-5-yl, 2,3-dihydro: 2-(3,4-dimethoxy-phenyl): 7-methoxy-3-methoxy-carbonyl-propan-1-oic acid methyl ester	Brazil (SA)	Pieters *et al.*, 1992
Benzofuran-5-yl, 2,3-dihydro: 2-(3,4-dimethoxy-phenyl): 7-methoxy-3-methoxy-carbonyl-propen-1-oic acid methyl ester	Brazil (SA)	Pieters *et al.*, 1992
Benzofuran-5-yl, 2,3-dihydro: 2-(4-hydroxy-3-methoxy-phenyl): 7-methoxy-3-methoxy-carbonyl-propen-1-oic acid methyl ester	Brazil (SA)	Pieters *et al.*, 1992
(±) *erythro*-Guaiacyl-glycerol-β-*O*-4'-dihydroconiferyl ether	Italy (LX)	Marino *et al.*, 2008
2-[4-(3-Hydroxypropyl)-2-methoxyphenoxy]-propane-1,3-diol	Italy (LX)	Marino *et al.*, 2008
Megastigmanes		
Blumenol B	Italy (LX)	Marino *et al.*, 2008
Blumenol C	Italy (LX)	Marino *et al.*, 2008
4,5-Dihydroblumenol A	Italy (LX)	Marino *et al.*, 2008
Steroids		
Daucosterol	Ecuador (B, SA)	Cai *et al.*, 1993a Chen *et al.*, 1994
Sitostenone, β	Ecuador (B, SA)	Cai *et al.*, 1993a
Sitosterol, β	Ecuador (B, SA)	Cai *et al.*, 1993a Chen *et al.*, 1994

Croton levatii Guill.

The trunk bark of *C. levatii* is reported to be used as perfume and as an aphrodisiac for dancers on the Ifate Island (Vanuatu) (Moulis *et al.*, 1992b).

Diterpenes

Crovatin	Ifate Island (SB)	Moulis *et al.*, 1992a
Levatin	Vanuatu (SB)	Moulis *et al.*, 1992b

Contd...

Table 7.1–*Contd...*

Compound Type Chemical Name	Place (Part Used)	References

Croton linearis Jacq

Used in traditional medicine of the Bahamas as a tea of the leaves prepared for easing menstrual pains and to eliminate worms (Elbridge, 1975).

Alkaloids

Crotonoside	Jamaica (EP)	Farnsworth *et al.*, 1969
Crotonosine	Jamaica (EP, BR)	Farnsworth *et al.*, 1969 Stuart and Graham, 1973
Hernovine	Jamaica (EP)	Farnsworth *et al.*, 1969
Hernovine, 10-O-methyl	Jamaica (EP)	Farnsworth *et al.*, 1969
Hernovine, *N*-methyl	Jamaica (EP)	Farnsworth *et al.*, 1969
Hernovine, *N*-methyl: 10-O-methyl	Jamaica (EP)	Farnsworth *et al.*, 1969
Jacularine	Jamaica (EP)	Farnsworth *et al.*, 1969
Linearisine	Jamaica (EP, BR)	Farnsworth *et al.*, 1969 Stuart and Graham, 1973
Linearisine, homo	Jamaica (EP)	Farnsworth *et al.*, 1969
Nuciferine, pro: (+)	Jamaica (EP)	Farnsworth *et al.*, 1969
Salutaridine, 8,14-dihydro	Jamaica (EP)	Farnsworth *et al.*, 1969
Salutaridine, nor: 8,14-dihydro	Jamaica (EP)	Farnsworth *et al.*, 1969
Wilsonirine	Jamaica (EP)	Farnsworth *et al.*, 1969

Diterpene

Croton diterpene 1	Jamaica (T, L)	Alexander *et al.*, 1991

Monoterpene

Angelic acid	Jamaica (EP)	Farnsworth *et al.*, 1969

Sesquiterpene

Vomifoliol	Jamaica (S, L)	Stuart *et al.*, 1976

Croton lobatus L.

This plant is used in Ivory Coast to sterilize females (Bouquet and Debrary, 1974). A medicinal plant used in western Africa in traditional folk medicine to cure malaria, pregnancy troubles, and dysentery (Attioua *et al.*, 2007).

Sesquiterpene

Vomifoliol	Guyana (S, L)	Stuart and Woo-Ming, 1975

Aliphatics

(*Z,Z,Z*)-9,12,15-octadecatrienoic acid methyl ester	Abidjan (S, L)	Chabert *et al.*, 2006
(*Z,Z*)-9,12-octadecadienoic acid (2E,6E,10E)-3,7,11,15-tetramethyl-2,6,10,14-hexadeca-tetraenyl ester	Abidjan (S, L)	Chabert *et al.*, 2006
Geranylgeraniol	Abidjan (S, L)	Chabert *et al.*, 2006

Contd...

Table 7.1–*Contd...*

Compound Type Chemical Name	Place (Part Used)	References
Lipids		
Lobaceride	Abidjan (S, L)	Chabert *et al.*, 2006
Polyciclics		
(*E*)-3-(4-Methoxy-phenyl)-2-phenyl-acrylic acid	Abidjan (S, L)	Chabert *et al.*, 2006
N-(2-Hydroxy-1-phenyl-propyl)-benzamide	Abidjan (S, L)	Chabert *et al.*, 2006
Steroids		
Cholestan-5,7-dien-3-ol	Abidjan (S, L)	Chabert *et al.*, 2006
Ergosterol	Abidjan (S, L)	Chabert *et al.*, 2006
3-Hydroxy-cholest-5-en-7-one	Abidjan (S, L)	Chabert *et al.*, 2006
Cholestan-3one	Abidjan (S, L)	Chabert *et al.*, 2006
Triterpenes		
Betulinic acid	Abidjan (S, L)	Chabert *et al.*, 2006

***Croton lucidus* L.**

Diterpene		
Crotonin	Jamaica (L)	Chan *et al.*, 1968

***Croton lutzelburguii* Pax. and Hoffm**

Benzoyl-Methylpolyols		
1-Benzoyloxy-4-para-hydroxy-benzoyloxy-2-3-5-trihydroxy-5-methylhexane	Brazil (L)	Barbosa *et al.*, 2004
1,5-Dibenzoyloxy-4-para-hydroxy-benzoyloxy-5-methylhex-2-ene	Brazil (L)	Barbosa *et al.*, 2004
3-Benzoyloxy-4-para-hydroxy-benzoyloxy-2-ethoxy-1-hydroxy-5-methylhexane	Brazil (L)	Barbosa *et al.*, 2004

***Croton lyratus* S.**

Diterpene		
Plaunotol	Japan (CT)	Morimoto, 1988

***Croton macrostachys* A. Rich.**

Occurs in the African Continent. It has a wide use in folk medicine to treat malaria (Kasa, 1991; Kokwaro, 1976), against parasites, hemorrhage on the delivery, dysentery (Kloos *et al.*, 1978; Kloos *et al.*, 1997), stomachic (Abede, 1986), sterilizer (Klauss and Adala, 1994), against conjunctivitis, venereal diseases and also as abortive (Klauss and Adala, 1994; Mazzanti *et al.*, 1987; Wilson and Marian, 1979).

Alicyclic		
Crotepoxide	Kenya (F) Ethiopia (F)	Addae-Mensah *et al.*, 1992a Kupchan *et al.*, 1968 Kupchan *et al.*, 1969
Diterpenes		
Cleroda-5,10-diene-19,6β,20,12-diolide-neo	Tanzania (R)	Kapingu *et al.*, 2000

Contd...

Table 7.1—*Contd...*

Compound Type Chemical Name	Place (Part Used)	References
Trachyloban-18-oic, acid	Tanzania (R)	Kapingu *et al.*, 2000
Trachyloban-19-oic, acid	Tanzania (R)	Kapingu *et al.*, 2000
Trachylobane,3α,18,19-trihydroxy	Tanzania (R)	Kapingu *et al.*, 2000
Trachylobane, 3α, 19-dihydroxy	Tanzania (R)	Kapingu *et al.*, 2000
Triterpenes		
Betulin	Kenya (SB)	Addae-Mensah *et al.*, 1992a
Lupeol	Kenya (SB)	Addae-Mensah *et al.*, 1992a
Taraxer-14-en-28-oic acid	Thailand (R)	Kapingu *et al.*, 2000
Taraxer-14-en-28-oic acid, 3β-acetoxy	Tanzania (R)	Kapingu *et al.*, 2000

Croton malambo Kartz

Croton malambo is a small tree that grows in west Venezuela region and in the northern part of Colombia. It is known as *palomatías, torco,* and *cáscara de lombrices* The aerial parts of the tree have an aromatic smell and a pungent, bitter taste with a calamus flavour.

An infusion of *C. malambo* bark is widely used in traditional medicine for treatment of diverse diseases such as diabetes, diarrhea, rheumatism, gastric ulcers, and as an anti-inflammatory and analgesic agent (Garcia, 1975).

Monoterpenes		
Pinene, β	Venezuela (EO, B)	Suárez *et al.*, 2005
Limonene	Venezuela (EO, B)	Suárez *et al.*, 2005
Linalil-oxide, trans	Venezuela (EO, B)	Suárez *et al.*, 2005
Norboneol	Venezuela (EO, B)	Suárez *et al.*, 2005
Terpinol, α	Venezuela (EO, B)	Suárez *et al.*, 2005
Verbenone	Venezuela (EO, B)	Suárez *et al.*, 2005
Phenylpropanoids		
Eugenol	Venezuela (EO, B)	Suárez *et al.*, 2005
Methyl eugenol	Venezuela (EO, B)	Suárez *et al.*, 2005
Methyl isoeugenol	Venezuela (EO, B)	Suárez *et al.*, 2005
Elemicine	Venezuela (EO, B)	Suárez *et al.*, 2005
3,4,5-Trimethoxy-benzaldehyde	Venezuela (EO, B)	Suárez *et al.*, 2005
Isoelemicine	Venezuela (EO, B)	Suárez *et al.*, 2005
Sesquiterpenes		
Cubebene, α	Venezuela (EO, B)	Suárez *et al.*, 2005
Gurjunene, α	Venezuela (EO, B)	Suárez *et al.*, 2005
Bergamotene, β	Venezuela (EO, B)	Suárez *et al.*, 2005
Bergamotene, α	Venezuela (EO, B)	Suárez *et al.*, 2005
Cedrene, β	Venezuela (EO, B)	Suárez *et al.*, 2005
Caryophyllene, β	Venezuela (EO, B)	Suárez *et al.*, 2005

Contd...

Table 7.1–*Contd...*

Compound Type Chemical Name	Place (Part Used)	References
Valencene	Venezuela (EO, B)	Suárez *et al.*, 2005
Veratral	Venezuela (EO, B)	Suárez *et al.*, 2005
Cadinene, γ	Venezuela (EO, B)	Suárez *et al.*, 2005
Bisabolene, β	Venezuela (EO, B)	Suárez *et al.*, 2005
Cadinene, α	Venezuela (EO, B)	Suárez *et al.*, 2005
Cadinene, β	Venezuela (EO, B)	Suárez *et al.*, 2005
Calacorene	Venezuela (EO, B)	Suárez *et al.*, 2005
Globulol	Venezuela (EO, B)	Suárez *et al.*, 2005
Humulene epoxide	Venezuela (EO, B)	Suárez *et al.*, 2005
Cadinol, γ	Venezuela (EO, B)	Suárez *et al.*, 2005
Asaraldehyde	Venezuela (EO, B)	Suárez *et al.*, 2005
Carotol	Venezuela (EO, B)	Suárez *et al.*, 2005
Cadinol, t	Venezuela (EO, B)	Suárez *et al.*, 2005
Cadinol, δ	Venezuela (EO, B)	Suárez *et al.*, 2005
Nootkatone	Venezuela (EO, B)	Suárez *et al.*, 2005

Croton matourensis Aubl

It is a tree widely spread in Northern Brazil, where it is popularly known as "maravuvuia". When cut, the bark exudates a reddish latex (Gottlieb *et al.*, 1981).

Diterpene

Maravuic acid	Brazil (B)	Schneider *et al.*, 1994 Schneider *et al.*, 1995

Monoterpenes

Cymene, *para*	Brazil (SB)	Gottlieb *et al.*, 1981
Phellandrene	Brazil (SB)	Gottlieb *et al.*, 1981
Pinene, α	Brazil (SB)	Gottlieb *et al.*, 1981

Phenylpropanoid

Elemicin	Brazil (SB)	Gottlieb *et al.*, 1981

Croton mayumbensis J. Leonard

Croton mayumbensis (Euphorbiaceae) is a tree up to 34 m growing in the rain forest of the Central African Republic. It is an African medicinal plant that barks and leaves are empirically used to treat microbial infections and human parasitic diseases (Lejolly, 1956).

Diterpenoids

Centrafricine I	Central African Republic (SB)	Yamale *et al.*, 2009

Contd...

Table 7.1–*Contd...*

Compound Type Chemical Name	Place (Part Used)	References

Croton megalocarpus Hutch.

Occurs in East Africa. Its properties as antihelmints, anticancer and against bronchitis are attributed to its diterpenes (Addae-Mensah *et al.*, 1989; Addae-Mensah *et al.*, 1992b; Weckert *et al.*, 1992). In folk medicine it is used for kidney stones, malaria and anti helmints (Johns *et al.*, 1994).

Alyphatics

Cyclopent-2-en-1-one, 3,4,5-trimethyl	Kenya (W)	Kinyanjui *et al.*, 2000

Carbohydrate

Sucrose	Kenya (B)	Addae-Mensah *et al.*, 1992b

Diterpenes

Chiromodine	Ghana (SB) Kenya (SB)	Weckert *et al.*, 1992 Addae-Mensah *et al.*, 1989
Chiromodine, epoxy	Kenya (B)	Addae-Mensah *et al.*, 1992b

Lactone

Pyran-2-one, 2(H)	Kenya (W)	Kinyanjui *et al.*, 2000

Oxygen Heterocycle

Furan,3-methyl	Kenya (W)	Kinyanjui *et al.*, 2000

Phenylpropanoids

Ferulic acid, *trans*: hexacosyl ester	Kenya (B)	Addae-Mensah *et al.*, 1992b
Ferulic acid, *trans*: octacosyl ester	Kenya (B)	Addae-Mensah *et al.*, 1992b
Ferulic acid, *trans*: tetracosyl ester	Kenya (B)	Addae-Mensah *et al.*, 1992b

Sesquiterpenes

Cedrol	Kenya (W)	Kinyanjui *et al.*, 2000

Steroid

Sitosterol, β	Kenya (SB)	Addae-Mensah *et al.*, 1989

Triterpenes

Aleuritolic acid, 3-*O*-acetyl	Kenya (B)	Addae-Mensah *et al.*, 1992b
Betulin	Kenya (SB)	Addae-Mensah *et al.*, 1989
Lupeol	Kenya (SB)	Addae-Mensah *et al.*, 1989
Lupeol, 3-*O*-aceto-acetyl	Kenya (B)	Addae-Mensah *et al.*, 1992b

Croton membranaceus Muell. Arg.

The root extract of *Croton membranaceus* Mull. Arg. (Euphorbiaceae) is used in formulations for the treatment and management of prostate and related cancers and measles in Ghana (Mshana *et al.*, 2000).

Alkaloid

Julocrotine	Ghana (R)	Aboagye *et al.*, 2000

Contd...

Table 7.1—*Contd...*

Compound Type Chemical Name	Place (Part Used)	References

Croton menthodorus Benth.

In folk medicine, the powdered seeds are used as antiinflamatory (Ortega *et al.*, 1996). In Ecuador, the leaves extract is used for dermatitis.

Alkaloids

Flavinantine, *O*-methyl	Not stated (AP)	Capasso *et al.*, 2000
Magnoflorine	Not stated (AP)	Capasso *et al.*, 2000

Flavonoids

Nicotiflorin	Not stated (AP)	Capasso *et al.*, 2000
Quercetin-3-*O*-(2-*O*-b-D-apiofuranosyl)-rutinoside	Not stated (AP)	Capasso *et al.*, 2000
Rutin	Not stated (AP)	Capasso *et al.*, 2000

Phenylpropanoid

Caffeic acid	Not stated (AP)	Capasso *et al.*, 2000

Croton micans Muell. Arg.

It grows in Northeastern Brazil, where it is popularly known as "marmeleiro" (Craveiro *et al.*, 1981a). The essential oils from the fresh leaves did not show molluscicidal activity (Rouquayrol *et al.*, 1980).

Monoterpenes

Myrcene	Brazil (EO, L)	Craveiro *et al.*, 1981a
Phellandrene, α	Brazil (EO, L)	Craveiro *et al.*, 1981a
Pinene, α	Brazil (EO, L)	Craveiro *et al.*, 1981a
Pinene, β	Brazil (EO, L)	
Sabinene	Brazil (EO, L)	Craveiro *et al.*, 1981a
Thujene, α	Brazil (EO, L)	Craveiro *et al.*, 1981a

Sesquiterpenes

Caryophyllene, β	Brazil (EO, L)	Craveiro *et al.*, 1981a
Elemene, β	Brazil (EO, L)	Craveiro *et al.*, 1981a
Elemene, δ	Brazil (EO, L)	Craveiro *et al.*, 1981a
Elemene, γ	Brazil (EO, L)	Craveiro *et al.*, 1981a
Germacrene B	Brazil (EO, L)	Craveiro *et al.*, 1981a
Humulene α	Brazil (EO, L)	Craveiro *et al.*, 1981a

Croton moritibensis Baill.

Croton moritibensis Baill, a native shrub from our region, where it is popularly known as "velame-preto" (Araújo-Júnior *et al.*, 2004).

Alkaloids

Harman	Brazil (AP)	Araújo-Júnior *et al.*, 2004
Harman, tetrahydro	Brazil (AP)	Araújo-Júnior *et al.*, 2004
Harman, tetrahydro: 2-ethoxy-carbonyl	Brazil (AP)	Araújo-Júnior *et al.*, 2004
Harman, tetrahydro: 6-hydroxy-2-methyl	Brazil (AP)	Araújo-Júnior *et al.*, 2004

Contd...

Table 7.1–*Contd...*

Compound Type Chemical Name	Place (Part Used)	References
Diterpenes		
Podocarpa-1,8,11,13-tetraen-3-one, 12-hydroxy-13-methyl	Brazil (AP)	Araújo-Júnior *et al.*, 2004
Podocarpa-8,11,13-trien-3-one, 12-hydroxy-13-methyl	Brazil (AP)	Araújo-Júnior *et al.*, 2004
Sonderianol	Brazil (AP)	Araújo-Júnior *et al.*, 2004

Croton mucronifolius Muell. Arg.

In Northeastern Brazil, where it is know by the common name "velaminho" is used for treatment of influenza, rheumatism and syphilis. Pharmacological studies with the leaves have shown antitumor and antibacterial activities (Lemos *et al.*, 1992).

Monoterpenes		
Ascaridol	Brazil (EO, L)	Lemos *et al.*, 1992
Camphene	Brazil (EO, L)	Lemos *et al.*, 1992
Camphor	Brazil (EO, L)	Lemos *et al.*, 1992
Carvacrol	Brazil (EO, L)	Lemos *et al.*, 1992
Cineole, 1,8	Brazil (EO, L)	Lemos *et al.*, 1992
Cymene, *para*	Brazil (EO, L)	Lemos *et al.*, 1992
Limonene	Brazil (EO, L)	Lemos *et al.*, 1992
Linalool	Brazil (EO, L)	Lemos *et al.*, 1992
Pinene, α	Brazil (EO, L)	Lemos *et al.*, 1992
Pinene, β	Brazil (EO, L)	Lemos *et al.*, 1992
Terpinen, 4-ol	Brazil (EO, L)	Lemos *et al.*, 1992
Terpinene, γ	Brazil (EO, L)	Lemos *et al.*, 1992
Terpineol, α	Brazil (EO, L)	Lemos *et al.*, 1992

Croton nepetaefolius Baillon

Croton nepetaefolius is an aromatic plant abundantly found in the Northeastern Brazil, where it is commonly known as "marmeleiro sabiá". The genus *Croton* is one of the most strongly represented in the flora of this region and includes a large number of aromatics species which are widely exploited in local folk medicine. Infusions of the bark and leaves are commonly used for their antispasmodic properties and to relieve flatulence and to increase appetite (Magalhaes *et al.*, 1988; Lahlou *et al.*, 1999).

Aliphatic		
Octacosan, 1-ol	Brazil (W)	Craveiro *et al.*, 1980
Benzenoids		
Acetophenone, 2-hydroxy-3,4,6-trimethoxy	Brazil (W)	Craveiro *et al.*, 1980
Cathecol, *n*-propyl	Brazil (EO, L)	Craveiro *et al.*, 1981a
Xanthoxylin	Brazil (EO, S, L, W)	Craveiro *et al.*, 1980 Moura *et al.*, 1990 Magalhaes *et al.*, 1998 Lahlou *et al.*, 1999

Contd...

Table 7.1–*Contd...*

Compound Type Chemical Name	Place (Part Used)	References
Diterpenes		
Croton casbene diterpene 1	Brazil (S)	Moura *et al.*, 1990
Monoterpenes		
Camphene	Brazil (EO, B)	Craveiro *et al.*, 1980
Camphor	Brazil (EO, B)	Craveiro *et al.*, 1980
Cineole, 1,8	Brazil (EO, L, AP)	Craveiro *et al.*, 1980 Craveiro *et al.*, 1981a Magalhaes *et al.*, 1998 Lahlou *et al.*, 1999 Lima-Accioly *et al.*, 2006
Limonene	Brazil (EO)	Lahlou *et al.*, 1999
Pinene, α	Brazil (EO, L, B, AP)	Craveiro *et al.*, 1980 Craveiro *et al.*, 1981a Lahlou *et al.*, 1999 Lima-Accioly *et al.*, 2006
Pinene, β	Brazil (EO, B, L)	Craveiro *et al.*, 1980 Craveiro *et al.*, 1981a Lahlou *et al.*, 1999
Sabinene	Brazil (EO, L, AP)	Craveiro *et al.*, 1980 Craveiro *et al.*, 1981a Lima-Accioly *et al.*, 2006 Magalhaes *et al.*, 1998
Terpineol, α	Brazil (EO, L, AP)	Craveiro *et al.*, 1980 Craveiro *et al.*, 1981a Magalhaes *et al.*, 1998 Lahlou *et al.*, 1999 Lima-Accioly *et al.*, 2006
Phenylpropanoids		
Elemicin	Brazil (EO, S, L, W, B, AP)	Craveiro *et al.*, 1980 Craveiro *et al.*, 1981a Moura *et al.*, 1990 Lima-Accioly *et al.*, 2006
Eugenol, methyl ether	Brazil (EO, L, W)	Craveiro *et al.*, 1980 Craveiro *et al.*, 1981a Magalhaes *et al.*, 1998 Lahlou *et al.*, 1999
Catechol, propyl	Brazil (EO, L)	Craveiro *et al.*, 1980
Sesquiterpenes		
Aromadendrene, allo	Brazil (EO, AP)	Lahlou *et al.*, 1999 Lima-Accioly *et al.*, 2006
Bergamoptene, α	Brazil (EO, W, B, S, AP)	Craveiro *et al.*, 1980 Lahlou *et al.*, 1999 Lima-Accioly *et al.*, 2006

Contd...

Table 7.1–*Contd...*

Compound Type Chemical Name	Place (Part Used)	References
Bicyclogermacrene	Brazil (EO, AP)	Lahlou *et al.*, 1999 Lima-Accioly *et al.*, 2006
Cadinene, δ	Brazil (EO, W, B)	Craveiro *et al.*, 1980
Caryophyllene	Brazil (EO, AP)	Lahlou *et al.*, 1999 Lima-Accioly *et al.*, 2006
Caryophyllene, β	Brazil (EO, L)	Craveiro *et al.*, 1980 Craveiro *et al.*, 1981a Magalhaes *et al.*, 1998
Caryophyllene, oxide	Brazil (EO, AP)	Lahlou *et al.*, 1999 Lima-Accioly *et al.*, 2006
Copaene, α	Brazil (EO, B, S,W)	Craveiro *et al.*, 1980
Cubebene, α	Brazil (EO, W, S, B)	Craveiro *et al.*, 1980
Elemene, β	Brazil (EO, L)	Craveiro *et al.*, 1980 Craveiro *et al.*, 1981a Lahlou *et al.*, 1999
Elemene, δ	Brazil (EO, L, AP)	Craveiro *et al.*, 1980 Craveiro *et al.*, 1981a Lahlou *et al.*, 1999 Lima-Accioly *et al.*, 2006
Elemene, γ	Brazil (EO, L)	Craveiro *et al.*, 1980 Craveiro *et al.*, 1981a
Germacrene A	Brazil (EO)	Lahlou *et al.*, 1999
Germacrene C	Brazil (EO)	Lahlou *et al.*, 1999
Humulene, α	Brazil (EO, L)	Craveiro *et al.*, 1980 Craveiro *et al.*, 1981a Lahlou *et al.*, 1999
Humulene	Brazil (EO, L)	Craveiro *et al.*, 1980
Santalene, α	Brazil (EO, W, B, S)	Craveiro *et al.*, 1980 Lahlou *et al.*, 1999
Spathulenol	Brazil (EO, AP)	Lahlou *et al.*, 1999 Lima-Accioly *et al.*, 2006

Croton nitens Sw.

Diterpenes

Crotonitenone	Jamaica (L, T)	Burke *et al.*, 1981

Croton nitrariaefolius Baillon

Found in the State of Rio Grande do Sul, Southern Brazil, where it is popularly know as "formigueirinha" (Siqueira *et al.*, 1984).

Monoterpenes

Cineole, 1,8	Brazil (EO)	Siqueira *et al.*, 1984
Limonene	Brazil (EO)	Siqueira *et al.*, 1984
Pinene, α	Brazil (EO)	Siqueira *et al.*, 1984
Pinene, β	Brazil (EO)	Siqueira *et al.*, 1984
Terpinene, α	Brazil (EO)	Siqueira *et al.*, 1984

Contd...

Table 7.1–*Contd...*

Compound Type Chemical Name	Place (Part Used)	References
	***Croton niveus* Jacq.**	

The classical pharmacognosy of this species has been reported. It is used as popular medicine in Curaçao. Aqueous extracts of the bark and of stems, leaves, bark and roots were slightly toxic to American cockroaches, and non-toxic to German cockroaches and milkweed bugs.

Diterpene

Nivenolide	Mexico (L)	Rojas and Rodriguez-Hahn, 1978
	***Croton oblongifolius* Roxb.**	

C. oblongifolius is widely distributed in Thailand. Its has been used as a traditional medicine for many applications such as for dysmenorrhea, as a purgative and to treat dyspepsia and dysenteries (Rao *et al.*, 1968; Sabnis *et al.*, 1983).

Diterpenes

Croblongifolin	Thailand (SB)	Roengsumran *et al.*, 2002
Crotocembranal, neo	Thailand (SB)	Roengsumran *et al.*, 1999b
Crotocembraneic acid	Thailand (SB)	Roengsumran *et al.*, 1998 Roengsumran *et al.*, 1999b
Crotocembraneic acid, neo	Thailand (SB)	Roengsumran *et al.*, 1998 Roengsumran *et al.*, 1999b
Crotohalimaneic acid	Thailand (SB)	Roengsumran *et al.*, 2004
Crotohalimaneic-12-benzoyloxy acid	Thailand (SB)	Roengsumran *et al.*, 2004
Crotohalimoneic acid	Thailand (SB)	Roengsumran *et al.*, 2004
Labda-7-*trans*,12,14-triene	Thailand (SB)	Roengsumran *et al.*, 1999a
Labda-7-*trans*,12,14-trien-17-al	Thailand (SB)	Roengsumran *et al.*, 1999a
Labda-7-*trans*-12,14-trien-17-oic acid	Thailand (SB)	Roengsumran *et al.*, 1999a Roengsumran *et al.*, 2002
Labda-7-*trans*-12,14-trien-17-ol	Thailand (SB)	Roengsumran *et al.*, 1999a
Labda-8(17)-*trans*,12,14-triene-2,3-diacetoxy	Thailand (SB)	Roengsumran *et al.*, 2001
Labda-8(17)-*trans*,12,14-triene-2,3-dihydroxy	Thailand (SB)	Roengsumran *et al.*, 2001
Labda-8(17)-*trans*,12,14-triene-2-acetoxy-3-hydroxy	Thailand (SB)	Roengsumran *et al.*, 2001
Labda-8(17)-*trans*,12,14-triene-3-acetoxy-2-hydroxy	Thailand (SB)	Roengsumran *et al.*, 2001
Plaunol A	Thailand (S)	Ogiso *et al.*, 1981
Plaunol E	Thailand (S)	Ogiso *et al.*, 1981
Cleistantha-4(18)-13(17)-15-trien-3-oic acid	Thailand (SB)	Israngkura *et al.*, 2000
Crovatin	Thailand (SB)	Roengsumran *et al.*, 2002
Hardwickiic acid	Thailand (SB)	Chaichantipyyuth *et al.*, 2005
Kaur-16-en-19-oic acid, ent: (-)	Thailand (SB)	Ngamrojnavanich *et al.*, 2003

Contd...

Table 7.1–*Contd...*

Compound Type Chemical Name	Place (Part Used)	References
Labda-13(16)-14-diene, ent: 6α, 7β, 8α-trihydroxy	Thailand (SB)	Chaichantipyyuth *et al.*, 2005
Labda-13(16)-14-diene, ent: 7β, 8α-dihydroxy	Thailand (SB)	Chaichantipyyuth *et al.*, 2005
Labda-cis-12,14,17-trien-18-oic acid	Thailand	Roengsumran *et al.*, 2002
Manoyl oxide, ent: 1,2-dehydro: 12α-hydroxy-3-oxo	Thailand (SB)	Chaichantipyyuth *et al.*, 2005
Manoyl oxide, ent: 1,2-dehydro: 3-oxo	Thailand (SB)	Chaichantipyyuth *et al.*, 2005
Manoyl oxide, ent: 1β, 3α-dihydroxy	Thailand (SB)	Chaichantipyyuth *et al.*, 2005
Manoyl oxide, ent: 1β-hydroxy-3-oxo	Thailand (SB)	Chaichantipyyuth *et al.*, 2005
Manoyl oxide, ent: 3α-hydroxy	Thailand (SB)	Chaichantipyyuth *et al.*, 2005
Manoyl oxide, ent: 3-oxo	Thailand (SB)	Chaichantipyyuth *et al.*, 2005
Nidorellol	Thailand (SB)	Roengsumran *et al.*, 2002
Oblongifoliol	Not stated (EP)	Farnsworth *et al.*, 1969
3-Hydroxycleistantha-13(17),15-diene	Thailand (SB)	Roengsumran *et al.*, 2009
3,4-Seco-cleistantha-4(18),13(17), 15-trien-3-oic acid	Thailand (SB)	Roengsumran *et al.*, 2009
Flavonoids		
Hyperoside	India (L)	Subramanian *et al.*, 1971
Quercetin	India (L)	Subramanian *et al.*, 1971
Rhamnetin, iso	India (L)	Subramanian *et al.*, 1971
Furanocembranoids		
Furanocembranoid 1	Thailand (SB)	Pudhom *et al.*, 2007
Furanocembranoid 2	Thailand (SB)	Pudhom *et al.*, 2007
Furanocembranoid 3	Thailand (SB)	Pudhom *et al.*, 2007
Furanocembranoid 4	Thailand (SB)	Pudhom *et al.*, 2007
Xanthones		
Shamixanthone	Thailand (L)	Pompakakul *et al.*, 2006
14-Methoxytajixanthone-25-acetate	Thailand (L)	Pompakakul *et al.*, 2006
Tajixanthone methanoate	Thailand (L)	Pompakakul *et al.*, 2006
Tajixanthone hydrate	Thailand (L)	Pompakakul *et al.*, 2006
***Croton ovalifolius* Vahl**		
Monoterpenes		
Camphor	Venezuela (EO, L)	Meccia *et al.*, 2000
Bornyl acetate	Venezuela (EO, L)	Meccia *et al.*, 2000
Terpineol, α	Venezuela (EO, L)	Meccia *et al.*, 2000
Borneol	Venezuela (EO, L)	Meccia *et al.*, 2000
Ocimene, (Z)-β	Venezuela (EO, L)	Meccia *et al.*, 2000
Ocimene, α	Venezuela (EO, L)	Meccia *et al.*, 2000

Contd...

Table 7.1 –*Contd...*

Compound Type Chemical Name	Place (Part Used)	References
Sesquiterpenes		
Bicyclogermacrene	Venezuela (EO, L)	Meccia *et al.*, 2000
Cadinol, γ	Venezuela (EO, L)	Meccia *et al.*, 2000
Caryophyllene, oxide	Venezuela (EO, L)	Meccia *et al.*, 2000
Caryophyllene, β	Venezuela (EO, L)	Meccia *et al.*, 2000
Germacrene B	Venezuela (EO, L)	Meccia *et al.*, 2000
Spathulenol	Venezuela (EO, L)	Meccia *et al.*, 2000
Cadinene, δ	Venezuela (EO, L)	Meccia *et al.*, 2000
Cyperene	Venezuela (EO, L)	Meccia *et al.*, 2000
α-copaene	Venezuela (EO, L)	Meccia *et al.*, 2000
β-bourbonene	Venezuela (EO, L)	Meccia *et al.*, 2000
α-humulene	Venezuela (EO, L)	Meccia *et al.*, 2000
Farnesal, (*E,E*)	Venezuela (EO, L)	Meccia *et al.*, 2000
Farnesol, (*Z,E*)	Venezuela (EO, L)	Meccia *et al.*, 2000
Cadinol, δ	Venezuela (EO, L)	Meccia *et al.*, 2000
Elemene, γ	Venezuela (EO, L)	Meccia *et al.*, 2000
Muurolene, γ	Venezuela (EO, L)	Meccia *et al.*, 2000
Muurolol, γ	Venezuela (EO, L)	Meccia *et al.*, 2000
Elemene, β	Venezuela (EO, L)	Meccia *et al.*, 2000
Alloaromadendrene	Venezuela (EO, L)	Meccia *et al.*, 2000

Croton palanostigma K.L.

Occurs in Northern Peru. It is another of the plants popularly know as "sangre de drago" or "sangre de dragon", due to its red latex. The latex is used to cure wounds, stomach and uterine ulcers (Ramirez *et al.*, 1988). It also has a citotoxic effect, which is attributed to the alkaloid taspine (Pieters *et al.*, 1992; Itokawa *et al.*, 1991b).

Alkaloid

Taspine	Peru (SA)	Itokawa *et al.*, 1991b Itokawa *et al.*, 1992

Croton parvifolius Muell. Arg.

Phenylpropanoids		
Elemicin	Argentina (EO)	Etcheves *et al.*, 1981
Eugenol, methyl ether	Argentina (EO)	Etcheves *et al.*, 1981
Sesquiterpene		
Humulene	Argentina (EO)	Etcheves *et al.*, 1981

Contd...

Table 7.1–*Contd...*

Compound Type Chemical Name	Place (Part Used)	References
***Croton penduliflorus* Hutch**		

Occurs in Nigeria, where it is used in folk medicine as a purgative (Shetty *et al.*, 1983).

Aliphatics

Arachidic acid	Nigeria (SE)	Asuzu *et al.*, 1988 Asuzu *et al.*, 1989a Asuzu *et al.*, 1989b
Palmitic acid	Nigeria (SE)	Asuzu *et al.*, 1988 Asuzu *et al.*, 1989a Asuzu *et al.*, 1989b
Stearic acid	Nigeria (SE)	Asuzu *et al.*, 1988 Asuzu *et al.*, 1989a Asuzu *et al.*, 1989b

Diterpene

Penduliflaworosin	Nigeria (RB)	Adesogan, 1981

***Croton plumieri* Urb.**		

Alkaloids

Crotonosine	Jamaica (L, S)	Stuart and Woo-Ming, 1969
Crotonosine, *N*-methyl: L	Jamaica (L, S)	Stuart and Woo-Ming, 1969
Linearisine	Jamaica (L, S)	Stuart and Woo-Ming, 1969
Salutaridine	Jamaica (L, S)	Stuart and Woo-Ming, 1969
Salutaridine, 8,14-dihydro	Jamaica (L, S)	Stuart and Woo-Ming, 1969
Salutaridine, nor: 8,14-dihydro	Jamaica (L, S)	Stuart and Woo-Ming, 1969

***Croton poilanei* Gagnep**		

In Tayland the bark is used for the treatment of stomach pain (Esser, 2003).

Diterpene

Poilaneic acid	Thailand (L)	Sato *et al.*, 1981

***Croton polyandrus* Spreng**		

It is common to the sandbanks and coastal vegetation in Northeastern Brazil, where it is popularly known as "croton de tabuleiro" (Araújo-Júnior *et al.*, 2002).

Diterpenes

Cordatin	Brazil (R)	Araújo-Júnior *et al.*, 2002
Hardwickiic, 12-oxo acid: methyl ester	Brazil (R)	Araújo-Júnior *et al.*, 2002
Labda-18-hydroxy-7,13-dien-15-oic acid: methyl ester	Brazil (R)	Araújo-Júnior *et al.*, 2002

***Croton pullei* var. *glabrior* Lanj.**		

Alkaloids

Crotonimide A	Brazil (S, SB)	Barbosa *et al.*, 2007
Crotonimide B	Brazil (S, SB)	Barbosa *et al.*, 2007

Contd...

Table 7.1–*Contd...*

Compound Type Chemical Name	Place (Part Used)	References

Croton pyramidalis D.S.

Occurs in the areas of Tuxtla and Vera Cruz in Mexico. It forms small groups of trees of the same species, what indicates that it inhibits the circunjacent flora (Rodriguez-Hahn *et al.*, 1981).

Diterpene

Pyramidolactone	Mexico (L, S)	Rodriguez-Hahn *et al.*, 1981

Flavonoid

Flavone, 3,5-dihydroxy-4',7-dimethoxy	Mexico (L, S)	Rodriguez-Hahn *et al.*, 1981

Croton regelianus Muell. Arg.

C. regelianus, popularly known as 'velame-de-cheiro', is a medicinal plant, particularly used to treat malignant tumors (Torres *et al.*, 2008).

Monoterpenes

Ascaridole	Brazil (L)	Torres *et al.*, 2008
Camphene	Brazil (L)	Torres *et al.*, 2008
Camphor	Brazil (L)	Torres *et al.*, 2008
Cineole, 1,8	Brazil (L)	Torres *et al.*, 2008
Cymene, p	Brazil (L)	Torres *et al.*, 2008
Isoascaridole	Brazil (L)	Torres *et al.*, 2008
Isoborneol	Brazil (L)	Torres *et al.*, 2008
Limonene	Brazil (L)	Torres *et al.*, 2008
Linalool	Brazil (L)	Torres *et al.*, 2008
Myrcene	Brazil (L)	Torres *et al.*, 2008
Phellandrene, α	Brazil (L)	Torres *et al.*, 2008
Pinene, α	Brazil (L)	Torres *et al.*, 2008
Pinene, β	Brazil (L)	Torres *et al.*, 2008
Terpinene, α	Brazil (L)	Torres *et al.*, 2008
Terpinene, γ	Brazil (L)	Torres *et al.*, 2008
Terpineol, 4	Brazil (L)	Torres *et al.*, 2008
Terpineol, α	Brazil (L)	Torres *et al.*, 2008
Terpinolene	Brazil (L)	Torres *et al.*, 2008
Thujene, α	Brazil (L)	Torres *et al.*, 2008
Tricyclene	Brazil (L)	Torres *et al.*, 2008

Croton reflexifolius H.B.K.

In the Huasteca Hidalguense traditional medicine of Mexico, the leaves of *Croton reflexifolius* H.B.K. (Euphorbiaceae), locally known as "huilocuahuil," is prepared as tea and used to treat cough, gastric ulcers, and diabetes (Estrada, 1985). *C. reflexifolius,* a shrub up to 3–5 m high, is commonly found in the states of Hidalgo, Sinaloa, Veracruz, Chiapas, and Yucat´an in Mexico (Martínez, 1987). In spite of the uses of this plant by the Huastecans, no report has been issued with a chemical or biological evaluation of *C. reflexifolius*.

Diterpenes

Polyalthic acid	Mexico (L)	Sánchez-Mendoza *et al.*, 2008

Contd...

Table 7.1–*Contd...*

Compound Type Chemical Name	Place (Part Used)	References

Croton rhamnifolius H.B.K.

Occurs in the State of Ceará, Northeastern Brazil. It is popularly known as "marmeleiro" and "velame" (Craveiro *et al.*, 1981a). The essential oil from the fresh leaves did not show molluscicidal activity (Rouquayrol *et al.*, 1980).

Monoterpenes

Car-3-ene	Brazil (EO, L)	Craveiro *et al.*, 1981a
Cineole, 1,8	Brazil (EO, L)	Craveiro *et al.*, 1981a
Cymene, *para*	Brazil (EO, L)	Craveiro *et al.*, 1981a
Linalool	Brazil (EO, L)	Craveiro *et al.*, 1981a
Phellandrene, α	Brazil (EO, L)	Craveiro *et al.*, 1981a
Sabinene	Brazil (EO, L)	Craveiro *et al.*, 1981a
Terpinen-4-ol	Brazil (EO, L)	Craveiro *et al.*, 1981a
Terpineol, α	Brazil (EO, L)	Craveiro *et al.*, 1981a

Sesquiterpenes

Aromadendrene	Brazil (EO, L)	Craveiro *et al.*, 1981a
Cadinene, α	Brazil (EO, L)	Craveiro *et al.*, 1981a
Caryophyllene, β	Brazil (EO, L)	Craveiro *et al.*, 1981a
Copaene, α	Brazil (EO, L)	Craveiro *et al.*, 1981a
Cubebene, α	Brazil (EO, L)	Craveiro *et al.*, 1981a
Elemene, γ	Brazil (EO, L)	Craveiro *et al.*, 1981a
Humulene, α	Brazil (EO, L)	Craveiro *et al.*, 1981a

Croton robustus Kurz

Croton robustus Kurz. or *Croton siamensis* Craib. (Euphorbiaceae) was known as "Plao Lueat" in Thailand. It is scarcely distributed in the northern part of Thailand. *C. robustus* is used as antianemic agent, and the barks and leaves are used to stop bleeding and to treat skin diseases. Up to date, there is no report on the chemical constituents and biological activity of this plant in the literature (Ngamrojnavanich *et al.*, 2003).

Diterpenes

Poilaneic acid	Thailand (SB)	Ngamrojnavanich *et al.*, 2003
Trachyloban-19-oic acid	Thailand (SB)	Ngamrojnavanich *et al.*, 2003
Trachyloban-19-ol	Thailand (SB)	Ngamrojnavanich *et al.*, 2003

Croton ruizianus Muell. Arg.

C. ruizianus is a small shrub growing in the Central sierra of Peru. The plant is known by vernacular name of "cabra-cabra", "upalu" and "matarracra". The infusion of its leaves is employed in traditional Peruvian medicine as vulnerary and antiespasmodic (Del-Castillo *et al.*, 1996).

Alkaloids

Crotsparine	Peru (L)	Del-Castillo *et al.*, 1996
Flavinantine	Peru (L)	Piacente *et al.*, 1998
Flavinantine, *O*-methyl	Peru (L)	Piacente *et al.*, 1998
Jacularine	Peru (L)	Del-Castillo *et al.*, 1996

Contd...

Table 7.1–*Contd...*

Compound Type Chemical Name	Place (Part Used)	References
Steroids		
Pregnan-4-one, 3β,14β,15β,16α-tetrahydroxy-3-*O*-β-D-glucopyranosyl (1-4)-β-D-oleandro-pyranosyl(1-4)-β-D-oleandropyranosyl (1-4)-β-D-digitoxopyranosyl (1-4)-β-D-oleandropyranosyl	Peru (L)	Piacente *et al.*, 1998
Pregnan-4-one, 3β,14β,15β,16α-tetrahydroxy-3-*O*-β-D-oleandropyranosyl(1-4)-β-D-oleandro-pyranosyl(1-4)-β-D-digitoxopyranosyl(1-4)-β-D-oleandropyranosyl	Peru (L)	Piacente *et al.*, 1998

Croton salutaris Casar

It is used as an febrifuge in malaria (Brandão *et al.*, 1985). It grows in the Brazilian states of Rio de Janeiro (Barnes and Soeiro, 1981) and São Paulo (Rao *et al.*, 1968).

Alkaloids		
Salutaridine	Brazil (L, T, EP)	Barnes and Soeiro, 1981
Salutarine	Brazil (L, T)	Farnsworth *et al.*, 1969 Barnes and Soeiro, 1981
Diterpenes		
Hexadeca-1-*cis*-6-*trans*-10,14-tetraene-5, 13-dione, 3,12-di-hydroxy-3,7,11,15-tetramethyl	Brazil (T)	Itokawa *et al.*, 1991a
Hexadeca-1-*trans*-10,14-triene-5-13-dione, 3,12-dihydroxy-3,7,11,15-tetramethyl	Brazil (T)	Itokawa *et al.*, 1991a
Hexadeca-1-*trans*,6-*trans*,10-14-tetraene-5, 13-dione, 3,12-dihydroxy: 3,7,11,15-tetramethyl	Brazil (T)	Itokawa *et al.*, 1991a
Podocarpa-9,11,13-trien-3-one-12-hydroxy-13-methyl	Brazil (T)	Itokawa *et al.*, 1991a
Sonderianol	Brazil (T)	Itokawa *et al.*, 1991a

Croton sarcopetalus Muell.

C. Sarcopetalus commonly known as "lecheron" is shrub that grows in Northwestern and Central Argentina (Heluani *et al.*, 1998; Heluani *et al.*, 2000).

Aliphatic		
Undecanoic acid	Argentina (EO, W)	Heluani *et al.*, 2000
Benzenoids		
Veratraldehyde	Argentina (R)	Heluani *et al.*, 2000
Diterpenes		
Junceic acid	Argentina (R)	Heluani *et al.*, 2000
Sarcopetaloic acid	Argentina (R)	Heluani *et al.*, 2000
Sarcopetalololide	Argentina (R)	Heluani *et al.*, 2000
Sarcopetalolide	Argentina (R)	Heluani *et al.*, 2000
Yucalexin A-16	Argentina (R)	Heluani *et al.*, 2000

Contd...

Table 7.1–*Contd...*

Compound Type Chemical Name	Place (Part Used)	References
Yucalexin B-6	Argentina (R)	Heluani *et al.,* 2000
Yucalexin P-4	Argentina (R)	Heluani *et al.,* 2000
Monoterpenes		
Borneol	Argentina (EO, W)	Heluani *et al.,* 2000
Borneol, iso	Argentina (EO, W)	Heluani *et al.,* 2000
Camphene	Argentina (EO, W)	Heluani *et al.,* 2000
Cineole, 1,8	Argentina (EO, W)	Heluani *et al.,* 2000
Cymene, *para*	Argentina (EO, W)	Heluani *et al.,* 2000
Limonene	Argentina (EO, W)	Heluani *et al.,* 2000
Linalool, oxide	Argentina (EO, W)	Heluani *et al.,* 2000
Linalool, dihydro	Argentina (EO, W)	Heluani *et al.,* 2000
Myrcenol, formate	Argentina (EO, W)	Heluani *et al.,* 2000
Pinene, α	Argentina (EO, W)	Heluani *et al.,* 2000
Pinene, β	Argentina (EO, W)	Heluani *et al.,* 2000
Sabinene	Argentina (EO, W)	Heluani *et al.,* 2000
Sabinene, hydrate	Argentina (EO, W)	Heluani *et al.,* 2000
Thujene, α	Argentina (EO, W)	Heluani *et al.,* 2000
Myarcene	Argentina (EO, W)	Heluani *et al.,* 2000
Phenylpropanoids		
Anethole	Argentina (EO, W)	Heluani *et al.,* 2000
Eugenol, methyl ether	Argentina (EO, W)	Heluani *et al.,* 2000
Eugenol, iso-*cis*	Argentina (EO, W)	Heluani *et al.,* 2000
Eugenol, iso-*trans*	Argentina (EO, W)	Heluani *et al.,* 2000
Eugenol, iso-*trans*, methyl ether	Argentina (EO, W, R)	Heluani *et al.,* 2000
Sesquiterpenes		
Bisabolene	Argentina (EO, W)	Heluani *et al.,* 2000
Ionone, β: dihydro	Argentina (EO, W)	Heluani *et al.,* 2000

Croton schiedeanus Scheacht

Croton schiedeanus Schlecht (Euphorbiaceae) is a tree which grows widely in south and central America. In Colombia, it is used in Folk Medicine for treating hypertension (Puebla *et al.,* 2005b).

Diterpenes

Cascarillone, (12R)-12-hydroxy	Colombia (AP)	Puebla *et al.,* 2003
Crotonin, *cis*-dehydro	Colombia (AP)	Puebla *et al.,* 2003
Crotonin, *trans*-dehydro	Colombia (AP)	Puebla *et al.,* 2003
Crotonin, 5β-hydroxy-*cis*-dehydro	Colombia (AP)	Puebla *et al.,* 2003
Cleroda-cis-3,13-dien-15,16-olide-20-oic acid, 19-nor: 2-oxo-16-hydroxy: methyl ester	Colombia (AP)	Puebla *et al.,* 2005a

Contd...

Table 7.1–*Contd...*

Compound Type Chemical Name	Place (Part Used)	References
Cleroda-cis-3,13-dien-18-oic acid, 15-hydroxy: (+)	Colombia (AP)	Puebla *et al.*, 2005a
Cleroda-cis-3,13-dien-18-oic acid-15, 16-olide, hydroxy: (-)	Colombia (AP)	Puebla *et al.*, 2005a
Floridolide A	Colombia (AP)	Puebla *et al.*, 2005a
Floridolide A, 15-methoxy: (+)	Colombia (AP)	Puebla *et al.*, 2005a
Haplopappic acid	Colombia (AP)	Puebla *et al.*, 2005a
Flavonoid		
Ayanin	Colombia (AP)	Guerrero *et al.*, 2002
Phenylbutanoid		
7,9-Dimethoxyrhododendrol and 2-Acetoxy-7,9-Dimethoxyrhododendrol	Colombia (NS)	Puebla *et al.*, 2005b

Croton selowii Baill.

Sesquiterpenes		
Caryophyllene, oxide	Brazil (L)	Palmeira *et al.*, 2004
Caryophyllene, *trans*	Brazil (L)	Palmeira *et al.*, 2004
Cubenol	Brazil (L)	Palmeira *et al.*, 2004
Eudesmol, α	Brazil (L)	Palmeira *et al.*, 2004
Eudesmol, β	Brazil (L)	Palmeira *et al.*, 2004
Eudesmol, γ	Brazil (L)	Palmeira *et al.*, 2004

Croton sonderianus Muell. Arg.

It is a shrub widespread in the Brazilian Northeast and known in the region as "marmeleiro preto". The bark in used in folk medicine for treatment of gastric diseases. Hexane or benzene extracts of its heartwood and roots have shown antifungal and antibacterial activity (Mc Chesney *et al.*, 1984; Silveira and Mc Chesney, 1994).

Diterpenes		
Annonene, *trans*	Brazil (R)	Mc Chesney and Silveira, 1990
Annonene, 6α-hydroxy	Brazil (R)	Silveira and Mc Chesney, 1994
Annonene, 6α,7α-dihydroxy	Brazil (R)	Silveira and Mc Chesney, 1994
Annonene, 6α,7β-diacetoxy	Brazil (R)	Silveira and Mc Chesney, 1994
Beyer-15-en-18-oic, acid: ent	Brazil (R)	Mc Chesney *et al.*, 1991a
Cascarillone, *trans*	Brazil (R)	Mc Chesney and Silveira, 1990
Hardwickiic acid,	Brazil (R)	Mc Chesney and Silveira, 1989 Mc Chesney and Silveira, 1990
Hardwickiic acid, (-)	Brazil (R)	Mc Chesney *et al.*, 1991b
Hardwickiic acid, 12-hydroxy	Brazil (R)	Mc Chesney and Silveira, 1989
Sonderianial	Brazil (R)	Mc Chesney and Silveira, 1989

Contd...

Table 7.1–*Contd...*

Compound Type Chemical Name	Place (Part Used)	References
Sonderianin	Brazil (HW, R)	Craveiro *et al.*, 1981b Mc Chesney and Silveira, 1989 Mc Chesney and Silveira, 1990 Mc Chesney *et al.*, 1991b Craveiro and Silveira, 1982
Sonderianol	Brazil (HW)	Craveiro and Silveira, 1982
Sonderianol, 3,4-seco	Brazil (HW)	Craveiro and Silveira, 1982
Trachylobanic acid, 3,4-seco	Brazil (R)	Mc Chesney *et al.*, 1991b
Annonene, 3,4-dihydro: 2,3,4-trihydroxy	Brazil (R)	Pessoa *et al.*, 2000
Coumarin		
Scopoletin	Brazil (HW)	Craveiro and Silveira, 1982
Monoterpenes		
Camphene	Brazil (EO, L)	Mc Chesney *et al.*, 1984
Camphor	Brazil (EO, L)	Mc Chesney *et al.*, 1984
Car-3-ene	Brazil (EO, L)	Craveiro *et al.*, 1981a
Cymene, *para*	Brazil (EO, L)	Craveiro *et al.*, 1981a
Limonene	Brazil (EO, L)	Mc Chesney *et al.*, 1984
Myrcene	Brazil (EO, L)	Craveiro *et al.*, 1981a Mc Chesney *et al.*, 1984
Phellandrene, α	Brazil (EO, L)	Craveiro *et al.*, 1981a
Pinene, α	Brazil (EO, L)	Craveiro *et al.*, 1981a Mc Chesney *et al.*, 1984
Pinene, β	Brazil (EO, L)	Mc Chesney *et al.*, 1984
Sabinene	Brazil (EO, L)	Craveiro *et al.*, 1981a
Terpinen-4-ol	Brazil (EO, L)	Mc Chesney *et al.*, 1984
Terpinene, γ	Brazil (EO,L)	Mc Chesney *et al.*, 1984
Terpinolene, α	Brazil (EO, L)	Craveiro *et al.*, 1981a
Thujene, α	Brazil (EO, L)	Craveiro *et al.*, 1981a
Sesquiterpenes		
Aromadendrene	Brazil (EO, L)	Craveiro *et al.*, 1981a
Cadinene, δ	Brazil (EO, L)	Mc Chesney *et al.*, 1984
Cadinene, γ	Brazil (EO, L)	Craveiro *et al.*, 1981a Mc Chesney *et al.*, 1984
Caryophyllene, β	Brazil (EO, L)	Craveiro *et al.*, 1981a Mc Chesney *et al.*, 1984
Copaene	Brazil (EO, L)	Mc Chesney *et al.*, 1984
Crotosondin	Brazil (R)	Craveiro *et al.*, 1983
Cyperene	Brazil (EO, L)	Mc Chesney *et al.*, 1984

Contd...

Table 7.1–*Contd...*

Compound Type Chemical Name	Place (Part Used)	References
Elemene, β	Brazil (EO, L)	Craveiro *et al.*, 1981a Mc Chesney *et al.*, 1984
Farnesene, β	Brazil (EO, L)	Mc Chesney *et al.*, 1984
Guaiazulene	Brazil (EO, L)	Mc Chesney *et al.*, 1984
Gurjunene, α	Brazil (EO, L)	Mc Chesney *et al.*, 1984
Humulene, α	Brazil (EO, L)	Craveiro *et al.*, 1981a
Marmelerin	Brazil (EO, L, R)	Mc Chesney *et al.*, 1984 Mc Chesney and Silveira, 1990
Muurolene, γ	Brazil (EO, L)	Mc Chesney *et al.*, 1984
Palustrol	Brazil (EO, L)	Mc Chesney *et al.*, 1984
Thujopsene	Brazil (EO, L)	Mc Chesney *et al.*, 1984

Croton sparsiflorus Morong.

C. sparsiflorus (sin. *C. bonpladianum* Bayl.) is used in Argentina as insecticide, vermifuge and antiseptic (Schmeda-Hirschmann and Rojas de Arias, 1992).

Alkaloids

Crotosparinine	Not stated (L, T)	Casagrande *et al.*, 1975
Crotosparinine, *N*-methyl	Not stated (L, T)	Casagrande *et al.*, 1975
Crotosparinine, iso	Not stated (L, T)	Casagrande *et al.*, 1975
Crotosparinine, iso-*N*-methyl	Not stated (L, T)	Casagrande *et al.*, 1975
Glaziovine, tetrahydro	Not stated (L, T)	Casagrande *et al.*, 1975
Mophinandien-7-one, 4,6-dihydroxy-3-methoxy	Not stated (L, T)	Tiwari *et al.*, 1981
Sinoacutine, nor	India (L)	Tiwari *et al.*, 1981
Sparsiflorine	India (L)	Casagrande *et al.*, 1975

Diterpenes

Phorbol 12-*O*-dodecanoyl-13-acetate	India (SE)	Upadhyay and Hecher, 1976
Phorbol 12-*O*-dodecanoyl-13-*O*-acetyl-20-linoleate	India (SE)	Upadhyay and Hecher, 1976

Flavonoid

Quercetin-3-*O*-α-rhamnosyl glycoside	India (L)	Satishi and Bhakuni, 1972

Sesquiterpene

Vomifoliol	India (L)	Satishi and Bhakuni, 1972

Steroid

Sitosterol, β	India (L, S)	Bhakuni *et al.*, 1971

Triterpenes

Taraxerol	Sri Lanka (R)	Bhakuni *et al.*, 1971
Ursolic acid	India (R)	Satishi and Bhakuni, 1972

Contd...

Table 7.1–*Contd...*

Compound Type Chemical Name	Place (Part Used)	References
Croton speciosus **Müll. Arg.**		
Alkaloid		
Turumiquirensine	Not stated (EP)	Farnsworth *et al.*, 1969
Croton stelluliferus **Hutch.**		
The essential oil from the bark showed antibacterial and antifungal activities (Martins *et al.*, 2000).		
Monoterpenes		
Camphene	Portugal (EO, B)	Martins *et al.*, 2000
Cymene, *para*	Portugal (EO, B)	Martins *et al.*, 2000
Limonene	Portugal (EO, B)	Martins *et al.*, 2000
Linalool	Portugal (EO, B)	Martins *et al.*, 2000
Phellandrene, α	Portugal (EO, B)	Martins *et al.*, 2000
Phellandrene, β	Portugal (EO, B)	Martins *et al.*, 2000
Pinene, α	Portugal (EO, B)	Martins *et al.*, 2000
Terpinen, 4-ol	Portugal (EO, B)	Martins *et al.*, 2000
Terpinene, γ	Portugal (EO, B)	Martins *et al.*, 2000
Terpinolene	Portugal (EO, B)	Martins *et al.*, 2000
Thujene, α	Portugal (EO, B)	Martins *et al.*, 2000
Sesquiterpenes		
Copaene, α	Portugal (EO, B)	Martins *et al.*, 2000
Kessane	Portugal (EO, B)	Martins *et al.*, 2000
Croton stenophyllus **Griseb**		
Alkaloids		
Salutaridine	Cuba (L, EP)	Sanchez and Sandoval, 1982a Sanchez and Sandoval, 1982b
Salutaridine, 8,14-dihydro	Cuba (L)	Sanchez and Sandoval, 1982
Croton steenkampianus		

Croton steenkampianus Gerstner (Euphorbiaceae), commonly known as "Marsh Fever-berry" and "Tonga Croton", is a shrub or tree endemic to restricted areas of central Africa and eastern parts of southern Africa (Pooley, 1993). Various medicinal uses of the genus *Croton* are reported in countries all over the world, and many species are used to treat bleeding, bleeding gums, chest complaints, coughs, fever, indigestion, malaria, and rheumatism (Pooley, 1993). Chemically, the genus contains very diverse compound types including alkaloids, flavonoids, and terpenoids (Gupta *et al.*, 2004 and Cai *et al.*, 1991).

Aliphtic		
2,6-Dimethyl-1-oxo-4-indanecarboxylic acid	African (L)	Adelekan *et al.*, 2008
Diterpenoids		
Steenkrotin A	African (L)	Adelekan *et al.*, 2008
Steenkrotin B	African (L)	Adelekan *et al.*, 2008

Contd...

Table 7.1–*Contd...*

Compound Type Chemical Name	Place (Part Used)	References
Croton stipuliformis J. Murillo		
Diterpenois		
ent-12,15-Dioxo-3,4-seco-4,8,13-labdatrien-3-oic acid	Chinchiná (L)	Ramos *et al.*, 2008
ent-12,15-Epoxy-3,4-seco-4,8,12,14-labdatetraen-3-oic acid	Chinchiná (L)	Ramos *et al.*, 2008
ent-15-nor-14-oxo-3,4-seco-4,8,12(*E*)-Labdatrien-3-oic acid	Chinchiná (L)	Ramos *et al.*, 2008
ent-12,15-dioxo-8,13-Labdadien-3a-ol	Chinchiná (L)	Ramos *et al.*, 2008
Croton sublyratus Kurz.		

Used in folk medicine of Thailand as antihelmintic, antiseptic and antiulcers, these properties are attributed to the diterpenes present in the plant (Vongchareonsathit and De-Eknamkul, 1998; Ogiso *et al.*, 1978).

Diterpenes

Geranyl-geraniol, 18-hidroxy	Thailand (S)	Kitazawa *et al.*, 1980 Ogiso *et al.*, 1981
Hexadeca-2-*trans*,6-*cis*,10-*trans*,14-tetraen-1-ol-7-hydroxymethyl-3,11,15-trimethyl	Thailand (S)	Ogiso *et al.*, 1978
Kaurane, 16β,17-dihydroxy: ent	Thailand (S)	Kitazawa and Ogiso, 1981
Manool, 13-epi: 3α-hydroxy: ent	Thailand (S)	Kitazawa and Ogiso, 1981
Plaunol A	Thailand (S)	Kitazawa *et al.*, 1979 Kitazawa *et al.*, 1980
Plaunol B	Thailand (S)	Kitazawa *et al.*, 1979 Kitazawa *et al.*, 1980 Ogiso *et al.*, 1981
Plaunol C	Thailand (S)	Kitazawa *et al.*, 1980 Ogiso *et al.*, 1981
Plaunol D	Thailand (S)	Kitazawa *et al.*, 1980 Ogiso *et al.*, 1981
Plaunol E	Thailand (S)	Kitazawa *et al.*, 1980 Ogiso *et al.*, 1981
Plaunolide	Thailand (S, EP)	Haruyama *et al.*, 1983 Takahashi *et al.*, 1983
Plaunotol	Thailand (EP, L)	Shibata *et al.*, 1996 Siriphol *et al.*, 1997 Vongchareonsathit and De-Eknamkul, 1998
Croton sublyratus furanoditerpene A	Thailand (S)	Tansakul and De-Eknamkul, 1998 Ogiso *et al.*, 1978
Croton sublyratus furanoditerpene B	Thailand (S)	Ogiso *et al.*, 1978
Geranyl-geraniol ester A	Thailand (L)	Kitazawa *et al.*, 1982

Contd...

Table 7.1–*Contd...*

Compound Type Chemical Name	Place (Part Used)	References
Geranyl-geraniol ester B	Thailand (L)	Kitazawa *et al.,* 1982
Geranyl-geraniol ester C	Thailand (L)	Kitazawa *et al.,* 1982
Geranyl-geraniol ester D	Thailand (L)	Kitazawa *et al.,* 1982
Geranyl-geraniol ester E	Thailand (L)	Kitazawa *et al.,* 1982
Geranyl-geraniol ester F	Thailand (L)	Kitazawa *et al.,* 1982
Geranyl-geraniol ester G	Thailand (L)	Kitazawa *et al.,* 1982
Planunol A	Thailand (S)	Ogiso *et al.,* 1981
Plaunol	Thailand (EP)	Mishima *et al.,* 1977
Plaunol monoacetate	Thailand (EP)	Mishima *et al.,* 1977
Steroids		
Sitosterol, β	Thailand (CT)	Eknamkul and Potduang, 2003
Stigmasterol	Thailand (CT)	Eknamkul and Potduang, 2003

Croton texensis Muell. Arg.

In the United States the leaves are used as a laxative (Hokanson and Cassady, 1976), in India the seeds are used to treat worm infestations (Selvanayahgam *et al.,* 1994).

Phenylpropanoid

Coumaric acid; *para* ethyl ester	U.S.A. (EP)	Hokanson and Cassady, 1976

Proteid

Protein	U.S.A. (SE)	Earle *et al.,* 1960

Croton tiglium L.

The seeds are used in Korea for the treatment of thorax and abdomen cancers, as contraceptives, abortive and emenagog (Woo *et al.,* 1981) in Thailand it is used as laxative (Mimmanhemin *et al.,* 1979).

Aliphatic

Triacontanoyl-hexacosanoate	Taiwan (SE)	Lin and Huang, 1977

Alkaloid

Guanosine, iso	China (SE) South Korea (SE)	Kim *et al.,* 1994 Liebich *et al.,* 1998
Crotonoside	Not stated (O, SE)	Farnsworth *et al.,* 1969

Diterpenes

Phorbol, 12-*O*-tiglyl-13-isobutyrate	Not stated (SE, O)	Marshall *et al.,* 1985 Erdelmeier *et al.,* 1988
Phorbol, 12-*O*-tetradecanoyl-13-acetate	U.S.A. (SE, O)	Kinghorn and Marshall, 1984
Phorbol, 12-*O*-dodecanoyl-13-acetate	Not stated (SE, O)	Marshall *et al.,* 1985 Erdelmeier *et al.,* 1988
Phorbol, 12-*O*-decanoyl-13-acetate	Not stated (SE, O)	Marshall *et al.,* 1985 Erdelmeier *et al.,* 1988

Contd...

Table 7.1–*Contd...*

Compound Type Chemical Name	Place (Part Used)	References
Phorbol-12-*O*-(α-methyl)butyryl-13-decanoate	Not stated (SE, O)	Marshall *et al.*, 1985
Phorbol, 12-*O*-acetyl-13-decanoate	Not stated (SE, O)	Marshall *et al.*, 1985 Erdelmeier *et al.*, 1988
Phorbol, 12-*O*-tiglyl-13-acetate	Not stated (SE, O)	Marshall *et al.*, 1985 Erdelmeier *et al.*, 1988
Phorbol,12-*O*-(2-methyl)butyryl-13-isobutyrate	Not stated (SE, O)	Erdelmeier *et al.*, 1988
Phorbol, 12- tiglate	Not stated (SE, O)	Marshall and Kinghorn, 1984
Phorbol, 12-*O*-acetyl-13-acetate	Not stated (SE, O)	Marshall and Kinghorn, 1984
Phorbol, 12-*O*-(2-methyl)butiryl-13-acetate	Not stated (SE, O)	Marshall and Kinghorn, 1984
Phorbol, 13-acetate	Not stated (SE, O)	Marshall and Kinghorn, 1984
Phorbol	U.S.A. (SE, O)	Marshall and Kinghorn, 1980 Kinghorn and Marshall, 1984 Farnsworth *et al.*, 1969 Edwards *et al.*, 1983 Cairnes *et al.*, 1981 Mishra *et al.*, 1986 Marshall and Kinghorn, 1981 Tseng *et al.*, 1977 Hickey *et al.*, 1981 Mirvish *et al.*, 1985
Phorbol, 12-*O*-acetyl-13-tigliate	Egypt (SE)	El Mekkawy *et al.*, 2000
Phorbol, 13-*O*-acetyl-20-linoleate	Egypt (SE)	El Mekkawy *et al.*, 2000
Phorbol, 13-*O*-tigloyl-20-linoleate	Egypt (SE)	El Mekkawy *et al.*, 2000
Phorbol, 4α-deoxy-12-*O*-tiglyl 13-isobutyrate	Not stated (SE, O)	Marshall and Kinghorn, 1984
Phorbol, 4α-deoxy-12-*O*-tiglyl 13-acetate	Not stated (SE, O)	Marshall and Kinghorn, 1984
Phorbol, 4α-deoxy-12-*O*-(2-methyl)butiryl-13-acetate	Not stated (SE, O)	Marshall and Kinghorn, 1984
Phorbol, 4α-deoxy-13-acetate	Not stated (SE, O)	Marshall and Kinghorn, 1984
Phorbol, 12-*O*-decanoyl-13-(2-methyl)butyrate	Egypt (SE)	El Mekkawy *et al.*, 2000
Phorbol, 12-*O*-tigloyl-13-(2-methyl)butyrate	U.S.A (SE, O) Egypt (SE)	Kinghorn and Erdelmeier, 1991 El Mekkawy *et al.*, 2000
Phorbol, 12-*O*-hexadecanoyl-13-acetate	Not stated (SE, O)	Erdelmeier *et al.*, 1988
Phorbol, 12-*O*-tiglyl-13-decanoate	Not stated (SE, O)	Erdelmeier *et al.*, 1988
Phorbol, 12-*O*-acetyl-13-dodecanoate	Not stated (SE, O)	Erdelmeier *et al.*, 1988
Phorbol, 12-*O*-tiglyl-13-octanoate	Not stated (SE, O)	Erdelmeier *et al.*, 1988
Phorbol, 4α-deoxy-5-hydroxy-12-*O*-(2-methylaminobenzoyl)-20-acetoxy	Not stated (SE, O)	Edwards *et al.*, 1983

Contd...

Table 7.1–*Contd...*

Compound Type Chemical Name	Place (Part Used)	References
Phorbol, 13-acetate-4,20-dideoxy-5-hydroxy-12-*O*-(2-methylaminobenzoyl)	Not stated (SE, O)	Edwards *et al.,* 1983
Phorbol, 4,20-dideoxy-5,13-diacetoxy-12-*O*-(2-methylaminobenzoyl)	Not stated (SE, O)	Edwards *et al.,* 1983
Phorbol, 4,20-dideoxy-5-hydroxy-12-*O*-(2-methylaminobenzoyl)	Not stated (SE, O)	Edwards *et al.,* 1983
Phorbol-13-acetate, 4,20-dideoxy-5-hidroxy-12-*O*-(*n*-deca-2,4,6-trieoyl)	Not stated (SE, O)	Edwards *et al.,* 1983
Phorbol, 4α-deoxy-5,13,20-triacetoxy-12-*O*-(2-methylaminobenzoyl)	Not stated (SE, O)	Edwards *et al.,* 1983
Phorbol, 4-deoxy-5,13,20-triacetoxy-12-*O*-(n-deca-2,4,6-trienoyl)	Not stated (SE, O)	Edwards *et al.,* 1983
Phorbol, 4-deoxy-5-hydroxy-12-*O*-(2-methylamino-benzoyl)-13-acetate	Not stated (SE, O)	Edwards *et al.,* 1983
Phorbol butyrate tiglate	Not stated (SE, O)	Bauer *et al.,* 1983
Phorbol caprylate acetate	Not stated (SE, O)	Bauer *et al.,* 1983
Phorbol caprylate tiglate	Not stated (SE, O)	Bauer *et al.,* 1983
Phorbol linoleate acetate	Not stated (SE, O)	Bauer *et al.,* 1983
Phorbol myristate acetate	Egypt (SE, O)	Hecker, 1968 El Mekkawy *et al.,* 2000 Pieters and Vlietinck, 1986 Erdelmeier *et al.,* 1988 Bauer *et al.,* 1983 Marshall *et al.,* 1985
Phorbol, 12-*O*-(2-methyl)-butyryl: 13-decanoate	Not stated (SE, O)	Erdelmeier *et al.,* 1988
Phorbol, 12-*O*-tiglyl: 13-(2-methyl)-butyrate	Not stated (SE, O)	Erdelmeier *et al.,* 1988
Phorbol, 4α	U.S.A. (SE, O)	Kinghorn and Marshall, 1984 Marshall and Kinghorn, 1981 Marshall and Kinghorn, 1980
Phorbol, 4α-deoxy: 5-hydroxy: 12-*O*-(*N*-deca-2,4,6-trienoyl)	Not stated (SE, O)	Edwards *et al.,* 1983
Phorbol, 4-deoxy-5,13,20-triacetoxy-12-*O*-(2-methyl-amino-benzoyl)	Not stated (SE, O)	Edwards *et al.,* 1983
Phorbol,5-hydroxy-4α-20-dideoxy-12-*O*-(2-methyl-amino-benzoyl)	Not stated (SE, O)	Edwards *et al.,* 1983
Phorbol,5-hydroxy-4α-deoxy-12-*O*-(2-methyl-amino-benzoyl)	Not stated (SE, O)	Edwards *et al.,* 1983
Phorbol, deoxy: tiglate acetate	Not stated (SE, O)	Bauer *et al.,* 1983
Phobol-12-tiglate-13-decanoate	Not stated (SE, O)	Kupchan *et al.,* 1976
Phorbol-13-acetate, 12-*O*-acetyl	Not stated (SE, O)	Marshall and Kinghorn, 1984

Contd...

Table 7.1—*Contd...*

Compound Type Chemical Name	Place (Part Used)	References
Phorbol-13-acetate, 12-*O*-tiglyl	Not stated (SE, O)	Marshall and Kinghorn, 1984
Phorbol-13-acetate, 4,20-dideoxy-5-hydroxy-12-*O*-(*N*-tetradecanoyl)	Not stated (SE, O)	Edwards *et al.,* 1983
Phorbol-13-acetate, 4-deoxy-5-hydroxy-12-*O*-(*N*-deca-2,4,6-trienoyl)	Not stated (SE, O)	Edwards *et al.,* 1983
Phorbol-13-decanoate, 12-*O*-acetyl	Egypt (SE)	El Mekkawy *et al.,* 2000
Phorbol-13-iso-butyrate, 12-*O*-(2-methyl-butyryl)	Not stated (SE, O)	Marshall and Kinghorn, 1984
Phorbol-13-iso-butyrate, 12-*O*-tiglyl	Not stated (SE, O)	Marshall and Kinghorn, 1984
Monoterpene		
Angelic acid	Not stated (SE, O)	Farnsworth *et al.,* 1969
Pyrazine		
2-(Furan-2-yl)-5-(2,3,4-trihydroxy-butyl)-1,4-diazine	China (L)	Wu *et al.,* 2007
Proteids		
Crotin I	China, Taiwan, Sri Lanka (SE)	Stirpe *et al.,* 1976 Lin and Huang, 1977 Chen and Pan, 1993
Crotin II	China, Sri Lanka (SE) Chen and Pan, 1993	Stirpe *et al.,* 1976
Crotin tiglium lectin (M.W. 88,000)	Taiwan (SE)	Fuh and Chen, 1982
Crotin tiglium lectin (M.W. 220,000)	Taiwan (SE)	Fuh and Chen, 1982
Sesquiterpenes		
Phorbol, 12-*O*-(2-methyl)butyryl-13-dodecanoate	Egypt (SE)	Erdelmeier *et al.,* 1988
Phorbol-13-dodecanoate,12-*O*-(2-methyl-butyryl)	Egypt (SE)	El Mekkawy *et al.,* 2000

Croton tonkinensis Gagnep.

The alkaloidal fraction obtained from the leaves showed antimalarial activity (Thuan *et al.,* 1991).

Alyphatics		
Nonacosan-2-ol	Vietnam (L)	Phan and Phan, 2004
Triacontan-1-ol	Vietnam (L)	Phan and Phan, 2004
Diterpene		
Kaur-16-en-18-ol-7β-hydroxy-15-oxo: *ent*	Vietnam (L)	Minh *et al.,* 2004
Kaur-16-en-15-one, 1 α-acetoxy-7β-14α-dihydroxy: *ent*	Vietnam (L)	Minh *et al.,* 2003 Minh *et al.,* 2004
Kaur-16-en-15-one, 18-acetoxy-7α, 14β-dihydroxy: *ent*	Vietnam (L)	Giang *et al.,* 2003

Contd...

Table 7.1–*Contd...*

Compound Type Chemical Name	Place (Part Used)	References
Kaur-16-en-15-one, 18-acetoxy-7α-hydroxy: *ent*	Vietnam (L)	Giang *et al.*, 2003
Kaur-16-en-15-on-18-oic-acid, *ent*	Vietnam (L)	Giang *et al.*, 2005
Kaur-16-en-15-one, 11α-18-diacetoxy-7β-hydroxy: *ent*	Vietnam (L)	Giang *et al.*, 2005
Kaur-16-en-15-one, 11α-acetoxy-7β, 14α-dihydroxy: *ent*	Vietnam (L)	Giang *et al.*, 2005
Kaur-16-en-15-one,18-acetoxy-11α-hydroxy: *ent*	Vietnam (L)	Phan *et al.*, 2005 Giang *et al.*, 2005
Kaur-16-en-15-one, 18-acetoxy-14(R)-hydroxy	Vietnam (L)	Giang *et al.*, 2005
Kaur-16-en-15-one, 18-acetoxy-7α-hydroxy	Vietnam (L)	Giang *et al.*, 2005
Kaur-16-en-15-one, 1β-14(R)-diacetoxy-7α-hydroxy	Vietnam (L)	Giang *et al.*, 2005
Kaur-16-en-15-one, 1β, 7α-diacetoxy-14(R)-hydroxy	Vietnam (L)	Giang *et al.*, 2005
Kaur-16-en-15-one, 1β-acetoxy-7α, 14β-dihydroxy:*ent*	Vietnam (L)	Giang *et al.*, 2003
Kaur-16-en-15-one, 7α, 14β-dihydroxy: *ent*	Vietnam (L)	Giang *et al.*, 2003
Kaur-16-en-15-one, 7β, 18-dihydroxy: 18-acetate: *ent*	Vietnam (L)	Phan *et al.*, 1999
Kaur-16-en-15-one, 7β-acetoxy-11α-hydroxy: *ent*	Vietnam (L)	Giang *et al.*, 2005
Kaur-16-en-15-oxo-18-oic acid, *ent*	Vietnam (L)	Phan *et al.*, 2005
Kaur-16-en-18-oic acid, 11α-acetoxy: *ent*	Vietnam (L)	Giang *et al.*, 2005
Kaur-16-en-18-ol, 7β-hydroxy-15-oxo: acetate: *ent*	Vietnam (L)	Son *et al.*, 2000
Kaur-16-ene, 15α,18-dihydroxy: *ent*	Vietnam (L)	Giang *et al.*, 2005
Kaur-16-ene, 18-hydroxy: *ent*	Vietnam (L)	Giang *et al.*, 2005 Phan *et al.*, 2005
Kaur-15-one, 16(S): 1α,14α-diacetoxy-7β-hydroxy-17-methoxy: e*nt*	Vietnam (L)	Giang *et al.*, 2005
Kaur-15-one, 18-acetoxy-7α-hydroxy	Vietnam (L)	Giang *et al.*, 2005
Kaur-8(14),16-diene-9,15-dione, 8,9-seco-	Thailand (L)	Thongtan *et al.*, 2003
7α,11β-diacetoxy: *ent*		
Kaur-8(14),16-diene-9,15-dione, 8,9-seco-7α-hydroxy-11β-acetoxy: *ent*	Thailand (L)	Thongtan *et al.*, 2003
Kaur-16-ene-9,15-dione, 8,9-seco-8,14-epoxy-7α-hydroxy-11β-acetoxy: *ent*	Thailand (L)	Thongtan *et al.*, 2003
Kaur-16-en-18-ol, 7β-hydroxy-15-oxo: acetate	Thailand (L)	Thongtan *et al.*, 2003

Contd...

Table 7.1–*Contd...*

Compound Type Chemical Name	Place (Part Used)	References
(1β,7α)-7-Hydroxy-9,15-dioxo-ent-8,9-secokaura-8(14),16-dien-1-yl acetate	China (L)	Chen *et al.*, 2006
(1β,7α)-8,14-Epoxy-7-hydroxy-9,15-dioxo-ent-8,9-secokaur-16-en-1-yl acetate	China (L)	Chen *et al.*, 2006
(1β,7α)-8,14-Epoxy-1,7-dihydroxy-ent-8,9-secokaur-16-ene-9,15-dione	China (L)	Chen *et al.*, 2006
(1β,7α)-1,7-Dihydroxy-ent-8,9-secokaura-8(14),16-diene-9,15-dione	China (AP)	Chen *et al.*, 2007
(1β,7α,14β)-1,7,14-Trihydroxy-ent-kaur-16-en-15-one 1,7,14-Triacetate	China (AP)	Chen *et al.*, 2007
14α-Hydroxykaur-16-en-7-one	Vietnam (EP)	Kuo *et al.*, 2007
14α-Acetoxy-17-formylkaur-15-en-18-ol	Vietnam (EP)	Kuo *et al.*, 2007
7α,10α-Epoxy-14β-hydroxygrayanane-1(5),16(17)-dien-2,15-dione	Vietnam (L)	Thuong *et al.*, 2009
7α,10α-epoxy-14β-hydroxygrayanane-1(2),16(17)-dien-15-one	Vietnam (L)	Thuong *et al.*, 2009
(1β,7α)-7-ethoxy-9,15-dioxo-ent-8,9-secokaura-8(14),16-dien-1-yl-acetate	China (L)	Yang *et al.*, 2009
Flavonoids		
Tiliroside	Vietnam (L)	Phan *et al.*, 2004
Vitexin	Vietnam (AP, L)	Minh *et al.*, 2004 Phan *et al.*, 2004
Vitexin, iso	Vietnam (AP, L)	Minh *et al.*, 2004 Phan *et al.*, 2004
***Croton triangularis* Muell. Arg.**		
Monoterpenes		
Camphene	Brazil (EO, L)	Lemos *et al.*, 1992
Cineole, 1,8	Brazil (EO, L)	Lemos *et al.*, 1992
Cubebene, α	Brazil (EO, L)	Lemos *et al.*, 1992
Pinene, α	Brazil (EO, L)	Lemos *et al.*, 1992
Pinene, β	Brazil (EO, L)	Lemos *et al.*, 1992
Camphor	Brazil (EO, L)	Lemos *et al.*, 1992
Phenylpropanoid		
Eugenol, methyl ether	Brazil (EO, L)	Lemos *et al.*, 1992
Sesquiterpenes		
Calarene	Brazil (EO, L)	Lemos *et al.*, 1992
Caryophyllene, β	Brazil (EO, L)	Lemos *et al.*, 1992
Copaene, α	Brazil (EO, L)	Lemos *et al.*, 1992
Humulene, α	Brazil (EO, L)	Lemos *et al.*, 1992
Muurolene, γ	Brazil (EO, L)	Lemos *et al.*, 1992

Contd...

Table 7.1–*Contd...*

Compound Type Chemical Name	Place (Part Used)	References

Croton trinitatis Mylls.

The aerial parts collected in Peru showed antibacterial and antiviral activities (Macrae *et al.*, 1988)

Sesquiterpene

| Vomifoliol | Guyana (L, S) | Stuart and Woo-Ming, 1975 |

Croton turumiquirensis Steyerm.

Alkaloid

| Magnoflorine | Venezuela (NS) | Burnell *et al.*, 1981 |

Croton urucurana Baillon

It grows in several areas of Brazil. Its bark is used for the treatment of cancer and for healing wounds. In Argentina, it is used in the treatments of diarrhea and infections of the urinary and respiratory systems (Peres and Anesini, 1994).

Diterpenes

Barbascoatic acid,12-epi-methyl ester	Brazil (SB)	Peres *et al.*, 1998a
Cleroda-3,13(16) trien-2-one, 15,16-epoxy	Brazil (SB)	Peres *et al.*, 1998a
Sonderianin	Brazil (B, SB)	Peres *et al.*, 1997 Peres *et al.*, 1998a Peres *et al.*, 1998b

Carbohydrate

| Croton fucoarabinogalactan | Paraguay (G) | Milo *et al.*, 2002 |

Flavonoids

| Catechin, (+) | Brazil (B, SB) | Peres *et al.*, 1997 Peres *et al.*, 1998b |
| Gallocatechin, (+) | Brazil (B, SB) | Peres *et al.*, 1997 Peres *et al.*, 1998b |

Steroids

Campesterol	Brazil (B,SB)	Peres *et al.*, 1997 Peres *et al.*, 1998b
Daucosterol	Brazil (B, SB)	Peres *et al.*, 1997 Peres *et al.*, 1998b
Sitosterol, β	Brazil (B, SB)	Peres *et al.*, 1997 Peres *et al.*, 1998b
Stigmasterol	Brazil (B, SB)	Peres *et al.*, 1997 Peres *et al.*, 1998b

Triterpene

| Aleuritolic acid, acetyl | Brazil (B, SB) | Peres *et al.*, 1997 Peres *et al.*, 1998b |

Contd...

Table 7.1–*Contd...*

Compound Type Chemical Name	Place (Part Used)	References
Croton verreauxii Baillon		

In Australia, this plant has been suspect of being the cause of death of pigs with violent vomiting (Farnsworth *et al.*, 1969).

Diterpenes

| Croverin | Not stated (AP) | Fujita *et al.*, 1980 |
| Croverin, dihydro | Not stated (AP) | Fujita *et al.*, 1980 |

Croton wilsonii Griseb.

Alkaloids

Hernovine	Jamaica (EP)	Farnsworth *et al.*, 1969
Hernovine, 10-*O*-methyl	Jamaica (EP)	Farnsworth *et al.*, 1969
Hernovine, *N*-methyl	Jamaica (EP)	Farnsworth *et al.*, 1969
Hernovine, *N*-ethyl: 10-*O*-methyl	Jamaica (EP)	Farnsworth *et al.*, 1969
Wilsonirine	Jamaica (EP)	Farnsworth *et al.*, 1969

Croton zambesicus Muell. Arg.

In Africa, the Masai utilize this species with *Grewia villosa* as a strengthening medicine (Farnsworth *et al.*, 1969).

Aliphatic

Octan-2-one	Cameroon (L, RB, SB)	Boyom *et al.*, 2002
Dodecan-2-one	Cameroon (L, RB)	Boyom *et al.*, 2002
Undecan-2-one	Cameroon (L, RB)	Boyom *et al.*, 2002

Diterpenes

Crotocorylifuran	Cameroon (SB)	Ngadjui *et al.*, 1999 Ngadjui *et al.*, 2002
Crotonadiol	Cameroon (SB)	Ngadjui *et al.*, 1999 Ngadjui *et al.*, 2002
Crotozambefuran A	Cameroon (SB)	Ngadjui *et al.*, 2002
Crotozambefuran B	Cameroon (SB)	Ngadjui *et al.*, 2002
Crotozambefuran C	Cameroon (SB)	Ngadjui *et al.*, 2002
Trachyloban-3β-ol: ent	Benin (L)	Block *et al.*, 2002
Trachyloban-18-oic acid, 7β-acetoxy	Cameroon (SB)	Ngadjui *et al.*, 1999 Ngadjui *et al.*, 2002
Trachyloban-7β,18-diol	Cameroon (SB)	Ngadjui *et al.*, 1999 Ngadjui *et al.*, 2002
ent-18-Hydroxy-trachyloban-3-one	Benin (L)	Block *et al.*, 2005
ent-Trachyloban-3-one	Benin (L)	Block *et al.*, 2005
Isopimara-7,15-dien-3β-ol	Benin (L)	Block *et al.*, 2005
ent-Kaurane-3β,16β,17-triol	Africa (F)	Mohamed *et al.*, 2009

Contd...

Table 7.1–*Contd...*

Compound Type Chemical Name	Place (Part Used)	References
ent-Trachylobane	Benin (L)	Block *et al.*, 2006
Sandaracopimaradiene	Benin (L)	Block *et al.*, 2006
Kaurene	Benin (L)	Block *et al.*, 2006
ent-Trachyloban-3-one	Benin (L)	Block *et al.*, 2006
ent-Trachyloban-3β-ol	Benin (L)	Block *et al.*, 2006
Isopimara-7,15-dien-3β-ol	Benin (L)	Block *et al.*, 2006
Flavonoids		
Vitexin	Africa (F)	Mohamed *et al.*, 2009
Monoterpenes		
Thujene, α	Nigeria (EO, L)	Usman *et al.*, 2009 Boyom *et al.*, 2002 Block *et al.*, 2006
Pinene, α	Nigeria (EO, L)	Usman *et al.*, 2009 Boyom *et al.*, 2002 Block *et al.*, 2006
Sabinene	Nigeria (EO, L)	Usman *et al.*, 2009 Boyom *et al.*, 2002 Block *et al.*, 2006
Pinene, β	Nigeria (EO, L)	Usman *et al.*, 2009 Boyom *et al.*, 2002 Block *et al.*, 2006
Myrcene	Nigeria (EO, L)	Usman *et al.*, 2009 Boyom *et al.*, 2002 Block *et al.*, 2006
Limonene	Nigeria (EO, L)	Usman *et al.*, 2009 Boyom *et al.*, 2002 Block *et al.*, 2006
Cineole, 1,8	Nigeria (EO, L)	Usman *et al.*, 2009 Block *et al.*, 2006
Ocimene, *cis*	Nigeria (EO, L)	Usman *et al.*, 2009
Terpinene, γ	Nigeria (EO, L, RB)	Usman *et al.*, 2009 Boyom *et al.*, 2002 Block *et al.*, 2006
Borneol	Nigeria (EO, L)	Usman *et al.*, 2009 Boyom *et al.*, 2002 Block *et al.*, 2006
Terpinen-4-ol	Nigeria (EO, L)	Usman *et al.*, 2009 Boyom *et al.*, 2002 Block *et al.*, 2006
Terpineol, α	Nigeria (EO, L)	Usman *et al.*, 2009 Boyom *et al.*, 2002
Neral	Nigeria (EO, L)	Usman *et al.*, 2009

Contd...

Table 7.1–*Contd...*

Compound Type Chemical Name	Place (Part Used)	References
Geranial	Nigeria (EO, L)	Usman *et al.*, 2009
Borneol acetate	Nigeria (EO, L)	Usman *et al.*, 2009
Camphene	Cameroon (L, RB, SB)	Boyom *et al.*, 2002 Block *et al.*, 2006
Phellandrene, α	Cameroon (L, RB, SB)	Boyom *et al.*, 2002
Terpinene, α	Cameroon (L, SB)	Boyom *et al.*, 2002
Ocimene, (Z)-β	Cameroon (L, SB)	Boyom *et al.*, 2002
Ocimene (E)-β	Cameroon (L, SB)	Boyom *et al.*, 2002
Linalool oxide, (Z)	Cameroon (SB)	Boyom *et al.*, 2002
Linalool oxide, (E)	Cameroon (SB)	Boyom *et al.*, 2002
Linalool	Cameroon (L, RB, SB)	Boyom *et al.*, 2002 Block *et al.*, 2006
Camphor	Cameroon (L, RB, SB)	Boyom *et al.*, 2002 Block *et al.*, 2006
Cymen-8-ol, p	Cameroon (L,RB, SB)	Boyom *et al.*, 2002
Myrtenol	Cameroon (L, RB, SB)	Boyom *et al.*, 2002 Block *et al.*, 2006
Bornyl formate	Cameroon (L, RB, SB)	Boyom *et al.*, 2002
Thymol	Cameroon (SB)	Boyom *et al.*, 2002
Bornyl acetate	Cameroon (L, RB, SB)	Boyom *et al.*, 2002
Geranylacetone	Cameroon (RB)	Boyom *et al.*, 2002
Cymene, p	Cameroon (L, RB, SB)	Boyom *et al.*, 2002 Block *et al.*, 2006
Sabinene hydrate, cis	Benin (L)	Block *et al.*, 2006
Pinocarveol, trans	Benin (L)	Block *et al.*, 2006
Pinocarvone	Benin (L)	Block *et al.*, 2006
Cyclosativene	Benin (L)	Block *et al.*, 2006
Phenylpropanoids		
Eugenol	Nigeria (EO, L)	Usman *et al.*, 2009
Elemicin	Nigeria (EO, L)	Usman *et al.*, 2009
Viridiflorol	Nigeria (EO, L)	Usman *et al.*, 2009
Torreyol	Nigeria (EO, L)	Usman *et al.*, 2009
Benzyl benzoate	Nigeria (EO, L)	Usman *et al.*, 2009
Sesquiterpene		
Betulinol	Cameroon (SB)	Ngadjui *et al.*, 2002
Copane, α	Nigeria (EO, L)	Usman *et al.*, 2009
Elemene, β	Nigeria (EO, L)	Usman *et al.*, 2009 Boyom *et al.*, 2002

Contd...

Table 7.1–*Contd...*

Compound Type Chemical Name	Place (Part Used)	References
Caryophyllene, β	Nigeria (EO, L)	Usman *et al.*, 2009 Boyom *et al.*, 2002 Block *et al.*, 2006
Ethyl cinamate	Nigeria (EO, L)	Usman *et al.*, 2009
Germacrene D	Nigeria (EO, L, RB)	Usman *et al.*, 2009 Boyom *et al.*, 2002
Bicyclogermacrene	Nigeria (EO, L, RB)	Usman *et al.*, 2009 Boyom *et al.*, 2002
Bisabolene, β	Nigeria (EO, L)	Usman *et al.*, 2009
Acetyl eugenol	Nigeria (EO, L)	Usman *et al.*, 2009
Elemene, δ	Cameroon (L, RB)	Boyom *et al.*, 2002
Cubebene, α	Cameroon (L, RB, SB)	Boyom *et al.*, 2002
Copaene, α	Cameroon (L, RB, SB)	Boyom *et al.*, 2002 Block *et al.*, 2006
Cubebene, β	Cameroon (L, RB)	Boyom *et al.*, 2002
Cyperene	Cameroon (L, RB)	Boyom *et al.*, 2002 Block *et al.*, 2006
Bergamotene	Cameroon (L, RB, SB)	Boyom *et al.*, 2002
Humulene, α	Cameroon (L, RB, SB)	Boyom *et al.*, 2002 Block *et al.*, 2006
aromadendrene	Cameroon (L, RB)	Boyom *et al.*, 2002 Block *et al.*, 2006
Selinene, β	Cameroon (RB)	Boyom *et al.*, 2002
Muurolene, α	Cameroon (L, RB, SB)	Boyom *et al.*, 2002
Cadinene, γ	Cameroon (L, RB, SB)	Boyom *et al.*, 2002
Cadinene, δ	Cameroon (L, RB, SB)	Boyom *et al.*, 2002 Block *et al.*, 2006
Calamenene, (Z)	Cameroon (L, RB)	Boyom *et al.*, 2002
Calacorene, (E)	Cameroon (L, RB, SB)	Boyom *et al.*, 2002
Germacrene B	Cameroon (L, RB)	Boyom *et al.*, 2002
Curcumone, *ar-*	Cameroon (L, RB)	Boyom *et al.*, 2002
Nerolidol, α	Cameroon (L, RB, SB)	Boyom *et al.*, 2002
Nerolidol, β	Cameroon (L)	Boyom *et al.*, 2002
Spathulenol	Cameroon (L, RB, SB)	Boyom *et al.*, 2002
Caryophyllene oxide	Cameroon (L, RB, SB)	Boyom *et al.*, 2002 Block *et al.*, 2006
Guaiol	Cameroon (L, RB)	Boyom *et al.*, 2002 Block *et al.*, 2006
Humulene oxide	Cameroon (L, RB, SB)	Boyom *et al.*, 2002 Block *et al.*, 2006

Contd...

Table 7.1–*Contd...*

Compound Type Chemical Name	Place (Part Used)	References
Longiborneol	Cameroon (RB, SB)	Boyom *et al.*, 2002
14-Hydroxy muurolene	Cameroon (L, RB, SB)	Boyom *et al.*, 2002
Cadinol, α	Cameroon (RB, SB)	Boyom *et al.*, 2002
Bourbonene, β	Benin (L)	Block *et al.*, 2006
Cedrene, β	Benin (L)	Block *et al.*, 2006
Muurolene, γ	Benin (L)	Block *et al.*, 2006
Cuparene	Benin (L)	Block *et al.*, 2006
Cadinen, *trans*-γ	Benin (L)	Block *et al.*, 2006
Steroids		
Daucosterol	Cameroon (SB)	Ngadjui *et al.*, 1999 Ngadjui *et al.*, 2002
Sitosterol, β	Cameroon (SB)	Ngadjui *et al.*, 1999 Ngadjui *et al.*, 2002
Triterpenes		
Lupeol	Cameroon (SB) Africa (F)	Ngadjui *et al.*, 1999 Ngadjui *et al.*, 2002 Mohamed *et al.*, 2009
Betulinic Acid	Africa (F)	Mohamed *et al.*, 2009
Betulin	Africa (F)	Mohamed *et al.*, 2009

Croton zehntneri Pax and K. Hoffm.

It grows in the State of Ceará, Northeastern Brazil. It smells like carnation flowers. Its bark and leaves extracts are used in perfumes and to aromatize food and beverages. In folk medicine it is used as a sedative and to treat gastric problems (Albuquerque *et al.*, 1995).

Aliphatics

Eicosane, *n*	Brazil (EO, S)	Craveiro *et al.*, 1978b
Heptadecane, *n*	Brazil (EO, S)	Craveiro *et al.*, 1978b
Crototropone	Brazil (R)	Bracher *et al.*, 2008

Monoterpenes

Borneol, iso	Brazil (EO, S, L)	Craveiro *et al.*, 1978b
Camphor	Brazil (EO, S)	Craveiro *et al.*, 1978b
Cineole, 1-8	Brazil (EO, S, W)	Craveiro *et al.*, 1978b
Cymene, *para*	Brazil (EO, L)	Craveiro *et al.*, 1981a
Geranial	Brazil (EO, L)	Craveiro *et al.*, 1981a
Linalool	Brazil (EO, L)	Craveiro *et al.*, 1981a
Myrcene	Brazil (EO, S, L, W)	Craveiro *et al.*, 1978b Craveiro *et al.*, 1981a
Neral	Brazil (EO, L)	Craveiro *et al.*, 1981a

Contd...

Table 7.1–*Contd...*

Compound Type Chemical Name	Place (Part Used)	References
Pinene, α	Brazil (EO, S, L)	Craveiro *et al.*, 1978b Craveiro *et al.*, 1981a
Pinene, β	Brazil (EO, S)	Craveiro *et al.*, 1978b
Phenylpropanoids		
Anethole	Brazil (EO, S, L)	Albuquerque *et al.*, 1995
Estragole	Brazil (EO, S, L, B, W)	Craveiro *et al.*, 1978b Craveiro *et al.*, 1981a Albuquerque *et al.*, 1995 Coelho-de-Sousa *et al.*, 1997
Eugenol	Brazil (EO, L)	Craveiro *et al.*, 1981a
Eugenol, methyl ether	Brazil (EO, S, L, W)	Craveiro *et al.*, 1978b
Eugenol, iso: methyl ether	Brazil (EO, S, W)	Craveiro *et al.*, 1978b
Safrole	Brazil (EO, L)	Craveiro *et al.*, 1978b
Anethole, trans	Brazil (EO, S, B, L)	Craveiro *et al.*, 1978b Craveiro *et al.*, 1981a Craveiro and Lemos, 1980
Anethole, cis	Brazil (B)	Craveiro and Lemos, 1980
Sesquiterpenes		
Caryophyllene	Brazil (EO, S, L, W)	Craveiro *et al.*, 1978b
Caryophyllene, β	Brazil (EO, L)	Craveiro *et al.*, 1981a
Elemene, γ	Brazil (EO, S, L, W)	Craveiro *et al.*, 1978b Craveiro *et al.*, 1981a
Farnesene, β	Brazil (EO, L)	Craveiro *et al.*, 1981a
Guayene, β	Brazil (EO, L)	Craveiro *et al.*, 1981a
Muurolene, γ	Brazil (EO, L)	Craveiro *et al.*, 1981a
Bergamoatene, α	Brazil (EO, L)	Craveiro *et al.*, 1981a

AP: Aerial parts; B: Bark; BR: Branches; C: Cortex; CT: Callus tissue; EO: Essential oil; EP: Entire plant; F: Fruits; FL: Flowers; HW: Heartwood; L: Leaves; LX: Latex; NS: Not specified; O: Oil; R: Roots; RB: Root bark; S: Stems; SA: Sap; SE: Seeds; SB: Stem bark; T: Twigs; TB: Trunk bark; TW: Trunk wood; W: Wood.

Conclusions

Of the some 700 species of plants belonging to genus *Croton*, only 108 have been chemically studied. The plants are distributed worldwide. It is important to state the variety of chemical classes of natural products found in the genus. They are rich in essential oils, in which can be found Phenylpropanoids, Monoterpenes and Sesquiterpenes. It is also possible to find alkaloids, diterpenes, triterpenes, flavonoids, lignoids and a few different compounds classified as miscellaneous.

Figure 7.1: Structure of compounds of the genus *Croton*.

Alicyclic

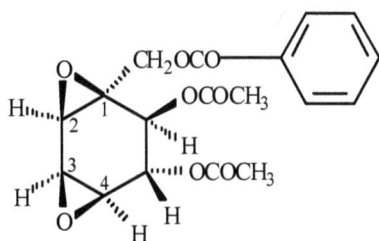

(1) Crotepoxide

Aliphatics

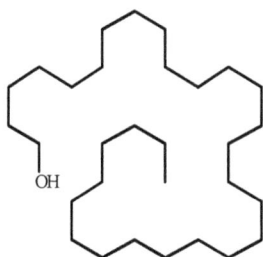

(2) Triacontan-1-ol

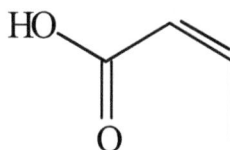

(3) Crotonic acid

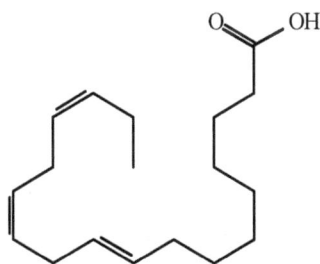

(4) Linolenic acid

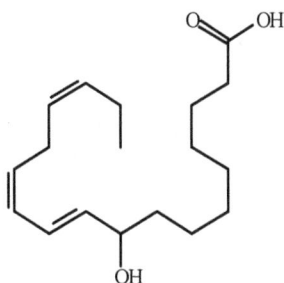

(5) Octadeca-10-*trans*-12-*cis*-15-*cis*-trienoic acid, 9-hydroxy

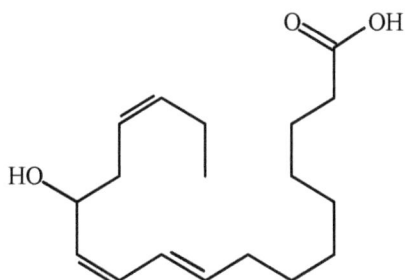

(6) Octadeca-9-*cis*-11-*trans*-15-*cis*-trienoic acid, 13-hydroxy

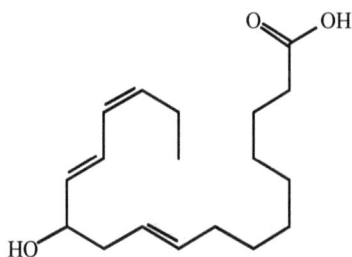

(6) Octadeca-9-*cis*-13-*trans*-15-*cis*-trienoic acid-12-hydroxy

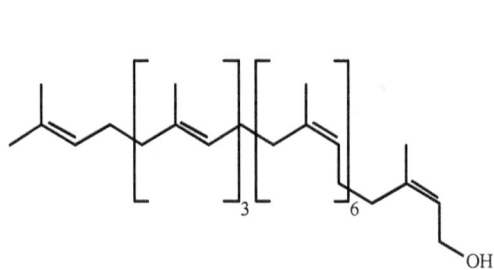

(8) Castaprenol 11

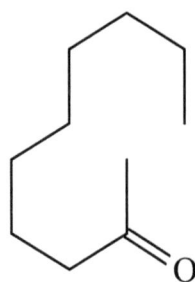

(9) Decan-2-one

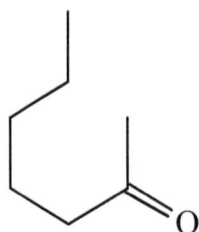

(10) Heptan-2-one

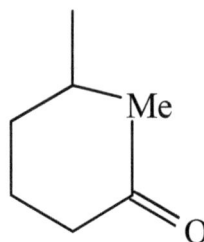

(11) Heptan-2-one, 6-methyl

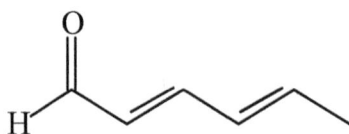

(12) Hexa-2,4-dien-1-al

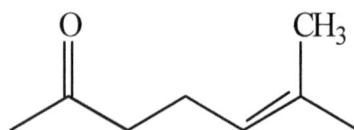

(13) Hept-5-en-2-one, 6-methyl

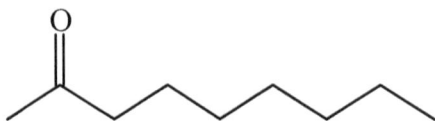

(14) Nonan-2-one

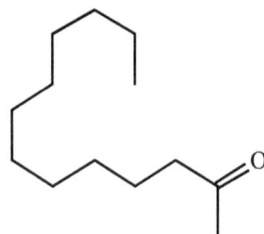

(15) Tridecan-2-one

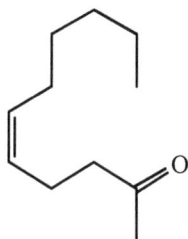

(16) Undec-*cis*-5-en-2-one

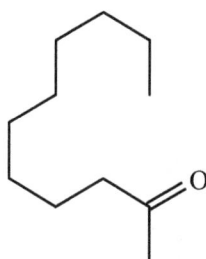

(17) Undecan-2-one

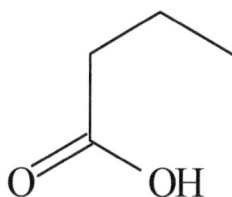

(18) Valeric acid

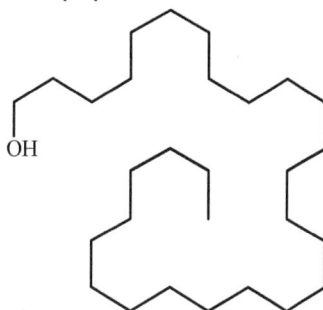

(19) Octacosan, 1-ol

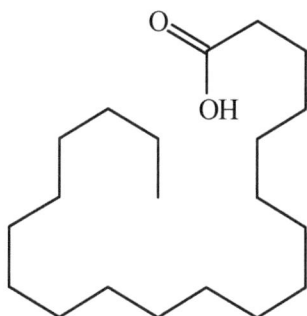

(20) Arachidic acid

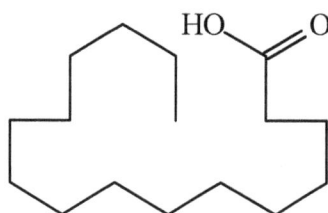

(21) Palmitic acid

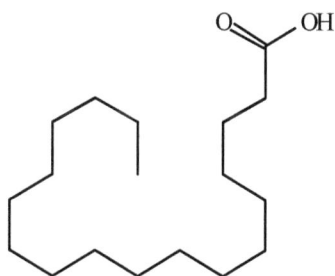

(22) Stearic acid

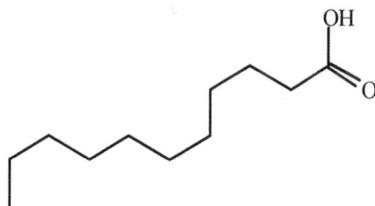

(23) Undecanoic acid

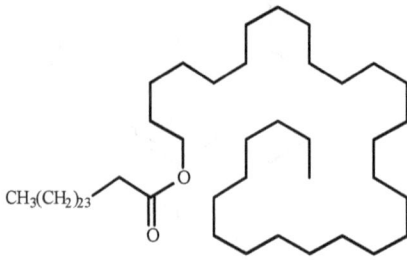

(24) Triacontanoyl-hexacosanoate

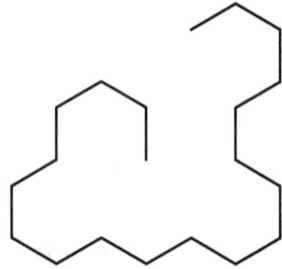

(25) Eicosane, *n*

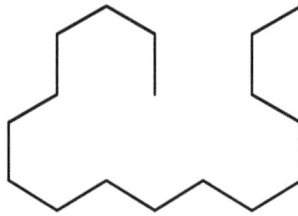

(26) Heptadecane, *n*

Alkaloids

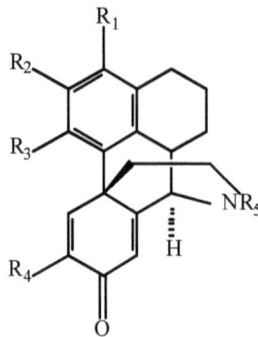

(27a) R_1=H R_2=OMe R_3=OH R_4=OMe R_5=Me - Salutaridine
(27b) R_1=H R_2=OMe R_3=OH R_4=OMe R_5=H - Salutaridine, nor
(27c) R_1=OH R_2=OMe R_3=H R_4=OMe R_5=Me - Salutaridine, iso
(27d) R_1=OMe R_2=OH R_3=H R_4=OMe R_5=Me - Flavinantine
(27e) R_1=OMe R_2=OMe R_3=H R_4=OMe R_5=Me - Flavinantine, *O*-methyl
(27f) R_1=OMe R_2=OH R_3=H R_4=OMe R_5=Me - Sebiferine

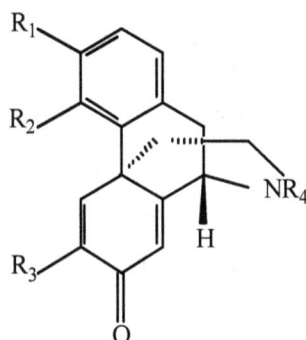

(28a) R_1=OMe R_2=OH R_3=OMe R_4=Me - Sinoacutine
(28b) R_1=OMe R_2=OH R_3=OMe R_4=H - Sinoaucutine, nor

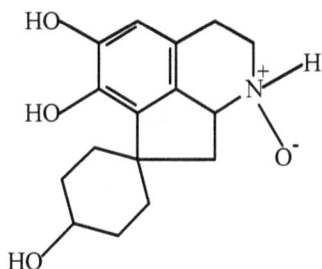

(29) Crotosinoline-*N*-oxide, 1,2,10-trihydroxy

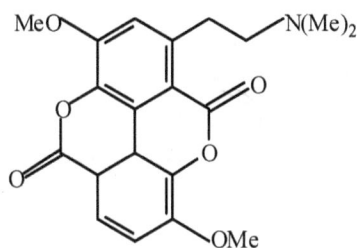

(30) Taspine

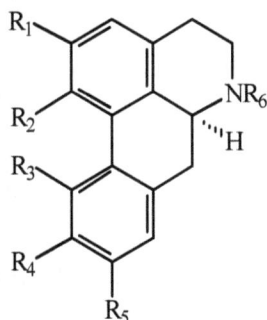

(31a) R_1=OMe R_2=OH R_3=H R_4=OMe R_5=OH R_6=Me - Boldine, iso
(31b) R_1=OMe R_2=OMe R_3=OH R_4=OMe R_5=H R_6=Me - Coridine, isso
(31c) R_1=OMe R_2=OH R_3=H R_4=OMe R_5=OMe R_6=Me - Thalyporphine
(31d) R_1=OMe R_2=OH R_3=H R_4=OH R_5=H R_6=H - Sparsiflorine
(31e) R_1=OMe R_2=OH R_3=H R_4=OMe R_5=OH R_6=H - Laurelliptine
(31f) R_1=OMe R_2=OH R_3=H R_4=OMe R_5=OH R_6=Me - Laurelliptine, *N*-methyl
(31g) R_1=OMe R_2=OH R_3=OMe R_4=OMe R_5=H R_6=Me - Corydine
(31h) R_1=OMe R_2=OMe R_3=H R_4=OMe R_5=OMe R_6=Me - Glaucine

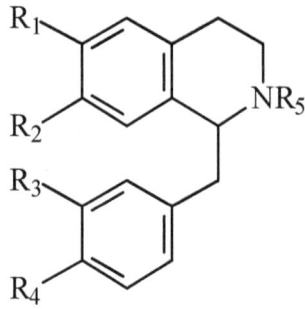

(32a) R₁=OMe R₂=OMe R₃=OH R₄=OMe R₅=Me - Laudanidine, (+)

(32b) R₁=OMe R₂=OH R₃=OH R₄=OMe R₅=Me - Reticuline, (-)

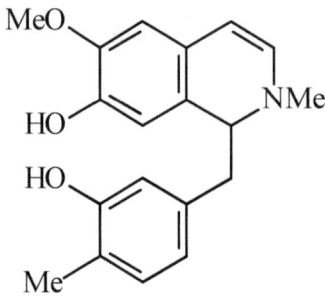

(33) Reticuline, 3,4-dehydro

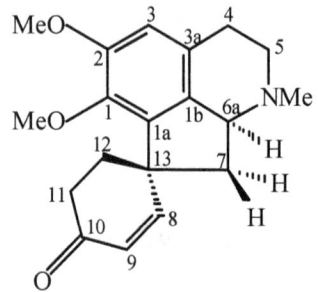

(34) Amuronine, (-)

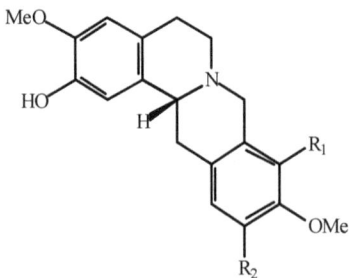

(35a) R₁=H R₂=OH - Coreximine
(35b) R₁=OH R₂=H - Scoulerine

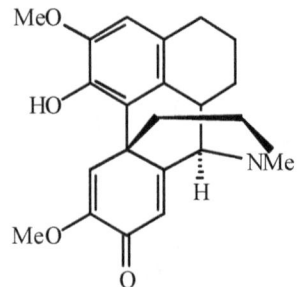

(36) Salutarine

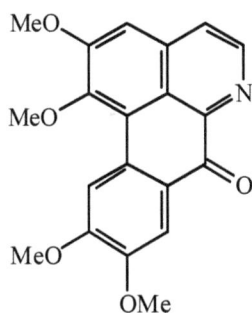

(37) Glaucine, oxo

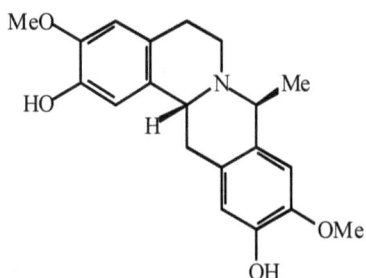

(38) Hemiargyrine

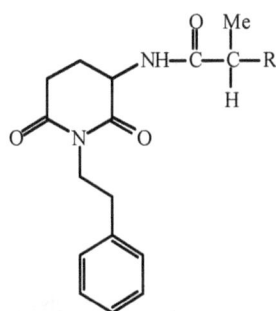

(39a) R=Me - Glutarimide, 2-[*N*-(2-methyl-propanoyl]-*N*-phenylethyl

(39b) R=CH$_2$CH$_3$ - Glutarimide, 2-[*N*-(2-methylbutanoyl]-*N*-phenylethyl

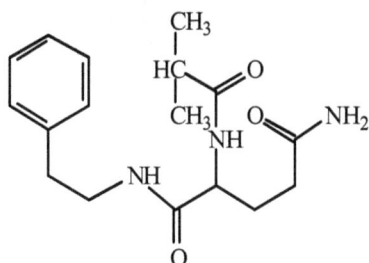

(40) Phenylethylamine, 2:*N*-[*N*-(2-methylpropanoyl]-L-glutaminoyl

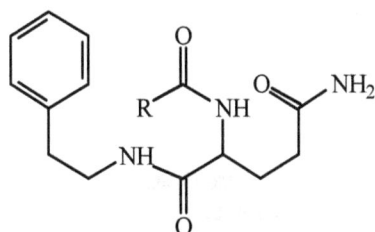

(41a) R=CH(CH$_3$)CH$_2$CH$_3$. Phenylethylamine, 2:*N*-[*N*-(2-methylbutanoyl]-glutaminoyl

(41b) R=CH(CH$_3$)CH$_3$. Phenylethylamine, 2:*N*-[*N*-2-methylpropionoyl]-glutaminoyl

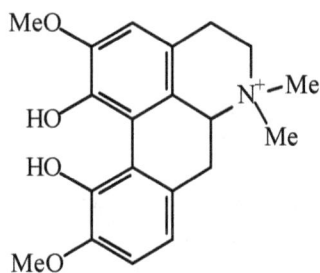

(42) Magnoflorine

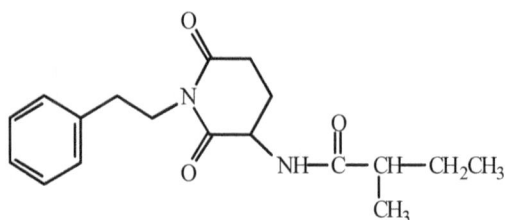

(43) Julocrotine

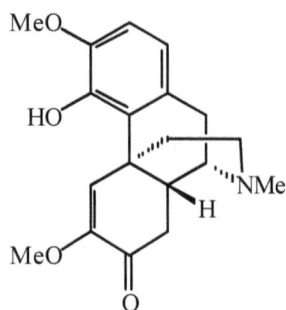

(44) Salutaridine, 8,14-dihydro

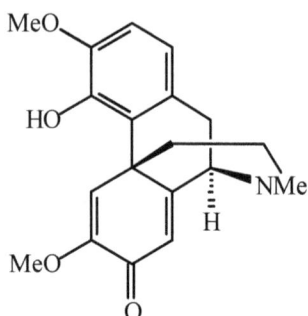

(45) Salutaridine, 8,14-dihydro

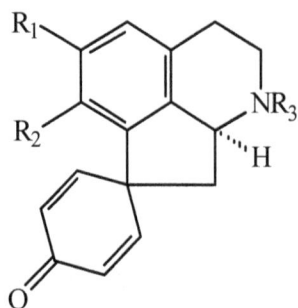

(46a) R_1=OMe R_2=OH R_3=H -
Crotsparine

(46b) R_1=OMe R_2=OH R_3=Me -
Jacularine

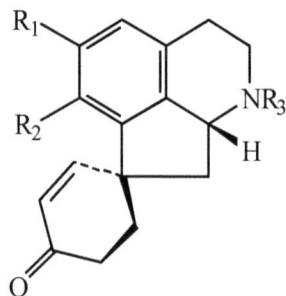

(47a) R_1=OMe R_2=OH R_3=H -
Crotosparinine

(47b) R_1=OMe R_2=OH R_3=Me -
Crotosparinine, *N*-methyl

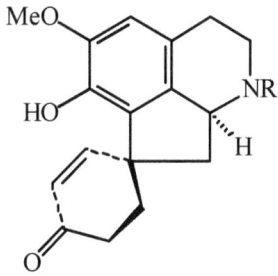

(48a) R=H - Crotosparinine, iso
(48b) R=Me - Crotosparinine, iso-*N*-methyl

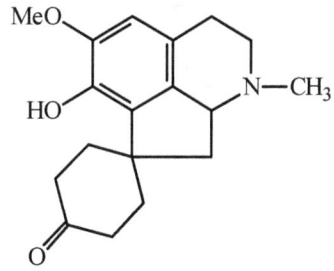

(49) Glaziovine, tetrahydro

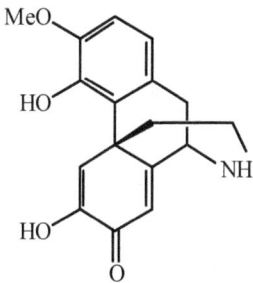

(50) Mophinandien-7-one,
4,6-dihydroxy-3-methoxy

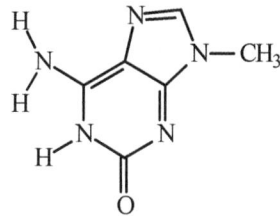

(51) Guanosine, isso

Benzenoids

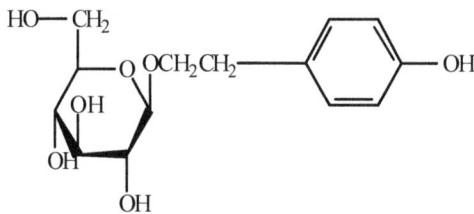

(52) Salidroside

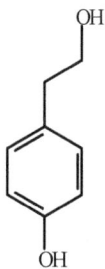

(53) Tyrosol

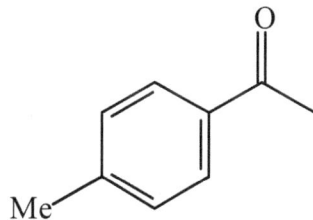

(54) Acetophenone, *para*-methyl

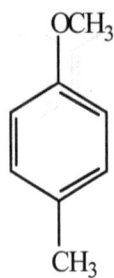

(55) Anisole, *orto*-methyl

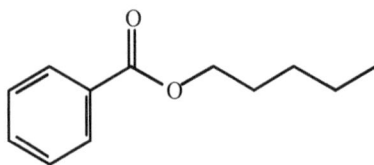

(56) Benzoic acid, pentyl ester

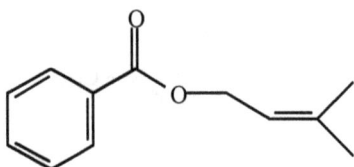

(57)

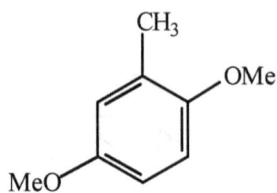

(58) Toluene, 2,5-dimethoxy

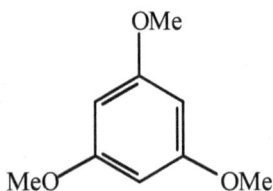

(59) Benzene, 1,3,5-trimethoxy

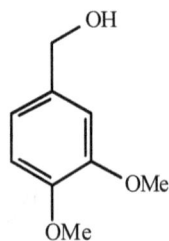

(60) Benzyl alcohol, 3,4-dimethoxy

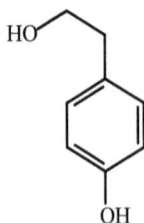

(61) Phenethyl alcohol, 4-hydroxy

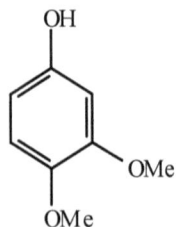

(62) Phenol, 3,4-dimethoxy

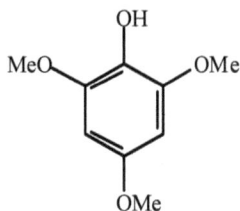

(63) Phenol, 2,4,6-trimethoxy

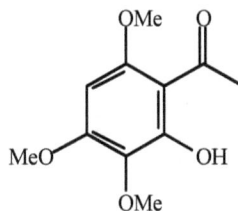

(64) Acetophenone, 2-hydroxy-3,4,6-trimethoxy

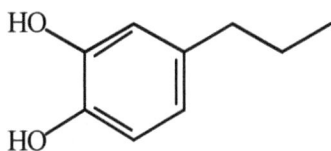

(65) Cathecol, *n*-propyl

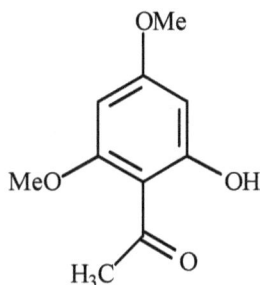

(66) Xanthoxylin

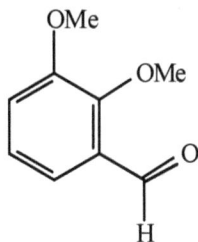

(67) Veratraldehyde

Carbohydrates

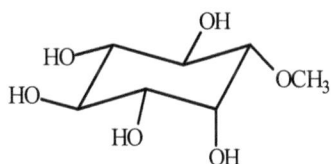

(68) Bornesitol, D: (-)

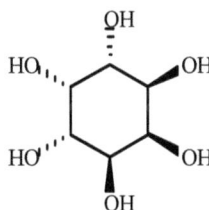

(69) Inositol, neo

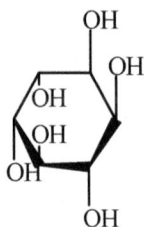

(70) Inositol, L: (-)

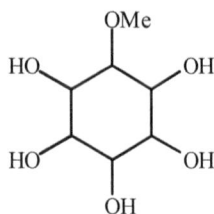

(71) Inositol, muco: 1-*O*-methyl

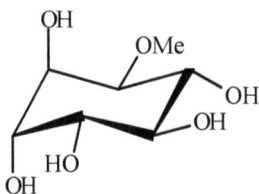

(72) Quebrachitol

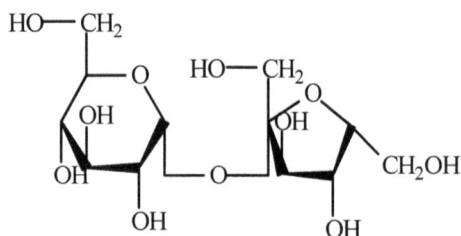

(73) Sucrose

Diterpenes

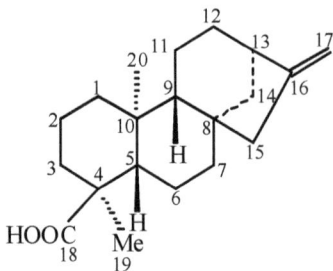

(74) Argyrophylic acid

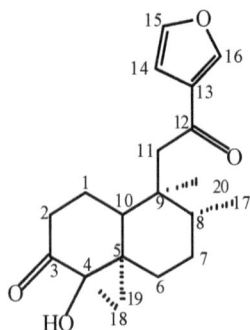

(75) Cleroda-13(16),14-diene, 3,12-
dioxo-15,16-epoxy-4-hydroxy

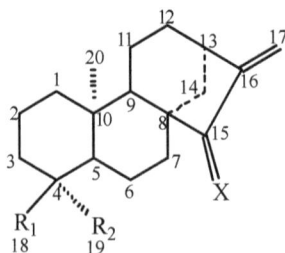

(76a) X= H, H R₁=COOH R₂=Me - Kaur-16-en-13-oxo-18-oic acid: ent

(76b) X= O R₁=COOH R₂=Me - Kaur-16-en-19-oic acid: ent

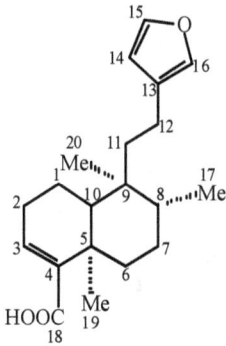

(77) Hardwickiic acid, (-)

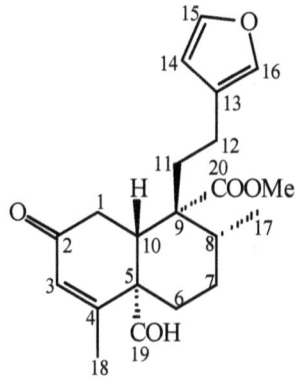

(78) Cajucarin A

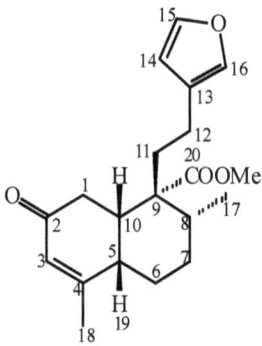

(79) Cajucarin B

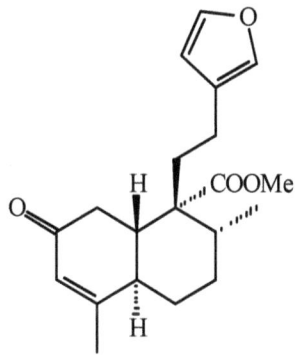

(80) Cajucarin B, *trans*

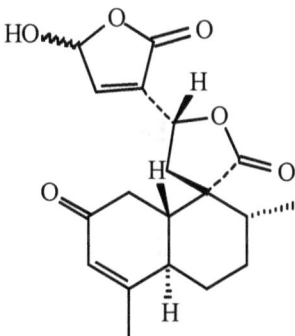

(81) Cajucarinolide

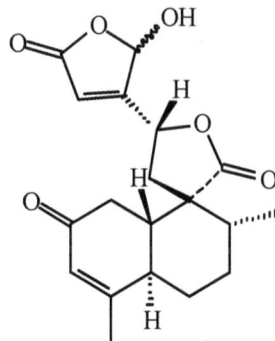

(82) Cajucarinolide, iso

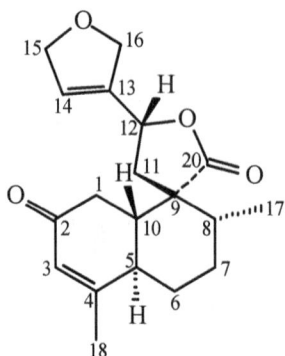

(83) Crotonin, dehydro

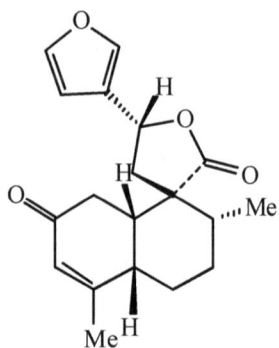

(84) Crotonin, *cis*-dehydro

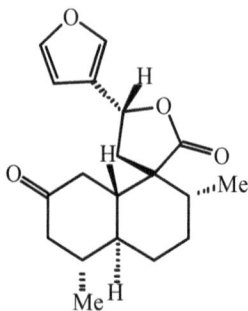

(85) Crotonin, *trans*

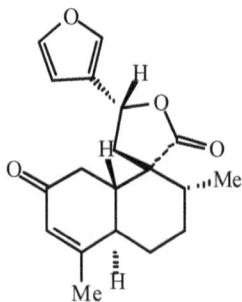

(86) Crotonin, *trans*-dehydro

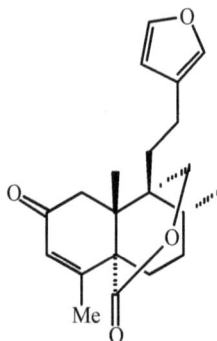

(87) Sacacarin

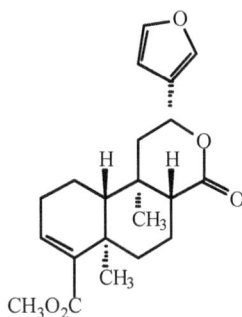

(88) Barbascoate methyl, (-)

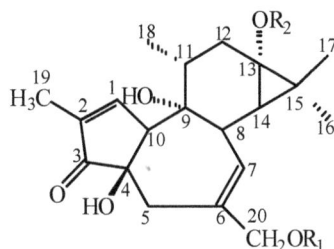

(89a) R$_1$=CO(CH$_2$)$_8$CH$_3$ R$_2$= CO(CH$_2$)$_8$CH$_3$. Phorbol, 12-deoxy:13,20-*O*-didecanoyl

(89b) R$_1$=CO(CH$_2$)$_{10}$CH$_3$ R$_2$= CO(CH$_2$)$_8$CH$_3$. Phorbol, 12-deoxy-13-*O*-decanoyl-20-*O*-dodecanoyl

(89c) R$_1$=CO(CH$_2$)$_{14}$CH$_3$ R$_2$= CO(CH$_2$)$_8$CH$_3$. Phorbol, 12-deoxy:13-*O*-decanoyl-20-*O*-hexadecanoyl

(89d) R$_1$= CO(CH$_2$)$_{12}$CH$_3$ R$_2$= CO(CH$_2$)$_8$CH$_3$. Phorbol, 12-deoxy:13-*O*-decanoyl-20 tetradecanoyl

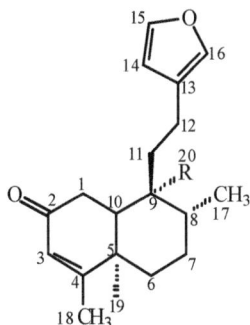

(90a) R=CH$_3$. Cleroda-3,13(16)-14-trien-15,16-epoxy-2-oxo: ent
(90b) R=CH$_2$OAc - Cleroda-3,13(16)-14-trien-15,16-epoxy-20-acetoxy-2-oxo: *ent*
(90c) R=CH$_2$OH - Cleroda-3,13(16)-14-triene,15,16-epoxy-20-hydroxy-2-oxo:*ent*

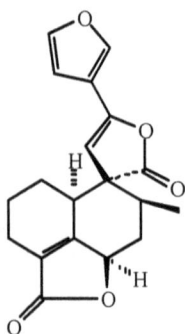

(91) Crotocaudin

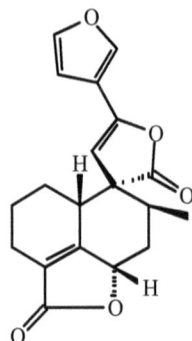

(92) Crotocaudin, iso

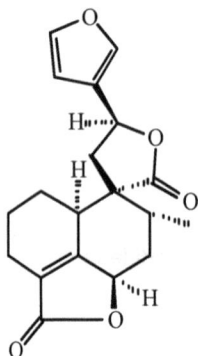

(93) Teucvidin

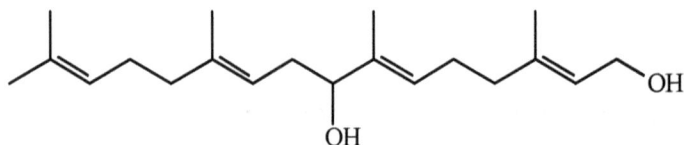

(94) Geranyl-geraniol, all-*trans*:8-hydroxy

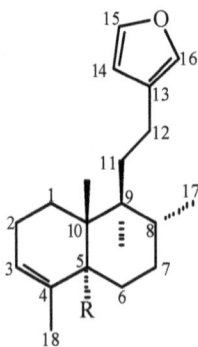

(95a) R=CO₂H - Hoffmanniaaldehyde
(95b) R=COH - Strigyllanoic acid B

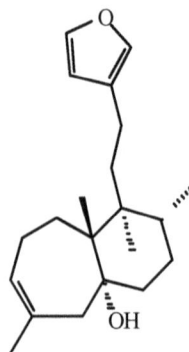

(96) Printziane,10-β-(H): 5,10-
dihydro;5α- hydroxy

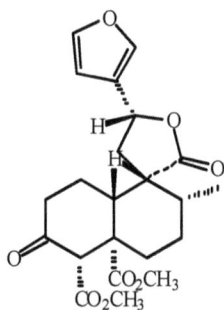

(97) Corylifuran

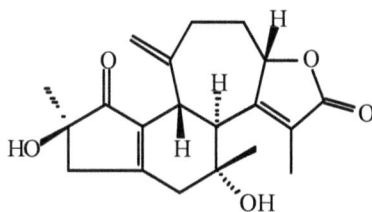

(98) Crotofolin A

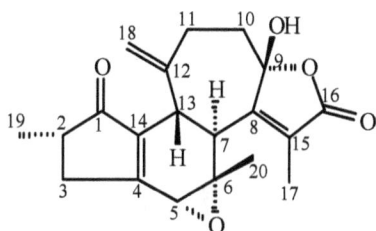

(99) Crotofolin E

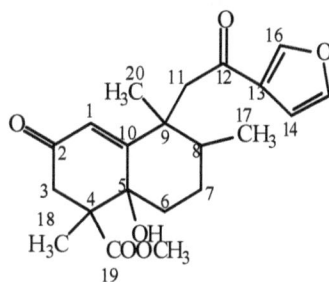

(100) Chettaphanin I

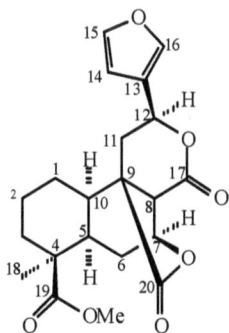

(101) Diasin

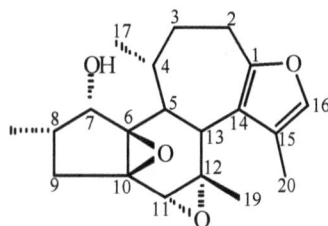

(102) Crotoxide A

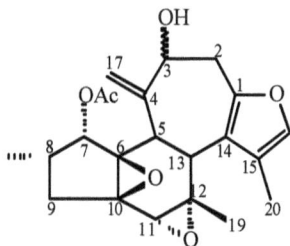

(103) Crotoxide B

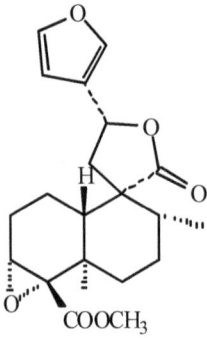

(104) Draconin

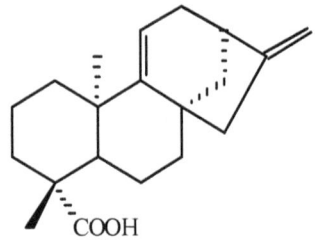

(105)

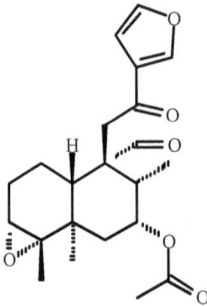

(106) Cascarillin B

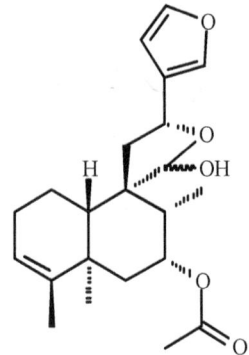

(107) Cascarillin C

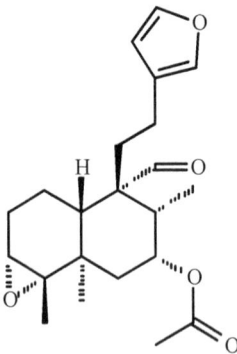

(108) Cascarillin D

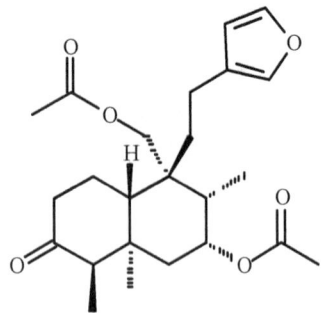

(109) Eluterin A

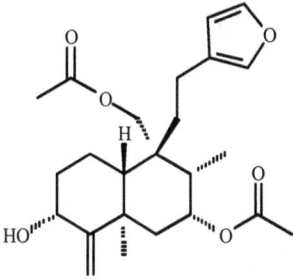

(110) Eluterin B

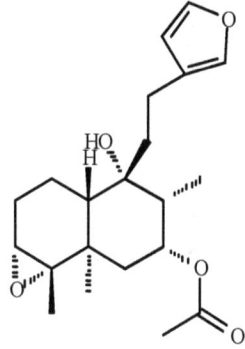

(111) Eluterin C

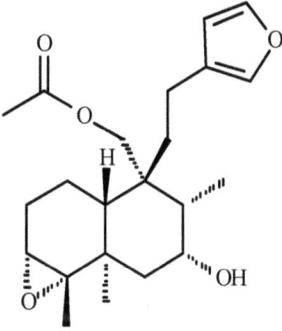

(112) Eluterin D

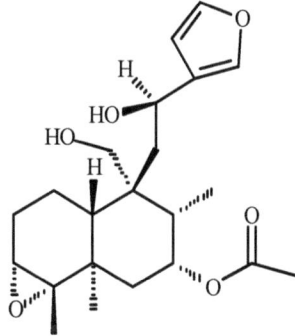

(113) Eluterin E

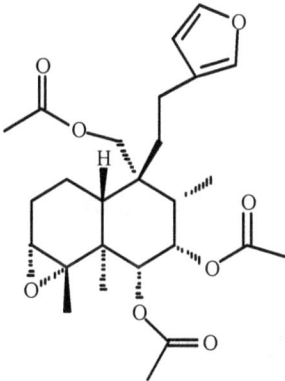

(114) Eluterin F

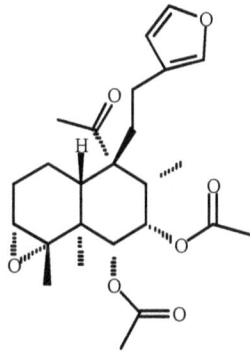

(115) Eluterin G

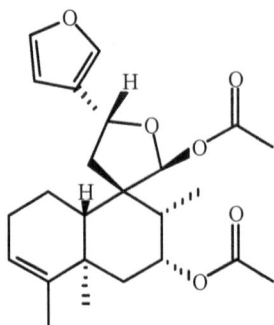

(116) Eluterin H

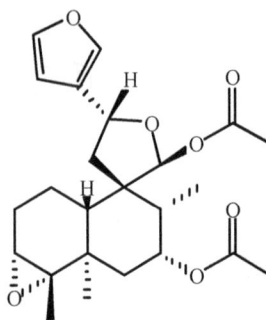

(117) Eluterin I

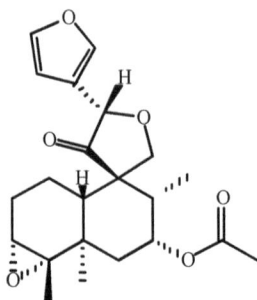

(118) Eluterin J

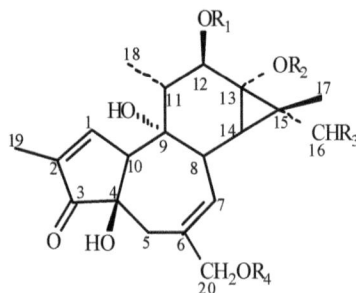

(119a) R_1=CO(CH$_2$)$_{14}$CH$_3$ R_2=COCH$_3$ R_3=HOH R_4=CO(CH$_2$)$_8$CH$_3$. Phorbol, 16-hydroxy-12-hexadecanoate-13-acetate-20-decanoate

(119b) R_1=CO(CH$_2$)$_{12}$CH$_3$ R_2=COCH$_3$ R_3=HOH R_4=H - Phorbol-13-acetate, 16-hydroxy-12-myristate

(119c) R_1=CO(CH$_2$)$_{14}$CH$_3$ R_2=COCH$_3$ R_3=HOH R_4=H - Phorbol-13-acetate, 16-hydroxy-12-*O*-Palmitate

(119d) R_1=CO(CH$_2$)$_{16}$CH$_3$ R_2=COCH$_3$ R_3=HOH R_4=H - Phorbol-13-acetate, 16-hydroxy-12-stearate

119e) R_1=CO(CH$_2$)$_{14}$CH$_3$ R_2=COCH$_3$ R_3=HOH R_4=CO(CH$_2$)$_8$CH$_3$. Phorbol-13-acetate, 16-hydroxy-20-decanoate-12-palmitate

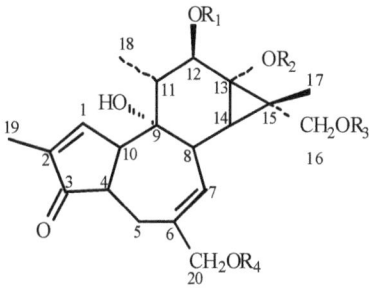

(120) R₁=CO(CH₂)₁₄CH₃ R₂=COCH₃ R₃=H
R₄=H - Phorbol, 4-deoxy-16-hydroxy-12-
O-hexadecanoate-13-*O*-acetate

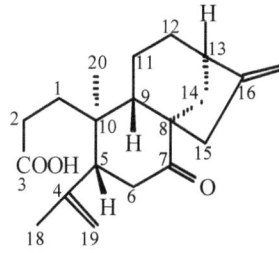

(121) Geayine

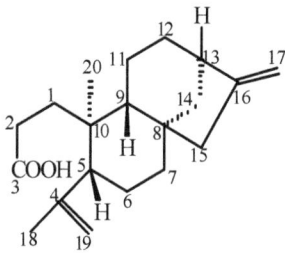

(122) Geayine, 7-deoxo

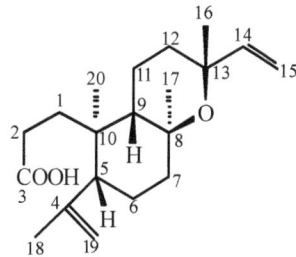

(123) Geayinine,iso

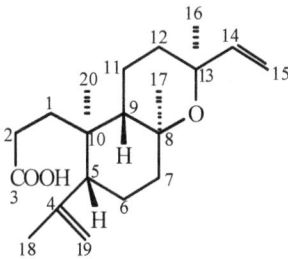

(124) Geayinine

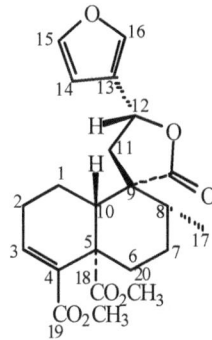

(125) Crotocorylifuran

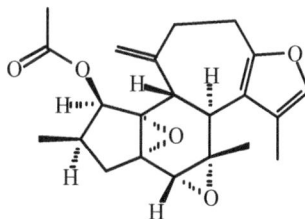

(126) Crotohaumanoxide

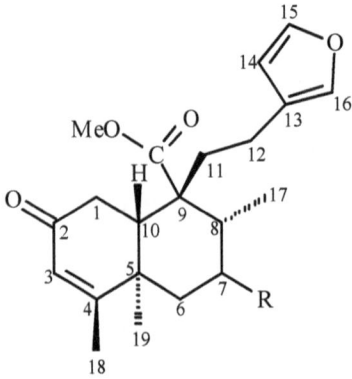

(127a) R= H,H - Cromiargyne

(127b) R= α-OAc - Cromiargyne, 7-acetoxy

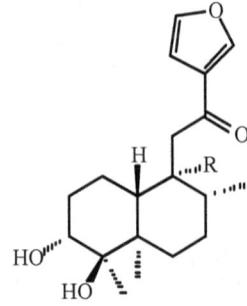

(128a) R=CH₃. Cleroda-13(16),14-dien, 9-al-3α,4β-dihydroxy-15,16, epoxy-12-oxo

(128b) R=CHO - Cleroda-13(16),14-dien, 3α,4β-dihydroxy-15,16, epoxy-12-oxo

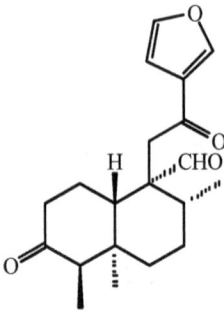

(129) Cleroda-13(16),14-dien-9-al,3,12-dioxo-15,16-epoxy

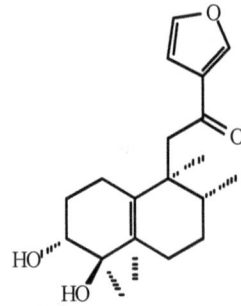

(130) Cleroda-5(10),13(16),14-triene,19-nor-3α, 4β-dihydroxy-15,16- epoxy-12-oxo

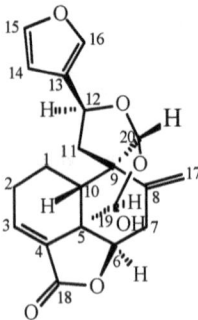

(131) Plaunol A

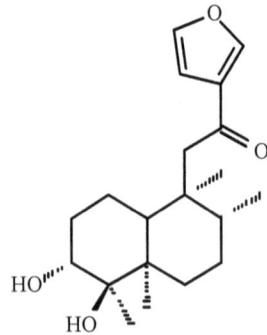

(132) Cleroda-13(16), 14-diene

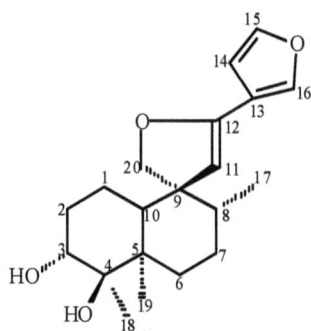

(133) Crotinsularin

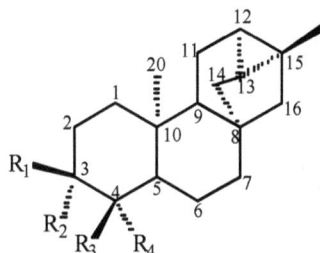

(134a) $R_1, R_2 = O$ $R_3 = CH_3$ $R_4 = CH_3$ -
Trachyloban-3-one: *ent*

(134b) $R_1 = H$ $R_2 = H$ $R_3 = CH_3$ $R_4 = COOH$ -
Trachyloban-19-oic acid: *ent*

(134c) $R_1 = H$ $R_2 = OH$ $R_3 = CH_3$ $R_4 = CH_2OH$
- Trachylobane, $3\alpha, 19$-dihydroxy: *ent*

(134d) $R_1 = H$ $R_2 = OH$ $R_3 = CH_3$ $R_4 = CH_3$ -
Trachyloban-3β-ol: *ent*

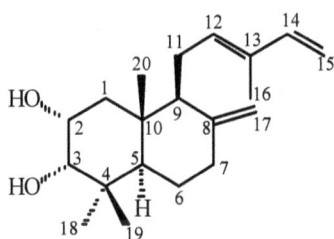

(135) Labda-8(17),12(13),14(15)-
triene,2α,3α-dihydroxy

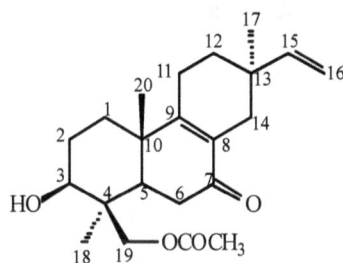

(136) Pimara-8(9),15-dien-7-one, 19-*O*-
acetyl-3β-hydroxy

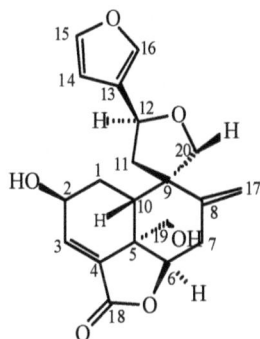

(137) Plaunol C

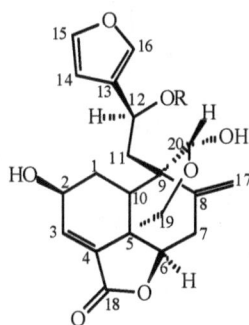

(138a) R=H - Plaunol D
(138b) R=Ac - Plaunol E

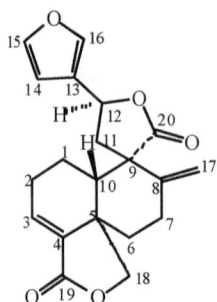

(139) Swassin

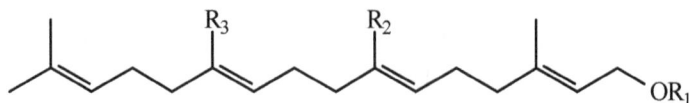

(140a) R$_1$=H R$_2$=Me R$_3$=CH$_2$OH - Hexadeca-2,6,1,1,14-tetraen-1-ol, [*E,E,Z*]-11-hydroxymethyl-3,7,15-trimethyl

(140b) R$_1$=H R$_2$=Me R$_3$=COH - Hexadeca-2,6,10,14-tetraen-1-ol, [*E,E,E*]-11-formyl-3,7,15-trimethyl

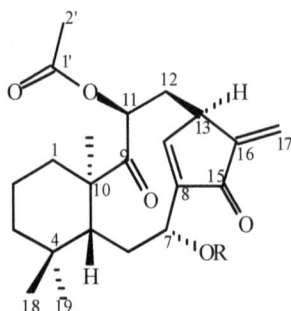

(141a) R=Ac
(141b) R=H

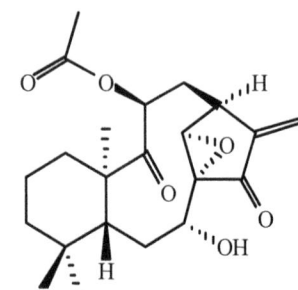

(142)

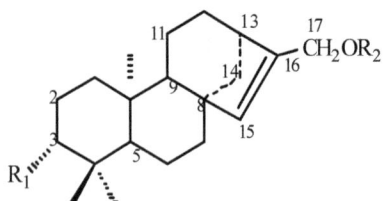

(143a) R$_1$=OAc R$_2$=H - Kaur-15-en-17-hydroxy-3β-yl acetate: *ent*

(143b) R$_1$=OH R$_2$=H - Kaur-15-en-3β,17-diol, *ent*

(143c) R$_1$=H R$_2$=H - Kaur-15-en-17-ol, *ent*

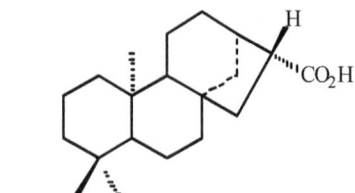

(144) Kauran-17-oic acid,16α-(H): *ent*

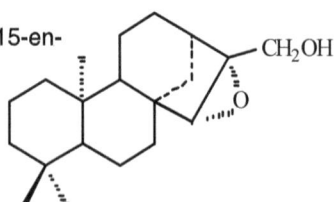

(145) Kauran-17-ol, 15β-16-epoxy: *ent*

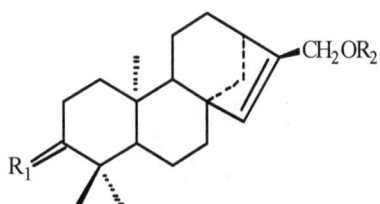

(146) R$_1$=αOH,H R$_2$=H - Kauran-
3β, 16β,17-triol: *ent*

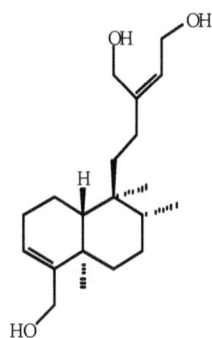

(147) Bincatriol

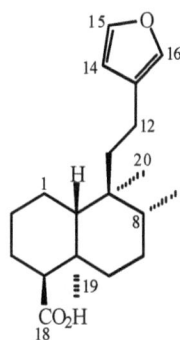

(148) Crolechinic acid

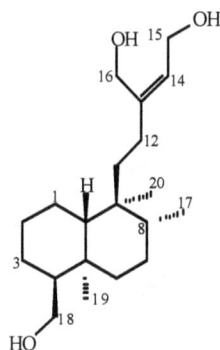

(149) Crolechinol

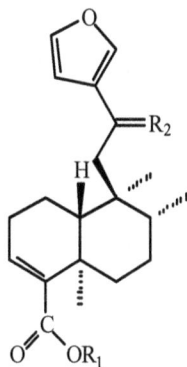

(150a) R$_1$=H R$_2$=H - Hardwickiic acid

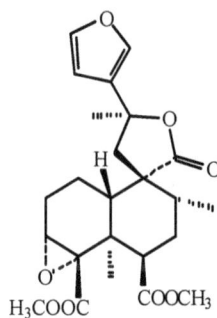

(151) Korberin A

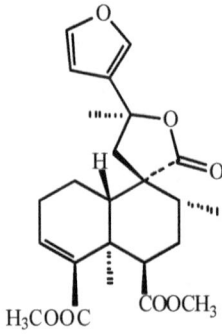

(152) Korberin B

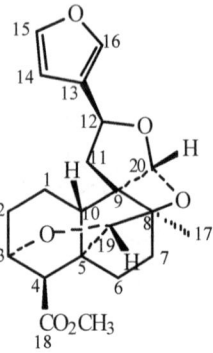

(153) Crovatin

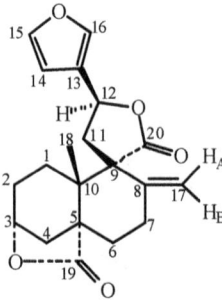

(154) Levatin

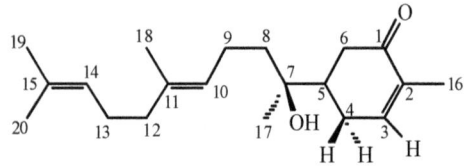

(155) Croton diterpene 1

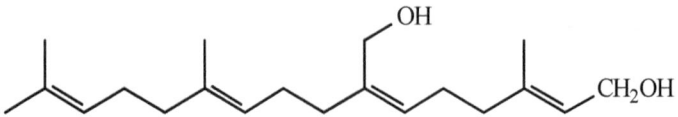

(156) Plaunotol

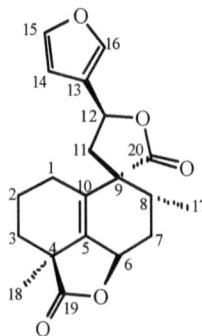

(157) Cleroda-5,10-diene-19,6β,20,12-diolide-neo

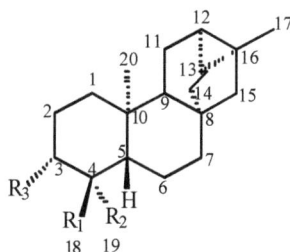

(158a) R$_1$=COOH R$_2$=CH$_3$ R$_3$=H - Trachyloban-18-oic, acid
(158b) R$_1$=CH$_3$ R$_2$=COOH R$_3$=H - Trachyloban-19-oic, acid
(158c) R$_1$=CH$_2$OH R$_2$=CH$_2$OH R$_3$=OH - Trachylobane,3α,18,19-trihydroxy
(158d) R$_1$=CH$_3$ R$_2$=CH$_2$OH R$_3$=OH - Trachylobane, 3α,19-dihydroxy
(158e) R$_1$=CH$_3$ R$_2$=CH$_2$OH R$_3$=H - Trachyloban-19-ol

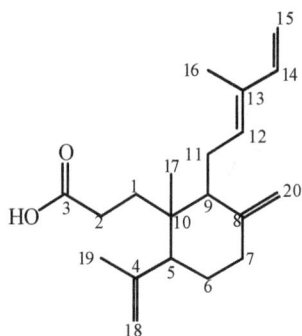

(159) Maravuic acid

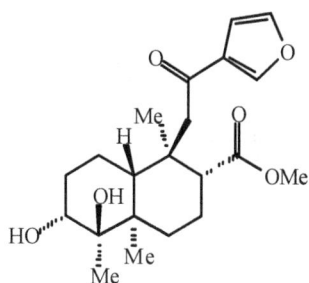

(160) Chiromodine

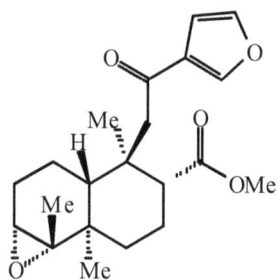

(161) Chiromodine, epoxy

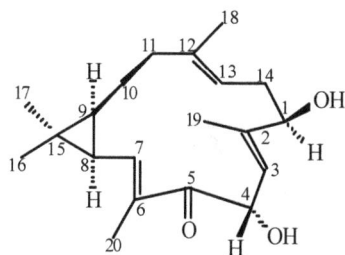

(162) Croton casbene diterpene1

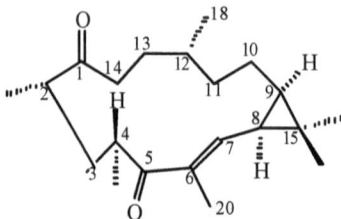

(163) Crotonitenone

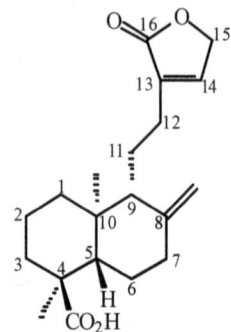

(164) Nivenolide

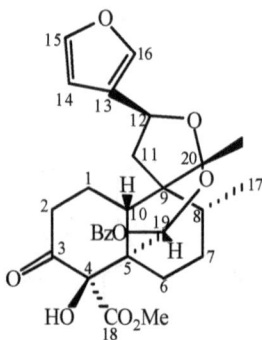

(165) Croblongifolin

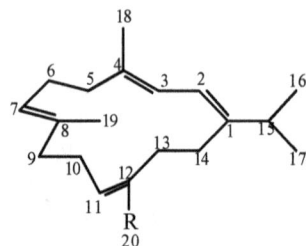

(166a) R=CHO, Δ^{11}=E - Crotocembranal, neo
(166b) R=COOH, Δ^{11}=Z - Crotocembraneic acid
(166c) R=COOH, Δ^{11}=E - Crotocembraneic
acid, neo

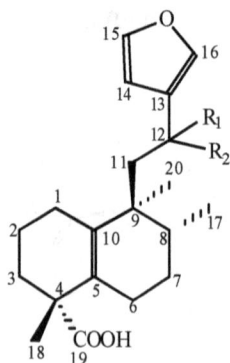

(167a) R$_1$=H R$_2$=H - Crotohalimaneic
acid

(167b) R$_1$=OBz R$_2$=H -
Crotohalimaneic-12-benzoyloxy acid

(167c) R$_1$= =O R$_2$= - - Crotohalimoneic
acid

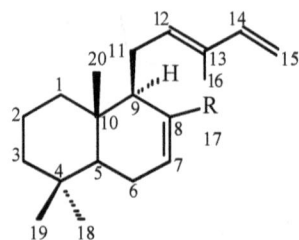

(168a) R=Me - Labda-7-*trans*,12,14-triene
(168b) R=COH - Labda-7-*trans*,12,14-trien-17-al
(168c) R=CO$_2$H - Labda-7-*trans*-12,14-trien-17-
oic acid
(168d) R=CH$_2$OH - Labda-7-*trans*-12,14-trien-
17-ol

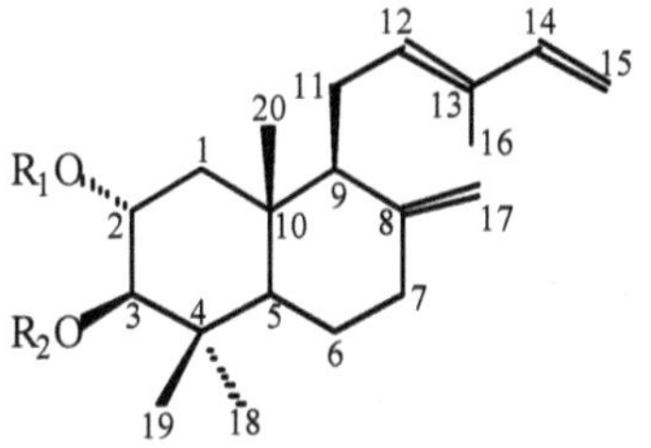

(169a) R₁=Ac R₂=Ac - Labda-8(17)-*trans*,12,14-triene-2,3-diacetoxy

(169b) R₁=H R₂=H - Labda-8(17)-*trans*,12,14-triene-2,3-dihydroxy

(169c) R₁=Ac R₂=H - Labda-8(17)-*trans*,12,14-triene-2-acetoxy-3-hydroxy

(169d) R₁=H R₂=Ac - Labda-8(17)-*trans*,12,14-triene-3-acetoxy-2-hydroxy

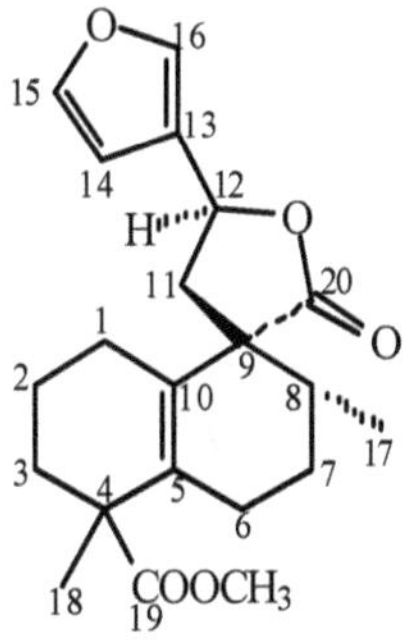

(170) Penduliflaworosin

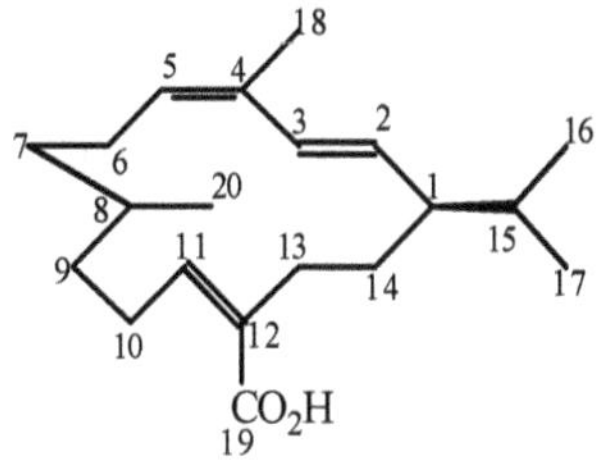

(171) Poilaneic acid

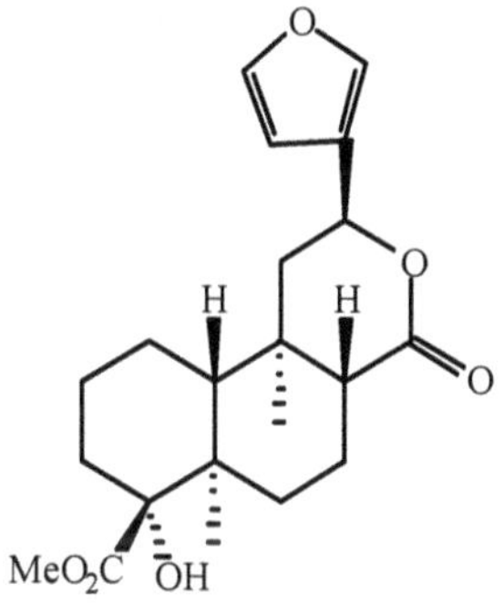

(172) Cordatin

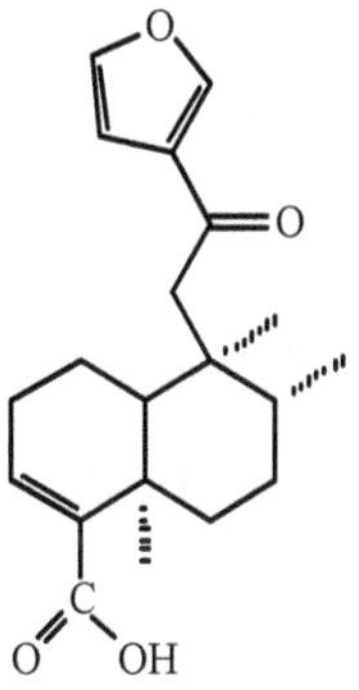

(173) Hardwickiic, 12-oxo acid: methyl ester

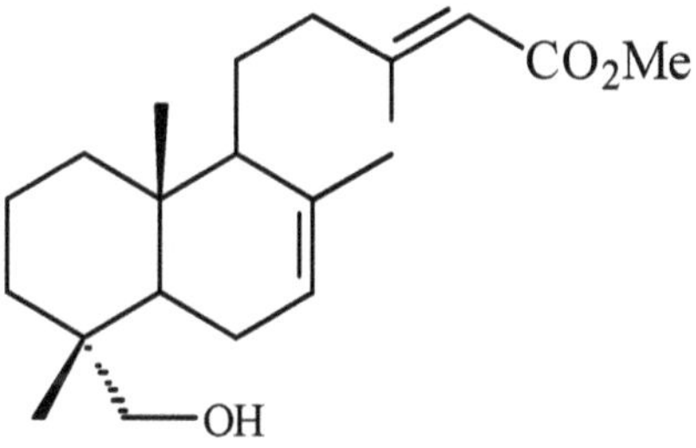

(174) Labda-18-hydroxy-7,13-dien-15-oic acid: methyl ester

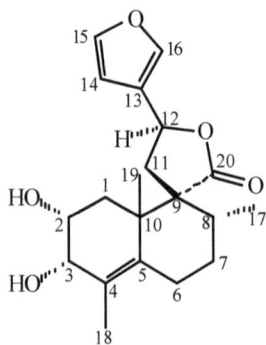

(175) Pyramidolactone

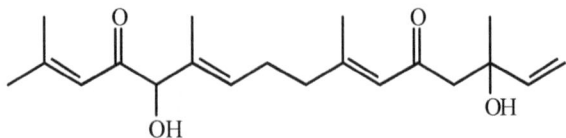

(176) Hexadeca-1-*cis*-6-*trans*-10,14-tetraene-5,13-dione, 3,12-di-hydroxy-3,7,11,15-tetramethyl

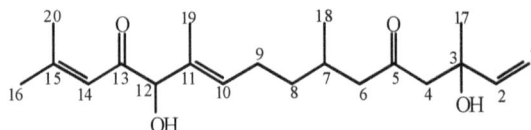

(177) Hexadeca-1-*trans*-10,14-triene-5-13-dione, 3,12-dihydroxy-3,7,11,15-tetramethyl

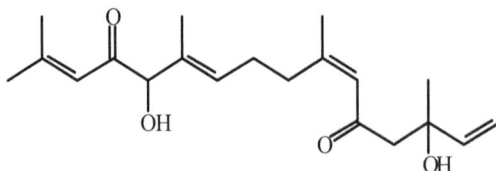

(178) Hexadeca-1-*trans*,6-*trans*,10-14-tetraene-5,13-dione, 3,12-dihydroxy: 3,7,11,15-tetramethyl

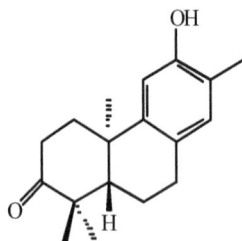

(179) Podocarpa-9,11,13-trien-3-one-12-hydroxy-13-methyl

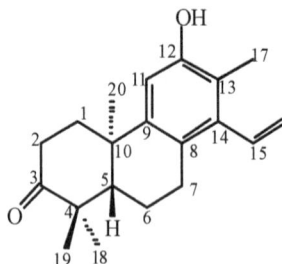

(180) Sonderianol

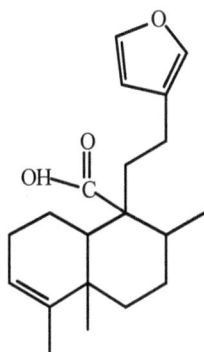

(181) Junceic acid

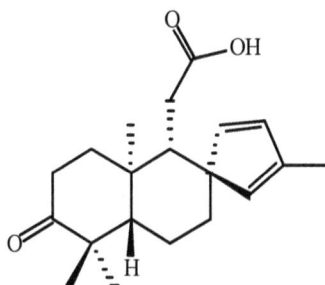

(182) Sarcopetaloic acid

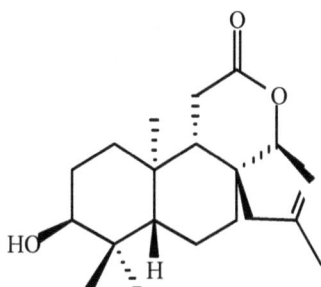

(183) Sarcopetalololide

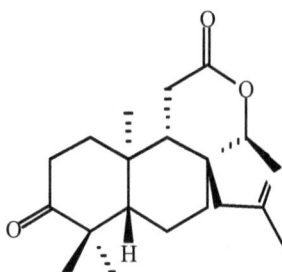

(184) Sarcopetalolide

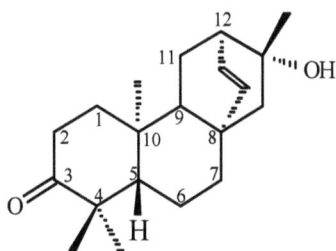

(185) Yucalexin A-16

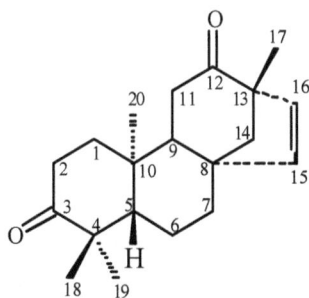

(186) Yucalexin B-6

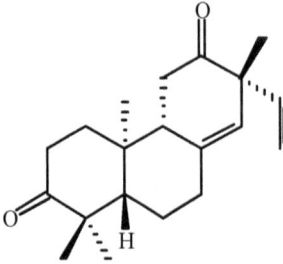

(187) Yucalexin P-4

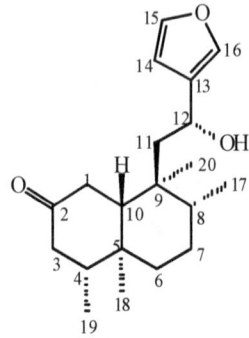

(188) Cascarillone, (12R)-12-hydroxy

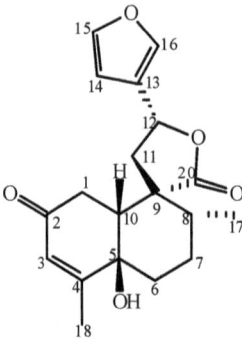

(189) Crotonin, 5β-hydroxy-*cis*-dehydro

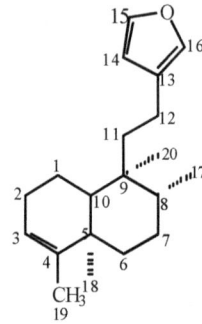

(190) Annonene, *trans*

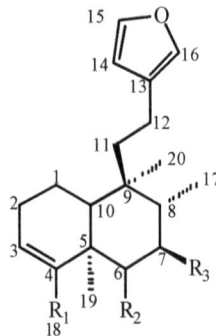

(191a) R₁=Me R₂=α-OH R₃=H - Annonene, 6α-hydroxy

(191b) R₁=Me R₂=α-OH R₃=OH - Annonene, 6α,7β-dihydroxy

(191c) R₁=Me R₂=α-OAc R₃=OAc - Annonene, 6α,7β-diacetoxy

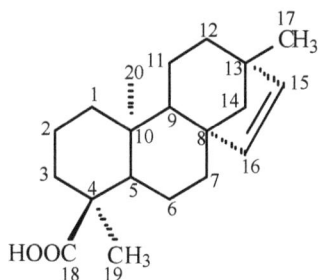

(192) Beyer-15-en-18-oic, acid: ent

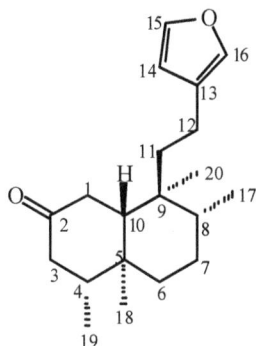

(193) Cascarillone, *trans*

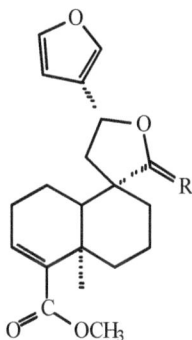

(194a) R=α-H, β-OH - Sonderianial
(194b) R=O - Sonderianin

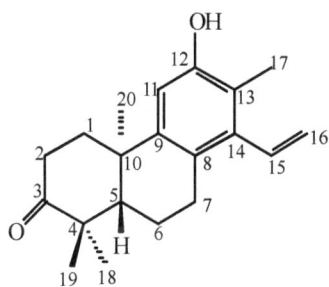

(195) Sonderianol

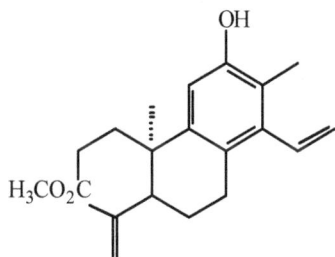

(196) Sonderianol, 3,4-seco

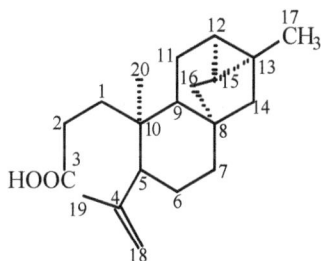

(197) Trachylobanic acid, 3,4-seco

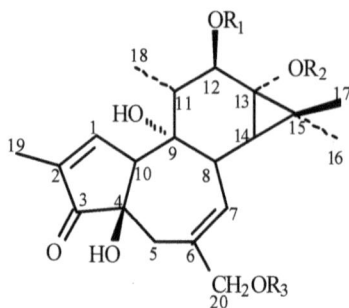

(198a) R₁=CO(CH₂)₁₀CH₃ R₂=Ac R₃=H - Phorbol 12-*O*-dodecanoyl-13-acetate

(198b) R₁=CO(CH₂)₁₀CH₃ R₂=Ac R₃=CO(CH₂)₅(CH₂CH=CH)₃CH₂CH₃. Phorbol 12-*O*-dodecanoyl-13-*O*-acetyl-20-linoleate

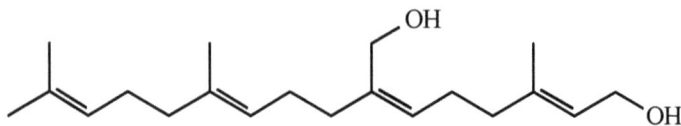

(199) Geranyl-geraniol, 18-hidroxy

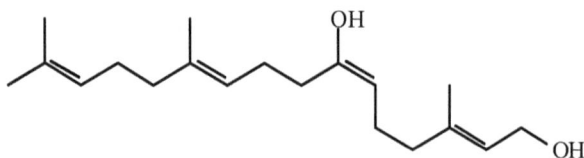

(200) Hexadeca-2-*trans*,6-*cis*,10-*trans*,14-tetraen-1-ol-7-hydroxymethyl-3,11,15-trimethyl

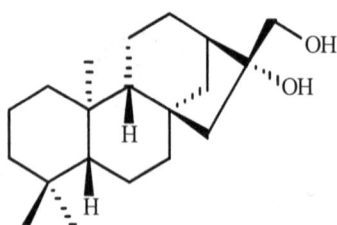

(201) Kaurene, 16β,17-dihydroxy: ent

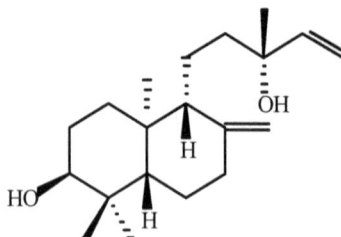

(202) Manool, 18 upi. 8α-hydroxy. ent

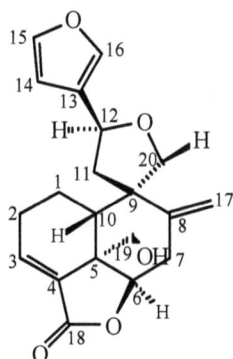

(203) Plaunol B

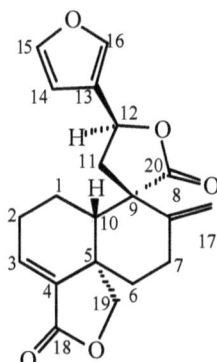

(204) Plaunolide

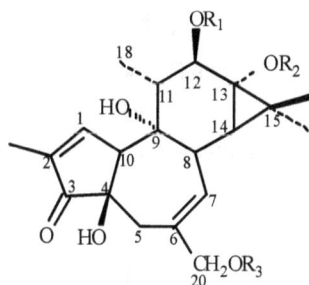

(205a) R_1=Tiglate R_2=Isobutyrate R_3=H - Phorbol, 12-*O*-tiglyl-13-isobutyrate

(205b) R_1=Tetradecanoate R_2=Acetate R_3=H - Phorbol, 12-*O*-tetradecanoyl-13-acetate

(205c) R_1=Dodecanoate R_2=Acetate R_3=H - Phorbol, 12-*O*-dodecanoyl-13-acetate

(205d) R_1=Decanoate R_2=Acetate R_3=H - Phorbol, 12-*O*-decanoyl-13-acetate

(205e) R_1=α-Methylbutyrate R_2=Decanoate R_3=H - Phorbol-12-*O*-(α-methyl)butyryl-13-decanoate

(205f) R_1=Acetate R_2=Decanoate R_3=H - Phorbol, 12-*O*-acetyl-13-decanoate

(205g) R_1=Tiglate R_2=Acetate R_3=H - Phorbol, 12-*O*-tiglyl-13-acetate

(205j) R_1=2-methylbutyrate R_2=Isobutyrate R_3=H .Phorbol,12-*O*-(2-methyl)butyryl-13-isobutyrate

(205k) R_1=Tiglate R_2=H R_3=H - Phorbol, 12- tiglate

(205l) R_1=Acetate R_2=Acetate R_3=H - Phorbol, 12-*O*-acetyl-13-acetate

(205m) R_1=2-methylbutyrate R_2=Acetate R_3=H - Phorbol, 12-*O*-(2-methyl)butiryl-13-acetate

(205n) R_1=H R_2=Acetate R_3=H - Phorbol, 13-acetate

(205q) R_1=H R_2=H R_3=H - Phorbol

(205r) R_1=Acetate R_2=Tiglate R_3=H - Phorbol, 12-*O*-acetyl-13-tigliate

(205s) R_1=H R_2=Acetyl R_3=Linoleate - Phorbol, 13-*O*-acetyl-20-linoleate

(205t) R_1=H R_2=Tiglate R_3=Linoleate - Phorbol, 13-*O*-tigloyl-20-linoleate

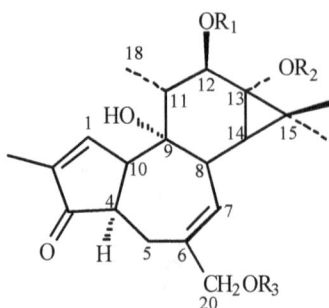

(206a) R$_1$=Tiglate R$_2$=Isobutyrate R$_3$=H - Phorbol, 4α-deoxy-12-*O*-tiglyl 13-isobutyrate

(206b) R$_1$=Tiglate R$_2$=Acetate R$_3$=H - Phorbol, 4α-deoxy-12-*O*-tiglyl 13-acetate

(206c) R$_1$=2-methylbutyrate R$_2$=Acetate R$_3$=H - Phorbol, 4α-deoxy-12-*O*-(2-methyl)butiryl-13-acetate

(206d) R$_1$=H R$_2$=Acetate R$_3$=H - Phorbol, 4α-deoxy-13-acetate

(206e) R$_1$=Acetate R$_2$=Acetate R$_3$=Acetate

(206f) R$_1$=Tiglate R$_2$=H R$_3$=H

(206g) R$_1$=2-methylbutyrate R$_2$=H R$_3$=H

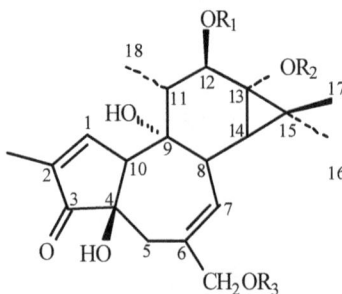

(207a) R$_1$=H R$_2$=Acetate R$_3$=C$_{18}$H$_{31}$O

(207b) R$_1$=H R$_2$=Tigloyl R$_3$=C$_{18}$H$_{31}$O

(207c) R$_1$=Acetate R$_2$=Tigloyl R$_3$=H

(207d) R$_1$=C$_{10}$H$_{19}$O R$_2$=2-methylbutyryl R$_3$=H - Phorbol, 12-*O*-decanoyl-13-(2 methyl)butyrate

(207e) R$_1$=Tigloyl R$_2$=2-methylbutyryl R$_3$=H - Phorbol, 12-*O*-tigloyl-13-(2-methyl)butyrate

(207f) R$_1$=Hexadecanoyl R$_2$=Acetate R$_3$=H - Phorbol, 12-*O*-hexadecanoyl-13-acetate

(207g) R$_1$=Tiglate R$_2$=Decanoate R$_3$=H - Phorbol, 12-*O*-tiglyl-13-decanoate

(207h) R$_1$=Acetate R$_2$=Dodecanoate R$_3$=H - Phorbol, 12-*O*-acetyl-13-dodecanoate

(207i) R$_1$=Tiglate R$_2$=Octanoate R$_3$=H - Phorbol, 12-*O* tiglyl 13 octanoate

(207j) R$_1$=2-methylbutyryl R$_2$= C$_{12}$H$_{23}$O R$_3$=H

(208) R$_1$=2-methylaminobenzoyl R$_2$=Acetate R$_3$=H - Phorbol, 4α-deoxy-5-hydroxy-12-O-(2-methylaminobenzoyl)-20-acetoxy

(209a) R$_1$=2-methylaminobenzoyl R$_2$=Acetate R$_3$=H
(209b) R$_1$=2-methylaminobenzoyl R$_2$=Acetate R$_3$=Acetate
(209c) R$_1$=2-methylaminobenzoyl R$_2$=H R$_3$=H
(209d) R$_1$=(n-deca-2,4,6-trieneyl) R$_2$=Acetate R$_3$=H

(210a) R$_1$=2-methylaminobenzoyl R$_2$=Acetate R$_3$=Acetate R$_4$=Acetate
210b) R$_1$=(n-deca-2,4,6-trieneyl) R$_2$=Acetate R$_3$=Acetate R$_4$=Acetate

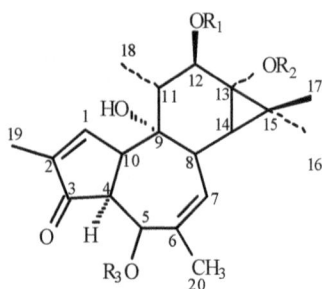

(211a) R$_1$=n-tetradecanoyl R$_2$=Acetate R$_3$=H R$_4$=H
(211b) R$_1$=2-methylaminobenzoyl R$_2$=Acetate R$_3$=H R$_4$=H

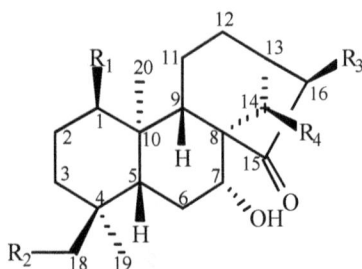

(212a) R$_1$=H R$_2$=OH R$_3$= =CH$_2$ R$_4$=H - Kaur-16-en-18-ol-7β-hydroxy-15-oxo: *ent*
(212b) R$_1$=H R$_2$=OAc R$_3$= =CH$_2$ R$_4$=H - Kaur-16-en-15-one, 1α-acetoxy-7β-14α-dihydroxy: *ent*
(212c) R$_1$=H R$_2$=OAc R$_3$=CH$_3$ R$_4$=H - Kaur-16-en-15-one, 18-acetoxy-7α,14β-dihydroxy: *ent*
(212d) R$_1$=OAc R$_2$=H R$_3$= =CH$_2$ R$_4$=OH - Kaur-16-en-15-one, 18-acetoxy-7α-hydroxy: *ent*
(212e) R$_1$=H R$_2$=H R$_3$=CH$_3$ R$_4$=OH - Kaur-16-en-15-on-18-oic-acid, *ent*
(212f) R$_1$=H R$_2$=OAc R$_3$= =CH$_2$ R$_4$=OH - Kaur-16-en-15-one, 11α-18-diacetoxy-7β-hydroxy: *ent*
(212g) R$_1$=H R$_2$=OAc R$_3$= =CH$_2$ R$_4$=H - Kaur-16-en-15-one, 11α-acetoxy-7β,14α-dihydroxy: *ent*
(212h) R$_1$=H R$_2$=OH R$_3$= =CH$_2$ R$_4$=H - Kaur-16-en-15-one,18-acetoxy-11α-hydroxy: *ent*

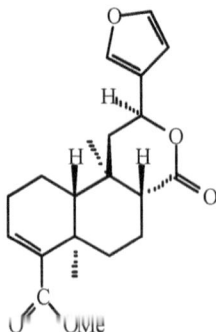

(213) Barbascoatic acid,12-epi-methyl ester

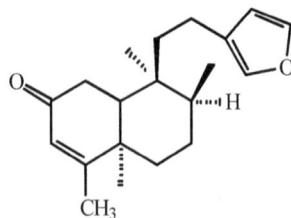

(214) Cleroda-3,13(16) trien-2-one, 15,16-epoxy

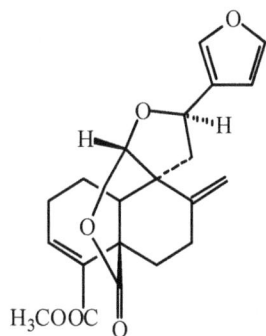

(215) Croverin

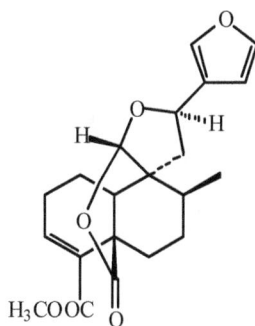

(216) Croverin, dihydro

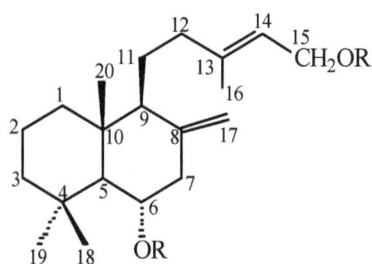

(217a) R=Ac - Crotocorylifuran
(217b) R=H - Crotonadiol

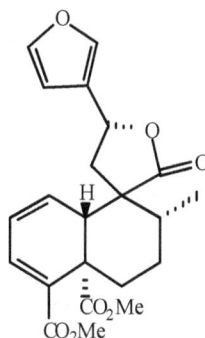

(218) Crotozambefuran A

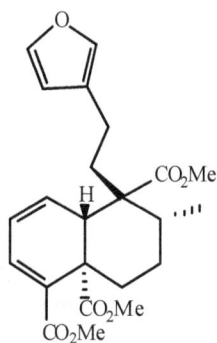

(219) Crotozambefuran B

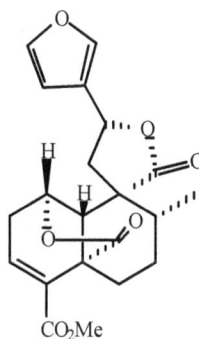

(220) Crotozambefuran C

(221)

(222) Isopimara-7,15-dien-3β-ol

(223a) R₁=COOH R₂=CH₃ R₃=H R₄=β-OAc - Trachyloban-18-oic acid, 7β-acetoxy

(223b) R₁=OH R₂=CH₃ R₃=H R₄=OH - Trachyloban-7β,18-diol

(223c) R₁=OH R₂=CH₃ R₃= =O R₄=H,H - *ent*-18-Hydroxy-trachyloban-3-one

Flavonoids

(224) Flavone, 5-hydroxy-7,4'-dimethoxy

(225a) R₁=CH₃ R₂=H R₃=CH₃ R₄=H - Kaempferol 3,7-dimethyl-ether

(225b) R₁=CH₃ R₂=CH₃ R₃=CH₃ R₄=H - Kaempferol 3,5,7-trimethyl-ether

(221)

(222) Isopimara-7,15-dien-3β-ol

(223a) R_1=COOH R_2=CH$_3$ R_3=H R_4=β-OAc - Trachyloban-18-oic acid, 7β-acetoxy

(223b) R_1=OH R_2=CH$_3$ R_3=H R_4=OH - Trachyloban-7β,18-diol

(223c) R_1=OH R_2=CH$_3$ R_3= =O R_4=H,H - *ent*-18-Hydroxy-trachyloban-3-one

Flavonoids

(224) Flavone, 5-hydroxy-7,4'-dimethoxy

(225a) R_1=CH$_3$ R_2=H R_3=CH$_3$ R_4=H - Kaempferol 3,7-dimethyl-ether

(225b) R_1=CH$_3$ R_2=CH$_3$ R_3=CH$_3$ R_4=H - Kaempferol 3,5,7-trimethyl-ether

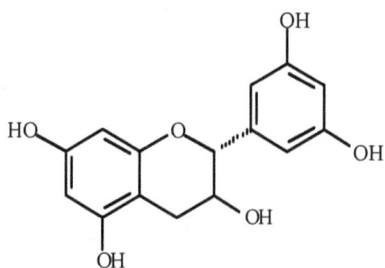

(226) Flavan, epi:3,3',5,5',7-pentahydroxy: (-)

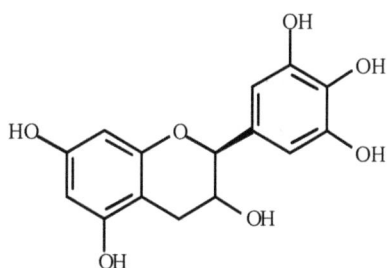

(227) Gallocatechin, (+)

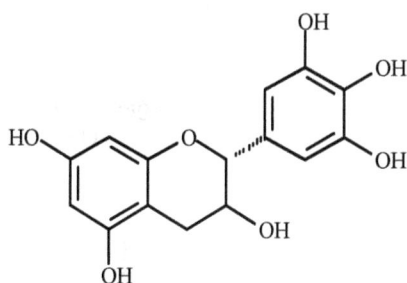

(228) Gallocatechin, epi (-)

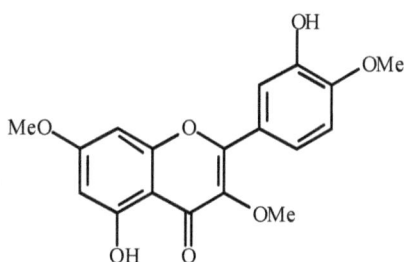

(229) Ayanin

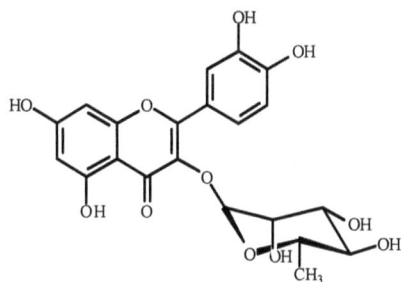

(230) Quercitrin

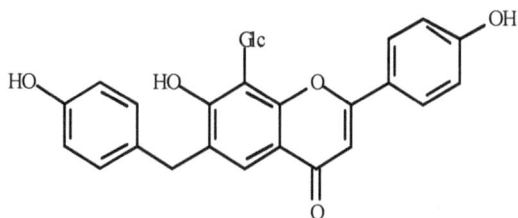

(231) Vitexin

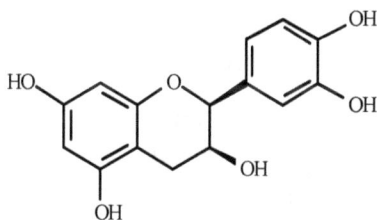

(232) Catechin, (+)

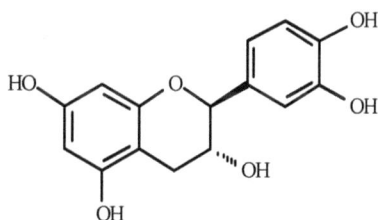

(233) Catechin, epi (-)

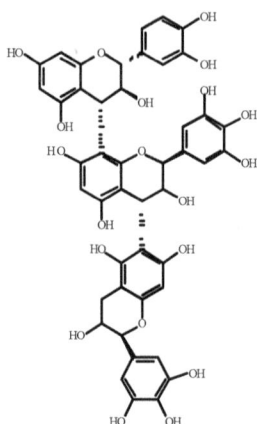

(234) Catechin (4α-8)-gallocatechin
(4α-6)-gallocatechin

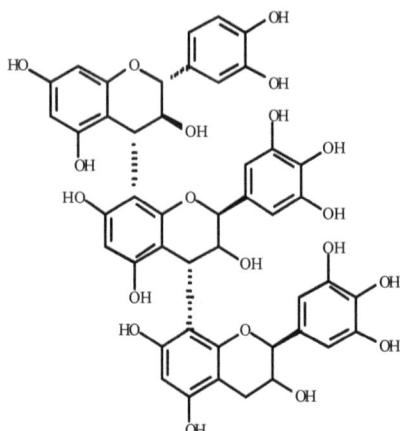

(235) Catechin (4α-8)-gallocatechin
(4α-8)-gallocatechin, (+)

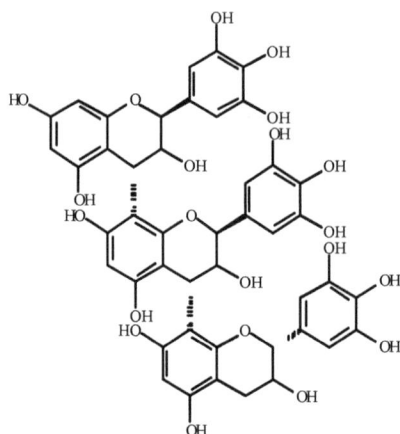

(236) Gallocatechin (4-α-8)-gallocatechin
(4-α-8)-epi-gallocatechin

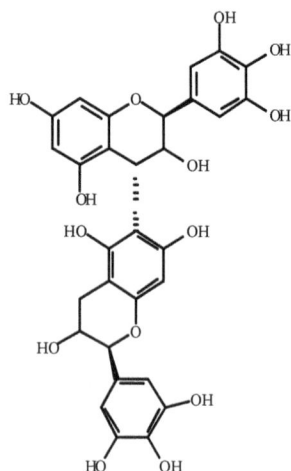

(237) Gallocatechin-(4-α-6)-epi-
gallocatechin

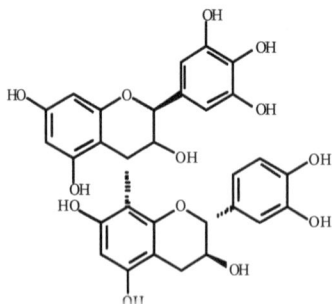

(238) Gallocatechin-(4-α-8)-epi-catechin

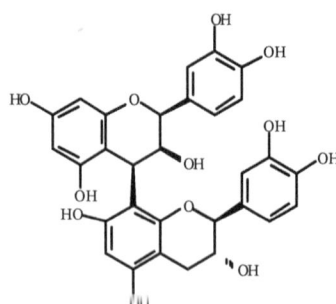

(239) Procyanidin, B-1

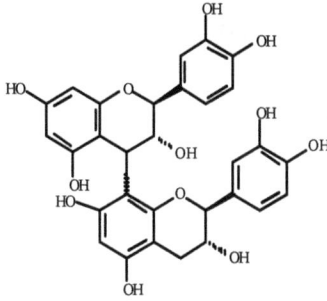

(240) Procyanidin, B-4

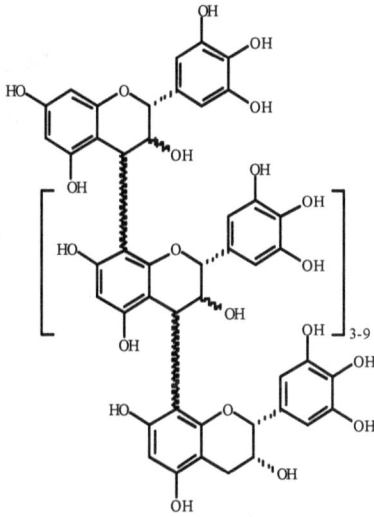

(241) SP-303

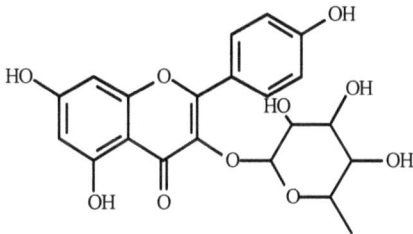

(242) Nicotiflorin

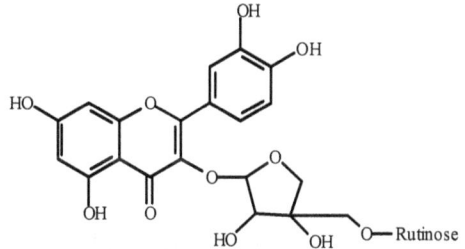

(243) Quercetin-3-*O*-(2-*O*-β-D-apiofuranosyl)-rutinoside

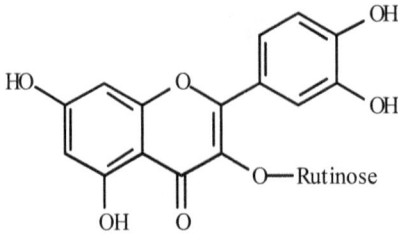

(244) Rutin

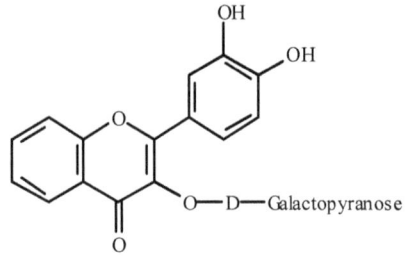

(245) Hyperoside

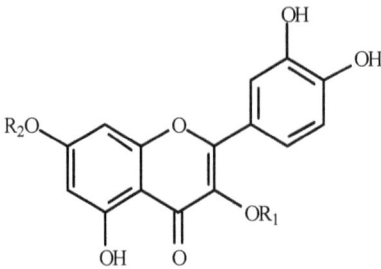

(246a) R₁=H R₂=H - Quercetin

(246b) R₁=CH₃ R₂=CH₃

(246c) R₁=Rhamnosyl R₂=H -
Quercetin-3-*O*-α-rhamnosyl glycoside

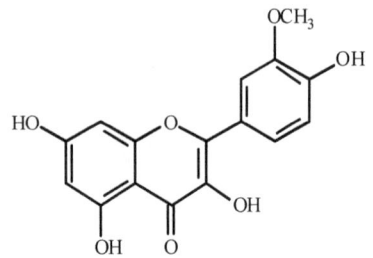

(247) Rhamnetin, iso

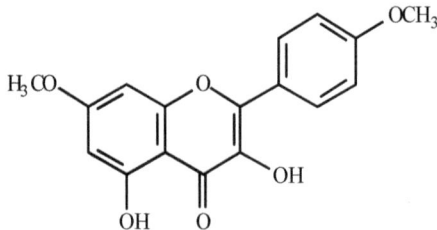

(248) Flavone, 3,5-dihydroxy-4',7-dimethoxy

Lignans

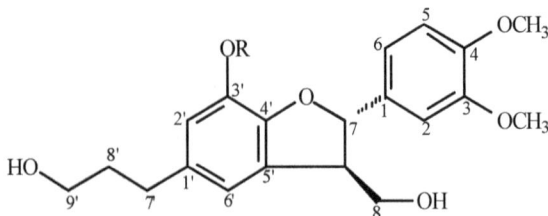

(249a) R=CH₃ Cedrusin, 3',4-*O*-dimethyl
(249b) R=H

Lipid

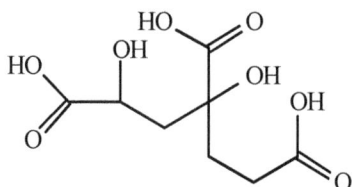

(250) Phorbic acid

Oxygen heterocycle

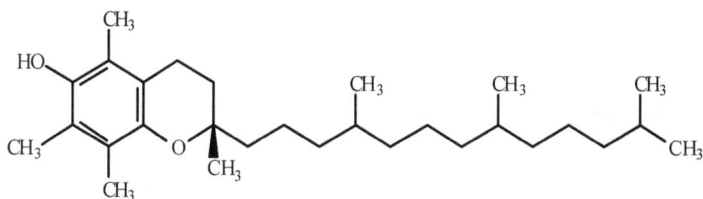

(251) Tocopherol, α

Peptides

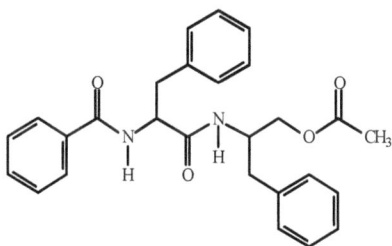

(252)

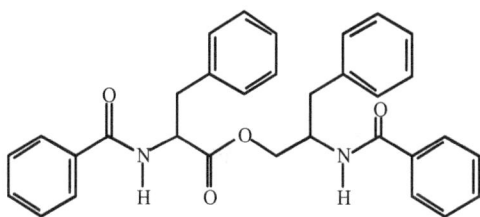

(253)

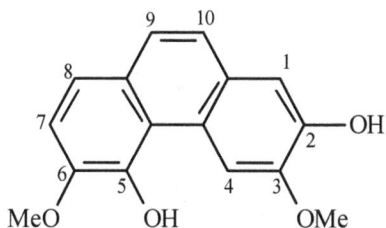

(254) Crotoflavol

Phenylpropanoids

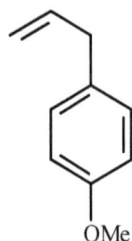

(255) Estragole

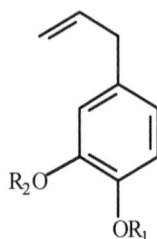

(256a) R_1=H R_2=CH_3. Eugenol
(256b) R_1=CH_3 R_2=CH_3. Eugenol, methyl ether

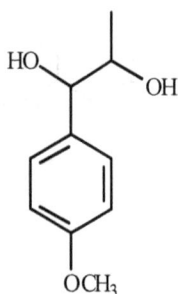

(257) Anethole

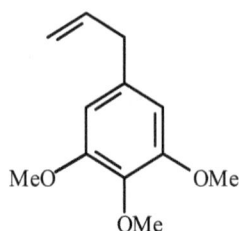

(258) Elemicin

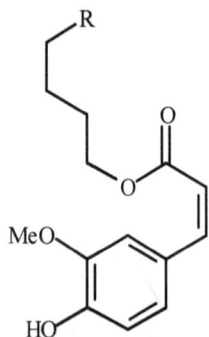

(259a) R=$(CH_2)_{21}CH_3$. Ferulic acid, *trans*: hexacosyl ester

(259b) R=$(CH_2)_{23}CH_3$. Ferulic acid, *trans*: octacosyl ester

(259c) R=$(CH_2)_{19}CH_3$. Ferulic acid, *trans*: tetracosyl ester

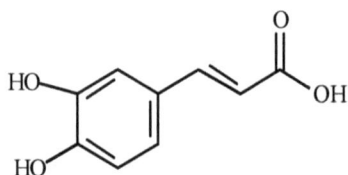

(260) Caffeic acid

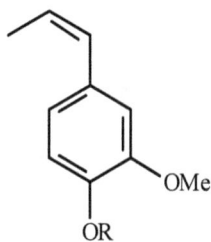

(261a) R=H - Eugenol, iso-*cis*
(261b) R=Me

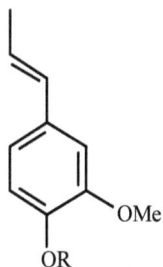

(262a) R=H - Eugenol, iso-*trans*
(262b) R=Me - Eugenol, iso-*trans*, methyl ether

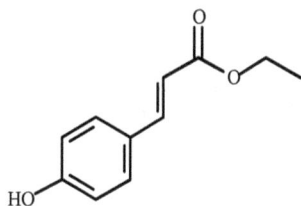

(263) Coumaric acid; *para* ethyl ester

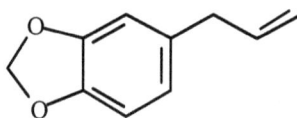

(264) Safrole

Proteids

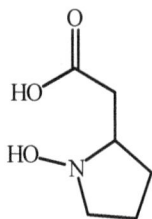

(265) Hygric acid, 4-hydroxy

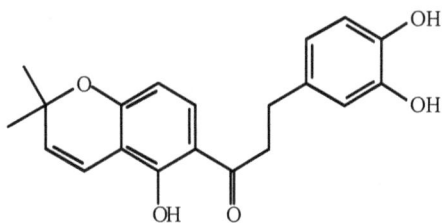

(266) Crotin I

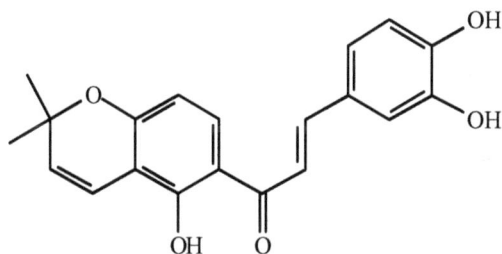

(267) Crotin II

Quinone

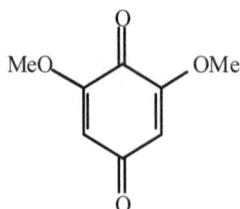

(268) Benzoquinone, 2,6-dimethoxy

Sesterpene

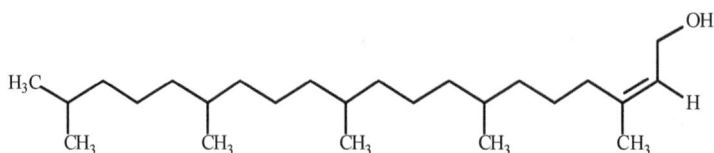

(269) Eicos-2-1-ol,*trans-* 2,3,7,11,15,19-pentamethyl

Steroids

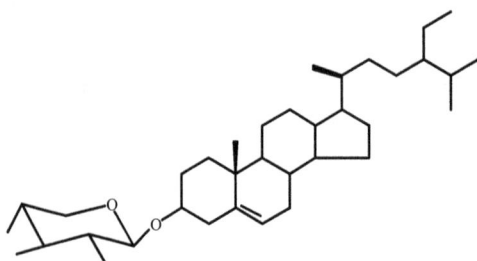

(270) Daucosterol

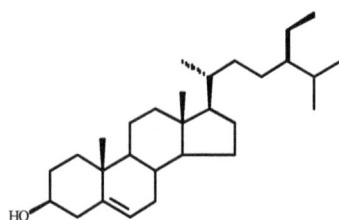

(271) Sitosterol, β

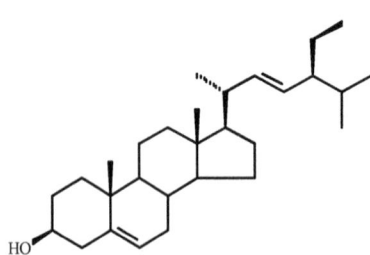

(272) Otigmasterol

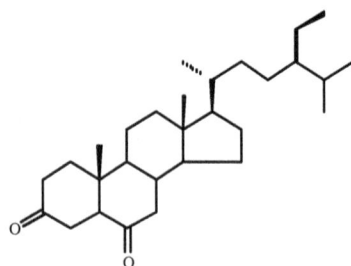

(273) Stigmastane-3,6-dione, 5α

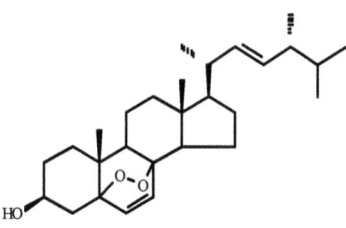

(274) Ergosterol, peroxide

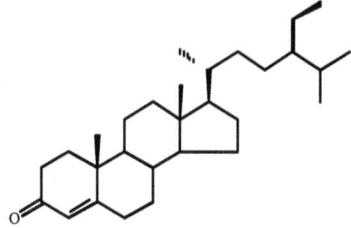

(275) Sitostenone, β

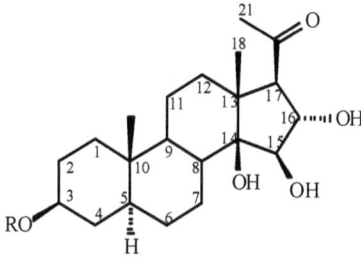

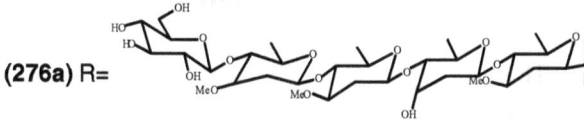

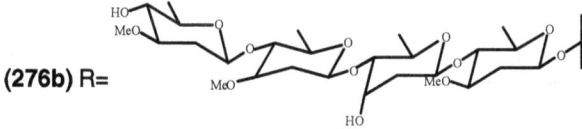

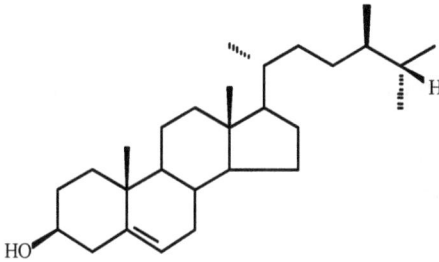

(276a) R=

(276b) R=

(277) Campesterol

Triterpenes

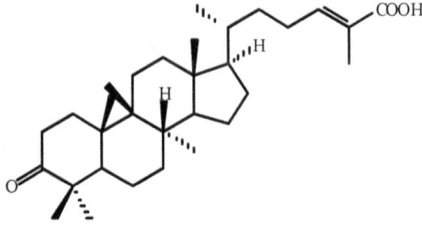

(278) Cycloart-24-en-26-oic-3-oxo acid

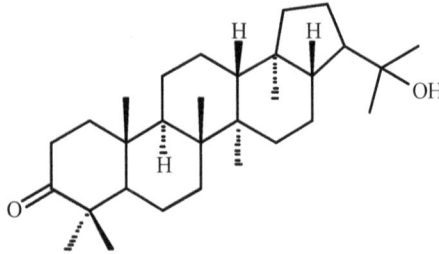

(279) Hopane, 22-hydroxy -3-oxo

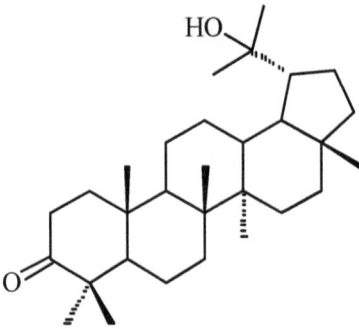

(280) Lupane, 20-hydroxy -3-oxo

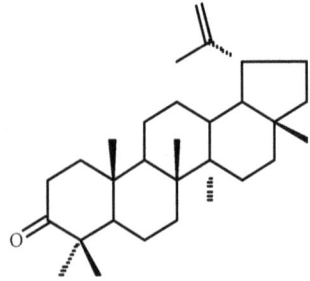

(281) Lupenone

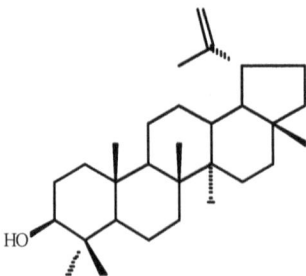

(282) Lupeol

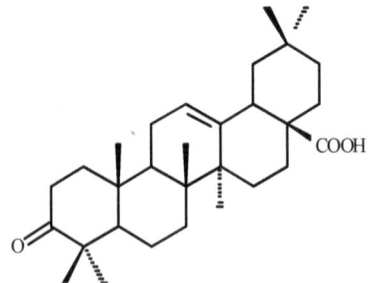

(283) Olean-12-en-28-oic-3-oxo acid

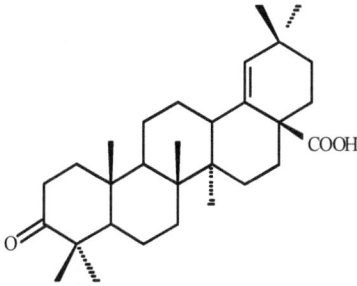

(284) Olean-18-en-28-oic-3-oxo acid

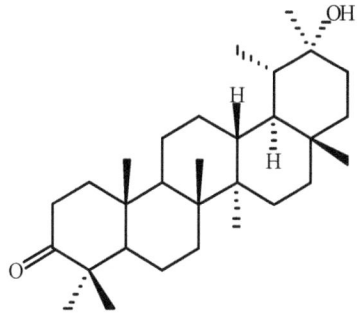

(285) Taraxastane, 3-oxo-20β-hydroxy

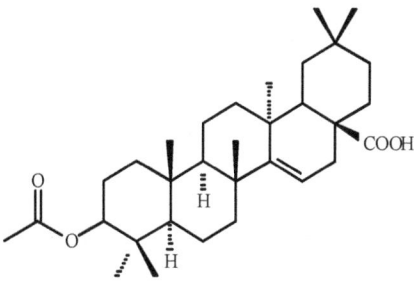

(286) Aleuritolic acid, acetyl

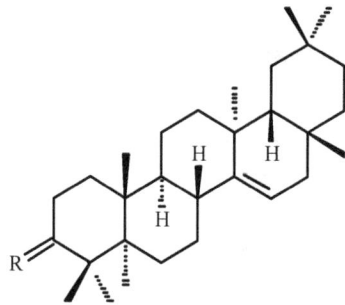

(287a) R=α-H, β-OH - Taraxerol
(287b) R=O - Taraxerone
(287c) R=α-H, β-OAc - Taraxeryl acetate

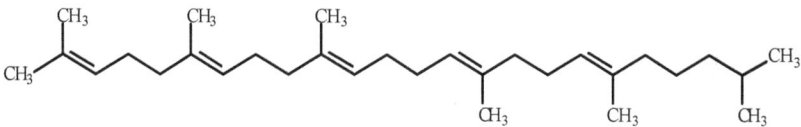

(288) Squalene

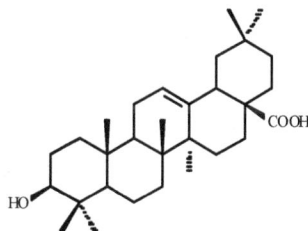

(289) Oleanolic acidOleanolic acid

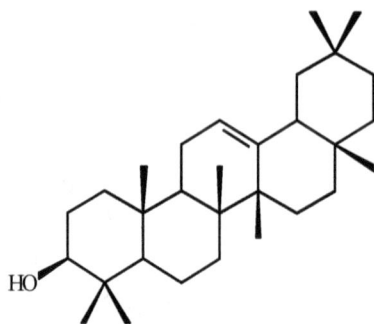

(290) Amyrin, β

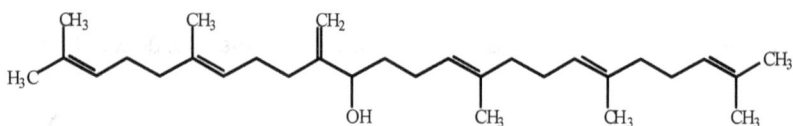

(291) *All-trans*-2,6,15,19,23-pentamethyl-tetracosa-2,6,10 (28),14,22,28-hexaene-11-ol

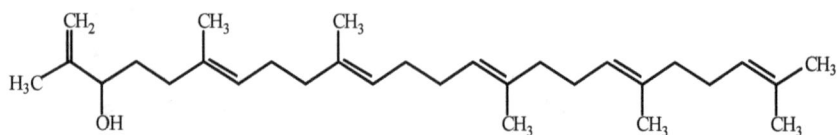

(292) *All-trans*-10-methylene-2,6,10,14,18,22-pentamethyl-tetracosa-1,6,10,14,18,22-hexaen-3-ol

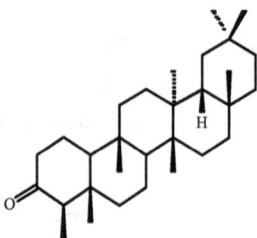

(293) Friedelin

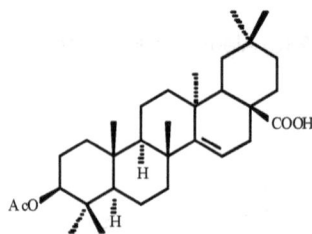

(294) Friedoolean-14-en-28-oic acid, 3-β-acetoxy

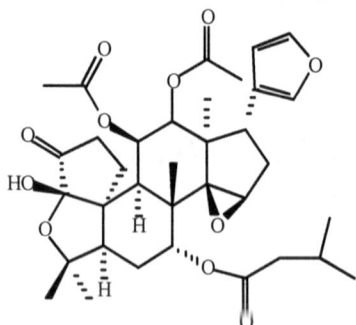

(295) Dumsin

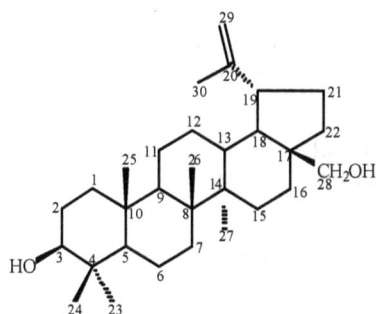

(296) Betulin

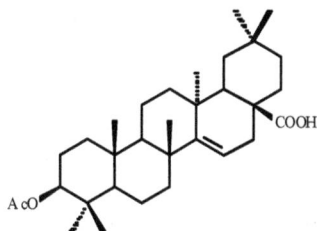

(297) Taraxer-14-en-28-oic acid

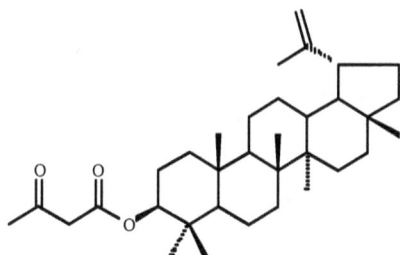

(298) Taraxer-14-en-28-oic acid

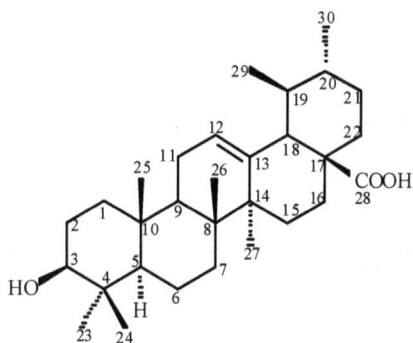

(299) Ursolic acid

Xanthone

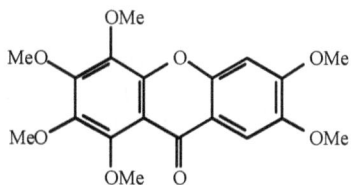

(300) Xanthone-1,2,3,4,6,7-hexamethoxy

Acknowledgements

The authors wish to express their sincere thanks to the College of Pharmacy, The University of Illinois at Chicago, Chicago, Illinois 60612-7231, U.S.A., for assistance with the computer aided NAPRALERT search of genus Croton and CNPq/CAPES/ Brazil for financial support.

References

Abede, W. (1986). A survey of prescriptions used in traditional medicine in Gonbar region, North Western Ethiopia : General Pharmaceutical Pratice. *J. Ethnopharcol.*, **18(2):** 147-165.

Aboagye, F. A., Sam, G. H. (2000). Massiot, E.; Lauaud, C.: Julocrotine, a glutarimide alkaloid from *Croton membranaceus. Fitoterapia*, **71**: 461-462.

Addae-Mensah, I., Waibel, R., Achenbach, H., Muriuki, G., Pearce, C., and Sanders, J. K. M. (1989). A clerodane diterpene and other constituents of *Croton megalocarpus. Phytochemistry,* **28(10)**: 2759-2761.

Addae-Mensah, I., Muriuki, G., Karanja, G., Wandera, C., Waibel, R., and Achenbach, H. (1992a). Constituents of the stem bark and twigs of *Croton macrostachys. Fitoterapia,* **63(1):** 81.

Addae-Mensah, I., Achenbach, H., Thoithi, G. N., Waibel, R., and Mwangi, J. W. (1992b). Epoxychiromodine and other constituents of *Croton megalocarpus. Phytochemistry,* **31(6):** 2055-2058.

Adelekan, A. M., Prozesky, E. A., Hussein, A. A., Urena, L. D., Van Rooyen, P. H., Liles, D. C., Meyer, J. J. M., and Rodriguez, B. (2008). Bioactive diterpenes and other constituents of *Croton steenkampianus. Journal Natural Products,* **71 (11):** 1919-1922.

Adesogan, E. K. (1981). The structure of penduliflaworosin, a new furanoid diterpene from *Croton penduliflorus. J Chem Soc Perkin Trans.,* **1981**: 1151-1153.

Agner, A. R., Maciel, M. A. M., Pinto, A. C., Pamplona, S. G. S. R., and Côlus, I. M. S. (1999). Investigation of genotoxic activity of *trans*-dehydrocrotonin, a clerodane diterpene from *Croton cajucara. Teratogen Carcinogen Mutagen,* **19(6):** 377-384.

Agner, A. R., Maciel, M. A. M., Pinto, A. C., and Colus, I. M. S. (2001). Antigenotoxicity of Trans-dehydrocrotonin, a clerodane diterpene from *Croton cajucara. Planta Medica,* **67 (9):** 815-819.

Albuquerque, M. M. F., Lyra, F. D. A., Melo, J. F., Lima, O. G., Delle-Monache, F., Diu, M. B. D. S., and Moreira, L. C. (1974). Antimicrobial substances of higher plants. Communication XLIV. Isolation of a diterpenic acid from *Croton aff argyrophylloides* (Euphorbiaceae). *Rev. Inst Antibiot Univ Fed Pernambuco Recife.,* **14**: 83.

Albuquerque, A. A. C., Sorenson, A. L., and Leal-Cardoso, J. H. (1995). Effects of essential oil of *Croton zehntneri*, and anethole and estragole on skeletal muscles. *J Ethnopharmacol.,* **49(1):** 41-49.

Alexander, I. C., Pascoe, K. O., Manchard, P., and Lawrence, A. D. W. (1991). An insecticidal diterpene from *Croton linearis. Phytochemistry,* **30(6):** 1801-1803.

Alvarenga, M. A., Gottlieb, O. R., Gottlieb, H. E., Magalhaes, M. T., and Da Silva, V. O. (1978). Diasin, a diterpene from *Croton diasii. Phytochemistry,* **17:** 1773-1776.

Amaral, A. C. F., and Barnco, R. A. (1997). Alkaloids from *Croton celtidifolius. Planta Med.,* **63(5):** 485.

Amaral, A. C. F., and Barnes, R. A. (1998a). Clerodane diterpenoids from *Croton hemiargyreus. Nat. Prod Lett.*, **12(1):** 41-46.

Amaral, A. C. F., and Barnes, R. A. (1998b). A tetrahydroprotoberberine alkaloid from *Croton hemiargyreus. Phytochemistry,* **47(7):** 1445-1447.

Appendino, G., Borrelli, F., Capasso, R., Campagnuolo, C., Fattorusso, E., Petruccil, F., and Taglialatela Scafati, O. (2003). Minor diterpenoids from cascarilla (*Croton eluteria* Bennet) and evaluation of the cascarilla extract and cascarillin. *J Agr Food Chem.*, **51 (24):** 6970-6974.

Aquino, R., Civatta, M. L., and De Simone, F. (1991). Catechins from *Croton draconoides. Fitoterapia,* **62(5):** 454.

Araujo, V. C., Correa, R. G. C., Gottlieb, O. R., Leado da Silva, M., Marx, M. C., Maia, J. G. S., and Magalhaes, M. I. (1972). Linalool-containing amazonian essential oils. *A Acad Brasil Cienc.*, **44:** 317.

Araujo-Junior, V. T., Navarro, P. A., Silva, M. S., Da Cunha, E. V. L., Agra, M. F., Gray, A. I., and Barbosa-Filho, J. M. (2002). Diterpenes From *Croton polyandrus.Ciencia,* **10(3):** 286-290.

Araujo-Junior, V. T., Da Silva, M. S., Da-Cunha, E. V. L., Agra, M. F., Da Silva-Filho, R. N., and Barbosa-Filho, J. M. (2004). Braz-Filho, R.: Alkaloids and diterpenes from *Croton moritibensis. Pharmaceutical Biol.*, **42(1):** 62-67.

Asuzu, I. U., Gray, A. I., and Waterman, P. G. (1988). The extration, isolation and identification of the purgative component of *Croton penduliflorus* seed oil. *J Ethnopharmacol.*, **23(2/3):** 267-271.

Asuzu, I. U., Shetty, S. N., and Anika, S. M. (1989a). Effects of the gut-stimulating principle in *Croton penduliflorus* seed oil on the central nervous system. *J Ethnopharmacol.*, **26(2):** 111-119.

Asuzu, I. U., Shetty, S. N., and Anika, S. M. (1989b). The toxic effects of a chronic administration of the gut-stimulating principle in *Croton penduliflorus* Hutch. seeds in mice. *Drug Chem. Toxicol.*, **12(1):** 85-93.

Attioua, B., Weniger, B., and Chabert, P. (2007). Antiplasmodial activity of constituents isolated from *Croton lobatus. Pharmaceutical Biology,* **45(4):** 263-266.

Bandara, B. M. R., Wimalasiri, W. R., and Bandara, K. A. N. P. (1987). Isolation and insecticidal activity of (-)- Hardwickiic acid from *Croton aromaticus. Planta Med.*, **53(6):** 575.

Bandara, B. M. R., and Wimalasiri, W. R. (1988). Diterpene alcohols from *Croton lacciferus. Phytochemistry,* **27(1):** 225-226.

Bandara, B. M. R., Wimalasiri, W. R., and Mac Leod, J. K. (1988). Ent-kauranes and oleananes from *Croton lacciferus. Phytochemistry,* **27(3):** 869-871.

Bandara, B. M. R., Wimalasiri, W. R., Wickremasinghe, W. A., and Bandara, K. A. N. P. (1990). Cyperenoic acid and (-)-Hardwickiic acid from the chloroform extract of the roots of *Croton aromaticus*: isolation and insecticidal properties. *J Natl Sci Counc Sri Lanka,* **18(2):** 119-126.

Banerjee, A., Nandi, G., and Kundu, A. B. (1988). Investigation of *Croton caudatus* Geisel: Isolation of stigmastan-3,6-dione-5-alfa. *J. Indial Chem Soc.*, **65(6):** 459.

Barbosa, P. R., Fascio, M., Martins, D., Guedes, M. L. S. and Roque, N. F. (2003). Triterpenes of *Croton betulaster* (Euphorbiaceae). *Biochem Syst Ecol.*, **31:** 307-308.

Barbosa, P. R., Fascio, M., Martins, D., and Roque, N. F. (2004). Benzoyl-methylpolyols from *Cróton* species (Euphorbiaceae). *Arkivoc.*, **VI:** 95-102.

Barbosa, P. S., Abreu, A. S., Batista, E. F., Guilhon, G. M. S. P., Mueller, A. H., Arruda, M. S. P., Santos, L. S., Arruda, A. C., and Secco, R. S. (2007). Glutarimide alkaloids and terpenoids from *Croton pullei* var. *glabrior* Lanj. *Biochemical Systematics and Ecology*, **35(12):** 887-890.

Barnes, R. A., and Soeiro, O. M. (1981). The alkaloids of *Croton salutaris*. *Phytochemistry*, **20:** 543-544.

Bauer, R., Tittel, G., and Wagner, H. (1983). Isolation and detection of phorbol esters in *Croton* oil with HPLC. A new method for diterpenester-screening in Euphorbiaceae. *Planta Med.*, **48(1):** 10-16.

Bettolo, R. M., and Scarpati, M. L. (1979). Alkaloids of *Croton draconoides*. *Phytochemistry*, **18:** 520.

Bhakuni, D. S., Gupta, N. C., Satish, S., Sharma, S. C., Shukla, Y. N., and Tandon, J. S. (1971). Chemical constituents of *Actinodaphne agustifolia*, *Croton sparsiflorus*, *Duabanga sonneratioides*, *Glyscosmis mauritiana*, *Hedyotes auricularia*, *Lyonia ovalifolia*, *Micromelum pubescens*, *Pyrus pashia* and *Rhododendron niveun*. *Phytochemistry*, **10:** 2247-2249.

Bhakuni, D. S., Jain, S., and Chaturvedi, R. (1979). The biosynthesis of nornuciferine I (2-methoxy-6(A)-alpha-aporphine-T-ol). *Tetahedron*, **35:** 2323-2326.

Bhakuni, D. S., and Jain, S. (1981). The biosynthesis of the alkaloids of *Croton sparsiflorus* Morong. *Tetrahedron*, **37:** 3175-3181.

Bhakuni, D. S. (1984). Biosynthesis, aberrant biosynthesis and biogenetic type synthesis of some alkaloids of Indian medicinal plants. Proc Fifth Asian Symposium on Medicinal Plants and Spices Seoul Korea August 5: 509-518.

Bittner, M., Silva, M., Aqueveque, P., Kufer, J., Jakupovic, J., and Murillo, R. (1997). Alkaloids and other contituents from *Croton chilensis*. *Bol Soc Chil Quim.*, **42(2):** 223-228.

Block, S., Stevigny, C., De Pauw-Gillet, M. C., de Hoffmann, E., Liabrees, G., Adjakidje, V., and Quetin-Leclercq, J. (2002). Ent-Trachyloban-3-beta-ol, a new cytotoxic diterpene from *Croton zambesicus*. *Planta Med.*, **68(7):** 647-649.

Block, S., Bacelli, C., Tinant, B., Meervelt, L. V., Rozenberg, R., Jiwan, J. L. H., Llabres, G., Gillet, M. C. P., and Leclerq, J. Q. (2004). Diterpenes from leaves of *Croton zambesicus*. *Phytohemistry*, **65:** 1165-1171.

Block, S., Brkic, D., Hubert, P., and Quetin-Leclercq, J. (2005). A validated method for the quantification of pimarane and trachylobane diterpenes in the leaves of

Croton zambesicus by capillary gas chromatography. *Phytochemical Analysis,* **16**: 342-348.

Block, S., Flamini, G., Brkic, D., Morelli, I., and Quetin-Leclercq, J. (2006). Analysis of the essential oil from leaves of *Croton zambesicus* Muell. Arg. Growing in Benin. *Flavour and Fragrance Journal,* **21**: 222-224.

Boonyrathanakornkit, L., Che, C. T., Fong, H. H. S., and Farnsworth, N. R. (1988). Constituents of *Croton crassifolius* roots. *Planta Med.,* **54(1)**: 61-63.

Boonyaratavej, S., and Roengsumran, S. (1988). X-ray structure of cyperenoic acid from *Croton crassifolius. J Nat Prod.,* **51(4)**: 769-770.

Borquez, J., Mancilla, A., Pedreros, S., Loyola, L. A., Morales, G., Wittke, O., and Brito, I. (1995). Isolation and structure of crotonic acid, a clerodane diterpenoid from *Croton chilensis. Bol Soc Chil Quim.,* **40(2)**: 157-162.

Bouquet, A., and Debrary, M. (1974). Medicinal plants of the Ivory Coast. *Trav Doc Orstom.,* **32**: 1.

Boyom, F. F., Keumedjio, F., Dongmo, P. M. J., Ngadjui, B. T., Zollo, P. H. A., Menut, C., and Bessiere, J. M. (2002). Essential oils from *Cróton zambesicus* Muell. Arg. Growing in Cameroon. *Flavour and Fragrance Journal,* **17**: 215-217.

Bracher, F., Randau, K. P., and Lerche, H. (2008). Crototropone, a new tropone derivative from *Croton zehntneri. Fitoterapia,* **79 (3)**: 236-237.

Braga, R. (1976). Plantas do Nordeste especialmente do Ceará. 3ª ed. Coleção Mossoroense, 362.

Brandão, M., Botello, M., and Krettli, E. (1985). Antimalarial experimental chemotherapy using natural products. *Cienc Cult.,* **37(7)**: 1152-1163.

Brito, A. R. M. S., Rodriguez, J. A., Hiruma-Lima, C. A., Haun, M., and Nunes, D. S. (1998). Antiulcerogenic activity of trans-dehydrocrotonin from *Croton cajucara. Planta Medica,* **64 (2)**: 126-129.

Burke, B. A., Chan, W. R., Prince, E. C., Manchand, P. S., Eickman, N., and Clardy, J. (1976). The structure of corylifuran, a clerodane-type diterpene from *Croton corylifolius* LAM. *Tetrahedron,* **32**: 1881-1884.

Burke, B. A., Chan, W. R., Pascoe, K. O., Blount, J. F., and Manchand, P. S. (1979). The structure of Crotofolin E, a novel tricyclic diterpene from *Croton corylifolius. Tetrahedron Letters,* **36**: 3345-3348.

Burke, B. A., Chan, W. R., Pascoe, R. O., Blount, J. F., and Manchand, P. S. (1981). The structure of crotonitenone, a novel casbane diterpene from *Croton nitens* Sw (Euphorbiaceae). *J. Chem Soc Perkin Trans.,* **1981**: 266-2669.

Burnell, R. H., Chapelle, A., and Bird, P.H. (1981). S-(+) Magnoflorine bromide: Isolation from *Croton turumiquirensis. J Nat Prod.,* **44(2)**: 238.

Cai, Y., Evans, F. J., Roberts, M. F., Phillipson, J. D., Zenk, M. H., and Gleba, Y. Y. (1991). *Phytochemistry,* **30**: 2033–2040.

Cai, Y., Evans, F. J., Roberts, M. F., Phillipson, J. D., Zenk, M. H., and Gleba, Y. Y. (1991). Polyphenolic compounds from *Croton lechleri*. *Phytochemistry*, **30(6):** 2033-2040.

Cai, Y., Chen, Z. P., and Phillipson, J. D. (1993a). Diterpenes from *Croton lechleri*. *Phytochemistry*, **32(3):** 755-760.

Cai, Y., Chen, Z. P., and Phillipson, J. D. (1993b). Clerodane diterpenoids from *Croton lechleri*. *Phytochemistry*, **34(1):** 265-268.

Cairnes, D. A., Mirvish, S. S., Wallcave, L., Nagel, D. L., and Smith, J. W. (1981). A rapid method for isolating phorbol from *Croton* oil. *Cancer Lett.*, **14:** 85-91.

Capasso, A., Piacente, S., De Tommassi, N., Ragucci, M., and Pizza, C. (2000). Constituents of *Croton menthodorus* and their effects of electrically induced contractions of the guinea-pig isolated ileum. *Phytother Res.*, **14(3):** 156-159.

Carlin, L., Vaisberg, A. J., and Hammond, G. B. (1996). Isolation of sinoaucutine from the leaves of *Croton lechleri*. *Planta Med.*, **62(1):** 90-91.

Carvalho, J. C. T., Silva, M. F. C., Maciel, M. A. M., Pinto, A. D. C., Nunes, D. S., Lima, R. M., Bastos, J. K., and Sarti, S. J. (1996). Investigation of anti-inflammatory and antinociceptive axtivities of trans-dehydrocrotonin, a 19-nor-clerodane diterpene from *Croton cajucara*. *Planta Medica*, **62 (5):** 402-404.

Casagrande, C., Canonica, L., and Severini-Ricca, G. (1975). Studies on proaporphine and aporphine alkaloids. VII. Stereochemistry of reduced proaporphines of *Croton sparsiflorus* and *C. linearis*. *J Chem Perkin Trans.*, I: 1659.

Catalan, C. A. N., Heluani, C. S., Kotowicz, C., Gedris, T. E., and Herz, W. (2003). A linear sesterpene, two squalene derivatives and two peptide derivatives from *Croton hieronymi*. *Phytochemistry*, **64:** 625-629.

Chabert, P., Attioua, B., and Brouillard, R. (2006). *Croton lobatus*, an African medicinal plant: Spectroscopic and chemical elucidation of its many constituents. *Bio Factors*, **27:** 69-78.

Chaichantipyyuth, C., Petsom, A.,Taweechotipatr, P., Muangsin, N., Chaichit, N., Puthong, S., Roengsumran, S., Kawahata, M., Watanabe, T., and Ishikawa, T. (2005). New labdane-type diterpenoids from *Croton oblongifolius* and their cytotoxic activity. *Heterocycles*, **65 (4):** 809-822.

Chan, W. R., Taylor, D. R., and Willis, C. R. (1968). Terpenoids from the Euphorbiaceae. Part 1. The structure of crotonin, a norditerpene from *Croton lucidus* L. *J Chem Soc C.*, **22:** 2781-2785.

Chan, W. R., Prince, E. C., Marchand, P. S., Springer, J. P., and Clardy, J. (1975). The structure of Crotofolin A, a diterpene with a new skeleton. *J. Am Chem Soc.*, **97(150):** 4437-4439.

Charris, J., Dominguez, J., De la Rosa, C., and Caro, C. (2000). (-)-Amuronine from the leaves of *Croton flavens* L. (Euphorbiaceae). *Biochem Syst Ecol.*, **28(8):** 795-797.

Chatterjee, A., Banerjee, A., and Bohlmann, F. (1977). Crotocaudin: a rearranged labdane type norditerpene from *Croton caudatus* Geisel. *Tetrahedron*, **33**: 2407-2414.

Chatterjee, A., Banerjee, A., and Bohlmann, F. (1978). Isocrotocaudin a new norclerodane–type diterpene from *Croton caudatus*. *Phytochemistry*, **17**: 1777-1779.

Chavez, P. I., Jolad, S. D., Hoffman, J. J., and Cole, J. R. (1982). Four new 12-deoxyphorbol diesters from *Croton californicus*. *J Nat Prod.*, **45(6)**: 745-748.

Chen, M. H., and Pan, K. Z. (1993). Isolation, characterization, crystallization of croton toxin from the seeds of *Croton tiglium* L. *Shen Huaxue Zazhi.*, **9(1)**: 104-108.

Chen, Z. P., Cai, Y., and Phillipson, J. D. (1994). Studies on the anti-tumor, antibacterial, and wound-healing properties of Dragon's Blood. *Planta Med.*, **60(6)**: 541-545.

Chen, W., Yang, X. D., Zhao, J. F.,Yang, J. H., Zhang, H. B., Li, Z. Y., and Li, Liang. (2006). Three new, 1-oxygenated *ent*-8,9-secokaurane diterpenes from *Croton kongensis*. *Helyetica Chimica Acta.*, **89**.

Chen, W., Li, Z., Yang, X., Zhao, J., Yang, J., and Li, L. (2006). Studies on chemical components of *Croton cascarilloides*. *Yunnan Daxue Xuebao, Ziran Kexueban.*, **28 (3)**: 247-250.

Chen, X. (2007). Chinese medicinal compositions containing *Croton* and *Phytolacca* and others for treating hepatic cirrhosis. *Faming Zhuanli Shenging Gongkai Shuomingshu.*, **10**.

Ciccio, J. F., and Segnini, M. (2002). Composition of the essential oil from leaves *Croton jimenezii* from Costa Rica. *J Essent Oil Res.*, **14(5)**: 357-360.

Coelho-de-Sousa, A. N., Barata, E. L., Magalhaes, P. J. C., Lima, C. C., and Leal-Cardoso, J. H. (1997). Effects of the essential oil of *Croton zehntneri*, and its constituent estragole on intestinal smooth muscle. *Phytother Res.*, **11(4)**: 299-304.

Costa, A. M. L., Silva, J. C. R., Campos, A. R., Rao, V. S. N., Maciel, M. A. M., and Pinto, A. C. (1999). Antioestrogenic effect of trans-dehydrocrotonin, a nor-clerodane diterpene from *Croton cajucara* Benth, in rats. *Phytother Res.*, **13(8)**: 689-691.

Craveiro, A. A., Monte, F. J., Matos, F. J. A., and Alencar, J. W. (1978a). Volatile constituentes of *Croton* aff *argyrophylloides*. *Rev Latinoamer Quim.*, **9(3)**: 98-99.

Craveiro, A. A., Andrade, C. H. S., Matos, F. J. A., and Alencar, J. W. (1978b). Anise-like flavor *Croton* aff *zehntneri*. *J Agr Food Chem.*, **26(3)**: 772-773.

Craveiro, A. A., Andrade, C. H. S., Matos, F. J. A., Alencar, J. W. and Dantas, T. N. C. (1980). Fixed and volatile constituents of *Croton* Aff *nepetaefolius*. *J Nat Prod.*, **43(6)**: 756-757.

Craveiro, A. A.,and Lemos, T. L. G. (1980). Interconversion of arylpropanoids in the essential oil of *Croton* Aff *zehntneri*. *J Nat Prod.*, **43(5)**: 634-636.

Craveiro, A. A., Rodrigues, A. S., Andrade, C. H. S.,Matos, F. J. A., Alencar, J. W., and Machado, M. I. L. (1981a). Volatile constituents of Brazillian Euphorbiaceae. Genus *Croton. J Nat Prod.*, **44(5):** 602-608.

Craveiro, A. A., Silveira, E. R., Braz-Filho, R., and Mascarenhas, I. P. (1981b). Sonderianin, a furanoid diterpene from *Croton sonderianus. Phytochemistry,* **20:** 852-854.

Craveiro, A. A., and Silveira, E. R. (1982). Two cleistanthane type diterpenes from *Croton sonderianus. Phytochemistry,* **21:** 2571-2574.

Craveiro; A. A., Silveira, E. R., and Mc Chesney, J. D. (1983). Crotosondin; a benzofuran sesquiterpene from *Croton sonderianus.* Abstr 24 th *Annual Meeting American Society of Pharmacognosy Univ Mississippi Oxford* July-24-28-1983: ABTR-35.

Craveiro, A. A., Alencar, J. W., Matos, F. J. A., and Machado, M. I. L. (1990). The essential oil of *Croton adenocalyx* A. DC. *J Essent Oil Res.,* **23:** 145-146.

Cronquist, A. (1981). An integrated system of classification of flowering plants. *New York: Columbia University Press,* 55.

Cuong, N. M., Sung, T. V., and Ahn, B. Z. (2002). Cytotoxic compounds from *Croton cascarilloides. Korean J Pharmacog.,* **33(3):** 207-210.

Del-Castillo, H. C. C., De Simone, F., and De Feo, V. (1996). Proaporphine alkaloids from *Croton ruizianus* Muell Arg. *Biochem Sist. Ecol.,* **24(5):** 463-464.

Di Stasi, L. C., Guimarães-Santos, E. M., Santos, C. M., and Hiruma, C. A. (1989). Plantas Medicinais na Amazônia. Ed. UNESP, São Paulo, 194p.

Dominguez, X. A., and Alcorn, J. B. (1985). Screening of medicinal plants used by Huastec Mayans of Northeastern Mexico. *Journal of Ethnopharmacol.,* **13(2):** 139-156.

Earle, F. R., Glass, C. A., Geisinger, G. C., Wolff, I. A., and Jones, Q. (1960). Search for new industrial oils. *J Amer oil Chem Soc.,* **37:** 440.

Edwards, M. C., Taylor, S. E., Williamson, E. M., and Evans, F. J. (1983). New phorbol and deoxyphorbol esters: Isolation and relative potencies in inducing platelet aggregation and erythema of skin. *Acta Pharmacol Toxicol.,* **53(3):** 177-187.

Eisenreich, W. J., and Bracher, F. (2000). Alkaloids from *Croton balsamifera. Archiv der Pharmazie,* **333** Supplement., 2/00 42.

Eisenreich, W. J., and Bracher, F. (2001). Crotoflavol, a new phenanthrene from *Croton flavens. Nat Prod Lett.,* **15(2):** 147-150.

Eisenreich, W. J., Hofner, G., and Bracher, F. (2003). Alkaloids from *Croton flavens* L. and their affinities to Gaba-receptors. *Nat Prod Research,* **17 (6):** 437-440.

Eknamkul, W., and Potduang, B. (2003). Biosyntehsis of beta-sitosterol and stigmasterol in *Croton sublyratus* proceeds via a mixed origin of isoprene units. *Phytochemistry,* **62(3):** 289-298.

El Babili, F., Moulis, C., Bon, M., Respaud, M. J., and Fourasté, I. (1998). Three furano- diterpenes from the bark of *Croton campestris. Phytochemistry,* **48(1):** 165-169.

Elbridge, J. (1975). Bush medicine in the exumas and Long Island, Bahamas: A Field Study. *Econ Bot.*, **29:** 307-332.

El Mekkawy, S., Meselhy, M. R., Nakamura, N., Hattori, M., Kawahata, T., and Otake, T. (2000). Anti-HIV-1 phorbol esters from the seeds of *Croton tiglium*. *Phytochemistry*, **53(4):** 457-464.

Erdelmeier, C. A. J., Van Leeuwen, P. A. S., and Kinghorn, A. D. (1988). Phorbol diester constituents of *Croton* oil: Separation by two-dimensional TLC and rapid purification utilizing Reversed-Phase Overpressure Layer Chromatography (RP-OPLC). *Planta Med.*, **54(1):** 71-75.

Esser, H. J. (2003). Flora of Thailand Euphorbiaceae. 28 *Croton*. (Accessed from: http: //nhncml.leidenuniv.nl/thaieuph/ThCroton.htm#Croton per cent 20columnaris).

Estrada E. (1985). Jardín Botánico de Plantas Medicinales Maximino Martínez. *Universidad Autónoma Chapingo*, **12**.

Etcheves, M. D. C. P., Gros, E. G., and Retamar, J. A. (1981). Chemical study of *Croton parvifolius*. *Essenze Deriv Agrum.*, **51:** 253-261.

Farias, R. A. F., Rao, V. S. N., Viana, G. S. B., Silveira, E. R., Maciel, M. A. M., and Pinto, A. C. (1997).

Hypoglicemic effect of *trans*-dehydrocrotonin, a nor-clerodane diterpene from *Croton cajucara*. *Planta Med.*, **63(6):** 558-560.

Farnsworth, M. R., Blomste, R. N., Messmer, W. M., King, J. C., Persinos, G. F., and Nilkes, J. D. A. (1969). Phytochemical and biological review on the genus *Croton*. *Lloydia*, **32(1):** 1-28.

Fattorusso, E., Taglialatela-Scafati, O., Campagnuolo, C., Santelia, F. U., Appendino, G., and Spagliardi, P. (2002). Diterpenoids from cascarilla (*Croton eluteria* Bennet). *J Agric Food Chem.*, **50(18):** 5131-5138.

Fischer, H., Machen, T. E., Widdicombe, J. H., Carlson, T. J. S., King, S. R., Chow, J. W. S., and Illek, B. (2004). A novel extract SB-300 from the stem bark latex of *Croton lechleri* inhibits CFTR-mediated chloride secretion in human. *J Ethnopharmacol.*, **93(2-3):** 351-357.

Freise, F. W. (1935). Essential oils from Brazilian Euphorbiaceae. *Perfum Essent Oil Rec.*, **26:** 219.

Fuh, L. F., and Chen, C. C. (1982). Isolation of lectins from *Croton tiglium* seeds. *Sheng Wu K'o Hsueh.*, **19:** 45-50.

Fujita, E., Nagao, Y., and Node, M. (1976). Diterpenoids of isodon and teucrium plants. *Heterocycles*, **5:** 793-838.

Fujita, E., Node, M., Nishide, K., Sai, M., Fuji, K., Mc Phail, A. T., and Lamberton, J. A. (1980). Structures of croverin (X-Ray-Analysis) and dihydrocroverin, two new diterpene lactones from *Croton verreauxii* Baill. *Chem Commun.*, **1980:** 920-921.

Garcia, H. (1975). Flora medicinal de Colombia. *Universidad Nacional de Bogota*.

Garcia, L., Guarin, D. L., and Tobar, M. C. (1986). Isolation of ayanin from *Croton glabellus* leaves. *Rev Colomb Cienc Quim Farm.*, **15**: 95-98.

García, A., Ramírez-Apan, T., Cogordan, J. A., and Delgado, G. (2006). Absolute configuration assignments by experimental and theoretical approaches of *ent*-labdane- and *cis-ent*-clerodane-type diterpenes isolated from *Croton glabellus*. *Can. J. Chem.*, **84**: 1593-1602.

Giang, P. M., Jin, H. Z., Son, P. T., Lee, J. H., Hong, Y. S., and Lee, J. J. (2003). Ent-kaurane diterpenoids from *Croton tonkinensis* inhibit LPS-induced NF-kappa B activation and NO production. *J Nat Prod.*, **66(9)**: 1217-1220.

Giang, P. M., Son, P. T., Hamada, Y., and Otsuka, H. (2005). Cytotoxic diterpenoids from Vietnamese medicinal plant *Croton tonkinensis* Gagnep. *Chem Pharm Bull.*, **53 (3)**: 296-300.

Gonzaâlez-Vaâzquez, R., Díaz, B. K., Aguilar, M. I., Diego, N., and Lotina-Hennsen, B. (2006).Pachypodol from *Croton ciliatoglanduliferus* Ort. As Water-Splitting Enzyme Inhibitor on Thylakoids. *J. Agric. Food Chem.*, **54**: 1217-1221.

Gottlieb, O. R.,Koketsu, M., Magalhães, M. T., Guilherme S. Maia, J., Mendes, P. H., Da Rocha, A. I., Da Silva, M. L., and Wilbrelg, V. C. (1981).Essential oil of Amazonia VII. *Acta Amazônica.*, **11**: 143-148.

Grynberg, N. F., Echevarria, A.,Lima, J. E., Pamplona, S. S. R., Pinto, A. C., and Maciel, M. A. M. (1999). Anti-tumor activity of two 19-nor-clerodane diterpenes, trans-dehydrocrotonin and trans-crotonin. From *Croton cajucara*. *Planta Medica*, **65(8)**: 687-689.

Graikou, K., Aligiannis, N., Skaltsounis, A. L., Chinou, I., Michel, S., Tillequin, F., and Litaudon, M. (2004). New diterpenes from *Croton insularis*. *J Nat Prod.*, **67(4)**: 685-688.

Guerrero, M. F., Puebla, P., Carbon, R., Martin, M. L., and Roman, L. S. (2002). Quercetin 3,7-dimethyl ether: Vasorelaxant isolated from *Croton scchiedeanus* Schleacht. *J Pharm Pharmacol.*, **54(10)**: 1373-1378.

Guadarrama, A. B. A., and Rios, M. Y. (2004). Three new sesquiterpenes from *Croton arboreous*. *J Nat Prod.*, **67(5)**: 914-917.

Gupta, M., Mazumder, U. K., Vamsi, M. L. M., Sivakumar, T., and Kandar, C. C. (2004). *J. Ethnopharmacol.*, **90**: 21–25.

Hagedorn, M. L., and Brown, S. M. (1991). The constituents of cascarilla oil (*Croton eluteria* Bennett). *Flavour Fragrance J.*, **6(3)**: 193-204.

Haruyama, H., Hata, T., Ogiso, A., Tamura, C. and Kitazawa, E. (1983). Structure of Plaunolide, a new furanoid diterpene from *Croton sublyratus*, C-20-H-20-0-5. *Acta Crystallogr Ser C.*, **39(2)**: 255-257.

Hecker, E. (1968). Cocarcinogenic principles from the seed oil of *Croton tiglium* and from other Euphorbiaceae. *Cancer Res.*, **28**: 2338-2349.

Hecker, E. (1984). Co-carcinogenic diterpene esters as principal risk factors in local life style esophageal cancer in Curaçao. *Act Pharmacol Toxicol.*, **55(52)**: 148-153.

Heluani, C. S., Catalano, C. A. N., Hernandez, L. R., Burgueno-Tapia, E. and Joseph-Nathan, P. (1998). ^{13}C NMR assignments and conformational evalution of diterpenes from *Croton sarcopetalus* Muell. *Magn Reson Chem.*, **36(12)**: 947-950.

Heluani, C. S., Catalan, C. A. N., Hermandez, L. R., Burgueno-Tapia, E. and Joseph-Nathan, P. (2000). Three new diterpenoids based on the novel sarcopetalane skeleton from *Croton sarcopetalus*. *J Nat Prod.*, **63 (2)**: 222-225.

Hernandez, J., Delgado, G. (1992).Terpenoids from aerial parts of *Croton dracco. Fitoterapia*, **63(4):** 377-378.

Hickey, T. A., Worobec, S. M. and West, D. P. (1981). Kinghorn, A. D.: Irritant contact dermatitis in humans from phorbol and related esters. *Toxicon*, **19**: 841-850.

Hiruma-Lima, C. A., Spadari-Bratfisch, R. C., Grassi-Kassisse, D. M. and Brito, A. R. M. S. (1999). Antiulcerogenic mechanisms of dehydrocrotonin, a diterpene lactone obtained from *Croton cajucara. Planta Med.*, **65(4):** 325-330.

Hokanson, G. C. and Cassady, J. M. (1976). Potential antitumor agents from higher plant. Part I: *Strychnos hennigs* II. Part II: *Centaurea solstittalis*. Part III: *Croton texensis. Diss Abstr Int.*, B **97**: 1265-1266.

Ichihara, Y.; Takeya, K., Hitotsuyanagi, Y., Morita, H., Okuyama, S., Suganuma, M., Fujiki, H., Motidome, M. and Itokawa, H. (1992). Cajucarinolide and isocajucarinolide: anti-inflammatory diterpenes from *Croton cajucara. Planta Med.*, **58(6):** 549-551.

Israngkura, A., Ayutthaya, N., Chaichantipyuth, C., Siriwat, K., Petsom, A. and Roengsumran, S. (2000). Cleistanthane diterpenoid from *Croton oblongifolius. Thai J Pharm Sci.*, **24**: 18.

Itokawa, H., Ichihara, Y., Kojima, H., Watanabe, K. and Takeya, K. (1989). Nor-clerodane diterpenes from *Croton cajucara. Phytochemistry*, **28(6):** 1667-1669.

Itokawa, H.; Ichihara, Y., Shimizu, M., Takeya, K. and Motidome, M. (1990). Cajucarins A and B, new clerodane diterpenes from *Croton cajucara*, and their conformations. *Chem. Pharm. Bull.*, **38(3):** 701-705.

Itokawa, H., Ichihara, Y., Takeya, K., Morita, H. and Motidome, M. (1991a).Diterpenes from *Croton salutaris. Phytochemistry*, **30(12):** 4071-4073.

Itokawa, H., Ichihara, Y.,Mochizuki, M., Enomori, T., Morita, H., Shirota, O., Inamatsu, M. and Takeya, K. (1991b). A cytotoxic substance from Sangre de Grado. *Chem Pharm Bull.*, **39(4):** 1041-1042.

Itokawa, H., Takeya, K., Watanabe, K., Morita, H., Ichihara, Y., Totsuka, N., Shirota, O., Izumi, H., Satoke, M., Yasuda, I., Sankama, U., Motidome, M. and Flores, F. A. (1992). Antitumor substances from South American plants. *J Pharmacobio Dyn.*, **15(1):** S-2.

Jogia, M. K., Andersen, R. J., Parkaniy, L., Clardy, J., Dublin, H. T. and Sinclair, A. R. E. (1989). Crotofolane diterpenoids from the African shrub *Croton dichogamus* PAX. *J. Org Chem.*, **54(7):** 1654-1657.

Johns, T., Mhoro, E. B., Sanaya, P. and Kimanani, E. K. (1994). Herbal remedies of the babatemi of Ngorongoro District, Tanzania: A quantitative appraisal. *J. Herbs Spices Med Plants,* **2(4):** 45-100.

Jones, K. (2003). Review of sangre de drago (*Croton lechleri*)- a South American tree sap in the treatment of diarrhea, inflamation, insect bites, viral infections, and wounds: tradicional uses to clinical research. *J Altern Compl Med.*, **9(6):** 877-896.

Kapoor, V. K., Chawla, A. S., Soni, P. K. and Bedi, K. L. (1986). Studies on some Indian seed oils. *Fitoterapia*, **42(3):** 188-190.

Kapingu, M. C., Guillaume, D., Mbwambo, Z. H., Moshi, M. J., Uliso, F. C. and Mahuannah, L. A. (2000). Diterpenoids from the roots of *Croton macrostachys*. *Phytochemistry*, **54(8):** 767-770.

Kasa, M. (1991). Antimalarial activities of local medicinal plants. *Trad Med New Lett.*, **2:** 1-3.

Kim, J. H., Lee, S. J., Haw, J. B., Moon, J. J. and Kim, J. B. (1994). Isolation of isoguanosine from *Croton tiglium* and its antitumor activity. *Arch Parm Res.*, **17(2):** 115-118.

Kinghorn, A. D. and Marshall, G. T. (1984). Purication of 12-O–tetradecanoyl-phorbol-13-acetate, phorbol and 4-alpha-phorbol from *Croton* Oil. *Patent-Us-4*, 468,328.

Kinghorn, A. D. and Erdelmeier, C. A. J. (1991). Analysis of *Croton* oil by reversed-phase overpressure layer chromatography. Modern methods of plant anaçysis, new series, essential oils and waxes (H. F. Linskens and J. F. Jackson, EDS.), Springerverlag, Germany **12:** 175-184.

Kinyanjui, T., Gitu, P. M. and Kamau, G. N. (2000). Potential antitermite compounds from *Juniperus procera* extracts. *Chemosphere*, **41(7):** 1071-1074.

Kitazawa, E., Ogiso, A., Takahashi, S., Sato, A., Kurabayashi, M., Kuwano, H., Hata, T. and Tamura, C. (1979). Plaunol A and B, new anti-ulcer diterpene lactones from *Croton sublyratus*. *Tetrahedron Lett.*, **13:** 1117-1120.

Kitazawa, E., Sato, A., Takahashi, S., Kuwano, H. and Ogiso, A. (1980). Novel diterpenelactones with anti-peptic ulcer activity from *Croton sublyratus*. *Chem Pharm Bull.*, **28 (1):** 227-234.

Kitazawa, E. and Ogiso, A. (1981). Two diterpene alcohols from *Croton sublyratus*. *Phytochemistry*, **20:** 287-289.

Kitazawa, E., Kurabayashi, M., Kasuga, S., Oda, O. and Ogiso, A. (1982).New esters of a diterpene alcohol from *Croton sublyratus*. *Ann Rep Sankyo Res Lab.*, **34:** 39-41.

Klauss, U. and Adala, H. S. (1994). Tradicional herbal eye medicine in Kenya. *World Health Forum*, **15(9):** 198-143.

Kloos, H., Tekle, A., Yohannes, L. W., Yosef, A. and Lemma, A. (1978). Preliminary studies of traditional medichial plants in nineteen markets in Ethiopia: Use patterns and public health aspects. *Ethiopian Med.*, **16:** 33.

Kloos, H. (1997). Preliminary studies of medicinal plants and plant products in markets of Central Ethiopia. *Ethno Medicine*, **4(1):** 63-104.

Kokwaro, J. (1976). Medicinal plants of East Africa. *East Afr Literature Bureau, Book.*

Kostova, I., Iossifova, T., Rostan, J., Vogler, B., Kraus, W. and Navas, H. (1999). Chemical and biological studies on *Croton panamensis* latex (Dragon's blood). *Pharm Pharmacol Lett.*, **9(1):** 34-36.

Krebs, H. C. and Ramiarantsoa, H. (1996). Clerodane diterpenes and other constituents of *Croton hovarum*. *Phytochemistry*, **41(2):** 561-563.

Krebs, H. C. and Ramiarantsoa, H. (1997). Clerodane diterpenes of *Croton hovarum*. *Phytochemistry*, **45(2):** 379-381.

Kubo, I., Hanke, F. J., Asaka, Y., Matsumoto, T. and He, C. H.; Clardy, J. (1990). Insect antifeedants from tropical plants I. structure of dumsin. *Tetrahedron*, **46(5):** 1515-1522.

Kubo, I., Asaka, Y. and Shibata, K. (1991). Insect growth inhibitory nor-diterpenes, *cis*-dehydrocrotonin and *trans*-dehydrocrotonin from *Croton cajucara*. *Phytochemistry*, **30(8):** 2545-2546.

Kuo, P. C., Shen, Y. C., Yang, M. L., Wang, S. H., Thang, T. D., Dung, N. X.,Chiang, P. C., Lee, K. H., Lee, E. J. and Wu, T. S. (2007). Crotonkinins A and B and Related Diterpenoids from *Croton tonkinensis* as Anti-inflammatory and Antitumor Agents. *Journal of Natural Products*, **70(12):** 1906-1909.

Kupchan, S. M., Hemingway, R. J., Coggon, P., Mc Phail, A. T. and Sim, G. A. (1968). Crotepoxide, a novel cyclohexane diepoxide tumor inhibitor from *Croton macrostachys*. *J Amer Chem Soc.*, **90:** 2982.

Kupchan, S. M., Hemingway, R. J. and Smith, R. M. (1969). Tumor inhibitors. Crotepoxide, a novel cyclohexane diepoxide tumor inhibitor from *Croton macrostachys*. *J Org Chem.*, **34:** 3898.

Kupchan, S. M., Uchida, I., Branfman, A. R., Dailey Jr, R. G. and Fei, B. Y. (1976). Antileukemic principles isolated from Euphorbiaceae plants. *Science*, **191:** 571-572.

Kutney, J. P., Klein, F. K., Knowles, G. and Stuart, K. L. (1971a). Alkaloids from *Croton* species XI-1. Peptidyl compounds from *Croton humilis* L. *Tetrahedron Lett.*, **35:** 3263-3266.

Kutney, J. P., Klein, F. K., Eigendorf, G., Mc Neill, D. and Stuart, K. L. (1971b). Alkaloids from *Croton* species. XII-1. Glutarimide peptides from *Croton humylis* L. *Tetrahedron Lett.*, **52:** 4973-4975.

Lahlou, S., Leal-Cardoso, J. H., Magalhães, P. J. C., Coelho-de-Souza, A. N. and Duarte, G. P. (1999). Cardiocascular effects of the essential oil of *Croton nepetaefolius* in Rats: Role of the Autonomic Nervous System. *Planta Med.*, **65(6):** 553-557.

Leão, I. M. S., Andrade, C. H. S., Pinheiro, M. L. B., Rocha A. F. I., Machado, M. I. L., Craveiro, A. A., Alencar, J. W. and Matos, F. J. A. (1998). Essential oil of *Croton lanjouwensis* Jablonski from Brazillian Amazonian Region: *J. Essent Oil Res.,* **10**: 643-644.

Lemos, T. L. G., Monte, F. J. Q., Matos, F. J. A., Alencar, J. W., Craveiro, A. A., Barbosa, R. C. S. B. and Lima, E. O. (1992). Chemical composition and antimicrobial activity of essential oils from Brazilian plants. *Fitoterapia,* **63(3):** 266-268.

Liebich, H. M., Lehmann, R., Di Stefano, C., Haring, H. U., Kim, J. H. and Kim, K. R. (1998). Analysis of traditional chinese anticancer drugs by capillary electrophoresis. *J Chromatogr* A, **795(2)**: 388-393.

Lima, D. P., Leoncio D'Albuquerque, I., Gonçalves de Lima, O., Lacerda, A. L., Maciel, G. M. and Martins, D. (1973). Substâncias antimicrobianas de plantas superiores XLIII. Primeiras observações sobre ação antimicrobianas de principio ativo isolado do caule do *Croton jacobinensis* Baill (Euphorbiaceae). *Rev. Inst Antibiot, Recife.,* **13**: 31-35.

Lima, S. G. (2000). Contribuição ao conhecimento químico de plantas do Nordeste do Brasil: *Combretum laxum* Jack. (Combretaceae)*; Croton adenocalyx* Baill. (Euphorbiaceae) e *Julocroton* sp. (Euphorbiaceae). Dissertação de Mestrado. Universidade Federal do Ceará.

Lima- Accioly, P. M., Lavor-Porto, P. R., Cavalcante, F. S., Magalhães, P. J. C., Lahlou, S., Morais, S. M. and Leal-Cardoso, J. H. (2006). Essential Oil of *Croton nepetaefolius* and its main constituent, 1,8-cineole, block excitability of rat sciatic nerve *in vitro*. *Clinical and Experimental Pharmacology and Physiology*, **33**: 1158-1163.

Lin, J. Y. and Huang, M. Y. (1977). Isolation of a hemolitic protein, Crotin I from *Croton tiglium*. *Taiwa Yao Hsueh Tsa Chih.,* **28**: 1071.

Lin, W. H., Fu, H. Z., Cheng, G. and Barnes, R.A. (2003). The alkaloids from leaves of *Croton hemiargyreus* var *gymnodiscus*. *J Chin Pharm Sci.,* **12 (3):** 117-122.

Lopes, D., Bizzo, H. R., Sobrinho, A. F. S. and Pereira, M. V. G. (2000). Linalool- rich essential oil from leaves of *Croton cajucara* Benth. *J Essent Oil Res.,* **12(6):** 705-708.

Luzbetak, D. J., Torrance, S. J., Hoffmann, J. J. and Cole, J. R. (1979). Isolation of (-)-Hardwickiic acid and 1-Triacontanol from *Croton californicus*. *J Nat Prod.,* **42(3):** 315-316.

Maciel, M. A. M., Pinto, A. C., Brabo, S. N. and Da Silva, M. N. (1998). Terpenoids from *Croton cajucara*. *Phytochemistry,* **49(3):** 823-828.

Maciel, M. A. M., Pinto, A. C., Arruda, A. C., Pamplona, S. G. S. R., Vanderlinde, F. A., Lapa, A. J., Echevarria, A., Grynberg, N. F., Côlus, I. M. S., Farias, R. A. F., Costa, A. M. L. and Rao, V. S. N. (2000). Ethnopharmacology, phytochemistry and pharmacology: a successful combination in the study of *Croton cajucara*. *Journal of Ethnopharmacol,* **70(1):** 41–55.

Macrae, W. D. Hudson, J. B and Towers, G. H. N. (1988). Studies on the pharmacological activity of amazonian Euphorbiaceae. *J Ethnopharmacol,* **22**: 143-172.

Magalhães, P. J. C., Criddle, D. N., Tavares, R. A., Melo, E. M., Mota, T. L. and Leal-Cardoso, J. H. (1988). Intestinal myorelaxant and antispasmodic effects of the essential oil of *Croton nepetaefolius* and its contituents cineole, methyl-eugenol and terpineol. *Phytother Res.,* **12(3):** 172-177.

Mahmood, N., Pizza, C., Aquino, R., De Tommasi, N., Piacente, S., Colman, S., Burke, A. and Hay, A.J. (1993).Inhibition of HIV infection by flavonoids. *Antiviral Res.,* **22(2/3):** 189-199.

Marino, S., Gala, F., Zollo, F., Vitalini, S., Fico, G., Visioli, F. and Iorizzi, M. (2008). Identification of minor secondary metabolites from the látex of *Croton lechleri* (Muell-Arg) and evaluation of their antioxidant activity. *Molecules,* **13:** 1219-1229.

Marshall, G. T. and Kinghorn, A. D. (1984). Short-chain phorbol ester constituents of *Croton* oil. *J Amer Oil Chem Soc.,* **61(7):** 1220-1225.

Marshall, G. T. and Kinghorn, A. D. (1980). Purification method for phorbol and 4-alpha-phorbol from *Croton* oil using droplet counter-current chromatography. Patent-Us-Appl Nov 12 1980.

Marshall, G. T. and Kinghorn, A. D. (1981). Isolation of phorbol and 4-alpha-phorbol from *Croton* oil by droplet counter-current chromatography. *J Chromatogr.,* **206:** 421-424.

Marshall, G. T., Klocke, J. A., Lin, L. J. and Kinghorn, A. D. (1985). Effects of diterpene esters of tigliang, daphnane, ingenane and lathyrane types of pink bollworm, pectinophora *Gossypiella saunders* (Lepidotera: Gelechidade). *J Chem Ecol.,* **11(2):** 191-206.

Martínez, M. (1987). Catálogo de Nombres Vulgares y Científicos de Plantas Mexicanas. *Fondo de Cultura Económica,* 215.

Martins, A. P., Salqueiro, L. R., Gonçalves, M. J., Vila, R.; Tomi, F., Adzet, T., Dacunha, P., Canigueral, S. and Casanova, J. (2000). Antimicrobial activity and chemical composition of the bark oil of *Croton stelluliferus*, an endemic species from S. Tome e Principe. *Plant Med.,* **66(7):** 647-650.

Mazzanti, G., Bolle, P., Martinoli, L., Piccinelli, D., Grweurina, I., Animati, F. and Mucne, Y. (1987). *Croton macrostachys,* a plant used in traditional medicine with purgative and inflammatory activity. *J Ethnopharmacol,* **19(2):** 213-219.

Mc Chesney, J. D., Silveira, E. R., Craveiro, A. A. and Shoolery, J. N. (1984).The use of carbon-carbon connectivity in the structure determination of marmelerin, a novel benzofuran sesquiterpene from *Croton sonderianus. J Org Chem.,* **49(26):** 5154-5157.

Mc Chesney, J.D. and Silveira, E.R. (1989). 12-hydroxyhardwickiic acid and sonderianal, neo-clerodanes from *Croton sonderianus. Phytochemistry,* **28(12):** 3411-3414.

Mc Chesney, J. D. and Silveira, E. R. (1990).Ent-clerodanes of *Croton sonderianus. Fitoterapia,* **61(2):** 172-175.

Mc Chesney, J. D., Clark, A. M. and Silveira, E. R. (1991a). Antimicrobial diterpenes of *Croton sonderianus.* II. Ent-beyer-15-en-18-oic acid. *Pharmaceut Res.,* **8(10):** 1243-1247.

Mc Chesney, J. D., Clark, A. M. and Silveira, E. R. (1991b). Antimicrobial diterpenes of *Croton sonderianus.* I. Hardwickiic acid and 3,4-secotrachylobanoic acids. *J Nat Prod.,* **54(6):** 1625-1663.

Meccia, G., Rojas, L. B., Rosquete, C. and Feliciano, A. S. (2000). Essential oil of *Croton ovalifolius* Vahl from Venezuela. *Flavour Fragrance J.,* **15(3):** 144-146.

Milanowski, D. J., Winter, R. E. K., Elvin-Lewis, M. P. F. and Lewis, W. H. (2002). Geografic distribution of three alkaloid chemotypes of *Croton lechleri. J Nat Prod.,* **65(6):** 814-819.

Milo, B., Risco, E., Vila, R.; Iglesias, J. and Canigueral, S. (2002). Characterization of a fucoarabinogalactan, the main polysaccharide from the gum exudate of *Croton urucurana. J Nat Prod.,* **65(8):** 1143-1146.

Mimmanhemin, S. and Apisariyakul, A. (1979). Isolation of active components from *Croton* seeds and their pharmacological activity on rat intestine. *Bull Chieng Mai Ass Med Sci.,* **12(3):** 125-133.

Minh, P. T.H., Ngoc, P. H., Quang, N. Q., Hashimoto, T., Takaoka. S. and Asakawa, Y. (2003). A novel ent-kaurane diterpenoid from the *Croton tonkinensis* Gagnep. *Chem Pharm Bull.,* **51(5):** 590-591.

Minh, P. T. H., Ngoc, P. H., Taylor, W. C. and Cuong, N. M. (2004). A new ent-kaurane diterpenoid from *Croton tonkinensis* leaves. *Fitoterapia,* (Article in press).

Mirvish, S. S., Salmasi, S., Lawson, T. A., Pour, P. and Sutherland, D. (1985). Test of catechol, tannic acid, bidens pilosa, *Croton* oil, and phorbol for cocarcinogenesis of esophageal tumors induced in rats by methyl-N-amylnitrosamine. *J Nat Cancer Inst.,* **74 (6):** 1283-1289.

Mishima, H., Ogiso, A. and Kobayashi, S. (1977). Diterpene alcohol from *Croton*-Plants. *Patent-Japan Kokai-* 77 70,010.

Mishra, N. C., Estensen, R. D. and Abdel-Monem, M. M. (1986). Isolation and purification of phorbol from *Croton* oil by reversed-phase column chromatography. *J Chromatogr.,* **369:** 435-439.

Mohamedi, I. E., El Nuri, E. B. E., Choudhary, M. I. and Khan, S. N. (2009). Bioactive Natural Products from Two Sudanese Medicinal Plants *Diospyros mespiliformis* and *Croton zambesicus. Rec. Nat. Prod.,* **3(4):** 198-203.

Mokkhasmit, M., Ngarmwathana, W., Swasdimongkol, K. and Permphiphat, U. (1971a). Pharmacological evaluation of thai medicinal plants. *J Med Ass Thailand,* **54(7):** 490-504.

Mokkhasmit, M., Swatdimondigkol, K. and Satrawaha, P. (1971b). Study on toxicity of thai medicinal plants. *Bull Dept Med Sci.,* **12(2/4):** 36-65.

Monte, F. J. Q., Dantas, E. M., Crovoiro, A. A., Andrade, C. H. S. and Diaz Filho, R. (1983). Diterpenes from *Croton argyrophylloides.* Annual Meeting American Society of Pharmacognosy Univ Mississippi Oxford **46:** 24-28.

Monte, F. J. Q., Andrade, C. H. S., Craveiro, A. A. and Braz-Filho, R. (1984). New tetracyclic diterpenes from *Croton argyrophylloides*. *J Nat Prod.*, **47(1):** 55-58.

Monte, F. J. Q., Dantas, E. M. G. and Braz-Filho, R. (1988). New diterpenoids from *Croton argyrophylloides*. *Phytochemistry*, **27(10):** 3209-3212.

Morales-Flores, F., Aguilar, M. I., King-Díaz, B., Santiago-Gómez, J. R. and Lotina-Hennsen, B. (2007). Natural diterpenes from *Cróton ciliatoglanduliferus* as photosystem II and photosystem I inhibitors in spinach chloroplasts. *Photosynth. Res.*, **91:** 71-80.

Morimoto, H. (1988). Plaunotol manufacture by plant tissue culture of *Croton species*. *Patent-Japan Kokai Tokkyo Koho*- 65 317 90: 51PP.

Morton, J. F. (1968). A survey of medicinal plants of Curacão. *Econ Bot.*, **22:** 87.

Moulis, C., Fouraste, I., Bon, M. and Jaud, J. (1992a). Crovatin, a furanoid diterpene from *Croton levatii*. *Phytochemistry*, **31(4):** 1421-1423.

Moulis, C., Fouraste, I. and Bon, M. (1992b). Levatin, an 18-norclerodane diterpene from *Croton levatii*. *J Nat Prod.*, **55(4):** 445-449.

Moura, V. L. A., Monte, R. F. O. and Braz-Filho, R. (1990). A new casbane-type diterpenoid from *Croton nepetaefolius*. *J Nat Prod.*, **53(6):** 1566-1571.

Mukherjee, R. and Axt, E. M. (1984). Cyclitols from *Croton celtidifolius*. *Phytochemistry*, **23(11):** 2682-2684.

Murillo, R. M.and Jakupovic, J. (2000). Clerodanes and secoclerodanes of the *Croton jimenezii. Ingenieria y Ciencia Quimica.*, **19(2):** 68-73.

Murillo, R. M., Jakupovic, J. and Rivera, J. (2002). Diterpenes and other constituents from *Croton dracco* (Euphorbiaceae). *Rev Biol Trop.*, **49(1):** 259-264.

NAPRALERT-NATURAL PRODUCTS ALERT.Chicago: Universidade de Illinois. *http: //www.uic.edupharmacy/depts/PCRP/NAPRALERT.htm.Consultado*em 2002.

Ngadjui, B. T., Folefoc, G. G., Keumedjio, F., Dongo, E., Sondengan, B. L. and Connolly, J. D. (1999). Crotonadiol, a labdane diterpenoid from the stem bark of *Croton zambesicus*. *Phytochemistry*, **51(1):** 171-174.

Ngadjui, B. T., Abegaz, B. M., Keumedjio, F., Folefoc, G. N. and Kapche, G. W. F. (2002). Diterpenoids from the stem bark of *Croton zambesicus*. *Phytochemistry*, **60(4):** 345-349.

Ngamrojnavanich, N., Tonsiengsom, S., Lertpratchya, P., Roengsumran, S., Puthong, S. and Petsom, A. (2003). Diterpenoids from the stem barks of *Croton robustus*. *Arch Pharm Res.*, **26(11):** 898-901.

Nihei, K., Hanke, F. J., Asaka, Y., Matsumoto, T. and Kubo, I. (2002). Insect antifeedants from tropical plants II: structure of zumsin. *J Agr Food Chem.*, **50(18):** 5048-5052.

Nihei, K. I., Asaka, Y., Mine, Y. and Kubo, I. (2005). Insect antifeedants from *Croton jatrophoides*: structures of zumketol, zumsenin, and zumsenol. *J Nat Prod.*, **68(2):** 244-247.

Nihei, K., Asaka, Y., Mine, Y., Yamada, Y., Iigo, M., Yanagisawa, T. and Kubo, I. (2006). Musidunin and musiduol, insect antifeedants from *Croton jatrophoides*. *Journal of Natural Products*, **69(6)**: 975-977.

Nordal, A., Krogh, A. and Ogner, G. (1965). The occurrence of phorbic acid in plants. *Acta Chem Scand.*, **19(7)**: 1705-1708.

Novoa, B. E., Céspedes, A., De Garcia, L. A. and Olarte, C. J. (1979). Quercitrina um flavonoide com atividad hipotensora obtenido del *Croton glabellus*. *Rev. Col Cien. Quim Farm.*, 7-13.

Ogiso, A., Kitazawa, E., Kurabayashi, M., Sato, A., Takahashi, S., Noguchi, H., Kuwano, H., Kobayashi, S. and Mishima, H. (1978). Isolation and structure of antipeptic ulcer diterpene from Thai medicinal plant. *Chem Pharm Bull.*, **26(10)**: 3117-3123.

Ogiso, A., Kitazawa, E., Mikuriya, I. and Promdej, L. (1981). Original plant of a Thai crude drug, Plau-noi. *Shoyakugaku zasshi.*, 35: 287-290.

Ortega, T., Carretero, M. E., Pascual, E., Villar, A. M. and Chiribiga, X. (1996). Anti-inflammatory activity of ethanolic extracts of plants used in traditional medicine in Ecuador. *Phytother Res.*, 10: S121-S1222.

Palazzino, G., Federici, E., Rasoanaivo, P., Galeffi, C. and Delle Monache, F. (1997). 3,4-seco diterpenes of *Croton geayi*. *Gaz. Chimica. Italian*, **127(6)**: 311-314.

Palmeira, S. F., Moura, F. D., Alves, V.D., Oliveira, F. M., Bento, E. S., Conserva, L. M. and Andrade, E. H. D. (2004). Neutral components from hexane extracts of *Croton sellowii*. *Flavour and Fragrance Journal*, **19(1)**: 69-71.

Paredes, A. L., Luque, R. S., Olarte, J. C. and Caile, A. J. (1985). Caracterizacion de la fraccion responsable del efecto hipotensor del *Croton glabellus*: *Rev. Col Cienc Quim Farm.*, **4(2)**:

Payo, A. H., Sandoval, D. L., Vélez, H. C. and Oquendo, M. S. (2001). Alcaloides en la especie Cubana *Croton micradenus* Urb. *Rev Cubana Farm.*, **35(1)**: 61-65.

Pereira, A. S., Amaral, C. F., Barnes, R. A., Cardoso, J. N. and Neto, F. R. A. (1999). Identification of isoquinoline alkaloids in crude extracts by High Temperature Gas Cromatography-Mass Spectrometry. *Phytochem Anal.*, **10(5)**: 254-258.

Peres, C. and Anesini, C. (1994). Inhibition of *Pseudomonas aerguinosa* by Argentineam medicinal. *Fitoterapia*, **65(2)**: 196-172.

Peres, M. T. L. P., Monache, F. D., Cruz, A. B., Pizzolatti, M. G. and Yunes, R. A. (1997). Chemical composition and antimicrobial activity of *Croton urucurana* Baillon (Euphorbiaceae). *J Ethnopharmacol.*, **56(3)**: 223-226.

Peres, M. T. L. P., Pizzolatti, M. G., Yunes, R. A. and Delle Monache, F. (1998a). Clerodane diterpenes of *Croton urucurana*. *Phytochemistry*, **49(1)**: 171-174.

Peres, M. T. L. P., Monache, F. D., Pizzolatti, M. G., Santos, A. R. S., Beirith, A., Calixto, J. B. and Yunes, R. A. (1998b). Analgesic compouds of *Croton urucurana* Baillon. Pharmacochemical criteria used in their isolation. *Phytother Res.*, **12(3)**: 209-211.

Persinos, G. J. (1972). Anti-inflammation compositions containing Taspine or Acid Salts there of and method of use. *Patent-Us-* 3 694: 557- 688.

Persinos, G. J., Farnsworth, N. R., Blomster, R. N. and Blake, D. A. (1974). Studies on South American plants II. Taspine, isolation and anti-inflammatory activity. *Lloydia*, **37(4)**: 644C.

Persinos, G. J., Blomster, R. N., Blake, D. A. and Farnsworth, N. R. (1979). South American plants II. Taspine, isolation and anti-inflammatory activity. *J Pharm Sci.*, **68(1)**: 124-126.

Pessoa, C., Silveira, E. R., Lemos, T. L. G., Wetmore, L. A., Moraes, M. O. and Leyva, A. (2000). Antiproliferative effect of compounds derived from plants of northeast Brazil. *Phytother Res.*, **14(3)**: 187-191.

Phan, T. S., Van, N. H., Phan, M. G. and Taylor, W. C. (1999). Contribution to the study on bioactive substances from *Croton tonkinensis* Gagnep (Euphorbiaceae). *Tap Chi Hoa Hoc.*, **37(4)**: 1-2.

Phan, M. G. and Phan, T. S. (2004). Two long chain alkyl alcohols from the leaves of *Croton tonkinensis* Gagnep, Euphorbiaceae. *Tap Chi Hoa Hoc.*, **42(1)**: 132.

Phan, M. G., Lee, J. J. and Phan, T. S. (2004). Flavonoid glucosides from the leaves of *Croton tonkinensis* Gagnep, Euphorbiaceae. *Tap Chi Hoa Hoc.*, **42(1)**: 125-128.

Phan, M. G., Otsuka, H. and Phan, T. S. (2005). Three minor ent-kaur-16-ene-type diterpenes from *Croton tonkinensis* Gagnep. *Tap Chi Hoa Hoc.*, **43(2)**: 263-264.

Piacente, S., Belisario, M. A., Del Castillo, H., Pizza, C. and De Feo, V. (1998). *Croton ruizianus*: platelet proaggreganting activity of two new pregnane glycosides. *J Nat Prod.*, **61(3)**: 318-322.

Pieters, L. A. and Vlietinck, A. J. (1986). Phorbol ester constituents of *Croton* oil. *Planta Med.*, **52(6)**: 465-468.

Pieters, L. A. C., Vanden Berghe, D. A. and Vlietinck, A. J. (1990). A dihydrobenzofuram lignan from *Croton erythrochilus*. *Phytochemistry*, **29(1)**: 348-349.

Pieters, L., De Bruyne, T., Mei, G., Lemiere, G., Vanden Berghe, D. and Vlietinck, A. J. (1992). *In vitro* and *in vivo* biological activity of South American Dragon's Blood and its constituents *Planta Med Suppl.*, **58(1)**: A582-A583.

Pieters, L., De Bruyne, T., Claeys, M., Vlietinck, A., Calomme, M. and Vanden Berghe, D. (1993). Isolation of a dihydrobenzofuran lignan from South American Dragon's blood (*Croton* spp.) as an inhibitor of cell proliferation. *J Nat Prod.*, **56(6)**: 899-906.

Pieters, L., De Bruyne, T., Van Poel, B., Vingerhoets, R., Totte, J., Vanden Berghe, D. and Vlietinck, A. J. (1995). *In vivo* wound healing activity of Dragon's Blood (*Croton* spp), a traditional South American drug, and its constituents. *Phytomedicine*, **2(1)**: 17-22.

Pooley, E. (1993). *The Complete Field Guide to Trees of Natal, Zululand and Transkei.* Natal Flora Publication Trust 222.

Pornpakakul, S., Liangsakul, J., Ngamrojanavanich, N., Roengsumran, S., Sihanonth, P., Piapukiew, J., Sangichien, E., Puthong, S. and Petsom, A. (2006). Cytotoxic activity of four xanthones from *Emericella variecolor*, an endophytic fungus isolated from *Croton oblongifolius. Arch Pharm Res.*, **29(2)**: 14-144.

Pudhom, K., Vilaivan, T., Ngamrojanavanich, N., Dechangyipart, S., Sommit, D., Petsom, A. and Roengsumran, S. (2007). Furanocembranoids from the stem bark of *Croton oblongifolius. Journal of Natural Products,* **70(4)**: 659-661.

Puebla, P., López, J. L., Guerrero, M., Carrón, R., Martin, M. L., Roman, L. S. and Feliciano, A. S. (2003). Neo-clerodane diterpenoids from *Croton schiedeanus. Phytochemistry,* **62**: 551-555.

Puebla, P., Correa, S. X., Guerrero, M., Carron, R. and San Feliciano, A. (2005a). New cis-clerodane diterpenoids from *Croton schiedianus. Chem Pharm Bull.,* **53(93)**: 328-329.

Puebla, P., Correa, S. X., Guerrero, M. F. and San Feliciano, A. (2005b). Phenylbutanoid derivatives from *Croton schiedeanus. Biochemical Systematics and Ecology,* **33(8)**: 849-854.

Quintyne-Walcott, S., Maxwell, A. R. and Reynolds, W. F. (2007). Crotogossamide, a cyclic nonapeptide from the latex of *Croton gossypifolius. J Nat Prod.,* **70(8)**: 1374-1376.

Ramos, F., Takaishi, Y., Kashiwada, Y., Osorio, C., Duque, C., Acuna, R. and Fujimoto, Y. (2008). Ent-3,4-seco-labdane and ent-labdane diterpenoids from *Croton stipuliformis* (Euphorbiaceae). *Phytochemistry,* **69(12)**: 2406-2410.

Ramirez, U. R., Mostacero, L. J., Garcia, A. E., Mejia, C. F., Palael, P. F., Medina, C. D. and Miranda, C. H. (1988). Vegetales empleados en medicina tradicional Norpervana. *Banco Arg. Del Peru Nac Univ Trujillo,* Trujillo, Peru Junho, 1988: 54 PD.

Rao, P. S., Sachdeu, G. P., Seshadri, T. R. and Singh, H. B. (1968). Isolation and constitution of Oblongifoliol, a new diterpene of *Croton oblongifolius.Tetrahedron Letters,* **45**: 4685-4686.

Ribeiro, E. M., Paulo, M. Q. and Souza Brito, A. R. M. S. (1993). Isolation of active substances from *Croton campestris* ST. HILL (Euphorbiaceae) leaves. *Rev. Bras Farm.,* **74(2)**: 36-41.

Rios, M. Y. and Aguilar-Guadarrama, A. B. (2006). Nitrogen-containing phorbol esters from *Cróton ciliatoglandulifer* and their effects on cyclooxygenases-1 and 2. *J. Nat. Prod.,* **69**: 887-890.

Risco, E., Ghia, F., Vila, R., Iglesias, J., Alvarez, E. and Caniqueral, S. (2003). Immunomodulatory activity and chemical characterisation of sangre de drago (Dragon's blood) from *Croton lechleri. Planta Med.,* **69(9)**: 785-794.

Rodriguez-Hahn, L., Rodriguez, J. J. and Romo, J. (1975). Isolation and structure of draconin. *Rev Latinoamer Quim.*, **6(3)**: 123-126.

Rodriguez-Hahn, L., Valencia, A., Saucedo, R. and Diaz, E. (1981). Aislamiento y estructura de los componentes quimicos de *Croton pyramidalis*. *Rev. Latinoamer Quim.*, **12(1)**: 16-19.

Roengsumran, S., Luangdilok, W., Petsom, A., Praruggamo, S. and Pengprecha, S. (1982). Struture of swassin, a new furanoid diterpene. *J Nat Prod.*, **45(6)**: 772-773.

Roengsumran, S., Achayindee, S., Petson, A., Pudhom, K., Singtothong, P., Surachetapan, C. and Vilaivan, T. (1998). Two new cembranoids from *Croton oblongifolius*. *J Nat Prod.*, **61(5)**: 652-654.

Roengsumran, S., Petson, A., Sommit, D. and Vilaivan, T. (1999a). Labdane diterpenoids from *Croton oblongifolius*. *Phytochemistry*, **50(3)**: 449-453.

Roengsumran, S., Singtothong, P., Pudhom, K., Ngamrochanavanich, N., Petsom, A. and Chaichantipyuth, C. (1999b). Neocrotocembranal from *Croton oblongifolius*. *J Nat Prod.*, **62(8)**: 1163-1164.

Roengsumran, S., Petsom, A., Kuptiyanuwat, N., Vilaivan, T., Ngamrojnavanich, N., Chaichantipyuth, C. and Phuthong, S. (2001). Cytotoxic labdane diterpenoids from *Croton oblongifolius*. *Phytochemistry*, **56(1)**: 103-107.

Roengsumran, S., Musikul, K., Petsom, A., Vilaivan, T., Sangvanish, P., Pompakakul, S., Puthong, S., Chaichantipyuth, C., Jaiboon, N. and Chaichit, N. (2002). Croblongifolin, a new anticancer clerodane from *Croton oblongifolius*. *Planta Med.*, **68(3)**: 274-277.

Roengsumran, S., Pornpakakul, S., Muangsin, N., Sangvanich, P., Nhujak, T., Singtothong, P., Chaichit, N., Puthong, S. and Petsom, A. (2004). New halimane diterpenoids from *Croton oblongifolius*. *Planta Med.*, **70(1)**: 87-89.

Roengsumran, S., Pata, P., Ruengraweewat, N., Tummatorn, J., Pompakakul, S., Sangvanich, P., Puthong, S. and Petsom, A. (2009). New cleistanthane diterpenoids and 3,4-seco-cleistanthane diterpenoids from *Croton oblongifolius*. *Chemistry of Natural Compounds*, **45(5)**: 641-646.

Rojas, E. T. and Rodriguez-Hahn, L. (1978). Nivenolide, a diterpene lactone from *Croton niveus*. *Phytochemistry*, **17**: 574-575.

Rosa, M. S. S., Mendonça Filho, R. R., Bizzo, H. R., Rodrigues, I. A., Soares, R. M. A., Souto Padron, T., Alviano, C. S. and Lopes, A. H. C. S. (2003). Antileishmanial activity of a linalool-rich essential oil from *Croton cajucara*. *Antimicrob Agents Chemother.*, **47(6)**: 1985-1901.

Rouquayrol, M. Z., Fonteles, M. C., Alencar, J. E., Matos, F. J. A. and Craveiro, A.A. (1980). Molluscicidal activity of essential oils from Northeastern Brazilian plants. *Rev Brasil Pesq Med Biol.*, **13**: 135-143.

Sabnis, S. D. and Bedi, S. J. (1983). Ethnobotanical studies in Dadra-Nagar Haveli and Daman Indian. *J. For.*, **6(1)**: 65-69.

Sanchez, V. and Sandoval, D. (1982a). Alkaloids in Cuba species of *Croton* L. Genus.II. Chemical of *C. stenophyllus*. *Rev Cubana Farm.*, **16**: 45-55.

Sanchez, V. and Sandoval, D. (1982b). Alkaloids in Cuban species of *Croton* L. genus, 2. Chemical study of *C. stenophyllus*. *Rev Cubana Farm.*, **16**: 46-55.

Sánchez-Mendoza, M. E., Reyes-Trejo, B., la Rosa, L., Rodríguez-Silverio, J., Castillo-Henke, C. and Arrietal, J. (2008). Polyalthic Acid isolated from *Cróton reflexifolius* hás relaxing effect in guinea pig tracheal smooth muscle. *Pharmaceutical Biology*, **46(10-11)**: 800-807.

Santos, H. S., Barros, F. W. A., Albuquerque, M. R. J. R., Bandeira, P. N., Pessoa, C., Braz-Filho, R., Monte, F. J. Q., Leal-Cardoso, J. H. and Lemos, T. L. G. (2009). Cytotoxic diterpenoids from *Croton argyrophylloides*. *Journal of Natural Products*, **72(10)**: 1884-1887.

Satishi, S. and Bhakuni, D. S. (1972).Constituents of Indian and Other Plants. *Phytochemistry*, **11(9)**: 2888-2890.

Sato, A., Ogiso, A. and Kuwano, H. (1980). Acyclic diterpenes from *Croton kerrii*. *Phytochemistry*, **19**: 2207-2209.

Sato, A., Kurabayashi, M., Ogiso, A. and Kuwano, H. (1981). Poilaneic acid, a cembranoid diterpene from *Croton poilanei*. *Phytochemistry*, **20(8)**: 1915-1918.

Schmeda-Hirschmann and Rojas de Arias, A. (1992). A screening method for natural products on triatomine bugs. *Phytother Res.*, **6(2)**: 68-73.

Schneider, C., Breitmaier, E., Bayma, J. C., França, L. F., Kneifel, H. and Krebs, H. C. (1994). Maravuic acid, a new seco-labdane diterpene from *Croton matourensis*. *Liebigs Ann chem.*, **1995(4)**: 709-710.

Schneider, C., Breitmaier, E., Bayma, J. D., De Franca, L. F., Kneifel, H. and Krebs, H. C. (1995). Maravuic acid, a new seco-labdane diterpene from *Croton matourensis*. *Liebigs Ann Chem.*, **4**: 709-710.

Selvanayahgam, Z. E., Gnanevendhan, S. G., Balakrishna, K. and Rao, R. B. (1994). Antisnake venon botanicals from ethnomedicine. *J Herbs Spices Med Plants*, **2(4)**: 45-100.

Shetty, S. N., Anika, S. M. and Asuzu, U. I. (1983). Investigations on*Croton penduliflorus* Hutch: I Observation on pharmacognostic, physicochemical and pharmacological characteristics. *J. Crude Drug Res.*, **21(2)**: 49-58.

Shibata, W., Murai, F., Akiyama, T., Siriphol, M., Matsunaga, E. andMorimoto, H. (1996). Micropropagation of *Croton sublyratus* a tropical medicinal importance. *Plant Cell Rep.*, **16(3/4)**: 147-152.

Siems, K., Dominguez, X. A. and Jakupovic, J. (1992). Diterpenes and other constituents from *Croton cortesianus*. *Phytochemistry*, **31(12)**: 4363-4365.

Silveira, E. R. and Mc Chesney, J. D. (1994). 6,7-Oxygenated neo-clerodane furan diterpenes from *Croton sonderianus*. *Phytochemistry*, **36(6)**: 1457-1463.

Simões, J. C., Silva, A. J. R., Serruya, H. and Bentes, M. H. S. (1979). Dehydrocrotonin, a norditerpenoid from *Croton cajucara* Benth. (Euphorbiaceae). *Cienc Cult.*, **31(10)**: 1140-1141.

Siqueira, N. C. S., Silva, G. A. A. B., Bauer, L., Alice, C. and Santana, B. (1984). Alguns componentes do óleo essencial de *Croton nitrariifolius* Baill-Euphorbiaceae (nota previa). *Rev Bras Farm.*, **65(1/3)**: 62-64.

Siriphol, M., Matsunaga, E., Shibata, W., Murai, F. and Toshiyuki, A. (1997). Cultivation of *Croton sublyratus* Kurz in Thailand. *Proc Int Symp Nat Med*, oct 28-30-1997, Kyoto, Japan: 58-64.

Son, P. T., Giang, P. M. and Taylor, W. C. (2000). An ent-kaurane diterpenoid from *Croton tonkinensis*. *Aust J Chem.*, **53(11/12)**: 1003-1005.

Spencer, C. F., Koniuszy, F. R., Rogers, E. F., Shavel, J. R. J., Easton, N. R., Kaczka, E. A. and Seeler, A. O. (1947). Survey of plants for antimalarial activity. *Lloydia*, **10**: 145-174.

Stirpe, F., Pession-Brizzi, A., Lorenzoni, E., Strocchi, P., Montanaro, L. and Sperti, S. (1976). Studies on the proteins from seeds of *Croton tiglium* and of *Jatropha curcas*. Toxic properties and inhibition of protein synthesis in vitro. *Biochem J.*, **156**: 1-6.

Stuart, K. L. and Woo-Ming, R. B. (1969). Alkaloids from *Croton plumier*. *Phytochemistry*, **8**: 777-780.

Stuart, K. L. (1970). Chemical and biological investigations of the *Croton* genus. *Rev. Latinoamer. Quim.*, 40-143.

Stuart, K. L., Mc Neill, D., Kutney, J. P., Eigendorf, G. and Klein, F. K. (1973). Isolation and synthesis of glutamine and glutarimide derivatives from *Croton humilis*. *Tetrahedron*, **29**: 4071-4075.

Stuart, K. L. and Graham, L. (1973). Alkaloid biosynthesis in *Croton flavens*. *Phytochemistry*, **12**: 1967-1972.

Stuart, K. L. and Woo-Ming, R. B. (1975). Vomifoliol in *Croton* and *Paulicourea* species. *Phytochemistry*, **14**: 594-595.

Stuart, K. L., Roberts, E. V. and Whittle, Y. G. (1976). A general method for vomifoliol detection. *Phytochemistry*, **15**: 332-333.

Suarez, A. I., Blanco, Z., Monache, F. D., Compagnone, R. S. and Arvelo, F. (2004). Three new glutarimide alkaloids from *Croton cuneatus*. *Nat Prod Res.*, **18(5)**: 421-426.

Suárez, A. I., Vasquez, L. J., Manzano, M. A. and Compagnone, R. S. (2005). Essential oil composition of *Cróton cuneatus* and *Croton malambo* growing in Venezuela. *Flavour and Fragrance Journal*, **20**: 611-614.

Subramanian, S. S., Nagarajan, S. and Sulochana, N. (1971). Flavonoids of some Euphorbiaceous plants. *Phytochemistry*, **10**: 2548-2549.

Sutthivaiyakit, S., Nareeboon, P., Ruangrangsi, N.,Ruchirawat, S., Pisitjaroenpong, S. and Mahidol, C. (2001). Labdane and pimarane diterpenes from*Croton joufra*. *Phytochemistry*, **56(8)**: 811-814.

Takahashi, S., Kurabayashi, M., Kitazawa, E., Haruyama, H. and Ogiso, A. (1983). Plaunolide, a furanoid diterpene from *Croton sublyratus*. *Phytochemistry*, **22(1)**: 302-303.

Tammami, B. Torrance, S. J. Fabela, F. V. Wiedhopf, R. M. and Cole, J. R. (1977). Preliminary investigation of *Croton californicus* var. Tenuis and *Uvaria kirkii*: a xanthone and benzyldihydrochalcone. *Phytochemistry*, **16**: 2040.

Tan, G. T., Pezzuto, J. M., Kinghorn, A. D. and Hughes, S. H. (1991). Evaluation of natural products as inhibitors of human immunodeficiency virus type 1 (HIV-1) reverse transcriptase. *J Nat Prod.*, **54**: 143-154.

Tansakul, P. and De-Eknamkul, W. (1998). Geranylgeraniol-18-hydroxylase: the last enzyme on the plaunotol biosynthetic pathway in *Croton sublyratus*. *Phytochemistry*, **47(7)**: 1241-1246.

Tchissambou, L., Chiaroni, A., Riche, C. and Khuong-Huu, F. (1990). Crotocorylifuran and Crotohaumanoxide, new diterpenes from *Croton haumanianus* J. Leonard: *Tetrahedron*, **46(15)**: 5199-5202.

Thongtan, J., Kittakoop, P., Ruangrungsi, N., Saenboonrueng, J. and Thebtaranonth, Y. (2003). New antimycobacterial and antimalarial 8,9-secokaurane diterpenes from *Croton kongensis*. *J Nat Prod.*, **66(6)**: 868-870.

Thuan, B. T. and Nhu, T. V. (1991). The chemical composition of the*Croton tonkinensis* Gagnep and its action on parasites of experimental malaria. *Tap Chi Duoc Hoc.*, **5**: 11-12.

Thuong, P. T., Dao, T. T., Pham, T. H. M., Nguyen, P. H., Le, T. V. T., Lee, K. Y. and Oh, W. K. (2009). Crotonkinensins A and B, diterpenoids from the Vietnamese medicinal plant *Croton tonkinensis*. *Journal of Natural Products*, **72(11)**: 2040-2042.

Tiwari, K. P., Choudari, R. N. and Paney, G. D. (1981). 3-Methoxy-4-6-dihydroxymorphinandien-7-one, an alkaloid from *Croton bomplandianun*. *Phytochemistry*, **20**: 863-864.

Torres, M. C. M., Assunção, J. C., Santiago, G. M. P., Andrade-Neto, M., Silveira, E. R., Costa-Lotufo, L. V., Bezerra, D. P., Marinho Filho, J. D. B.,Viana, F. A. and Pessoa, O. D. L. (2008). Larvicidal and nematicidal activities of the leaf essential oil of *Cróton regelianus*. *Chemistry and Biodiversity*, 5.

Tseng, S. S., Van Duuren, B. L. and Solomon, J. J. (1977). Synthesis of 4A alpha-phorbol 9-myristate 9A-acetate and related esters. *J Org Chem.*, **42**: 3645-3649.

Ubillas, R., Jolad, S. D., Bruening, R. C., Kernan, M. R., King, S. R., Sesin, D. F., Barrett, M., Stoddart, C. A., Flaster, T., Kuo, J., Ayala, F., Meza, E., Castanel,

M., Mc Meekin, D., Rozhon, E., Tempesta, M. S., Barnard, D., Huffman, J., Smee, N., Sidwell, R., Soike, K., Brazieri, A., Safrin, S., Orlando, R., Kenny, P. T. N., Berova, N. and Nakanish, K. (1994). SP-303, an Antiviral Oligomeric Proanthocyanidin from The Latex of *Croton lechleri* (Sangue de Grado). *Phytomedicine*, **1(2)**: 17-106.

Upadhyay, R. R. and Hecher, E. (1976). A New criptic irritant and cocarcinogen from seeds of *Croton sparsiflorus*. *Phytochemistry*, **15**: 1070-1072.

Usman, L. A., Olawore, N. O., Oladosu, I., Hamid, A. A. and Elaigwu, S. E. (2009). Constituent of leaf oil of *Croton zambesicus* Muell. Arg growing in North Central Nigeria. *Middle-East Journal of Scientific Research*, **4(4)**: 242-244.

Vaisberg, A. J., Milla, M., Planas, M. D. C., Cordova, J. L., De Augusti, E. R., Ferreira, R., Mustiga, M. D. C., Carlin, L. and Hammond, G. B. (1989). Taspine is the cicatrizant principle in Sangre de Grado extracted from *Croton lechleri*. *Planta Med.*, **55(2)**: 140-143.

Vigor, C., Fabre, N., Fouraste, I. and Moulis, C. (2001). Three clerodane diterpenoids from *Croton eluteria* Bennett. *Phytochemistry*, **57(8)**: 1209-1212.

Vigor, C., Fabre, N., Fouraste, I. and Moulis, C. (2002). Neoclerodane diterpenoids from *Croton eluteria*. *J Nat Prod.*, **65(8)**: 1180-1182.

Vongchareonsathit, A. and De-Eknamkul, W. (1998). Rapid TLC-Densometric analysis of plaunotol from *Croton sublyratus* leaves. *Planta Med.*, **64(3)**: 279-280.

Weber, J. and Hecker, E. (1977). Cocarcinogens of the diterpene ester type from *Croton flavens* and esophageal cancer in Curacão. *Experientia*, **34**: 679-682.

Webster, G. L. (1994). Systematics of the Euphorbiaceae. In: *Annals of the Missouri Botanic Garden. California*, EUA. N.81, v.1, p44.

Weckert, E., Hummer, K., Addae-Mensah, I. and Achenbach, H. (1992). The absolute configuration of Chiromodine. *Phytochemistry*, **31(6)**: 2170-2172.

Williams III, L., Evans, P. E. and Bowers, W. S. (2001). Defensive chemistry of an aposematic bug, *Pachycoris stallii* uhler and volatile compounds of its host plant *Croton californicus* Muell-Arg. *Journal of Chemical Ecology*, **27(2)**.

Wilson, R. S., Neubert, L. A. and Huffman, J. C. (1976). The chemistry of Euphorbiaceae. A new diterpene from *Croton californicus*. *J Amer Soc.*, **98**: 3669.

Wilson, R. T. and Marian, W. G. (1979). Medicine and magic central tigre a contribution to the ethnobotany of the Ethiopean Plateau. *Econ Bot.*, **33**: 29-34.

Woerdenbag, H. J., Bos, R., Van Meeteren, H. E., Baarslag, J. J. J., De Jong Van Der Berg, L. T. W., Pras, N., Do Rego Kuster, G., Petronia, R. R. L. and Vos, G. I. (2000). Essential oil of *Croton flavens* L., a medicinal plant from Curaçao. *J Essent Oil Res.*, **12(6)**: 667-671.

Woo, W. S., Lee, E. B., Shin, K. H., Kang, S. S. and Chi, H.J. (1981). A review of research in plants for fertility regulation in Korea. *Korean J Pharmacog.*, **12(3)**: 153-170.

Wu, X. A., Zhao, Y. M. and Yu, N. J. (2007). A novel analgesic pyrazine derivative from leaves of *Croton tiglium* L. *Journal of Asian Natural Products Research*, **9 (5)**: 451-455.

Yamale, S. C., Koudou, J., Samb, A., Heitz, A. and Teulade, J. C. (2009). Structural elucidation of a new furoclerodane from stem barks of *Croton mayumbensis* J. Leonard. *International Journal of Physical Sciences*, **4(3)**: 96-100.

Yang, X., Chen, S., Lin, Q. and Deng, S. (2009). Terpenoid from roots of *Croton crassifolius*. *Guangxi Zhiwu.*, **29(2)**: 272-274.

Natural Products: Research Reviews Vol. 1 (2012)
Editor: V.K. Gupta
Published by: DAYA PUBLISHING HOUSE, NEW DELHI

Pages 371–408

8

Anti-Amyloidogenic Effect of Natural Products: Implications for the Prevention and Therapeutics of Alzheimer's Disease

N. Suganthy[1], S. Karutha Pandian[1] and K. Pandima Devi[1]*

ABSTRACT

Alzheimer's disease (AD) is a debilitating neurodegenerative disorder in the elderly affecting millions of individuals throughout the world. Because of the severity and increasing prevalence of the disease in the population, it is urgent that better treatments be developed. AD is a complex disease, involving multiple factors such as the production of aggregation-prone amyloid beta (A β) peptides, the formation of fibrillarly tangles of microtubule-associating proteins, Tau, and the polymorphism of cholesterol binding protein, APOE4. Despite the major efforts aimed at elucidating the molecular basis and physiopathology of AD, there is still no effective treatment available for this devastating disorder. Evidence gathered over the last two decades has implicated the abnormal accumulation of Aβ, in particular the longer more amyloidogenic form Aβ 42, as a potential causative agent in the disease. Given its central role in the neuropathology of AD, the A β is the main focus of AD therapy. As growing number of studies suggest that natural extracts and phytochemicals have a positive impact on brain aging, this review focuses on natural compounds with outstanding profile against Aβ induced AD. This review highlights phytoconstituents which prevent amyloidogenesis in three different mechanisms (i) by inhibiting the enzymes involved in proteolytic processing of the amyloid precursor protein and production of A β (ii) by clearing preformed A β and (iii) by inhibiting the amyloid aggregation thereby preventing cerebral amyloid deposition and

1   Department of Biotechnology, Alagappa University, Karaikudi – 630 003, Tamil Nadu, India.

*   *Corresponding author:* E-mail: devikasi@yahoo.com

neurotoxicity. Significant progress has been achieved in these directions, opening up new perspectives towards the development of effective approaches for the treatment or prevention of AD.

Keywords: Secretase inhibitors, Metal chelators, Beta amyloid peptide, Antioxidants, Natural products.

Introduction

Alzheimer's disease (AD) is a progressive and fatal neurodegenerative disorder characterized by impairment of memory and cognitive function. Initially mild cognitive impairment and deficits in short-term and spatial memory appear (Selkoe, 2001), but the symptoms become more severe with disease progression, eventually culminating in loss of executive function. The risk of AD dramatically increases with aging, affecting 7-10 per cent of individuals over age 65, and about 40 per cent of persons over 80 years of age, and it is predicted that the incidence of AD will increase 3-fold within the next 50 years if no therapy intervenes (Sisodia, 1999). In developed societies where life expectancy has been considerably extended, this devastating disease actually represents a major public health concern, being estimated that 22 million people worldwide will develop this progressive neurodegenerative disorder by 2025 (Sleeger and Van Duijn, 2001). In United States, AD is the seventh leading cause of all deaths and the prevalence has been estimated to be 5.3 million cases in 2010, which will increase by almost 3-fold *i.e.* 11-16 million by 2050 (Maslow, 2010). Current data from developing countries suggest that prevalence of AD are high (>or=5 per cent) in certain Asian and Latin American countries, but consistently low (1-3 per cent) in India and sub-Saharan Africa. The annual incidence of AD for those aged 65 years and older was found to be 1.3 per cent (0.8-1.8) in India and in china it is 3.1 per cent (0.2-1.9) (Kalaria *et al.*, 2008).

Despite the strong progresses made in AD research in the last decades, no treatment with a strong disease-modifying effect is currently available. The neuropathological features of AD are senile plaques, neurofibrillary tangles of hyperphosphorylated tau protein and neurotransmitter deficit (Nordberg, 2004). Beta amyloid (Aβ 1-42) peptide is the major component of senile plaques and it is considered to have a causal role in the neuronal damage and cell death in AD (Behl, 1999). Experimental data from *in vitro* and *in vivo* studies indicate that Aβ deteriorates a variety of neuronal and glial functions, thereby ultimately leading to apoptotic death in these cells (Behl, 1999). In particular, increasing evidence support that Aβ caused neuronal cell death is mediated via oxidative stress (Markesbery, 1997; Chauhan and Chauhan, 2006).

Risk Factors Leading to AD

Genetics of Alzheimer's Disease

☆ *Early Onset Familial AD (FAD)*: Autosomal inherited mutation of the genes encoding amyloid precursor protein (APP), presenilin 1 (PSEN1) and presenilin 2 (PSEN2) leads to FAD (Figure 8.1). All these genes are involved

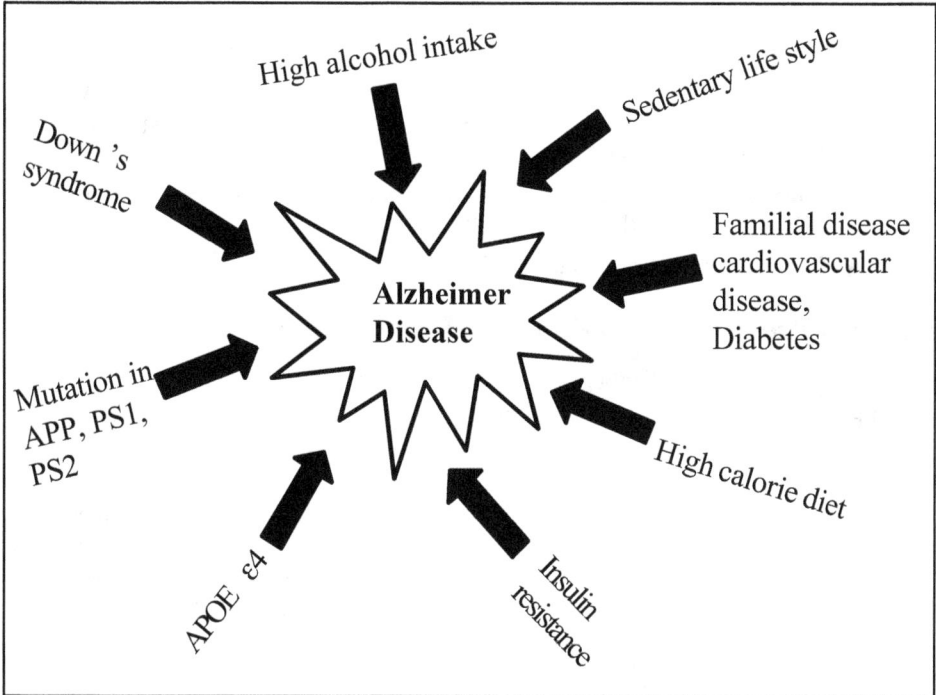

Figure 8.1: Etiological causatives of AD.

with β-amyloid processing and accounts for less than 5 per cent of total AD (Hardy, 1997; Tilley *et al.*, 1998). Mutations in the APP gene on chromosome 21 either increases total Aβ levels or just Aβ42 alone (Jankowsky *et al.*, 2004). Mutations in the PS-1 gene (chromosome 14) and the much smaller number of known mutations in the PS-2 gene (chromosome 1) results in an increase in the production of Aβ42 (Tomita *et al.*, 1997).

☆ Down's syndrome (DS, trisomy 21), an extra copy of chromosome 21 is assumed to results in an increased APP expression and a corresponding increase in Aβ levels (Mori *et al.*, 2002).

☆ *Sporadic late-onset AD (LOAD)*: Main risk factor is e4 allele of the apolipoprotein E gene (APOE). E4 enhances Aβ aggregation and reduces Aβ clearance. In addition, data suggest that E4 might increase the risk of AD by enhancing amyloidogenic processing of APP, promoting cerebrovascular pathology, increasing oxidative stress and impairing neuronal plasticity (Mahley *et al.*, 2006).

☆ Other genes associated with AD risk include those encoding lipoprotein receptors, α1 antichymotrypsin, α2 macroglobulin, and butyrylcholi-nesterase (Tilley *et al.*, 1998). Single-nucleotide polymorphisms in the urokinase-type plasminogen activator gene on chromosome 10q are associated with elevated Aβ42 and increased risk of LOAD (Ertekin-Taner *et al.*, 2005).

Other Risk Factors Associated with AD

☆ Family history of the disease such as cardiovascular disease, diabetes, hypertension, heart disease, high alcohol intake and prior head injury

☆ Low education level, consumption of high-calorie, high-fat diets and a sedentary lifestyle (Mayeux, 2003).

☆ Low dietary folate intake, as an apparent consequence of increased levels of homocysteine (Mattson, 2003).

☆ High intake of lipids and metals such as copper and iron (Bush, 2003).

Amyloid Precursor Protein (APP)-Processing Pathways

Non-amyloidogenic Pathway

APP is widely expressed in cells throughout the body where the amount produced is influenced by the developmental and physiological state of the cells. It has important roles in regulating neuronal survival, neurite outgrowth, synaptic plasticity and cell adhesion. APP is transported along axons to presynaptic terminals where it accumulates at relatively high levels, which can result in Aβ deposition at synapses. One possible function of APP is that it acts as a cell surface receptor that transduces signals within the cell in response to an extracellular ligand. APP is an integral membrane protein with a single membrane spanning domain, a large extracellular glycosylated N terminus and a shorter cytoplasmic C terminus-Aβ is located at the cell surface (or on the lumenal side of ER and Golgi membranes), with part of the peptide embedded in the membrane. APP is produced in several different isoforms ranging in size from 695 to 770 amino acids. The most abundant form in brain (APP695) is produced mainly by neurons, and differs from longer forms of APP in that it lacks a kunitztype protease inhibitor sequence in its ectodomain. The immature form of APP undergoes several post-translational modifications including N-glycosylation, O-glycosylation, and Tys sulfation to give the mature form of APP (Selkoe, 2001). Following these steps, the routes of APP metabolism become more complex and result in different pathways leading to proteolytic processing of the precursor by at least three proteolytic enzymes (Nunan and Small, 2000). APP can be processed by both amyloidogenic and nonamyloidogenic pathways (Figure 8.2). Near the cell surface or in a secretory vesicle a protease, *i.e.* α-secretase, cleaves APP in the extracellular domain and releases the ectodomain sAPPα (soluble APPα) into the extracellular space. This proteolytic cleavage occurs within the Aβ sequence, therefore preventing the formation of amyloidogenic fragments constitutes the non-amyloidogenic pathway. Production of sAPP α increases in response to electrical activity and activation of muscarinic acetylcholine receptors, suggesting that neuronal activity increases α-secretase cleavage of APP. sAPPα regulates neuronal excitability and enhances synaptic plasticity, learning and memory, possibly by activating a cell surface receptor that modulates the activity of potassium channels and also activates the transcription factor NF-kB.

Amyloidogenic Pathway

Amyloidogenic processing of APP involves sequential cleavages by beta secretase (BACE1) and γ-secretase at the N and C termini of Aβ, respectively. The generation of

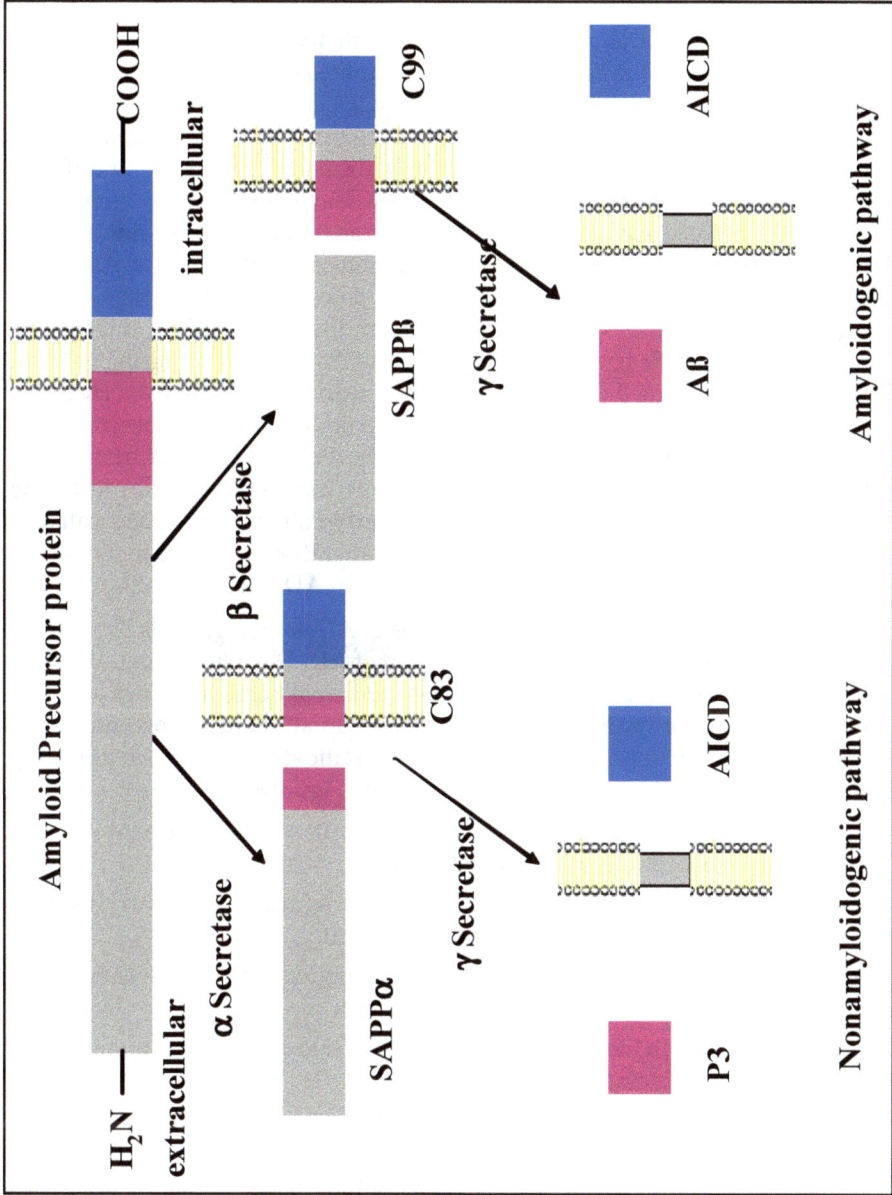

Figure 8.2: Schematic diagram representing cleavage of Amyloid Precursor Protein (APP).

Aβ is initiated by endoprotease, BACE1 which cleaves at the N-terminus of APP (Seubert *et al.*, 1993). Since the enzymatic activity of BACE1 is a prerequisite for the generation of β-amyloid peptides, inhibition of BACE1 appears to be a promising pharmacological target to reduce the production of Aβ peptides and thereby the formation of β -amyloid plaques in AD patients. Normally level of BACE1 does not change with aging, but in AD, however, increase in both BACE1 protein concentration and enzymatic activities have been reported (Fukumoto *et al.*, 2002). The 99-amino-acid C-terminal fragment of APP generated by BACE1 cleavage is internalized and further processed by γ-secretase to produce Aβ40/42 in endocytic compartments. γ-secretase, which cleaves APP within a transmembrane region, involves four different proteins, presenilin, nicastrin, Aph-1 and Pen-2. Cleavage of C99 by γ-secretase liberates an APP intracellular domain (AICD) that translocates in to the nucleus and regulates the gene expression, including the induction of apoptotic genes. As APP is axonally transported, Aβ1-42 therefore likely accumulates at synapses in high amount in AD. Aβ1-42 can have multiple adverse effects on the functions and integrity of both pre- and postsynaptic terminals including oxidative stress, impairing calcium homeostasis and perturbing the functions of mitochondria and the ER. Abnormalities in axons may result from adverse effects of Aβ 1-42 on tau, and microtubules resulting in neurofibrillary tangle formation and cell death. Presynaptic disturbances in synaptic vesicle trafficking and axonal transport may also contribute to the dysfunction and death of neurons in AD. Oxidative stress, perturbed calcium regulation and mitochondrial impairment are major alterations involved in functional and structural abnormalities in synapses and axons in AD (Mattson, 2004).

As Aβ is the prime factor in the pathogenesis of AD, targeting Aβ is likely to be considered a therapeutic strategy for the treatment of AD. This review highlights the role of natural ingredients from the dietary and medicinal herbs in reduction of Aβ accumulation directly or indirectly which could be an alternative preventive and therapeutic intervention in AD. The potential therapeutic strategies which prevent β amyloid plaque formation (Citron, 2004) are shown in Figure 8.3

1. *β and γ-secretase inhibitors*: Prevents formation of Aβ 1-42 fragment which is responsible for β amyloid plaque formation

2. *α secretase activators*: Enhances the non amyloidogenic pathway thereby preventing the formation of Aβ 1-42 fragment

3. Amyloid antiaggregant therapies: Prevents aggregation of Aβ 1-42 fragment the precursor in β amyloid plague formation.

4. Clearance of Aβ: Neutralizing or removing the toxic aggregate or misfolded forms of Aβ protein

5. Antioxidants: Scavenges ROS and RNS the major causative agent for oxidative injury in neurons caused by Aβ.

6. *Metal chelators*: Chelates Cu^{2+} and Zn^{2+} which plays a major role in beta amyloid plaque aggregation.

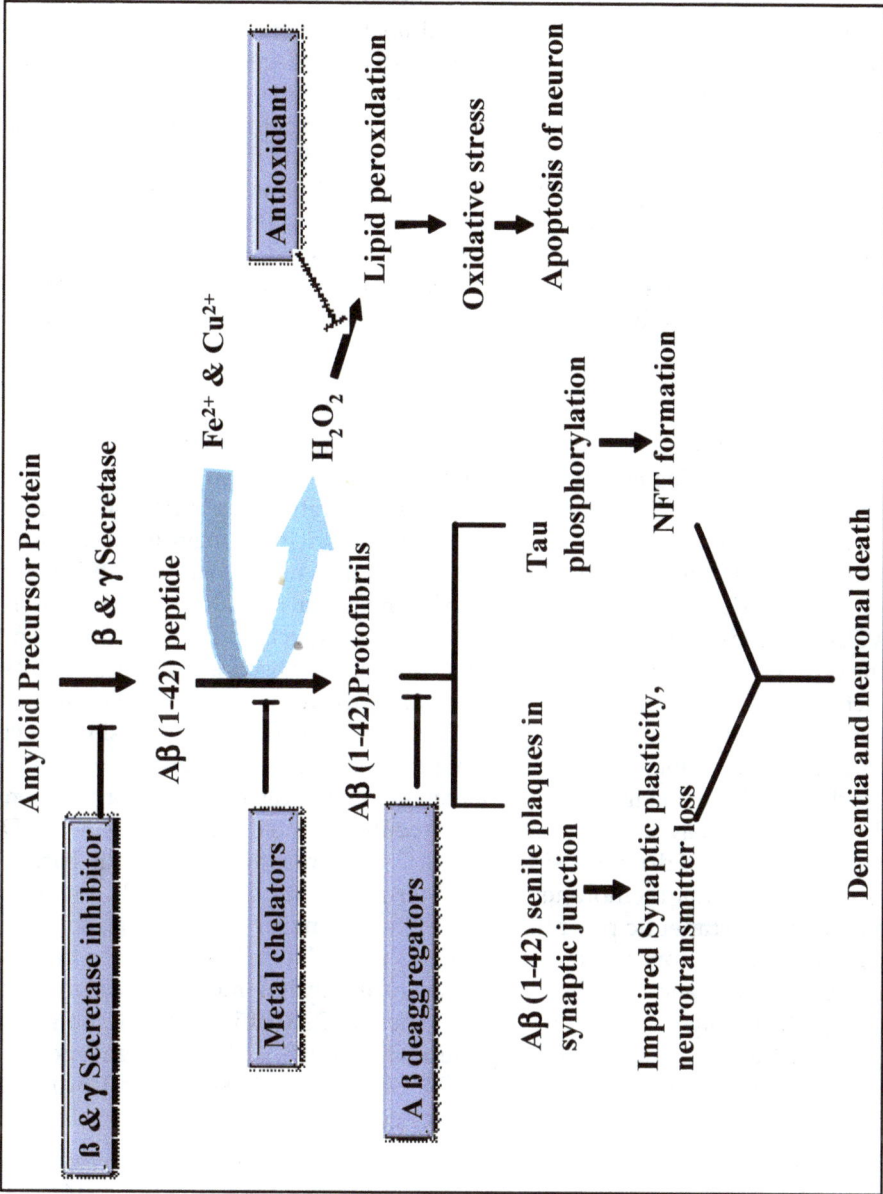

Figure 8.3: Potential therapeutic strategies in AD.

Traditional Medicine in the Treatment of Dementia

Nature is a rich source of biological and chemical diversity. The unique and complex structures of natural products cannot be obtained easily by chemical synthesis. Herbal medicine is a traditional medicinal or folk medicine practice based on the use of plants and plant extracts. In traditional practices numerous plants have been used to treat cognitive disorders, including neurodegenerative diseases and different neuropharmacological disorders. An ethnopharmacological approach has provided leads to identify plants and potential new drugs that are relevant for the treatment of cognitive disorders, including AD, and may aid the discovery of a more varied and efficacious selection of drugs for AD treatment. Nowadays, herbal medicine has received much attention and is recommended as a natural alternative to maintain one's health. This review focuses on the recently reported medicinal plants with pharmacological activities relevant to the treatment of dementia including β and γ secretase inhibitory activity, metal chelating activity and Amyloid antiaggregant properties.

β-secretase Inhibitors from Natural Source

β-Secretase, a novel transmembrane aspartic protease is also known as BACE 1(the β-site APP-cleaving enzyme). This enzyme cleaves an easily accessible site at the luminal side of β-APP, and its activity is the rate-limiting step in Aβ 1-42 peptide production (Vassar *et al.*, 1999). BACE1 activity is present in the majority of cells and tissues of the body with maximal activity in neural tissues (Zhao *et al.*, 1996). As BACE1 plays a prime role in the intial step of pathological cascade of AD, inhibition of BACE1 may reduce the production of Aβ peptides expecting that it might slow or halt the progression of AD. There is strong evidence that absence of BACE1 inhibits production of amyloid and the C99 stubs without any major side-effect (Dewachter and van Leuven, 2002). Therefore, the BACE1 inhibitors could be a promising target for developing anti-dementia drug. Compounds that alter the proteolytic cleavage of APP, including those that inhibit β- or γ-secretase activity, can reduce the production of Aβ peptides, and may have interesting therapeutic potential in the treatment of AD (Owens *et al.*, 2003). All drugs considered for AD must cross the blood–brain–barrier (BBB) and the plasma membrane (Dewachter and Van Leuven, 2002). Enzyme inhibitors with therapeutic potential are preferably smaller than 700 Da, so large peptide-based inhibitors are not viable drug candidates. Thus, the secondary metabolites of plants and microbes which have relatively low-molecular weight and high lipophilicity might be good drug candidates for BACE1 inhibitors (Jeon *et al.*, 2003). Several transition-state analog peptide inhibitors of BACE1 have been reported with relatively low Ki values (Schmidt, 2003; Ghosh *et al.*, 2000). Natural product inhibitors, however, have rarely been investigated.

1. Naturally occurring flavonols such as myricetin, quercetin, kaempherol and morin along with one flavones apigenin directly inhibited BACE-1 activity in a concentration dependent manner. IC$_{50}$ values of myricetin, quercetin, kaempherol and morin was found to be $2.8 \pm 0.3, 5.4 \pm 0.5, 14.7 \pm 1.3, 21.7 \pm 1.1$ and 38.5 ± 2.6 mM respectively (Shimmyo *et al.* 2008)

2. *Aloe vera* (Liliaceae) is a well-known pharmaceutical herb that has long been used in the traditional Chinese systems of medicine (TCM) for the treatment of various diseases. Chromone glycosides isolated from *A. vera* showed significant inhibitory activity against BACE1 with IC_{50} values of 39.0 and 20.5×10^{-6} M, as well as inhibition of Aβ (1-42) production by 7.4 and 12.3 per cent, respectively, in B103 neuroblastoma cells (Lv *et al.*, 2008).

3. *Punica granatum* (Lythraceae) commonly termed as pomegranate has been used in traditional Indian medicine for the treatment of diarrhea, dysentery and intestinal parasites. In the course of screening for anti-dementia agents from natural products, two BACE1 inhibitors were isolated from the husk of pomegranate by activity-guided purification. They were identified as ellagic acid and punicalagin with IC_{50} values of $3.9 \times 10^{(-6)}$ and $4.1 \times 10^{(-7)}$ M and Ki values of $2.4 \times 10^{(-5)}$ and $5.9 \times 10^{(-7)}$ M, respectively. Kinetic analysis showed the type of inhibition as noncompetitive type inhibition. Moreover ellagic acid and punicalagin were less inhibitory to α secretase and other serine proteases such as chymotrypsin, trypsin, and elastase, thus indicating that they were relatively specific inhibitors of BACE1 (Kwak*et al.*, 2005).

4. *Panax notoginseng* (Araliaceae) is a Chinese herb widely used in TCM to improve learning and memory function. In TCM it is termed as notoginseng "the miracle root for the preservation of life." Clinically it has been used as anti-inflammatory agent, immune booster (Gao*et al.*, 1996), hepatoprotective (Yoshikawa, 2003) and possibly reverse dementia by regenerating injured brain cells (Tohda *et al.*, 2002). Ginsenoside Rg1 the major active components of *P. notoginseng* were assessed under *in vitro* condition for its ability to protect against Aβ induced neurotoxicity. The results showed that ginsenoside could inhibit BACE1 activity *in vitro* and also protect the PC12 cells against injuries caused by exposure of PC12 cells at 5 to 50 mM Aβ for 48 h. The cell death, LDH release, NO release, ROS production, lipid peroxidation, intracellular calcium elevation, and apoptosis are associated events induced by Aβ that can be rescued by ginsenoside in PC12 cells. So ginsenoside Rg 1 can be used as promising drug for AD, and the mechanism is related to BACE1 inhibition and protection against Aβ-induced cytotoxicity (Wang and Du., 2009).

5. Magnolol, honokiol, and obovatol are well known bioactive constituents of the bark of *Magnolia officinalis* (Magnoliaceae), which has been used as TCM for treatment of neurosis, anxiety, stroke, fever and headache (Song*et al.*, 1989). Magnolol, obovatol, and honokiol possess anti-inflammatory (Choi *et al.*, 2007) and anti-oxidative activity (Lin *et al.*, 2006). Honokiol was known to promote potassium-induced release of acetylcholine in a rat hippocampal slice (Tsai *et al.*, 1995) and to enhance neurite sprouting (Fukuyama *et al.*, 1992). Ethanol extract of *M. officinalis* were assessed for BACE1 inhibitory activity under *in vitro* condition. 4-O-methylhonokiol the bioactive constituent of ethanol extract of *M. officinalis* inhibited BACE1 in concentration dependent manner with IC_{50} value of 10.3 µM. In addition

western blot analysis showed that 4-O methylhonokiol decreased the expression of BACE1 in both cortex and hippocampus region of the mice brain dose dependently (Lee *et al.*, 2010).

6. *Eisenia bicyclis* (Laminariaceae) a common perennial brown algae possess rich source of phlorotannins. This alga is frequently consumed as foodstuff and also as raw material for sodium alginate preparation. Phlorotannins isolated from *Ecklonia* and *Eisenia* species, possess wide range of pharmacological properties including anti-diabetic, anti-inflammatory, antioxidant and acetylcholinesterase inhibitory activity. In addition phlorotannins from *Ecklonia* species showed memory enhancing activity. As a part of anti-AD remedy BACE1 inhibitory activity of *E. bicyclis* and its phlorotannins was evaluated. Methanolic fraction of *E. bicyclis* showed significant BACE1 inhibitory activity with IC_{50} value of 4.87 mg/ml. Bioactive guided fractionation showed the presence of phlorotannins such as phloroglucinol, dioxinodehydroeckol, eckol, phlorofurofucoeckol-A, dieckol, triphloroethol A, 7-phloroethol which showed significant BACE1 inhibitory activity with IC_{50} value of 36.47, 5.35, 12.20, 2.13, 2.21, 11.68 and 8.59 mM when compared to standard quercetin (IC_{50} value 10.82 mM). (Jung *et al.*, 2010).

7. *Sanguisorbae radix* (Leguminosae) In the course for screening anti-dementia agents from natural products, two BACE1 inhibitors were isolated from the ethyl acetate soluble fraction of *S. radix*. Bioactive guided fractionation of ethyl acetate fraction showed the presence of 1,2,3-trigalloyl-4,6-hexahydroxydiphenoyl-13-D-glucopyranoside (Tellimagrandin II) and 1,2,3,4,6-pentagalloyl-13-D-glucopyranoside. Both the compounds showed noncompetitive type of inhibition against BACE1 with IC_{50} value of 3.10×10^6 and 3.7×10^6 M respectively. Ki values of the respective compounds were found to be 6.84×10^6 M and 5.13×10^6 M. They were less inhibitory to α secretase and other serine proteases suggesting that they were relatively specific inhibitors of BACE1 (Lee *et al.*, 2005).

8. *Coptidis rhizoma* (Ranunculaceae) also known as 'Huang Lian' is widely used in the treatment of various diseases in TCM due to their anti-diabetic, relaxant, pyretic, antibacterial, and antiviral effects. *C. rhizoma* and its isolated alkaloids have been reported to exhibit cognitive enhancing, anti-depressing, and cholinesterase-inhibitory effects. Six protoberberine alkaloids were assessed for its BACE1 inhibitory activity. Protoberberine such as epiberberine and groenlandicine showed significant BACE1 inhibitory activity with IC_{50} values of 8.55 and 19.68 mM, respectively. Both the compounds exhibited noncompetitive type inhibition with Ki value of 10 and 21.2mM. Protoberberine alkaloids due to their low molecular weight and high lipophilicity can readily cross the blood brain barrier, so they are supposed to act as promising therapeutic drugs for AD (Asai *et al.*, 2007).

9. In the course of searching for BACE1 inhibitors from natural products, the ethyl acetate soluble fraction of rhizome of *Smilax china* showed potent inhibitory activity. The active compounds were identified as a trans/cis-

resveratrol mixture, oxyresveratrol, veraphenol, and cis-scirpusin A. They were shown to non-competitively inhibit BACE1 with the Ki values of 5.4×10^{-6}, 5.4×10^{-6}, 3.4×10^{-6}, and 5.4×10^{-6} M and IC_{50} values of 1.5×10^{-5}, 7.6×10^{-6}, 4.2×10^{-6}, and 1.0×10^{-5} M, respectively. The active compounds were less inhibitory to α-secretase (TACE) and other serine proteases suggesting that they were relatively specific inhibitors of BACE1 (Jeon *et al.*, 2007).

10. *Sophora flavescens* (Fabaceae) one of the most ubiquitous traditional herbal medicines in East Asia, with an array of biological activities has been assessed for BACE1 inhibitory activity. Chloroform root extract of *S. flavescens* showed positive results and the bioactive guided fractionation showed the presence of eight flavanones of which sophoraflavanone G, kurarinone, leachianone A and (2S)-20-methoxy kurarinone showed significant inhibitory activity against BACE1 with IC_{50} values of 5.2, 3.3, 8.4, 2.6, and 6.7 mM respectively. Enzyme kinetic analysis showed the type of inhibition as noncompetitive type. Structure activity relationship shows that lavandulyl group present in the position 8 of flavanone ring was responsible for the BACE1 inhibitory activity. In addition, lipophilic nature of these lavandulyl flavanones makes them as a promising drug for the treatment of AD (Hwang *et al.*, 2008)

11. *Angelica dahurica* (Umbelliferae) is a perennial herb used as traditional folk medicine for the treatment of headache, bleeding, menstrual disorder and neuralgia (Kimura *et al.*, 1996). In the course for screening antidementia agents from natural products, BACE1 inhibitors were isolated from the root extract of *A. dahurica*. They were identified as furanocoumarins, isoimperatorin, imperatorin, (+)-oxypeucedanin, (+)-byakangelicol and (+)-byakangelicin, which inhibited BACE1 in a dose dependent manner. Among them isoimperatorin and (+)-oxypeucedanin showed significant inhibitory activity against BACE1 with IC_{50} values of 91.8 ± 7.5 and 104.9 ± 2.4 mM, respectively. BACE1 inhibitory activity of imperatorin was due to presence of prenyloxy group at the C-8 position. Furanocomarins due to its low molecular weight can readily cross the blood brain barrier and reach the site to action *i.e.* brain (Marumoto and Miyazawa, 2010).

12. *Aralia cordata* (Araliaceae) is widely used in TCM for the treatment of rheumatism, lumbago and lameness (Perry, 1980). As reports regarding anti-alzhiemer effect are limited, diterpenoids isolated from *A. cordata* were assessed for BACE1 inhibitory activity. Of the fourteen diterpenoids isolated, 16α-hydroxy-17-isovaleroyloxy-*ent*-kauran-19-oic acid exhibited good inhibitory activity with an IC_{50} value of 18.58 m M, while 7-oxo-*ent*-pimara-8(14),15-diene-19-oic acid; 17-hydroxy-*ent*-kaur-15-en-19-oic acid exhibited moderate inhibitory activity with an IC_{50} values of 24.10 and 23.40 mM respectively (Jung *et al.*, 2009).

13. *Camellia sinensis* (Theaceaes): In the course of searching for BACE1 inhibitors from natural products, methanolic extract of green tea leaves exhibited potent BACE1 inhibitory activity. Bioactive guided fractionation showed that ethyl acetate soluble fraction, rich in catechin content, showed potent

inhibitory activity. (-)-Epigallocatechin gallate, (-)-epicatechin gallate, and (-) gallocatechin gallate showed potent BACE1 ihibitory activity with IC_{50} values of 1.6 ×10^{-6}, 4.5×10^{-6}, and 1.8×10^{-6} M, respectively. (-)-Catechin gallate, (-)-gallocatechin, and (-)-epigallocatechin significantly inhibited BACE1 activity with IC_{50} values of 6.0×10^{-6}, 2.5×10^{-6} and 2.4×10^{-6}M, respectively. However, (+)-catechin, (-)-catechin, (+)-epicatechin, and (-)-epicatechin exhibited about ten times less inhibitory activity. Inhibitory activity of catechin moiety is due to the presence of pyrogallol moiety on C-2 and C-3 catechin skeleton. Type of inhibition was observed as noncompetitive type (Jeon *et al.*, 2003).

14. *Polygala tenuifolia* (Polygalaceae): An extract from *Polygala tenuifolia*, which is called as tenuigenin has been widely used in TCM to improve memory and intelligence for more than 2000 years. Recent report shows that tenuigenin plays an important role in cognitive improvement and neuroprotection and it has been used for the treatment of AD in TCM nowadays. As reports regarding BACE1 inhibitory activity is lacking, tenuigenin was assessed for BACE1 inhibitory activity in different doses. Tenuigenin inhibited BACE1 in a dose-dependent manner, with the IC_{50} value of 0.25mg/ml. Tenuigenin inhibited the expression BACE1 alone without affecting the secretion of SAPPα and full length APP. Taken together, these results suggest that Tenuigenin may be worthy of future study as an anti-AD drug (Jia *et al.*, 2004).

15. A new isoflavone, neocorylin isolated from the seed extract of *Psoralea corylifolia* showed significant BACE1 inhibitory activity under *in vitro* condition (Choi *et al.*, 2008).

16. *Perilla frutescens* (Lamiaceae): In the course for screening natural products as BACE1 inhibitors, methanolic extract of *Perilla frutescens* showed the high inhibitory activity. Bioactive guided fractionation showed that the active compounds rosamarinic acid and luteolin exhibited significant inhibitory activity with IC_{50} values of 5.0×10^{-7} M and 2.1×10^{-5}M, respectively. They inhibited BACE1 in a non-competitive manner. Ki values of rosamrinic acid and luteolin was found to be 6.2×10^{-5} M and 3.9×10^{-5} M, respectively. They were less inhibitory against other enzymes such asα-secretase (TACE), acetylcholine esterase (AchE), chymotrypsin and elastase, indicating that they were relatively specific inhibitors of BACE1 (Choi *et al.*, 2008).

γ-secretase Inhibitors from Plants

γ-secretase determines the ratio of Aβ 1-40 to Aβ 1-42 and has several unusual properties including its ability to cut substrates in the middle of the transmembrane domain in a water-free environment. γ-secretase is a multiprotein complex and presenilin proteins are the catalytic site of this complex. Presenilins are involved in the cleavage of the notch receptor and the blocking of this pathway in the embryo is lethal. Recently it has been found that high concentration of certain non-steroidal anti-inflammatory drugs (NSAIDs) modulates the γ secretase cleavage such that Aβ

42 is reduced with out affecting the Notch signaling (Weggen *et al.*, 2001). *In vivo* studies showed that γ-secretase inhibitors reduce amyloid burden in animal models and several synthetic drugs have already entered phase I clinical trials (Dovey *et al.*, 2001), but no reports regarding g secretase inhibitors from natural source is available.

α-secretase Activators

α-secretase pathway stimulation leads to a reduction of the APP substrate that is available for Aβ 1-42 formation, and it was demonstrated early that this pathway can be stimulated through cell-surface receptors. Nowadays much research has been focused on the therapeutic potential of natural compounds which stimulate α secretase thereby promoting non-amyloidogenic pathway. Results suggest that medications and dietary regiments (low cholesterol diet) which enhance the nonamyloidogenic pathway of APP processing to be a valuable approach to Alzheimer's disease therapy.

1. Green tea promotes the nonamyloidogenic α-secretase pathway (Levites *et al.*, 2003). Intraperitoneal injection of EGCG daily for 60 days (20mg/Kg) in 12 months old AD induced mice increased the α-secretase activity there by promoting non-amyloidogenic pathway and reducing the Aβ 1-42 generation.

2. Cryptotanshinone (CTS), the active component of the medicinal herb *Salvia miltiorrhiza* increased the expression of α secretase in cultured neuronal cells treated with Aβ1-42 thereby increasing the level of SAPPα. CTS induced ADAM10 increase and promoted non-amyloidogenic α-secretase processing of APP in SweAPP cortical neurons (Mei *et al.*, 2010).

3. Berberine, an isoquinoline alkaloid isolated from *Coptidis rhizoma*, a major herb widely used in TCM has been assessed for its anti-amyloidogenic effect. Result showed that berberine increased the expression of α-secretase activity and downregulated β-secretase activity in H4 neuroglioma cells thereby promoting non-amyloidogenic pathway. The compound can act as an effective drug in the treatment of AD, as berberine can cross the blood brain barrier (Asai *et al.*, 2007).

4. *Ginkgo biloba* extract (EGb761) a effective ROS scavenger and neuroprotector at the concentration of 150mg/Kg increased the activity of α-secretase thereby promoting the secretion of SAPPα. EGb761 effectively mitigates the Aβ induced toxicity in neuronal cells at the same concentration used for *in vivo* studies. Antioxidative, anitamyloidogenic and neuroprotective effect of EGb761 makes it effective drug in the treatment of AD (Colciaghi *et al.*, 2004).

5. Monascus-fermented red mold rice (RMR), which act as multiple cholesterol-lowering agents, antioxidants, and anti-inflammatory agents has been studied for its effect on APP metabolism in cholesterol treated neuroblastoma IMR32 cell. Ethanol extract of RMR increased the α-secretase activity and suppress the β-secretase activity thereby promoting the secretion of SAPPα. Moreover it has been proven to improve the cognitive deficit induced by Aβ 40 infusion in rats (Lee *et al.*, 2010).

Antioxidants as an Anti-dementia Agent

The central nervous system is particularly vulnerable to oxidative insult on account of the high rate of oxygen utilization, the relatively poor concentration of classical antioxidants and related enzymes, and the high content of polyunsaturated lipids which are highly susceptible to oxidations. In addition, the presence of high concentration of redox-active transition metals induces the catalytic generation of reactive oxygen species (ROS). Thus, it is not surprising that oxidative stress is a common point for neurodegenerative disease (Sayre *et al.*, 2008). In AD the Aβ 1- 42 fragment produced from APP by amyloidogenic pathway acts as a source of oxidative stress which plays a significant role in AD pathogenesis (Butterfield *et al.*, 2001). A considerable number of oxidative stress markers are found in the brain of AD patients associated with neuritic plaques and NFT. Interactions of Aβ oligomers in the presence of Fe^{2+} or Cu^{2+} generates H_2O_2 which leads to membrane-associated oxidative stress resulting in lipid peroxidation (Bush *et al.*, 2003; Smith *et al.*, 1997) and the consequent generation of 4-hydroxynonenal (4HNE), a neurotoxic aldehyde that covalently modifies proteins on cysteine, lysine and histidine residues (Mattison, 1997). Some of the proteins oxidatively modified by this Aβ-induced oxidative stress include membrane transporters, receptors, GTP binding proteins ('G proteins') and ion channels (Hong *et al.*, 2000). Oxidative modifications of tau by 4HNE and other ROS can promote its aggregation and may thereby induce the formation of neurofibrillary tangles. Aβ can also cause mitochondrial oxidative stress and dysregulation of Ca^{2+} homeostasis, resulting in impairment of the electron transport chain, increased production of superoxide anion radical and decreased production of ATP. Superoxide is converted to H_2O_2 by the activity of superoxide dismutases (SOD) and superoxide can also interact with nitric oxide (NO) via nitric oxide synthase (NOS) to produce peroxynitrite (ONOO*). Interaction of H_2O_2 with Fe^{2+} or Cu^{2+} generates the hydroxyl radical (OH*), a highly reactive oxyradical and potent inducer of membrane-associated oxidative stress that contributes to the dysfunction of the ER. By disturbing cellular ion homeostasis and energy metabolism, relatively low levels of membrane-associated oxidative stress can render neurons vulnerable to apoptosis (Chauhan, 2006; Mattson, 2004). Moreover oxidative stress induced by Aβ peptide leads to dysfunction and degeneration of synapses leading to death of neurons in the hippocampal gyri. Hence for slowing the progression of AD and minimizing neuronal degeneration, antioxidants may be used as potential agents (Wang *et al.*, 2006). Clinical studies revealed that high dose of antioxidants such as vitamin E and NADH (Sano *et al.*, 1997; Birkmayer, 1996) had beneficial effect in the treatment of AD. As synthetic antioxidants possess severe side effects, the search for antioxidants from natural sources has received much attention and efforts have been put into identifying compounds that can act as suitable antioxidants to replace synthetic ones. In addition, these naturally-occurring antioxidants can be formulated to give nutraceuticals that can help to prevent oxidative damage from occurring in the body. Epidemiological investigations revealed that high consumption of food rich in polyphenols such as fruits, vegetable juices; green tea, wine, curry spice turmeric and Mediterranean diet were inversely associated with AD incidence (Dai *et al.*, 2006). As it is well known that polyphenols are excellent antioxidants both as ROS scavengers and transition metal chelators (Rice –Evans *et al.*, 2006), the anti-AD effects of these foods were naturally linked to their antioxidant potential. Thus, it is interesting to note that some antioxidants derived from these foods go beyond modulating ROS.

Antioxidant Compounds Derived from Natural Source

Flavonoids are the most commonly occurring polyphenols in fruit, vegetable juices and green tea. Pharmacological effects of flavonoids are due to presence of catechols and pyrogallol group. Hydroxyl group present in the catechol moiety acts as ROS neutralizer, potent chelator of transition metal ions and as hydrogen bond donor and acceptor. (Taniguchi *et al.*, 2005). Therefore presence of catechol moiety plays a key role in exerting the multifunction of flavonoids against diverse targets in AD.

1. Flavonoids such as quercetin, isoquercitrin, rutin, gossypetin, myricetin, xanthones, which are widely distributed in fruits and vegetable juices acts as potent ROS scavenger (Porat *et al.*, 2006; Bruhlmann *et al.*, 2004). These flavonoids can pass through the blood brain barrier which makes them effective in the treatment of AD.

2. Green tea polyphenols such as - (-) epicatechin (EC), (-)-epicatechin-3-gallate (ECG) and (-)-epigallocatechin-3-gallate (EGCG) serve as powerful antioxidants against free radicals such as DPPH radicals (Nanjo al., 1996), superoxide anion, lipid free radicals and OH radicals (Zhao *et al.*, 1989; Guo *et al.*, 1996). In the central nervous system, there is also some evidence to show that oral administration of green tea polyphenols and flavonoid-related compounds has preventive effects on iron-induced lipid peroxide accumulation and age-related accumulation of neurotoxic lipid peroxides in rat brain (Inanami *et al.*, 1998). In addition the ability of these catechins to cross the blood brain barrier makes green tea polyphenols as effective antioxidant to combat AD.

3. Nicotine present in cigrattes acts as potent antioxidant activity which is higher than vitamin C (Liu *et al.*, 2003).

4. Curcumin, a yellow-orange pigment extracted from curry spice turmeric (*Curcuma longa*), has long been used as a food additive in India. Epidemological findings shows that AD prevalence is only 1 per cent in people of >65 years of age in rural India. Many pharmacological effects have been identified for this pigment such as antioxidant and anti-inflammatory activity (Aggarwal *et al.*, 2007). Interestingly curcumin-Cu^{2+} complex may become more active than the parent curcumin in scavenging ROS by catalyzing the dismutation of superoxide anion radicals or by donating a proton or electron (Barik *et al.*, 2005).

5. Ginkgo-Biloba extract (EGb761) is a well known plant extract obtained from green leaves of *Ginkgo biloba*, which contains flavonoids like ginkgolides and biolobalids. EGb 761 exhibits broad range of biochemical and pharmacological activities such as antioxidant and free radical scavenging as well as nootropic and neurotrophic activities in the hippocampal region. EGb and its flavonoid fraction exhibits protective

effect against Aβ peptides by strongly inhibiting Aβ 25-35 fragment and NO induced ROS accumulation and the total extract protected cells against the harmful effects of H_2O_2 a purported mediator of Aβ toxicity (Bastianetto *et al.*, 2001).

Antioxidants from Marine Natural Products

Natural products isolated from marine resources have also been shown to have a great potential in drug discovery. With the ocean covering 70 per cent of the Earth's surface, and with the uniqueness of the environmental conditions present in the oceans, it is easily understandable why the ocean can be considered as a very promising source of natural drugs – or synthetic derivatives thereof – for the future. Marine organisms are potentially prolific sources of highly bioactive secondary metabolites that might represent useful leads in the development of new pharmaceutical agents (Iwamoto *et al.*, 2001). During the last four decades, numerous novel compounds have been isolated from marine organisms and many of these substances have been demonstrated to possess interesting biological activities. At present, marine plants (marine algae and mangroves) are becoming the focus for targeting effective antioxidants towards oxidative stress in human body due to the presence of diverse natural products with unique structures possibly caused by extreme marine environment (El Gamal, 2010).

Seaweeds

Seaweeds have been used in food and traditional remedies in Asian countries. It has been reported that seaweeds serve as an important source of bioactive natural products. Many metabolites isolated from marine algae have been shown to possess bioactive effects (Faulkner, 2002). In fact, the discovery of metabolites with biological activities, from macro algae has increased significantly in the past three decades. On the other hand, seaweeds have recently received significant attention for their potential as natural antioxidants (Chandini *et al.*, 2008; Duan *et al.*, 2006).

Marine algae particularly brown algae have gained interest in the field of antioxidants due to the presence of phlorotannins. Several prenyl toluquinones isolated from the brown alga *Cystoseira crinite* showed potent radical scavenging activity (Fisch *et al.*, 2003). The plastiquinones isolated from brown alga *Sargassum micracanthum* displayed significant antioxidant activity (Iwashima *et al.*, 2005). The sargachromanols A–P the monoterpenoids of the chromene class, isolated from S. *siliquastrum* showed significant antioxidant potential. S. *thunbergii* afforded a novel chromene, sargothunbergol A, which acts as a free radical scavenger (Seo *et al.*, 2007). The known compounds taondiol, isoepitaondiol, stypodiol, stypoldione and sargaol, isolated from brown alga *Taonia atomaria* exhibited free radical-scavenging activity (Nahas *et al.*, 2007). (2R)-2-(2,3,6-tribromo– 4,5-dihydroxybenzyl) cyclohexanone isolated from red alga *Symphyocladia latiussula* acts as potent free radical scavenger with IC_{50} value of lg/ml (Choi *et al.*, 2000). S. *fusiforme* a brown algae exhibited potent antioxidant activity which is due to presence of uronic acid (Zhou *et al.*, 2008). Phlorotannins present in *Ecklonia stolonifera* and *E. cava* used in Korean traditional medicine showed potent antioxidant activity (Kang *et al.*, 2004a). Sulphated polysaccharides of *Porphyra haitanesis* showed excellent antioxidant activity (Zhang *et al.*, 2010). Different solvent fraction of *S. marginatum, Padina tetrastomatica* and *Turbinaria conoides* were assessed for antioxidant activity and the result showed that ethyl acetate fraction of S *marginatum* showed higher radical scavenging activity (Chandini *et al.*, 2008). Methanol and aqueous fraction of *Kappaphycus alvarezii*, showed significant antioxidant activity, which was correlated to its high polyphenolic content (Suresh kumar *et al.*, 2008). The edible seaweeds *Fucus vesiculosus* and *P. gymnospora* showed beneficial antioxidant activity which is due to presence of fucoidan (Rocha de souza *et al.*, 2007). Methanolic extract of edible seaweed *Gelidiella acerosa* showed excellent antioxidant activity and its activity was correlated with its total polyphenolic content (Devi *et al.*, 2008). *In vitro* antioxidant activities of Indian red seaweeds *Euchema kappaphycus, Gracilaria edulis* and *Acanthophora spicifera* were evaluated. Result showed that methanolic extract of all the three seaweeds showed antioxidant activity in a concentration dependent manner (Ganesan *et al.*, 2008).

Mangroves

Mangroves are integral primary producers in estuarine systems, and provide habitat and breeding grounds for many commercial fish species (Laegdsgraad and Johnson, 1995). Mangroves have substantial medicinal values and are well documented in Arab pharmacopoeia and in an Australian traditional medicinal system termed as Bush medicine. Mangroves comprising of intertidal marine plants forms a dominant ecosystem predominantly bordering the margins of tropical coastlines around the world and they possess unique adaptation to combat environmental stress *e.g.* high salinity, high temperature, low nutrient and excessive radiation. An inevitable consequence of this process results in the production of ROS and accordingly the antioxidant enzymes were upregulated in them.

Methanolic fraction of *R. mucronata* leaf showed significant antioxidant activity, which was correlated to the presence of tannins (Suganthy *et al.*, 2009). Bioactive guided fractionation showed that catechin present in the bark extract of *R. apiculata* showed significant antioxidant activity (Rahim *et al.*, 2008). Methanolic extract of *A. ilicifolius* exhibited potent ferric reducing capacity and peroxyl radical scavenging activity (Mani Senthil Kumar *et al.*, 2008). Pyroligneous acid isolated from dichloromethane extract of *R. apiculata* showed significant free radical scavenging activity (Loo *et al.*, 2006). Ethanolic extract of bark and twigs of *Laguncularia racemosa* showed significant antioxidant activity. Bioactive guided fractionation showed that antioxidant activity is due to presence of (2R, 3R) pinobanksin- 3-caffeoylate (Shi *et al.*, 2009). *Aegiceras corniculatum* (Aegicerataceae) widely distributed along the coastlines of tropical and subtropical regions were evaluated for its antioxidant activity. Result showed that n-hexane, ethyl acetate and methanol extracts of *A.corniculatum* stem showed prominent antioxidant activity. These extracts contain a plethora of chemicals that can scavenge free radicals, chelate metal ions, inhibit lipid peroxidation, diminishes the respiratory burst in cells and also exert a protective effect against oxidative damage by •CCl_3 in rat liver and by •OH in mouse (Roome *et al.*, 2008). Activity is related to the presence of alkyl substituted benzoquinone (*n*-hexane extract), flavonoids and triterpenes (ethyl acetate fraction) and tannins, saponins (methanolic extract).

Antioxidants from Plant Source

Plants possess rich source of antioxidants as an inbuilt defense mechanism against oxidative insult (Spiteller, 1993). Plants such as *Aeseclus hippocastanum, Allium nutans, Artemisia* spp., *Guiera senegalensis, Hamamelis virginiana, Rosmarinus officinalis, Salvia officinalis, Taraxacum officinale* and *Thymus vulgaris* and their phytochemicals like cinnamic acids, coumarins, diterpenes, flavonoids, lignans, monoterpenes, phenylpropanoids, tannins and triterpenes exhibit excellent antioxidant property (Aruoma *et al.*, 1996; Bouchet *et al.*, 1998; Cuvelier *et al.*, 1996; Deans *et al.*, 1993; Masaki *et al.*, 1995; Stajner *et al.*, 1999; Youdim and Deans, 1999). Curcumin from *Curcuma longa* reduced lipid peroxidation in rat brain following oral administration to rats with ethanol induced brain injury (Rajakrishnan *et al.*, 1999), and *Bacopa monniera*, which is reported to have cognition enhancing effects, induced a dose related increase in superoxide dismutase, catalase and glutathione peroxidase activities in the rat frontal cortex, striatum and hippocampus (Bhattacharya *et al.*, 2000b). In another study, essential oil from *Thymus vulgaris* maintained higher PUFA levels in various tissues, including the brain in rats, indicating protective antioxidant effects (Youdim and Deans, 1999). Several species of *Salvia* exhibited excellent antioxidant activity. *S. officinalis* inhibits lipid peroxidation due to presence of antioxidant compounds like caffeic acid, carnosic acid, carnosol and rosmarinic acid (Wang *et al.*, 2000). *S. lavandeulafolia* exhibited antioxidant effects due to the presence of the major components 1,8-cineole, β pinene and α pinene in its essential oil (Perry *et al.*, 2001). The antioxidant effect of root extract of *S. miltiorrhiza* is due to the presence of dihydrotanshinone, tanshinone I, methylene tanshinquinone and cryptotanshinone, salvianolic acids A and B (compounds found to protect against memory impairment induced by cerebral ischaemia in mice) (Du *et al.*, 2000; Guanhua and Juntian, 1997), rosmariquinone (also known as miltirone) and several other phenolic compounds (Huang and Zhang, 1992; Kang *et al.*, 1997). Such antioxidant compounds may be useful in AD therapy. Crude extract of root of *Withania somnifera* exhibited excellent antioxidant and anti-inflammatory activities which may be relevant in AD therapy. Compounds responsible for antioxidant activity include the withanolides and glycowithanolides, which reduced lipid peroxidation in brain of rodents. Moreover glycowithanolides and sitoindosides (VII–X) enhanced the activities of catalase and glutathione peroxidase in rat frontal cortex and striatum (Bhattacharya *et al.*, 2001). *Gardenia jasminodes* traditionally used as anti-inflammatory agent, reduced the oxidative stress and cytotoxicity of Aβ in PC 12 cells (Choi *et al.*, 2007). *Bacopa monnieri* (Brahmi) is a traditional Ayurvedic medicinal plant and is used in India as a nerve tonic. Aqueous extract of aerial parts of brahmi lowers the level of ROS lowering the oxidative stress induced damage in the neurons (Limpeanchob *et al.*, 2008).

Metal Chelators (Indirect Antioxidants)

The brain is a specialized organ that concentrates metals (Cu^{2+}, Zn^{2+}, Fe^{3+}) in the neocortex. All these metals play an important role in the cortical physiology and Zn^{2+}, Cu^{2+} are normally released during neurotransmission. The metalloprotein Aβ 1-42 has a high affinity for these metal ions, hence accumulation of these ions play an major role in the aggregation of Aβ 1-42. When Cu^{n} and Fe^{n} readily bind to the Aβ 1-42 peptide it promotes the release of H_2O_2, the proxidant molecule which induces the production of ROS thereby leading to oxidative stress damage to neurons in the neocortex (Bush, 2002). Transition metal ions such as Cu^{2+} and Fe^{3+} are important catalysts for the generation of highly reactive OH radicals via the Fenton reaction in both *in vivo* and *in vitro* systems. These ions can be rendered catalytically silent by ligands like secondary antioxidants, which bind to the metal ions and alters its redox potentials. The Secondary antioxidants sequesters Cu^{2+} and Fe^{3+} by "wrapping" themselves around these ions. There by these ligands help, intercept and suppress the formation of radicals via catalysis from fuelling a chain reaction (Aruoma *et al.*, 1987). Nowadays research has been focused on the use of metal chelators for AD therapy (Huang *et al.*, 1999), since they can interact with Aβ 1-42 and slow down the pathogenesis of AD. In 1999 Cherny and co-workers showed that Cu/Zn chelators solubilises Aβ plaques from brains of patients with AD after death. Clioquinol is a hydrophobic Cu/Zn chelator that freely crosses the blood–brain barrier and was used as an oral antiamoebic treatment for many years before being withdrawn from the market in the 1970s because of its association with subacute myelooptic neuropathy. This effect is now believed to be preventable with Vitamin B12 supplementation (Yassin *et al.*, 2000). Many natural products have been reported as metal chelators, some of which are given below

1. Flavonoids derived from fruits, vegetable juices and green tea acts as good chelators of pro-oxidant transition metal ions. Metal chelating activity of flavonoids is due to the presence of hydroxyl group in the phenolic moiety such as catechol and pyrogallol (Taniguchi *et al.*, 2005).

2. Xanthones are special kind of flavonoids with multipotent metal chelating activity, which is attributable to the catechol moiety.

3. Curcumin is good metal chelator, which readily chelates Cu^{2+}, and Curcumin- Cu^{2+} is more active than the parent compound in scavenging ROS by catalyzing the dismutation of superoxide anion radical or by donating the electron (Barik *et al.*, 2005).

4. Green tea polyphenols such as EC, EGC, EGCG and flavaonoids such as quercetin, gossypetin, myricetin, quercitrin, isoquercitrin and rutin, from natural sources act as very good metal chelators (Rice-Evans *et al.*, 1996; Taniguchi *et al.*, 2005).

5. TCM phytochemicals such as Bellidifolin, Isogntisin, Swerchirin, Glycyrrhisoflavone and Morin also showed effective metal chelating activity. Nicotine prevents aggregation of Aβ peptides through regulating metal (copper and zinc) homeostasis. Copper (II)-nicotine chelates hold SOD-like activity, which may play a role in the neuroprotective effects of nicotine (Zhang *et al.*, 2006).

6. Chloroform and methanol extract of roots and stems of *Rhubarb ribes* (Polygonaceae) were assessed for its metal chelating activity. Methanol extract of stem showed the highest Fe^{2+} chelating activity (93.71±0.80 per cent) when compared with standard quercitin and positive control EDTA. Metal chelating activity might be due to presence of stilbenes (Ozturk *et al.*, 2007).

7. *Ferula assafoetida* (Umbelliferae) which is rich in essential oil were assessed for metal chelating activity. Result showed that methanolic extract of aerial parts of *F. assafoetida* showed effective Fe^{2+} chelating activity in concentration dependent manner with IC_{50} value of 0.57µg/ml. Chelating activity of the extract is due to the presence of secondary antioxidants (Dehpour *et al.,* 2009).

8. Stem bark extract of *Spondias pinnata* (Anacardiaceae) showed Fe^{2+} chelating activity of 51 per cent (45µg/ml) when compared to positive control EDTA. Chelating activity is due to presence of flavonoids (Hazra *et al.,* 2008).

9. *Mucuna pruriens* (Fabaceae) a tropical plant exhibits neuroprotective effect especially in the treatment of Parkinson's disease by restoring the level of dopamine in the substantia nigra. *Mucuna pruriens* cotyledon powder showed significant metal chelating effect, which might be due to presence of polyphenolic content (Dhanasekaran *et al.,* 2008).

10. *Azadirachta indica* (Meliaceae) a traditional medicinal plant in India rich in azadirachtin and nimbin were assessed for its metal chelating activity. Result showed that methanolic bark extract showed significant Fe^{2+} chelating effect of 74.2 per cent than the leaf extract (Ghimeray *et al.,* 2009).

11. *Hizikia fusiformis* the most common edible brown seaweeds (*Sargassaceae*) were assessed for its metal chelating activity. Aqueous fractions of chloroform (IC_{50}= 0.131 ± 0.002 mg/ml) and ethyl acetate (IC_{50} = 0.18 ± 0.07 mg/ml) exhibited significantly (P<0.05) higher chelating effects compared with the effects of a tocopherol (IC_{50} = 1.72 ± 0.2 mg/ml). Iron binding capacity, suggested their ability as a peroxidation protector which is due to rich polyphenolic content present in it (Karawita *et al.,* 2005).

Amyloid Antiaggregant Therapies

The hallmark pathology of AD has been the deposition of Aβ in the form of senile plaques (Selkoe, 2001). Aβ can self-assemble to form dimers, soluble oligomers, and protofibrils and diffuse plaques through multistep-nucleated polymerization (Lambert *et al.,* 1998). The self-assembling evaluation of Aβ *in vitro* will provide an opportunity to screen molecules for anti-amyloidogenic property. The prevention of the formation of oligomers and the fibrils from soluble monomers is of therapeutic significance for AD drug discovery (Smith *et al.,* 2007). Over the past decades researchers have identified several peptidic inhibitors that inhibit these specific protein–protein interactions. Soto and colleagues (2001) have designed short synthetic peptides homologous to the central hydrophobic region of Aβ that disrupts β-sheet stabilization. Recently, the same group has shown a reduction in aggregated amyloid in transgenic mice overproducing APP that were treated with a five-residue β-sheet breaker peptide (iAβ5p). The peptide did not induce antibody production and could cross the blood–brain barrier. Clinical trials of antiaggregants as potential therapies for AD are now possible. Nonpetidic inhibitors mostly organic compounds such as congo red has been used as antiaggregants but unfortunately its carcinogenicity and inability to cross the blood brain barrier has hindered it from therapeutics. Though natural products as alternatives for AD discovery are a current trend, only few reports on antiamyloidogenic properties of plants crude extracts or pure compounds are available, however with unclear mechanisms of their therapeutic potential.

1. Salvianolic acid B (Sal B), a potent antioxidant component from *Salvia miltiorrhizae* a Chinese herb used in TCM for the treatment of coronary heart disease. Salvianolic acid (10-10nmol/l) completely prevented the Aβ fibril formation within 30 hours in PC12 cell lines. Moreover result showed that Sal B (1μmol/L) significantly attenuated the toxicity induced by Aβ 1-42 in aged PC12 cells (Tang and Zhang, 2001). Sal B acts both as Aβ deaggregator and attenuator of Aβ toxicity.

2. *Ginkgo biloba* extract - EGb 761 contains flavonol-*O*-glycosides, 6 per cent terpenoids (known as ginkgolides A, B, C, M, J and bilobalide), 5–10 per cent organic acids, and >0.5 per cent proanthocyanidins. EGb 761 exhibits a broad range of pharmacological activities such as antioxidant and free radical scavenging as well as nootropic and/or neurotrophic activities in the hippocampal formation. *G. biloba* extract was able to inhibit Aβ1–42 fibril formation and also disrupt Aβ fibrinilolysis. Recent studies revealed that EGb 761 increased gene expression for transthyretin, a protein that may play a neuroprotective role by Aβ sequestration. Multipotent action of EGb 761 makes it an effective drug in AD therapy.

3. *Caesalpinia crista* (Fabaceae) has been traditionally used as anthelmintic, anitmalarial, antipyretic and anti-inflammatory agent. The aqueous leaf extract of *C. crista* has been used as mental stress relaxant health drink by forest dwellers. This tempted the researchers to investigate the antiamylodiogenic effect of *C. crista*. Result showed that aqueous extract of *C. crista* not only prevented the Aβ fibril formation but also disaggregated the Aβ fibrils and the antiamyloidogenic property was attributed to the water-soluble polyphenols (Ramesh *et al.*, 2010).

4. MegaNatural-AZ (MN) a water-soluble polyphenolic extract from *Vitis vinifera* grape seeds consists of catechin and epicatechin, in monomeric (8 per cent), oligomeric (75 per cent), and polymeric (17 per cent) forms. MN, which is a proven antioxidant, was assessed for its antiamyloidogenic property. Result showed that MN (25mM) inhibited protofibril formation, pre-protofibrillar oligomerization and destabilized the already formed fibrils. MN readily binds to the Aβ fibrils preventing its interaction with cell surface receptors thereby preventing its cytotoxicity. In transgenic mouse model, the inhibitory effect of MN on Aβ assembly was coupled with attenuation of AD type cognitive deterioration and reduction in cerebral amyloid deposition (Wang *et al.*, 2008), which suggested that MN is worthy of consideration as a therapeutic agent for AD (Onto *et al.*, 2008).

5. EGCG, the most abundant polyphenolic extract from green tea, directly bind to the natively unfolded polypeptides and redirects their aggregation down an off-folding pathway, resulting in the formation of unstructured, innocuous and highly stable oligomers. EGCG not only reverse the fibrillation

pathway but also disrupts the preformed amyloid fibrils. In all, the high efficacy of EGCG as a generic anti-aggregate makes it an excellent candidate for the design and synthesis of more potent fibrillation blockers (Hudson *et al.*, 2009).

6i. Ellagic acid (EA), a dimeric derivative of gallic acid, present in fruits and nuts was assessed for its effect on Aβ aggregation. Recent research has shown that Aβ non-fibrillar aggregates or soluble oligomers, rather than mature amyloid fibrils, are the pathogenic components that drive neurodegeneration and neuronal death. Based on this hypothesis it was tested whether EA decreases the levels of Aβ oligomers and it was found that EA conjugated with Aβ by aromatic stacking or hydrophobic forces and promoted fibrillation thereby decreasing the levels of pathogenic Aβ oligomers. Supporting data shows that intake of pomegranate juice rich in EA decreases soluble Aβ42 levels and attenuates cognitive deterioration in AD (Feng *et al.*, 2009).

7. Galantamine isolated from *Galanthus woronowi* is currently prescribed as a drug treatment for AD because of its activity as a moderate AChE inhibitor. Galanthamine was assessed for its antiamyloidogenic effect and the result showed that galanthamine inhibited both Aβ1-40 and Aβ1-42 aggregation, into oligomers of β-amyloid in a concentration dependent manner (25–1000 μM). In addition galanthamine also acted as a deaggregator of Aβ fibrils and inhibited Aβ induced cell mediated cytotoxicity (Matharu *et al.*, 2009).

8. The organosulfur compound S-allyl-l-cysteine (SAC) is, a natural constituent of garlic. Though SAC is a water-soluble compound, it is sufficiently hydrophobic to cross the blood brain barrier and gain access to the CNS. Since SAC showed a protective effect against Aβ induced neurotoxicity, it was assessed whether SAC has antiaggregation property also. The result showed that due to the compact structure of SAC, it specifically binds to Aβ and destabilizes the b sheet rich conformation in fibrils. Binding could also be induced by hydrophobic interactions between allyl chain of SAC and the hydrophobic region of Aβ, thus blocking association between Aβ molecules. These interactions could also be reinforced by the H-bond between the –OH group of the carboxylic moiety of SAC and donor/acceptor groups of Aβ. Thus, SAC with potent antioxidant motifs, could bind specifically to Aβ and inhibit fibrillar Aβ formation and destabilize preformed fibrillar Aβ. SAC also act as a breaker of the preformed Aβ fibrils. Chauhan (2006) have recently demonstrated anti-amyloidogenic, anti-inflammatory and anti-tangle effects of aged garlic extract and its constituents namely SAC and diallyl- disulfide in AD transgenic model Tg2576 which shows that SAC can act as effective drug in the treatment of AD.

9. Resveratrol (trans-3,4,5-trihydroxystilbene), a polyphenolic phytoalexin and a main ingredient of polyphenols in wine, mitigates or delays the onset of neurodegenerative disease and prevent learning impairment in transgenic AD models (Kim *et al.*, 2007; Karuppagounder *et al.*, 2009).

Resveratrol exert a direct protective effect against A β42 via its antioxidation or anti-inflammatory activity (Albani *et al.*, 2009). Reports on effect of resveratrol on Aβ42 oligomerization showed that resveratrol (100mM) showed 90 per cent effective inhibition against the Aβ aggregation when compared to other polyphenols such as curcumin and catechin. Structure activity relationship showed that resveratrol contains two 3,4-methoxyhydroxyphenyl ring stilbene, which may be quite suitable for binding to b-sheet rich conformation of A β42 oligomers, thereby inhibiting fibrilliar formation and destabilizing fibrils to small aggregates. Resveratrol also intercalates into the β-sheet of Aβ fibrils and increase fibrillar Aβ disaggregation. As reseveratrol possess antioxidant and antiamyloidogenic activity this natural compound could be a promising molecule for the development of therapeutics for AD (Feng *et al.*, 2009).

10. Tannins (TA) are water-soluble polyphenols which differ from most other natural phenolic compounds in their ability to precipitate proteins such as gelatin from solution. The effect of polymeric polyphenol tannic acid on the extension and destabilization of fibrillar Aβ (1-42) (fAβ) was assessed and compared with other polyphenols such as myricetin, morin and catechin. TA dose-dependently inhibited fAβ formation from Aβ (1–42), as well as their extension. Moreover, it dose-dependently destabilized preformed fAβs. The effective concentrations (EC$_{50}$) of TA for the formation, extension and destabilization of fAβs were in the order of 0–0.1 mM. Although the exact mechanism by which TA inhibits fAβ formation and extension unclear, it could act as a key molecules for the development of therapeutics for AD (Ono *et al.*, 2004).

11. Curcumin (Cur), a major component of the yellow curry spice turmeric is used in traditional diet and herbal medicine in India. It is proposed that because of this reason the frequency of AD in India is roughly one quarter of that in the US (Ganguli *et al.*, 2000). Curcumin prevents lipid peroxidation in brain and protects it from Aβ induced neuronal toxicity. Assessment of effect of curcumin on Aβ fibril formation and extension showed that curcumin inhibited fAβ formation, extension and destabilization of preformed fAβ with IC$_{50}$ value of 0.63, 0.52 and 0.32 mM respectively. Structure activity relationship showed that curcumin has a 3,4-methoxyhydroxyphenyl rings symmetrically bound by a short carbohydrate chain. This compact symmetric structure specifically binds to free Aβ and subsequently inhibit polymerization of Aβ into fAβ. Alternatively, this structure might be suitable for specific binding to fAβ and subsequent destabilization of the β-sheet-rich conformation of Aβ molecule in f Aβ. The antioxidant and antiamyloidogenic molecule of curcumin makes this a key molecule in the development of therapeutic against AD (Ono *et al.*, 2004).

12. The effect of walnut extract on Aβ fibrillization was assessed by Thioflavin T fluorescence spectroscopy and electron microscopy. Methanolic extract of walnut not only inhibited Aβ fibril formation in a concentration and time-

dependent manner but it was also able to defibrillize Aβ preformed fibrils. Over 90 per cent inhibition of Aβ fibrillization was observed with 50mg/ ml of methanolic extract of walnut and maximum defibrillization (91.6 per cent) was observed when preformed Aβ fibrils were incubated with 100 mg/ml of extract. Antiamyloidogenic effect of walnut might be due to presence of polyphenolic compound (flavonoids) with molecular wt less than 10 kDa. These results suggest that walnuts may reduce the risk or delay the onset of Alzheimer's disease by maintaining Aβ in the soluble form (Chauhan *et al.*, 2001).

13. Rosmarinic acid (RA) is an ester of caffeic acid and 3, 4-dihydroxyphenylactic acid commonly found in the Lamiaceae family which possess several biological activities such as antioxidant and anti inflammatory activity. RA dose-dependently inhibited fAβ formation from Aβ(1–42), their extension, and destabilized preformed fAβs with IC_{50} value of 1.10, 0.80 and 0.60 mM respectively. RA has two 3, 4- dihydroxyphenyl rings symmetrically bound by a short carbohydrate chain. This compact structure might be quite suitable for specifically binding to free Aβ and subsequently inhibiting the polymerization of Aβ into fAβ. Alternatively, this structure might be suitable for specific binding to fAβ and subsequent destabilization of the β-sheet-rich conformation of Aβ molecules in fAβ (Ono *et al.*, 2004).

14. Polyphenols comprise a chemical class with over 8000 members, many of which are found in high concentrations in wine, tea, nuts, berries, cocoa, and other plants. A substantial body of evidence suggests that polyphenols have ROS scavenging activity. Many dietary polyphenolic compounds have been reported to have inhibitory effect on amyloid fibril formation. Wine related polyphenols such as Quercitin (Qur), gossypetin, myricetin (Myr), morin (Mor), kaempferol (Kmp), catechin (Cat) and epicatechin (epi-Cat) were assessed for its anitamyloidogenic effect. All these polyphenols inhibited the formation, extension and destabilization of fAβ at pH 7.5 at 37°C *in vitro* in dose dependent manner. The anti-amyloidogenic and fibril-destabilizing activity of the polyphenols was reported to be in the order: Myr=Mor=Qur>Kmp>Cat=epi-Cat. Myr, Mor, Qur and Kmp have no chirality and the hydroxyphenyl and benzopyran rings of these molecules are located on the same plane by the rotation of the hydroxyphenyl ring. On the other hand, Cat and epi-Cat have a chirality and the two rings cannot be located on the same plane. This difference in the three-dimensional structure of polyphenols may affect greatly the anti-amyloidogenic and fibril destabilizing activity. Second, the numbers of hydroxyl groups in Myr, Mor, Qur and Kmp are 6, 5, 5 and 4, respectively. These numbers may also affect the activity; the more hydroxyl groups in the molecule renders it to have high anti-amyloidogenic activity. However, there are few reports on the effects of the polyphenols on the fAβ burden *in vivo* (Onto *et al.*, 2003).

15. *Crocus sativus stigmas* (Iridaceae), one of the widely known spices (saffron) consists of unusually polar carotenoids which exhibits anti-cancer activity and acts as a memory enhancer. *In vitro* evaluation of the antiamyloidogenic effect of methanol: water extract of *C. sativus* stigmas showed that the carotenoid constituent, *trans*-crocin-4, the digentibiosyl ester of crocetin, inhibited Aβ fibrillogenesis at lowest concentration. Moreover these carotenoids inhibited Aβ induced toxicity also (Papandreou *et al.*, 2006).

Conclusions

The current review focuses on the pharmacological applications of natural products and their phytoconstituents for anti-amyloidogenic effect and other activities relevant to the treatment of AD. It is clear that dietary antioxidants/herbal extracts can significantly contribute to the modulation of the complex mechanisms of neurodegenerative diseases. Developing countries tend to retain traditional herbal medical practices and thus offer an invaluable resource for new anti-dementia therapies. One of the long-term controlled clinical trials in progress on dementia prevention is based on the Asian traditional tree medicine *Gingko biloba*. Preliminary data have indicated significant effects on dementia progression, but the most recent Cochrane analysis concluded that evidence of predictable and clinically significant benefit of *G biloba* and standardised extract (EGb 761) for people with dementia is inconsistent and unconvincing. Huperzine A and its derivative, ZT-1, are currently developed anti-AD drug. A plethora of pharmacognostic practices, including those for cognitive care, still exist in countries such as Africa, South America, India, and in other aboriginal cultures. Other relevant phytotherapeutics including combinations of traditional Chinese medicinal herbs (*yi-gan san* and *ba wei di huang wan*), sage (*Salvia officinalis* and *Salvia lavandulaefolia*), and lemon balm (*Melissa officinalis*), which have shown positive benefits on behavioral symptoms and cognition, need to be explored in wider studies. Several species of medicinal plants have been reported for activities that are relevant to AD under both *in vitro* and *in vivo* condition (*e.g.*, anti-cholinesterase, anti-amyloid, anti-inflammatory, antioxidant, neuroprotective, and memory enhancing). The most frequently reported are blueberry, cannabis, club moss, curcumin, garlic, ginseng, green tea, pomegranate, and rhubarb. Only few plant-derived drugs have been approved currently for clinical use because most herbal medicines are complex mixtures of chemical components and have diverse biological and pharmacological actions. Some of the phyto-constituents such as honokiol, magnolol, flavonoids, glycowithanolides, ferulic acid, galanthamine, Huperzine A, curcumin, salvianolic acids A and B, EGCG and ellagic acid for which, behavioral effects and pharmacological properties have been well characterized can serve as a good drug candidate for the treatment of AD. Further investigations of these drugs may ultimately result in clinical use. Also, some of the herbal constituents with well-defined chemical structures may offer templates and models for synthesis of analogous drugs with higher efficacy and less adverse effects. One such example is rivastigmine a synthetic chemical analogue of physostigmine (*Physostigma venenosum*). Although many of these herbal preparations and its phytoconstituents have shown positive therapeutic potentials in animal models, the clinical efficacy of most herbal extracts and herbal mixtures is still in its infancy. Knowledge of active constituents opens up a possibility for development of standardized products, which would help secure a more reliable medication for patients. Over all the information collected in this review on a large number of herbal extracts and its constituents which possess antiamyloidogenic effects may be useful in the search for novel pharmacotherapies for the treatment of AD.

Future Therapeutic Perspectives

To date the complex pathophysiology of AD is not yet clearly clarified. Further research will provide better understanding of the molecular pathways involved and thereby leads to the development of additional pharmacological test systems, in which activities may yet be observed. Most of the herbal drugs such as *Ginko biloba, Withania somnifera* and *Bacopa monneri* which showed positive effect on cognitive function require further studies regarding the compounds responsible for their activity. Traditional medical systems such as Ayurveda emphasize health maintenance and disease prevention over curative treatments. Hence, preclinical and clinical research into protective and preventive effects of herbals drugs should be carried out in the future. Several herbal plants which possess both supporting activity relevant to AD, needs further investigation and development for treatment of AD. In some cases the active constituents have been isolated and structures have been identified. Mechanism of action of these compounds can be elucidated using QSAR studies. Elucidation of the relationships between the structures and anti-amyloidogenic activities of these compounds is expected to lead to the development of more potent anti-amyloidogenic compounds for prevention and treatment of AD. In addition long term clinical trials on herbal extracts and its constituents will provide a possible role of natural product based preventive therapy in AD. Initiatives are thus needed to continue to protect, assess, and standardise traditional herbal medicines for the treatment of AD.

References

Aggarwal, B.B., Sundaram, C., Malani, N., and Ichikawa, H. (2007). Curcumin: the Indian solid gold. *Advances in Experimental Medicine and Biology*, **595**: 1-75.

Albani, D., Polito, L., Batelli, S, De Mauro, S., Fracasso, C., Martelli, G., Colombo, L., Manzoni, C., Salmona, M., Caccia, S., Negro, A., and Forloni, G. (2009). The SIRT1 activator resveratrol protects SK-N-BE cells from oxidative stress and against toxicity caused by alpha-synuclein or amyloid-beta (1-42) peptide. *Journal of Neurochemistry*, **110**: 1445–1456.

Aruoma, O.I., Spencer, J.P., Rossi, R., Aeschbach, R., Khan, A., Mahmood, N., Munoz, A., Murcia, A., Butler, J., and Halliwell, B. (1996). An evaluation of the antioxidant and antiviral action of extracts of rosemary and Provencal herbs. *Food and Chemical Toxicology*, **34**: 449–456.

Asai, M., Iwata, N., Yoshikawa, A., Aziaki, Y., Ishiura, S., Saido, T.C., and Maruyama, K. (2007). Berberine alters the processing of Alzheimer's amyloid precursor protein to decrease Aβ secretion. *Biochemical and Biophysical Research Communications, 352. 490 502.*

Barik, A., Mishra, B., Shen, L., Mohan, H., Kadam RM, Dutta S, Zhang H.Y., and Priyadarsini K.I. (2005). Evaluation of new copper-curcumin complex as superoxide dismutase mimic and its free radical reactions. *Free Radical Biology and Medicine,* **39**: 811-822.

Bastianetto, S., Ramassamy, C., Dore, S., Christen, Y., Poirier, J., and Quirion, R. (2001). The ginkgo biloba extract (EGb 761) protects hippocampal neurons against cell death induced by β-amyloid. *European Journal of neuroscience,* **12**: 1882-1890.

Bayer, M., Deng, Z.W., Kubbutat, M.H.C., Waejen, W., Proksch, P., Lin, W.H., Shi, C., and Xu, M.H. (2010). Phenolic compounds and their anti-oxidative properties and protein kinase inhibition from the Chinese mangrove plant Laguncularia racemosa. *Phytochemistry,* **71**: 435–442

Behl, C. (1999). Alzheimer's disease and oxidative stress: implications for novel therapeutic approaches. *Progress in Neurobiology,* **57**: 301–323.

Bhattacharya, A., Ghosal, S., and Bhattacharya, S.K. (2001). Antioxidant effect of *Withania somnifera* glycowithanolides in chronic foot shock stress-induced perturbations of oxidative free radical scavenging enzymes and lipid peroxidation in rat frontal cortex and striatum. *Journal of Ethanopharmacology,* **74**: 1-6.

Bhattacharya, S.K., Bhattacharya, A., Kumar, A., and Ghosal, S. (2000b). Antioxidant activity of *Bacopa monniera* in rat frontal cortex, striatum, and hippocampus. *Phytotheraphy Research,* **14**: 174-179.

Birkmayer, J.G.D. (1996). Coenzyme nicotinamide adenine dinucleotide. New therapeutic approach to improving dementia of the Alzheimers type. *Annals of Clinical and Laboratory Science,* **26**: 1-9.

Bouchet, N., Barrier, L., and Fauconneau, B. (1998). Radical scavenging activity and antioxidant properties of tannins form *Guiera senegalensis* (*Combretaceae*). *Phytotheraphy Research,* **12**: 159-162.

Bruhlmann, C., Marston, A., Hostettmann, K., Carrupt, P.-A., and Testa, B. (2004). Screening of non-alkaloidal natural compounds as acetylcholinesterase inhibitors. *Chemical and Biodiversity,* **1**: 819–829.

Bush, A. I., Masters, C. L., and Tanzi, R. E. (2003). Copper, beta-amyloid, and Alzheimer's disease: tapping a sensitive connection. *Proceedings of the National Academy of Sciences U.S.A* **100**: 11193–11194.

Butterfield, D.A., Drake, J., Pocernich, C., and Castegna, A. (2001). Evidence of oxidative damage in Alzheimer's disease brain: central role for amyloid beta-peptide. *Trends in Molecular Medicine,* **7**: 548–554.

Chandini, S.K., Ganesan, P., and Bhaskar, N. (2008). In vitro antioxidant activities of three selected brown seaweeds of India. *Food Chemistry,* **107**: 707–713.

Chauhan, N., Wang, K.C., Wegiel, J., and Malik M.N. (2004). Walnut Extract Inhibits the Fibrillization of Amyloid Beta-Protein, and also Defibrillizes its Preformed Fibrils. *Current Alzheimer Research,* **1**: 183-188.

Chauhan, N.B. (2006). Effect of aged garlic extract on APP Processing and tau phosphorlyation in Alzheimer's transgenic model Tg2576. *Journal of Ethnopharmacology,* **108**: 385–394

Chauhan, N.B. (2006). Effect of aged garlic extract on APP processing and tau phosphorylation in Alzheimer's transgenic model Tg2576. *Journal of Ethnopharmacology,* **108**: 385-394.

Chauhan, V., and Chauhan, A. (2006). Oxidative stress in Alzheimer's disease. *Pathophysiology,* **13**: 195-208.

Choi J.S., Park H.J., Jung H.A., Chung H.Y., Jung J.H., and Choi W.C. (2000). A cyclohexanonyl bromophenol from the red alga *Symphyocladia latiuscula. Journal of Natural Products* **63**: 1705-1706.

Choi M.S., Lee S.H., Cho H.S., Kim Y., Yun Y.P., Jung HY, Jung J. K., Lee B.C., Pyo H.B., and Hong J.T. (2007) Inhibitory effect of obovatol on nitric oxide production and activation of NF-kappaB/MAP kinases in lipopolysaccharide- treated RAW 264.7cells. *European Journal of Pharmacology,* **556**: 181–189.

Choi Y.H., Yon G.H., Hong K.S., Yoo D.S., Choi C.W., Park W.K., Kong J.Y., Kim Y.S., and Ryu S.Y. (2008). In vitro BACE-1 Inhibitory Phenolic Components from the Seeds of *Psoralea corylifolia. Planta Medicine,* **74**: 1405-1408.

Choi, J.S., Park, H.J., Jung, H.A., Chung, H.Y., Jung, J.H. and Choi, W.C. (2000). A cyclohexanonyl bromophenol from the red alga *Symphyocladia latiuscula. Journal of Natural Products,* **63**: 1705–1706.

Choi, S.H., Hur, J.M., Yang, E.J., Jun, M., Park, H.J., Lee, K.B., Moon, E., and Soon, K.S. (2008). β-secretase (BACE1) inhibitors from *Perilla frutescens* var. *Acuta. Archives of Pharmacal Research* **31**: 183-187.

Choi, S.J., Kim, M.J., Heo, H.J., Hong, B., Chao, H.Y., Kim, Y.J., Kim, H.K., Lim, S.T., Jun, W.J., kim, E.K., and Shin, D.H. (2007). Ameliorating effect of *Gardenia jasminoides* extract on amyloid beta peptide induced neuronal cell deficit. *Molecules and cells,* **24:** 113-118.

Citron, M. (2004). Strategies for disease modification in Alzheimer's disease. *Nature Reviews Neuroscience,* **5**: 677-685.

Colciaghi, F., Borroni, B., Zimmermann, M., Bellone, C., Longhi, A., Padovani, A., Cattabeni, F., Christen, Y., and Lucaa, M.D. (2004). Amyloid precursor protein metabolism is regulated toward alpha-secretase pathway by *Ginkgo biloba* extracts. *Neurobiology of Disease,* **16**: 454– 460.

Cuvelier, M.E., Richard, H., and Berset, C. (1996). Antioxidative activity and phenolic composition of pilot-plant and commercial extracts of sage and rosemary. *Journal of American Oil Chemist Society,* **73**: 645-652.

Dai, Q., Borenstein, A.R., Wu, Y., Jackson, J.C., and Larson, E.B. (2006). Fruit and vegetable juices and Alzheimer's disease: The Kame project. *The American Journal of Medicine,* **119:** 751-759.

de Souza R.M.C., Marques, C.T., Dore, C.M.G., da Silva, F., Rocha O.H.A., and Leite, E.L. (2007). Antioxidant activities of sulfated polysaccharides from brown and red seaweeds. *Journal of Applied Phycology*, **19**: 153-160.

Dehpour, A.A., Ebrahimzadeh, M.A., Fazel, N.Z., and Mohammad N.S. (2009). Antioxidant activity of the methanol extracts of *Ferula assafoetida* and its essential oil composition *Grasastaceites* **60**: 405-412.

Devi, K.P., Suganthy, N., Kesika, P., and Pandian, S.K. (2008). Bioprotective properties of seaweeds: *In vitro* evaluation of antioxidant activity and antimicrobial activity against food borne bacteria in relation to polyphenolic content. *BMC Complemenentary and Alternative Medicine*, **8**: 38. Doi: 10.1186/1472-6882-8-38.

Dewachter, I., and van Leuven, F. (2002). Secretase as targets for the treatment of Alzheimer's disease: The prospects. *Lancet Neurology*, **1**: 409-416.

Dhanasekaran, M., Tharakan, B., and Manyam, B.V. (2008). Antiparkinson Drug – *Mucuna pruriens* shows Antioxidant and Metal Chelating Activity. *Phytotheraphy Research* **22**: 6–11.

Dovey, H.F., John, V., Anderson, J.P., Chen, L.Z., de Saint Andrieu, P., Fang, L.Y., Freedman, S.B., Folmer, B., Goldbach, E., Holsztynska, E.J., Hu, K.L., Johnson-Wood, K.L., Kennedy, S.L., Kholodenko, D., Knops, J.E., Latimer, L.H., Lee, M., Liao, Z., Lieberburg, I.M., Motter, R.N., Mutter, L.C., Nietz, J., Quinn, K.P., Sacchi, K.L., Seubert, P.A., Shopp, G.M., Thorsett, E.D., Tung, J.S., Wu, J., Yang, S., Yin, C.T., Schenk, D.B., May, P.C., Altstiel, L.D., Bender, M.H., Boggs, L.N., Britton, T.C., Clemens, J.C., Czilli, D.L., Dieckman-McGinty, D.K., Droste, J.J., Fuson, K.S., Gitter, B.D., Hyslop, P.A., Johnstone, E.M., Li, W.Y., Little, S.P., Mabry, T.E., Miller, F.D., and Audia, J.E. (2001). Functional gamma-secretase inhibitors reduce beta-amyloid peptide levels in brain. *Journal of Neurochemistry*, **76**: 173–181.

Du, G.H., Aiu, Y., and Zhang, J.T. (2000). Salvianolic acid B protects the memory functions against transient cerebral ischemia in mice. *Journal of Asian Natural Product Research*, **2**: 145-152.

Duan, X.J., Zhang, W.W., Li, X.M., and Wang, B.G. (2006). Evaluation of antioxidant property of extract and fractions obtained from a red alga, *Polysiphonia urceolata*. *Food Chemistry*, **95**: 37–43.

El Gamal, A.A. (2010). Biological importance of marine algae. *Saudi Pharmaceutical Journal*, **18**: 1–25.

Ertekin-Taner, N., Ronald, J., Feuk, L., Prince, J., Tucker, M., Younkin, L., Hella, M., Jain, S., Hackett, A., Scanlin, L., Kelly, J., Kihiko-Ehman, M., Neltner, M., Hershn, L., Kindy, M., Markeshery, W., Hutton, M., de Andrade, M., Petersen, R.C., Petersen, R.C., Graff-Radford, N., Estus, S., Brookes, A.J., and Younkin S.G. (2005). Elevated amyloid β protein (Aβ42) and late onset Alzheimer's disease are associated with single nucleotide polymorphisms in the urokinase-type plasminogen activator gene. *Human Molecular Genetics*, **14**: 447-460.

Faulkner, D.J. 2002. Marine natural products. *Natural Product Report,* **17**: 1-55.

Feng Y., Yang, S., Du, X., Zhang, X., Sun, X., Zhao, M., Sun, G., and Liu, R. (2009). Ellagic acid promotes Ab42 fibrillization and inhibits Ab42-induced neurotoxicity. *Biochemical and Biophysical Research Communications,* **390**: 1250–1254.

Feng, Y., Wang, X., Yang, S., Wang, Y., Zhang, X., Du, X., Sun, X., Zhao, M., Huang, L., and Liu, R. (2009). Resveratrol inhibits beta-amyloid oligomeric cytotoxicity but does not prevent oligomer formation. *NeuroToxicology,* **30**: 986–995.

Fisch, K.M., Bohm, V., Wright, A.D. and Konig G.M. (2003). Antioxidative monoterpenoids from the brown alga *Cystoseira crinite. Journal Natural Products,* **66**: 968–975.

Fukumoto, H., Cheung, B. S., Hyman, B. T. and Irizarry, M. C. (2002). Beta-secretase protein and activity are increased in the neocortex in Alzheimer's disease. *Archives of neurology,* **59**: 1381 -1389.

Fukuyama, Y., Otoshi, Y., Miyoshi, K., Nakamura, K., Kodama, M., and Nagasawa, M. (1992). Neurotrophic sesquiterpeneneolignans from *Magnolia obovata*: structure and neurotrophic activity. *Tetrahedron,* **48**: 377–392.

Ganesan, P., Chandini, S.K., and Bhaskar, N. (2008).Antioxidant properties of methanol extract and its solvent fractions obtained from selected Indian red seaweeds. *Bioresource Technology,* **99** : 2717–2723

Ganguli, M., Chandra, V., Kamboh, M.I., Johnston, J.M., Dodge, H.H., Thelma, B.K., Juyal, R.C., Pandav, R., Belle, S.H., and DeKosky, S.T. **(2000)** Apolipoprotein E polymorphism and Alzheimer's disease: the Indo-US Cross-National Dementia Study. *Archives of Neurology,* **57**: 824-830.

Ganguli, M., Chandra, V., Kamboh, M.I., Johnston, J.M., Dodge, H.H., Thelma, B.K., Juyal, R.C., Pandav, R., Belle, S.H., and DeKosky, S.T. (2000). Apolipoprotein E polymorphism and Alzheimer disease: the Indo-US Cross-National Dementia Study. *Archives of Neurology,* **57**: 824–830.

Gao, H., Wang, F., Lien, E.J., and Trousdale, M.D. (1996). Immunostimulating polysaccharides from *Panax notoginseng. Pharmacal Research,* **13**: 1196-1200.

Ghimeray, A.K., Jin, C.W., Ghimire, B.K., and Cho, D.H. (2009). Antioxidant activity and quantitative estimation of azadirachtin and nimbin in *Azadirachta Indica Azadirachta Juss* grown in foothills of Nepal. *African Journal of Biotechnology,* **8**: 3084-3091.

Ghosh, A.K., Hong, L., and Tang, J. (2002). Beta-secretase as a therapeutic target for inhibitor. *Current Medicinal Chemistry,* **9**: 1135–1144.

Guanhua, D., and Juntain, Z. (1997). Protective effects of salvianolic acid A against impairment of memory induced by cerebral ischemia-reperfusion in mice. *Chinese Medical Journal (Beijing),* **110**: 65-68.

Guo, Q., Zhao, B.L., Li, M.F., Shen, S.R., and Xin, W.J. (1996). Studies on protective mechanisms of four components of green tea Polyphenols (GTP) against lipid peroxidation in synaptosomes. *Biochimica Biophysica et Acta,* **1304**: 210–222.

Hardy, J. (1997). The Alzheimer's family of diseases: many etiologies, one pathogenesis. *Proceedings of the National Academy of Sciences U. S. A*, **94**: 2095-2097.

Hazra B., Biswas S., and Mandal N. (2008). Antioxidant and free radical scavenging activity of *Spondias pinnata*. *BMC Complementary and Alternative Medicine*, **8**: 63 doi: 10.1186/1472-6882-8-63.

Hong, G.Z., Arzu, E., Butterfield, D.A., and Mattson M.P. (2000). Beneficial Effects of Dietary Restriction on Cerebral Cortical Synaptic Terminals: Preservation of Glucose and Glutamate Transport and Mitochondrial Function after Exposure to Amyloid [beta]-Peptide, Iron, and 3-Nitropropionic Acid. *Clinical Neurochemistry and Disease*, **75**(1): 314-320.

Huang, X., Moir, R.D., Tanzi, R.E., Bush, A.I., and Rogers, J.T. (2004). Redoxactive metals, oxidative stress, and Alzheimer's disease pathology. *Annals of New York Academic Science*, **1012**: 153–163.

Huang, Y.S., and Zhang, J.T. (1992). Antioxidative effect of three water soluble components isolated from *Salvia miltiorrhiza in vitro*. *ACTA Pharmacologica Sinica*, **27**: 96-100.

Hudson, S.A., Ecroyd, H., Dehle, F.C., Musgrave, I.F., and Carver, J.A. (2009). (-)-Epigallocatechin-3-Gallate (EGCG) Maintains κ-Casein in Its Pre-Fibrillar State without Redirecting Its Aggregation Pathway. *Journal of Molecular Biology*, **392**: 689-700.

Hwang, E.M., Ryu, Y.B., Kim, H.Y., Kim, D.G., Hong, S.G., Lee, J.H., Curtis-Long, M.J., Jeon, S.H., Park, J.Y., and Park, K.H. (2008). BACE1 inhibitory effects of lavandulyl flavanones from *Sophora flavescens*. *Bioorganic and Medicinal Chemistry*, **16**: 6669–6674.

Inanami, O., Watanabe, Y., Syuto, B., Nakano, M., Tsuji, M., and Kuwabara, M. (1998). Oral administration of (-) catechin protects against ischemia-reperfusion-induced neuronal death in the gerbil. *Free Radical Research*, **29**: 359–365.

Iwamoto, C., Yamada, T., Ito, Y., Minoura, K., and Numata, A. (2001). Cytotoxic cytochalasans from a Penicillium species separated from a marine alga. *Tetrahedron*, **57**: 2997-3004.

Iwashima, M., Mori, J., Ting, X., Matsunaga, T., Hayashi, K., Shinoda, D., Saito, H., Sankawa, U., and Hayashi, T. (2005). Antioxidant and antiviral activities of plastoquinones from the brown alga *Sargassum micracanthum*, and a new chromene derivative converted from the plastoquinones. *Biological and Pharmaceutical Bulletin*, **28**: 374-377.

Jankowsky, J.L., Fadale, D.J., Anderson, J., Xu, G.M., Gonzales, V., Jenkins, N.A., Copeland, N.G., Lee, M.K., Younkin, L.H., Wagner, S.L., Younkin, S.G. and Borchelt, D.R. (2004). Mutant presenilins specifically elevate the levels of the 42 residue beta-amyloid peptide *in vivo*: evidence for augmentation of a 42-specific gamma secretase. *Human Molecular Genetics*, **13**: 159-170.

Jeon, S.Y., Bae, K., Seong, Y.H., and Song, K.S. (2003). Green tea catechins as a BACE1 (beta-secretase) inhibitor. *Bioorganic Medicinal Chemistry Letters,* **17**: 3905-3908.

Jeon, S.Y., Kwon, S.H., Seon, Y.H., Bae, K., Hur, J.M., Lee, Y.Y., Suh, D.Y., and Song, K.S. (2007). β-secretase (BACE1)-inhibiting stilbenoids from *Smilax Rhizoma. Phytomedicine,* **14**: 403–408.

Jia, H., Jiang, Y., Ruan, Y., Zhang, Y., Ma, X., Zhang, J., Beyreuther, K., Tu, P., and Zhang, D. (2004). Tenuigenin treatment decreases secretion of the Alzheimer's disease amyloid beta-protein in cultured cells. *Neuroscience Letters,* **367**: 123–128.

Jung, A.H., Ju, E.L., Kim, S., Kang, S.S., Lee, J.H., Min, B.S., and Choi, J.S. (2009). Cholinesterase and BACE1 inhibitory diterpenoids from *Aralia cordata. Archives of Pharmacal Research,* **32**: 1399-1408.

Jung, H.A., Oh, S.H., and Choi, J.S. (2010). Molecular docking studies of phlorotannins from *Eisenia bicyclis* with BACE1 inhibitory activity. *Bioorganic and Medicinal Chemistry Letters,* **20**: 3211–3215.

Kalaria, R.N., Maestre, G.E., Arizaga, R., Friedland, R.P., Galasko, D., Hall, K., Luchsinger, J.A., Ogunniyi, A., Perry, E.K., Potocnik, F., Prince, M., Stewart, R., Wimo, A., Zhang, Z., and Antuono, P. (2008). Alzheimer's disease and vascular dementia in developing countries: prevalence, management, and risk factors. *Lancet Neurology,* **7**: 812–826.

Kang, H.S., Chung, H.Y., Jung, J.H., Kang, S.S., and Choi, J.S. (1997). Antioxidant effect of *Salvia miltiorrhiza. Archieves of Pharmacal Research,* **20**: 495-500.

Kang, H.S., Kim, H.R., Byun, D.S., Son, B,W., Nam, T.J., and Choi, J.S. (2004). Tyrosinase Inhibitors Isolated from the Edible Brown Alga *Ecklonia stolonifera. Archives of Pharmacal Research,* **27**: 1226-1232.

Kang, T. H., Jeong, S. J., Ko, W. G., Kim, N. Y., Lee, B. H., Inagaki, M., Miyamoto, T., Higuchi R., and Kim Y. C. (2000). Cytotoxic lavandulyl flavanones from *Sophora flavescens. Journal of Natural Product,* **63**: 680-681.

Karawita, R., Siriwardhana, N., Lee, K.W., Heo, M.S., Yeo, I.K., Lee, Y.D., and Jeon, Y.J. (2005). Reactive oxygen species scavenging, metal chelation, reducing power and lipid peroxidation inhibition properties of different solvent fractions from *Hizikia fusiformis. European Food Research and Technology,* **220**: 363–371.

Karuppagounder, S.S., Pinto, J.T., Xu, H., Chen, H.L., Beal, M.F., and Gibson, G.E. (2009). Dietary supplementation with resveratrol reduces plaque pathology in a transgenic model of Alzheimer's disease. *Neurochemistry International,* **54**: 111–118.

Kenjiro, O., Hasegawa, K., Naiki H., and Yamada M. (2004). Anti-amyloidogenic activity of tannic acid and its activity to destabilize Alzheimer's β-amyloid fibrils in vitro. *Biochimica et Biophysica Acta,* **1690**: 193– 202.

Kim, H.J., Lee, K.W., and Lee, H.J. (2007). Protective effects of piceatannol against beta-amyloid-induced neuronal cell death. *Annals of New York Academy Science,* **1095**: 473–482.

Kumar, S., Ganesan, K., and Subba rao, P.V. (2008). Antioxidant potential of solvent extracts of *Kappaphycus alvarezii*: An edible seaweed. *Food Chemistry*, **107**: 289-295.

Kwak, H.M., Jeon, S.Y., Sohng, B.H., Kim, J.G., Lee, J.M., Lee, K.B., Jeong, H.H., Hur, J.M., Kang, Y.H., and Song, K.S. (2005). β-Secretase (BACE1)Inhibitors from Pomegranate (*Punica granatum*) Husk. *Archives of Pharmacal Research*, **28**: 1328-1332.

Lambert, M.P., Barlow, A.K., Chromy, B.A, Edward, C., Freed, R., Liosatos, M., Mogan, T.E., Rozovsky, I., Trommer, B., Viola, K.L., Wals, P., Zhang, C., Finch, C.E., Krafft, G.A. and Klein, W.L. (1998). Diffusible, non-fibrillar ligands derived from Abeta (42) are potent central nervous system neurotoxins. *Proceedings of National Academic Science U.S.A.*, **95**: 6448–6453.

Lee, C.L., Kwo, T.F., Wu, C.L., Wang, J.J., and Pan, T.M. (2010). Red mold rice promotes neuroprotective sAPPα secretion instead of Alzheimer's risk factors and amyloid beta expression in hyperlipidemic A beta40-infused rats. *Journal of Agricultural and food chemistry*, **58**: 2230-2238.

Lee, H.J., Seong, Y.H., Bae, K.H., Kwon, S.H., Kwak, H.M., Nho, S.K., Kim, K.A., Hur, J.M., Lee, K.B., Kang, Y.H., and Song, K.H. (2005). β -Secretase (BACE1) inhibitors from *Sanguisorbae Radix*. *Archives of Pharmacal Research*, **2**: 799-803.

Lee, J.W., Lee, Y.K., Lee, B.J., Nam, S.J., Lee, S.I., Kim, Y.H., Kim, K.H., Oh, K.W. and Hong, J.T. (2010). Inhibitory effect of ethanol extract of *Magnolia officinalis* and 4-O-methylhonokiol on memory impairment and neuronal toxicity induced by beta-amyloid. *Pharmacology Biochemistry and Behavior*, **95**: 31–40.

Levites, Y., Weinreb, O., Maor, G., Youdim, M.B., and Mandel, S. (2001). Green tea polyphenol (-)-epigallocatechin-3-gallate prevents N-methyl-4-phenyl- 1,2,3,6-tetrahydropyridine-induced dopaminergic neurodegeneration. *Journal of Neurochemistry*, **8**: 1073–1082.

Limpeanchob, N., Jaipan, S., Rattanakaruna, S., Phrompittayarat, W., and Ingkaninan, K. (2008). Neuroprotective effect of *Bacopa monnieri* on beta-amyloid-induced cell death. *Journal of Ethnopharmacology*, **120**: 112-117.

Liu, Q., Tao, Y., and Zhao, B.L. (2003). ESR study on scavenging effect of nicotine on free radicals. *Applied Magnetic Resonance*, **24**: 105–112.

Loo, A.Y., Jain, K., and Darah, I. Antioxidant activity of compounds isolated from the pyroligneous acid, *Rhizophora apiculata*. *Food Chemistry*, **107**: 1151-1160.

Lv, L., Yang, Q.Y, Zhao, Y., Yao, C.S., Sun, Y., Yang, E.J., Song, K.S., Mook-Jung, I., and Fang W.S. (2008). BACE1 (beta-secretase) inhibitory chromone glycosides from *Aloe vera* and *Aloe nobilis*. *Planta Medicine*, **74**: 540-545.

Magdalini, A.P, Charalambos, D. K., Moschos, G. P., Efthimiopoulos, S., Cordopatis, P., Margarity, M., and Lamari, F.N. (2006). Inhibitory Activity on Amyloid-β Aggregation and Antioxidant Properties of *Crocus sativus* Stigmas Extract and Its Crocin Constituents. *Journal of food and agricultural chemistry*, **54**: 8762–8768.

Mahley, R.W., Weisgraber, K.H., and Huang, T. (2006). Apolipoprotein E4: A causative factor and therapeutic target in neuropathology, including Alzheimer's disease. *Proceedings of National Academic Science U.S.A,* **103**: 5644–5651.

Markesbery, W. R. (1997). Oxidative stress hypothesis in Alzheimer's disease. *Free Radical Biology and Medicine,* **23**: 134-147.

Marumoto, S., and Miyazawa, M. (2010). β-secretase inhibitory effects of furanocoumarins from the root of *Angelica dahurica. Phytotherapy Research,* **24**: 510–513.

Maslow, K. (2010). Alzheimer's Association Report Alzheimer's disease facts and figures. *Alzheimer's and Dementia,* **6**: 158–194.

Matharu, B., Gibson, G., Parsons R., Huckerby T.N., Moore S.A., Cooper L.J., Millichamp, R., Allsop, D., and Austen, B. (2009). Galantamine inhibits β-amyloid aggregation and cytotoxicity. *Journal of the Neurological Sciences,* **280**: 49–58.

Mattson, M. P. (1997). Cellular actions of beta-amyloid precursor protein and its soluble and fibrillogenic derivatives. *Physiological Reviews,* **77**: 1081–1132.

Mattson, M.P. (2003). Gene-diet interactions in brain aging and neurodegenerative disorders. *Annals of Internal Medicine,* **139**: 441–444.

Mattson, P. (2004). Pathways towards and away from Alzheimer's disease. *Nature,* **430**: 631-639.

Mayeux, R. (2003). Epidemiology of neurodegeneration. *Annual Review of Neuroscience,* **26**: 81–104.

Mehmet, O., Aydogmus, F.O., Duru, M.E., and Topcu, G. (2007). Antioxidant activity of stem and root extracts of *Rhubarb (Rheum ribes)*: An edible medicinal plant. *Food Chemistry,* **103**: 623–630.

Mei, Z., Situ, B., Tana, X., Zheng, S., Zhang, F., Yan, P., and Liu, P. (2010). Cryptotanshinione upregulates α-secretase by activation PI3K pathway in cortical neurons. *Brain Research,* **1348**: 165-173.

Mori, C., Spooner, E.T., Wisniewski, K.E., Wisniewski, T.M., Yamaguchi, H., Saido, T.C., Tolan, D.R., Selkoe, D.J., and Lemere C.J. (2002). Intraneuronal Aβ42 accumulation in Down syndrome brain. *Amyloid,* **9**: 88 - 102.

Nahas, R., Abatis, D., Anagnostopoulou, M.A., Kefalas, P., Vagias, C., Roussis, V. (2007). Radical-scavenging activity of Aegean Sea marine algae. *Food Chemistry,* **102:** 577–581.

Nanjo, F., Goto, K., Seto, R., Suzuki, M., Sakai, M., Hara, Y. (1996). Scavenging effect of tea catechins and their derivatives on 1,1- diphenyl-2-picrylhdrazyl radical. *Free Radical Biology and Medicine,* **21**: 895– 902.

Nordberg, A. (2004). PET imaging of amyloid in Alzheimer's disease. *Lancet Neurology,* **3**: 519–527.

Nunan, J., and Small, D.II. (2000). Regulation of APP cleavage by alpha-, beta- and gamma-secretases. *FEBS Letters,* **483**: 6–10.

Ono, K., Hasegawa, K., Naiki, H., and Yamada M. (2004). Anti-amyloidogenic activity of tannic acid and its activity to destabilize Alzheimer's b-amyloid fibrils*in vitro*. *Biochimica et Biophysica Acta*, **1690**: 193– 202.

Ono, K., Yoshike, Y., Takashima, A., Hasegawa, K., Naiki, H., and Yamada, M. (2003). Potent anti-amyloidogenic and fibril-destabilizing effects of polyphenols *in vitro*: implications for the prevention and therapeutics of Alzheimer's disease. *Journal of Neurochemistry*, **87**: 172-181.

Onto, K., Margaret, M., Ho, C.L., Wang, J., Zhao, W., Pasinetti, G.M., and Teplow D.B. (2008). Effects of Grape Seed-derived Polyphenols on Amyloid β-Protein Self-assembly and Cytotoxicity. *The Journal of Biological Chemistry*, **283**: 32176-32187.

Owens, A.P., Nadin, A., Talbot, A.C., Clarke, E.E., Harrison, T., Lewis, H.D., Reilly, M., Wrigley, J.D., and Castro, J.L. (2003). High affinity, bioavailable 3-amino-1,4-benzodiazepine-based gamma-secretase inhibitors. *Bioorganic and Medicinal Chemistry Letters*, **17**: 4143-4145.

Papandreou, M.A., Kanakis, C.D., Polissiou, M.G., Efthimiopoulos, S., Cordopatis, P., Margarity, M., and Lamari, F.N. (2006). Inhibitory activity on amyloid-beta aggregation and antioxidant properties of *Crocus sativus* stigmas extract and its crocin constituents. *Journal of Agricultural and Food Chemistry*, **54**: 8762–8768.

Park, P.J., Shahidi, F., Jeon, Y.J. (2004). Antioxidant activities of enzymatic extracts from and edible seaweed *Sargassum horneri* using ESR spectroscopy. *Journal of Food Lipids*, **11**: 15–27.

Perry, N.S., Houghton, P.J., Sampson, J., Theobald, A.E., Hart, S., Lis-Balchin, M., Hoult, J.R., Evans, P., Jenner, P., Milligan, S., and Perry, E.K. (2001). *In-vitro* activity of *S. lavandulaefolia* (Spanish sage) relevant to treatment of Alzheimer's disease. *Journal of Pharmacy and Pharmacology*, **53**: 1347-56.

Porat, Y., Abramowitz, A., Gazit, E. (2006). Inhibition of amyloid fibril formation by polyphenols: structural similarity and aromatic interactions as a common inhibition mechanism. *Chemical Biology and Drug Design*, **67**: , 27-37.

Rahim, A.A., Rocca, E., Steinmetz, J., Kassim, M.J., Sani-Ibrahim, M., and Osman, H. (2008). Antioxidant activities of mangrove *Rhizophora apiculata* bark extracts. *Food Chemistry*, **107**: 200-207.

Rahim, A.A., Rocca, E.,Steinmetz, J.M., Kassim, J.M., Ibrahim, S., and Osman, H. (2008). Antioxidant activities of mangrove *Rhizophora apiculata* bark extracts. *Food Chemistry*, **107**: 200–207.

Rajakrishnan, V., Viswanathan, P., Rajasekharan, K.N., and Menon, V.P. (1999). Neuroprotective role of curcumin from Curcuma longa on ethanol-induced brain damage. *Phytotheraphy Research*, **13**: 571-574.

Ramesh, B.N., Indi, S.S., and Rao, K.S.J. (2010). Anti-amyloidogenic property of leaf aqueous extract of *Caesalpinia crista*. *Neuroscience Letters*, **475**: 110–114.

Rice-evans, C.A., Miller, N.J., and Paganga, G. (1996). Structure-antioxidant activity relationships of flavonoids and phenolic acids. *Free Radical Biology and Medicine*, **20**: 933-956.

Roome, T., Dar, A., Ali, S., Naqvi, S., and Choudhary, M.I. (2008). A study on antioxidant, free radical scavenging, anti-inflammatory and hepatoprotective actions of *Aegiceras corniculatum* (stem) extracts. *Journal of Ethnopharmacology*, **118**: 514–521.

Sayre, L.M., Perry, G., and Smith, M.A. (2008). Oxidative stress and neurotoxicity. *Chemical Research in Toxicology*, **21**: 172-188.

Schmidt, B. (2003). Aspartic proteases involved in Alzheimer's disease. *Chem Bio Chem.*, **4**: 366–378.

Selkoe, D.J. (2001). Alzheimer's disease results from the cerebral accumulation and cytotoxicity of amyloid beta protein. *Journal of Alzheimer's Disease*, **3**: 75–80.

Selkoe, D.J. (2001). Alzheimer's disease: genes, proteins, and therapy. *Physiological Reviews*, **81**: 741–766.

Senthil kumar, K.T.M., Gorain, B., Roy, D.K., Zothanpuia, S.S.K., Pal, M., Biswas, P., and Roy, A. (2008). Anti-inflammatory activity of *Acanthus ilicifolius*. *Journal of Ethanopharmacology*, **120**: 7-12.

Seo, Y., Park, K.E., and Nam Bull, T.J. (2007). Isolation of a new chromene from the brown alga *Sargassum thunbergii*. *Korean Chemical Society*, **28**: 1831–1833.

Seubert, P., Oltersdorf, T., Lee, M.G., Barbour, R., Blomquist, C., Davis, D.L., Bryant, K., Fritz, L.C., Galasko, D., Thal, L.J., Lieberburg, I., and Schenk D.B. (1993). Secretion of amyloid precursor protein cleaved at the amino terminus of the amyloid peptide. *Nature*, **361**: 260–263.

Shimmyo, Y., Kihara, T., Akaike, A., Niidome, T., and Sugimoto, H. (2008). Flavonols and flavones as BACE-1 inhibitors: Structure–activity relationship in cell-free, cell-based and in silico studies reveal novel pharmacophore features. *Biochimica et Biophysica Acta.*, **1780**: 819–825.

Sisodia S.S. (1999). Series Introduction: Alzheimer's disease: perspectives for the new millennium. The *Journal of Clinical Investigation*, **104**: 1169-1170.

Sleegers, K., and van Duijn, C.M. (2001). Alzheimer disease: Genes, pathogenesis and risk prediction. *Community Genetics*, **4**: 197-203.

Smith, D.G., Cappai, R., Barnham, K.J. (2007). The redox chemistry of the Alzheimer's disease amyloid beta peptide. *Biochimica Biophysica Acta*, **1768**: 1976–1990.

Smith, M.A., Harris, P.L.R., Sayre, L.M., Perry, G. (1997). Iron accumulation in Alzheimer disease is a source of redox-generated free radicals. *Proceedings of National Academy of Science USA*, **94**: 9866–9868.

Son, J. K., Park, J. S., Kim, J. A., Kim, Y., Chung, S. R., and Lee, S. H. (2003). Prenylated flavonoids from the roots of *Sophora flavescens* with tyrosinase inhibitory activity. *Planta Medicine*, **69**: 559-561.

Song, W.Z., Cui, J.F., and Zhang, G.D. (1989). Studies on the medicinal plants of Magnoliaceae tu-hou-po of Manglietia. *Yao Xue Xue Bao*, **24**: 295–299.

Soto, C. (2001). Protein misfolding and disease; protein refolding and therapy. *FEBS Letters*, **498**: 204–207.

Suganthy, N, Kesika, P, Pandian, S.K., and Devi, K.P. (2009). Mangrove plants extracts: Radical scavenging activity and the battle against food borne pathogens. *Forschende Komplementarmedizine*, **16**: 41-48.

Tang, M.K., and Zhang, J.T. (2001). Salvianolic acid B inhibits fibril formation and neurotoxicity of amyloid beta-protein *in vitro*. *Acta Pharmacologica Sinica*, **22**: 380–384.

Taniguchi, S., Suzuki, N., Masuda, M., Hisanaga S., Iwatsubo, T., Goedert, M., and Hasegawa, M. (2005). Inhibition of Heparin-induced Tau Filament Formation by Phenothiazines, Polyphenols and Porphyrins. *The Journal of Biological chemistry*, **280**: 7614-7623.

Tilley, L., Morgan, K., and Kalsheker, N. (1998). Genetic risk factors in Alzheimer's disease. *Molecular Pathology*, **51**: 293-304.

Tohda, C., Matsumoto, N., Zou, K., Meselhy, M.R., and Komatsu K. (2002). Axonal and dendritic extension by protopanaxadiol-type saponins from ginseng drugs in SK-N-SH cells. *Japanese Journal of Pharmacology*, **90**: 254-262.

Tomita, T., Maruyama, K., Saido, T.C., Kume, H., Shinozaki, K., Tokuhiro, S., Capell, A., Walter, J., Grunberg, J., Haass, C., Iwatsuboti, T., and Obata, K. (1997). The presenilin 2 mutation (N141I) linked to familial Alzheimer disease (Volga German families) increases the secretion of amyloid beta protein ending at the 42nd (or 43rd) residue. *Proceedings of National Academic Science U.S.A.*, **94**: 2025–2030.

Tsai, T.H., Westly, J., Lee, T.F., Chen, C.F., and Wang, L.C.H. (1995). Effects of honokiol and magnolol on acetylcholine release from rat hippocampal slices. *Planta Medicine*, **61**: 477–479.

Vassar, R., Bennett, B. D., Babu-Khan, S., Khan, S., Mendiaz, A., Denis, P., Teplow, D. B., Ross, S., Amarante, P., Loeloff, R., Luo, Y., Fisher, S., Fuller, J., Edenson, S., Lile, J., Jarosinski, M. A., Biere, A. L., Curran, E., Burgess, T., Louis, J. C., Collins, F., Treanor, J., Rogers, G., and Citron M. (1999). Beta secretase cleavage of Alzheimer's amyloid precursor protein by the transmembrane aspartic protease BACE1. *Science*, **286**: 735-741.

Wang, C.-N., Chi, C.-W., Lin, Y.-L., Chen, C.-F., and Shiao, Y.-J. (2001). The neuroprotective effects of phytoestrogens on amyloid h protein-induced toxicity are mediated by abrogating the activation of caspase cascade in rat corticalneurones. *Journal of Biological chemistry*, **276**: 5287 –5295.

Wang, J., Ho, L., Zhao, W., Ono, K., Rosensweig, C., Chen, L., Humala, N., Teplow, D.B., and Pasinetti, G.M. (2008). Grape-derived polyphenolics prevent Abeta oligomerization and attenuate cognitive deterioration in a mouse model of Alzheimer's disease. *Journal of Neuroscience*, **28**: 6388–6392.

Wang, J.Y., Wen, L.L., Huang, Y.N., Chen, Y.T., Ku, M.C. (2006). Dual Effects of Antioxidants in Neurodegeneration: Direct Neuroprotection against Oxidative

Stress and Indirect Protection via Suppression of Gliamediated Inflammation. *Current Pharmaceutical design,* **12:** 3521-3533.

Wang, M., Kikizaki, H., Zhu, N., Sang, S., Nakatani, N., and Ho, C.T. (2000). Isolation and structural elucidation of two new glycosides from sage (*Salvia officinalis*). *Journal of Agricultural Food Chemistry*, **48:** 235-238.

Wang, Y.H., and Du, G.H. (2009). Ginsenoside Rg1 inhibits beta- secretase activity in vitro and protects Aβ induced cytotoxicity in PC12 cells. *Asian Natural Product Research,* **11:** 604-612.

Weggen, S., Eriksen, J.L., Das, P., Sagi, S.A., Wang, R., Pietrzik, C.U., Findlay, K.A., Smith, T.E., Bulter, M.T., Kang, D.E., Sterling, N.M., Golde, T.E., and Koo, E.H. (2001). A subset of NSAIDs lower amyloidogenic Aβ42 independently of cyclooxygenase activity. *Nature,* **414:** 212–216.

Yassin, M.S., Ekblom, J., Xilinas, M., *Gottfries,* C.G., and Oreland, L. (2000). Changes in uptake of vitamin B (12) and trace metals in brains of mice treated with clioquinol. *Journal of Neurological Science,* **173:** 40–44.

Yoshikawa, M., Morikawa, T., Kashima, Y., Ninomiya, K., and Matsuda H. (2003). Structures of new dammarane-type Triterpene Saponins from the flower buds of *Panax notoginseng* and hepatoprotective effects of principal Ginseng Saponins. *Journal of Natural Products,* **66:** 922-927.

Youdim, K.A., and Deans, S.G. (1999). Dietary supplementation of thyme (*Thymus vulgaris L.*) essential oil during the lifetime of the rat: its effects on the antioxidant status in liver, kidney and heart tissues. *Mechanisms of Ageing and Development,* **109:** 163-175.

Zhang, J., Liu, Q., Chen, Q., Liu, N.Q., Li, F.L., Lu, Z.B., Qin, C., Zhu, H., Ying, H.Y., He, W., and Lu, A.B. (2006). Nicotine attenuates the β-amyloid neurotoxicity through regulating metal homeostasis. *FASEB Journal,* **20:** E399-E408.

Zhang, Z., Zhang, Q., Wang, J., Song, H., Zhang, H., and Niu, X. (2010). Regioselective syntheses of sulfated porphyrans from *Porphyra haitanensis* and their antioxidant and anticoagulant activities *in vitro. Carbohydrate polymers,* **79:** 1124-1129.

Zhao, B.-L., Li, X.-J., He, R.-G., Cheng, S.-J., and Xin, W.-J. (1989). Scavenging effect of extracts of green tea and natural antioxidants on active oxygen radicals. *Cell Biophysics,* **14:** 175–181.

Zhao, J., Paganini, L., Mucke, L., Gordon, M., Refolo, L., Carman, M., Sinha, S., Oltersdorf, T., Lieberburg, I., and McConglogue, L. (1996). Betasecretase processing of the beta amyloid precursor protein in transgenic mice is efficient in neurons but inefficient in astrocytes. *The Journal of Biological Chemistry,* **271:** 31407-31411.

Zhou, J., Hu, N., Wu, Y., Pan, Y.-J., and Sun, C.-R. (2008). Preliminary studies on the chemical characterization and antioxidant properties of acidic polysaccharides from *Sargassum fusiforme. Journal Zhejiang University Science B,* **9:** 721-727.

Natural Products: Research Reviews Vol. 1 (2012)   *Pages* **409–437**
Editor: **V.K. Gupta**
Published by: **DAYA PUBLISHING HOUSE, NEW DELHI**

9

An Overview of *Acmella oleracea*: A Multipurpose Species

Ana Claudia F. Amaral[1]*, José Luiz P. Ferreira[1,2],
Aline de S. Ramos[1], Deborah Q. Falcão[2], Cristina B. Viana[1],
Jane S. Inada[1], Sílvia L. Basso[3,4] and Jefferson Rocha de A. Silva[4,5]

ABSTRACT

Acmella is a genus of Asteraceae family with about 30 annual or perennial species and 9 additional infraspecific taxa. Acmella oleracea Murr. is popularly known in Brazil as "jambú", among other common names, and is a typical herb from north region, where it is widely used in the local cuisine. It is also grown widely in the USA, Northern and Southern America, Haiti, India and South Africa. A. oleracea is used in folk medicine for toothache, throat and mouth complaints, rheumatism and fever. In industry, extracts, fractions and isolated compounds have applications in oral preparations, flavor and refresher. In addition, cosmetics compositions have active components of this species to treat wrinkles, fine lines, laxity and

1 Laboratório de Plantas Medicinais e Derivados, Depto de Produtos Naturais, Farmanguinhos – FIOCRUZ – Rua Sizenando Nabuco, 100 – Manguinhos/RJ, 21041-250, Brazil.

2 Faculdade de Farmácia – UFF- R Mario Viana, 523-Niterói, 24241-000, RJ, Brazil.

3 Laboratório de Produtos Naturais, FUNTAC, Av. das Acácias, Lote 01, zona A, Distrito Industrial, Rio Branco, AC, 69917-100, Brazil.

4 Programa de Pós-graduação em Biotecnologia- PPGBIOTEC-UFAM – Setor Sul/Av. Rodrigo Otávio, 3000 - Japiim - Manaus/AM, 69077-000, Brazil.

5 Laboratório de Cromatografia – Depto. de Química - UFAM – Setor Sul/Av. Rodrigo Otávio, 3000 - Japiim - Manaus/AM, 69077-000, Brazil.

* *Corresponding author*: E-mail: acamaral@fiocruz.br; jrocha_01@yahoo.com.br; Phone: +552139772575

mottled pigmentation. These properties have been associated to the principal alkamide called spilanthol found in whole plant. Other alkamides, amines, terpenoids, amino acids, phenolics and carboxilic acids have been isolated from this species. Some pharmacological studies with this species have shown activities such as: anti-inflammatory, antiprotozoal, antifungical, analgesical and antibacterial. This work presents an overview of this important species.

Keywords: Asteraceae, *Acmella oleracea*, Botany, Alkamides, Pharmacology, Patents.

Introduction and Ethnobotanical Considerations

This work is a review about *Acmella oleracea* Murr. (Asteraceae family), species whose versatility and importance can be observed by the various uses in folk medicine, commerce and cuisine. This species is popularly known as "jambú, agrião-do-norte, agrião-do-pará, jambuassu, jabuaçú, erva maluca, jaburama, botão de ouro" and other common names, and is a typical herb from north of Brazil, where it is used in the local cuisine (Coutinho *et al.*, 2006). People make a famous beverage named "tucupi" with a thick milky liquid from the paste of *Manihot esculenta* Crantz ("mandioca") and leaves of *A. oleracea*. This liquid is used as spice in famous food as "Pato no tucupi" and "Tacacá" and to cure symptoms of the ethanol hangover (Carlini *et al.*, 2006). Other countries also cultivate this species and it is also known for many common names, for example in Thai as "phak-kratt huawaen", India as "akarkara" or toothache plant. It has a long history of use in folk medicine, *e.g.* for toothache, throat and mouth complaints, rheumatism and fever (Molina-Torres *et al.*, 1999). *A. oleracea* is one of the ingredients of "Malarial", an improved traditional medicine of Mali used for the treatment of malaria (Wu *et al.*, 2008). In many tribes of India, the leaves are also eaten raw or as a vegetable and the entire plant is local use against snakebite and rheumatic fever (Sandberg *et al.*, 2005). In Ayurvedic medicine, flower heads and roots are used in treatment of scabies, psoriasis, scurvy, toothache, infections of gums and throat, paralysis of tongue and remedy for stammering in children (Rani and Murty, 2006). This species has found applications in pharmaceuticals as oral preparations, flavor or refresher. In addition, the extract, fractions and alkamides are active components of cosmetics to skin and body (Nigrinis *et al.*, 1986; Prachayasittikul *et al.*, 2009). These properties have been associated to the principal alkamide called spilanthol or affinin (N-isobutyl- 2E,6Z,8E-decatrienamide) described at the first time by Geber in 1903 (Martin and Becker, 1984; Ramsewak*et al.*, 1999). It is known for the pungent taste, local anesthetic and for causing itching and salivating. The medicinal and commercial values of this plant have stimulated the market demand in several countries (Rani and Murty, 2006). Despite its long and wide application as folk medicine and other areas, the taxonomy of *A. oleracea* and its congeners has been disorderly. In all works published about of this species, the authors used several botanic synonyms and classification. In this review, based on current botanical classification, we used *Acmella oleracea* as unique scientific name. This works integrates information on botany, chemical and pharmacology of *A. oleracea*.

Geographic Distribution

Asteraceae is the second largest family of Angiospermae, comprising about 1,535 genera with over 20,000 species that occur predominantly in American mainland, particularly in Latin America. In Brazil, about 4,000 species are spread over arid, semiarid and mountain regions (Hind, 1993).

Acmella oleracea is largely grown for ornamental and medicinal purposes and spread mainly in the tropical and subtropical regions throughout the world. Not apparently known from the wild, although where naturalized it occurs in moist places on lake margins. Summarizing the various accounts of the species, and judging by a look at herbarium material, it is grown widely in the USA, Northern and Southern America, Haiti, India, South Africa, and appears to have become naturalized in Eastern Africa. Although Jansen (1985a,b) suggested the probable wild origin for the species is Peru or Brazil, more fieldwork is needed to confirm this (Hind and Biggs, 2003).

In Brazil *A. oleracea* is well-known all over the Amazon region, mainly in the Northern region, being a species that grows well in warm and moist climate by means of seeds propagation or cuttings. In Brazil, thanks to its large importance, this species can be found in the wild or cultivated.

Botanic Considerations

Taxonomic Classification

Acmella is a pantropical genus with about 30 annual or perennial species and 9 additional infraspecific taxa. Many species have become naturalized and invasive. Based on the cladistic analysis of morphological and cytological characters of *Spilanthes* Jacq. (sensu Moore, 1907), Jansen (1981; 1985a; 1985b) recircumscribed the genus and restored the generic status of *Acmella*, which had long been subsumed to a section under *Spilanthes* by earlier taxonomists. Additionally, he suggested that *Acmella oleracea* was mostly likely derived through cultivation from *Acmella alba* (L'Her.) R.K. Jansen in central Peru (Chung *et al.*, 2008).

Table 9.1: Differential characteristics between the genera *Acmella* and *Spilanthes* (Jansen, 1981; Bringel, 2007).

Characteristics	Acmella	Spilanthes
Capitula	Radiate or discoid	Discoid
Shape of corolla	Fauce smoothly widened after tube	Usually fauce is abruptly widened after tube
Transversal section of the cypselae	The outer ones are 3-angular, with 3 vascular bundles and the inner ones are 2- angular with 2 vascular bundles.	Usually 4-angular, rhombic, 2, 4 or 8 vascular bundles
Pappus	1-several bristles	1, 2(3) robust appendages
Chromosomic number	n=16	n=12, 13, 24, 26 ou 39

Kingdon: Plantae

(Unranked): Angiosperm

(Unranked): Eudicots

(Unranked): Asterids

Order: Asterales

Family: Asteraceae

Tribe: Heliantheae

Genus: Acmella

Species: *Acmella oleracea* (L.) R.K. Jansen

(Panero *et al.*, 1999)

(Wikipedia, the free encyclopedia, http://en.wikipedia.org/wiki/ Acmella_oleracea, 2010)

Botanical Synonymy

It is obvious that this plant is potentially masquerading under many names upon trade and certainly on www. Some sites appear authoritative, which gives rise to an inherent problem as people are used to rely upon the information provided.

Several sites provide the "accepted" "scientific name" as *Spilanthes acmella* (L.) Murr., also listing several synonyms, mostly referring to *species* in other genera. However, without exception, all those sites that illustrate the plant clearly show *Acmella oleracea* (Hind and Biggs, 2003).

Bidens acmella (L.) Lam.; *Bidens ocymifolia* Lam.; *Pyrethrum acmella* (L.) Medik.; *Spilanthes acmella* (L.) Murr.; *Spilanthes ocymifolia* (Lam.) A.H. Moore; *Verbesina acmella* L. (Hind and Biggs, 2003); *Bidens fervida* Lam.; *Bidens fusca* Lam. *Cotula pyretharia* L.; *Isocarpha pyrethraria* (L.) Cass.; *Pyrethrum spilanthus* Medik.; *Spilanthes acmella* var. *oleracea* (L.) C.B. Clarke ex Hook. f.; *Spilanthes fusca* Lam.; *Spilanthes oleraceae* L.; *Spilanthes oleracea* var, *fusca* (Lam.) DC.; *Spilanthes radicans* Schrad. ex DC. (PIER http:// www.hear.org/Pier/species/acmella_oleracea.htm, 2007; Lorenzi and Matos, 2008; Tropicos http://mobot.mobot.org/W3T/Search/vast.html, 2010); *Spilanthes arrayana* Gardn.; *S. melampodioides* Gardn.; *S. pseudo-acmella* (L.) Murr.; *Acmella linnaei* Cass.; *Spilanthes acmella* (L.) Murr. var. *typica*; *uliginosa* Sw. and *oleracea* L. (Kissmann and Groth, 1992).

Morphological Description

Prostate to ascending annual herbs (apparently up to 60 or 90 cm tall), or sometimes short-lived perennial herbs (if protected and kept warm during the winter months). Stems not usually rooting at nodes, usually decumbent to ascending, green to red, glabrous. Leaves opposite, petiolate, petiole 20-64 mm long, flattened, grooved on upper surface, narrowly winged, sparsely pilose; lamina broadly ovate to deltoid, 53-106 mm long, 40-79 mm wide, base truncate, attenuate in upper leaves, hairs sparse on both surfaces, mainly on upper midrib, hairs eglandular, uniseriate, base multicellular, slightly swollen, brown, tip unicellular, long, slender, white, lamina margin dentate, apex short acuminate to acute. Inflorescences of solitary, terminal and axillary pedunculate, capitula; peduncles 3.5-12.5 mm long, ebracteolate, hollow, glabrous to sparsely pilose, hairs eglandular; capitula 10.5-23.5 mm high, 11-17 mm diameter, pedunculate, homogeneous, discoid, florets yellow with distinct purple to red paleae visible in immature capitulum; involucre shallowly campanulate, 3-7 mm high, 9-15 mm diameter; phyllaries triseriate, imbricate, green, lanceolate, apices purple to red, margins entire, ciliate, apices acute, outer phyllaries 5-6, 5.8-7.3 mm long, middle phyllaries 5-6, 5.8-7.3 mm long, inner phyllaries 5-6, 5.5-6.5 mm long; receptacle oblanceolate or conical, white, spongy, 8.3-21.5 mm high, 3.5-8.5 mm diameter, paleaceous; paleae 5.3-6.2 mm long, 1-1.2 mm wide, white, top 0.5 mm purple to red, glabrous, except at tip, hairs translucent, uniseriate, short, paleae base with right-angled narrow keel, apex acute. Florets hermaphrodite, numerous (400-620), fertile; corolla-tube 2.7-3.3 mm long, green, glabrous, constricted into a tube at base; tube 0.5-0.7 mm long, 0.2-0.4 mm diameter throat inflated, 2.2-2.6 mm long, 0.5-1.0 mm diameter; corolla-lobes (4)-5, 0.5-0.6 mm long, yellow, papillose inside; anther cylinder included within corolla throat; filaments white, attached to base of corolla throat, flattened, lacking obvious anther collar; anthers 5; apical anther appendages triangular, with thickened apices, wider than long; basal anther appendages short, triangular, entire; pollen bright orange fading to pale yellow; style base with distinct nodes glabrous; style shaft glabrous; style arms frequently 3 in this material (other herbarium material of cultivated specimens at Kew have 2 style arms), apices truncate, papillose. Achenes 2.0-2.5 mm long, 0.9-1.1 mm wide, black, compressed between two marginal ribs, ribs ciliate along entire length, setuliferous on faces, setulae of "twin-hairs" with excentric, scarcely divided apices; carpopodium narrowly ovate, thickened portion drying amber with an elongated, narrow, white "tail"; pappus persistent, of 2 unequal bristles, finely and inconspicuously barbellate, longer 0.5-1.5 mm long, shorter 0.3-1.3 mm long, tips straight (Hind and Biggs, 2003; Chung et al., 2008).

Under a microscope, the mesophyll in transversal section showed 2 to 3 layers of palisade parenchyma and 5 to 6 of sponge parenchyma composed of round cells. The epidermis tissue, with predominant diacytic stomata is unistratified on both sides with smaller cells alongside the abaxial surface, except for the vascular bundles region, where they are larger. A sheath of parenchymatic nature enwraps the bundles entirely and elongates till reaching the abaxial epidermis, which from this moment on is slightly bent. The biconvex midrib presents an epidermic sheath with pluricellular tector trichomes, usually curved and glandular trichomes of two types: with pluricellular or subsessile pedicle, both supporting a glandular unicellular head. Inside there is an angular collenchyma, fundamental parenchyma and a vascular bicollateral system. The concave-convex petiole presents two lateral wings and is encased by an epidermal tissue with trichomes, angular collenchyma and several vascular bundles arranged in semi-circles, immersed in fundamental parenchyma and protected by an endodermis and a parenchymatic pericycle (Ferreira et al., 2007).

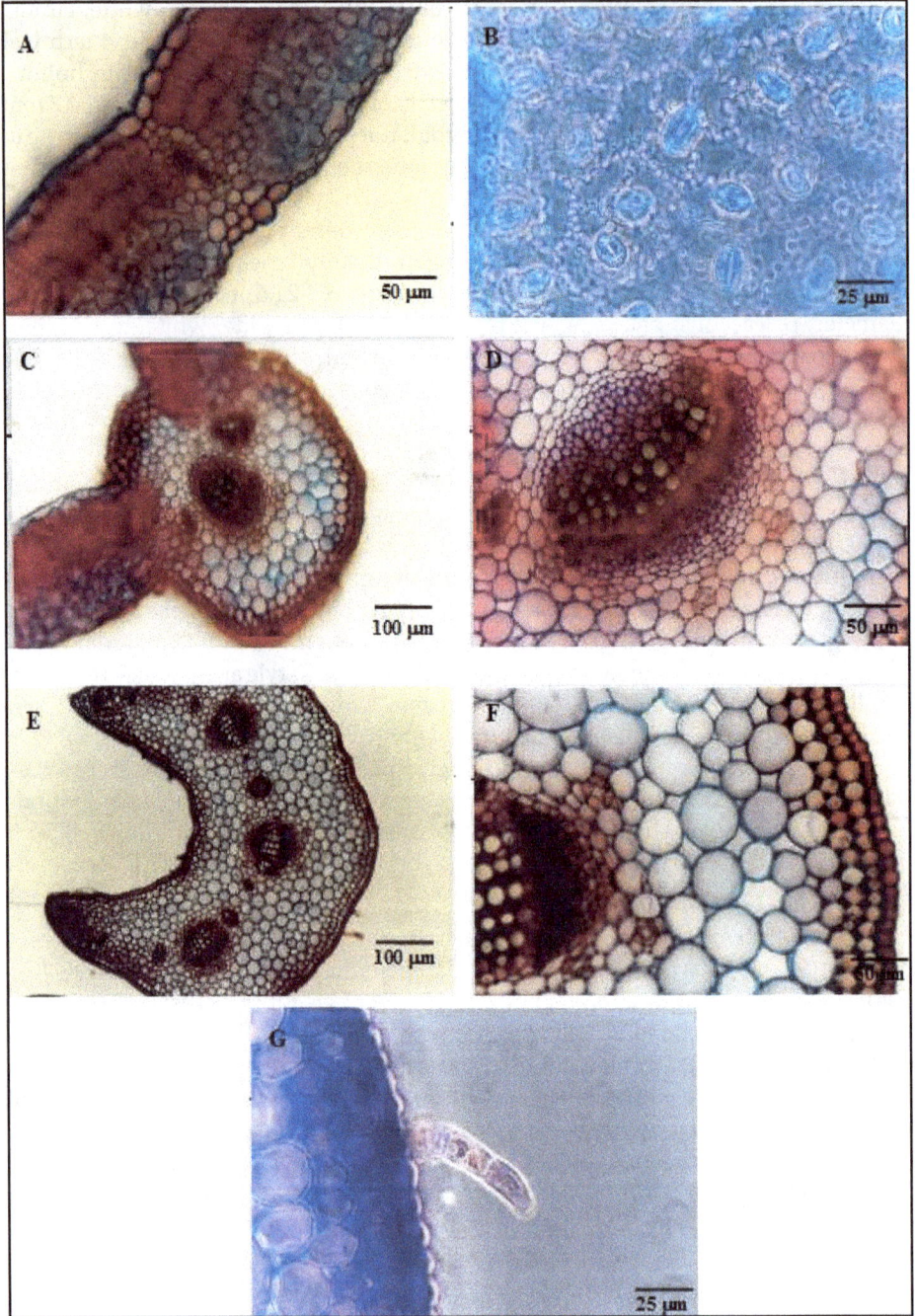

Figure 9.1: Leaf transversal section of *A. oleracea* (A; C-G); paradermic section (B); A: dorsiventral mesophyll. B: diacytic stomata of the abaxial epidermis of lamina leaf. C: biconvex midrib. D: vascular bundles in the midrib region. E: concave convex petiole. F: angular collenchyma adjacent to abaxial region in the petiole. G: multicellular tector trichomes of the petiole.

Chemical Constituents

The compounds isolated from *A. oleracea* are shown in the figures below (Greger *et al.*, 1985; Lemos *et al.*, 1991; Nagashima and Nakatani, 1992; Nakatani and Nagashima, 1992; Barreto *et al.*, 1993; Mondal *et al.*, 1998; Ley *et al.*, 2006a,b).

Biological Considerations

Extracts and Fractions

Human Leukocyte Elastase Inhibitory Activity

The oleoresin from *A. oleracea* was evaluated by the capacity to inhibit *in vitro* the enzyme human leukocyte elastase (HLE). The inhibitory property was performed, briefly, by measuring the HLE activity using N-(methoxysuccinyl)-Ala-Ala-Pro-Val 4-nitroanilide as substrate. Enzyme was preincubated for 5 min, in the presence of the oleoresin or vehicle (DMSO) for control. The hydrolysis was started by the addition of substrate in DMSO and release of *p*-nitroanilide was monitored by change in absorbance at 405 nm. Oleoresin from *A. oleracea* showed no HLE inhibitory activity and it was considered inactive above 20 ppm (Baylac and Racine, 2004).

Pancreatic Lipase Inhibitory Activity

The ethanol (70 per cent) extract of dried flower buds of *A. oleracea* presented pancreatic lipase inhibitory activity in a concentration related manner under *in vitro* conditions. Human pancreatic lipase (PL) and the extract in different concentrations (0.75, 1.0 and 2.0 mg/mL) were incubated for 5 min at 37 °C. Activator-reagent solution was added and incubated again for 3 min at 37 °C. The recorded rate of increase in absorbance at 550 nm due to the formation of a quinone diimine dye was used to determine the PL activity in the extract. Lipase inhibitory activity of 40 per cent was observed after treatment with 2.0 mg/mL of *A. oleracea* ethanolic extract (Ekanem *et al.*, 2007).

Antifungal Activity

Effect of different concentrations (0.1, 0.4. 1.0.1.6, and 2.0 mg.) of *A. oleracea* dried flower heads petroleum ether extract was tested against four different fungi. The samples were dissolved in dimethyl sulfoxide (DMSO). The maximum zone of inhibition was found at 2.0 mg concentration and the high inhibition zones were observed in *Fusarium oxysporium* (2.3 cm) and *F. moniliformis* (2.1 cm), followed by *Aspergillus niger* (2.0) and *A. paraciticus* (1.8 cm) (Rani and Murty, 2006).

Anti-inflammatory Activity

Hexane, chloroform, ethyl acetate and butanol extracts from dried flowers of *A. oleracea* were bioassayed for the inhibitory effect on nitric oxide (NO) production in LPS-stimulated macrophages, RAW 264.7 cells. The effect of extracts was determined by using the Griess reaction to measure the level of nitrite at 550 nm. Concentrations were calculated against a sodium nitrite standard curve. LPS-induced (1 µg/mL) RAW 264.7 cells with or without extracts (80 µg/mL) were incubated for 24 h to study the inhibition of NO production by the extracts. The hexane and chloroform extracts strongly suppressed the production of NO to 28 and 15 per cent, respectively, whereas

Amides

Spilanthol

N-propyl-acetamide

Homospilanthol

Undeca-2E,7Z,9E-trienoic acid isobutylamide

(2Z)-N-isobutyl-2-nonene-6,8-diynamide

N-phenethyl-2,3-epoxy-6,8-nonadiynamide

(2E,4Z)-N-isobutyl-2,4-undecadiene-8,10-diynamide

(2E)-N-isobutyl-2-undecene-8,10-diynamide

(2E)-N-(2-methylbutyl)-2-undecene-8,10-diynamide

(2E, 7Z)-N-isobutyl-2,7-tridecadiene-10,12-diynamide

(2E, 7Z)-N-isobutyl-2,7-decadienamide

(2E,6Z,8E)-N-(2-methylbutyl)-2,6,8-decatrienamide

(2E,6Z,8E,10Z)-N-isobutyl-dodeca-2,4,8,10-tetraenamide

Amines

2-phenylethylamine

isobutylamine

2-methylbutylamine

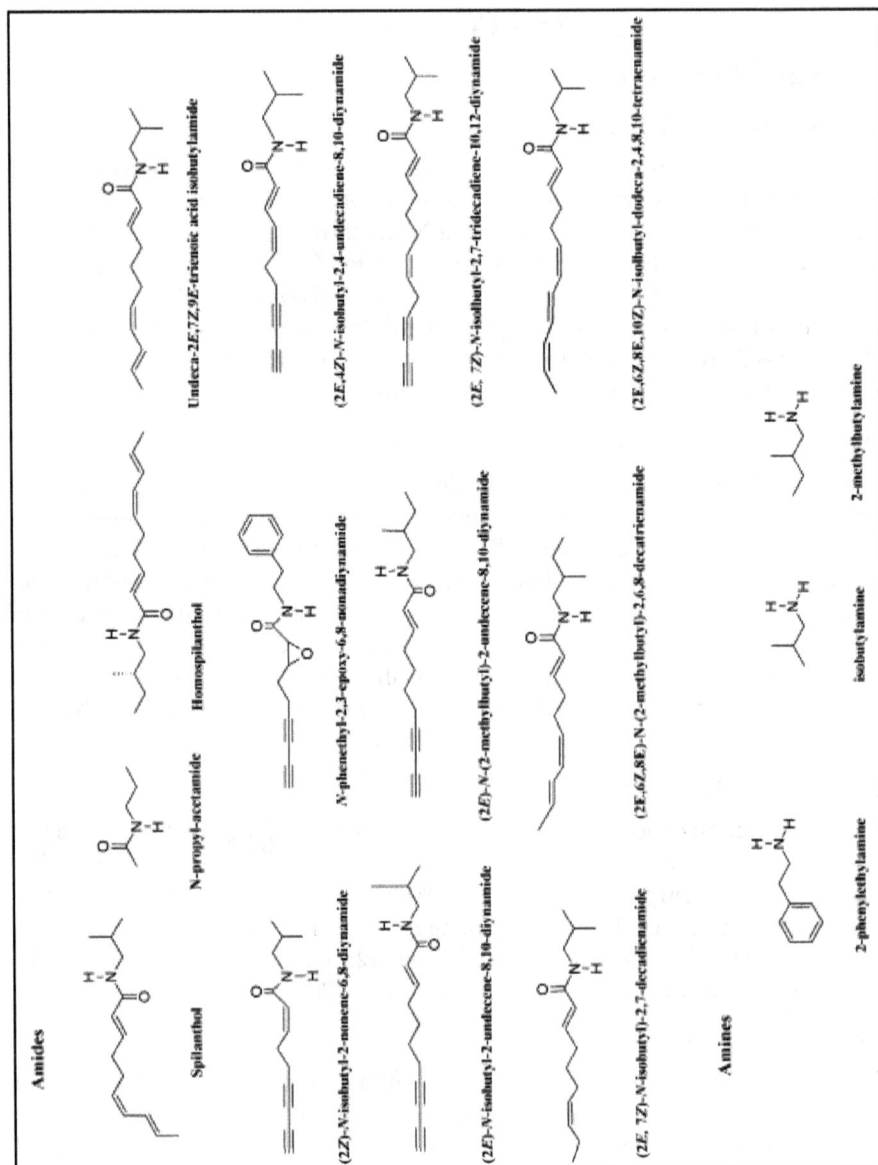

Figure 9.2: Amides and amines.

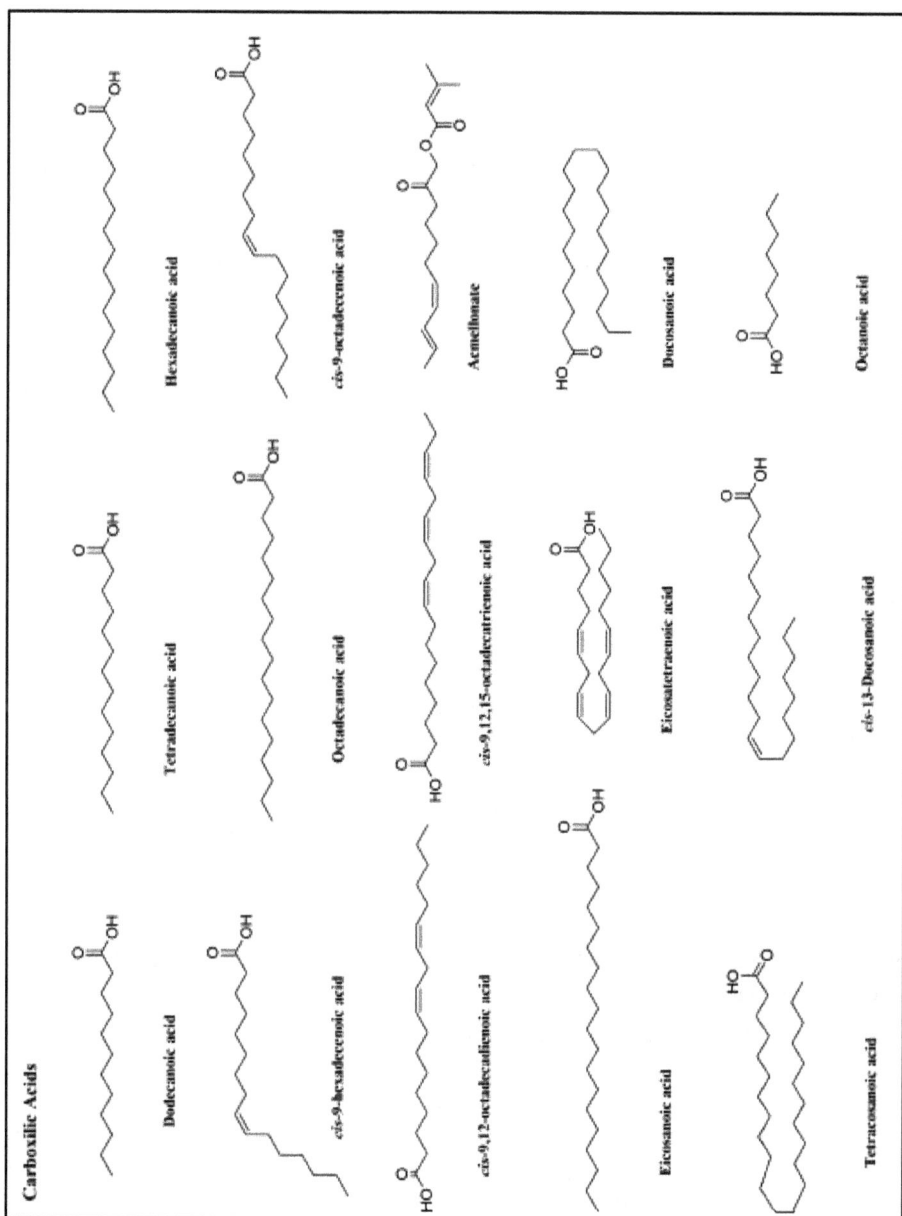

Figure 9.3: Carboxilic acids.

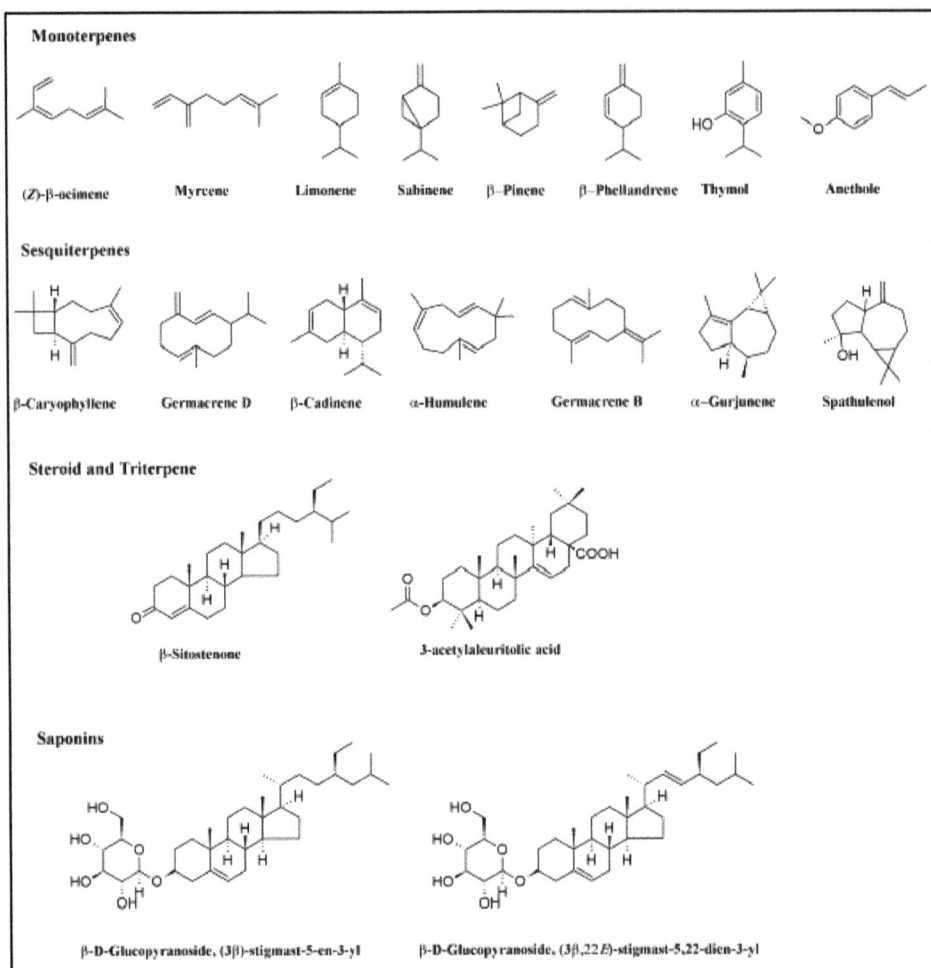

Figure 9.4: Monoterpenes, sesquiterpenes, triterpene, steroid and saponins.

ethyl acetate and butanol extracts reduced less NO production to 64 and 77 per cent, respectively (Wu *et al.*, 2008).

Aqueous extract from fresh aerial parts of *A. oleracea* was evaluated *in vivo* for anti-inflammatory action by carrageenan-induced rat paw edema. Acute inflammation was produced by subplantar injection of 1 per cent suspension of carrageenan with 2 per cent gum acacia, in the right hind paw of the rats, one hour after oral administration of the drugs. The paw volume was measured at '0' and '3' hours after the carrageenan injection. The difference between the two readings was taken as the volume of edema and the percentage anti-inflammatory activity was calculated. Aspirin (100 mg/kg, p.o.) suspended in 2 per cent gum acacia was used as the standard drug. The aqueous extract of *A. oleracea* was administered as a suspension in 2 per cent gum acacia to the animals in doses of 100, 200 and 400 mg/kg. In the

Phenolics

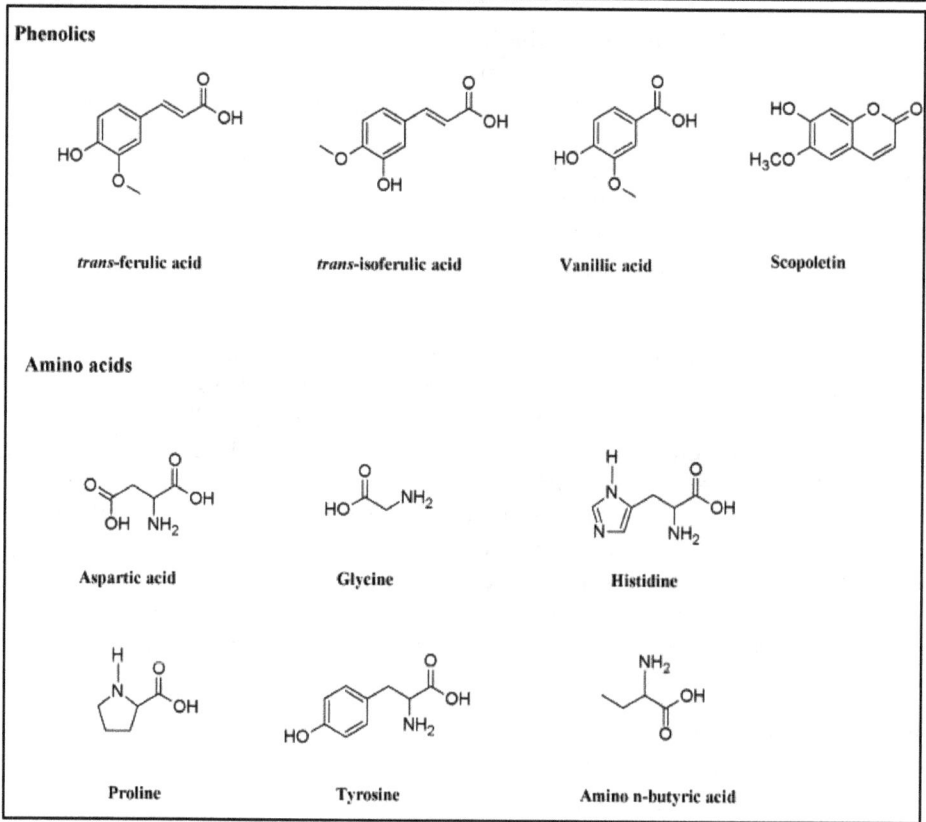

| trans-ferulic acid | trans-isoferulic acid | Vanillic acid | Scopoletin |

Amino acids

| Aspartic acid | Glycine | Histidine |

| Proline | Tyrosine | Amino n-butyric acid |

Figure 9.5: Phenolics and amino acids.

acute inflammation model, the extract produced dose-dependent inhibition of paw edema, showing 52.6, 54.4 and 56.1 per cent inhibition of paw edema, respectively, at the end of three hours (Chakraborty *et al.*, 2004).

Analgesic Activity

Aqueous extract from fresh aerial parts of *A. oleracea* was evaluated *in vivo* for analgesic activity using acetic acid-induced writhing test in albino mice and tail flick response in albino rats. In both methods aqueous extract of *A. oleracea* was administered as a suspension in 2 per cent gum acacia in doses of 100, 200 and 400 mg/kg. Acetic acid-induced writhing test was performed using Aspirin in doses of 50, 100 and 150 mg/kg, suspended in 2 per cent gum acacia, as the standard drug. The samples were autoclaved at 121°C for 30 min and administered subcutaneously. Writhing was induced 30 min later by intraperitoneal injection of 10 mL/kg of 0.6 per cent acetic acid in distilled water. The number of writhes was counted for 30 min immediately after the acetic acid injection. The aqueous extract of *A. oleracea* suppressed the acetic acid-induced writhing response significantly in a dose-dependent manner and the percentage of protection from writhing was 46.9, 51.0 and 65.6 per cent, respectively. In tail flick method pethidine (5 mg/kg) acted as the standard drug. The samples were administered intraperitoneally and the tail flick latency was assessed by the analgesiometer. The aqueous extract of *A. oleracea* in the above doses produced a dose-dependent action, it increased the pain threshold significantly after 30 min, 1, 2 and 4 h of administration. However, the analgesic activity of *A. oleracea* was found to be more significant on the acetic acid-induced model than the tail flick model and thus it appears that the test drug inhibits predominantly the peripheral pain mechanism (Chakraborty *et al.*, 2004).

Diuretic Activity

The diuretic property of the cold water extract from *A. oleracea* fresh flowers was investigated. Different concentrations of extract (500, 1000, 1500 mg/kg) or vehicle or furosemide (13 mg/kg) were orally administered to hydrated male albino rats and their urine output was monitored at several intervals of time (1–5 h). *A. oleracea* extract showed strong diuretic action. The highest dose significantly and markedly increased the urine output. The onset of this diuretic action was extremely prompt (within 1 h) and lasted up to 5 h. The peak effect was evident between 1 and 2 h. Further, the intensity of diuresis induced by the aqueous extract in the first hour was almost similar to that of furosemide. The extract also caused marked increase in urinary Na^+ and K^+ levels and a reduction in the osmolarity of urine suggesting that it is mainly acting as a loop diuretic (Ratnasooriya *et al.*, 2004).

Vasorelaxant Activity

Effects of dried aerial parts from *A. oleracea* extracts (hexane, chloroform, ethyl acetate and methanol) on vascular function of rat (male Sprague-Dawley) thoracic aorta precontracted with phenylephrine (PE) were investigated under various conditions; in the presence or absence of inhibitors, namely, NG-nitro-L-arginine methyl ester (L-NAME) and indomethacin (INDO) and under removal of functional endothelial cells. In addition, the effects of acetylcholine (ACh) as a positive control and sodium nitroprusside (SNP) a negative were studied. The extracts showed maximal vasorelaxations in a dose-dependent manner, but their effects are less than acetylcholine-induced nitric oxide (NO) vasorelaxation. Significant reduction of vasorelaxations was observed in both L-NAME and INDO. In the presence of L-NAME plus INDO, synergistic effects were observed, leading to loss of vasorelaxation of both ACh and the extracts. The tested extracts exhibited vasorelaxation *via* partially endothelium-induced NO and prostacyclin in a dose-dependent manner. Significantly, the ethyl acetate extract demonstrated immediate vasorelaxation (ED_{50} 76.1 ng/mL) and the chloroform extract showed the highest activity (ED_{50} 4.28 ng/mL) (Wongsawatkul *et al.*, 2008).

Human Cytochrome P450$_{2E1}$ Inhibitory Activity

The ethanol extract from fresh *Acmella oleracea* was examined with regard to its ability to inhibit cytochrome P450$_{2E1}$ mediated oxidation of *p*-nitrophenol *in vitro*. The 95 per cent ethanol extract from whole flowering *A. oleracea* showed modest inhibition of P450$_{2E1}$. It was required in human liver microsomes 8 per cent of extract (40 µL in a 500 µL reaction) to achieve 40 per cent inhibition using a *p*-nitrophenol concentration of 0.020 mM. Interestingly, 60 per cent inhibition was observed in the expressed $P450_{2E1}$ using the same conditions (Raner *et al.*, 2007).

Antiprotozoal Activity

Antimalarial

Decoction from dried capitula flowers of *A. oleracea* showed a similar activity against the chloroquine sensitive (FCC2 chloroquine sensitive) and resistant strains (FZR chloroquine resistant) of *P. falciparum* (IC_{50} 0.18 mg/mL and 0.20 mg/mL, respectively). The *in vivo* activity of this extract was evaluated in mice (*Mus musculus* OF1) infected by intraperitoneal route with *P. berghei*. Mice were given a single dose of 200 mg/kg *per os* each day. The activity in mice was evaluated daily by quantifying the parasitemia on smears of blood of each mouse and by measuring the survival time of the treated mice compared with the controls (not treated infected mice). None of the treated mice recovered; however, parasitaemia was slowed after treatment with the extract (IC_{50} 400 µg/mL) (Gasquet *et al.*, 1993).

Trypanocidal

Hydroalcoholic extract from *A. oleracea* (1.0 µg/mL) was evaluated on cell growth and differentiation of *Herpetomonas samuelpessoai* (ATCC 30252), a non-pathogenic trypanosomatid used as biological model for its similar antigens to *Trypanosoma cruzi*. Cell growth was estimated by counting in Neubauer's chamber after 72 h incubation and cell differentiation was examined by light microscope. *A. oleracea* hydroalcoholic extract showed no antiprotozoan activity demonstrating 36.7 per cent of stimulating effect on *H. samuelpessoai* growth. In contrast, *A. oleracea* extract stimulated cell differentiation in this flagellate. The percentages of promastigote, paramastigote and opisthomastigote forms observed in untreated cells were 87 per cent, 11 per cent and 2 per cent, respectively. Proportion of promastigote forms was 19 per cent when treated with *A. oleracea* extract (Holetz *et al.*, 2002b).

Antileishmanial

Crude extracts (dichloromethane, methanol and aqueous) from *A. oleracea* blossoms were assayed for their antileishmanial effects against *Leishmania major* on both extracellular and intracellular forms of the parasite. Survival (per cent) of free-living *L. major* promastigotes was measured with extracts at 35 and 17.5 µg/mL. Survival (per cent) of intracellular *L. major* was performed with RAW 264.7 parasitized mouse macrophages at the same concentrations. Amphotericin B was used as a reference compound. All *A. oleracea* extracts showed no significant activity against *L. major* in the employed models (Ahua *et al.*, 2007).

Anti-giardial

Chloroform, methanol and aqueous extracts from whole dried *A. oleracea* were evaluated for their anti-giardial activities. The extracts (31.25 to 1,000 µg/mL) and a standard drug, metronidazole (0.625 to 20 µg/mL), were incubated with 2×10^5 trophozoites of *Giardia intestinalis* per millilitre of growth medium in 96-well tissue culture plates under anaerobic conditions for 24 h. All tested extracts showed no significant activity (MIC $\geq$ 1,000 µg/mL; $IC_{50} \geq$ 500 µg/mL) and were considered inactive by the authors (Sawangjaroen *et al.*, 2005).

Anti-amoebic

Chloroform, methanol and aqueous extracts from whole dried *A. oleracea* were evaluated for their anti-amoebic activities against *Entamoeba histolytica* strain HTH-56:MUTM and strain HM1:IMSS growing *in vitro*. The extracts (1,000 µg/mL) were incubated with 2×10^5 *E. histolytica* trophozoites/mL of medium at 37°C under anaerobic conditions for 24 h. Metronidazole was selected as standard drug. *A. oleracea* extracts showed no significant activity (Sawangjaroen *et al.*, 2006).

Antibacterial and Antifungal Activities

Hydroalcoholic extract from *A. oleracea* leaves (2.0 mg/mL) was evaluated for its antimicrobial activity against *Escherichia coli* ATCC 25922, *Pseudomonas aeruginosa* ATCC 15442, *Bacillus subtilis* ATCC 6623, *Staphylococcus aureus* ATCC 25923, *Candida albicans*, *C. krusei*, *C. parapsilosis* and *C. tropicalis*. The minimum inhibitory concentrations (MICs) of extract and reference antibiotics (tetracycline, vancomycin, penicillin, and nistatin) were determined by microdilution techniques in Mueller-Hinton broth for bacteria and RPMI-1640 for fungi. The extract was considered inactive for all the microorganisms tested (Holetz *et al.*, 2002a).

Ethanol tincture made from fresh *A. oleracea* was prepared and applied to blank diffusion disks (25 µL). These disks were desiccated and used in Kirby-Bauer disk diffusion susceptibility tests on three bacteria: *Staphylococcus aureus, Pseudomonas aeruginosa* and *Escherichia coli*. The efficacy of the tincture was compared to that of commercially prepared antibiotic diffusion disks. Ten micrograms of ampicillin was used on *E. coli*, 30 µg of cefotaxime was used against *P. aeruginosa* and 30 µg of vancomycin was used on *S. aureus*. No inhibition was observed with *A. oleracea* tincture (Romero *et al.*, 2005).

Crude extracts (hexane, chloroform, ethyl acetate and methanol) and some fractions from aerial parts of *A. oleracea* were tested for antimicrobial activity against 27 strains of microorganisms using the agar dilution method and ampicillin as control for growth inhibition. The hexane and chloroform extracts completely inhibited the growth of *Saccharomyces cerevisiae* ATCC 2601 with minimum inhibitory concentration (MIC) 256 µg/mL. The chloroform extract also completely exhibited antigrowth activity against *Streptococcus pyogenes* II with MIC 256 µg/mL. Fractions from the chloroform extract exhibited activity against *Corynebacterium diphtheriae* NCTC 10356 with MIC 64-256 µg/mL and one fraction from the same extract completely inhibited the growth of *Bacillus subtilis* ATCC 6633 (MIC 128 µg/mL) and *Bacillus cereus* with MIC 256 µg/mL. Antigrowth property of ethyl acetate and methanol extracts were evaluated at 256 µg/mL, but no activitiy was observed. However, some fractions of ethyl acetate and methanol extracts inhibited the growth of *C. diphtheriae* NCTC 10356 with MIC 64-128 µg/mL and 128-256 µg/mL, respectively. In addition, methanol fractions also showed antimicrobial activity against *B. subtilis* ATCC 6633 with MIC 128-256 µg/mL, *Micrococcus lutens* ATCC 10240, *Staphylococcus epidermidis* ATCC 12228 and *B. cereus* with MIC 128-256 µg/mL (Prachayasittikul *et al.*, 2009).

Cell Lines Activity

The crude extracts (hexane, chloroform, ethyl acetate and methanol) from *A. oleracea* were tested against the KB and HuCCA-1 cell lines. The KB cell lines were originally derived from epidermoid carcinoma of the floor of the oral cavity and the HuCCA-1 cells were established from chlolangiocarcinomas experimentally induced in hamsters. Cells were incubated at 37°C in 96-well culture plates with different concentrations of tested extracts (0.001–10 µg/mL) or taxol (0.012–1.2 µg/mL), used as control for 48 h. The results showed that all the extracts exhibited ED_{50} values greater than 10 µg/mL and were consequently considered to be inactive (Prachayasittikul *et al.*, 2009).

Antioxidant Activity

The radical scavenging activity of the hexane, chloroform, ethyl acetate and butanol extracts from dried flowers of *A. oleracea* were assayed by DPPH and 2,2'-azinobis(3-ethylbenzothiazoline-6-sulfonate (ABTS) radical scavenging assays. The ethyl acetate extract from *A. oleracea* had the highest radical scavenging capacity, with EC_{50} values of 1.38μmol of vitamin C equiv/mg of dried extract and 3.32μmol of Trolox equiv/mg of dried extract, as determined by DPPH and ABTS assays, respectively. Chloroform and butanol extracts exhibited similar radical scavenging abilities according to both assays (Wu *et al.*, 2008).

The antioxidant activity of hexane, chloroform, ethyl acetate and methanol extracts from dried aerial parts of *A. oleracea* was also evaluated by Wongsawatkul *et al.* (2008) using DPPH and SOD assays. The results were in accordance with those previous described. In DPPH assay at 200 μg/mL, the ethyl acetate and methanol extracts displayed comparable activity and the highest radical scavenging activity (47.90 and 47.76 per cent) with IC_{50} 216 and 223 μg/mL. The chloroform extract exhibited 29.82 per cent radical scavenging activity. The hexane extract produced some activity (4.90 per cent radical scavenging).

Larvicidal Activity

Hexane, ethyl acetate and methanol extracts from freshly harvested flowers of *A. oleracea* were bioassayed against *Aedes aegyptii* larvae and *Helicoverpa zea* (corn earworm) neonates. The mosquitocydal assay was carried out with 10-15 mosquito larvae, *A. aegyptii*, placed in 980 mL of distilled water and 20 mL of DMSO solution containing test extracts. The test concentrations were 250 µg/mL for crude extracts. A control was prepared with distilled water and DMSO solution to which larvae were added. The mortality was recorded at 2, 4, 12 and 24 h intervals. *Corn earworm* assay was conducted briefly, placing the *H. zea* eggs in an incubator at 27°C for 24-36 h. The insect diet was an agar (1.4 per cent) in water solution added with the dry diet for *Corn earworm* until the total diet weight reached 5 g. DMSO (25 mL) or DMSO solutions containing test extracts were then mixed with the diet separately. Three to four drops of this diet was then dispensed into polystyrene vials and stored at 4°C for 24 h. To each vial one neonate larvae was placed using a fine point, sterilized brush. The treatment and control vials were held in a growth chamber at a photoperiod of 16 h day and 8 h night with day temperature of 28°C and night temperature of 24°C. Each treatment had fifteen replicates. The larvae were weighed on an analytical balance after six days. The hexane extract showed the highest activity. The fractionation of this bioactive extract furnished six fractions (A-F) and fraction E (1.65 g/600 mL) was the biologically active (Ramsewak *et al.*, 1999).

Crude hexane extract from dried flower heads of *Acmella oleracea* was investigated for the larvicidal efficacy against malaria (*Anopheles stephensi* and *A. culicifacies*, species C) and filaria vectors (*Culex quinquefasciatus*) following the standard World Health Organization larval susceptibility test method. Different dilutions of the extract were made (50, 25, 12.5, 6.25, 3.125 and 1.5625 ppm) through serial dilution in sterile water with triton (10 μL/L of water). Results were scored after 24 h of continuous exposure to the test solution and were expressed as percent mortality. Pure water and hexane without the dissolved plant extract served as the positive control, whereas pure water alone served as the control. The tested extract showed variable mortality against all the vectors. It induced complete lethality (100 per cent mortality) at minimum doses, the respective LC_{50} and LC_{90} values being 4.57 and 7.83 (*A. stephensi*), 0.87 and 1.92 (*A. culicifacies*) and 3.11 and 8.89 ppm (*C. quinquefasciatus*) (Pandey *et al.*, 2007).

The same evaluation was carried out by the same group using plant material obtained from *in vitro* micropropagation of *A. oleracea*. Hexane extracts were obtained from different parts of the fully matured plants (flower heads, stem, leaves, and roots) and evaluated for their larvicidal activity as previously described. A hundred percent larval mortality has been achieved within 24 h of exposure of both the mosquito vectors (*A. stephensi* and *C. quinquefasciatus*) with different concentrations of root, stem, leaf, and flower head extracts. However, the mortality was dose-dependent. Different plant part extracts of *A. oleracea* exhibited different rates of lethality. Among the aforesaid extracts tried, root extract proved to be the most toxic against the mosquito larvae of both the species. One hundred percent mortality was obtained at 6.25 ppm (*A. stephensi*) and 3.125 ppm (*C. quinquefasciatus*), respectively. The respective lethal doses (LC_{50} and LC_{90}) required for achieving 50 per cent and 90 per cent mortalities are 2.71 and 4.26 ppm against *Anopheles* larvae and 1.19 and 2.06 ppm against *Culex* larvae. Incidentally, a steep increase in LC_{50} and LC_{90} values were observed with the extracts of flower > leaf > stem against both the larvae. The root extract was 6.9-fold more effective than flower extract in the case of *A. stephensi* and 2.61-fold against *C. quinquefasciatus* mosquito larvae. Similarly, leaf and stem extract has also exhibited 100 per cent lethality, but their doses were 22.57- and 32.07-fold (*A. stephensi*) and 5.37- and 9.72-fold (*C. quinquefasciatus*) less effective than the root extract (Pandey and Agrawal, 2009).

Insecticidal Activity

The toxicity of leaf and flower heads extracts (petroleum distillate, chloroform and methanol) of *A. oleracea* against adult male American cockroach, *Periplaneta americana*, was studied by Kadir *et al.* (1989). Extracts were applied topically to the dorsal thoracic surface of each insect (10 μg/g insect). Mortality was scored after 24 h exposure period. The control insects were treated with the organic solvent (5μL) only. The tested extracts were toxic against adults of *P. americana*. The flower heads extracts showed a high acute toxicity of greater than 50 per cent mortality, the petroleum distillate extract showed the highest insecticidal effect (95-100 per cent). The leaf extracts, however, showed a lower percent mortality of less than 30 per cent.

N-Alkylamides

Analgesic Activity of Spilanthol

Spilanthol was tested to evaluate its influence on increasing GABA concentration in brain. A protocol using brain slices (temporal cortex) from adult albino female mice and spilanthol at 1×10^{-4} M, 1×10^{-6} M or 1×10^{-8} M was carried out six times. Experiments without spilanthol were the negative control and potassium chloride (KCl 47 mM) was used as positive control. The influence of spilanthol was observed only at 1×10^{-4} M and 1×10^{-6} M. At 1×10^{-6} M, GABA released increased from 125 to 194 pmol GABA/μg protein 1 min after stimulation and achieved 252 pmol GABA/μg protein 5 min after stimulation. At 1×10^{-4} M, an abrupt increase until 521 pmol GABA/ μg protein was observed immediately after spilanthol addition, followed by a decreased to 280 pmol GABA/μg protein 1 min after stimulation and a slight increase in the following 5 min (Rios *et al.*, 2007).

Anti-inflammatory Activity

The inhibitory effect on nitric oxide production on LPS-stimulated murine macrophages-like RAW 264.7 cells was used to evaluate the anti-inflammatory activity of spilanthol (Wu *et al.*, 2008). Macrophages produced nitric oxide to mediate inflammation when activated by LPS. LPS-induced (1 μg/mL) macrophages with various concentrations of spilanthol (20-360 μM) were incubated for 24 h to study the inhibition of nitric oxide production. 360 μM of spilanthol suppressed only 20 per cent of the nitric oxide production in comparison with control. Followed that test, a western blot analysis was performed on whole cell lysates to determine whether the inhibition of nitric oxide production resulted from the diminished levels of LPS-induced inducible nitric oxide synthase (iNOS) protein expression and if inducible cyclooxygenase (COX-2), highly expressed during inflammation, was suppressed. Spilanthol was tested at 90 μM and 180 μM. In both responses, it was observed a dose-dependent inhibition. The expression of COX-2 protein was suppressed by 180 μM spilanthol. Lower levels of proinflammatory cytokines (IL-1β, IL-6 and TNF-R) in the supernatant of culture were also verified by ELISA kits. In a reverse transcription polymerase chain reaction (RT-PCR) analysis, an inhibitory dose-dependent effect of spilanthol on LPS-induced iNOS and COX-2 mRNA production was observed at 45 μM, 90 μM and 180 μM, suggesting that the administration of spilanthol restrained iNOS and COX-2 production at the transcriptional and translational levels (Wu *et al.*, 2008).

Blockage of Muscular Contractions

The nerve-muscle co-culture model was used to study the influence of purified spilanthol on muscle contraction frequency. After 21 days of culture, muscle fibres formed contracted spontaneously and the number of contractions was counted using automates counting software. Carisprodol (1 mM) was used as a positive control for reversible blockage of muscle contractions. The number of contractions was counted for each measurement period: before incubation (pre-incubation frequency), during incubation and during the contractile activity recuperation phase after elimination of spilanthol. At the concentrations 40×10^{-5} per cent and 160×10^{-5} per cent, spilanthol blocked muscle contractions after 5 min of incubation. After 6 h, the fibers were washed out and recuperation of contractile activity was not observed even 24 h after elimination of the substance. Thus, spilanthol showed the ability to inhibit contractile activity, suggesting the same anti-wrinkle potential as botulinum toxin (Demarne and Passaro, 2008).

Antimicrobial Activity

Spilanthol was assayed for antibacterial activity against *Escherichia coli, Pseudomonas solanacearum, Bacillus subtilis* and antifungical activity against the yeast *Saccharomyces cerevisiae* (Molina-Torres *et al.*, 1999). The study was based on the evaluation of growth inhibition in liquid cultures by determining turbidity at 650 nm. Flasks containing 15 mL of potato-dextrose broth and different concentrations of spilanthol (25-150 µg/mL) were inoculated with 0.3 mL of the microorganism culture. At 25 µg/mL, the inhibition of *E. coli* growth was close to 90 per cent after 8 h of incubation. Higher concentrations of spilanthol (50 and 75 µg/mL) almost completely inhibited growth. The effect on growth of *P. solanacearum* was less inhibitory and it was necessary 150 µg/mL of spilanthol to total inhibition. At 50 µg/mL, *B. subtilis* growth was reduced and it was observed total inhibition at 150 µg/mL, *S. cerevisiae* growth also was inhibited at 25 µg/mL of spilanthol and total inhibition was observed with higher concentrations tested.

In a further study, Molina-Torres *et al.* (2004) used the same protocol to test spilanthol against fungi – *Rhizoctonia solani* groups AG3 and AG5, *Sclerotium rolfsii, S. cerevisiae, Sclerotium cepivorum, Fusarium* sp. and *Verticillium* sp. – phytopathogenic Chromista *Phytophthora infestans*, and bacteria – *E. coli, Erwinia carotovora* and *B. subtilis*. The mycelium dry weight was determined after 10 days of incubation to evaluate fungi growth. *S. cerevisiae* and bacterial growth were estimated as turbidity at 650 nm. Spilanthol was assayed at 50, 75 and 150 µg/mL. *S. rolfsii* and *S. cepivorum* were more sensitive, with a mycelial growth inhibition of 100 per cent and 94 per cent, respectively, at the minimum concentration assayed. The growth of *E. coli, S. cerevisiae, P. infestans, R. solani* AG-3 and AG-5 were almost totally inhibited at spilanthol concentration of 75 µg/mL. The growth of *B. subtilis* was totally inhibited at 150 µg/mL of spilanthol, but this concentration only inhibited less than 50 per cent of the concentration of *Verticillium* sp. and *Fusarium* sp. Only *E. carotovora* was not sensitive in the tested conditions.

Sensory Activity

Spilanthol and homospilanthol were evaluated for their trigeminal sensory properties (burning, pungency, tingling, scratching, numbing, warming, mouth-watering and cooling) by trained panelists (Ley *et al.*, 2006b). Solutions containing 11 per cent sucrose and 0.03 mg/mL of the test substance were sipped for 10 to 20 s and then spat out. Estimates intensity ratings for descriptors were 1 (low) to 9 (high). Spilanthol was considered tingling (intensity 8) and mouth-watering (intensity 7) while homospilanthol was considered tingling (intensity 5) and numbing (intensity 5).

Human Cytochrome P450$_{2E1}$ Inhibitory Activity

The ability of spilanthol to inhibit cytochrome P450$_{2E1}$ was evaluated based on the oxidation of p-nitrophenol *in vitro*, using human liver microsomes and expressed cytochrome P450$_{2E1}$. At 0.067 mM, spilanthol inhibited P450$_{2E1}$ activity by 55 per cent in human liver microsome and 75 per cent in expressed P450$_{2E1}$ samples (Raner *et al.*, 2007).

Transmucosal Permeation

A commercial mouth gel containing spilanthol and water, glycerine, PVP, polyacrylic acid, propyleneglycol, hydroxypropyl methyl cellulose, benzyl alcohol, sodium hydroxide, sodium benzoate, xanthan gum, aromaparfum (cinnamon-mint), potassium sorbate, aroma (spilanthol), sodium hyaluronate and sucralose was tested by Franz diffusion cell (FDC) experiments using porcine buccal mucosa to evaluate the permeation of spilanthol through buccal mucosa. Spilanthol content in the formulation was 62.6 µg/mL. Permeation of spilanthol was determined using 5 mL static FDC and 0.5 g of the gel, in diffusion experiments performed in sixfold. The amount of spilanthol found in the mucosa at the end of the FDC experiments was (mean ± SEM) 0.72 ± 0.01 µg. This corresponded to a spilanthol concentration of 2.64 × 10^4 ng/mL, (or 1.19 × 10^5 nM), in the buccal mucosal tissue. After 5 h, only 4.464 ± 0.318 per cent of the applied dose was cumulatively found in the receptor chamber. The calculated permeability coefficient K_p was 19.516 ± 1.752 × 10^{-3} cm/h. These findings showed that spilanthol permeated the buccal mucosa and suggested that systemic and local mucosa effects with topical spilanthol formulations could be expected (Boonen *et al.*, 2010).

Larvicidal Activity

The undeca-2E-en-8,10-diynoic acid isobutylamide was assayed against *Aedes aegyptii* larvae. At 6.25 µg/mL, this alkylamide caused 30 per cent mortality of the larvae (Ramsewak *et al.*, 1999).

Spilanthol was tested against eggs, various instar larvae and pupae of *Anopheles*, *Culex* and *Aedes* mosquito. Aqueous solutions of spilanthol (1, 3, 5 and 10 ppm) were added to sets of 25 eggs of each species. Instar larvae in different stages and pupae were treated with 1-10 ppm spilanthol aqueous solution containing 0.1 per cent albumin and dried liver powder. Within 2-3 h, abnormal movements by larvae were observed. Between 10-15 h, slow upside-down movements and feeble jerks with lesser frequency suggested acute toxicity. Spilanthol was deleterious to various larval stages of mosquitoes at 4-7.5 ppm concentration (LC$_{50}$ 4 to 5 ppm and LC$_{95}$ 6 to 7 ppm. The study of the eggs showed no hatching at 3 ppm concentration of spilanthol and eggs became decolorized within 21 hrs. At 4-5 ppm concentration, spilanthol killed pupae within 3-5 h (Saraf and Dixit, 2002).

Insecticidal Activity

The toxicity of spilanthol against adult male American cockroach, *Periplaneta americana*, was studied by Kadir *et al.* (1989). The substance, diluted in 5 µL acetone, was applied topically to the dorsal thoracic surface of each insect, at 2 to 40 µg/g insect. Mortality was scored after 48 h. It was observed that spilanthol penetrated the integument of insects and caused tremors almost immediately. Mean lethal dose value (LD_{50}) was 2.46 µg/g. In comparison with conventional insecticides also assayed, spilanthol was 3.8, 2.6 and 1.3 times more toxic than lindane, bioresmethrin and carbaryl, respectively. In the electrophysiological investigation, the body tissues, guts and the trachea associated with the nervous system were partly dissected out and the abdominal cavity was filled with fresh insect saline and mineral oil. Spilanthol, diluted in 1 µL acetone, was applied to the circus, at doses corresponding to about LD_{10} (0.8 µg), LD_{25} (1.5 µg) and LD_{40} (2.0 µg). The electrical activity conducted by the abdominal nerve were amplified with a Universal Coupler and displayed in an oscilloscope. At 2.0 µg, the lowest dose that resulted in a significant effect, the spontaneous activity in the abdominal connectives increased, achieving a maximum in 5 min and remaining high for about 10 min. Thus, frequency decreased rapidly to zero. Control assays with acetone showed no abnormalities on nerve activity. The results indicated that the toxic action of spilanthol is related to interference with the nervous system.

Molluscicidal Activity

The molluscicidal activity of spilanthol was studied by Johns *et al.* (1982). Aqueous solutions of spilanthol (50 mg/L and 150 mg/L) were added to freshwater snails (*Physa occidentalis*) and *Leptocercous cercariae* of the Echinostome released by the mollusc. At 50mg/L, snails were inactive after 1 h and dead within 18h (LD_{50} = 100 µM). At 150 mg/L, snails movement ceased after 30 min and cercarial showed immobility after 5 s and convulsion after 1 min.

The undeca-2*E*-en-8,10-diynoic acid isobutylamide was assayed against *Helicoverpa zea* (corn earworm) neonates. At 250 µg/mL, this alkamide reduced the weight of corn earworm by 79 per cent (Ramsewak *et al.*, 1999).

Others Constituents

Crude extracts (chloroform, ethyl acetate and methanol) of aerial parts of *A. oleracea* and their fractions were tested for antioxidant activity using the 2,2-diphenyl-1-picrylhydrazyl (DPPH) (333.33 µg/mL) and superoxide dismutase (SOD) (300 µg/mL) assays. All tested fractions exhibited antioxidant activity in both assays. Particularly, fractions of the methanol extract displayed very potent antioxidant properties with 84.69-96.05 per cent radical scavenging activity (DPPH assay). The fraction of the ethyl acetate extract with the highest antioxidant activity in the DPPH assay (82.46 per cent) also produced the highest SOD activity (81.50 per cent). From these fractions were isolated *trans*-isoferulic acid, scopoletin, vanillic acid, stigmasteryl-3-O-β-D-glucopyranoside and a mixture of stigmasteryl-3-O-β-D-glucopyranoside and β-sitosteryl-3-O-β-D-glucopyranoside. These substances exhibited no antioxidant activity (Prachayasittikul *et al.*, 2009).

Toxic Activity

Extracts and Fractions

Macrophage Cytotoxicity

Crude extracts (dichloromethane, methanol and aqueous) from *A. oleracea* blossoms were assayed for their toxicity against RAW macrophages. Survival (per cent) of RAW 264.7 mouse macrophages were measured after addition of extracts at 35 and 17.5 µg/mL. All *S. oleracea* extracts showed no significant toxicity against RAW macrophages with 80.1-97.2 per cent of survival when exposed to the highest tested concentration (Ahua *et al.*, 2007).

The cytotoxic effects of *A. oleracea* extracts on macrophages, RAW 264.7, were also assayed by Wu *et al.* (2008). The growth inhibitory effect of hexane, chloroform, ethyl acetate and butanol extracts from dried flowers of *A. oleracea* was evaluated by MTT assay at different concentrations (20-100 µg/mL). After 24 h of incubation, hexane and chloroform extracts (80 µg/mL) reduced cell viability to about 75 and 81 per cent, respectively, whereas ethyl acetate or butanol extracts did not significantly alter cell viability (91 and 93 per cent, respectively). The results indicated that both hexane and chloroform extracts exhibited cytotoxic effects at high concentrations.

Aqueous extract from fresh aerial parts of *A. oleracea* was evaluated *in vivo* for acute toxicity. No adverse effect or mortality was detected in albino rats up to 3.0 g/kg (*p.o.*) of extract during the 24 h observation period (Chakraborty *et al.*, 2004).

Antimutagenic Activity

An 80 per cent methanol extract (0.5 g/25mL) from *A. oleracea* leaves (50 µL in methanol) was assayed for its antimutagenic property against Trp-P-1 in *Salmonella typhimurium* TA98 by the Ames preincubation method using S9 mix. The activity was evaluated by the amount of extract which suppressed 90 per cent of the mutagenesis (ED_{90}). *A. oleracea* extract showed no significant antimutagenic activity ($ED_{90} > 50 µL$/plate) (Nakahara *et al.*, 2002).

Convulsive Activity

To evaluate the convulsive activity of the hexane extract from *A. oleracea*, male Wistar rats were injected intraperitonialy with 50 to 150 mg/kg of the extract. The electroencephalogram (EEG) and animals behavior were observed for periods as long as 2 h. Following the lower doses (50 and 75 mg/kg) only minor behavioral changes such as grooming and wet dog shakes were observed. Higher doses (100 to 150 mg/kg) induced full tonic-cronic convulsions in dose-dependent manners which were accompanied by typical electrographic seizures in the EEG. These results showed that the hexane extract from *A. oleracea* is able to induce generalized convulsions in rats and indicated that it can be used as a tool in the development of new models of epilepsy (Moreira *et al.*, 1989).

Compounds

Macrophage Cytoxicity

Wu *et al.* (2008) studied cytotoxic effects of spilanthol on murine macrophages-like, RAW 264.7 cells. Spilanthol was used in the tests at 20-360 µM. At low concentrations, it was not observed significant cytotoxicity. Cell viability was higher than 90 per cent at 180 µM. However, significant cytotoxicity was observed at 360 µM (cell viability < 90 per cent).

Arrhythmic Activity

The influence of spilanthol on the electrical activity of isolated heart of rabbits was studied using the experimental electrocardiogram with Langendorff perfusion

(Herdy and Carvalho, 1984). Spilanthol was added to perfusion liquid, initially at 40 mg/L, gradually diminishing to 0.1 mg/L. Effects were compared to procainamide, at 10 mg/L to 400 mg/L doses. Spilanthol increased the automaticity, with supraventricular and ventricular arrhythmias. The main difference between spilanthol and procainamide was the enhancement on automaticity. The results suggested that spilanthol can not be used as an antiarrythmic drug, but can generate arrhytmias to test new drugs.

Patents

The several applications and commercial value of *A. oleracea* stimulated the market in several countries which resulted in large number of patents listed below.

Skin Preparation

JP60215610 – Plant (Michio and Fumiichirou, 1985);

JP2004189660 – Spilanthol (Hiroshi *et al.*, 2004);

JP2009073800 - Extract (Ikeda, 2009);

US 2004028643 - Extract (Chiba *et al.*, 2004);

US2004052735 - (Nakatsu *et al.*, 2004);

US2007041922 – Extract (Reinhart and Helman, 2007);

US2007048245 – Extract (Belfer, 2007);

US2008069912 – Extract (Frederic and Ghislaine, 2008);

US 2009280078 – Extract (Belfer, 2009).

Flavor Agent

JP2001178395 – Spilanthol (Norifumi *et al.*, 2001);

JP2006296357 – Extract (Toshio *et al.*, 2006A);

JP2006296356 – Extract, Essential Oil and Spilanthol (Toshio *et al.*, 2006B);

JP2006223104 - Extract, Essential Oil and Spilanthol (Toshio *et al.*, 2006C);

JP2010004767 – Spilanthol (Toshio *et al.*, 2010);

US2008242740 - Alkamides (Jakob *et al.*, 2008);

US2008050500 – Spilanthol; Spilanthol containing plant extract or essential oil (Muranishi *et al.*, 2008);

US2009124701 - Alkamides (Kathrin *et al.*, 2009);

US2009155445 – Jambu oleoresin, Spilanthol (Anh and Catherine, 2009).

Insecticidal and Anti-swarm

JP2002363012 – Extract (Taiji and Hisashi, 2002).

Hair Preparation

JP9194334 – Extract (Tomoko *et al.*, 1997).

Sensorial Uses

JP7090294 – Essential oil (Toshiya and Tetsuo, 1995).

Medicine Composition with Plants to Cure Cold

CN1079394 – Plant (Fang, 1993).

Oral Preparation

GB1438205 – Essential oil (No author, 1976);

JP54067040 - (Akinori and Youji, 1979);

JP60075424 – Spilanthol and Extract (Fumio *et al.*, 1985);

JP60041606 – Spilanthol (Michio *et al.*, 1985);

JP61155315 – Spilanthol and Oleoresin (Fumio *et al.*, 1986);

JP62198611 – Spilanthol (Hideaki *et al.*, 1987);

JP2008115115 – Spilanthol (Yukio *et al.*, 2008);

JP2010037318 – Spilanthol (No author, 2010);

US3720762 - Spilanthol or Essential oil (Hatasa and Lioka, 1971);

US2010184863 - Spilanthol (Louis *et al.*, 2010).

Process

BRPI0500886 – Extracts and isolation of compounds (Delarcina *et al.*, 2005).

Detergent

JP6072858 - Spilanthol (Hiroyasu, 1994).

Acknowledgments

The authors thank CNPq and Ministry of Health for the financial supports.

References

Ahua, K. M., Ioset, J. R., Ioset, K. N., Diallo, D., Mauël, J. and Hostettmann, K. (2007). Antileishmanial activities associated with plants used in the Malian traditional medicine. *Journal of Ethnopharmacology*, **110:** 99–104.

Akinori, T. and Youji, Y. (1979). Composition for intraoral use. *Japanese Patent Application* **54067040**.

Anh, L. and Catherine, M. (2009). Tingling and salivating compositions. *U.S. Patent Application* **2009155445**.

Barreto, M. B., Nartins Neto, J. S., Bezerra, A. M. E. and Souza Brasil, N. Baruah, R. N. and Leclercq, P. A. (1993). Characterization of the essential oil from flower heads of *Spilanthes acmella. Journal of Essential Oil Research.*, **5:** 693-695.

Baylac, S. and Racine, P. (2004). Inhibition of human leukocyte elastase by natural fragrant extracts of aromatic plants. *The International Journal of Aromatherapy*, **14:** 179–182.

Belfer, W.A. (2007). Cosmetic composition to accelerate repair of functional wrinkles. *U.S. Patent Application*, **2007048245**.

Belfer, W.A. (2009). Cosmetic composition to accelerate repair of functional wrinkles. *U.S. Patent Application*, **2009280078**.

Boonen, J., Baert, B., Burvenich, C., Blondeel, P., Saeger, D.S. and Spiegeleer, B.D. (2010). LC–MS profiling of N-alkylamides in*Spilanthes acmella* extract and the LC-MS profiling of *N*-alkylamides in *Spilanthes acmella* extract and the transmucosal behaviour of its main bio-active spilanthol. *Journal of Pharmaceutical and Biomedical Analysis*, **53**: 243-249

Carlini, E. A., Rodrigues, E., Mendes, F. R., Tabach, R. and Gianfratti, B. (2006). Treatment of drug dependence with Brazilian herbal medicines. *Brazilian Journal of Pharmacognosy*, **16** (Supl.): 690-695.

Chakraborty, A., Devi, R. K. B., Rita, S., Sharatchandra, K. and Singh, T. I. (2004). Preliminary studies on antiinflammatory and analgesic activities of *Spilanthes acmella* in experimental animal models. *Indian Journal of Pharmacology*, **36**: 148-150.

Chiba, K.; Sone, T.; Miyazaki, K.; Hanamizu, T.; Nishisaka, F.; Matsumoto, S. and Aiyama, R. (2004) Compositions for retarding skin aging. *U.S. Patent Application*, **2004028643**.

Chung, K.F., Kono, Y., Wang, C.M., and Peng, C.I. (2008). Notes on *Acmella* (Asteraceae: Heliantheae) in Taiwan. *Botanical Studies*, **49**: 73-82.

Coutinho, L. N., Aparecido, C. C. and Figueiredo, M. B. (2006). Galhas e deformações em Jambu (*Spilanthes oleraceae*) causadas por *Tecaphora spilanthes* (Ustilaginales). *Summa Phytopathologica*, **32** (3): 283-285.

Delarcina, J.R., S.; Cagnon, J.R., Silva, A. R. and Fukusama, V.E.N. (2005). A process of preparing jambu extract, use of said extract, cosmetic compositions comprising thereof and cosmetic products comprising said cosmetic compositions. *BR. Patent Application*, **PI0500886**.

Demarne, F. and Passaro, G. (2008). Use of an *Acmella oleracea* extract for the botox-like effect thereof in an anti-wrinkle cosmetic composition. *US Patent* **US 2008/ 0069912 A1**.

Ekanem, A. P., Wang, M., Simon, J. E. and Moreno, D. A. (2007). Antiobesity Properties of two African Plants (*Afromomum meleguetta* and *Spilanthes acmella*) by Pancreatic Lipase Inhibition. *Phytotherapy Research*, **21**: 1253–55.

Fang, L. (1993). Cold tea with traditional Chinese medicine. *Chinese Patent Application*, **1079394**.

Ferreira, J.L.P., Simas, N.K., Sampaio, T.R., de Araújo, R.B., and Amaral, A.C.F. (2007). Parâmetros anatômicos do pecíolo e da lâmina foliar de *Spilanthes oleracea*. *In:* 58° Congresso Nacional de Botânica, São Paulo, Brasil.

Frederic, D. and Ghislaine, P. (2008). Use of an *Acmella oleracea* Extract for the Botox-Like Effect Thereof in an Anti-Wrinkle Cosmetic Composition. *U.S. Patent Application*, **2008069912**.

Fumio, Y., Keiichi, Y., Norifumi, T. and Hideaki, S. (1986) *Composition for oral purpose Japanese Patent Application*, **61155315**.

Fumio, Y., Michio, U. and Seiji, K. (1985). Composition for oral cavity. *Japanese Patent Application*, **6007542**.

Gasquet, M., Delmas, F., Timon-David, P., Keita, A. Guindo, M., Koita, N., Diallo, D. and Doumbo, O. (1993). Evaluation *in vitro* and *in vivo* of a traditional antimalarial, "Malarial 5". *Fitoterapia*, **64**: 423-426.

Greger, H., Hofer, O. and Werner, A. (1985). New amides from *Spilanthes oleracea. Monatshefte fuer Chemie*, **116**: 273-277.

Hatasa, S. and Lioka, I. (1971). Spilanthol-containing compositions for oral use. *U.S. Patent Application*, **3720762**.

Herdy, G.V.H. and Carvalho, A.P. (1984). Ação do espilantol (extraído do jambu) sobre a atividade elétrica do coração do coelho, eletrocardiograma experimental. *Arquivos Brasileiros de Cardiologia*, **43**: 315-320.

Hideaki, S., Fumio, Y., Yukari, W. and Norifumi, T. (1987). Composition for oral cavity application. *Japanese Patent Application*, **62198611**.

Hind, D.J.N. (1993). New compositae from the Serra do Grão- Mogol (Mun. Grão-Mogol, Minas Gerais, Brazil) and the surrounding area. *Kew Bulletin*, **49**: 511-522.

Hind, N., and Biggs, N. (2003). *Acmella oleracea*. Compositae. *In*: Royal Botanic Gardens, Kew, Blackwell Publishing Ltd, UK, 31-39.

Hiroshi, Y.; Mutsuko, T.; Takashi, S. and Masashi, K. (2004). External preparation for skin and application thereof. *Japanese Patent Application*, **2004189660**.

Hiroyasu, K. (1994). Detergent for body and hair of head. *Japanese Patent Application*, 6072858.

Holetz, F.B., Pessini, G.L., Sanches, N.R., Cortez, D.A.G., Nakamura, C.V. and Dias Filho, B.P. (2002a). Screening of Some Plants Used in the Brazilian Folk Medicine for the Treatment of Infectious Diseases. *Memórias do Instituto Oswaldo Cruz*, **97**: 1027-1031.

Holetz, F.B., Ueda-Nakamura, T., Dias Filho, B.P., Cortez, D.A.G., Mello, J.C.P. and Nakamura, C.V. (2002b). Effect of plant extracts used in folk medicine on cell growth and differentiation of *Herpetomonas samuelpessoai* (Kinetoplastida, Trypanosomatidae) cultivated in defined medium. *Acta Scientiarum*, **24**: 657-662.

Ikeda, T. (2009). Cosmetics containing extracts of aerial parts of *Spilanthes acmella*. *J.P. Patent Application*, **2009073800A**.

Jakob, L., Kathrin, F., Gerhard, K., Gerald, R. (2008). Aroma compositions of alkamides with hesperetin and/or 4- hydroxydihydrochalcones and salts thereof for enhancing sweet sensory impressions. *U.S. Patent Application*, **2008242740**.

Jansen, R.K. (1981). Systematics of *Spilanthes* (Compositae: Heliantheae). *Systematic Botany*, **6**: 231-257.

Jansen, R.K. (1985a). Systematic significance of chromosome numbers in *Acmella* (Asteraceae). *American Journal of Botany*, **72:** 1835-1841.

Jansen, R.K. (1985b). The Systematics of *Acmella* (Asteraceae: Heliantheae). *Systematic Botany Monographs*, **8:** 1-115.

Johns, T., Graham, K., Towers, G.H.N. (1982). Molluscicidal activity of affinin and other isobutylamides from the Asteraceae. *Phytochemistry*, **21:** 2737-2738.

Kadir, H.A., Zakaria, M.B., Kechil, A.A. and Azirun, M.S. (1989). Toxicity and electrophysiological effects of *Spilanthes acmella* Murr. extracts on *Periplaneta americana* L. *Pesticide Science*, **25:** 329-35.

Kathrin, L.; Jakob, L.; Gerald, R.; Gunter, K. and Gerhard, K. (2009). Use of alkamides for masking an unpleasant flavor. *U.S. Patent Application*, **2009124701.**

Kissmann, K.G. and Groth, D. (1992). *Spilanthes acmella* (L.) Murr. *In:* Plantas Infestantes e Nocivas, Vol. II, BASF Brasileira S.A., Brasil, pp.346-348.

Lemos, T.L.G., Pessoa, O.D.L., Matos, F.J.A., Alencar, J.W. and Craveiro, A.A. (1991). The essential oil of *Spilanthes acmella* Murr. *Journal of Essential Oil Research*, **3:** 369-70.

Ley, J. P., Blings, M., Krammer, G., Reinders, G., Schmidt, C.-O. and Bertram, H.-J. (2006a). Isolation and synthesis of acmellonate, a new unsaturated long chain 2-ketol ester from *Spilanthes acmella*. *Natural Product Research, Part A: Structure and Synthesis.* **20:** 798-804.

Ley, J.P., Krammer, G., Looft, J., Reinders, G., Bertram, H-J. (2006b). Structure-activity relationships of trigeminal effects for artificial and naturally occurring alkamides related to spilanthol. *Flavour Science: Recent Advances and Trends*, **43:** 21-24.

Lorenzi, H. and Matos, F.J. (2008). Plantas Medicinais no Brasil. Nativas e Exóticas. Instituto Plantarum de Estudos da Flora LTDA, Nova Odessa.

Louis, L., Michael L., Kenya, I.; Shigeru, T., Hideo, U., Kenji, Y., Jennifer, M., Carter, G. and Amrit, M. (2010). Synthetic spilanthol and use thereof. *U.S. Patent Application*, **2010184863.**

Martin, R. and Becker, H. (1984). Spilanthol-related amides from *Acmella ciliata*. *Phytochemistry*, **23:** 1781-1783.

Michio, M., Fumiichirou, H. (1985). Bath preparation. *Japanese Patent Application.* **60215610.**

Michio, U., Fumio, Y. and Seiji, K. (1985). Dentifrice composition. *Japanese Patent Application*, **60041606.**

Molina-Torres, J., García-Cháves, A. and Ramírez-Chávez, E. (1999). Antimicrobial properties of alkamides present in flavouring plants traditionally used in Mesoamerica: affinin and capsaicin. *Journal of Ethnopharmacology*, **64:** 241-248.

Molina-Torres, J., Salazar Cabrera, C.J., Armenta-Salina, C. and Ramírez-Chávez, E. (2004) Fungistatic and bacteriostatic activities of alkamides from *Heliopsis longipes* roots. affinin and reduced amides. *Journal of Agricultural and Food Chemistry*, **52:** 4700-4704.

Mondal, A.K., Parui, S. and Mandal, S. (1998). Analysis of the Free Amino Acid content in pollen of nine Asteraceae species of Known allergenic activity. *Annals of Agricultural Environmental Medicine,* **5:** 17–20.

Moreira, V. M., Maia, J.G., de Souza, J.M., Bortolotto, Z.A. and Cavalheiro, E.A. (1989). Characterization of convulsions induced by a hexanic extract of *Spilanthes acmella* var. *oleracea* in rats. *Brazilian Journal of Medical and Biological Research,* **22(1):** 65-7.

Muranishi, S., Miyake, K., Uesugi, T., Mori, Y. and Miyazawa, T. (2008). Additive for Carbonated Beverage. *U.S. Patent Application,* **2008050500.**

Nagashima, M. and Nakatani, N. (1992). LC-MS and structure determination of pugent alkamides from *Spilanthes acmella* L flowers. *Food Science Technology,* **25:** 417-421.

Nakahara, K., Trakoontivakorn, G., Alzoreky, N.S., Ono, H., Onishi-Kameyama, M. and Yoshida, M. (2002). Antimutagenicity of Some Edible Thai Plants, and a Bioactive Carbazole Alkaloid, Mahanine, Isolated from *Micromelum minutum. Journal of Agricultural and Food Chemistry,* **50:** 4796-4802.

Nakatani, N. and Nagashima, M. (1992). Pungent alkamides from *Spilanthes acmella* L. var. oleracea Clarke. *Bioscience Biotechnology and Biochemistry,* **56:** 759-762.

Nakatsu, T., Mazeiko, P. J., Lupo, A.T., Green, B.C., Manley, C.H., Spence, D.J. and Ohta, H. (2004). Sensate composition imparting initial sensation upon contact. *U.S. Patent Application,* **2004052735.**

Nigrinis, L. S. O., Caro, J. O. and Olarte, E. N. (1986). Estudio Fitofarmacologico de la fraccion liposoluble de las flores de la *Spilanthes americana* (Mutis) parte I: Estudo Fitoquimico. *Revista Colombiana de Ciencias Quimico-Farmacêuticas,* **15:** 37-47.

No authors. (1976). Dentifrice composition. *Great Britain Patent Application,* **1438205.**

No authors. (2010). No Title. *Japanese Patent Application,* **2010037318.**

Norifumi, T., Makoto, W. and Yoko, H. (2001). Xylitol-containing composition. *Japanese Patent Application,* **2001178395.**

Pandey, V. and Agrawal, V. (2009). Efficient micropropagation protocol of *Spilanthes acmella* L. possessing strong antimalarial activity. *In Vitro Cellular* and *Developmental Biology - Plant,* **45:** 491–499.

Pandey, V., Agrawal, V., Raghavendra, K. and Dash, A.P. (2007). Strong larvicidal activity of three species of *Spilanthes* (Akarkara) against malaria (*Anopheles stephensi* Liston, *Anopheles culicifacies,* species C) and filaria vector (*Culex quinquefasciatus* Say). *Parasitology Research,* **102:** 171–4.

Panero, J.L., Jansen, R., and Clevinger, J.A. (1999). Phylogenetic relationships of subtribe Ecliptinae (Asteraceae: Heliantheae) based on chloroplast DNA restriction site data. (1999). *American Journal of Botany,* **86:** 413-427.

PIER http://www.hear.org/Pier/species/acmella_oleracea.htm, July 19/2007.

Prachayasittikul, S., Suphapong, S., Worachartcheewan, A., Lawung R., Ruchirawat, S. and Prachayasittikul, V. (2009). Bioactive Metabolites from *Spilanthes acmella* Murr. *Molecules*, **14**: 850-867.

Ramsewak, R.S., Erickson, A.J. and Nair, M. G. (1999). Bioactive N-isobutylamides from the flower buds of *Spilanthes acmella*. *Phytochemistry*, **51**: 729-732.

Raner, G.M., Cornelious, S., Moulick, K., Wang, Y., Mortenson, A. and Cech, N.B. (2007). Effects of herbal products and their constituents on human cytochrome P450(2E1) activity. *Food and Chemical Toxicology*, **45**: 2359-2365.

Rani, S. A. and Murty, S. U. (2006). Antifungal potential of flower head extract of *Spilanthes acmella* Linn. *African Journal of Biomedical Research*, **9**: 67 -69.

Ratnasooriya, W.D., Pieris, K.P.P., Samaratunga, U. and Jayakody, J.R.A.C. (2004). Diuretic activity of *Spilanthes acmella* flowers in rats. *Journal of Ethnopharmacology*, **91**: 317–320.

Reinhart, G.M., Helman, M.D.(2007). Compositions for Treating Keratinous Surfaces. *U.S. Patent Application*, **2007041922**.

Rios, M.Y., Aguilar-Guadarrama, A.B., Gutiérrez, M.C. (2007). Analgesic activity of affinin, an alkamide from *Heliopsis longipes* (Compositae). *Journal of Ethnopharmacology*, **110**: 364-367.

Romero, C.D., Chopin, S.F., Buck, G., Martinez, E., Garcia, M. and Bixby, L. (2005). Antibacterial properties of common herbal remedies of the southwest. *Journal of Ethnopharmacology*, **99**: 253–257.

Saraf, D.K., Dixit, V.K. (2002). Spilanthes acmella Murr.: Study on its extract spilanthol as larvicidal compound. *Asian Journal of Experimental Sciences*, **16**: 9-19.

Sandberg, F., Perera-Ivarsson, P. and El-Seedi, H. R. (2005). A Swedish collection of medicinal plants from Cameroon. *Journal of Ethnopharmacology*, **102**: 336–343.

Sawangjaroen, N., Phongpaichit, S., Subhadhirasakul, S., Visutthi, M., Srisuwan, N. and Thammapalerd, N. (2006). The anti-amoebic activity of some medicinal plants used by AIDS patients in southern Thailand. *Parasitology Research*, **98**: 588–592.

Sawangjaroen, N., Subhadhirasakul, S., Phongpaichit, S., Siripanth, C., Jamjaroen, K. and Sawangjaroen, K. (2005). The *in vitro* anti-giardial activity of extracts from plants that are used for self-medication by AIDS patients in southern Thailand. *Parasitology Research*, **95**: 17–21.

Taiji, O. and Hisashi, T. (2002). Insecticidal and anti-swarm agent for plant and method for its use. *Japanese Patent Application*, **2002363012**.

Tomoko, O., Koji, T., Yuka, M. and Ryoichi, Y. (1997). Hair tonic and hair nourishing agent. *Japanese Patent Application*, **9194334**.

Toshio, M., Takeshi, Y., Katsuyuki, M. and Shuichi, M. (2010). Taste improver for potassium salt and potassium salt-containing food and drink. *Japanese Patent Application*, **2010004767**.

Toshio, M., Tomoko, M., Shuichi, M. and Kazuyuki, M. (2006B). Food or drink flavor-reinforcing agent. *Japanese Patent Application*, **2006296356**.

Toshio, M., Tomoko, M., Shuichi, M. and Kazuyuki, M. (2006C). Taste improver for highly sweet sweetener. *Japanese Patent Application*, **2006223104**.

Toshio, M., Tomoko, M., Shuichi, M. and Kazuyuki, M. (2006A). Taste-reinforcing agent and spice containing the taste-reinforcing agent and food or drink containing the spice. *Japanese Patent Application*, **2006296357**.

Toshiya, S. and Tetsuo, G. (1995). Essential oil having high spilanthol content, production thereof and composition for oral cavity blended with essential oil having high spilanthol content. *Japanese Patent Application*, **7090294**.

Tropicos http://mobot.mobot.org/W3T/Search/vast.html, August/2010.

Wikipedia, the free encyclopedia, http://en.wikipedia.org/wiki/Acmella_oleracea, August/2010.

Wongsawatkul, O., Prachayasittikul, S., Isarankura-Na-Ayudhya, C., Satayavivad, J., Ruchirawat, S. and Prachayasittikul, V. (2008). Vasorelaxant and Antioxidant Activities of *Spilanthes acmella* Murr. *International Journal of Molecular Sciences*, 9: 2724-44.

Wu, L., Fan, N., Lin, M., Chu, I., Huang, S., Hu, C. and Han, S. (2008). Anti-inflammatory Effect of Spilanthol from *Spilanthes acmella* on Murine Macrophage by Down-Regulating LPS-Induced Inflammatory Mediators. *Journal of Agricultural and Food Chemistry*, **56**: 2341–2349.

Yukio, Y., Takashi, H. and Chiharu, G. (2008). Tooth-cleaning agent composition. *Japanese Patent Application*, **2008115115**.

Natural Products: Research Reviews Vol. 1 (2012)
Editor: V.K. Gupta
Published by: DAYA PUBLISHING HOUSE, NEW DELHI

Pages 439–460

10

Effect of Natural Plant Based and Non-Steroidal Anti-Inflammatory Drugs on the Immune System

Amit Gupta[1]* Anamika Khajuria[2],
Jaswant Singh[2], Surjeet Singh[2] and V.K. Gupta[2]

ABSTRACT

Generally, the inflammatory process involves a series of events that can be elicited by numerous stimuli such as infectious agents, ischaemia, antigen–antibody interaction and thermal or physical injury. Inflammation is usually associated with pain as a secondary process resulting from the release of algesic mediators Therapy of inflammatory diseases is usually directed at the inflammatory processes. Through years of ingenious synthesis and structural modifications, which usually accompany design and development of new drug substances, many non-steroidal anti-inflammatory agents have been prepared and marketed. These have been of immense help in the management of various inflammatory conditions like rheumatism, arthritis and pain. Moreover, in the face of rising costs of orthodox medicines, phytomedicinal treatment of diseases has become the order of the day in all over the world, chiefly because of their ready affordability and availability, especially in the rural set-ups where the greater percentage of the people are poor and merely subsisting. In this review, we discuss both the effect of natural plant based and non-steroidal anti-inflammatory drugs on the immune system.

Keywords: Inflammation, Ischaemia, Non-steroidal anti-inflammatory agents, Rheumatism, Arthritis.

---

1   School of Pharmacy, National Taiwan University Hospital, Taipei

2   Division of Pharmacology, Indian Institute of Integrative Medicine (CSIR), Canal, Road, Jammu   180 001, India

*   *Corresponding author*: E-mail: amitrrl@yahoo.com; amitrrl@gmail.com

Introduction

A number of natural products are used in the traditional medical systems in many countries. Alternative medicine for treatment of various diseases is getting more popular. Many medicinal plants provide relief of symptoms comparable to that obtained from allopathic medicines. In recent times, focus on plant research has increased all over the world and a large body of evidence has collected to show immense potential of medicinal plants showing anti-inflammatory activity. The present review aims to compile data generated through the research activity using modern scientific approaches and innovative scientific tools. To facilitate the readers to look at their areas of interest more easily, the data in the present review have been organized in two sections *i.e.*, natural plant based and non-steroidal anti-inflammatory drugs.

Anti-inflammatory Activity of Natural Products

The severe side effects of steroidal and nonsteroidal anti-inflammatory drugs evoked us to search for new anti-inflammatory agents from natural botanical sources which may have minimal drawbacks. Our review of 'folk medicine' indicated that many plants possessing an anti-inflammatory activity were sometimes consumed by humans. These were the roots of *Glycyrrhiza glabra* L., family Leguminosae (Liquorice), the seeds of *Trigonella foenum groecum* L., family Leguminosae (Fenugreek), and the fruits of *Coriandrum sativum* L., family Umbelliferae (Coriander). The anti-inflammatory activity of liquorice has been studied and still current research is being carried out by some authors (Amagaya *et al.*, 1984; Akamatsu *et al.*, 1991). The majority of literature deals only with the activity of the aqueous extract and its main component, glycyrrhizin. The anti-inflammatory activity of fenugreek seeds have been reported (Ammar *et al.*, 1992); the authors noticed only the topical anti-inflammatory activity of the aqueous alcoholic extract on treating Aphthus ulcers and attributed its activity to the presence of triterpenoid saponins. Scanty literature concerning the anti-inflammatory effect of coriander fruits were available, only the topical anti-inflammatory activity of glycerol-ethanol extract of coriander fruits was studied and proved to be successful (Asylgaraj *et al.*, 1993). Following are the plants showed anti-inflammatory activity:-

1. *Cymbopogon citratus*, an herb known worldwide as lemongrass, is widely consumed as an aromatic drink, and its fresh and dried leaves are currently used in traditional cuisine. However, little is known about the mechanism of action of *C. citratus*, namely, the anti-inflammatory effects of its dietary components. Because nitric oxide (NO), produced in large quantities by activated inflammatory cells, has been demonstrated to be involved in the pathogenesis of acute and chronic inflammation, we evaluated the effects of the infusion of dried leaves from *C. citratus*, as well as its polyphenolic fractions-flavonoid, tannin and phenolic acid-rich fractions (FF, TF, and PAF, respectively) on the NO production induced by lipopolysaccharide (LPS) in a skin-derived dendritic cell line (FSDC). *C. citratus* infusion significantly inhibited the LPS-induced NO production and inducible NO synthase (iNOS) protein expression. All the polyphenolic fractions tested also reduced the iNOS protein levels and NO production stimulated by

LPS in FSDC cells, without affecting cell viability, with the strongest effects being observed for the fractions with mono- and polymeric flavonoids (FF and TF, respectively). It also showed the anti-inflammatory properties of FF are mainly due to luteolin glycosides. In conclusion, *C. citratus* has NO scavenging activity and inhibits iNOS expression and should be explored for the treatment of inflammatory diseases, in particular of the gastrointestinal tract (Figuerinha *et al.,* 2010).

2. Plant flavonoids show anti-inflammatory activity *in vitro* and *in vivo* (Kim *et al.*, 2004). Although not fully understood, several action mechanisms are proposed to explain *in vivo* anti-inflammatory action. One of the important mechanisms is an inhibition of eicosanoid generating enzymes including phospholipase A2, cyclooxygenases, and lipoxygenases, thereby reducing the concentrations of prostanoids and leukotrienes. Recent studies have also shown that certain flavonoids, especially flavone derivatives, express their anti-inflammatory activity at least in part by modulation of proinflammatory gene expression such as cyclooxygenase-2, inducible nitric oxide synthase, and several pivotal cytokines. Due to these unique action mechanisms and significant *in vivo* activity, flavonoids are considered to be reasonable candidates for new anti-inflammatory drugs. To clearly establish the therapeutic value in inflammatory disorders, *in vivo* anti-inflammatory activity and action mechanism of varieties of flavonoids need to be further elucidated. The effect of flavonoids on eicosanoid and nitric oxide generating enzymes and the effect on expression of proinflammatory genes. As natural modulators of proinflammatory gene expression, certain flavonoids have a potential for new anti-inflammatory agents.

3. Methanolic extract of *Solanum trilobatum* Linn. (MEST) belonging to the family of Solanacea was evaluated by hot plate and acetic acid induced writhing methods to assess its analgesic activity (Pandurangana *et al.*, 2008). The extract was also showed anti-inflammatory activity by subjecting into carrageenan, and cotton pellet induced granuloma tests and showed its effect on acute and chronic phase inflammation models in rats, as well as analgesic activity in mice. It was found that the extract caused an inhibition on the writhing response induced by acetic acid in a dose dependent manner. It was also indicated that the MEST showed significant antinociceptive action in hot plate reaction time method in mice.

4. The ethanolic extract of the leaf of *Vitex leucoxylon* showed significant inhibition of carrageenin paw oedema and granulation tissue formation in rats (Makwana *et al.*, 1994).

5. The aqueous suspension of dried latex of *Calotropis procera* (Arka) showed anti-inflammatory property when tested in the carrageenin and formalin induced rat paw oedema models (Kumar and Basu, 1994).

6. The roots and leaves of *Butea frondosa* (Palash) were evaluated for ocular anti inflammatory activity in rabbits. The results showed that the gel formulation of *Butea frondosa* leaves, prepared using a commercially available,

pluronic F-127, reduced the intra-ocular pressure, decreased leucocytosis and miosis and was comparable to flubiproten gel (Mengi and Deshpande, 1995).

7. The triglyceride fraction of oil of *Ocimum sanctum* (Tulsi) offered higher protection against carrageenin induced paw oedema in rats and acetic acid induced writhing in mice, as compared to the fixed oil (Singh *et al.*, 1996). Fixed oil of *Ocimum sanctum* and linolenic acid were found to possess significant anti-inflammatory activity against PGE2, leukotriene and arachidonic acid induced paw oedema. The anti-inflammatory activity of linolenic acid present in the fixed oil of *Ocimum sanctum* was probably due to blockade of both, the cyclo-oxygenase and lipo-oxygenase pathways of arachidonic acid metabolism (Singh and Majumdar, 1997).

8. Alcoholic extract of *Ochna obtusata* stem bark demonstrated potent anti-inflammatory effects in the rat paw oedema and cotton pellet granuloma models (Sivaprakasam *et al.*, 1996).

9. All extracts of the root of *Pongamia pinnata* showed significant anti-inflammatory activity (compared to phenylbutazone) in carrageenin and PGE1 induced oedema models. Possible mechanism of action could be prostaglandin inhibition, especially by EE and AE. The BE was effective in carrageenin but not the PGE1 model of inflammation. The anti-inflammatory property appears to reside mainly in the intermediate polar constituents and not in lipophilic or extremely polar constituents (Singh and Pandey, 1996). The PEE and CE of the seeds of *Pongamia pinnata* showed potent acute anti-inflammatory effect whereas the aqueous suspension showed pro-inflammatory effects. Further studies have shown that maximum anti-inflammatory effect was seen in the bradykinin induced oedema model with the direct EE (Singh and Pandey, 1996). Possible mechanism of action could be inhibition of prostaglandin synthesis and decreased capillary permeability. PEE and AE inhibited histamine and 5- hydroxytryptamine induced inflammation probably by their lipophilic constituents preventing the early stages of inflammation. However, the fractions were not effective against Freund's adjuvant arthritic model. The latter finding indicates that the plant may not be effective in rheumatoid arthritis.

10. All extracts of *Abies pindrow* Royle leaf showed anti-inflammatory effect in various animal models of inflammation such as carrageenin induced paw oedema, granuloma pouch and Freund's adjuvant arthritis. Chemical analysis indicated the presence of glycosides and steroids in the PEE and BE and terpenoids and flavonoids in the AE and EE. Flavonoids and terpenoids are polar substances effective in acute inflammation whereas glycosides and steroids are non-polar substances effective in chronic inflammation ((Singh and Pandey, 1997).

11. The methanolic extracts of the flowers of *Michelia champaca* Linn. (Champaka), *Ixora brachiata* Roxb (Rasna) and *Rhynchosia cana* Willd were found to possess significant anti-inflammatory activity against cotton pellet induced subacute inflammation in rats. The latter two drugs showed higher

activity as compared to *Michelia champaca*. They also reduced the protein content, acid phosphatase, glutamate pyruvate transaminase and glutamate oxaloacetate transaminase activities in the liver and serum. These properties are probably due to the presence of flavonoids in the flowers of these plants (Vimala *et al.*, 1997).

12. The water soluble part of the alcoholic extract of *Azadirachta indica* exerted significant anti-inflammatory activity in the cotton pellet granuloma assay in rats. Levels of various biochemical parameters studied in cotton pellet exudate were also found to be decreased *viz.*, DNA, RNA, lipid peroxide, acid phosphatase and alkaline phosphatase suggesting the mechanism for the anti-inflammatory effect of *Azadirachta indica* (Chattopadhyay, 1998).

Non-steroidal Anti-inflammatory Drugs

The majority of clinically important medicines belong to steroidal or non-steroidal anti-inflammatory chemical therapeutics for treatment of various inflammatory diseases. The non-selective non-steroidal anti-inflammatory drugs are a class of drug used widely (Dubois *et al.*, 2004). Non-steroidal anti-inflammatory drugs are one of the classes of drugs most prescribed worldwide (Dubois *et al.*, 2004). However, concern over the overall non-steroidal anti-inflammatory drug consumption has arisen due to issues around their toxicity (Dubois *et al.*, 2004) and also because of their high patterns of utilization, often in inappropriate population groups (Barozzi and Tett, 2007). Non-steroidal anti-inflammatory drugs have been used to decrease pain and inflammation for major and minor musculoskeletal disorders for years (Green, 2001; Hawkins and Hanks, 2000).

Non-steroidal anti-inflammatory drugs are drugs with analgesic and antipyretic (fever-reducing) effects and which have, in higher doses, anti-inflammatory effects (reducing inflammation). The term non-steroidal is used to distinguish these drugs from steroids, which (among a broad range of other effects) have a similar eicosanoid-depressing, anti-inflammatory action. Inflammatory cells produce a highly complicated mixture of growth and differentiation cytokines as well as biologically active arachidonate metabolites. In addition, they possess the ability to generate and release a spectrum of reactive oxygen species (ROS), reactive nitrogen species (RNS) and free radicals during oxidative burst. Among inflammatory cells, polymorphonuclear leukocytes (PMNs) are particularly adept at generating and releasing ROS and RNS, including superoxide, hydrogen peroxide, nitric oxide, hypochlorous acid, singlet oxygen and hydroxyl radical (Ramos *et al.*, 1992 and Steineck *et al.*, 1992). The excessively produced reactive oxygen species can injure cellular biomolecules such as nucleic acids, proteins, carbohydrates, and lipids, causing cellular and tissue damage, which in turn augments the state of inflammation (Cochrane, 1991). Therefore, compounds that have scavenging activities toward these radicals and/or suppressive activities on lipid peroxidation may thus be expected to have therapeutic potentials for several inflammatory diseases (Trenam *et al.*, 1992).

In one study, non-steroidal anti-inflammatory drugs with cyclooxygenase (COX) inhibitory activity are commonly used in various inflammatory diseases. In the previous study, to examine the immunomodulatory effects of well known non-steroidal anti-inflammatory drugs at clinically available doses, macrophage and T cell-mediated immune responses such as tumor necrosis factor (TNF)-alpha release and nitric oxide (NO) production, cell-cell adhesion, phagocytic uptake and lymphocyte proliferation were investigated. Non-steroidal anti-inflammatory drugs tested significantly enhanced TNF-α release from lipopolysaccharide (LPS)-activated RAW264.7 cells at certain concentrations (fenoprofen, indomethacine, piroxicam, ace-dofenac, diclofenac and sulindac) or in a dose-dependent manner (aspirin and phenyibuta-zone). Of non-steroidal anti-inflammatory drugs, phenylbutazone and aspirin most potently attenuated NO production, although sulindac was the only compound with cytoprotective activity against LPS-induced cytotoxicity. Most non-steroidal anti-inflammatory drugs used displayed weak or no modulatory effects on phagocytic uptake and CD29- or CD43-mediated cell-cell adhesion. Interestingly, however, phenylbutazone itself triggered cell-cell clustering under normal culture conditions and enhanced the phagocytic activity. Aspirin and phenylbutazone also dose-dependently attenuated CD4+ T cell proliferation stimulated by concanavalin A (Con A) and CD8+ CTLL-2 cell proliferation induced by interleukin (IL)-2. Sulindac only blocked CTLL-2 cell proliferation. These results suggest that non-steroidal anti-inflammatory drugs may differentially exert immunomodulatory effects on activated macrophages and lymphocytes, and some of the effects may enforce NSAID's therapeutic effect against inflammatory symptoms (Cho, 2007).

Enzymes as Non-steroidal Anti-inflammatory Drugs

The anti-inflammatory activity of proteolytic enzymes has been attributed to several mechanisms. They can reduce the swelling on mucous membranes, decrease capillary permeability, dissolve blood clot-forming fibrin deposits and microthrombi. Enzymes reduce the viscosity of the blood and thus improve circulation. This consequently increases the supply of oxygen and nutrients to traumatized tissue, at the same time transporting harmful waste products away from it. Bromelain has been shown to inhibit platelet aggregation by using *in vitro* and *in vivo* models (Metzig *et al.*, 1999).

Proteolytic enzymes also help break down plasma proteins and cellular debris at the site of an injury into smaller fragments facilitating their passage through the lymphatic system and resulting in more rapid resolution of swelling. The net result is relief of pain and discomfort. Comparative studies in animal models for anti-inflammatory action of proteolytic enzymes with standard drugs such as phenylbutazone, hydrocortisone, indomethacin, and acetylsalicylic acid (aspirin) revealed that the enzymes were on par with the standard drugs and at times even superior (Netti *et al.*, 1972).

Immune System

Dendritic Cells: Essential Components of the Immune System

The immune system has multiple pathways for recognizing and responding to microbial components and other disease-related stimuli. The dendritic-cell lineage of white blood cells controls this intricate system. After nearly 35 years of research, there are many perspectives from which to appreciate the powerful influence of dendritic cells within the broad reach of immunology. There are different perspectives of dendritic cells are:

Natural selection, dendritic cells help the immune system defend against more than a thousand different forms of infection. They capture microbial proteins and lipids, and present them to lymphocytes, thereby launching lymphocyte responses.

Physiological perspective, resistance to infection is not a single automatic response. Instead, dendritic cells select from a host of rapid, short-lived innate reactions and from the more slowly acquired and longer-lived adaptive responses.

Cellular perspective, dendritic cells are best known for their role in initiating T cell immunity.

Medical perspective, dendritic cells influence many clinical conditions. In addition to providing resistance to some diseases, dendritic cells can investigate autoimmune inflammation and allergy and transplant rejection, and they can be exploited by several infections and tumors.

Dendritic cells are the most potent antigen-presenting cells in the body, and their unique ability to stimulate a primary T cell response places them at the center of the immune response (Banchereau *et al.*, 2000). Immature dendritic cells differentiate from bone marrow progenitors or from monocytes and then either stay in the blood stream or migrate into the peripheral tissues. Immature dendritic cells, such as Langerhans cells in the skin, survey incoming pathogens. They are equipped with receptors to become activated when exposed to pathogen-associated molecular patterns (Medzhitov and Janeway, 1997). Their capacity to recognize pathogens and become activated therefore represents the first critical event in the initiation of the immune response. An encounter with a pathogen leads to dendritic cell maturation and migration through lymphatic vessels to T cell areas of secondary lymphoid organs. Antigen presentation by dendritic cells activates specific naive T cells to express CD40 ligand (CD154) (Caux *et al.*, 1994), which, in turn, activates dendritic cells, achieving their terminal differentiation, as assessed by the up-regulation of MHC I and II molecules and of co-stimulatory molecules CD80/CD86 and by the production of cytokines, such as IL-12 and IL-1α/β, which all participate in T cell stimulation (Cella *et al.*, 1996; Koch *et al.*, 1996; Ridge *et al.*, 1998) and in the development of adaptive immunity (Cella *et al.*, 1996; Ridge *et al.*, 1998). Mature dendritic cells modulate T cell responses through the secretion of various cytokines such as IL-12, promoting a Th1-type cellular immune response (Macatonia *et al.*, 1995), or IL-4 following thymic stromal lymphopoietin activation, promoting a Th2-type humoral immune response (Soumelis *et al.*, 2002). Finally a dendritic cells apoptosis program can be triggered at the end of the maturation process so that mature dendritic cells do not produce an overstimulation of the immune system (De Smedt *et al.*, 1998). Dendritic cells frequently operate under stress conditions induced by tissue damage, infectious pathogens, or inflammatory reactions. Oxidative stress enhances the production of inflammatory cytokines by dendritic cells (Verhasselt *et al.*, 1998).

Macrophages

The innate arm of the immune system functions as a first line of defense against pathogens that enter from the external environment. One key component of this system is the macrophage, a professional phagocyte that is essential for host defense.

Origin

Macrophages differentiate through monoblasts, promonocytes, and monocytes from hematopoietic stem cells that originate from the bone marrow. Under physiological conditions, monocytes are released from the bone marrow and then move to various tissues and organs such as the liver and lungs, where they differentiate into various types of macrophages such as Kupffer cells and alveolar macrophages due to M-CSF and possibly GM-CSF.

Function

Phagocytosis, a primary function of the macrophage, is initiated by the engagement of any number of numerous receptors on the cell surface. Although a complex series of events ensues, the end result is the digestion of the pathogen and the subsequent presentation of pathogen-derived peptides to the adaptive arm of the immune system. This level of participation in the immune system is not constrained to specific sites since macrophages are strategically situated in all tissues throughout the body, with characteristics and functions specific to their location. For example, Baroni *et al.* (1987) have demonstrated that differences exist between macrophages that populate different lymphoid tissues. Despite the heterogeneity of tissue-specific macrophage populations, they still share some common characteristics such as their phagocytic ability and their non-specific esterase activity. They also play an important role in the immune system as antigen-presenting cells and regulatory cells.

Macrophages and Apoptotic Cells

Macrophages are professional phagocytic cells and constitute the first line of defense against foreign pathogens and environmental particles. Moreover, clearance of apoptotic cell corpses by macrophages is essential to maintain normal tissue homeostasis and plays an important role in the resolution of inflammation and prevention of autoimmune responses (Witasp *et al.*, 2008; Savill *et al.*, 2002; Savill and Fadok, 2000). Discrimination between living, dying and dead cells is an essential requirement for appropriate clearance of apoptotic cells. Failure to clear dying cells may sometimes reflect an imbalance between the number of dead cells and the local availability of functional scavenger phagocytes. Factors that increase apoptosis (such as viral infections, local inflammation, and neonatal tissue remodeling) and factors that decrease the phagocytic ability of macrophages (such as phagocytic exhaustion and the cytokine milieu) may contribute to the accumulation of uncleared apoptotic bodies and breaking of immune tolerance (Maderna and Godson, 2003).

Apoptosis is a physiological program of cell suicide that directs engulfment and safe destruction of cell corpses by healthy neighbouring cells or professional phagocytic scavengers such as macrophages (Fadok and Chimini, 2001). The processes of apoptosis and phagocytosis work in concert to perform central roles in biological functions such as embryogenesis; mature tissue homeostasis; elimination of infected, aged, and injured cells; cellular immunity; and resolution of inflammation (Cohen 1993; Han *et al.*, 1993; Hopkinson-Woolley *et al.*, 1994).

The ability of professional phagocytes to clear apoptotic cells in a non-inflammatory manner relies on the expression by the dying cells of eat-me signals and other specific ligands of dying cells. The expression of these apoptotic signals is common to all physiological forms of cell death, regardless of the stimulus that triggered cell death (Cocco and Ucker, 2001; Cvetanovic and Ucker, 2004). Cells that die by necrosis are also recognized by professional phagocytes; however, necrotic corpuscles do not downregulate inflammatory responses and, instead, usually promote inflammation. It is not clear how macrophages recognize necrotic cells (Krysko *et al.*, 2006). While it was suggested that some macrophage receptors involved with engulfment of apoptotic cells also contribute to the uptake of necrotic cells (Bottcher *et al.*, 2006), recognition of apoptotic and necrotic cells may take place through distinct mechanisms (Cocco and Ucker, 2001). Recent studies indicate that the interaction of macrophages with dying cells initiates internalization of the apoptotic or necrotic targets, and that internalization can be preceded by "zipper"-like and macropinocytotic mechanisms, respectively (Krysko *et al*, 2006).

B Cells

B cells are the least efficient antigen presenting cells. Unlike the other two APC's, they possess specific antigen receptors, surface immunoglobulins. B cells ingest soluble proteins by pinocytosis. They also possess specific uptake receptors in surface Immunoglobulins. B cells present antigen via MHC-II. But these cells do not express co-stimulatory molecules. In order to do so, they need to be activated by Th cells. The role of B cells as APC's *in vivo* is not very well understood.

Examples of Non-steroidal Inflammatory Drugs Functions that are Currently Under Investigation

Macrophages are a first line of defense against microbial invaders and malignancies by nature of their phagocytic, cytotoxic, and intracellular killing capacities (Adams and Hamilton, 1984). Macrophage activation by lipopolysaccharide (LPS), the major component of gram-negative bacteria cell wall, results in the release of several inflammatory mediators such as nitric oxide (NO) and the proinflammatory cytokines, tumor necrosis factor-α (TNF-α), interleukin (IL)-6, and IL-12 (Schimmer and Parker, 2001). NO is a highly reactive molecule produced from a guanidine nitrogen of NO synthase (NOS). Three isoforms of NOS have been identified and are classified into two major categories, namely, constitutive and inducible NOS. Neuronal and endothelial NOSs, which are constitutively expressed, are activated by calcium and calmodulin and are called constitutive NOSs (Nathan, 1992). Of the three NO synthases, inducible NOS (iNOS), the high-output isoform, is the most widely expressed in various cell types after its transcriptional activation (Xie *et al.*, 1992). Most importantly, iNOS is highly expressed in LPS-activated macrophages, and this contributes to the pathogenesis of septic shock (Petros *et al.*, 1994). The physiologic or normal production of NO from phagocytes is beneficial for the host defense against micro organisms, parasites, and tumor cells (Thiermermann and Vane, 1990). However, overproduction of NO can be harmful and may result in septic shock, neurologic disorders, rheumatoid arthritis, and autoimmune diseases (Evans, 1995, O Shea *et al.*, 2002; Thiermermann and Vane, 1990). Therefore, inhibition of NO production is an important therapeutic target in the development of anti-inflammatory agents. Many anti-inflammatory drugs (non-steroidal) have been reported to act either as inhibitors of free radical production or as radical scavengers. Compounds with antioxidant properties could be expected to offer protection in rheumatoid arthritis and inflammation and to lead to potentially effective drugs.

Indomethacin

Indomethacin (IMN), a nonsteroidal anti-inflammatory drug widely used to treat arthritic diseases, causes gastric lesions (Langman *et al.*, 1991) by inhibiting prostaglandin biosynthesis. Decreased prostaglandin level impairs almost all aspects of gastroprotection and increases acid secretion to aggravate the ulcer (Milller, 1983). Other major factors include indomethacin-induced microvascular injury (Takeuchi *et al.*, 1990), neutrophil infiltration (Wallace *et al.*, 1990), induction of proinflammatory TNF-α expression (Souza *et al*, 2004), nitric oxide imbalance and apoptosis (Souza *et al.*, 2004; Fujii *et al.*, 2000) and extracellular matrix damage by modulation of matrix metalloproteinases −9 and −2 (Swarnakar *et al.*, 2005).

Reactive oxygen species (ROS) also play a vital role in indomethacin-induced gastric damage (Yoshikawa *et al.*, 1993; Kusuhara *et al.*, 1999). Indomethacin causes gastric erosions with increased lipid peroxidation and decreased glutathione peroxidase activity (Yoshikawa *et al.*, 1993). Treatment with superoxide dismutase and catalase inhibits the lesions suggesting involvement of ROS in gastric damage (Yoshikawa *et al.*, 1993). Cultured gastric mucosal cells exposed to H_2O_2 undergo oxidative injury, which is protected by catalase and desferrioxamine (Hiraishi *et al.*, 1993). Indomethacin also induces apoptosis by DNA fragmentation in cultured gastric mucosal cells by generating ROS, which is blocked by some antioxidants (Kusuhara *et al.*, 1999).

Acetaminophen

Acetaminophen (APAP) is the leading cause of drug induced liver disease in the United States resulting in over 56,000 emergency room visits and approximately 500 deaths each year (Nourjah *et al.*, 2006). One of the problems associated with APAP toxicity is the wide availability of the drug. APAP is present in more than 180 over the counter (OTC) products, which increases the probability of accidental overdose. Acute overdose of APAP leads to severe hepatic centrilobular necrosis (Golden *et al.*, 1981). Rapid treatment with *N*-acetylcysteine (NAC) is currently the clinical treatment for APAP overdose.

The toxicity of APAP is mediated through its biotransformation into *N*-acetyl-p-benzoquinoneimine (NAPQI) by cytochrome P450 2E1, 3A4, and 1A2 (Corcoran *et al.*, 1980; Dahlin *et al.*, 1984; Patten *et al.*, 1993). NAPQI is a strong electrophile that rapidly adducts sulfhydryl groups like those found on reduced glutathione (GSH) (Streeter *et al.*, 1984). GSH depletion by NAPQI precedes APAP toxicity (Larrauri *et al.*, 1987). In addition to adducting proteins, NAPQI also induces mitochondrial dysfunction leading to a severe energy debt and the formation of reactive oxygen species (ROS) that induce further damage in the hepatocytes (Andersson *et al.*, 1990). The current treatment for APAP overdose is *N*-acetylcysteine (NAC). NAC functions by replenishing cellular stores of cysteine which is involved in the rate-limiting step in the formation of GSH.

N omega-nitro-L-arginine Methyl Ester (L-NAME): Inhibitor of Nitric Oxide Function as Anti-inflammatory Agent

Effect of L-NAME on nitric oxide production was observed in dendritic cells of untreated mice. Dendritic cells were cultured in RPMI-FBS (10 per cent) and contents of the nitric oxide were measured in the supernatant as shown in Figure 10.1. Increasing doses of L-NAME registered a significant decrease in the production of nitric oxide as compared to control. We also evaluated the *in vitro* effect of L-NAME with apoptotic EL4 cells on nitrite secretion by dendritic cells obtained from untreated

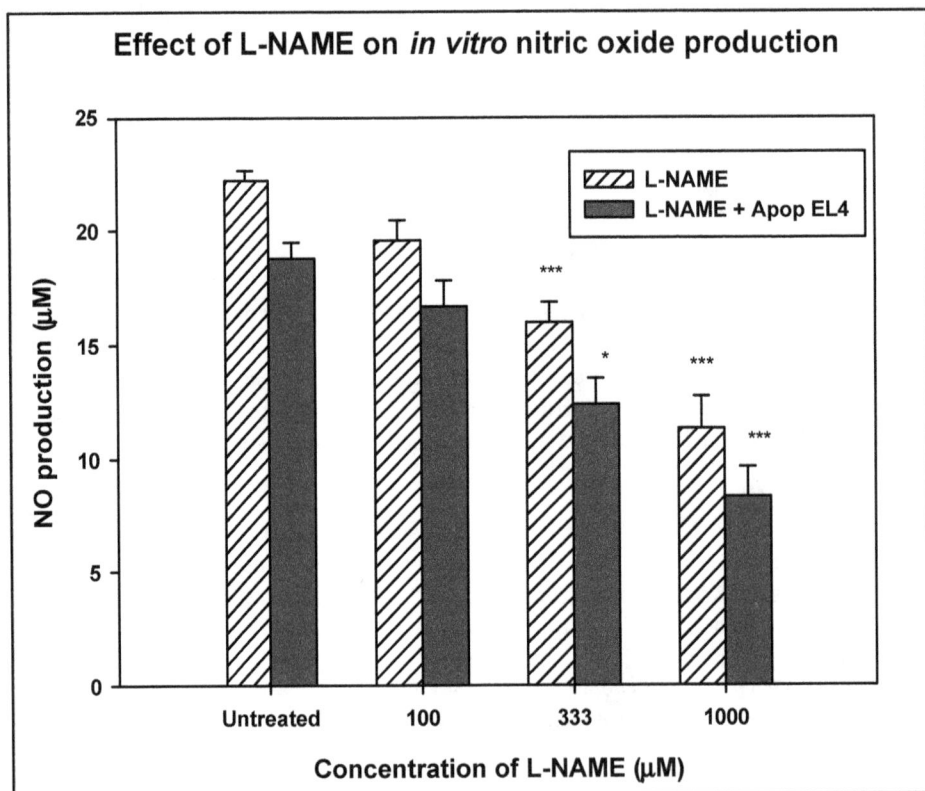

Figure 10.1: Effect of L-NAME on nitric oxide production.
Dendritic cells (10^5 cells/well) cocultured with or without apoptotic EL4 cells (10^5 cells/well) for 24 h incubation. Centrifuge the plate and collect the supernatant for the estimation of nitric oxide through Griess assay. Values are expressed as Mean ± SE of six set of experiments. The difference between the untreated and drug treated is determined by one way ANOVA analysis (Bonferroni multiple comparison test). * P < 0.05; ***P < 0.001.

mice. The results showed that dendritic cells were incubated with apoptotic EL4 cells or with different concentrations of L-NAME showed drastic decrease in nitric oxide production as compared to control.

For *in vitro* experiments, bone marrow dendritic cells were incubated with OVA (0.1 mg/ml) for 24 h and subsequently cocultured with apoptotic EL4/L121 cells and B3Z cells as shown in Figure 10.2. To determine B3Z activation, the medium was removed after 24h of coculture with B3Z cells and supernatant were incubated with

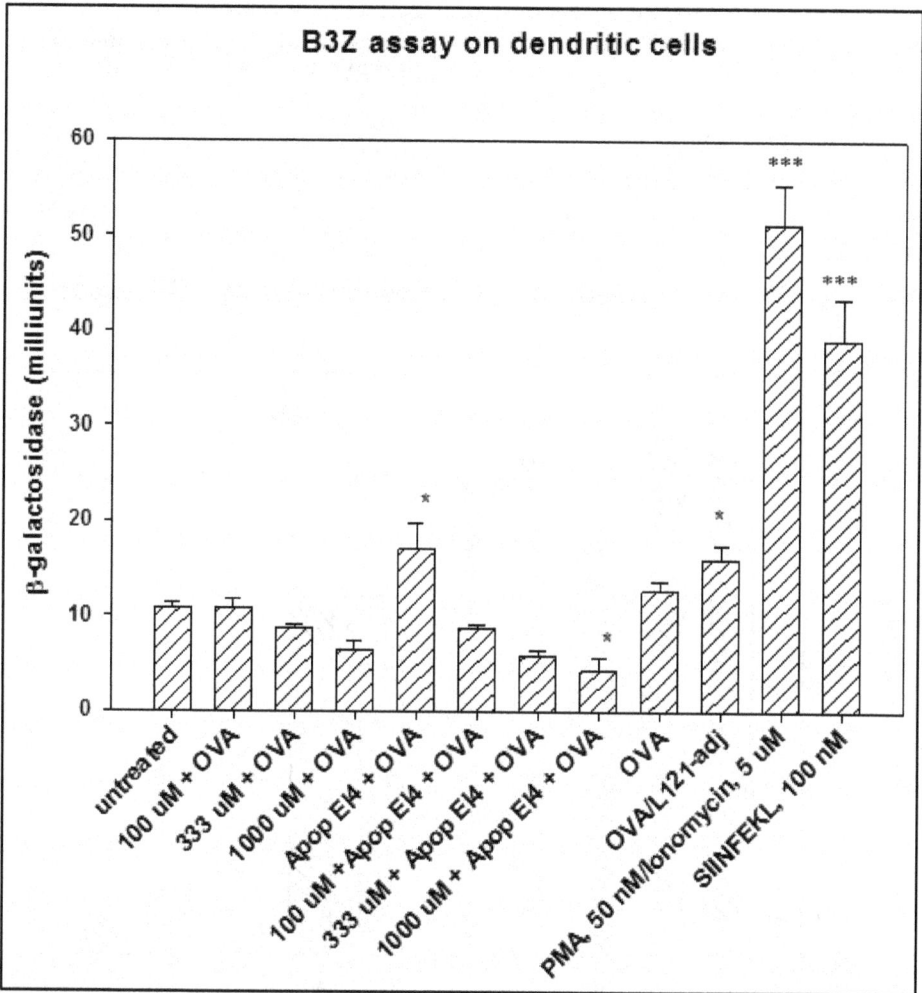

Figure 10.2: Effect of L-NAME on B3Z assay.

Dendritic cells (10^5 cells/well) cocultured with or without apoptotic EL4 cells (10^5 cells/well) for 24 h incubation. Wash the plate and then add B3Z cells (10^5 cells/well) again incubate for 18-24 h incubation. Wash and Centrifuge the plate and transfer the supernatant into another plate and add substrate into the wells. The results are compared with the standard curve of beta galactosidase. The difference between the untreated and drug treated is determined by one way ANOVA analysis (Bonferroni multiple comparison test). * P < 0.05; ***P < 0.001.

0.3 mM chlorophenol red-β-D-galactopyranoside (CPRG). After 5 h incubation, B3Z activation was analyzed by Elisa reader. The results showed that that dendritic cells were incubated with apoptotic cells EL4/L121 adjuvant emulsion resulted in a significantly increased cross-presentation of OVA while L-NAME was associated with decreased cross-presentation of OVA to MHC I as compared to control. PMA/ Ionomycin and SIINFEKL used as standard showed a drastic increase in cross presentation of OVA to MHC-I presentation as compared to control.

The MTT assays were used to examine the effects of L-NAME on J744A.1 cell viability. L-NAME caused a dose-dependent decrease in cell viability as compared to control. The survival and suppressive dose of L-NAME is up to 1 mM and 10 mM (Figure 10.3).

For *in vivo* experiment, animals were immunized on day 0 and 7 with OVA or with OVA/L121-adj or with or without L-NAME (10 mg/kg subcutaneously). On day 10, macrophages and serum were collected for the estimation of nitric oxide in

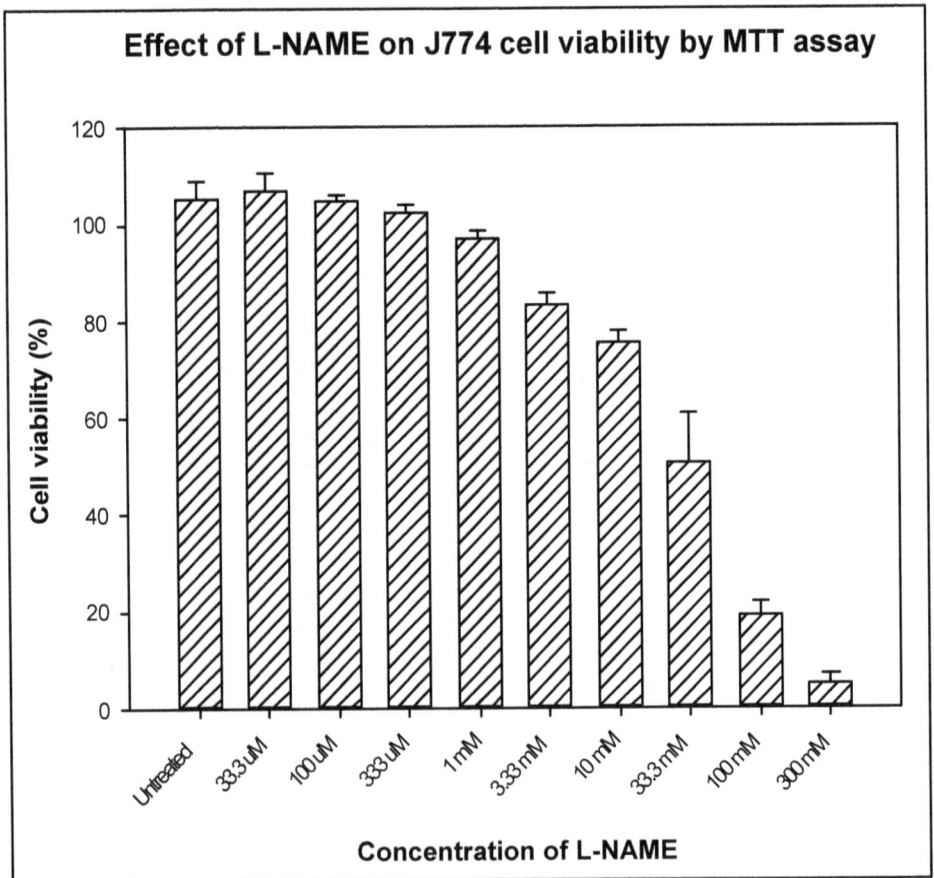

Figure 10.3: Effect of L-NAME on J774A cell viability by MTT assay.
J774A cells were treated for 24 h with various concentrations of L-NAME, then examined using the MTT assay. The data are presented as the mean ± S.E.

macrophages and estimation of cytokines (TNF alpha, IL-6 and IL-1 beta) in serum. The results showed that L-NAME/L121-adj showed drastic increase in nitric oxide production from macrophages and also cytokines in serum as compared to control (Figures 10.4 and 10.5).

Conclusions

In summary, persistent inflammatory pain is a condition that requires treatment. But there are two types of treatment available *i.e.* plant based and non-steroidal anti-inflammatory drugs. Currently available non-steroidal anti-inflammatory drugs can be used to treat mild to moderately severe inflammatory pain. Non-steroidal anti-

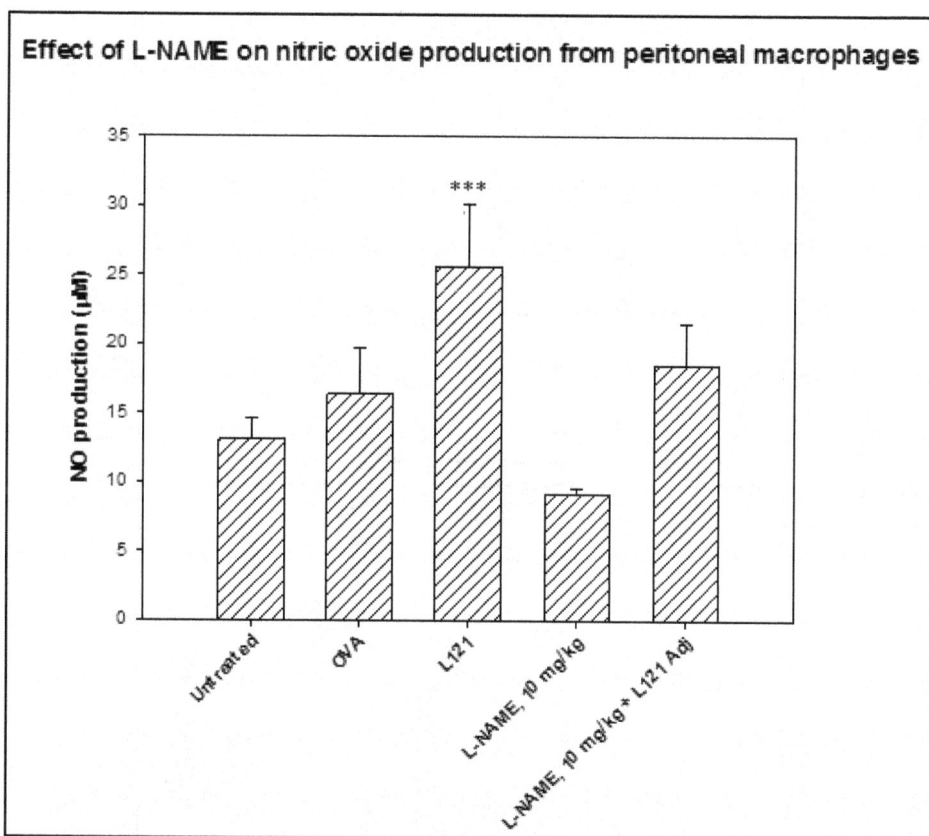

Figure 10.4: Effect of L-NAME on nitric oxide production from the peritoneal macrophages. Mice were sacrificed and peritoneal exudates cells isolated. The cells were washed twice with RPMI and centrifuged at 1200 rpm for 6 min and finally adjusted to 106 cells/ml in RPMI-FBS (10%). 100 ml of cell suspension was dispensed in triplicates in 96-well culture plates. Macrophages were allowed to adhere to the bottom of the wells at 37°C for 24 h in CO_2 incubator. After 24 h, centrifuge the plate and collect the supernatant for the estimation of nitric oxide through Griess assay. Values are expressed as Mean ± SE of three set of experiments. The difference between the untreated and drug treated is determined by one way ANOVA analysis (Bonferroni multiple comparison test).***$P < 0.001$.

Figure 10.5: Estimation of cytokines *i.e.* IL-1beta in serum.

On day 10, Blood was collected from the retro-orbital plexus of mice and serum was isolated for the estimation of cytokines *i.e.* IL-1beta. Values are expressed as Mean ± SE of four set of experiments. The difference between the untreated and drug treated is determined by one way ANOVA analysis (Bonferroni multiple comparison test). * $P < 0.05$; ***$P < 0.001$.

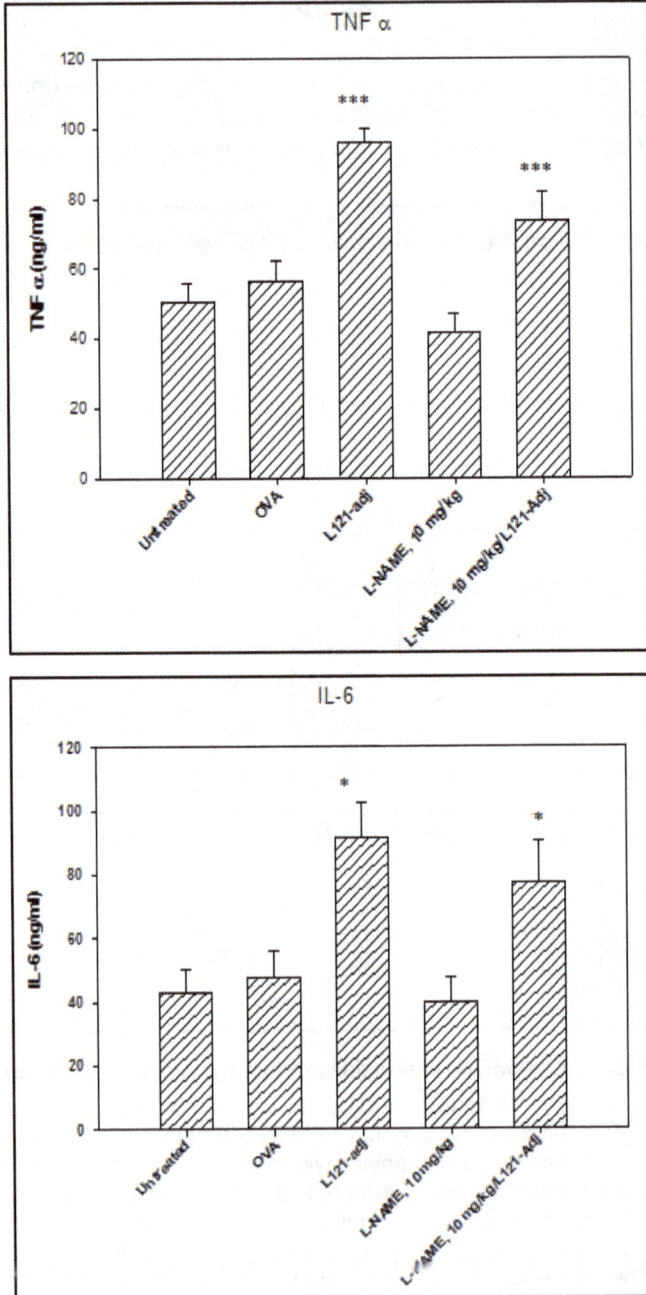

Contd...

Figure 10.5–*Contd...*

inflammatory drugs vary in their mechanisms of action and efficacy. Therefore, a number of different Non-steroidal anti-inflammatory drugs should be considered in order to achieve the most efficacious therapy for the equine patient. Non-steroidal anti-inflammatory drugs which act preferentially on COX-2 may be safer to use, particularly for long-term therapy. In Non-steroidal anti-inflammatory drugs, nitric oxide appears to play a role in inflammatory pain and has been recognized as one of the most versatile players in the immune system. It is involved in the pathogenesis and control of infectious diseases, tumors, autoimmune processes and chronic degenerative diseases.

In this review, it will mention about plant based and non-steroidal anti-inflammatory drugs, both of them played an important role in case of inflammation which may be correlated with nitric oxide production and its hypothesis will be supported by B3Z assay in case of non-steroidal anti-inflammatory drugs. Depending upon the time course of inflammation, suppression or enhancement of NO formation may influence the pain experience. But the main advantage is that non-steroidal anti-inflammatory drugs are used to control acute and chronic pain as well as to manage oncologic and neurologic diseases in human and veterinary patients.

References

Adams, D.O., and Hamilton, T.A. (1984). The cell biology of macrophage activation. *Annual Review of Immunology,* **2:** 283 – 318.

Akamatsu, H., Komura, J., Asada, Y., and Niwa, Y. (1991). Mechanism of anti-inflammatory action of glycyrrhizin: effect on neutrophil functions including reactive oxygen species generation. *Planta Med.*, **57**: 199 – 221.

Amagaya, S., Sugishita, E., Ogihara, Y., Ogawa, S., Okada, K., and Aizawa, T. (1984). Comparative studies of the anti-inflammatory activities of the stereoisomers of glycrrhetinic acid. *J Pharmacobio-Dyn.*, **7**: 923 - 928.

Ammar, N., Gaafar, S., and Khalil, R. (1992). Anti-inflammatory effect of natural steroidal sapogenins on oral aphthus ulcers. *Egyptian Dental Journal,* **38**: 89-98.

Andersson, B.S., Rundgren, M., Nelson, S.D., and Harder, S. (1990). N-acetyl-p-benzoquinone imine-induced changes in the energy metabolism in hepatocytes. *Chem Biol Interact.*, **75**: 201 – 211.

Asylgaraj, S.Y., Irina, I.V., Wlantseva, G.G., Sherman, B.L., and Margarita, V.P. (1993). Caries prophylactis and anti-inflammatory toothpaste. *Izobreteniya,* **8**: 204.

Banchereau, J., Briere, F., Caux, C., Davoust, J., Lebecque, S., Liu, Y. J., Pulendran, B., and Palucka, K. (2000). Immunobiology of dendritic cells. *Annu Rev Immunol.*, **18**: 767 – 811.

Baroni, CD., Vitolo, D., Remotti, D., Biondi, A., Pezzella, F., and Ruco, L.P. (1987). Immunohistochemical heterogeneity of macrophage subpopulations in human lymphoid tissues. *Histopathology*, **11** (10): 1029 – 1042.

Barozzi, N., and Tett, S. (2007). What happened to the prescribing of other COX-2 inhibitors, paracetamol and non-steroidal anti-inflammatory drugs when rofecoxib was withdrawn in Australia? *Pharmacoepidemiol Drug Saf.*, **16**: 1184 - 1191.

Bottcher, A., Gaipl, U.S., Furnrohr, B.G., Herrmann, M., Girkontaite, I., Kalden, J.R., and Voll, R.E. (2006). Involvement of phosphatidylserine, alphavbeta3, CD14, CD36, and complement C1q in the phagocytosis of primary necrotic lymphocytes by macrophages. *Arthritis Rheum.*, **54**: 927–938.

Caux, C., Massacrier, C., Vanbervliet, B., Dubois, B., Van Kooten, C., Durand, I., and Banchereau, J. (1994). Activation of human dendritic cells through CD40 cross-linking. *J. Exp. Med.*, **180**: 1263– 1272.

Cella, M., Scheidegger, D., Palmer-Lehmann, K., Lane, P., Lanzavecchia, A., and Alber, G. (1996). Ligation of CD40 on dendritic cells triggers production of high levels of interleukin-12 and enhances T cell stimulatory capacity: T-T help via APC activation. *J. Exp. Med.*, **184**: 747– 752.

Chattopadhyay, R.R. (1998). Possible biochemical mode of anti-inflammatory action of *Azadirachta indica* A. Juss in rats. *Indian J Exp Biol.*, **36**: 418 - 420.

Cho, J.Y. (2007). Immunomodulatory effect of nonsteroidal anti-inflammatory drugs (NSAIDs) at the clinically available doses. *Archives of pharmacal research,* **30(1)**: 64-74.

Cocco, R.E., and Ucker, D.S. (2001). Distinct modes of macrophage recognition for apoptotic and necrotic cells are not specified exclusively by phosphatidylserine exposure. *Mol Biol Cell.*, **12**: 919 – 930.

Cochrane, C.G. (1991). Cellular injury by oxidants. *The American Journal of Medicine*, **91**: 23S – 30S.

Cohen, J.J. (1993). Apoptosis. *Immunol Today*, **14**: 126–130.

Corcoran, G.B., Mitchell, J.R., Vaishnav, Y.N., and Horning, E.C. (1980). Evidence that acetaminophen and N-hydroxyacetaminophen form a common arylating intermediate, N-acetyl-p-benzoquinoneimine. *Mol Pharmacol.*, **18**: 536 – 542.

Cvetanovic, M., and Ucker, D.S. (2004). Innate immune discrimination of apoptotic cells: repression of proinflammatory macrophage transcription is coupled directly to specific recognition. *J Immunol.*, **172**: 880 – 889.

Dahlin, D.C., Miwa, G.T., Lu, A.Y., and Nelson, S.D. (1984). N-acetyl-p-benzoquinone imine: a cytochrome P-450-mediated oxidation product of acetaminophen. *Proc. Natl Acad. Sci. USA.*, **81**: 1327 – 1331.

De Smedt, T., Pajak, B., Klaus, G. G., Noelle, R. J., Urbain, J., Leo, O., and Moser, M. (1998). Antigen-specific T lymphocytes regulate lipopolysaccharide-induced apoptosis of dendritic cells *in vivo*. *J. Immunol.*, **161**: 4476 – 4479.

Dubois, R., Melmed, G., Henning, J., and Bernal, M. (2004). Risk of upper gastrointestinal injury and events in patients treated with cyclooxygenase (COX)-1/COX-2 nonsteroidal anti-inflammatory drugs (NSAIDs), COX-2 selective NSAIDs, and gastroprotective cotherapy – An appraisal of the literature. *J Clin Rheumatol.*, **10**: 178 - 189.

Evans, C.H. (1995). Nitric oxide: what role does it play in inflammation and tissue destruction? *Agents and Actions Supplements*, **47**: 107 – 116.

Fadok, V.A., and Chimini, G. (2001). The phagocytosis of apoptotic cells. *Semin Immunol.*, **13**: 365 – 372.

Figuerinha, A., Cruz, M.T., Francisco, V., Lopes, M.C., and Batista, M.T. (2010). Anti-Inflammatory Activity of *Cymbopogon citratus* Leaf Infusion in Lipopolysaccharide-Stimulated Dendritic Cells: Contribution of the Polyphenols. *J Med Food.*, **10.1089/jmf.2009.0115.pdf**

Fujii, Y., Matsura, T., Kai, M., Matsui, H., Kawasaki, H., and Yamada, K. (2000). Mitochondrial/cytochrome c release and caspase-3 like protease activation during indomethacin -induced apoptosis in rat gastric mucosal cells. *Proc Soc Exptl Biol Med.*, **224**: 102 – 108.

Golden, D.P., Mosby, E.L., Smith, D.J., and Mackercher, P. (1981). Acetaminophen toxicity. Report of two cases. *Oral Surg. Oral Med. Oral Pathol.*, **51**: 385 – 389.

Green, G.A. (2001). Understanding NSAIDs: From Aspirin to COX-2. *Clin Cornerstone.*, **3**: 50 – 59.

Han, H., Iwanaga, T., and Fujita, T. (1993). Species-differences in the process of apoptosis in epithelial cells of the small intestine: an ultrastructural and cytochemical study of luminal cell elements. *Arch Histol Cytol.*, **56**: 83 – 90.

Hawkins, C., and Hanks, G. (2000). The gastroduodenal toxicity of nonsteroidal anti-inflammatory drugs. A review of the literature. *J Pain Symptom Manage.*, **20**: 140 - 151.

Hiraishi, H., Yajima, N., Yamaguchi, N., Ishida, M., Katoh, Y., Hirada, T., Terano, A., and Ivey, K.J. (1993). Antioxidant protection against oxidant-induced damage in cultured gastric mucosal cells. *Gastroenterol Jpn.*, **28**: 132 – 138.

Hopkinson-Woolley, J., Hughes, D., Gordon, S., and Martin, P. (1994). Macrophage recruitment during limb development and wound healing in the embryonic and foetal mouse. *J Cell Sci.*, **107 (5)**: 1159 – 1167.

Koch, F., Stanzl, U., Jennewein, P., Janke, K., Heufler, C., Kampgen, E., Romani, N., and Schuler, G. (1996). High level IL-12 production by murine dendritic cells: upregulation via MHC class II and CD40 molecules and downregulation by IL-4 and IL-10. *J Exp Med.*, **184**: 741–746.

Kim, H.P., Son, K.H., Chang, H.W., and Kang, S.S. (2004). Anti-inflammatory plant flavonoids and cellular action mechanisms. *J Pharmacol Sci.*, **96 (3)**: 229 - 245.

Krysko, D.V., D'Herde, K., and Vandenabeele, P. (2006). Clearance of apoptotic and necrotic cells and its immunological consequences. *Apoptosis*, **11**: 1709 – 1726.

Kumar, V.L., Basu, N. (1994). Anti-inflammatory activity of the latex of *Calotropis procera*. *J Ethnopharmacol.*, **44**: 123 - 125.

Kusuhara, H., Kimatsu, H., Sumichika, H., and Sugahara, K. (1999). Reactive oxygen species are involved in the apoptosis induced by nonsteroidal antiinflammatory drugs in cultured gastric cells. *Eur J Pharmacol.*, **383**: 331 – 337.

Langman, M.T.S., Brooks, P., Hawkey, C.J., Silverstein, F., and Yeomanus, N. (1991). Nonsteroidal anti-inflammatory drug associated ulcer epidemiology, causation and treatment. *J Gastroenterol Hepatol.*, **6**: 442 – 449.

Larrauri, A., Fabra, R., Gomez-Lechon, M.J., Trullenque, R., and Castell, J.V. (1987). Toxicity of paracetamol in human hepatocytes. Comparison of the protective effects of sulfhydryl compounds acting as glutathione precursors. *Mol Toxicol.*, **1**: 301 – 311.
http: //www.sciencedirect.com/science?_ob=RedirectURL&_method= outwardLink&_partnerName=655&_originPage=article&_zone=art_page&_ targetURL=http per cent 3A per cent 2F per cent 2Fwww.scopus.com per cent 2Finward per cent 2Frecord.url per cent 3Feid per cent 3D2-s2.0-0023402269 per cent 26partnerID per cent 3D10 per cent 26rel per cent 3DR3.0.0 per cent 26md5 per cent 3Daa422990ba17ec257b8f7aab27431eb6 &_acct=C000051951 &_version=1 &_userid=7760972&md5=254754232 12356467dce70239b59e929

Macatonia, S. E., Hosken, N. A., Litton, M., Vieira, P., Hsieh, C. S., Culpepper, J. A., Wysocka, M., Trinchieri, G., Murphy, K. M., and O'Garra, A. (1995). Dendritic cells produce IL-12 and direct the development of Th1 cells from naive CD4+ T cells. *J Immunol.*, **154**: 5071–5079.

Maderna, P., and Godson, C. (2003). Phagocytosis of apoptotic cells and the resolution of inflammation. *Biochim Biophys Acta.*, **1639**: 141 – 151.

Makwana, H.G., Ravishankar, B., Shukla, V.J., *et al.* (1994). General pharmacology of *Vitex leucoxylon* Linn leaves. *Indian J Physiol Pharmacol*, **38**: 95 – 100.

Mengi, S.A., and Deshpande, S.G. (1995). Evaluation of ocular anti-inflammatory activity of *Butea frondosa*. *Indian J Pharmacol.*, **27**: 116 - 119.

Medzhitov, R., and Janeway, C. A., Jr. (1997). Innate immunity: the virtues of a nonclonal system of recognition. *Cell*, **91**: 295– 298.

Metzig, C., Grabowska, E., and Eckert, K., *et al.* (1999). Bromelain proteases reduce human platelet aggregation in vitro, adhesion to bovine endothelial cells and thrombus formation in rat vessels *in vivo*. *In vivo*, **13:** 7 - 12.

Milller, T. (1983). Protective effects of prostaglandins against gastric mucosal damage: current knowledge and proposed mechanism. *Am J Physiol.*, **235**: G601 – G623.

Nathan, C. (1992). Nitric oxide as a secretory product of mammalian cells. *The FASEB Journal*, **6**: 3051 – 3064.

Netti, C., Bandi, C., and Pecile, A. (1972). Anti-inflammatory action of proteolytic enzymes of animal, vegetable or bacterial origin administered orally compared with that of known antiphlogistic compounds. *Il Farmaco.*, **27:** 453 – 466.

Nourjah, P., Ahmad, S.R., Karwoski, C., and Willy, M. (2006). Estimates of acetaminophen (paracetamol)-associated overdoses in the United States. *Pharmacoepidemiol. Drug Saf.*, **15**: 398 – 405.

O'Shea, J.J., Ma, A., and Lipsky, P. (2002). Cytokines and autoimmunity. *Nature Reviews Immunology*, **2**: 37 – 45.

Pandurangan, A., Khosa, R.L., and Hemalatha, S (2008). Evaluation of Anti-inflammatory and Analgesic Activity of Root Extract of *Solanum Trilobatum* Linn. *Iranian Journal of Pharmaceutical Research*, **7 (3):** 217 – 221.

Patten, C.J., Thomas, P.E., Guy, R.L., Lee, M., Gonzalez, F.J., Guengerich, F.P., and Yang, C.S. (1993). Cytochrome P450 enzymes involved in acetaminophen activation by rat and human liver microsomes and their kinetics. *Chem Res Toxicol.*, **6:** 511 – 518.

Petros, A., Bennet, D., and Vallance, P. (1994). Effects of a nitric oxide synthase inhibitor in humans with septic shock, *Cardiovascular Research*, **28:** 34 – 39.

Ramos, C.L., Pou, S., Britigan, B.E., Cohen, M.S., and Rosen, G.M. (1992). Spin trapping evidence for myeloperoxidase-dependent hydroxyl radical formation by human neutrophils and monocytes. *The Journal of Biological Chemistry*, **267**: 8307 – 8312.

Ridge, J. P., Di Rosa, F., and Matzinger, P. (1998). A conditioned dendritic cell can be a temporal bridge between a CD4+ T-helper and a T-killer cell. *Nature*, **393**: 474– 478.

Savill, J., and Fadok, V. (2000). Corpse clearance defines the meaning of cell death. *Nature*, **407**: 784 – 788.

Savill, J., Dransfield, I., Gregory, C., and Haslett, C. (2002). A blast from the past: Clearance of apoptotic cells regulates immune responses. *Nat Rev Immunol.*, **2**: 965 – 975.

Schimmer, B.P., and Parker, K.L. (2001). Adrenocorticotropic hormone; adrenocortical steroids and their synthetic analogs; inhibitors of the synthesis and actions of adrenocortical hormones. In: J.G. Hardman, L.E. Limbird and A. Goodman Gilman, Editors. *The Pharmacological Basis of Therapeutics* (10th ed.) *McGraw-Hill New York*, 1649 – 1677.
http: //www.sciencedirect.com/science?_ob=RedirectURL&_method= outward Link&_partnerName=6 55&_originPage=article&_zone=art_page&_ target URL=http per cent 3A per cent 2F per cent 2Fwww.scopus.com per cent 2Finward per cent 2Frecord.url per cent 3Feid per cent 3D2-s2.0-0026644780 per cent 26partnerID per cent 3D10 per cent 26rel per cent 3DR3.0.0 per cent 26md5 per cent 3D2b34ec1bd33494e9a6b035562588747a &_acct=C000051951 &_version=1 &_userid=7760972&md5=e32796e806375 af175757798a9f91e8f

Singh, S., Majumdar, D.K., Yadav, M.R. (1996). Chemical and pharmacological studies on fixed oil of *Ocimum sanctum. Indian J Exp Biol.*, **34:** 1212 - 1215.

Singh, S., and Majumdar, D.K. (1997). Evaluation of anti-inflammatory activity of fatty acids of *Ocimum sanctum* fixed oil. *Indian J Exp Biol.*, **35**: 380 - 383.

Singh, R.K., and Pandey, B.L. (1996). Anti-inflammatory potential of *Pongamia pinnata* root extracts in experimentally induced inflammation in rats. *J Basic Appl Biomed.*, **4**: 21 - 24.

Singh, R.K., and Pandey, B.L. (1996). Anti-inflammatory activity of seed extracts of *Pongamia pinnata* in rat. *Indian J Physiol Pharmacol.*, **40**: 355 - 358.

Singh, R.K., and Pandey, B.L. (1997). Further study of anti-inflammatory effects of *Abies pindrow. Phytother Res.*, **11**: 535 - 537.

Sivaprakasam, P., Viswanathan, S., Thirugnanasambantham, P., Reddy, M.K., and Vijayasekaran, V. (1996). Pharmacological screening of *Ochna obtusata*. *Fitoterapia*, **67**: 117 - 120.

Soumelis, V., Reche, P. A., Kanzler, H., Yuan, W., Edward, G., Homey, B., Gilliet, M., Ho, S., Antonenko, S., Lauerma, A., Smith, K., Gorman, D., Zurawski, S., Abrams, J., Menon, S., McClanahan, T., de Waal-Malefyt, R., Bazan, F., Kastelein, R. A., and Liu, Y. J. (2002). Human epithelial cells trigger dendritic cell mediated allergic inflammation by producing TSLP. *Nat Immunol.*, **3**: 673 – 680.

Souza, M.H.L.P., Lemos, H.P., Oliviera, R.B., and Cunha, F.Q. (2004). Gastric damage and granulocyte infiltration induced by indomethacin in tumor necrosis factor receptor 1 (TNFR-1) or inducible nitric oxide synthase (iNOS) deficient mice. *Gut.*, **53**: 791 – 796.

Steineck, M.J., Khan, A.U., and Karnovsky, M.J. (1992). Intracellular singlet oxygen generation by phagocytosing neutrophils in response to particles coated with a chemical trap. *The Journal of Biological Chemistry*, **267**: 13425 – 13433.

Streeter, J., Dahlin, D.C., Nelson, S.D., and Baillie, T.A.(1984). The covalent binding of acetaminophen to protein. Evidence for cysteine residues as major sites of arylation *in vitro. Chem Biol Interact.*, **48**: 349 – 366.

Swarnakar, S., Ganguly, K., Kundu, P., Banerjee, A., Maity, P., and Sharma, A.V. (2005). Curcumin regulates expression and activity of matrix metalloproteinases 9 and 2 during prevention and healing of indomethacin-induced gastric ulcer. *J. Biol. Chem.*, **280:** 9409 – 9415.

Takeuchi, K., Okada, M. Ebara, S., and Osano, H. (1990). Increased microvascular permeability and lesion formation during gastric hypermotility caused by indomethacin and 2-deoxy-D-glucose in rat. *J Clin Gastroenterol.*, **12:** S76 – S84.

Thiemermann, A., and Vane, J. (1990). Inhibition of nitric oxide synthesis reduces the hypotension induced by bacterial lipopolysaccharides in the rat *in vivo*. *European Journal of Pharmacology*, **182:** 591 – 595.

Trenam, C.W., Blake, D.R., and Morris, C.J. (1992). Skin inflammation: reactive oxygen species and the role of iron. *The Journal of Investigative Dermatology*, **99:** 675 – 682.

Verhasselt, V., Goldman, M., and Willems, F. (1998). Oxidative stress up-regulates IL-8 and TNF-α synthesis by human dendritic cells. *Eur J Immunol.*, **28:** 3886–3890

Vimala, R., Nagarajan, S., Alam, M., Susan, T., and Joy, S. (1997). Anti-inflammatory and antipyretic activity of *Michelia champaca* Linn, (white variety), *Ixora brachiata* Roxb and *Rhynchosia Cana* (Willd.) D.C. flower extract. *Indian J Exp Biol.*, **35:** 1310 - 1314.

Xie, W., Cho, H.J.J., Mumford, R.A., Swiderek, K.M., Lee, T.D., Ding, A., Troso, T., and Nathan, C. (1992). Cloning and characterization of inducible nitric oxide synthase from mouse macrophages. *Science.*, **256:** 225 – 228.

Wallace, J.L., Keenan, C.M., and Granger, D.N. (1990). Gastric ulceration induced by nonsteroidal anti-inflammatory drugs is a neutrophil dependent process. *Am J Physiol.*, **259:** G462 – G467.

Witasp, E., Kagan, V., and Fadeel, B. (2008). Programmed cell clearance: molecular mechanisms and role in autoimmune disease, chronic inflammation, and anti-cancer immune responses. *Curr Immunol Rev.*, **4:** 53 – 69.

Yoshikawa, T., Naito, Y., Kishi, A., Tomii, T., Kaneko, T., Iinuma, S., Ichikawa, H., Yasuda, M., Takahashi, S., and Kondo, M. (1993). Role of active oxygen, lipid peroxidation and antioxidants in the pathogenesis of gastric mucosal injury induced by indomethacin in rats. *Gut.*, **34:** 732 – 737.

Natural Products: Research Reviews Vol. 1 (2012)   Pages 461–470
Editor: V.K. Gupta
Published by: DAYA PUBLISHING HOUSE, NEW DELHI

11

Inorganic Composition of Medicinal Plants

C.C. Silva[1]*, J.E.M. Gai[1], L.A.A. Freitas[1], L.S. Freire[1]
and V.F. Veiga-Junior[2]

ABSTRACT

Studies on inorganic chemicals in plants are geared mostly towards nutrition, which seeks to determine the macro and micronutrients important for species of commercial interest. The absorption of certain inorganic chemicals by plants is of great importance to their development. In phytotherapics, these studies are especially important, since the intake of medicines containing inorganic substances, at doses slightly above the limit of human tolerance, can cause severe poisoning. This work represents a broad survey of studies conducted in this area of research.

Keywords: Inorganic elements, Medicinal plant nutrition, Medicinal plant toxicity.

Introduction

According to World Health Organization (WHO), medicinal plants are those whose roots, leaves, seeds, bark, or other components that have therapeutic activity, purgative, tonic or any other pharmacological activity when administered to higher animals. In general, the main studies of medicinal plants are those that describe their

1   Universidade do Estado do Amazonas, Escola Superior de Tecnologia, Coordenação de Engenharia Química, Av. Darcy Vargas 1200, Manaus, Brazil. 69057-020.

2   Universidade Federal do Amazonas, Instituto de Ciências Exatas, Departamento de Química, Av. General Rodrigo Octávio Jordão Ramos 6200, Setor Norte, Manaus, Brazil. CEP. 69077-040.

*   *Corresponding author*: E-mail: ccsilva@uea.edu.br

organic constituents in the search for active ingredients. Studies on inorganic chemical composition in plants are geared mostly towards nutrition, seeking the determination of important macro and micronutrients for species of commercial interest, like soybeans, corn and wheat (Yannarelli *et al.*, 2007).

In phytotherapics these studies are especially important, since the intake of medicines containing inorganic substances can cause various levels of poisoning. In Chinese and Ayurvedic medicines, preparations containing metals such as mercury, lead and even arsenic are common. In these ancient processes, metals are added in minimum quantities with prescription and custom administration (Veiga Jr. *et al.*, 2005). Currently, however, the presence of metals in the preparation of phytotherapics from plants grown on a large scale can involve excessive absorption of heavy metals from the soil as larvicides and pesticides. Continued use of teas can cause the accumulation of these metals, which is very rarely discussed in literature, except for studies conducted in countries that are major consumers of herbal medicines, such as Germany and the United States.

Studies on the accumulation of metals originated from medicinal plants are uncommon (Scherz and Kirchhoff, 2006). The usual methodology for this type of analysis is atomic absorption, but the analysis through this technique presents the limitation of allowing only the observation of the constituents previously selected. In Rodushkin *et al.* (2008) we have a real sense of what it means to scan samples in search of any element that may be present in its composition. For each component to be studied there are dozens of dilutions, concentrations, standard solutions, etc. Researchers tend to simplify the problem, just watching for specific constituents (Lopes *et al.*, 2002), as Ca, Cu, Fe, Mg, Mn and Zn, or other nutrients relevant to the investigation in particular. But with that question, they neglect other elements that are not often discussed, but with greater potential nutritional value or toxicity, such as V, Rb and Ti.

Modern methods have been developed using multi-elemental analysis, allowing the observation of unusual bio-elements, such as tungsten, selenium, cobalt, and others. X-ray Fluorescence (XRF), for example, is a multi-elementary technique in which a unique quantitative analysis provides data about all constituents present with an atomic number between 11 and 92; that is, between sodium and uranium (Leyden, 1984; Butin, 1970). This also answers another question: "Which element may be present here?" And the interpretation of this answer may be of extreme importance for the populations that use these plants in their day-to-day lives. Some studies of medicinal plants are already underway with this method (Rüdiger *et al.*, 2009).

Despite the need of determination of those metals in phytotherapics, cosmetics, food, and perfumes, data concerning these are rarely observed in the literature. For products of the Amazonian biodiversity, in particular, studies that describe the inorganic composition, and that are able to guide the methodologies of quality control to be developed, are even rarer. Finally, it suggests that the chemical studies to determine the toxicological profile of each phytotherapies species and that describe the mineral bioavailability, are performed in parallel to agronomic studies, determining optimal growing conditions by ensuring uniformity of its features and greater security its use.

Plant Inorganic Composition

Several studies report the necessity of some chemical elements to plants (*i.e.*, nitrogen and phosphor) for the species of higher economical interest, such as soy and corn. One of the reference works in this area was carried out more than eight decades ago, in 1924, by Miller (1938). However, in most modern books of vegetal physiology, the mineral or inorganic physiology is not mentioned. Studies performed with plants of non intensive utilization, like the ones utilized for medicinal, cosmetics or perfume purposes are ever rarer.

The mineral nutrition (*i.e.*, the absorption of certain chemical elements by plants) is of great importance to its development. These components constitute raw material indispensable for the functioning of plant cells together with water, light and carbon dioxide (Epstein and Bloom, 2006; Geiz and Taiz, 2009). The nutrition of vegetable organism occurs through root and leaf systems.

Essential chemical elements are those indispensable to the life cycle of the plant and cannot be replaced by other elements because they are essential to their metabolism. In an environment with a deficiency of these minerals, the plant cell does not develop properly. They are required in certain metabolic pathways, such as enzyme activations or catalysis of some reactions. C, O, H, P, Ca, P, Mg, S, Cl, Mn, B, Zn, Fe, Cu, Ni and Mo are the 17 essential elements. All other chemical elements are considered non-essential or useful, since they can be replaced, depending on their availability in the soil (Meyer *et al.*, 1983; Prado, 2008; Salisbury and Ross, 1992; Geiz and Taiz, 2009). Independent of essentiality, minerals may be at higher or lower in concentration in the plant to be classified as macro or micronutrients. N, P, K, Ca, Mg and S are examples of macronutrients, whereas Si, Al, Fe and Mo are examples of micronutrients (Fernandes, 2006).

Several elements present in the earth show important functions for the living organisms, being significant for their development, even when in small quantities. To the terrestrial vegetables the main substrate is the soil. It can be defined as the soil layer that covers the earth's surface, with its ability to support living organisms. This results from the action of environmental and climatic conditions (Lepsch, 1993) and has five main components: minerals, water with dissolved salts, gases, organic matter, and living organisms, which are interested in particular: minerals and water. The inorganic elements exist in two forms in soil: insoluble complexes released by dissolution, and in simple form (ionic) readily assimilated, released by ion exchange. They are found in both cations and anions, depending only on the local position of the minerals found there (Brady, 1974; Epstein and Bloom, 2006).

Plant cells may preferentially absorb certain ions, which are needed in greater quantity. For example, cations are preferentially absorbed compared to anions (Larcher, 2000). While there is abundance of one element in the soil, it is necessary that it be present in the way that botanical species are able to absorb it. Approximately 98 per cent of these bio-elements are associated with inorganic materials in forms difficult to solubilize, or even embedded in minerals (Larcher, 2000). One element is bioavailable when it is in ionic form. If Ca^{+7} ions are not present in the soil, for example, but another type of cation with similar charge and size, are present, then the plant can absorb it.

The number of inorganic nutrients in a plant mainly depends on the elements present in the soil. The concentration of elements in a plant will vary according to many factors, as plant species, age, distribution of roots, the physical and chemical nature of the soil, the proportion and distribution of the elements, methods of cultivation, and the general climatic conditions (Miller, 1938). The proportion between the various bio-elements in the vegetable is directly related to the plant part that is being assessed. Teas made from the bark of a plant can result in the opposite effect to what would be obtained if it was made from the leaves of the same vegetable. The medicinal benefits obtained through consumption of a product derived from the extraction of an adult plant are not the same as that generated by a younger plant. There is a limit to absorptive capacity and selection of bio-elements. The efficiency of inorganic components acquisition from the root and the preference for a particular bio-element are genetic features that generate specific binding affinities of a particular ion or group of them. The rate of absorption also depends on the adjustability of the root, depending on the amount of available ions (Larcher, 2000).

In addition to the bioavailability of chemical species in a plant and chemical seasonal variability (Yariwake *et al.*, 2005), the chemical conditions of the rhizosphere and the metabolic adjustments that the plant performs in relation to these, determine the nutrient supply and directly influence the applicability of a medicinal plant (Larcher, 2000).

Plant cells can not totally exclude unnecessary or toxic salt, even if they prejudice their full development (Larcher, 2000). Medicinal plants that come from places where the soil is exposed to heavy metals, for example, are subject to contain these inorganic components in their structures, which in turn may be ingested. In the studies of Silva *et al.* (2010), an attempt to relate environmental pollution, where *Cymbopogon citratus* Stapf was present, their chemical composition resulted in the concentration of certain organic components being changed. If these properties were modified its inorganic composition probably was also changed.

There are mechanisms capable to eliminate at least some of these contaminants. In the work of Morita *et al.* (2008), a detailed study of Al absorption by roots of *Camellia sinensis* (L.) Kuntze, widely used in tea form, shows that the plant itself can promote the elimination of excess metal, through the complexation of Al and subsequent elimination. For the same purpose there are certain peptides (phytochelatins) which are produced by plants when they are in areas with high concentrations of metals (Cd, Pb, Zn, etc). They form chelate complexes with metal and then the plant eliminates them. In the article by Zhang *et al.* (2008), they were studied in medicinal plants, showing another defense mechanism.

The Medicinal Plants Consumption

In cases of chemical profiling study, all plant parts, preferably *in natura*, should have their levels of inorganics evaluated, as well as the composition with the lowest degree of contamination and volatile loss. But, there are several forms of medicinal plant consumption. Among the most commonly used are: infusions, syrups, juices, plasters, "bottled" (composition of various plants in alcoholic or water solution) and *in natura*. Same plant parts are used as medicine: fruits, leaves, bark, roots or flowers (Agra *et al.*, 2008). Research with a toxicity focus on the ingested form by consumers is more interesting. Once the plant has to be consumed provides a certain amount of metals, these can be extracted during the process of preparing for consumption, as infused. These values are more interested to be studied in the case of medical consumption.

Inorganic Composition and Medicinal Benefits

With the evolution of chemistry and biology, particularly in the seventeenth century, a new outlook was launched on the plants used as medicine for thousands of years. Their molecular compositions became to be questioned. This outlook has survived to modern times in which species identification becomes essential for the study of new herbal drugs. The molecular composition provides the chemical profile of that species, collected in a specific place and time, and under the action of particular environmental factors. Knowing that the addition of all these factors directly influences the biological activity expected for the medicinal plant in question, the study of inorganic constituents becomes relevant.

Trace metals play an important role in the management of diabetes mellitus. Kar *et al.* (1999) selected 30 hypoglycemic herbs from indigenous folk medicines, Ayurvedic, Unani and Siddha systems of medicines and activities for glucose tolerance were compared between the organic parts of each plant with the same activity when given only the ash (consisting mainly of inorganic salts that were present in plants). In certain inorganic samples, more pronounced action was noticed than their corresponding organic parts. The differences in the concentration of the elements are attributed to soil composition and the climate in which the plant grows. In their study, Ravi *et al.* (2004) have shown that by increasing the concentration of certain inorganic elements in the constitution of plants, without going beyond the values stipulated by the WHO, there was an increase in biological activity of *Eugenia jambolana* Lam in battle of diabetes mellitus.

In work published by Pereira and Felcman (1998), 16 different medicinal plants have had their contents of Si, Mn, Fe, Cu and Zn determined. It was demonstrated that the presence of these elements enhances their action as medicines. The authors suggest that these metals when in contact with the injury, is complexed with biological agents clearly showing that the biological activity can, in many cases, be related to the inorganic composition of the plant. In another work (Singh and Garg, 1997), 8 samples of fruits, 5 to 6 leaves and roots, used in Ayurvedic medicine, were studied to determine the concentration of 20 inorganic elements. All concentrations varied widely from plant to plant and also from one part of the plant to another. The researchers suggested that more detailed studies should be conducted to identify the functions of these elements in the activities of these medicinal plants. Along with these studies, we can mention several others (Ražiæ *et al.*, 2005; Rajasekaran *et al.*, 2005).

Martins *et al.* (2009) assessed the minerals present in Amazonian medicinal plants. Only the chemical elements Co, Mg, Fe, Cu and Zn were analyzed. The authors concluded that drinking tea composed of these plants could complement the diets of people lacking these elements. A study of plant species with anti inflammatory and analgesic effects were achieved (Delaporte *et al.*, 2005) where the ashes of the leaves were analyzed for K, N, P, Mn, S and B, in addition to the others studied by Martins *et al.* (2009). They concluded that the levels of the studied elements were within the parameters identified as non-toxic by the WHO. However, they do not take in account other elements such as Pb and Cd that are already in very low concentration in the Amazon soils have a greater possibility to be present naturally.

Rüdiger *et al.* (2009) studied the species from burceraceae oleoresin in which all the chemical elements with atomic numbers between 11 (Na) and 92 (U) were analyzed, almost the entire periodic table of elements. They detected the presence of metals such as Cr, Mn, Cu and Br, in the oleoresins. Their quantities vary from species to species, and this is a work for both phytochemical profile and for toxicity studies.

In many countries there are specific laws that control the presence of metals in natural products that are commercialized. Bioactive oils or plant extracts commercialized as personal hygiene products, cosmetics and perfumes must be free of As, Pb, Cr, Cd, Te, Ta, and Sb. In food, there are maximum limits for the following elements: As, Pb, Cd, Sn, Cu, and Hg. Quality control tests for heavy metals must be indicated for phytotherapic registration as well (ANVISA, 2009). The work of most national agencies responsible for herbal medicines control is based on the norms of the WHO.

Toxicity

Until today most of the research on herbal medicines has left aside the search for human body inorganic substances, primarily to the identification of organic species of interest (Riet-Correa and Medeiros, 2001). Many of these nutrients cause diseases and intoxications in concentrations just a bit higher than the necessary concentration for a healthy life. Heavy metals, for example, show high toxicity, even at very low concentrations. Some forms cause severe intoxication processes, such as chrome. Its accumulation in the organism can lead to the formation of carcinomas. Copper in excess originate diseases like the Menks Syndrome and the Wilson disease, both with cellular degeneration in humans (Butis and Ashwood, 1970).

The indistinct use of herbal medicines is based on the popular belief that plant-based medicines are free of health risks (Lanini, 2009). This marketing is supported by advertisements that promise benefits inherently safe and many consumers feel encouraged by the natural character of such products (Veiga Jr, 2005).

The ease of obtaining these products and their commercialization in large public places such as pharmacies and open markets, coupled with the poor performance of government bodies responsible for the following inspection, promotes further use (Veiga Jr, 2005). Although there has been some increase in the ease of access to the allopathic medicine in recent decades, there are basic obstacles in its use by the underserved populations, ranging from access to hospital care centers to achievement of medical examinations and medicines (Silveira, 2008).

In addition, one should be aware of the specific characteristics of each organism. Depending on the dose, such products may have beneficial or toxic effects (Sanchez, 1998). If used concomitantly with allopathic medicines they can minimize, enhance, or oppose their effects (Pinn, 2001).

The presence of metals in medicinal plants must be studied due the possibility of intoxication on ingesting their extracts from infusions or teas, and even through cutaneous absorption on treating wounds (Lopes *et al.*, 2002; Gomes *et al.*, 2004). In Brazil, the studies about metal detection in vegetables are not common. In a recent study, samples of boldo (*Peumos boldus* Mol.) from Brazil, Chile, and Argentina, were analyzed using the atomic absorption spectrometry allowing the detection of iron, manganese, copper and nickel and absence of plumb, chrome and cobalt. Much more research on medicinal plants popularly used is necessary, not only for the confirmation of the activities described by traditional use, but also for the security to be established (Schwanz *et al.*, 2008).

Based on personal evidence or reports from third parties, medicinal plants and their derivatives are used for multiple purposes. They are often combined, even with conventional medicine and do not generally have attributed side effects or adverse reactions to its use. But there are several reports in the literature of cardiac, liver, renal, hematological and intestinal complications caused by phytomedicines (Turolla, 2006).

Despite such scientific evidence, belief in the provision of security in its use is not easily contradicted, because this information hardly comes within range of the public who make use of such medicine. This is enhanced by the majority of individuals with low education and little access to scientific information. Further aggravating this factor, the press often publishes and disseminates misinformation in the media of mass communication (Silveira, 2008). The herbal drugs are used, in most cases, by self-medication and part of the medicine prescribed by doctors does not come with well known toxicity profiles. Often the supposed pharmacological properties advertised have no scientific validity to back them up (Veiga Jr, 2008).

There is also the possibility of contamination and/or tampering during its production and through the soil contamination, as well as the risk in the consumption of herbal medicines imported from Asian countries. Their formulations often contain heavy metals (gold, copper, tin, lead, mercury, iron, silver and zinc, arsenic and thallium) in concentrations that often exceed the safe values for consumption (Veiga Jr, 2005). The misidentification of the species (by the merchant or the supplier) and the lack of regulation in marketing are also risk factors (Turolla, 2006) for the occurrence of adverse reactions, poisonings and other complications arising from their use. This has not only immediate effects, but is also easily correlated with their intake (Silveira, 2008), as well as effects that settle over time as asymptomatic.

Currently there are few means of enforcement to ensure a strict quality control of phytomedicines. By the standardization of the inorganic compounds and the creation of an international system, pharmacovigilance of medicinal plants can identify the unknown side effects, identify and quantify risk factors, standardize terms, disseminate experiences, among others. This would allow their safe and effective use from the view of the new rules of regulatory agencies (Silveira, 2008).

References

Agra, M. F., Silva, K. N., Diniz, I. J. L., Freitas, P. F., and Barbosa Filho, J. M. (2008). Survey of medicinal plants used in the region Northeast of Brazil. *Brasilian Journal of Pharmacognosy*, **18(3):** 472-508.

ANVISA RDC 48 (16-3-2004); Portaria 685 (27-8-1998); RDC 48 (16-3-2006), in http://www.anvisa.gov.br/e-legis, accessed in January 2009.

Brady, N. C. (1983). Natureza e Propriedades dos solos. Freitas Bastos, Rio de Janeiro.

Burtis, C. A., and Ashwood, E. R. (1970). In Fundamentos de Química Clínica. Guanabara Koogan, Rio de Janeiro.

Butin, E. P. (1970). Principles and Pratice of X-Ray Spectrometric Analisis. Plenum Publishing Corporation, New York.

Cordeiro, C. H. G., Chung M. C., and Sacramento, L. V. S. (2005). Interações medicamentosas de fitoterápicos e fármacos: *Hypericum perforatum* e *Piper methysticum*. *Brasilian Journal of Pharmacognosy*, **15**: 272-278.

Deloporte, R. H., Guzen, K. P., Takemura, O. S., and Mello, J. C. P. (2005). Estudo mineral das espécies vegetais *Althermantera brasiliana* (L.) Kuntze e *Bouchea fluminensis* (*Vell.*) *Molde. Brasilian Journal of Pharmacognosy*, **15 (2):** 133-136.

Epstein, E., and Bloom, A. J. (2006). Nutrição Mineral de Plantas. Editora Planta, Londrina.

Fernandes, M. S. (2006). Nutrição mineral de plantas. Sociedade Brsileira de Ciência do Solo, Viçosa.

Gomes, M. R., Soledad, C., Olsina, R. A., Silva, M., and Martinez, L. D., J. (2004). Metal content monitoring in *Hypericum perforatum* pharmaceutical derivatives by atomic absorption and emission spectrometry. *Journal of Pharm Biomed Anal*, **34:** 569-576.

Kar, A., Choudhary, B. K., and Bandyopadhyay,N. G. (1999). Preliminary studies on the inorganic constituents of some indigenous hypoglycaemic herbs on oral glucose tolerance test. *Journal of Ethnopharmacology*, **64:** 179-184.

Lanini, J., Duarte-Almeida, J. M., Nappo, S., and Carlini, E. A. (2009). "O que vem da terra não faz mal" – relatos de problemas relacionados ao uso de plantas medicinais por raizeiros de Diadema/SP. *Brasilian Journal of Pharmacognosy*, **19 (1A):** 121-129.

Larcher, V. (2000). Ecofisiologia Vegetal, Rima, São Carlos.

Lepsch, I. F. (1993). Solos: formação e conservação. Melhoramentos, São Paulo.

Leyden, D. E. (1984) Fundamentals of X-Ray Spectrometry as Applied to Energy Dispersive Thecniques, Tracor X-Ray. Mountain View, California.

Lopes, M. F. G., Almeida, M. M. B., Nogueira, C. M. D., Magalhães, C. E. C., and Morais, N. M. T. (2002). Estudo mineral de plantas medicinais. *Brasilian Journal of Pharmacognosy*, **12**: 115-116.

Martins, A. S., Alves, C. N. A., Lameira, O. A., Santos, A. S., and Muller, R. S. C. (2009). Evaluation de minerais em plantas medicinais amazônicas. *Brasilian Journal of Pharmacognosy,*19 **(2B):** 621-625.

Meyer, B., Anderson, D., Bohning, R., and Fratianne, D. (1983). Introdução à Fisiologia Vegetal. Atlantida, Coimbra.

Miller, E. (1938). Plant Physiology. McGraw-Hill Book Company, New York.

Morita, A., Yanagisawa, O., Takatsu, S., Maeda, S., and Hiradate, S. (2008). Mechanisms for the detoxification of aluminium in roots of tea plant (*Camellia sinensis* (L.) Kuntze). *Phytochemistry*, **69**: 147-153.

Pereira, C. E., and Felcman, J. (1998). Correlation between five minerals and the healing effect of Brazilian medicinal plants. *Biological Trace Element Research,* **65(3):** 251–259.

Pinn, G. (2001). Adverse effects associated with herbal medicine. *Australian Family Physician*, **30:** 1070-1075.

Prado, R. M. (2008). Nutrição de plantas Unesp, São Paulo.

Rajaserakan, S., Sivagnanam, K., and Subramanian, S. (2005). Mineral contents of *Aloe vera* leaf gel and their role on streptozotocin-induced diabets in rats. *Biological Trace Element Research*, **108**: 185-195.

Ravi, K., Sekar, D. S., and Subramanian, S. (2004). Hypoglycemic activity of inorganic constituents in Eugenia jambolana seed on streptozotocin-induced diabets in rats. *Biological Trace Element Research*, **99**: 145-155.

Ražić, S., Đogo, S., Slavkovć, L., and Popović, A. (2005). Inorganic analysis of herbal drugs. Part I. metal determination in herbal drugsoriginating from medicinal plants of the family *Lamiaceae. J. Serb. Chem. Soc.* **70(11):** 1347-1355.

Riet-Correa F., and Medeiros R. M. T. (2001). Intoxicações por plantas em ruminantes no Brasil e no Uruguai: importância econômica, controle e riscos para a saúde pública. *Pesquisa Veterinária Brasileira*, **21:** 38-42.

Rodushkin, I., Engström, E., Sörlin, D., and Baxter, D. (2008). Levels of inorganic constituents in raw nuts and seeds on the Swedish market. *Science of The Total Environment*, **392 (2-3):** 290-304.

Rüdiger, A. L., SILVA, C. C., and Veiga Junior, V. F. (2009). EDXRF Analysis of Amazonian Burseraceae Oleoresins. *Journal of the Brazilian Chemical Society (Online)*, **20:** 1077-1081.

Salisbury, F. B., and Ross, C. W. (1992). Plant Physiology. Wadsworth Publishing Co, Belmont.

Sanchez, P. (1998). Plantas ornamentais tóxicas. Remédios venenosos da toxidez a letalidade. Site do grupo Plantamed. 1998. Available at: http: //www.plantastoxicas.hpg.com.br. Acessado em: 10 jan. 2005.

Scherz, H., and Kirchhoff, E. (2006). Trace elements in foods: zinc contents of raw foods - a comparison of data originating from different geographical regions of the world. *Journal of Food Composition and Analysis,* **19:** 420-433.

Schwanz, M., Ferreira, J. J., Fröehlich, P., Zuanazzi, J. A. S., Henriques, A. T., and Braz. J. (2008). Anaçose de metais pesados em amostras de *Peumus boldus* Mol. (Monimiacea). *Brasilian Journal of Pharmacognosy,* 18: 98-101.

Silva, M. A. L., Marques, G. S., Santos, T. M. F., Xavier, H. A., Higino, J. S., and Melo, A. F. M. (2010). Avaliação da composição química de Cymbopogon citratus Stapf cultivado em ambientes com diferentes níveis de poluição e a influência na composição do chá. *Acta Scientiarum. Health Sciences.* 32(1): 67-72.

Silva, S. A., Ribeiro, S. G., Bender, A. E., Timm, F. C., Garcias, G. L., and Martino-Roth, M. G. (2008). Estudo da atividade mutagênica das plantas, *Euphorbia milii* Des Moulins e *Ricinus communis* L através do teste de *Allium cepa*. *Brasilian Journal of Pharmacognosy,* 19 (2): 418-422.

Silveira, P. F., Bandeira, M. A. M., and Arrais, P. S. D. (2008). Farmacovigilância e reações adversas às plantas medicinais e fitoterápicos: uma realidade. *Brasilian Journal of Pharmacognosy,* 18: 618-626.

Singh, V., and Garg, A. N. (1997). Availability of essential trace elements in ayuvedic indian medicinal herbs using instrumental neutron activation analysis. *Appl. Radiat. Isot,* 48 (1): 97-101.

Taiz, L., and Zeiger, E. (2009). Fisiologia Vegetal. Artmed, Porto Alegre.

Turolla, M. S. R., and Nascimento, E. S. (2006). Informações toxicológicas de alguns fitoterápicos utilizados no Brasil. *Brazilian Journal of Pharmaceutical Sciences,* 42 (2): 289-306.

Veiga Jr, V. F. (2008) Estudo do consumo de plantas medicinais na Região Centro-Nortedo estado do Rio de Janeiro: aceitação pelos profissionais de saúde e modo de uso pela população. *Brasilian Journal of Pharmacognosy,* 18: 308-313.

Veiga Jr, V. F., Pinto, A. C., and Maciel, M. A. M. (2005). Plantas medicinais: cura segura? *Química. Nova,* 28: 519-528.

Veiga Jr, V.F., and Mello, J.C.P. (2008). As monografias sobre plantas medicinais. *Brasilian Journal of Pharmacognosy,* 18: 464-471.

Yanarelli, G. G., Fernández-Alvarez, A. J., Santa-Cruz, D. M., and Tomaro, M. L. (2007). Glutathione reductase activity and isoforms in leaves and roots of wheat pants subjected to cadmium stress. *Phytochemistry,* 68: 505-512.

Yariwake, J. H., Lanças, F. M., Cappelaro, E. A., Vasconcelos, E. C., Tiberti, L. A., Pereira, A. M. S., and França, S. C. (2005). Variabilidade sazonal de constituintes químicos (triterpenos, flavonóides e polifenóis) das folhas de *Maytenus aquifolium* Mart. (Celastraceae). *Brasilian Journal of Pharmacognos,* 15: 162 – 168.

Yunes, R., Pedrosa, R.C., and Filho, V.C. (2001). Fármacos e fitoterápicos: a necessidade do desenvolvimento da indústria de fitoterápicos e fitofármacos no Brasil. *Química Nova.* 24 (1): 147-152.

Zhang, Z., Gao, X., and Qiu, B. (2008). Detection of phytochelatins in the hyperaccumulator *Sedum alfredii* exposed to cadmium and lead. *Phytochemistry,* 69: 911-918.

Natural Products: Research Reviews Vol. 1 (2012)
Editor: V.K. Gupta
Published by: DAYA PUBLISHING HOUSE, NEW DELHI

Pages 471–499

12

Medicinal Plants for Management of Anxiety

Richa Shri[1]*

ABSTRACT

Anxiety disorders are among the most common mental, emotional, and behavioural problems. The conventional method of management of anxiety disorders with psychotherapy and pharmacotherapy. Common limitations of antianxiety drug therapy include co-morbid psychiatric disorders and increase in dose leading to intolerable side effects. These limitations have prompted scientists to investigate plants which are commonly employed in traditional and alternative systems of medicine for sleep disorders and related diseases with a view to find safer drugs. This article aims to review critically literature published mainly within this millennium on the plants/plant extracts/plant products that have shown some potential as anxiolytic agents in experimental animals as well as in clinical trials.

Keywords: Anxiety disorders, Medicinal plants, Anxiolytic agents.

Introduction

Anxiety is a normal, emotional, reasonable and expected response to real or potential danger. However, if the symptoms of anxiety:

☆ are prolonged, irrational, disproportionate and/or severe,

☆ occur in the absence of stressful events or stimuli,

☆ or interfere with everyday activities,

then, these are called Anxiety Disorders (DSM IV-TR, 2000).

1   Department of Pharmaceutical Sciences and Drug Research, Punjabi University, Patiala - 147002, Panjab, India.

*   E-mail: rshri587@hotmail.com

Anxiety disorders are among the most common mental, emotional, and behavioural problems (Kessler *et al.*, 2005a, b; Kessler and Wang, 2008). These affect one-eighth of the total population worldwide, and have become a very important area of research interest in psychopharmacology (Eisenberg *et al.*, 1998; Dopheide and Park, 2002; W.H.O., 2004).

Management of Anxiety

If anxiety becomes excessive, out of proportion and interferes with everyday life, management becomes necessary. Anxiety disorders are the most prevalent of psychiatric disorders, yet less than 30 per cent of individuals who suffer from anxiety disorders seek treatment (Lepine, 2002). People with anxiety disorders can benefit from a variety of treatments and services. Following an accurate diagnosis, possible treatments include (Moquin *et al.*, 2009; NIMH, 2006; Barlow, 2001; Baldessarini, 2001; Kessler *et al.*, 2001):

Management of anxiety

❖ **Psychological Treatments**

❖ **Pharmacotherapy**

- **Benzodiazepines** (*e.g.*, Diazepam, Lorazepam)
- **Non-benzodiazepines**:
  Beta blockers *e.g.*, Propanolol, Atenolol
  Selective serotonin reuptake inhibitors (SSRIs) *e.g.*, Citalopram, Fluvoxamine, Paroxetine
  Tricyclic antidepressants (TCAs) *e.g.*, Nortriptyline Amitriptyline Imipramine
  Monoamine oxidase inhibitors (MAOIs), *e.g.*, Selegilene, Isocarboxid
  Newer atypical antidepressants like Azaspirones *e.g.*, Buspirone

❖ **Combination of psychotherapy and pharmacotherapy**

❖ **Complementary and Alternative therapies**
  e.g., Relaxation techniques, **herbal medicines**, massage, chiropractic, spiritual healing, nutritional supplements

---

Common limitations of antianxiety drug therapy include co-morbid psychiatric disorders (Regier *et al.*, 1998) and increase in dose leading to intolerable side effects like the following:

Class of Antianxiety Drugs	Common Side Effects/Limitations
Benzodiazepines	Potentially habit-forming, can cause drowsiness, confusion and disorientation. Can produce withdrawal symptoms, discontinuation should be done slowly.
Beta blockers	Should not be used with pre-existing medical conditions, such as asthma, congestive heart failure, diabetes, vascular disease, hypothyroidism and angina pectoris
Selective serotonin reuptake inhibitors	Nausea, nervousness, headaches, sleepiness, sexual dysfunction, dizziness, weight gain
Monoamine oxidase inhibitors	Strict dietary restrictions are required and potential drug interactionsmay occur, changes in blood pressure, moderate weight gain, reduced sexual response, insomnia
Azaspirones	Nausea, headaches, dizziness, drowsiness, constipation, diarrhea, dry mouth

Cates *et al.*, 1996; Baldessarini, 2001.

The limitations of conventional therapy have prompted scientists to investigate plants which are commonly employed in traditional and alternative systems of medicine for sleep disorders and related diseases with a view to find effective and safer drugs (Spinella, 2001; Chung *et al.*, 2005; Kumar, 2006).

Plants Used for Management of Anxiety

The World Health Organisation estimates that 80 per cent of the world population relies on herbal medicine (Eisenberg *et al.*, 1998). Various plants have been investigated for their anxiolytic effects (Carlini, 2003; Sarris *et al.*, 2011) and many have shown marked antianxiety activity. Table 12.1 summarises the reports on plants investigated for their anxiolytic potential on experimental animals in the current millennium.

It is evident from Table 12.1 that many plants have anxiolytic potential in the pre-clinical stage. Of these *Passiflora incarnata* L., *Valeriana* spp. and *Piper methysticum* have demonstrated significant antianxiety activity (Carlini, 2003). Some of these leads have been followed by clinical trials. Results of some clinical trials have shown that plants may be effective clinically also. In a review (Ernst, 2006) eight monoherbal preparations containing *Scutellaria laterifolia, Centella asiatica, Paullinia cupana, Piper methysticum, Bacopa monniera, Cymbopogan citratus, Passiflofa incarnata* and *Valeriana officinalis* were reported to be subjected to randomised clinical trials to study their effect in alleviation of anxiety. According to the reported data, *Piper methysticum* (Pittler *et al.*, 2002) and *Bacopa monniera*, (Stough *et al.*, 2001) are associated with anxiolytic activity in humans. In another trial on generalized anxiety disorder (GAD) in hospital based clinical set-up, *Ocimumn sanctum* significantly attenuated generalized anxiety disorders and also attenuated its correlated stress and depression (Bhattacharyya *et al.*, 2008).

Only for *Piper methysticum* or kava independent replications were available. Kava has been shown beyond reasonable doubt to have anxiolytic effects in humans (Ernst, 2006). Table 12.2 summarizes some findings of clinical trials.

Table 12.1: Various plants investigated for their anxiolytic effects in experimental animals.

Plant/Active Constituent	Plant Part/Extract	Dose (mg/kg)	Model Employed for Evaluation of Anxiety	Activity Observed	References
Abies pindrow (Pinaceae)	Ethanolic extract of leaf	50–100	EPM, OFT, elevated zero maze	Dose dependent anxiolytic effect	Singh et al., 1998; Kumar et al., 2000a
Adiantum tetraphyllum (Adiantaceae)	Ethanol extract of leaves	50, 150, 300	EPM	Suppressed certain components of anxiety and fear	Bourbonnais-Spear et al., 2007
Aegle marmelos (Rutaceae)	Methanol extract of leaves	75, 150 and 300	EPM, tail suspension test	Anxiolytic and antidepressant activities	Kothari et al., 2010
Aethusa cynapium	New fatty acid-trideca-7,9,11-trienoic acid from methanol extract	25	EPM	Significant anxiolytic activity	Shri et al., 2010
Albizzia julibrissin (Mimosaceae)	Aqueous extract of stem bark	100–200	EPM	Anti anxiety effect via serotonergic system	Kim et al., 2004
Albizzia lebeck (Mimosaceae)	Saponin rich n-butanol fraction of leaves	25	EPM	Anxiolytic	Une et al., 2001
Aloysia polystachya (Verbenaceae)	Hydro-ethanolic extract obtained from the aerial parts	1, 10,100	EPM	Anxiolytic without sedative side effect	Helli'on-Ibarrola et al., 2006
Apocynum venetum (Apocynaceae)	Ethanol extract of leaves and kaempferol	30, 125	EPM	Anxiolytic activity mediated via the GABAergic system.	Grundmann et al., 2007 and 2009
Angelica archangelica (Apiaceae)	Essential oil	21, 42	Social interaction test of anxiety and the Hole-board test of exploration	Useful against various types of anxiety-related disorders and social failure	Min et al., 2005

Contd...

Table 12.1–*Contd...*

Plant/Active Constituent	Plant Part/Extract	Dose (mg/kg)	Model Employed for Evaluation of Anxiety	Activity Observed	References
Aniba riparia (Lauraceae)	Riparin I, riparin III from unripe fruit	25, 50	Open field, EPM and Hole board tests	Riparin III presents anxiolytic effects in the plus maze and hole board tests which are not influenced by the locomotor activity in the open field test	de Sousa *et al.,* 2005; de Melo *et al.,* 2006
Annona cherimolia (Annonaceae)	Hexane extract of leaves	6.25, 12.5, 25, 50	Mouse avoidance exploratory behavior and the burying behavior tests	Involvement of the GABA(A) receptor complex in the anxiolytic-like actions	Lopez-Rubalcava *et al.,* 2006
Artemisia copa (Asteraceae)	Aqueous extract from aerial parts	0.5-1.5 g/kg,	Exploration in the hole-board, pentobarbital-induced hypnosismarble-burying test and anticonvulsant activity on convulsions induced by pentylenetetrazol.	Contains sedative principles with potential anxiolytic and anticonvulsant activities	Miño *et al.,* 2010
Avena sativa (Poaceae)	Tincture of immature seeds	1-5ml	EPM	Useful in acute and chronic anxiety	Abascal and Yarnell, 2004
Azadirachta indica (Meliaceae)	Aqueous extract of leaves	500 for 15 days 10–800	EPM, OFT, Morris water maze test	Significantly reduced hypoperfusion induced functional disturbances like a propensity towards anxiety and disturbances of learning/memory	Jaiswal *et al.,* 1994; Yanpallewar *et al.,* 2005
Bacopa monnieri (Scrophulariaceae)	Standardised plant extract containing 25.5% bacosides	5, 10, 20	Vogel rat model	Mild anxiolytic and nootropic	Ghosal and Bhatta-charya, 1980; Bhattacharya *et al.,* 2000a; bacopin.com, 2001

Contd...

Table 12.1–*Contd...*

Plant/Active Constituent	Plant Part/Extract	Dose (mg/kg)	Model Employed for Evaluation of Anxiety	Activity Observed	References
Byrsocarpus coccineus (Connaraceae)	Aqueous extract of leaves	50 and 100	Y–maze, EPM, hexobarbitone sleeping time and hole board models	Anxiolytic and sedative	Akindele and Adeyemi, 2010
Calotropis gigantea (Asclepidiaceae)	Alcoholic extract of peeled roots	250,500	EPM	Anxiolytic, sedative	Argal and Pathak, 2006
Cannabis sativa (Cannabinaceae)	Cannabidiol (CBD)	2.5, 5, 10	Vogel test	Inhibits the anxiogenic activity of high doses of Δ^9-tetrahydro-cannabinol and induces anxiolytic-like effects in the rat Vogel conflict test	Moreira *et al.*, 2006; Zuardi *et al.*, 2006
Casimiroa edulis (Rutaceae)	Hydroalcoholic extract of leaves	25, 35	EPM, forced swimming	Sedative principles with potential anxiolytic and antidepressant properties	Molina-Hernandez *et al.*, 2004; Mora *et al.*, 2005
Cassia siamea (Leguminosae)	Aqueous extract, Barakol	1, 6, 12 p.o. 10, i.p.	EPM	Anxiolytic properties similar to diazepam but it increases exploratory behaviour	Thongsaard *et al.*, 1996
Cecropia glazioui (Moraceae)	Aqueous extract and two semipurified fractions	0.25–1 g/kg	EPM	After repeated administration of aqueous extract and butanol fraction, the frequency of entries in the open arms of EPM was increased threefold.	Rocha *et al.*, 2002
Centella asiatica (Apiaceae)	Hexane, ethyl acetate and methanol extracts; and asiaticoside.	Extracts 200 Asiaticoside 3, 5,10	EPM	Methanol and ethyl acetate extracts as well as the pure asiaticoside exhibited anxiolytic activity. Furthermore, the asiaticosides do not have sedative effects in rodents.	Wijeweera *et al.*, 2006

Contd...

Tab 12.1–Contd...

Plant/Active Constituent	Plant Part/Extract	Dose (mg/kg)	Model Employed for Evaluation of Anxiety	Activity Observed	References
Clerodendrum philippinum (Verbenaceae)	Ethanolic extract of the flowers	125 and 250	EPM	Anxiolytic	Lalitha et al., 2010
Cissus sicyoides (Vitaceae)	Hydroalcoholic extract of aerial parts	300, 600, 1000	EPM, hole board test	Anxiolytic and anticonvulsant action probably due to the action of flavonoid(s), linalool, and alpha-tocopherol	de Almeida et al., 2009
Citrus aurantium (Rutaceae)	Volatile oil; hydro-ethanolic leaf extracts and hexane, dichloro-methane, and aqueous fractions	0.5–1g/kg	EPM, light-dark box, marble-burying test	Anxiolytic	Carvalho-Freitas and Costa, 2002; Pultrini et al., 2006
Citrus limon (Rutaceae)	Methanol extract of leaves	100	EPM	Anxiolytic	Shri and Siddana, 2008
Citrus paradisi (Rutaceae)	Methanol extract of leaves	100,200,400	EPM	Anxiolytic and antidepressant	Gupta et al., 2010
Citrus sinensis (Rutaceae)	Volatile oil	100, 200 or 400 microl	EPM	Anxiolytic	Faturi et al., 2010
Clitoria ternatea (Fabaceae)	Root methanol extract	100–400	EPM	Anxiolytic, antidepressant and antidementia actions	Rai et al., 2002; Jain et al., 2003
Coccasia esculenta (Asteraceae)	Hydroalcoholic extract of leaves	100, 200, 400	Behavior despair and EPM tests	Anxiolytic and antidepressant activities	Kalariya et al., 2010
Convolvulus pluricaulis (Convolvulaceae)	Ethanolic extract of petals	100, 200 and 400	EPM	Anxiolytic	Sharma et al., 2009
	Ethanol extract of the aerial parts	100	EPM, OFT	Anxiolytic and antioxidant	Nahata et al., 2009
Coriandrum sativum (Apiaceae)	Aqueous extract of dried seeds	10, 25, 50, 100	EPM	Anxiolytic effect and potential sedative and muscle relaxant effects	Emamghoreishi et al., 2005

Contd...

Table 12.1–Contd...

Plant/Active Constituent	Plant Part/Extract	Dose (mg/kg)	Model Employed for Evaluation of Anxiety	Activity Observed	References
Crocus sativus (Iridaceae)	Crocins (water soluble carotenoids)	50	Light/dark test	Significantly increased the latency to enter the dark compartment and prolonged the time spent in the lit chamber in the rats indicating anxiolytic-like effects	Pitsikas *et al.,* 2008
Curcuma longa (Zingiberaceae)	Curcumin	10 and 20	EPM, OFT, light/dark test and social interaction test	Antianxiety effect with the possible involvement of inducible NOS	Gilhotra and Dhingra, 2010
Davilla rugosa (Polypodiaceae)	Stem hydroalcoholic extracts	7.5–60	EPM, OFT	Anxiolytic	Guaraldo *et al.,* 2000
Drymaria cordata	Hydroethanolic extract of leaves	25,50 and 100	EPM, OFT, light/dark and hole board test	Anxiolytic	Barua *et al.,* 2009
Eclipta alba (Asteraceae)	Aqueous and hydro-alcoholic extracts; hydrolyzed fraction of aqueous extract	150, 300, 30	EPM, holeboard model	Anxiolytic activity along with significant protection against stress induced alterations	Thakur and Mengi, 2005
Echium amoenum (Boraginaceae)	Ethanol extract of flowers	50	EPM	Anxiolytic effect with lower sedative activity than that of diazepam	Rabbani *et al.,* 2004
Echinacea species (Asteraceae)	Five different *Echinacea* preparations	3-8	EPM, social interaction and shock-induced social avoidance tests	Safe and efficient anxiolytic potential	Haller *et al.,* 2010
Elaeocarpus sphaericus	Methanolic extract of fruits	200	EPM	Anxiolytic with no significant effects on total sleeping time induced by ketamine.	Shah *et al.,* 2010

Contd...

Table 12.1–*Contd...*

Plant/Active Constituent	Plant Part/Extract	Dose (mg/kg)	Model Employed for Evaluation of Anxiety	Activity Observed	References
Equisetum arvense (Equisetaceae)	Ethanolic extract of stems	100	EPM	Anxiolytic effect with lower sedative activity than that of diazepam attributed to the flavonoid content	Singh *et al.,* 2011
Erythrina mulungu (Fabaceae)	Water-alcohol extract	100, 200, 400	Elevated T-maze, the light/dark transition, cat odor test	Anxiolytic-like effects particularly those that have been shown to be sensitive to low doses of benzo-diazepines	Onusic *et al.,* 2002
Euphorbia hirta (Euphorbiaceae)	Lyophilised aqueous extract	100	Staircase test	Anxiolytic and sedative	Lanhers, 1990
Euphorbia neriifolia (Euphorbiaceae)	Hydroalcoholic extract of leaves	100, 200, 400	EPM	Anti-anxiety, anti-psychotic and anti-convulsant activity	Bigoniya and Rana, 2005
Evolvulus alsinoides (Convolvulaceae)	Ethanol extract of the aerial parts	100	EPM, OFT	Anxiolytic and antioxidant	Nahata *et al.,* 2009
Flavonoids	Apigenin, chrysin	25	Dark-light model	Anxiolytic effect could be linked to an activation of the $GABA_A$ receptor	Zanoli *et al.,* 2000
Galphimia glauca (Malpighiaceae)	Galphinine A,B,E, methanol extract	15, 125, 250, 500, 1000, 2000	EPM, light-dark test	Anxiolytic activity of the compounds was due to the presence of free hydroxyl groups at C-4, C-6, and C-7 and the presence of the double bond in the A ring	Herrera-Ruiz *et al.,* 2006a; 2006b
Gastrodia elata (Orchidaceae)	Aqueous extract of the rhizome; its phenolic constituents, 4-hydroxy benzyl alcohol and 4-hyro-xybenzaldehyde	5,10,25, 50,100	EPM	Significantly increased the percent-age of time spent and arm entries into open arms with no changes in the locomotor and myo-relaxant effects via the GABAergic system	Jung *et al.,* 2006a

Contd...

Table 12.1–Contd...

Plant/Active Constituent	Plant Part/Extract	Dose (mg/kg)	Model Employed for Evaluation of Anxiety	Activity Observed	References
Gelsemium sempervirens (Loganiaceae)	A fraction (F9.4) derived from the methanol extract	10	EPM	Anxiolytic effect	Dutt et al., 2010
Ginkgo biloba (Ginkgoaceae)	Leaf ginkgolic acid conjugates (GAC)	0.3–0.6	EPM, OFT, SI, novelty feeding	Anti anxiety effect	Satyan et al., 1998; Kuribara et al., 2003
Glycine max (Fabaceae)	Phyto-600 diet, containing 600 mg of phytoestrogens/g of diet	200–600 µg/kg	EPM	Anxiolytic effect by activation of GABA receptors or the enhanced action of GABA	Lund and Lephart, 2001
Glycyrrhiza glabra (Fabaceae)	Hydroalcoholic extract of roots and rhizomes	10–300	EPM, foot shock induced aggression	Increased duration of occupancy in the open arms of the EPM	Ambawade et al., 2001
Hibiscus sabdariffa (Malvaceae)	Aqueous, hydroalcoholic and ethanol extracts and fractions of dried calyx	5–300	EPM	Anxiolytic and sedative effects which become more pronounced with administration of repeated doses of the extracts.	Fakeye et al., 2008
Hippeastrum vittatum (Amaryllidaceae)	Isoquinoline alkaloid – montanine	10, 30, 60	EPM	Anxiolytic, antidepressive and anticonvulsive effects.	da Silva et al., 2006
Hypericum perforatum (Hypericaceae)	Aerial parts	62.5–500	EPM, OFT, SI, Black/white chamber test, elevated zero and T-maze, cat odor test, novelty feeding, defense test	Enhancement of GABA$_A$/BZD receptor complex	Kumar et al., 2000b, 2000c; Coleta et al., 2001; Flausino et al., 2002; Vandenbogaerde et al., 2002; Beijamini and Andreatini, 2003a, 2003b
Ipomoea stans (Convolvulaceae)	Ethyl acetate extract of root	2.5–20.0	EPM	Anxiolytic effect	Herrera-Ruiz et al., 2007

Contd...

Table 12.1–Contd...

Plant/Active Constituent	Plant Part/Extract	Dose (mg/kg)	Model Employed for Evaluation of Anxiety	Activity Observed	References
Justicia pectoralis (Acanthaceae)	Aqueous standardized extract	50, 100, 200	EPM, light/dark, open field, rota rod and pentobarbital sleep time	Anxiolytic-like effect, without sedative effects.	Venâncio et al., 2011
Kielmeyera coriacea (Guttiferae)	Leaf hydroalcoholic extract	120	EPM, OFB	Significant antianxiety effect	Audi et al., 2002
Magnolia officinalis (Magnoliaceae)	Bark extract and active constituents: honokiol and magnolol	Honokiol: 0.1–20 Magnolol: 0.1–80	EPM	Honokiol acts on $GABA_A$ and $GABA_C$ receptors	Kuribara et al., 1998, 1999, 2000a, 2000b; Maruyama et al., 1998; Squires et al., 1999; Maruyama and Kuribara, 2000; Ai et al., 2001
Matricaria recutita (Asteraceae)	Apigenin from flowers	Apigenin: 1–25	EPM	Anxiolytic	Viola et al., 1995; Zanoli et al., 2000
Melissa officinalis (Lamiaceae)	Aqueous extract of aerial parts	5–100	EPM, OFT	Acting at ACh receptors	Coleta et al., 2001
Nauclea latifolia (Rubiaceae)	Decoction of roots	16, 40, 80, 160 and 360	Hole board test	Anxiolytic	Taiwe et al., 2010
Nelumbo nucifera (Nelumbonceae)	Neferine	25, 50, 100	EPM	Antianxiety, muscle relaxant and anti-convulsant effects	Sugimoto et al., 2008
Nepeta persica (Lamiaceae)	Hydroalcoholic extract of aerial parts	50,100	EPM	Anxiolytic	Rabbani et al., 2008
Ocimum gratissimum (Lamiaceae)	Petroleum ether and methanol extract of leaves	200, 400	OFT	Anticonvulsant and anxiolytic	Okoli et al., 2010
Palisota hirsute (Commelinaceae)	Ethanol extract of leaves	30–300	EPM,OFT, Light and dark test	Anxiolytic and antidepressant	Woode et al., 2010

Contd...

Table 12.1–*Contd...*

Plant/Active Constituent	Plant Part/Extract	Dose (mg/kg)	Model Employed for Evaluation of Anxiety	Activity Observed	References
Passiflora alata (Passifloraceae)	Methanol extract of aerial parts and chrysin	75, 150 Chrysin: 1	EPM, staircase test	Enhancing GABAA/BZD receptors	Wolfman *et al.,* 1994; Soulimani *et al.,* 1997; Petry *et al.,* 2001
Passiflora alata (Passifloraceae) and *Valeriana officinalis* (Valeriaceae)	Combined aqueous-ethanol extract of the two plants containing 0.02 per cent valerenic acid and 0.13 per cent flavonoids (apigenin)	20 (for 15 days)	EPM and OFT	Anxiolytic and sedative effects	Otobone *et al.,* 2005
Passiflora incarnata (Passifloraceae)	Aqueous extract	400	EPM	Enhancing GABAA/BZD receptors	Dhawan *et al.,* 2001a, b, 2002, 2003
Piper methysticum (Piperaceae)	Ethanol (96 per cent) extract of roots containing 30% kavapyrones	120–240	EPM	Anxiolytic behaviour comparable to the effects of diazepam by modulation of $GABA_A$ receptors	Jussofie *et al.,* 1994; Pepping, 1999; Smith *et al.,* 2001; Bilia *et al.,* 2002a; Rex *et al.,* 2002; Feltenstein *et al.,* 2003
Rollinia mucosa (Annonaceae)	Hexane extract of leaves	1.62 to 6.25 mg/kg	Avoidance exploratory behavior paradigm	CNS depressant effects, presumably through an interaction with the GABA/benzodiazepine receptor complex	Rosa *et al.,* 2010
Rosa (Rosaceae)	Oil inhalation	1.0%, 2.5%, 5.0% w/w	EPM	Anxiolytic activity	de Almeida *et al.,* 2004
Rubus brasiliensis (Rosaceae)	Aqueous, butanolic extract and wax fractions	50, 100, 150	EPM	Anxiolytic activity through $GABA_A$ receptor	Nogueira *et al.,* 1998a, b
Salvia reuterana (Lamiaceae)	Hydroalcoholic extract of aerial parts	100	EPM	Increased the time spent and arm entries in the open arms	Rabbani *et al.,* 2005

Contd...

Table 12.1–*Contd...*

Plant/Active Constituent	Plant Part/Extract	Dose (mg/kg)	Model Employed for Evaluation of Anxiety	Activity Observed	References
Salix aegyptiaca (Salicaceae)	Flower extract	100 and 200	EPM	Anxiolytic	Rabbani *et al.,* 2010
Sapindus mukorossi Gaern. (Sapindaceae)	Methanolic extract	200 and 400	EPM, Y-maze, Hole-board, Acto-photometer, and Marble-burying behavior models	Anxiolytic	Chakraborty *et al.,* 2010
Scutellaria lateriflora (Lamiaceae)	Flavonoid baicalin and its aglycone baicalein from hydroalcoholic extract of the herb	Baicalin 40 Baicalein 31	EPM and open field	Anxiolytic activity as the flavonoids bind to the benzo-diazepine site of the GABA$_A$ receptor	Awad *et al.,* 2003
Sesbania grandiflora (Fabaceae)	Benzene: ethyl acetate fraction of the acetone soluble part of a petroleum ether extract	50, 100	EPM	Triterpene containing fraction exhibits a wide spectrum of anti-convulsant and anxiolytic activity	Kasture *et al.,* 2002
Sinapic acid	Sinapic acid	4	EPM	Anxiolytic-like effects mediated via GABA$_A$ receptors	Yoon *et al.,* 2007
Sonchus oleraceus (Asteraceae)	Hydroethanolic and dichloromethane extracts	30–300	EPM, OFT	Anxiolytic effect similar to clonaze-pam (0.5 mg/kg, p.o.).	Vilela *et al.,* 2009
Souroubea sympetala (Marcgraviaceae)	Betulinic acid-(BA) enriched extract	6-8	EPM	Significant anxiolysis	Mullally *et al.,* 2011
Sphaeranthus indicus (Asteraceae)	Hydroalcoholic extract of aerial parts	100, 200, 500	EPM,OFT	Anxiolytic and antidepressant	Galani *et al.,* 2010
Stachys lavandulifolia (Lamiaceae)	Hydroalcoholic extract and essential oil from aerial parts	100	EPM	Anxiolytic effect with relatively lower sedative activity than diazepam. The essential oil did not have any significant effects on the mice behaviour.	Rabbani *et al.,* 2003

Contd...

Table 12.1 *–Contd...*

Plant/Active Constituent	Plant Part/Extract	Dose (mg/kg)	Model Employed for Evaluation of Anxiety	Activity Observed	References
Tetrapleura tetraptera (Mimosaceae)	Aridanin	5 and 10 mg/kg, i.p.	EPM	Anxiolytic effect through interaction with GABAA- benzodi-azepine receptor complex.	Aderibigbe et al., 2010
Tilia americana var. *mexicana* (Malvaceae)	Methanol extract	25	EPM	Anxiolytic	Herrera-Ruiz et al., 2008
Turnera aphrodisiaca (Turneraceae)	Methanol extract of aerial parts	10, 25, 75	EPM	Significant anti-anxiety activity	Kumar and Sharma, 2005a,b
Uncaria rhynchophulla (Rubiaceae)	Aqueous extract of stem	200	EPM	Anxiolytic	Jung et al., 2006
Valeriana officinalis (Valeriaceae)	Methanol and ethanol extracts	500–1000	EPM	Anxiolytic	Hattesohl et al., 2008
Vitex negundo (Verbenaceae)	Ethanol extract of roots	100 and 200	EPM and light-dark exploration test	Anxiolytic	Adnaik et al., 2009
Withania somnifera (Solanaceae)	Root glycol-withanolides	20–50	EPM, SI, novelty feeding	Anxiolytic and antioxidant action	Mehta et al., 1991; Bhattacharya et al., 2000b
Zingiber officinale (Zingiberaceae)	Benzene fraction of petroleum ether extract of dried rhizomes	15, 30	EPM	Decreased occupancy in the closed arm of the EPM	Vishwakarma et al., 2002
Ziziphus jujuba (Rhamnaceae)	Ethanol extract of ripe fruits containing spinosin and jujubo-sides Sanjoinine A	0.5–2 g/kg	EPM, black/white chamber, OFT, hole board test	Jujuboside A inhibits hippocampal hyper-activity. Anxiolytic-like effects may be mediated by GABA-ergic transmission	Peng et al., 2000; Shou et al., 2002 Han et al., 2009

EPM: Elevated plus maze; OFT: Open field test; SI: Social interaction.

Table 12.2: Results of clinical trials of some plants with anxiolytic potential.

Plant	Plant Extract/Preparation	Type of Trial	Type of Anxiety Disorder	Result of clinical trial
Bacopa monniera (Stough et al., 2001)	B. monniera extract – Keenmind ® (300mg for 12 weeks)	Placebo-controlled RCT	GAD	Reduction in state anxiety in the experimental group compared to the control group.
Centella asiatica (Bradwejn et al., 2000; Jana et al., 2010)	Gotu kola (500 mg/capsule, twice daily, after meal)	Double blind, RCT	GAD	Attenuated anxiety related disorders and also significantly reduced stress phenomenon and its correlated depression
Ocimum amoenum (Sayyah et al., 2009)	E. amoenum aqueous extract (500 mg/day)	6-week, double blind, parallel-group trial	OCD	The extract showed a significant superiority over placebo in reducing obsessive and compulsive and anxiety symptoms. There was no significant difference between the two groups in terms of adverse effects.
Ginkgo biloba (Woelk et al., 2007)	Ginkgo biloba special extract EGb 761 (480 and 240 mg/day for 4 weeks)	Double-blind, placebo-controlled RCT	GAD	It was safe and well tolerated and may thus be of particular value in elderly patients with anxiety related to cognitive decline.
Ocimum sanctum (Bhattacharyya et al., 2008).	Lyophilised ethanol extract (500 mg/capsule, twice daily, p.o. after Meal)	Programmed clinical trail	GAD in hospital based clinical set-up	Significantly attenuated GAD and also attenuated its correlated stress and depression
Lavandula spica (Woelk and Schläfke, 2010)	Silexan - new oral lavender oil capsule preparation	Multi center, double blind, RCT	GAD	Silexan appears to be an effective and well tolerated alternative to benzodiazepines for amelioration of generalised anxiety
Passiflora incarnata (Akhondzadeh et al., 2001).	Passion flower tincture (45 drops/day)	Double blind, RCT	GAD	No significant differences in terms of anxiety levels as compared to oxazepam treatment. However fewer adverse effects were reported.
Piper methysticum (Pittler et al., 2002)	Mono-preparation of kava	Placebo controlled RCT	GAD	Significant reduction of anxiety.

Contd...

Table 12.2—*Contd...*

Plant	Plant Extract/Preparation	Type of Trial	Type of Anxiety Disorder	Result of clinical trial
Piper methysticum (Kava) Aqueous extract (Sarris *et al.*, 2009)	Five Kava tablets per day were prescribed containing 250 mg of kavalactones/day.	3-week placebo-controlled, double-blind crossover trial	GAD	*Piper methysticum* (Kava) has been withdrawn in European, British, and Canadian markets due to concerns over hepatotoxic reactions. The WHO recently recommended research into "aqueous" extracts of Kava. The aqueous Kava preparation produced significant anxiolytic and antidepressant activity and raised no safety concerns at the dose and duration studied.
Rhodiola rosea (Panossian *et al.*, 2010)	*R. rosea* extract SHR-5	RCT	GAD	Encouraging results in mild to moderate depression, and generalized anxiety.
Salvia officinalis (Kennedy *et al.*, 2006)	Dried sage leaf (300, 600 mg)	Double blind, placebo-controlled		Dose-dependent reduction of anxiety and increased 'alertness', calmness and contentedness
Trifolium pratense (Lipovac *et al.*, 2010)	Two capsules, daily of MF11RCE (80 mg red clover isoflavones	RCT	Anxiety and depressive symptoms among post-menopausal women	Significant reduction in anxiety and depression
Valeriana officinalis (Andreaatini *et al.*, 2002)	Valerian extract containing valepotriates (mean daily dose: 81.3 mg)	Double blind, RCT	GAD	Improvement in valerian treated, diazepam and placebo groups. No significant differences between the groups.
Melissa officinalis and *Valeriana officinalis* (Kennedy *et al.*, 2006)	A standardized product containing *M. officinalis* and *V. officinalis* extracts (600 mg, 1200 mg, 1800 mg)	Double-blind, placebo-controlled RCT	GAD	The combination possesses anxiolytic properties that deserve further investigation.

GAD: Generalized anxiety disorder; OCD: Obsessive compulsive disorder; RCT: Randomized controlled trial.

Conclusions

Plants have been used for their anti anxiety activity for centuries; the use of complementary and alternative medicine has increased over the past decade. The use of natural products for anxiety disorders is on the rise in modern medicine also and hence a critical evaluation of their safety and efficacy is required (Awad *et al.*, 2007). This review was therefore aimed at summarising the evidence for the anxiolytic efficacy of such treatments. Numerous plants have been explored for their antianxiety potential using experimental animal models. Some plants have demonstrated efficacy clinically also. The ultimate goal of all researchers and physicians involved in health care is to develop strategies that minimize risks and maximize benefits (Saeed *et al.*, 2007; Kinrys *et al.*, 2009). The availability of natural treatments for management of anxiety, that are supported by clinical evidence will help physicians collaborate with patients using or seeking natural remedies to maximize the potential for benefit and minimize the potential for harm.

References

Abascal, K., and Yarnell, E. (2004). Nervine herbs for treating anxiety. *Journal of Alternative and Complementary Therapies*, **10(6):** 309-315.

Aderibigbe, A.O., Iwalewa, E.O., andAdesina, S.K. (2010). Anxiolytic effect of Aridanin isolated from *Tetrapleura tetraptera* in mice. *Bioresearch Bulletin*, **1:** 1-6 1.

Akhondzadeh, S., Naghavi, H.R., Vazirian, M., Shayeganpour, A., Rashidi, H., and Khani, M. (2001). Passionflower in the treatment of generalized anxiety: a pilot double-blind randomized controlled trial with oxazepam. *J Clin Pharm Ther.*, **26(5):** 363-367.

Adnaik, R.S., Pai, P.T., Sapakal, V.D., Naikwade, N.S., and Magdum, C.S. (2009). Anxiolytic activity of *Vitex negundo* Linn. in experimental models of anxiety in mice. *International Journal of Green Pharmacy*, **3(3):** 243-247.

Akindele, A.J., and Adeyemi, O.O. (2010). Anxiolytic and sedative effects of*Byrsocarpus coccineus* Schum. and Thonn. (Connaraceae) extract. *International Journal of Applied Research in Natural Products*, **3(1):** 28-36.

Ai, J., Wang, X., and Nielsen, M. (2001). Honokiol and magnolol selectively interact with GABAA receptor subtypes *in vitro*. *Pharmacology*, **63:** 34–41.

Ambawade, S., Kasture, V.S., and Kasture, S.B. (2001). Anxiolytic activity of*Glycyrrhiza glabra* linn. *Journal of Natural Remedies*, **1:** 130-134.

Andreatini, R., Sartor, V.A., Seabra, M.L.V., and Leite, J.R. (2002). Effect of valepotriates (valerian extract) in generalized anxiety disorder: a randomized placebo-controlled pilot study. *Phytotherapy Research*, **16(7):** 650–654.

Argal, A., Pathak, A.K. (2006). CNS activity of *Calotropis gigantea* roots. *Journal of Ethnopharmacology*, **106(1):** 142-145.

Audi, E.A., Otobone, F., Martins, J.V., and Cortez, D.A. (2002). Preliminary evaluation of *Kielmeyera coriacea* leaves extract on the central nervous system. *Fitoterapia* **73:** 517–519.

Awad, R., Arnason, J.T., Trudeau, V., Bergeron, C., Budzinski, J.W., Foster, B.C., and Merali, Z. (2003). Phytochemical and biological analysis of skullcap (*Scutellaria lateriflora* L.): a medicinal plant with anxiolytic properties. *Phytomedicine*, **10(8)**: 640-9.

Awad, R., Levac, D., Cybulska, P., Merali, Z., Trudeau, V.L., and Arnason, J.T. (2007). Effects of traditionally used anxiolytic botanicals on enzymes of the gamma-aminobutyric acid (GABA) system. *Can J Physiol Pharmacol.*, **85(9)**: 933-42.

Baldessarini, R.J. (2001). Drugs and the treatment of psychiatric disorders. In: .Hardman, J.G and. Limbird, L.E (Eds.), Goodman and Gilman's The Pharmacological Basis of Therapeutics, 10th ed., The McGraw-Hill Companies, New York, USA, pp 399-427 and 447-477.

Barlow, D.H. (2001). Clinical handbook of psychological disorders, (3rd ed.). New York: Guilford.

Barua, C., Roy, J. D., Buragohain, B., Barua, A.G., Borah, P., and Lahkar, M. (2009). Anxiolytic effect of hydroethanolic extract of *Drymaria cordata* L Willd. *Indian Journal of Experimental Biology*, **47(12)**: 969-973.

Beijamini, V., and Andreatini, R. (2003a). Effects of *Hypericum perforatum* and paroxetine in the mouse defense test battery. *Pharmacology, Biochemistry and Behavior*, **74**: 1015–1024.

Beijamini, V., and Andreatini, R. (2003b). Effects of *Hypericum perforatum* and paroxetine on rat performance in the elevated Tmaze. *Pharmacological Research*, **48**: 199–207.

Bhattacharya, S.K., Bhattacharya, A., Kumar, A., and Ghosal, S. (2000a). Antioxidant activity of *Bacopa monniera* in rat frontal cortex, striatum and hippocampus. *Phytotherapy Research*, **14**: 174–179.

Bhattacharya, S.K., Bhattacharya, A., Sairam, K., and Ghosal, S. (2000b). Anxiolytic-antidepressant activity of *Withania somnifera* glycowithanolides: an experimental study. *Phytomedicine*, **7**: 463–469.

Bhattacharyya, D., Sur, T.K., Jana, U., and Debnath, P.K. (2008). Controlled programmed trial of *Ocimum sanctum* leaf on generalized anxiety disorders. *Nepal Med Coll J.*, **10(3)**: 176-9.

Bigoniya, P., and Rana, A.C. (2005). Psychopharmacological profile of hydro-alcoholic extract of *Euphorbia neriifolia* leaves in mice and rats. *Indian Journal of Experimental Biology*, **43(10)**: 859-62.

Bilia, A.R., Gallori, S., and Vincieri F.F. (2002). Kava-kava and anxiety: growing knowledge about the efficacy and safety. *Life Science*, **70**: 2581-97.

Bourbonnais-Spear, N., Awad, R., Merali, Z., Maquinc, P., Cal, V., and Arnason, J.T. (2007). Ethnopharmacological investigation of plants used to treat susto, a folk illness. *Journal of Ethnopharmacology*,**109**: 380–387.

Bradwejn, J., Zhou, Y., Koszycki, D., and Shlik, J. (2000). A double-blind, placebo-controlled study on the effects of Gotu Kola (*Centella asiatica*) on acoustic startle response in healthy subjects. *J Clin Psychopharmacol.*, **20(6)**: 680-4.

Carlini, E.A. (2003). Plants and the central nervous system. *Pharmacol., Biochem. and Behavior,* **75**: 501-512.

Carvalho-Freitas, M.I., and Costa, M. (2002). Anxiolytic and sedative effects of extracts and essential oil from *Citrus aurantium* L. *Biological and Pharmaceutical Bulletin,* **25(12):** 1629-33.

Cates, M., Wells, B.G., and Thatcher, G.W. (1996). Anxiety Disorders. In: HerFindal, E.T. and Gourley, D.R. (editors), Textbook of Therapeutics: Drug and Disease Management, 6th edition, Williams and Wilkins, Maryland, USA, pp1073-93.

Chakraborty, A., Amudha, P., Geetha, M., and Singh, N.S. (2010). Evaluation of anxiolytic activity of methanolic extract of *Sapindus mukorossi* gaertn. in mice. *International Journal of Pharma and Bio Sciences,* **1(3):**

Chung, L.Y., Goh, S.H., and Imiyabir, Z. (2005). Central nervous system receptor activities of some Malaysian plant species. *Pharm Biol.,* **43(3):** 280–8.

Coleta, M., Campos, M.G., Cotrim, M.D., and Proenca da Cunha, A. (2001). Comparative evaluation of *Melissa officinalis* L, *Tilia europaea* L, *Passiflora edulis* Sims and *Hypericum perforatum* L. in the elevated plus maze anxiety test. *Pharmacopsychiatry,* **34:** S20–S21.

da Silva, A.F.S., de Andrade, J.P., Bevilaqua, L.R.M., deSouza, M.M., Izquierdo, I., Henriques, A.T., and Zuanazzi, J.A.S. (2006). Anxiolytic, antidepressant and anticonvulsant - like effects of the alkaloid montanine isolated from *Hippeastrum vittatum. Pharmacology, Biochemistry and Behavior,* **85:** 148–154.

de Almeida, R.N., Motta, S.C., de Brito Faturi, C., Catallani, B., and Leite, J.R. (2004). Anxiolytic-like effects of rose oil inhalation on the elevated plus-maze test in rats. *Pharmacol Biochem Behav.,* **77(2):** 361-4.

de Almeida, E.R., Rafael K.R., Couto, G.B., and Ishigami, A.B. (2009). Anxiolytic and anticonvulsant effects on mice of flavonoids, linalool, and alpha-tocopherol presents in the extract of leaves of *Cissus sicyoides* L. (Vitaceae). *J Biomed Biotechnol.,* 274-740.

de Melo, C.T., Monteiro, A.P., Leite, C.P., de Araujo, F.L., Lima, V.T., Barbosa-Filho, J.M., de Franca Fonteles, M.M., de Vasconcelos, S.M., de Barros Viana, G.S., and de Sousa, F.C. (2006). Anxiolytic-like effects of (O-methyl)-N-2,6-dihydroxybenzoyl-tyramine (riparin III) from *Aniba riparia* (Nees) Mez (Lauraceae) in mice. *Biol Pharm Bull.,* **29(3):** 451-4.

de Sousa, F.C.F., Monteiro, A.P., de Melo, C.T.V., de Oliveira, G.R., Vasconcelos, S.M.M., de França Fonteles, M.M., Gutierrez, S.J.C., Barbosa-Filho, J.M., and Viana, G.S.B. (2005). Antianxiety effects of riparin I from *Aniba riparia* (Nees) Mez (Lauraceae) in mice. *Phytotherapy Research,* **19(12):** 1005 – 1008.

Dhawan, K., Kumar, S., and Sharma, A. (2001a). Anti-anxiety studies on extracts of *Passiflora incarnata* L. *Journal of Ethnopharmacology,* **78:** 165 –170.

Dhawan, K., Kumar, S., and Sharma, A. (2001b). Anxiolytic activity of aerial and underground parts of *Passiflora incarnata. Fitoterapia,* **72(8):** 922-6.

Dhawan, K., Kumar, S., and Sharma, A. (2002). Comparative anxiolytic activity profile of various preparations of *Passiflora incarnata* linneaus: a comment on medicinal plants' standardization. *Journal of Alternative and Complementary Medicine*, **8**: 283–291.

Dhawan, K., Dhawan, S., and Chhabra, S. (2003). Attenuation of benzodiazepine dependence in mice by a tri-substituted benzoflavone moiety of *Passiflora incarnata* Linneaus: a non-habit forming anxiolytic. *Journal of Pharmacy and Pharmaceutical Sciences*, **6**: 215–222.

Diagnostic and Statistical Manual for Mental Disorders (DSM-IV)-Text Revision (2000). American Psychiatric Association, USA.

Dopheide, J., and Park, S. (2002). The Psychopharmacology of Anxiety. *Psychiatric Times*, **19(3)**: 66.

Dutt, V., Dhar, V.J., and Sharma, A. (2010). Antianxiety activity of *Gelsemium sempervirens*. *Pharm Biol.*, **48(10)**: 1091-106.

Eisenberg, D.M., Davis, R.B., Ettner, S.L., Appel, S., Wilkey, S., and Van Rompay, M. (1998). Trends in alternative medicine use in the United States. *J Am Med Assoc.*, **280**: 1569–1575.

Emamghoreishi, M; Khasaki, M; and Aazam, M.F. (2005). *Coriandrum sativum*: evaluation of its anxiolytic effect in the elevated plus-maze. *Journal of Ethnophrmacology*, **96**: 365–370.

Ernst, E. (2006). Prevalence of use of Complementary and Alternative Medicine: a systematic review. *Bulletin of World Health Organization*, **78 (2):**

Fakeye, T.O., Pal, A., and Khanuja, S.P. (2008). Anxiolytic and sedative effects of extracts of *Hibiscus sabdariffa* Linn (family Malvaceae). *Afr J Med Sci.*, **37(1)**: 49-54.

Faturi,C.B., Leite, J.R., Alves, P.B., Canton, A.C., and Teixeira-Silva, F. (2010). Anxiolytic-like effect of sweet orange aroma in Wistar rats. *Prog Neuropsychopharmacol Biol Psychiatry*, **30**: 605-9.

Flausino, O.A., Zangrossi, H., Salgado, J.V., and Viana, M.B. (2002). Effects of acute and chronic treatment with *Hypericum perforatum* L. (LI 160) on different anxiety-related responses in rats. *Pharmacology, Biochemistry and Behavior*, **71**: 251–257.

Feltenstein, M.W., Lambdin, L.C., Ganzera, M., Ranjith, H., Dharmaratne, W., Nanayakkara, N.P., Khan, I.A., and Sufka, K.J. (2003). Anxiolytic properties of *Piper methysticum* extract samples and fractions in the chick social-separation-stress procedure. *Phytotherapy Research*, **17**: 210–216.

Galani, V.J., and Patel, B.G. (2010). Effect of hydroalcoholic extract of *Sphaeranthus indicus* against experimentally induced anxiety, depression and convulsions in rodents. *International Journal of Ayurveda Research*, **1(2)**: 87-92.

Ghosal, S., and Bhattacharya, S.K. (1980). Anxiolytic activity of a standardized extract of *Bacopa monniera* in an experimental study. *Phytomedicine*, **5**: 133–148.

Gilhotra, N., and Dhingra, D. (2010). GABAergic and nitriergic modulation by curcumin for its antianxiety-like activity in mice. *Brain Res.*, **1352**: 167-75.

Grundmann, O., Nakajima, Jun-Ichiro, Seo, S., and Butterweck, V. (2007). Anti-anxiety effects of *Apocynum venetum* L. in the elevated plus maze test. *Journal of Ethnopharmacology*, **110**: 406–411.

Guaraldo, L., Chagas, D.A., Konno, A.C., Korn, G.P., Pfiffer, T., and Nasello, A.G. (2000). Hydroalcoholic extract and fractions of *Davilla rugosa* Poiret : effects on spontaneous motor activity and elevated plus-maze behavior. *Journal of Ethnopharmacology*, **72(1)**: 61-67.

Gupta, V., Bansal, P., Kumar, P., and Shri, R. (2010). Anxiolytic and Antidepressant Activities of Different Extracts from *Citrus paradisi* Var. Duncan. *Asian Journal of Pharmaceutical and Clinical Research.*, **3(2)**: 98-100.

Haller, J., Hohmann, J., and Freund, T.F. (2010). The effect of *Echinacea* preparations in three laboratory tests of anxiety: comparison with chlordiazepoxide. *Phytother Res.*, **24(11)**:1605-1613.

Han, H., Ma, Y., Eun, J.S., Li, R., Hong, J.T., Lee, M.K., and Oh, K.W. (2009). Anxiolytic-like effects of sanjoinine A isolated from *Zizyphi spinosi* Semen: possible involvement of GABAergic transmission. *Pharmacol Biochem Behav.*, **92(2)**: 206-13.

Hattesohl, M., Feistel, B., Sievers, H., Lehnfeld, R., Heggera, M., and Winterhoff, H. (2008). Extracts of *Valeriana officinalis* L. show anxiolytic and antidepressant effects but neither sedative nor myorelaxant properties. *Phytomedicine*, **15**: 2–15.

Helli´on-Ibarrola, M.C.; Ibarrola, D.A.; Montalbetti, Y.; Kennedy, M.L.; Heinichen, O.; Campuzanoa, M.; Tortoriello, J.; Fern´andez, S.; Wasowski, C.; Marder, M.; De Limad, T.C.M.; and Morae, S. (2006). The anxiolytic-like effects of *Aloysia polystachya* (Griseb.) Moldenke (Verbenaceae) in mice. *Journal of Ethnopharmacology*, **105**: 400–408.

Hernández, E. A., Martínez, A.L., González-Trujano, M.E., Moreno, J., Vibrans, H. and Soto-Hernández M. (2006). Pharmacological evaluation of the anxiolytic and sedative effects of *Tilia americana* L. var. mexicana in mice. *Journal of Ethnopharmacology*, **109(1)**: 140-145.

Herrera-Ruiz, M., Gonzalez-Cortazar, M., Jimenez-Ferrer, E., Zamilpa, A., Alvarez, L., Ramirez, G., and Tortoriello, J. (2006a). Anxiolytic effect of natural galphimines from *Galphimia glauca* and their chemical derivatives. *Journal of Natural Products*, **69(1)**: 59-61.

Herrera-Ruiz, M., Jimenez-Ferrer, J.E., De Lima, T.C., Aviles-Montes, D., Perez-Garcia, D., Gonzalez-Cortazar, M., and Tortoriello, J. (2006b) Anxiolytic and antidepressant-like activity of a standardized extract from *Galphimia glauca*. *Phytomedicine*, **13(1-2)**: 23-8.

Herrera-Ruiz, M., Guti´errez,C., Jim´enez-Ferrer, J.E., Tortoriello, J., Mir´on, G., and Le´on, I. (2007). Central nervous system depressant activity of an ethyl acetate extract from *Ipomoea stans* roots. *Journal of Ethnopharmacology*, **112**: 243–247.

Herrera-Ruiz, M., Román-Ramos, R., Zamilpa, A., Tortoriello, J., and Jiménez-Ferrer, J.E. (2008). Flavonoids from *Tilia americana* with anxiolytic activity in plus-maze test. *Journal of Ethnopharmacology*, **118(2)**: 312-7.

Jain, N.N., Ohal, C.C., Shroff, S.K., Bhutada, R.H., Somani, R.S., Kasture, V.S., and Kasture, S.B. (2003). *Clitoria ternatea* and the CNS. *Pharmacology, Biochemistry and Behavior*, **75**: 529–536.

Jaiswal, A.K., Bhattacharya, S.K., and Acharya, S.B. (1994). Anxiolytic activity of *Azadirachta indica* leaf extract in rats. *Indian Journal of Experimental Biology*, **32**: 489–491.

Jana, U., Sur, T.K., Maity, L.N., Debnath, P.K., and Bhattacharyya, D. (2010). A clinical study on the management of generalized anxiety disorder with*Centellaasiatica*. *Nepal Med Coll J.*, **12(1)**: 8-11

Jung, J.W., Yoon, B.H., Oh, H.R., Ahn, J.H., Kim, S.Y., Park, S.Y., and Ryu, J.H. (2006a). Anxiolytic-like effects of *Gastrodia elata* and its phenolic constituents in mice. *Biological and Pharmaceutical Bulletin*, **29(2)**: 261-5.

Jung, J.W., Ahn, N.Y., Oh, H.K., Lee, B.K., Lee, K.J., Kim, S.Y., Cheong, J.H. and Ryu, T.H. (2006b). Anxiolytic effects of the aqueous extract of *Uncaria rhynchophylla*. *Journal of Ethnopharmacology*, **108**: 193-197.

Jussofie, A., Schmiz, A., and Hiemke, C. (1994). Kavapyrone enriched extract from *Piper methysticum* as modulator of the GABA binding site in different regions of rat brain. *Psychopharmacology (Berl)*, **116**: 469–474.

Kalariya, M., Parmar, S., and Sheth, N. (2010). Neuropharmacological activity of hydroalcoholic extract of leaves of *Colocasia esculenta*. *Pharm Biol.*, **48(11)**:1207-1212.

Kasture, V.S., Deshmukh, V.K., and Chopde, C.T. (2002). Anxiolytic and anticonvulsive activity of *Sesbania grandiflora* leaves in experimental animals. *Phytotherapy Research*, **16(5)**: 455-60.

Kennedy, D.O., Pace, S., and Haskell, C. (2006). Effects of cholinesterase inhibiting sage (*Salvia officinalis*) on mood, anxiety and performance on a psychological stressor battery. *Neuropsychopharmacology*, **31(4)**: 845-852.

Kessler, R.C., Berglund, P., Demler, O., Jin, R., Merikangas, K.R., and Walters, E.E. (2005a). Lifetime prevalence and age-of-onset distributions of DSM-IV disorders in the National Comorbidity Survey Replication. *Archives of General Psychiatry*, **62(6)**: 593–602.

Kessler, R.C., Chiu, W.T., Demler, O., and Walters, E.E. (2005b). Prevalence, severity, and comorbidity of 12month DSM-IV disorders in the National Comorbidity Survey Replication. *Archives of General Psychiatry*, **62**: 617-709.

Kessler, R.C., Soukup, J., Davis, R.B., Foster, D.F., Wilkey, S.A., Van Rompay, M.I., and Eisenberg, D.M. (2001). The use of complementary and alternative therapies to treat anxiety and depression in the United States. *American Journal of Psychiatry*, **158 (2)**: 289–294.

Kessler, R.C., and Wang, P.S. (2008). The descriptive epidemiology of commonly occurring mental disorders in the United States. *Annual Review of Public Health*, **29**: 115–129.

Kim, W.I., Jung, J.W., Ahn, N.Y., Oh, H.R., Lee, B.K., Oh, J.K., Cheong, J.H., Chun, H.S., and Ryu, H. (2004). Anxiolytic-like effects of extracts from *Albizzia julibrissin* bark in the elevated plus-maze in rats. *Life Sciences*, **75(23)**: 2787-2795.

Kinrys, G., Coleman, E., and Rothstein, E. (2009). Natural remedies for anxiety disorders: potential use and clinical applications. *Depress Anxiety*, **26(3)**: 259-65.

Kothari, S., Minda, M., and Tonpay, S.D. (2010). Anxiolytic and antidepressant activities of methanol extract of *Aegle marmelos* leaves in mice. *Indian J Physiol Pharmacol.*, **54(4)**: 318-328.

Kumar, V., Jaiswal, A.K., Singh, P.N., and Bhattacharya, S.K. (2000b). Anxiolytic activity of Indian *Hypericum perforatum* Linn: an experimental study. *Indian Journal of Experimental Biology*, **38**: 36–41.

Kumar, V., Singh, P.N., Muruganandam, A.V., and Bhattacharya, S.K. (2000c). Effect of Indian *Hypericum perforatum* Linn on animal models of cognitive dysfunction. *Journal of Ethnophamacology*, **72**: 119–128.

Kumar, S. and Sharma, A. (2005a). Anti-anxiety Activity Studies on Homoeopathic Formulations of *Turnera aphrodisiaca*. *Evidenc Based Complementary and Alternative Medicine*, **2**: 117–119.

Kumar S, and Sharma A. (2005b). Anti-anxiety activity studies of various extracts of *Turnera aphrodisiaca* Ward. *Journal of Herbal Pharmacotherapy*, **5(4)**: 13-21.

Kumar, V. (2006). Potential medicinal plants for CNS disorders: an overview. *Phytotherapy Research*, **20**: 1023–35.

Kumar, V., Jaiswal, A.K., Singh, P.N., and Bhattacharya, S.K. (2000b). Anxiolytic activity of Indian *Hypericum perforatum* Linn: an experimental study. *Indian Journal of Experimental Biology*, **38**: 36–41.

Kumar, V., Singh, P.N., Muruganandam, A.V., and Bhattacharya, S.K. (2000c). Effect of Indian *Hypericum perforatum* Linn on animal models of cognitive dysfunction. *Journal of Ethnophamacology*, **72**: 119–128.

Kumar, V., Singh, R.K., Jaiswal, A.K., Bhattacharya, S.K., and Acharya, S.B. (2000a). Anxiolytic activity of Indian *Abies pindrow* Royle leaves in rodents: an experimental study. *Indian Journal of Experimental Biology*, **38**: 343–346.

Kuribara, H., Stavinoha, W.B., and Maruyama, Y. (1998). Behavioural pharmacological characteristics of honokiol, an anxiolytic agent present in extracts of *Magnolia* bark, evaluated by an elevated plus-maze test in mice. *J. Pharm. Pharmacol.*, **50(7)**: 819-826.

Kuribara, H., Kishi, E., Hattori, N., Yuzurihara, M., and Maruyama, Y. (1999). Application of the elevated plus-maze test in mice for evaluation of the content of honokiol in water extracts of magnolia. *Phytotherapy Research*, **13**: 593–596.

Kuribara, H., Kishi, E., Hattori, N., Okada, M., and Maruyama, Y. (2000a). The anxiolytic effect of two oriental herbal drugs in Japan attributed to honokiol from Magnolia bark. *The Journal of Pharmacy and Pharmacology*, **52**: 1425–1429.

Kuribara, H., Kishi, E., Kimura, M., Weintraub, S.T., and Maruyama, Y. (2000b). Comparative assessment of the anxiolytic-like activities of honokiol and derivatives. *Pharmacology, Biochemistry and Behavior*, **67**: 597–601.

Kuribara, H., Weintraub, S.T., Yoshihama, T., and Maruyama, Y. (2003). An anxiolytic-like effect of *Ginkgo biloba* extract and its constituent, ginkgolide-A, in mice. *Journal of Natural Products*, **66(10)**: 1333-7.

Lalitha, K.G., Sathish,R., Gayathri, R., Karthikeyan, S., Kalaiselvi,P., Muthuboopathi, P., and Venkatachalam, T. (2010). Pharmacological Evaluation of *Clerodendrum philippinum* Schauer Flowers for Antianxiety and Central Nervous System Depressant Activity. *International Journal of Pharmaceutical Research*, **2(2)**: 13-15.

Lanhers, M.C., Fleurentin, J., Cabalion, P., Rolland, A., Dorfman, P., Misslinb, R., and Pelt, J.M. (1990). Behavioral effects of *Euphorbia hirta* l.: sedative and anxiolytic properties. *Journal of Ethnopharmacology*, **29**: 189- 198.

Lépine, J.P. (2002). The epidemiology of anxiety disorders: prevalence and societal costs. *Journal of Clinical Psychiatry*, **63** (Suppl 14): 4-8.

Lipovac, M., Chedraui, P., Gruenhut, C., Gocan, A., Stammler, M., and Imhof, M. (2010). Improvement of postmenopausal depressive and anxiety symptoms after treatment with isoflavones derived from red clover extracts. *Maturitas*, **65**: 258–261

Lopez-Rubalcava, C., Pina-Medina, B., Estrada-Reyes, R., Heinze, G., and Martinez-Vazquez, M. (2006). Anxiolytic-like actions of the hexane extract from leaves of *Annona cherimolia* in two anxiety paradigms: possible involvement of the GABA/benzodiazepine receptor complex. *Life Sciences*, **78(7)**: 730-7.

Lund, T.D.; Lephart, E. D. 2001. Dietary soy phytoestrogens produce anxiolytic effects in the elevated plus-maze. *Brain Research*, **913**: 180–184.

Maruyama, Y., Kuribara, H., Morita, M., Yuzurihara, M., and Weintraub, S.T. (1998). Identification of magnolol and honokiol as anxiolytic agents in extracts of saiboku-to, an oriental herbal medicine. *Journal of Natural Products*, **61**: 135–138.

Maruyama, Y., and Kuribara, H. (2000). Overview of the pharmacological features of honokiol. *CNS Drug Reviews*, **6**: 35–44.

Mehta, A.K., Binkley, P., Gandhi, S.S., and Ticku, M.K. (1991). Pharmacological effects of *Withania somnifera* root extract on GABAA receptor complex. *The Indian Journal of Medical Research*, **94**: 312–315.

Min, L., Chen, S.W., Li.,J., Wang, R., Li, Y.L., Wang, W.J., and Mi, X.J. (2005). The effects of *Angelica* essential oil in social interaction and hole-board tests. *Pharmacology Biochemistry and Behavior*, **81(4)**: 838-42.

Miño, J.H., Moscatelli, V., Acevedo, C., and Ferraro, G. (2010). Psychopharmacological effects of *Artemisia copa* aqueous extract in mice. *Pharm Biol.*, **48(12)**: 1392-1396.

Molina-Hernandez, M., Tellez-Alcantara, N.P., Garcia, J.P., Lopez, J.I., and Jaramillo, M.T. (2004). Anxiolytic-like actions of leaves of *Casimiroa edulis* (Rutaceae) in male Wistar rats. *Journal of Ethnopharmacology,* **93(1):** 93-8.

Moquin, B., Blackman, M.R., Mitty, E., and Flores, S. (2009). Complementary and Alternative Medicine (CAM). *Geriatric Nursing,* **30(3):** 196-203.

Mora, S., Diaz-Veliz, G., Lungenstrass, H., Garcia-Gonzalez, M., Coto-Morales, T., Poletti, C., De Lima, T.C., Herrera-Ruiz, M., and Tortoriello, J. (2005). Central nervous system activity of the hydroalcoholic extract of *Casimiroa edulis* in rats and mice. *Journal of Ethnopharmacology,* **97(2):** 191-7.

Moreira, F.A., Aguiar, D.C., and Guimarães, F.S. (2006). Anxiolytic-like effect of cannabidiol in the rat Vogel conflict test. *Progress in Neuro-Psychopharmacology and Biological Psychiatry,* **30:** 1466–1471.

Mullally, M., Kramp, K., Cayer, C., Saleem, A., Ahmed, F., McRae, C., Baker, J., Goulah, A., Otorola, M., Sanchez, P., Garcia, M., Poveda, L., Merali, Z., Durst, T., Trudeau, V.L., and Arnason, J.T. (2011). *Anxiolytic activity of a supercritical carbon dioxide extract of Souroubea sympetala* (Marcgraviaceae). *Phytother Res.,* **25(2):** 264-270.

Nahata, A., Patil, U.K., and Dixit, V.K. (2009). Anxiolytic activity of *Evolvulus alsinoides* and *Convulvulus pluricaulis* in rodents. *Pharmaceutical Biology,* **47(5):** 444-451

NIMH. (2006). National Institute of Mental Health. Anxiety disorders: NIH publication number 06-3879. US department of health and human service.

Nogueira, E., Rosa, G.J., Haraguchi, M., and Vassilieff, V.S. (1998a). Anxiolytic effect of *Rubus brasilensis* in rats and mice. *Journal of Ethnophamacology,* **61:** 111–117.

Nogueira, E., Rosa, G.J., and Vassilieff, V.S. (1998b). Involvement of GABA(A)-benzodiazepine receptor in the anxiolytic effect induced by hexanic fraction of *Rubus brasiliensis. Journal of Ethnophamacology,* **61:** 119–126.

Okoli, C.O., Ezike, A.C., Agwagah, O.C., and Akah, P.A. (2010). Anticonvulsant and anxiolytic evaluation of leaf extracts of *Ocimum gratissimum*, a culinary herb. *Pharmacognosy Research,* **2(1):** 36-40.

Onusic, G.M., Nogueira, R.L., Pereira, A.M., and Viana, M.B. (2002). Effect of acute treatment with a water-alcohol extract of *Erythrina mulungu* on anxiety-related responses in rats. *Brazilian Journal of Medical and Biological Research,* **35(4):** 473-7.

Otobone, F.J., Martins, J.F.C., Trombelli, M.A., Andreatini, R., and Audi, E.A. (2005). Anxiolytic and sedative effects of a combined extract of *Passiflora alata* Dryander and *Valeriana officinalis* L. in rats. *Acta Scientiarum. Health Science,* **27(2):** 145-150.

Panossian, A., Wikman, G., and Sarris, J. (2010). Rosenroot (*Rhodiola rosea*): traditional use, chemical composition, pharmacology and clinical efficacy. *Phytomedicine : international journal of phytotherapy and phytopharmacology,* **17(7):** 481-93.

Peng, W.H., Hsieh, M.T., Lee, Y.S., Lin, Y.C., and Liao, J. (2000). Anxiolytic effect of seed of *Ziziphus jujuba* in mouse models of anxiety. *Journal of Ethnophamacology,* **72:** 435–441.

Pepping, J. (1999). Kava: Piper methysticum.Am. J. Health- syst Pharm, 56, 957–63. *Through Pharmacology, Biochemistry And Behavior*, **75**: 501 - 512.

Petry, R.D., Reginatto, F., de-Paris, F., Gosmann, G., and Salguciro, J.B. (2001). Comparitive pharmacological study of hydroethanol extracts of *Passiflora alata* and *P. edulis* leaves. *Phytotherapy Reseaech*, **15**: 162-4.

Pittler, M.H., and Ernst, E. (2002). Kava extract for treating anxiety. *Cochrane Database Syst. Rev.* 2003; **(1):** Cd003383.

Pitsikas, N., Boultadakisa, A., Georgiadou, G., Tarantilis, P.A., and Sakellaridis, N. (2008). Effects of the active constituentsof*Crocus sativus* L., crocins, in ananimal model of anxiety. *Phytomedicine*, **15**: 1135–1139.

Pultrini, A. M., Galindo, L.A., and Costa, M. (2006). Effects of the essential oil from *Citrus aurantium* L. in experimental anxiety models in mice. *Life Sciences*, **78(15):** 1720-5.

Rabbani, M., Sajjadi, S.E., and Zarei, H.R. (2003). Anxiolytic effects of *Stachys lavandulifolia* Vahl on the elevated plus-maze model of anxiety in mice. *Journal of Ethnopharmacology*, **89 (2-3):** 271-6.

Rabbani, M., Sajjadi, S.E., Vaseghi, G. and Jafarian, A. (2004). Anxiolytic effects of *Echium amoenum* on the Elevated Plus Maze model of anxiety in mice. *Fitoterapia*, **75(5):** 457-464.

Rabbani, M., Sajjadi, S.E., Jafarian, A, and Vaseghi, G. (2005). Anxiolytic effects of *Salvia reuterana* Boiss. on the elevated plus-maze model of anxiety in mice. *Journal of Ethnopharmacology*, **101**: 100–103.

Rabbani, M., Sajjadi, S.E. and Mohammadi A. (2008). Evaluation of the anxiolytic effect of *Nepeta persica* Boiss. in mice. *Evidenc Based Complementary and Alternative Medicine*, **5(2):** 181-6.

Rabbani, M., Sajjadi, S.E., and Rahimi, F. (2010). Anxiolytic Effect of Flowers of *Salix aegyptiaca* L. in Mouse Model of Anxiety. *Journal of Complementary and Integrative Medicine*, **7(1)**.

Rai, K.S., Murthy, K.D., Karanth, K.S., Nalini, K., Rao, M.S., and Srinivasan, K.K. (2002). *Clitoria ternatea* root extract enhances acetylcholine content in rat hippocampus. *Fitoterapia*, **73**: 685–689.

Regier, D.A., Rae, D.S., Narrow, W.E., 1998. Prevalence of anxiety disorders and their comorbidity with mood and addictive disorders. *British Journal of Psychiatry Supplement*, **(34)**: 24-8.

Rex, A., Morgenstern, E., and Fink, H. (2002). Anxiolytic-like effects of kava-kava in the elevated plus maze test–a comparison with diazepam. *Progress in Neuro-Psychopharmacology and Biological Psychiatry*, **26**: 855–860.

Rocha, F.F., Lapa, A.J., and De Lima T.C.M. (2002). Evaluation of the anxiolytic-like effects of *Cecropia glazioui* Sneth in mice. *Pharmacology, Biochemistry and Behavior*, **71**: 183– 190.

Rosa, E-R, Carolina, L-R.,Luisa, R., Gerardo, H., Esquinca, G., Rosa, A., and Mariano, M.V. (2010). Anxiolytic-like and sedative actions of *Rollinia mucosa*: *Possible involvement of the GABA/benzodiazepine receptor complex*, **48(1)**: 70-75.

Saeed, S.A., Bloch, R.M., Antonacci, D.J. 2007. Herbal and dietary supplements for treatment of anxiety disorders. *Am Fam Physician*, **76(4)**: 549-56.

Sarris, J., Kavanagh, D; J., Byrne, G., Bone, K. M., Adams, J., and Deed, G. (2009).The Kava Anxiety Depression Spectrum Study (KADSS): a randomized, placebo-controlled crossover trial using an aqueous extract of *Piper methysticum*. *Psychopharmacology*, **205(3)**: 399-407.

Sarris, J., Panossian, A., Schweitzer, I., Stough, C., and Scholey, A. (2011). Herbal medicine for depression, anxiety and insomnia: A review of psychopharmacology and clinical evidence. *European Neuropsychopharmacology* (accepted, 2011).

Satyan, K.S., Jaiswal, A.K., Ghosal, S., and Bhattacharya, S.K. (1998). Anxiolytic activity of ginkgolic acid conjugates from Indian*Ginkgo biloba*. *Psychopharmacology (Berl)*, **136**: 148–152.

Sayyah, M., Boostani, H., Pakseresht, S., and Malaieri, A. (2009). Efficacy of aqueous extract of *Echium amoenum* in treatment of obsessive–compulsive disorder. *Progress in Neuro-Psychopharmacology and Biological Psychiatry*, **33(8)**: 1513-1516.

Shah, G., Shri, R., Mann, A., Rahar, S., and Panchal, V. (2010). Anxiolytic effects of *Elaeocarpus sphaericus* fruits on the elevated plus-maze model of anxiety in mice *International Journal of PharmTech Research*, **2(3)**: 1781-1786.

Sharma, K., Arora, V., Rana, A.C., and Bhatnagar, M. (2009). Anxiolytic effect of *Convolvulus pluricaulis choisy* petals on elevated plus maze model of anxiety in mice. *Journal of Herbal Medicine and Toxicology*, **3 (1)**: 41-46.

Shou, C., Feng, Z., Wang, J., and Zheng, X. (2002). The inhibitory effects of jujuboside A on rat hippocampus *in vivo* and *in vitro*. *Planta Medica*, **68**: 799–803.

Shri, R., Bhutani, K.K., and Sharma, A. (2010). A new anxiolytic fatty acid from *Aethusa cynapium*. *Fitoterapia*, Aug 31. [Epub ahead of print].

Shri, R., and Siddana, J.K. (2008). Anxiolytic Effects of Leaf Extracts and Essential Oil of *Citrus limon* L. *Pharmacos*.

Singh, N., Kaur, S., Bedi, P.M., and Kaur, D. (2011). Anxiolytic effects of *Equisetum arvense* Linn. extracts in mice. *Indian J. Exp. Biol.*, **49(5)**: 352-356.

Singh, R.K., Nath, G., Goel, R.K., and Bhattacharya, S.K. (1998). Pharmacological actions of *Abies pindrow* Royle leaf. *Indian Journal of Experimental Biology*, **36**: 187–191.

Smith, K.K., Dharmaratne, H.R., Feltenstein, M.W., Broom, S.L., Roach, J.T., Nanayakkara, N.P., Khan, I.A., and Sufka, K.J. (2001). Anxiolytic effects of kava extract and kavalactones in the chick social separation-stress paradigm. *Psychopharmacology*, **155(1)**: 86-90.

Soulimani, R., Younos, C., Jarmouni, S., Bousta, D., Misslin, R., and Mortier, F. (1997). Behavioural effects of *Passiflora incarnata* L and its indole alkaloid and flavonoid derivatives and maltol in the mouse. *Journal of Ethnophamacology*, **57**: 11–20.

Spinella, M. (2001). in The Psychopharmacology of Herbal Drugs, The MIT press. London, England, 195 - 232.

Stough, C., Lloyd, J., Clarke, J., Downey, L.A., Hutchison, C.W., Rodgers, T. and Nathan, P.J. (2001). The chronic effect of an extract of *Bacopa monnieri* on cognitive function in healthy normal subjects. *Human Psychopharmacology*, **16**: 345-351.

Sugimoto, Y., Furutani, S., Itoh, A., Tanahashi, T., Nakajima, H., Oshiro, H., Sun, S., and Yamada, J. (2008). Effects of extracts and neferine from the embryo of *Nelumbo nucifera* seeds on the central nervous system. *Phytomedicine*, **15(12)**: 1117-24.

Taiwe, G.S., Ngo Bum, E., Dimo, T., Talla, E., Weiss, N., and Dawe, A. (2010). Antidepressant, myorelaxant and anti-anxiety-like effects of *Nauclea latifolia* smith (rubiaceae) roots extract in murine models. *Int. J. Pharmacol.*, **6**: 364-371.

Thakur, V.D., and Mengi, S.A. (2005). Neuropharmacological profile of *Eclipta alba* (Linn.) Hassk. *Journal of Ethnopharmacology*, **102**: 23–31

Thongsaard, C. Deachapunya, S., Boyd,P.A., Bennetts, G. W., and Marsdens, A. (1996). Barakol: a potential anxiolytic extracted from *Cassia siamea*. *Pharmacology Biochemistry and Behavior*, **53(3)**: 753-758.

Une, H.D., Sarveiya, V.P., and Pal, S.C. (2001). Nootropic and anxiolytic activity of saponins of *Albizziz lebbek* leaves. *Pharmaology, Biochemistry and Behaviour*, **69(3/4)**: 439-444.

Vandenbogaerde, A., Zanolli, P., Puia, G., Truzzi, C., Kamuhabwa, A., De Witte, P., Merlevede, W., and Baraldi, M. (2002). Effects of acute and chronic treatment with *Hypericum perforatum* L. (LI 160) on different anxiety-related responses in rats. *Pharmacology Biochemistry and Behavior*, **71(1)**: 251-257.

Venâncio, E.T., Rocha, N.F., Rios, E.R., Feitosa, M.L., Linhares, M.I., Melo, F.H., Matias, M.S., Fonseca, F.N., Sousa, F.C., Leal, L.K., and Fonteles, M.M. (2011). Anxiolytic-like effects of standardized extract of *Justicia pectoralis* (SEJP) in mice: Involvement of GABA/benzodiazepine in receptor. *Phytother Res.*, **25(3)**: 444-450.

Vilela, F.C., Soncini, R., and Giusti-Paiva, A. (2009). Anxiolytic-like effect of *Sonchus oleraceus* L. in mice. *Journal of Ethnopharmacology*, **124(2)**: 325-7

Viola, H., Wasowski, C., Levi de Stein, M., Wolfman, C., Silveira, R., Dajas, F., Medina, J.H., and Paladini, A.C. (1995). Apigenin, a component of *Matricaria recutita* flowers, is a central benzodiazepine receptors-ligand with anxiolytic effects. *Planta Medica*, **61**: 213–216.

Vishwakarma, S.L., Pal, S.C., Kasture, V.S., and Kasture, S.B. (2002). Anxiolytic and antiemetic activity of *Zingiber officinale*. *Phytotherapy Research*, **16(7)**: 621-6.

Woelk, H., Arnoldt, K.H., Kieser, M., and Hoerr, R. (2007). *Ginkgo biloba* special extract EGb 761 in generalized anxiety disorder and adjustment disorder with anxious mood: a randomized, double-blind, placebo-controlled trial. *J Psychiatr Res.*, **41(6)**: 472-80.

Woelk, H., and Schläfke, S. (2010). A multi-center, double-blind, randomised study of the Lavender oil preparation Silexan in comparison to Lorazepam for generalized anxiety disorder. *Phytomedicine*, **17(2):** 94-9.

Woode, E., Boakye-Gyasi, E., Amidu, N., Ansah, C., and Duwiejua, M. (2010). Anxiolytic and Antidepressant Effects of a Leaf Extract of *Palisota hirsuta* K. Schum. (Commelinaceae) in Mice. *International Journal of Pharmacology*, **6(1):** 1-17.

WHO. (2004).The World Health Organization: The World Health Report 2004: Changing History, Annex Table 3: Burden of disease in DALYs by cause, sex, and mortality stratum in WHO regions, estimates for 2002. Geneva.

Wijeweeraa, P., Arnasona, J.T., Koszyckib, D., and Merali, Z. (2006). Evaluation of anxiolytic properties of Gotukola – (*Centella asiatica*) extracts and asiaticoside in rat behavioral models. *Phytomedicine*, **13:** 668–676.

Wolfman, C., Viola, H., Paladini, A., Dajas, F., and Medina, J.H. (1994). Possible anxiolytic effects of chrysin, a central benzodiazepine receptor ligand isolated from *Passiflora coerulea*. *Pharmacology, Biochemistry and Behavior*, **47:** 1–4.

Yanpallewar S., Rai, S., Kumar, M., Chauhan,S., and Acharya, S. B. (2005). Neuroprotective effect of *Azadirachta indica* on cerebral post-ischemic reperfusion and hypoperfusion in rats. *Life Sciences*, **76:** 1325–1338.

Yoon, B.H., Jung, J.W., Lee, J.J., Cho, Y.W., Jang, C.G., Jin C, Oh, T.H., and Ryu, J.H. (2007). Anxiolytic-like effects of sinapic acid in mice. *Life Sciences*, **81(3):** 234-40.

Zanoli, P., Avallone, R., and Baraldi, M. (2000). Behavioral characterisation of the flavonoids apigenin and chrysin. *Fitoterapia*, **71(1):** S117-S123.

Zuardi, A.W., Crippa, J.A., Hallak, J.E., Moreira, F.A., and Guimarães, F.S. (2006). Cannabidiol, a *Cannabis sativa* constituent, as an antipsychotic drug. Brazilian *Journal of Medical and Biological Research*, 421-9.

Index

(-) epicatechin (EC) 385

(-) gallocatechin gallate 382

(-)-4-Epi-lyoniresinol-3a-O-β-D-glucopyranoside 135

(-)-epicatechin gallate 382

(-)-epicatechin-3-gallate (ECG) 385

(-)-Epigallocatechin gallate 382

(-)-epigallocatechin-3-gallate (EGCG) 385

(-)-Lyoniresinol -2α-O-β-D-glucopyranoside 136

(-)-Lyoniresinol -3α-O-β-D-glucopyranoside 136

(+)-4-(20-hydroxy-30-methylbut-30-enyloxy)-8H-[1,3]dioxolo[4,5-h]chromen-8-one 145

(+)-byakangelicin 381

(+)-byakangelicol 381

(+)-Lyoniresinol-3α-O-β-D-glucopyranoside 137

(+)-oxypeucedanin 381

(2E)-N-(2-methylbutyl)-2-undecene-8,10-diynamide 416

(2E)-N-isobutyl-2-undecene-8,10-diynamide 416

(2E, 7Z)-N-isobutyl)-2,7-decadienamide 416

(2E, 7Z)-N-isolbutyl-2,7-tridecadiene-10,12-diynam 416

(2E,4Z)-N-isobutyl-2,4-undecadiene-8,10-diynamide 416

(2E,6Z,8E)-N-(2-methylbutyl)-2,6,8-decatrienamide 416

(2E,6Z,8E,10Z)-N-isolbutyl-dodeca-2,4,8,10-tetraen 416

(2S)-20-methoxy kurarinone 381

(2Z)-N-isobutyl-2-nonene-6,8-diynamide 416

(Z)-β-ocimene 418

1,8-cineole 17, 388

1,8-dihydroxy-3,5-dimethoxyxanthone 15

17-hydroxy-ent-kaur-15-en-19-oic acid 381

1-Hydroxypiniresinol 186

1-Methyl-2-(3'-methyl-but-2'-enyloxy)-anthraquin 147

2-Isopropenyl-4-methyl-1-oxa-cyclopenta [b]anthrace 147

2-methylbutylamine 416

2-phenylethylamine 416

3-acetylaleuritolic acid 418

3-Methylbut-2-en-1-ol 152

3-n-butylphthalide 172

3-octanone 169

4-hydroxynonenal (4HNE) 384

4-Methoxy-1-methyl-2-quinolone 137

4-O-methylhonokiol 379

6-Hydroxy-1-methoxy-3-methylanthra-
quinone 146

7,8-Dimethoxy-1-hydroxy-2-methyl-
anthraquinone 147

7-oxo-ent-pimara-8(14),15-diene-19-oic
acid 381

7-phloroethol 380

B Cells 447

α(1)-acid glycoprotein 81, 83, 90

α1 antichymotrypsin 373

α2 macroglobulin 373

α-amyrin 150

α-bisabolol 171

α-gurjunene 418

α-humulene 418

α-phellandrene 149

α-pinene 17, 388

α-secretase 374, 375

α-secretase activators: 376

β- and γ-secretase inhibitors 376

β-amyloid 373

β-cadinene 418

β-carotene 153

β-caryophyllene 418

β-cell K+ permeability 11

β-D-Glucopyranoside,(3β)-stigmast-5-en-
3-yl 418

β-D-Glucopyranoside,(3β,22E)-stigmast-
5,22-dien-3-yl 418

β-endorphin 81, 90, 91

β-phellandrene 418

β-pinene 388, 418

β-secretase (TACE) 382

β-secretase 375, 378

β-sitostenone 418

β-sitosterol 154

β-sitosterol-β-D-glucoside 154

β-sitosteryl-3-O-β-D-glucopyranoside 428

γ-fagarine 139

γ-glutamylcysteines 101

γ-secretase 375

A

Aβ deteriorates 372

Abies pindrow 442, 474

Abnoba VISCUM 80

Acalypha 222

Acanthophora spicifera 387

Acetaminophen 448

Acetovanillion 16

Acetylcholine esterase (AchE) 382

Acetylcholinesterase 380

Acetylsalicylic acid 444

Achillea millefolium L. 196

Acmella alba 411

Acmella oleracea 409

Acute lymphoblastic leukemia (human)
86

AD pathogenesis 384

Adaptive immunity 445

Adenosine 101

Adiantum tetraphyllum 474

AE inhibited histamine 442

Aedes aegyptii 423

Aegeline 138

Aegiceras corniculatum 387

Aegle marmelos 127, 474

Aeseclus hippocastanum 388

Aethusa cynapium 474

Affinin 410

Alanine 131

Albizzia julibrissin 474

Albizzia lebeck 474

Alicyclics 222

Aliphatics 222, 223

Alkaloids 162, 222, 223

Alkamides 410

Alliin 100

Allium cepa 100

Allium nutans 388

Allium sativum 100

Allo-imperatorin 143

Alloimperatorin methyl ether 144

Allopathic medicines 440, 466

Aloe vera 379

Aloysia polystachya 474

Alternative medicine 440

Alzheimer's disease 107, 372

Amides 416

Amines 80, 416

Amino acids 80, 419

Amino *n*-butyric acid 419

Amygdalitis 18

Amyloid antiaggregant therapies 376

Amyloid precursor protein (APP) 372, 375

Amyloidogenic pathway 374

Anacardium occidentale 198

Analgesic 443

Analgesic properties 130

Anethole 418

Angelica archangelica 474

Angelica dahurica 381

Angiogenesis 10, 71

Aniba riparia 475

Annona cherimolia 475

Anopheles stephensi 424

Anti-amoebic 422

Anti-angiogenesis 80

Anti-CTLA-4 antibody 46

Anti-diarrhoeal 130

Anti-giardial 421

Anti-hyperglycemic 110, 130

Anti-hyperlipidaemic 130

Anti-hypertensive 109

Anti-inflammatory 100, 102, 130, 380

Anti-inflammatory activity 415, 440

Anti-metastatic effect 89

Anti-rheumatic 16

Anti-spermatogenic 130

Antiasthmatic 110

Antibacterial 10, 102, 422

Anticarcinogenic 107

Antidementia 477

Antidepressant 7, 477

Antidesma 222

Antidiabetic 2, 100, 380

Antifungal 102

Antifungal Activities 415, 422

Antihelminthic 130

Antihypertensive 6, 7

Antileishmanial 421

Antimalarial 421

Antimicrobial 10, 17, 100,130

Antimutagenic 107, 429

Antinociceptive 441

Antioxidant 6, 100, 170, 380, 391, 423

Antiplatelet 111

Antiprotozoal 421

Antipyretic 17, 130, 443

Antithrombotic 102, 111

Antitumor 102

Anxiety disorders 471

Anxiolytic 477, 484

Anxiolytic agents 471

Aph-1 376

Apigenin 6

Apo2.7 81

Apocynum venetum 474

Apoptosis 80, 379

Apoptosis of neuron 377

Apoptotic cells 446

Apoptotic genes 376

Applied chemistry 162

Aralia cordata 381

Archangelica officinalis 178

Arctium minus (Hill) Bernh 3

Arrhythmic activity 429

Artemisia annua L. 199

Artemisia copa 475

Artemisia sieberi 193

Artemisia spp. 388

Arthritis 439

Arthrospira maxima 200

Ascorbic acid 152

Aspartic acid 132, 419

Aspergillus 104

Aspergillus niger 415

Aspirin 444

Astaxanthin 181

Asteraceae 410

Asymptomatic 467

Atherosclerosis 6

Aurapten 142

Australian traditional medicinal system 387

Autoimmune inflammation 445

Autoimmune processes 454

Autoimmunity 71

Avena sativa 475

Ayurvedic medicines 410, 462

Ayurvedic system of medicines 129

Azadirachta indica 53, 199, 390, 443, 475

Azadirachtin A 186

Azaspirones 473

B

B16-BL6 melanoma 84

Baccatin III 186

Baccharis dracunculifolia 196

BACE1 inhibitors 378

Bacillus cereus 170, 172, 177, 422

Bacillus subtilis 105, 422

Bacopa monniera 388, 485

Baicalein 187

Baicalin 187

Balantidium entozoon 103

Bellidifolin 389

Benzenoids 222, 223

Benzodiazepines 473

Benzoyl-methylpolyols 223

Beta amyloid peptide 372

Beta blockers 473

Bifidobacteria 113

Biliary disorders 9

Bio-elements 462, 464

Bioactivity 161

Biomedical functions 10

Biotin 101

Borago officinalis 202

Borneol 17

Botryococcus braunii 200

Bradykinin induced oedema 442

Bunium persicum Boiss. 196

Bush medicine 387

Butea frondosa 441

Butyrylcholi-nesterase 373

Byrsocarpus coccineus 476

C

C. krusei 422

C. parapsilosis 422

C. tropicalis 422

CaCx 66

CaCx-DC 66

Caesalpinia crista 391

Caffeic acid 7, 18, 388

Calendulae 197

Calotropis gigantea 476

Calotropis procera 441

Calu-1 (human lung carcinoma) 84

CAMbase 82

Camellia sinensis (L.) Kuntze 381, 464

CAMPHENE 169

Camphor 17

Cancer 6

Cancer Therapy 79

Candida 104

Candida albicans 105, 177, 422

Cannabis sativa 476

Capsiacin 179

Capsicum annuum 179, 197, 203

Carbohydrates 222, 223

Carboxilic Acids 417

Carcinogenesis 42

Carcinomas 466

Cardiotoxicity 107

Cardiovascular disease 3, 6

Carnosic acid 18, 388

Carnosol 18, 178, 388

Carotene 177

Carrageenan 441

Carsonic acid 178

Carum carvi 171

Carvacrol 162

Carvone 162

Casimiroa edulis 476

Cassia siamea 476

Catechin 172, 394

CD14+ monocytes 55

CD4+CD25+Foxp3+ Treg cells 57

CD56+ lymphocytes 55

CD8+ T cells 81

CEA delivery system 66

Cecropia glazioui 476

Centaurium erythraea Rafin 14

Centella asiatica 476, 485

Cephalosporium sacchari 156

Ceratocystis paradoxa 156

CERVARIX 44

Chamazulene 171

Chamomile 191

Chamomilla recutita 171

Charaka 127

Chemokines 42, 66

Chemotherapeutic drugs 85

Chlorella vulgaris 200

Chlorogenic acid 7

Chromene 386

Chronic degenerative diseases 454

Chymotrypsin 379, 382

CINEOLE 169

Cineole 148

Cis-13-Docosanoic acid 417

Cis-9-octadecenoic acid 417

Cis-9,12-octadecadienoic acid 417

Cis-9,12,15-octadecatrienoic acid 417

Cis-scirpusin A 381

Cissus sicyoides 477

Cistus ladanifer L. 5

Citral 148

Citrus aurantium 477

Citrus limon 477

Citrus paradisi 477

Citrus sinensis 477

Clerodendrum philippinum 477

Clitoria ternatea 477

Cobalt 462

Cocarcinogenic properties 222

Colds and catarrh 129

Colocasia esculenta 477

Colon 26-M3.1 carcinoma (mice) 86

Consumption 129

Convolvulus pluricaulis 477

Convulsive activity 429

Coptidis rhizoma 380, 383

Coriandrum sativum L. 194, 440, 477

Corn earworm 423

Corynebacterium diphtheriae 422

COX 104

Crataegi folium 197

Cravacrol 171, 178

Crithidia 103

Crocus sativus 478

Croton 221, 222

Croton adenocalyx A. DC 224

Croton affinis var. *mucronifolius* 224

Croton arboreous Millsp. 225

Croton argyrophylloides Muell. Arg. 226

Croton aromaticus L. 226

Croton betulaster Muell. Arg. 227

Croton bonplandianus Baill. 227

Croton cajucara Benth. 228

Croton californicus Müll. Arg 230

Croton campestris St. Hill. 230

Croton cascarilloides Raeusch. 231

Croton caudatus Geiseler 232

Croton celtidifolius Baillon 232

Croton chilensis Muell. Arg. 233

Croton ciliatoglanduliferus Ort. 233

Croton columnaris Airy. Slaw. 234

Croton cortesianus H.B.K. (Kunth) 234

Croton corylifolius LAM 235

Croton crassifolius Geiseler 235

Croton cuneatus Klotzsch. 235

Croton diasii Pires ex Secco and P.E. Berry 237

Croton dichogamus Pax 237

Croton discolor Willd. 237

Croton dracco Schlecht (Schltdl. e Cham.) 237

Croton draconoides Muell. Arg. 238

Croton echinocarpus Muell. Arg. (Baill.) 238

Croton eluteria Bennet 238

Croton erythrochilus Muell. Arg. 242

Croton essequiboensis Klotzsch 242

Croton flavens L. 243

Croton geayi Leandri. 244

Croton glabellus L. 244

Croton gossypifolius Vahl 245

Croton gratissimus Burch. 245

Croton gubouga S. Moore 245

Croton haumanianus J Léonard 245

Croton hemiargyreus Muell. Arg. 245

Croton hieronymi Griseb. 246

Croton hovarum Leandri 247

Croton humilis L. 248

Croton hutchinsonianus Hosseus 249

Croton insularis Baill. 249

Croton jacobinensis Baill. 249

Croton jatrophoides Pax 250

Croton jimenezii Standl et. (&) Valerio 250

Croton joufra Roxb 253

Croton kerrii A. Shaw 253

Croton lacciferus Linn. 253

Croton lanjouwensis Jablonski 254

Croton lechleri L. (Müll. Arg.) 255

Croton levatii Guill. 257

Croton linearis Jacq 258

Croton lobatus L. 258

Croton lucidus L. 259

Croton lutzelburguii Pax. and Hoffm 259

Croton macrostachys A. Rich. 259

Croton malambo Kartz 260

Croton matourensis Aubl 261

Croton mayumbensis J. Leonard 261

Croton megalocarpus Hutch. 262

Croton membranaceus Muell. Arg. 262

Croton menthodorus Benth. 263

Croton micans Muell. Arg. 263

Croton moritibensis Baill. 263

Croton mucronifolius Muell. Arg. 264

Croton nepetaefolius Baillon 264

Croton nitens Sw. 266

Croton nitrariaefolius Baillon 266

Croton niveus Jacq. 267

Croton oblongifolius Roxb. 267

Croton ovalifolius Vahl 268

Croton palanostigma K.L. 269

Croton parvifolius Muell. Arg. 269

Croton penduliflorus Hutch 270

Croton plumieri Urb. 270

Croton poilanei Gagnep 270

Croton polyandrus Spreng 270

Croton pullei var. *glabrior* Lanj. 270

Croton pyramidalis D.S. 271

Croton reflexifolius H.B.K. 271

Croton regelianus Muell. Arg. 271

Croton rhamnifolius H.B.K. 272

Croton robustus Kurz 272

Croton ruizianus Muell. Arg. 272

Croton salutaris Casar 273

Croton sarcopetalus Muell. 273

Croton schiedeanus Scheacht 274

Croton selowii Baill. 275

Croton sonderianus Muell. Arg. 275

Croton sparsiflorus Morong. 277

Croton speciosus Müll. Arg. 278

Croton steenkampianus 278

Croton stelluliferus Hutch. 278

Croton stenophyllus Griseb 278

Croton stipuliformis J. Murillo 279

Croton sublyratus Kurz. 279

Croton texensis Muell. Arg. 280

Croton tiglium L. 222, 280

Croton tonkinensis Gagnep. 283

Croton triangularis Muell. Arg. 285

Croton trinitatis Mylls. 286

Croton turumiquirensis Steyerm. 286

Croton urucurana Baillon 286

Croton verreauxii Baillon 287

Croton wilsonii Griseb. 287

Croton zambesicus Muell. Arg. 287

Croton zehntneri Pax and K. Hoffm. 291

Cryptococcus 104

Cryptotanshinone 388

Culex quinquefasciatus 424

Cuminum cyminum L. 177, 190

Curcuma longa L. 201, 385, 478

Curcumin 385

CXCR3B 66

Cyclitoles 80

Cyclooxygenase 102, 441, 442

Cyclooxygenase (COX) inhibitory activity 443

Cyclooxygenase-2 441

Cymbopogon citratus 190, 440, 464

Cystoseira crinite 386

Cytisus multiflorus (L' Hér.) Sweet 10

Cytokines 80, 441

Cytoplasmic C terminus-Aβ 374

Cytotoxic 85

Cytotoxic T cells 44

Cytotoxicity 80, 379

D

d-Limonene 148

Dancus carrota 178

Datura arobrea 180

Daucus carota L. 195

Davilla rugosa 478

DCNLGPCEA vaccine 66

Decaffeination 168

Decursinol 146

Dementia 377

Dendritic cell line (FSDC) 440

Dendritic cells 42, 444

Diabetes 374

Diabetes mellitus 6, 465

Diarrhea 9

Dichloromethane 186

Dictamine 138

Dieckol 380

Diethylether 186

Dihydrotanshinone 388

Dioxinodehydroeckol 380

Dipole-dipole interactions 163

Diterpenes 222, 223

Diuretic activity 420

Dizziness 473

DLD-1 cells 84

Docosahexaenoic acid (DHA) 187

Docosanoic acid 417

Dodecanoic acid 417

Down's syndrome 373

DPPH assay 423

DPPH radicals 385

Drymaria cordata 478

Drypetes 222

Dunaliella salina 200

Dyslipidaemia 10

Dysregulated Immune Functions 41

E

E. histolytica 103

Echinacea species 478

Echium amoenum 478, 485

Ecklonia stolonifera 386

Eckol 380

Eclipta alba 478

Eczema 130

Ehrlich's carcinoma (EC) 53

Eicosanoic acid 417

Eicosatetraenoic acid 417

Eisenia bicyclis 380

Elaeis guineensis 197

Elaeocarpus sphaericus 478

Elastase 379, 382

Ellagic acid 379

Ellagitannins 6

Embryogenesis 446

Entamoeba histolytica 103, 422

Epicatechin 172, 394

Epidemiologic problem 3

Epirosemanol 178

Epoxyaurapten 137

Equisetum arvense 479

Erwinia carotovora 426

Erythraea centaurium 14

Erythrina mulungu 479

Escherichia coli 105, 156, 422

Essential amino acids 102

Essential chemical elements 463

Essential oils 168, 178

Esters 163

Eucalyptus citriodora 190

Eucalyptus globulus 194

Euchema kappaphycus 387

Eugenia caryophyllata 194

Eugenia jambolana 465

Euphorbia hirta 479

Euphorbia neriifolia 479

Euphorbiaceae 236

Evolvulus alsinoides 479

Ex vivo 66

Excerpta Medica (EMBASE) 81

Extraction 161

F

F-98 anaplastic glioma 84

F-98 glioma (rat) 86

Fatty acids 101, 163

Fenchone 173

Fenoprofen 444

Fenugreek seeds 440

Ferula assafoetida 390

Ferulic ethyl ester 16

Ferulic methyl ester 16

Flavones 18

Flavonoid glycosides 18

Flavonoids 80, 222, 223

Foeniculum vulgare Mill. 173, 192, 195, 203

Folk medicine 9, 100, 410, 440

Food industry 168

Fractionation 162

Fraxinus japonica 186

Free radicals 385, 443

Freund's adjuvant arthritis 442

Fructan 4, 101

Fructooligosaccharides 113

Fucus vesiculosus 387

Furanocembranoids 223

Furanocoumarins 381

Furosemide 420

Fusarium oxysporium 104, 415

Fusarium sp. 426

G

Galanthus woronowi 392

Galphimia glauca 479

GARDASIL 44

Gardenia jasminodes 388

Garlic 100

Gastrodia elata 479

Gelidiella acerosa 387

Gelsemium sempervirens 480

Geranium robertianum L. 15

Germacrene B 418

Germacrene D 418

Giardia intestinalis 103, 421

Giardia lamblia 103

Ginkgo biloba 383, 385, 480, 485

Ginkgo-Biloba extract (EGb761) 385

Ginsenosides 163

Gluconeogenesis 19

Glutamic acid 132

Glycine 132, 419

Glycine max L. 200, 480

Glycolipids 102

Glycoproteins 81

Glycosylated N terminus 374

Glycrrhizic acid 163

Glycyrrhisoflavone 389

Glycyrrhiza glabra L. 440, 480

Glycyrrhizia radix 180

Glycyrrhizin 440

Gossypetin 394

Gracilaria edulis 387

Guanidine 3

Guiera senegalensis 388

H

H4 neuroglioma cells 383

Hamamelis virginiana 388
Haplopine 138
Haptoglobin 81, 83
Headaches 473
HeLa (human cervix carcinoma) 84
Helicobacter pylori 106
Helicoverpa zea 423
Helixor 80
Hepatic centrilobular necrosis 448
Hepatitis B virus 44
Hepatobiliary problems 14
Hepatocarcinoma cells 84
Hepatocellular carcinoma 44
Hepatocytes 19
Hepatoprotective 17, 379
Herbal medicine 165
Herbs 168
Herpetomonas samuelpessoai 421
Heterocycle 223
Hexadecanoic acid 417
Hibiscus sabdariffa 480
Hierochloe odorata 203
High density lipoproteins (HDL) 109
Hippeastrum vittatum 480
Hippocampus region 380
Histidine 133, 419
Hizikia fusiformis 390
HL-60 84
HL-60 human leukemia 84
HLA-ABC 66
HLA/DR-antigens 81
HNSCC-lymphocytes 66
Homeopathic 129
Homeostasis 446
Homospilanthol 416
Honokiol 379
Human A253 cells 84

Human colon cancer (COLO) 84
Human colon cancer HT29 84
Human Cytochrome P4502E1 Inhibitory Activity 420
Human hepatocarcinoma SK-Hep1 cells 84
Human Leukocyte Elastase Inhibitory Activity 415
Human lung carcinoma (A549) 84
Human lung carcinoma (Calu-1) 84
Human melanoma cells 84
Human myeloleukemic cells 84
Human myeloleukemic U937 cells 84
Human tumor models 84
Humulon 174
Humulus lupulus 174
Hydrocortisone 444
Hydrodistillation 161
Hyoscamine 180
Hypercholesterolemia 11
Hyperforin 181
Hyperglycaemia 3
Hyperhomocysteinemia 111
Hypericin 7, 181
Hypericum androsaemum L. 6
Hypericum perforatum L. 181, 200, 480
Hyperphosphorylated tau protein 372
Hypertension 6, 11, 374
Hypocholesterolemic 102
Hypolipidemic effects 156

I

IFN-γ 57, 81
IL-1 beta 452
IL-12 55
IL-4 81
IL-6 81, 452
Immune escape 44
Immunoglobulins 447

Immunology 445

Immunomodulatory 80

Immunomodulatory effect 88

Immunostimulation 80

Immunotherapy 42

Imperatorin 143, 381

In vitro 61, 66, 83, 85, 104, 111, 372, 382, 441

In vivo 46, 66, 85, 111, 372, 421, 441

Indomethacin 420, 444, 448

Inflammation 439

Inorganic elements 461

Inorganic nutrients 464

INOS 425

INOS expression 441

Insecticidal activity 424, 427

Insecticides 188

Insomnia 473

Integriquinolone 139

Inulin 4

Ipomoea stans 480

Iscador 80, 82

Ischaemia 439

Ischemia 6

Isoamyl acetate 152

Isobutylamine 416

Isoepitaondiol 386

Isogntisin 389

Isoimperatorin 381

Isolation 161

Isoleucine 133

Isorosemanol 178

Ixora brachiata Roxb 442

J

Jatropha 222

Jaundice 129

Juniperus communis L. 192

Juniperus virginiana L. 196

Jurkat T-cells 84

Justicia pectoralis 481

K

Kaempferol 6, 7, 394

Kappaphycus alvarezii 387

KATP channels 11

Kielmeyera coriacea 481

Kinetic analysis 379

Klebsiella 105

Km values 130

Korean traditional medicine 386

Kurarinone 381

L

L-Arginine 132

L-Cystine 132, 135

L-NAME 450

L-selectin 67

Lactobacilli 113

Lactone 223

Laguncularia racemosa 387

LAK cytotoxicity 81

Larvicidal activity 423, 427

Laryngitis 18

Laurus nobilis L. 194, 202

Lavandula angustifolia 192

Lavandula spica 485

LCMS 161

Leachianone A 381

Lectins 80

Leishmania 103

Leishmania major 421

Leptocercous cercariae 428

Leptomonas 103

Leucocytosis 442

Leukemia 85

Leukemic B- and T-cell lines 84

Leukotrienes 441

Lewis lung tumour (mice) 86

Lignans 222, 223

Limonene 167, 418

Limonoids 223

Linalool 167, 168

Linalys acetate 169

Linoleic acid 131

Lipids 80, 222, 223

Lipo-oxygenase 442

Lipopolysaccharide (LPS) 440

Lipoxygenases 102, 441

Lippia alba Mill. 195, 199

Liquorice 163

Listeria monocytogenes 172

LOX 104

Lung carcinoma 85

Lupeol 150

Lupinus albus L. 11

Lupulon 174

Luteolin 7

Lycopene 168

M

Macaranga 222

Maclura pomifera 197

Macrophage cytoxicity 429

Macrophages 42, 44, 446

Magnolia officinalis 379, 481

Magnolol 379

Malignant glioma 86

Mangiferin 7

Mangroves 387

Manihot 222

Manihot esculenta 410

Marchantia convoluta 193

Marmelide 129

Marmenol 144

Marmesin 145

Marmin 144

Matricaria recutita 481

Matricariae 197

Mauritia flexuosa 199

MB49 urinary bladder carcinoma (mice) 86

MCF-7 breast carcinoma 84

Medicinal and aromatic plants 2

Medicinal chemistry 162

Medicinal herbs 376

Medicinal plant nutrition 463

Medicinal plant toxicity 463

Medicinal plants 471

Medicinal properties 99, 127

Mediculture 162

MEDLINE 81

Megastigmanes 223

Melanoma 85

Melissa officinalis 190, 481, 486

Menks syndrome 466

Menstrual disorder 381

Mentha piperita 202

Mentha pulegium L. 190

Mentha spicata 202

Metal chelators 372, 376, 388

Methionine 133

Methylene tanshinquinone 388

Michelia champaca Linn. 442, 443

Microalgae 200

Micrococcus 105

Micrococcus lutens 422

Micronutrients 462

Microorganism 162

MIP1α 67

MIP1β 67

Mistletoe lectins 79

Mitochondrial dysfunction 10

ML lectins 81

MO/Mf primed T cells 67

Modifiers 167

Molluscicidal Activity 428

Molt-4 84

Monarda citriodora 190

Monoamine oxidase inhibitors 473

Mononuclear cells 53

Monoterpenes 222, 223, 292, 418

Montanine 140

Morin 378, 389, 394

MPC-11 plasmacytoma tumour 88

Mucuna pruriens 390

Mus musculus 421

MV3 melanoma (human) 86

Myeloid Derived Suppressor Cells (MDSCs) 46

Myelooptic neuropathy 389

Myrcene 418

Myricetin (Myr) 378, 394

Myristica fragrans 173

N

N omega-nitro-L-arginine Methyl Ester 449

N-fructosyl arginine 107

N-fructosyl lysine 107

N-glycosylation 374

N-p-cis-Coumaryltyramine 140

N-p-trans-Coumaryltyramine 140

N-phenethyl-2,3-epoxy-6,8-nonadiyna-mide 416

N-propyl-acetamide 416

Nannochloropsis gaditana 200

NAPRALERT 222

Natural killer cells (NK) 81, 90

Natural product inhibitors 378

Natural products 161, 372, 440

Nauclea latifolia 481

Nausea 87, 473

Necrosis 447

Neem leaf glycoprotein 42

Nelumbo nucifera 481

Nepeta persica 481

Nephropathy 3, 8

Nepta tuberose 198

Nervousness 473

Neuralgia 381

Neurodegenerative 6

Neurodegenerative disease 104

Neurodegenerative disorder 372

Neurofibrillary tangles 372

Neurologic disorders 448

Neurotransmitter deficit 372

Neurotrophic 385

NFkB 102, 374

NHL tumours 89

Niacin 18, 133

Nicastrin 376

Nicotinamide 18

Nicotine 385

Nicotinic acid 101

Nigella sativa 193

Nimbin 186

Nitric oxide synthase (NOS) 384

NK cells 42, 80

NK-T cells 44

NO production 440

Nocturnal seminal emission 129

Non-amyloidogenic Pathway 374

Non-hodgkin lymphoma (mice) 86

Non-steroidal anti-inflammatory agents 439

Nootropic 385

Notch signaling 383

Nutraceuticals 168

O

O-glycosylation 374

O-isopentinylhalfordinol 140

O-methylhalfordinol 141

Obovatol 379

Ocimum basilicum 202

Ocimum gratissimum 481

Ocimum sanctum 442, 485

Octadecanoic acid 417

Octanoic acid 417

OH radicals 385

Olea europaea L. 198, 200

Oleic acid 18

Oleoresin 415

Oncology 81

Onion 100

Opalina dimidicita 103

Opalina ranarum 103

Organosulphur compounds 100

Origanum majorana 170, 171, 201

Origanum vulgare 193

Oxidative stress 3, 372, 445

Oxygen 223

Oxygen heterocycles 222

Oxyresveratrol 381

P

P-CYMENE 169

P-Cymene 149

P-nitroanilide 415

P. falciparum 421

P38MAP kinase 67

P38MAPK pathway 55

Padina tetrastomatica 387

Palisota hirsute 481

Palmitic acid 131

Panax notoginseng 379

Pancreatic beta-cell dysfunction 3

Pancreatic lipase inhibitory activity 415

Pandanus amaryllifolius 200

Paprika 163

Passiflora alata 482

Passiflora incarnata 482, 485

Pathogenesis 6, 376, 440

Pathogens 445

Pectin 101

Pelargonium sp. 190, 198

Peptides 80, 222

Periplaneta americana 424

Petasites hybridus 171

Peumus boldus M. 199, 203, 467

Peyer´s patch 80

Phaffia rhodozyma 180, 200

Phagocytosis 446

Pharmaceuticals 168

Phenanthrene 222

Phenolic acid 440

Phenolics 177, 419

Phenylalanine 133

Phenylbutanoid 223

Phenylbutazone 444

Phenylpropanes 80

Phenylpropanoids 222, 223, 292

Phlorofurofucoeckol-A 380

Phloroglucinol 380

Phlorotannins 380, 386

Phorbol esters 222

Phospholipase A2 441

Phospholipids 102

Phosphorylation 67

Phyllantus 222

Physa occidentalis 428

Physalospora tucumanensis 156

Phytochelatins 464

Phytochemicals 15, 127

Phytochemistry 248

Phytohaemagglutinin – PHA 88

Phytophthora infestans 426

Phytosterols 80, 177

Phytotherapics 462

Phytotherapy 2

Pigments 163

Pimpinella anisum L. 178, 201

Piper methysticum 482, 485, 486

Piper nigrun L. 191

Piroxicam 444

Plasmacytoma (mice) 86

Plastiquinones 386

Platonia insignis 202

Polyciclics 223

Polygala cyparissias 191

Polygala tenuifolia 382

Polymorphonuclear leukocytes (PMNs) 443

Polysaccharides 80, 163

Pongamia pinnata 442

Porphyra haitanesis 386

Porphyromonas gingivalis 105

PPAR-γ 16

PPAR-γ activation 18

Prenyl toluquinones 386

Presenilin 376

Presenilin 1 (PSEN1) 372

Presenilin 2 (PSEN2) 372

Prevotella intermedia 105

Proanthocynidin 153

Proinflammatory genes 441

Proinflammatory messengers 102

Proline 134, 419

Prophylactic vaccine 44

Prostaglandin inhibition 442

Prostaglandins 101

Prostanoids 441

Proteids 222, 223

Proteus 105

Pseudomonas 105

Pseudomonas aeruginosa 177, 422

Pseudomonas solanacearum 426

Psoralea corylifolia 382

Psoralen 141

Psoriasis 410

Psychiatric disorders 472

Psychopharmacology 472

Pterospartum tridentatum (L.) Willk. subsp. triden 13

Punica granatum 379

Punicalagin 379

Pyrazine 223

Pyrethrins 163

Pyrethrum 197

Q

Quantification 161

Quercetins 6, 7, 378, 394

Quinoids 223

Quinone 222

R

RANTES 67

RAW 264.7 84

Reactive oxygen species (ROS) 384, 443, 448

Regulatory T cells 42, 44

Reperfusion 6

Reverse transcription polymerase chain reaction 425

Rheumatic problems 6

Rheumatism 17, 410, 439

Rheumatoid arthritis 448

Rhizoctonia solani 426

Rhizosphere 464

Rhodiola rosea 486

Rhodotorula 104

Rhynchosia cana Willd 442

Ridolfia segetum 194

Rollinia mucosa 482

Romanian Mentha 193

ROS scavenger 385

Rosa canin L. 196

Rosa damascena 199

Rosmanol 178

Rosmarinic acid 18, 388

Rosmarinus officinalis L. 17, 388

Rosmariquinone 388

Rubus brasiliensis 482

Rutaretin 143

Rutin 155

S

S-alk(en)yl-L-cysteine sulphoxides (ACSOs) 101

S-allyl-cysteine sulphoxide (ACSO) 101

S-allyl-l-cysteine (SAC) 392

S-methyl cysteine sulphoxide (MCSO) 101

S-propyl cysteine sulphoxide (PCSO) 101

S-trans-prop-1-enyl cysteine sulphoxide (PeCSO) 101

Sabinene 418

Saccharomyces cerevisiae 105, 422, 426

Safrole 173

Salix aegyptiaca 483

Salmonella 105

Salmonella enteritidis 105

Salmonella typhimurium 429

Salvia miltiorrhizae 383, 391

Salvia officinalis L. 18, 171, 190, 388, 486

Salvia reuterana 482

Salvia triloba L. 201

Salvianolic acids 388, 391

Sanguisorbae radix 380

Santalum spicatum 195

Santonin 186

Sapindus mukorossi 483

Saponins 418

SAPPα 374, 375

SAPPβ 375

Sargaol 386

Sargassum micracanthum 386

Sargothunbergol A 386

Scabies 410

Scavenger phagocytes 446

Schizandra chinensis 191, 201, 203

Schizonepeta tenuifolia 173

Sclerotium cepivorum 426

Sclerotium rolfsii 156, 426

Scoparone 142

Scopolamine 180

Scopoletin 142, 419

Scurvy 9, 129, 410

Scutellaria baicalensis 192

Scutellaria lateriflora L. 200, 483

Scutellariae radix 199

Secale cereale 197

Secondary metabolites 162

Secretase inhibitors 372

Sedanenolide 172

Sedanolide 172

Sedative 7

Selective serotonin reuptake inhibitors 473

Selenium 462

Sensory activity 426

Septic shock 448

Serine 134

Sesamum indicum L. 199

Sesbania grandiflora 483

Sesquiterpenes 177, 222, 223, 292, 418

Sesterpenes 222, 223

Sexual dysfunction 473

Shark liver 200

Siddha systems of medicines 465

Signal transduction gateway 71

Silybum marianum 200

Sinapic acid 483

Single-nucleotide polymorphisms 373

Skimmianine 139

Skimmiarepin-A 151

Skimmiarepin-C 151

Skimmin 146

Skin problems 129

Sleepiness 473

Smilax china 199, 380

SOD assays 423

Solanum trilobatum Linn. 441

Solvent extraction 161

Sonchus oleraceus 483

Sonication 161

Sophora flavescens 381

Sophoraflavanone G 381

Sorghum 198

Souroubea sympetala 483

Spathulenol 418

Sphaeranthus indicus 483

Spices 168

Spilanthes americana 177, 195, 196

Spilanthol 410, 416

Spirulina pacifica 197

Spondias pinnata 390

Sporadic late-onset AD 373

Stachys lavandulifolia 483

Staphylococcus aureus 105, 156, 172, 177, 422

Staphylococcus coagulans 172

Staphylococcus epidermidis 422

Stearic acid 131

Steroid 418

Steroidal 440

Steroids 162, 223, 224

Sterols 163

Stigmasteryl-3-O-β-D-glucopyranoside 428

Streptococcus mutans 105, 106

Streptococcus pyogenes 422

Streptococcus sobrinus 105

Stypodiol 386

Stypoldione 386

Suboptimally Matured Dendritic Cells (iDCs) 46

Supercritical fluid 161

Supercritical fluid extraction 161

Suppressor cells 44

Sweat inhibitor 18

Swerchirin 15, 389

Swertiamarin 15

Symphyocladia latiussula 386

Synthetic drugs 2

Syringic acid 16

T

T-cell leukemia 83

T-cell stimulation 445

T-cells 42, 81

T-lymphocytes 81

Tamoxifen 108

Tanacetum parthenium 171

Tannic acid 18

Tannin 440

Tanshinone I 388

Tanshinones 163

Taondiol 386

Taonia atomaria 386

Taraxacum officinale 388

Tau phosphorylation 377

Taxol 186

Terpenoids 162, 163, 222

Tetracosanoic acid 417

Tetradecanoic acid 417

Tetrapleura tetraptera 484

Theobroma cacao 203

Therapeutic cancer vaccine 71

Therapeutic drugs 380

Therapeutic efficacy 71

Therapeutic outcome 71

Thiamin 152

Thiols 80

Threonine 134

Thujones 18

Thymbra spicata 177, 194

Thymol 171, 177, 178, 418

Thymus vulgaris L. 171, 195, 199, 388

Tilia americana 484

TNF-R 425

TNFα 81

Toothache 410

Torulopsis 104

Traditional medical systems 440

Traditional medicine 127, 378, 410

Traditional methods 168

Traditional therapy 2

Traditional use 467

Tragia 222

Trans-3,4,5-trihydroxystilbene 392

Trans-anethole 173

Trans-ferulic acid 419

Trans-isoferulic acid 419

Transferrin 83, 90

Transmucosal permeation 427

Trichomonas vaginalis 103

Trichophyton 104

Trichosporon 104

Trifolium pratense 486

Triglycerides 178

Trigonella foenum groecum L. 440

Tripanosoma brucei 103

Triphloroethol A 380

Triterpenes 80, 223, 224

Triterpenoid saponins 440

Trypanocidal 421

Trypsin 379

Tuberculosis 9

Tumor necrosis factor (TNF) 444

Tumors 454

Tungsten 462

Turbinaria conoides 387

Turnera aphrodisiaca 484

Type 2 diabetes mellitus 2

Type 2 Helper (TH2) T-cells 46

Tyrosine 134, 419

Tys sulfation 374

U

U-266 plasmacytoma 84

U937 human monoblastic leukemia 84

Ulcers 130

Umbelliferone 141

Unani 465

Uncaria rhynchophulla 484

Undeca-2E,7Z,9E-trienoic acid isobutyla-
    mide 416

Uronic acid 386

Ursolic acid 18

Urtica dioica L. 203

Urticaria 130

V

Vaccinium myrtillus L. 8

Valeriana officinalis 482, 484, 486

Valine 134

Value addition 161

Vander wall forces 163

Vanillic acid 419, 428

Vanillin 149

Vasorelaxant Activity 420

Veraphenol 381

Verbenone 17

Very low density lipoproteins (VLDL) 109

Vibrio cholerae 156

Vindoline 180

Viscotoxins 80

Viscum album 85

Viscum album coloratum 86

Vitamin D3 162

Vitex leucoxylon 441

Vitex negundo 484

Vitis vinifera 202

Volatile oils 163

W

Wax 177

Weight gain 473

Wilson disease 466

Witchcraft 2

Withania somnifera 388, 484

Withdrawal symptoms 473

Wogonin 187

World Health Organization (WHO) 463

X

X-ray Fluorescence (XRF) 462

Xanthones 7, 223, 224

Xanthotoxin 141

Xanthotoxol 142

Xenobiotic 100

Xylopia aromatica 201

Y

Yajurveda 127

Z

Zataria multiflora Boiss. 178, 190

Zingiber officinale 484

Ziziphus jujuba 484